Onutt & Family

January 2004

The Science of Decision Making:
A Problem-Based Approach Using Excel

Eric V. Denardo
Yale University

John Wiley & Sons, Inc.

Acquisitions Editor *Wayne Anderson*
Marketing Manager *Katherine Hepburn*
Senior Production Editor *Christine Cervoni*
Senior Designer *Karin Kincheloe*
Illustration Editor *Anna Melhorn*

This book was set in *Times Roman* by *TechBooks* and printed and bound by *Hamilton*. The cover was printed by *Brady Palmer*.

This book is printed on acid free paper. ∞

Library of Congress Cataloging-in-Publication Data:
Denardo, Eric V.
The science of decision making: a problem-based approach using Excel / by Eric V. Denardo.
p. cm.
Includes index.
ISBN 0-471-31827-2 (alk. paper)
1. Microsoft Excel for Windows. 2. Electronic spreadsheets. 3. Operations Research.
I. Title
HF5548.4.M523 D46 2001
005.369—dc21 2001024566

Printed in the United States of America

10 9 8 7 6 5 4 3 2 1

This book is dedicated to my late parents, James Dante Denardo (1902–1950) and Inga Frederiksen Denardo Rockefeller (1902–2000); it was made possible by their nurture and priorities.

Preface

The proximate cause of this text was a request by Richard Brodhead, Dean of Yale College, that I develop a course that uses spreadsheets to introduce undergraduates to mathematical reasoning. Dean Brodhead had been convinced of the value of such a course by Donald Brown, Professor of Economics, who in turn had been sold on the idea by Wayne Winston at a National Science Foundation conference on the teaching of mathematics. Wayne had earned his Ph.D. in operations research at Yale in 1975. It is a small world.

SPREADSHEET-BASED LEARNING

At the time Dean Brodhead made his request, I was teaching a problem-based introduction to decision science. I was delighted to incorporate spreadsheets in that course. It was clear that spreadsheets would let students grapple with larger and more realistic problems and would help them explore the ways in which the models of decision science approximate reality.

Spreadsheet-based instruction proved to be a far better idea than I had imagined, however. Spreadsheets let me cover more material, probe topics in greater depth, and teach more effectively. As a result, enrollment mushroomed.

What accounts for the interest explosion? There seems to be a natural synergy between decision science and spreadsheets. Decision science is a problem-solving methodology, and spreadsheets are a problem-solving tool. Decision science puts spreadsheets to work, and spreadsheets help students grasp the methods of decision science. Each makes the other more useful, more potent.

PROBLEM-BASED LEARNING

This text is problem-based; almost without exception, examples are used to introduce ideas and methods. The advantages of this "inductive" style are widely accepted. Namely, it makes the ideas easier to grasp, it shows how to use them, and it points to their limitations.

When constructing examples, I have striven to present problems that are easy to state but that raise important issues and provide insight. I have included examples that are brief cases rather than trifles.

The problems at the end of each chapter are of two types. Some of them ask the student to review the material in that chapter, whereas others use the ideas introduced within the chapter in related ways. The latter are more challenging and can form a basis for classroom discussion.

SCOPE

This text contains material that belongs in a modern problem-based introduction to the science of decision making. When compared with more venerable introductions to operations research, this book includes some material that is traditionally excluded, and it excludes some material that is traditionally included.

On one hand, this text includes an introduction to probability, a topic that is omitted or abridged in many introductions to operations research. There are three good reasons to include it:

- The models of probability are especially easy to grasp when presented in a decision-making context, with spreadsheets.
- Probability models are themselves an important facet of decision science.
- Probability models dovetail with discussions of inventory theory, queueing, simulation, and game theory.

In brief, including probability models seems to be coherent and natural.

This text also emphasizes links between operations research and economic reasoning. Much of the content of this text has been assimilated by economics, and economists have contributed to it. Yet the connections between the fields are often suppressed or omitted. Emphasizing these connections helps students to learn economics. It also provides apt illustrations of many of the ideas developed in this book. For instance, a linear program is the ideal environment in which to learn about opportunity cost. Marginal analysis, mean-variance tradeoff, efficiency, and arbitrage make lovely applications of ideas in various chapters.

On the other hand, this text excludes nearly all of the algorithms that are a vital part of operations research. In my view, these algorithms now belong in more advanced courses. With spreadsheets, introductory courses should focus on formulating problems, computing their solutions, and making sense of them. A spreadsheet add-in called Solver gives the student easy access to algorithms that solve linear, nonlinear, and integer programs.

USES

This book is designed to serve as the primary resource for a variety of survey courses. It lays the foundation for traditional one-semester courses:

- An introduction to deterministic models of decision making.
- An introduction to stochastic models.
- An introduction to both.

Spreadsheets preserve the subjects of these courses but change their focus. In an earlier era, we were limited to problems whose solutions could easily be described in terms of the common mathematical functions. Now we can grapple with any problem whose solution can be computed on a spreadsheet. Spreadsheets enable students to tackle larger and more realistic problems, to learn more, and to learn more easily.

Although this book starts with the fundamentals, segments of it probe deeply enough to be used in higher level introductions that are geared for students with advanced backgrounds. When compared with traditional intermediate-level texts, this book emphasizes problem solving and deemphasizes mathematical manipulation. In addition, this text probes linear programming deeply enough to serve as the primary resource for an intermediate-level introduction to that subject.

By building links to probability theory, economics, and management, this text can also serve as a secondary resource for several different courses. It lays the foundation for an introduction to financial engineering; it reinforces many of the ideas of intermediate microeconomics; it investigates important models of operations management; and it forms the probability "half" of an introduction to statistics.

TARGET AUDIENCE

This text is aimed primarily at undergraduate and master's level students in the arts and sciences. For instance:

- Engineers may enjoy its problem-solving methodology, which applies to a range of issues in engineering and management.
- Economists may find its content to be directly relevant to their studies.
- Humanists may enjoy developing a new set of problem-solving skills, a facility for quantitative reasoning.

Professional master's students will find this material germane to several of their courses, although they may consider portions of it a bit too technical.

The mathematical prerequisite for this text is a firm grasp of high school mathematics, including a glimpse of calculus. As concerns college-level mathematics, this text is self-contained.

NO ALGORITHMS?

This text makes one important exception to the "no algorithms" rule. The penultimate chapter presents George Dantzig's simplex method. Why include an algorithm? The simplex method is itself so fundamental that many instructors will wish to teach it. This method is the centerpiece of the modern theory of optimization, and it plays a pivotal role in computer science. It is also true that a spreadsheet makes the simplex method especially easy to learn. In addition, the simplex method is the key to duality, which is the subject of the final chapter. Duality enlarges the scope of decision science, the range of problems that decision science solves. Duality is used to study the role of arbitrage in financial economics and the role of general equilibrium in microeconomics, for instance.

Instructors who wish to probe the simplex method in an introductory course or to study it more deeply will find the relevant materials in Chapters 17–19. In addition, the starred sections in Chapter 5 introduce algorithms that improve on the simplex method for particular network optimization problems.

ACKNOWLEDGMENTS

Spreadsheets were a recent innovation to this course. Its genesis lay in an earlier teaching experience at Yale's School of Management (SOM). For the first dozen years of SOM's existence, its curriculum included a problem-based introduction to models of decision making in the public and private sectors. Students had found that course to be useful, interesting, challenging, doable, and integrative—indeed, a highlight of their education. This text evolved from material taught there.

The SOM course and related courses had been developed by a team whose members included Kurt Anstreicher, Peter Cramton, Ron Dembo, Ludo Van der Heyden, Ed Kaplan, Offer Kella, Jon Lee, Janny Leung, Uriel Rothblum, Matthew Sobel, Arthur Swersey, and myself. Each of us contributed materials, Art Swersey being the de facto leader. Art crafted the notes on which the course rested. I owe a great debt to my colleagues on that team, and I owe an equal debt to the SOM students who took those courses. What I learned at SOM set the foundation for this text.

Each of us who uses spreadsheets in decision science owes special thanks to Sam Savage, Linus Schrage, and Kevin Cunningham, who pioneered the integration of spreadsheets with optimization, and to Dan Fylstra and Leon Lasdon, who developed Solver. Solver enormously enhances the power of spreadsheet computation. It is central to nearly every chapter of this book, including those that focus on probability. Thanks are also due to Michael Middleton, who wrote two Excel add-ins that are on the CD that accompanies this text. One of these add-ins solves decision trees, and the other runs simulations.

This text has benefited from the feedback of colleagues who reviewed drafts. Many or all of the chapters were read by Richard Francis, Carl Harris (deceased), Wallace Hopp, and Steven Robinson. Individual chapters were read by Ravindra Ahuja, Jerome Bracken, Ed Kaplan, A. David Paltiel, Alvin Roth, Christopher Tang, and Ward Whitt. I am sincerely grateful for their comments, which markedly improved both the content and the presentation.

Drafts of this text have been used by hundreds of Yale College undergraduate students. Their insights and comments have improved this text in myriad ways. One undergraduate, Jeremy Brandman '03, reviewed several chapters, each with uncanny acuity. At Wiley, I enjoyed unflinching support and guidance from Wayne Anderson, my acquisitions editor, valuable help in chapter design from Judith Goode, and the good-natured determination of my production editor, Christine Cervoni.

Eric V. Denardo

Contents

---

*Within parentheses is the prerequisite chapter or chapters.

Part A

Introduction

Chapter 1

The Science of Decision Making

1.1. INTRODUCTION

A substantial literature has developed about quantitative approaches to decision making. The roots of this literature are centuries old, but much of it emerged only during the past half century, in tandem with the digital computer. This literature developed so rapidly that researchers in different fields gave it several different names, with different nuances. In 1951, Morse and Kimball provided the following definition.[1]

> ***Operations research** is a scientific method of providing executive departments with a quantitative basis for decisions regarding operations under their control.*

Their definition remains useful, with one emendation. The method they describe applies to decisions made by executives, individuals, and all sorts of groups.

In modern parlance, the term *operations research* emphasizes the underlying mathematics and its applications to decisions faced in engineering and in military affairs. The term *management science* emphasizes applications of particular interest to managers. *Decision analysis* concentrates on decisions taken in the presence of uncertainty, with particular attention given to attitudes toward risk, emphasizing connections to psychology.

Operations research is a widely accepted name for this field. Yet, as the title of this book indicates, we feel that a more transparent abbreviation of Morse's definition would be **the science of decision making** or, more briefly, **decision science**. Our usage is somewhat idiosyncratic. In the economics literature, "decision science" focuses on quantitative models of competition between several decision makers. But, for us, "decision science" embraces the entire discipline, without any particular emphasis.

A key insight of decision science is how systems work. These include manufacturing systems, transportation systems, health care systems, systems of inventories, telecommunications systems, and systems for national defense. It may seem surprising that a single family of tools could relate to such dissimilar systems, but it is so.

[1] The quotation is the first sentence of *Methods of Operations Research* by Philip M. Morse and George E. Kimball (New York: John Wiley & Sons, Inc., 1951).

A Problem-Based Approach

The exposition in this book is problem-based; never is a technique introduced in an abstract setting. In each instance, we begin with a problem that requires solution and a decision that needs to be taken. Each problem introduces you to a mathematical method that will help you reach a decision.

If you have been schooled in the more traditional theory-first mode, our approach may seem strange. Let us mention some of its advantages. A well-selected problem provides context. The context (a decision that needs to be made) suggests why the mathematical model was devised, what it can accomplish, and, just as important, what its limitations are. The context can help you to grasp the model and to understand the mathematics, thereby providing ease of access. It's also true that focusing on a specific problem circumvents the thickets of specialized terminology and mathematical notation that an abstract discussion can require.

Spreadsheet Computation

In nearly every instance, the problems that we pose require more computation than can easily be done with pencil and paper. These computations are executed on spreadsheets, which have the following advantages:

- *Realism:* Spreadsheets help you grapple with complex, realistic problems that are too large to solve by hand.
- *Ubiquity:* Nearly every desktop computer arrives with a spreadsheet package. Developing your facility with spreadsheets will help you to harness the enormous power of personal computers. The time that you invest learning how to do spreadsheet computation may pay dividends for decades to come.
- *Power:* Modern spreadsheet programs are potent. A spreadsheet lets you use all of the elementary functions (the logarithm, the square root, and so forth) and all of the standard probability distributions (the binomial, the normal, and so forth).
- *Add-Ins:* Spreadsheet programs accommodate "add-ins" that enhance their power for particular purposes. We'll use a few of these.
- *Scope:* A spreadsheet provides easy access to a broad range of problems and solution methods.
- *Interactive learning:* A spreadsheet lets you develop a model gradually, and it provides instant feedback, with lots of detail. You try something out on a spreadsheet, but the answer looks fishy. You think about it, see something amiss, revise it, and look again. After a few tries, aha, that's it! In this way, the interactive use of a spreadsheet can help you to develop a method of analysis that you might not have found by another route.
- *A mathematical language:* A spreadsheet lets you express mathematical ideas directly rather than in mathematical notation, which can be difficult to grasp.
- *Mastery:* Spreadsheets help you to deepen your understanding of the material, to grasp the underlying mathematics, and to achieve mastery.

In sum, spreadsheets enable you to learn more and to learn more quickly. As we have said, decision science is a problem-solving methodology, and spreadsheets are a problem-solving tool, each enhancing the other. Used together, they represent a sea change in teaching and learning.

Prerequisites

The prerequisites for this text are a sound knowledge of high school mathematics and a familiarity with menu-driven software that uses the "mouse." This text is otherwise self-contained. When we venture beyond basic mathematics, we develop what we use. We also explain how to use spreadsheets.

To make full use of this book, you will need to have access to a personal computer (PC or Mac) whose software includes a spreadsheet program. Several different spreadsheet packages are available, and they are remarkably similar. Microsoft Excel is the most widely used package, and by a wide margin. Our explanations of spreadsheet operations are therefore presented for Excel. If you use a different spreadsheet package, you will find it necessary to adapt our explanations somewhat.

1.2. A BIT OF HISTORY

This section describes a few milestones in the history of decision science. The history is interesting in itself, but it also acquaints you with the scope of decision science, the ways in which it is used, and the professions to which it relates.

Telecommunication

In the early days of telephony, each telephone conversation was carried on its own pair of wires. If there were 25 pairs of wires between two central offices, as many as 25 conversations could occur simultaneously. If more than 25 calls were requested, the excess would need to be *blocked*, that is, denied access to the telephone network.

In the decade prior to 1917, a Danish mathematician, A. K. Erlang, built a mathematical model of the telephone traffic between a pair of central offices. His model took the form of a system of equations that prescribe the blocking rate in terms of three parameters—the traffic intensity (call rate), the mean call duration time, and the number of pairs of wires.

Erlang's insight was stunning. Today, nearly a century later, the mathematics that he introduced remains fundamental to the design and operation of telecommunication systems. Some of his ideas are introduced in Chapter 13.

Western Electric

Within the United States, AT&T emerged after 1900 as a near monopoly on the telephone system. By the 1920s, virtually all of the hardware (telephones, cable, and central offices) used by AT&T and its subsidiaries was made by AT&T's manufacturing subsidiary, Western Electric. Corporate policy was that Western Electric's hardware should be as reliable as possible and as inexpensive as possible. Western Electric aimed to build inexpensive equipment that would perform flawlessly for decades.

This wasn't easy, but with help from AT&T's researchers, Western Electric did it. How? By the 1920s, Western Electric was fully committed to the **design for efficient manufacture**, to the **continuous improvement** of its product, and to **quality control**. This methodology has remained the hallmark of the telephone gear made by AT&T's manufacturing subsidiary, which was eventually spun off as Lucent Technologies. The switches (central offices) that Lucent makes are large computers and software systems whose downtime is measured in *seconds per year.*

The design for efficient manufacture, continuous improvement of product, and quality control were American innovations. They took place within AT&T, a monopoly. Western's pioneers in this activity included Walter A. Shewhart (the father of quality control) and Joseph Juran. W. Edwards Deming would shortly become a disciple of Shewhart. After World War II, Western's production and quality methods were taught to the Japanese in an effort to help them rebuild their war-ravaged economy. The Japanese embraced these methods, improved them, welded them to corporate strategy, and deployed them worldwide, with brilliant success.

Military Engagement

In the popular press, battlefield success tends to be associated with heroics. In military planning, it's more quantitative, a matter of logistics and tactics. Quantitative models have long provided insight into military engagements. They played an important role in the planning for Operation Desert Storm in 1991, for instance.

In the Western literature, F. W. Lanchester published in 1916 what was to become a landmark model of battlefield operations. He used differential equations to compare the losses of opposing forces in a simplified military engagement. As has so often been the case, it was eventually discovered that a Russian had done similar mathematics a bit earlier, in this case, M. Osipov, in 1915.

Quantitative models of military engagements go back to the nineteenth century. Let us quote briefly from a Russian author, who wrote in the 1850s:

> *The spirit of an army is the factor when multiplied by the mass that gives the resulting force. . . .*

The author was Leo Tolstoy, the book, *War and Peace.* Tolstoy was thinking deeply and quantitatively about the determinants of success in battle. In his model, the outcome of a military engagement was determined by the ratio of opposing forces.

World War II

The terms *operations research* and *operational research* were coined in the United Kingdom just prior to World War II. Initially, they described research on how to make *operational* use of radar in the defense of Britain. The field coalesced during the war. Americans gravitated toward the term *operations research.* The British, who speak better English, preferred *operational research.* Taken within context, the name made sense; its focus was research on military operations.

Here are some issues that the early operations researchers tackled. How shall Britain's aircraft be launched to intercept incoming German fighters? Should bombing raids against Germany be made during daylight or at night? And what should be the targets of these bombing raids? How shall trans-Atlantic convoys be configured to minimize losses due to German U-boats?

Let us consider, briefly, the last of these questions, convoy size. A British operations research team asked themselves, "What data might be relevant to this issue?" They guessed that:

- A U-boat carried limited firepower, so it might do about the same damage to a small convoy as to a large one.
- The chance of encountering a U-boat during a trans-Atlantic crossing might be roughly independent of the size of the convoy.
- If so, larger convoys would get a given volume of materiel to England with fewer encounters and with the same damage per encounter.
- Reducing the number of convoys would also allow for more effective utilization of U-boat killing escorts.

- On the other hand, large convoys would need to rendezvous, which would take time and increase risk. And very large convoys could overwhelm English ports of debarkation, which would add to exposure.

From these speculations, you can probably visualize the methodology that these early operations researchers used. They guessed what data were relevant, obtained these data, built a mathematical model that evaluated alternative convoy sizes, tested out one or two promising candidates, compared what happened to what the model predicted, and revised the model if necessary.

Systems Analysis and The RAND Corporation

At the start of World War II, the United States military was technologically obsolete as illustrated by its lack of tanks. The tank had proved its worth during World War I, but budget cuts had forced tanks out of inventory. During the 1930s, George Patton commanded cavalry (horses)!

As World War II drew to a close, far-sighted people realized that the government must keep abreast of technology—indeed, government should *foster* the development of technology and its use for national security. In 1947, The RAND Corporation (a new entity) was chartered with a remarkably brief contract from the Air Force. RAND was to do research in the national interest.

RAND quickly became a leading proponent of a methodology called systems analysis. Briefly put, **systems analysis** meshes judgment with analysis. The analysis stems from mathematical models. The judgment reflects qualitative insight, experience, and an instinct for what is possible. An example of systems analysis follows.

A Nuclear Deterrent?

By the early 1950s, the United States was experiencing major diplomatic troubles with its allies. Hoping to deter the U.S.S.R. from initiating a nuclear exchange, the United States had ringed it with American bombers, which were based on allied soil and were prepared to deliver nuclear bombs to Soviet targets. The allies on whose soil these bombers were based had become fearful that they could be targets. RAND was asked how their concerns could be assuaged.

Almost immediately, it became clear to the researchers that they had been asked the wrong question. The Russians were developing intermediate-range missiles that would be able to reach these bases so quickly as to destroy the bombers before they could take off. Ringing the U.S.S.R. with bombers would not inhibit a nuclear attack. To the contrary, it could provoke one.

The search began for an alternative, for a viable nuclear deterrent. The RAND researchers soon hit upon the idea of developing a long-range bomber (the B-52) that could be based far from the U.S.S.R., with aerial refueling (by the KC-135) en route to targets. This became a principal deterrent to nuclear war. And—by the way—Boeing manufactured a version of its KC-135 for civilian use. The commercial version was the Boeing 707, which launched the jet age in air travel.

The Energy Crisis of 1979

In 1979, a worldwide energy crisis developed. Consumers experienced a severe gasoline shortage—short hours at gasoline stations, purchase limits, stockouts, and long waits in line for what gasoline there was. In this year, the price of crude oil tripled to $32.00 per barrel (in 1979 dollars). The United States was a principal importer of oil, much of it from the Middle East. Pundits predicted that oil imports would remain high for the indefinite future and advised consumers simply to get used to paying more and more for it. The price of crude oil was expected to increase almost without limit, perhaps going to $100.00 per barrel. But none of this occurred.

What data are relevant to this issue? Gasoline and diesel fuel accounted for *one-eighth* of U.S. energy consumption. Shifting to more efficient vehicles would reduce the U.S. dependence on imported oil, but the shift would be gradual, and its effect would be relatively minor. By contrast, process heat counted for *one-fifth* of U.S. energy consumption. At $32.00 per barrel, oil was *half again* as expensive as natural gas, on an energy-equivalent basis. And oil was *three times* the price of coal.

From these and similar data and from simple linear programs (which are studied in this text), it was easy to glean this insight: Tripling the price of crude oil had provided the major (nonautomotive) users of energy ample financial incentive to invest in plants that consumed energy more efficiently and to convert from oil to natural gas or, if possible, to coal. The incentives were so strong that these investments would occur immediately.

They did occur, throughout the economy. Oil consumption dropped, and within a few years, oil was in excess supply worldwide. As a result, its price plummeted, and it stayed low for the rest of the twentieth century. In this and many cases, a quick look at deftly selected data can help to bring an issue into focus.

The Laser Scanner and Shared Data Banks

Many applications of decision science require counting, for example, of the number of sales or of the movement of material through a production system. Until recently, counting was done manually and intermittently. It was expensive, time-consuming, and error-prone. Today, the **laser scanner** counts continuously, effortlessly, and accurately. In addition, modern computer software lets users access information in a **shared data bank**. In conjunction, the laser scanner and shared data banks have greatly improved our ability to control production and inventory.

Supply Chain Management

The supply chain is the sequence of organizations that supply goods to consumers. It consists of retailers, distributors and producers. Until recently, these were separate organizations that had difficulty coordinating their activities.

Laser scanners and shared data banks have enabled the participants in the supply chain to share information and coordinate activities. The process by which they coordinate their activities—**supply chain management**—is causing a revolution (the catch phrase is "reengineering") of production and distribution systems.

Airline Deregulation and Yield Management

Prior to deregulation of the U.S. airline industry, the federal government regulated ticket prices for each airline that flew between the states. These airlines could compete on the basis of schedule and service but not price. An airline was exempt from federal regulation if all of its flights took off and landed within a single state. In California, one airline (PSA) did a thriving business within state borders, at prices far below those that the interstate airlines could charge.

Lawyers and economists argued that deregulating the airline industry would encourage the formation of new airlines, motivate existing airlines to become more efficient, and thereby decrease the fares that consumers pay. Things have not worked out quite that way. Consumers who can preplan their trips do pay markedly lower fares. On the other hand, only the largest airlines have survived deregulation and for reasons that were evident to operations researchers.

In the current deregulated era, each airline is free to set the terms and prices at which it sells seats. An airline's goal is to maximize the net revenue that it can earn from each flight. The methods by which the airlines set prices and terms are known collectively as **yield management**. An effective program of yield management is essential to profitability in the modern era. In Chapter 12, a yield management problem is related to inventory control, that is, to the control of the inventory of unsold seats. In Chapter 15, simulation is used to study a yield management problem that is somewhat more complicated.

AIDS and Needle Exchange

AIDS (acquired immune deficiency syndrome) is transmitted through unprotected sexual intercourse, the sharing of contaminated needles by intravenous drug users (IDUs), the fetuses of infected mothers, and, very rarely, transfusions of contaminated blood.

Conceivably, the exchange of new needles for used ones might reduce the rate of transmission among IDUs. But would it? And by how much? To find out, one could interview the intravenous drug users, but the information obtained from interviews would be suspect.

In a needle exchange experiment run by the city of New Haven, Connecticut, the needles themselves were "interviewed." Participating IDUs were given condoms, bleach kits (that sterilize needles), information about drug treatment programs, and offers of help to enter these programs. Used needles were exchanged for new ones on a one-for-one basis. Each new needle was coded with a unique number, and each participating IDU was uniquely identified. From each needle that was returned, one could tell:

- Whether this needle had been issued by the needle exchange program or had been acquired elsewhere.
- Whether this needle had been returned by the IDU to whom it had been issued.
- How long this needle had been in circulation.
- Whether the fluid in this needle was HIV positive or negative.

From these data, Edward Kaplan and Elaine O'Keefe built a mathematical model and provided strong evidence that needle exchange reduces the annualized rate of HIV infection by participating IDUs by at least 33% per year. A simplified version of their analysis, with appropriate references, appears in Chapter 13 as an illustration of Markov chains.

The Early Years

In the early post–World War II era, RAND and other not-for-profit "think tanks" had been the focal point of systems analysis and decision science. RAND's staff included Richard Bellman (dynamic programming), George Dantzig (mathematical programming), David Fulkerson (networks), Harry Markowitz (simulation and portfolio theory), Lloyd Shapley (game theory), and many others. IDA (the Institute for Defense Analysis) housed Al Blumstein (criminology). By the 1960s, decision scientists had begun to migrate from the think tanks to the universities. Systems analysis tended to remain within the think tanks, however.

Recap

Collectively, these illustrations hint at the range of application of decision science. It encompasses telecommunications, manufacturing, quality control, military logistics, tactics and operations, energy issues, supply chains, yield management, and public health. In addition, the convoy sizing issue has hinted at the methodology of systems analysis.

1.3. WHAT CAN YOU LEARN FROM THIS TEXT?

This book introduces you to the methods of decision science. It aims to show you how each method can be used and what its limitations are. It does not pretend to make you an expert on any aspect of this discipline. Our primary goals are to:

- Acquaint you with valuable problem-solving tools and techniques.
- Improve your problem-solving ability.
- Suggest the uses of decision science in various professions.

Let us elaborate on each of these goals.

Tools and Techniques

A battery of tools has been brought to bear on decision science. Each of these tools is a topic within applied mathematics, and different tools apply in different situations. To see which tools apply, we categorize the issue that is under study:

- Does uncertainty play a major role or a minor one?
- Are the stakes small, or do they threaten the life of the organization?
- Is there a single decision maker, or are there several?

An issue is said to be **deterministic** if uncertainty plays a relatively minor role. Many resource allocation issues are deterministic. Deterministic issues are especially suitable for analysis by the modern theory of constrained optimization, which encompasses linear programming, nonlinear programming, and integer programming. This text devotes several chapters to these subjects.

An issue is said to be **stochastic** if uncertainty plays a central role. Several chapters of this text develop probability models of uncertain systems. Other chapters concern themselves with decision making in the face of uncertainty. Chapters on inventory, Markov chains, queueing, and simulation describe particular classes of the stochastic model.

An issue is said to be **operational** if it affects the day-to-day affairs of the organization but not its existence. An issue is said to be **strategic** if it concerns existential issues. For strategic issues, one must be especially attentive to the risk of bad outcomes. This text devotes one chapter to strategic issues.

In many decision-making situations, there is a single decision maker or decision-making group. In others, there are several decision makers who can compete or cooperate. **Game theory** is the study of mathematical models of conflict or cooperation among decision makers. This text devotes one chapter to game theory.

These tools are native to operations research, but they have been assimilated by economics, engineering, computer science, and mathematics. In its current stage of development, decision science is truly interdisciplinary. Decision science can help you to learn each of the disciplines that it touches.

Consider, for example, economics. Many of the ideas in this text play key roles in microeconomics. For instance, a linear program is an ideal setting in which to learn about shadow prices, opportunity costs, and marginal profit, which are important concepts in economics. Decision science anchors these economic concepts to specific problem settings.

Mathematics also illustrates this point. Each of the tools in this text is a topic in applied mathematics. Studying these tools in a problem-based setting with spreadsheets lets you learn the relevant math without the nomenclature that a more general presentation would entail. For beginners, this nomenclature can get in the way of understanding. This nomenclature is extremely valuable nonetheless, for it enables succinct and precise statement of complicated facts. Our approach is not to avoid the mathematics but to use spreadsheets to introduce it and to make the mathematics more accessible. In these and other ways, studying decision science deepens your mathematical capability.

Problem Solving

Each of us has a unique problem-solving style. For each of us, problem solving is a blend of qualitative and quantitative techniques. Studying this text can hone your ability to reason quantitatively, deepen your insight into problems, and alter the way you perceive the world.

If you fashion yourself as a math-phobe, the "word problems" in high school math may have seemed contrived, unrelated to issues of importance. We think that you will find the problems in this text to be relevant and germane to the decisions that you will face as individuals and as members of organizations.

Professions

Decision science relates to many professions, as can be demonstrated by the applications mentioned earlier. Erlang's work relates to telecommunications. Quality control and continuous improvement are important parts of industrial engineering. Supply chain management and yield management are components of operations management. Needle exchange is an application to medical decision making. The convoy escort problem and the nuclear deterrent problem relate to military affairs. Sprinkled throughout this text are applications to other fields, including transportation, manufacturing, logistics, chemical engineering, consulting, and finance. Thus, studying this book can help prepare you for a variety of fields of endeavor.

Operations research is itself a profession that saves millions of dollars for large companies. It requires good people skills and a deeper understanding of the methods than can be provided in an introductory text.

1.4. MODELS

This book is about models of decision making. In classical usage, a "model" is an ideal, as in a model citizen or a model of feminine beauty. That is not what we mean here, however. For us, a **model** is an approximation, a simplification that aims to preserve the essential features of the situation that it represents.

Our use of "model" is that of physics. Euclid's postulates are a model of the space in which we live. Newton's laws are a model of the motion that is due to gravity. Both of these models describe everyday events with spectacular accuracy, but neither is exactly correct. Albert Einstein showed that both models are false; special relativity and general relativity provide more accurate models of space–time and of gravity. Newton's laws remain important because they apply broadly, though not universally.

The Breadth of a Model

Physics is rife with models, that is, with approximations of reality. The **breadth** of a model describes the range of situations to which it applies. Newton's laws have enormous breadth. When NASA planned the first manned spacecraft trip to the moon, it needed to learn how to train astronauts to perform in a zero-gravity environment. But NASA could count on Newton's laws to describe the motion of the spacecraft.

The models that we encounter in engineering have narrower breadth. A model of the behavior of bridges (in gales, during earthquakes, and under heavy loads) rests not only on Newton's laws, but also on engineering approximations and on rules of thumb that reflect decades of experience. For a broad range of bridge designs, these rules of thumb can be used with confidence. If a bridge design is sufficiently radical, the rules of thumb may need to be re-tested, or extra margins of safety added.

The models that we encounter in the social sciences can have still narrower scope. A well-built model may describe a particular facet of a particular economy at a particular moment in time, but with less precision and with less certitude than the model of a bridge.

The methods developed in this text have broad applicability, but the models themselves tend to lack breadth. Typically, our models describe *specific* situations. We must ask ourselves whether such a model is valid. Do the conclusions to which the model leads apply to the situation that is being modeled? Does the model help us to make better decisions?

Sensitivity Analysis and Robustness

A **sensitivity analysis** checks whether the conclusions reached within the model continue to hold true when the assumptions that underlie the model are perturbed. A model is said to be **robust** if a sensitivity analysis indicates that its conclusions remain valid for a range of circumstances that are broad enough to give us confidence that it encompasses the situation that is being modeled. Sensitivity analysis is a vital part of decision science.

If our model of a decision-making situation is robust, the conclusions that we draw from the model can be used with confidence. If our model fails to be robust, we have learned something unpleasant. Either we did a poor job of modeling, or the relative desirability of the alternative actions depends precariously on factors that cannot be specified.

1.5. MODELING

Modeling is the art of building good models. In this text, we have attempted to illustrate the art of modeling. Accordingly, we have tried to model important issues, to present intelligent models, and to strive not just for numerical answers but for insight. We've also tried to acquaint you with the various tools and techniques that you can deploy when you do build a model.

One important facet of modeling is absent from this text. Typically, there are several possible models of a situation, with each model approximating it differently. Each focuses on certain features and suppresses others. The art of modeling includes the choice of which model to build. That's an advanced subject, one that is best understood after you know what each type of model can accomplish.

The First Guideline to Modeling

An early step in modeling—and, in our opinion, the most important step—is the following one.

> **First guideline to modeling.** When confronted with an issue, learn to ask yourself, "What data are relevant, and how can I find these data?"

Few of us seem to have an innate instinct for numbers, and instead most of us must acquire it. We need to train ourselves to ask what numerical data might be relevant to a particular issue. Nothing will enhance your problem-solving ability more than developing an eye for relevant data.

The earlier discussion of milestones in decision science has illustrated the first guideline. For instance:

- In the convoy sizing example, operations researchers asked themselves how a U-boat finds a convoy and what happens if it does.
- In the energy crisis example, a decision scientist asked herself or himself how energy is consumed, what the possibilities for substitution are, what the possibilities for conservation are, and what will be the effects of tripling the price of crude oil.
- In the needle exchange example, the decision scientists asked themselves what they could learn from the needles themselves and whether it was enough to predict the effect of needle exchange on the AIDS infection rate.

To ask these questions, one needs to be savvy. These questions reflect the judgment and practical know-how that is the hallmark of a systems analysis.

Data and Myth

While it is a good idea to be sensitive to data, it's also useful to check the facts, particularly the numbers that everyone in the organization believes to be true. These numbers may be myths—justifiers of organizational practice. They may once have been true. An example follows, and it's a true story.

In the recent past, four distinguished professors surveyed American adults about their sexual practices. Based on their surveys, these researchers reported that over the course of their lives, modern American males will have sex with an average (that's the mean, not the median) of about 13 different partners, while modern American women will have sex with an average of fewer than 4 different partners. Perhaps these researchers had been persuaded by a myth. Or, in more trenchant language, exactly who are the other 9 sex partners per American male?

The Second Guideline to Modeling

No model can be an exact replica of reality. As hard as we try to capture the essential features of a problem, we are bound to paint a stylized and somewhat distorted picture of the situation. In any model, complex interactions are simplified, and detail is summarized or omitted. Data whose values are uncertain are replaced by fixed numbers.

A good model is an intelligent approximation. If the model is skillfully built, its simplifications have value. They clear away the detail that is unimportant and focus on the main effects, which will stand out starkly. This embodies our second guideline to model-building, which is given the acronym KISS (in U.S. Army lingo, Keep It Simple, Stupid).

> **Second guideline to modeling (KISS).** Begin with a simple model—one in which the main effects may stand out starkly. Use sensitivity analysis to see what facets of the model should be expanded.

An elaborate model is enticing, for it may seem to be "realistic." But an elaborate model can require data that are hard to come by; it can take a great deal of time and money to build; it can be very difficult to debug; it can obscure the main effects; and it can be plain wrong in ways that are buried in its detail. Start simple.

Third Guideline to Modeling

Our third guideline to modeling is an admonition to modelers: Treat modeling as a tool, not a religion. Remember that models are coarse representations of reality. Early in the history of digital computation, Richard Hamming wrote, "The purpose of computation is insight, not numbers."[2] Hamming warned computer users not to stop thinking when the computer program coughs up an answer, but to try to figure out what, if anything, that answer tells us.

The models in this book are *quantitative*; they manipulate numbers and give numerical answers. The insights for which we strive are largely *qualitative.* These insights explain why the model gives the answers that it does.

To an extent, these insights arise from sensitivity analyses, which ascertain the validity of the conclusions we reach within the model. These insights may also arise from remembering that the mathematical model is only part of the process of systems analysis. The other part of systems analysis is to apply experience, judgment, and common sense to the issue at hand.

A good model can be enormously helpful, but it can never be the answer. We summarize:

> **Third guideline to modeling.** No model supplies a complete answer. Use a model interactively, and aim for insight that helps you to bring an issue into focus.

The needle exchange experiment illustrates this point. The model indicates that needle exchange programs can reduce the AIDS incidence rate among participating intravenous drug users by at least 33%. Whether needle exchange should be legalized is a different issue that this model can help bring into focus.

The Systems Analysis Methodology

Every introduction to modeling, however brief, should mention the methodology that has developed for doing an effective systems analysis. The **methodology** of systems analysis consists of these steps:

- Identify the objectives and identify plausible alternatives.
- Ask yourselves what data might be relevant.
- Try to obtain these data.
- Build a mathematical model that measures the effectiveness of the alternatives.

---

[2] The quote appears on page v of *Numerical Methods for Scientists and Engineers* by R.W. Hamming (New York: McGraw-Hill, 1962).

- Test one or more of the most attractive alternatives and check the predictions of the model.
- If necessary, modify the model and try again.
- Throughout the modeling process, involve the decision makers.

Involving the decision makers in the modeling activity is crucial because their experience and knowledge are essential. Initially, it will help to frame the study. If the study is a good one, the decision makers will learn from it. They will also be far more likely to accept its recommendations if they participated in the study itself.

An Example

A familiar setting is now used to illustrate the preceding guidelines.

Problem A (The Fire Drill)

Ms. See, the principal of a brand new elementary school, sounded the alarm for a fire drill. It took a very long time to empty the building.

It was evident to Ms. See that the schoolchildren were at risk. She puzzled over what course of action to pursue. The school has three exterior doors. The main door is served by a wide corridor, the side door by a narrower corridor, and the rear door by a dog-legged corridor. An undergraduate course prompted Ms. See to ask herself what data might be relevant and how to find them. It occurred to her to measure the rate at which school occupants can file through each door.

By experimenting with the classrooms adjacent to each of these doors, she determined that:

- About 1.5 minutes elapse between the announcement of an emergency and the first departure through each door.
- Students and staff can file through the doors at these rates—approximately 55 per minute through the main door, 30 per minute through the side door, and 25 per minute through the rear door.

Ms. See also observed that, on a typical day, about 450 persons are inside the building.

From these simple data, she determined that the aggregate rate at which people can leave the building is 110 per minute because $110 = 55 + 30 + 25$; that the minimum time until the building empties is approximately 5.6 minutes because

$$1.5 = \frac{450}{110} = 1.5 + 4.1 = 5.6;$$

and that to get out as quickly as possible, the building's population should be allocated as follows.

$$\frac{55}{110} = 0.50, \text{ or } 50\% \text{ to the main door,}$$

$$\frac{30}{110} = 0.27, \text{ or } 27\% \text{ to the side door,}$$

$$\frac{25}{110} = 0.23, \text{ or } 23\% \text{ to the rear door.}$$

These calculations enabled Ms. See to allocate classrooms to exit routes in a way that is approximately optimal. After presenting her plan at a faculty meeting, she could sound a second fire drill, test her routing plan and, if necessary, refine it.

The fire-drill problem illustrates the virtue of thinking quantitatively, of asking yourself what data are relevant to the issue at hand. Asking this question can bring an issue into focus. Each of us has experienced fire drills, but how many of us have ever thought to model them?[3]

1.6. REVIEW

This book introduces you to decision science, whose concern is quantitative models of decision making. The field took shape during World War II, when it was given the name operations research. Its basic insight centers on how systems work. It has become part of several disciplines, and it is relevant to several professions.

This text is problem-based; that is, each method is introduced via a problem that requires solution. The problems can make the methods easier to grasp. Collectively, the problems can reveal the range of applicability of decision science. In nearly every case, problems are solved on a spreadsheet.

This book introduces a variety of tools, each of which is a topic in applied mathematics. The presentation relies on a firm understanding of high school mathematics; it is otherwise self-contained. For instance, the chapters on probability constitute an introduction to probability, which is presented from a decision-making perspective.

We need to recall that our models are approximations. We must check whether the model is valid—whether the conclusions to which a model leads hold for a range of circumstances that is broad enough to encompass the situation that it represents. If so, the model has provided insight. This insight will be primarily qualitative; it will explain why something is true.

Being a survey, this book probes a large number of topics and studies no model exhaustively. Generally, we have included only the material that we feel to be appropriate to an introduction. We have ruthlessly omitted everything that we feel to be relevant only to an advanced discussion, no matter how fascinating it may be.

Linear programs are an exception to this rule. This book probes linear programs beyond the beginning level, for two reasons. First, linear programs are themselves fundamental; they are crucial to the modern theory of constrained optimization. Second, our pedagogy (problem-based instruction with solution on spreadsheets) seems to be particularly well suited to a deeper understanding of linear programs.

---

[3] The fire-drill example was kindly suggested by one of this book's reviewers, Richard L. Francis, who had done research on emergency evacuation planning.

Chapter 2

Getting Started with Spreadsheets

2.1. PREVIEW

This chapter gets you started, but it does not begin with a spreadsheet. Instead, it begins by introducing you to an optimization problem that is known as a linear program. From a practical viewpoint, the importance of linear programs would be hard to overstate. They and their generalizations are used throughout decision science. Linear programs also form a nice setting in which to learn about spreadsheets.

After introducing linear programs, this chapter shows how to use a spreadsheet add-in called *Solver* to compute the solutions to linear programs. This chapter continues by describing the other things that Solver can do and by introducing the spreadsheet features that will prove handy later on.

What is a Linear Program?

If you've taken a course in calculus, you have probably encountered optimization problems that you solved by differentiation. A linear program is a different type of optimization problem, and it is solved by methods other than differentiation.

When we speak of a linear program, the word "program" means agenda or schedule, as in a program of activities or a concert program, and the word "linear" means that the relationships are linear.

Before describing a linear program, we introduce two preliminary definitions. A **linear expression** appears below; its variables are A, B, and C, and the dependence on these variables is linear.

$$A - 3.4\,B + 2\,C$$

A **linear constraint** requires a linear expression to bear one of three relations to a number that are indicated below:

$$2\,A - 3\,B = 6 \quad \text{and} \quad A - 3.4\,B + 2\,C \leq -2 \quad \text{and} \quad C \geq 0.$$

Thus, in a linear constraint, the linear expression can be equal to the number, less than or equal to the number, or greater than or equal to the number. A **linear program** maximizes or minimizes a linear expression, subject to finitely many linear constraints.

What distinguishes a linear program from classical mathematics is the inequalities. Commodities exist in nonnegative quantities. Yet the requirement that variables be nonnegative played a minor role in mathematics before the advent of linear programming.

Are Linear Programs Important?

Yes! A staggering diversity of problems can be posed as linear programs. Linear programs are now used routinely in industry and government for long-range planning and for the control of day-to-day operations. Linear programming influences the management of forests, the operation of oil refineries, the scheduling of airlines, the planning for future energy needs, and many other activities. Several chapters of this text are devoted to linear programs and their generalizations. The term **linear programming** encompasses the art of formulating problems as linear programs and the science of solving them. Linear programming may be the most significant development in applicable mathematics in the twentieth century.

2.2. WHAT CAN YOU LEARN FROM THIS CHAPTER?

This chapter prepares you for the rest of the text by introducing the following three subjects:

- Linear programs.
- The fundamentals of Excel.
- A spreadsheet add-in called *Solver*, whose uses include the computation of solutions to linear programs.

These subjects complement each other. The spreadsheet operations that help us solve linear programs have many other uses, even in chapters that have nothing to do with linear programs. Solver will be particularly handy. Why? Spreadsheet functions cannot solve equations; Solver can.

2.3. A RESOURCE ALLOCATION PROBLEM

The efficient allocation of resources is important in many fields, including engineering, management, and economics. Engineers face the problem of designing and operating systems in ways that maximize quality and minimize cost. Managers have to grapple with the problem of organizing activities in a way that maximizes profit. For economists the efficient allocation of scarce resources is *the* central issue.

Resource allocation issues take many forms. Listed here are three types of resource allocation problems that may concern a manufacturing facility:

- Produce a required set of outputs as inexpensively as possible.
- Invest a given budget in capacity expansion in the way that makes the largest improvement in profitability.
- Manufacture the most profitable mix of products.

Linear programming provides insight into each of these resource allocation issues. To introduce linear programs, we turn our attention to the following resource allocation issues.

Problem A (Recreational Vehicles)

A single plant is used to assemble three different models of a recreational vehicle: Standard, Fancy and Luxury. The plant contains an Engine shop, a Body shop, and three finishing shops, one for each model. The Engine shop assembles engines and drive trains for all three models. The Body shop assembles bodies for all three models. In each model's finishing shop, its engine and body are joined. Table 2.1 contains the relevant data. The capacities in this table are expressed in hours per week. The capacity of the Engine shop is 120 hours per week, for instance. The manufacturing times in this table are expressed in hours per vehicle. Manufacturing each Standard model vehicle requires three hours in the Engine shop, for instance. The "contributions" in Table 2.1 are profit per vehicle manufactured; making each Standard model vehicle earns a profit of $840, for instance. The plant manager wishes to learn what production mix maximizes the profit that she can obtain from this plant.

Fixed Cost, Variable Cost, and Contribution

The word "contribution," as used in this text, takes its meaning from accounting. When one contemplates taking an action, a **variable cost** is an expense that is incurred if the action is taken and only if the action is taken and a **fixed cost** is a cost that is incurred whether or not that action is taken. When we are allocating this year's production capacity, the variable cost normally includes the parts, labor, and electricity that will be consumed during production, and the fixed cost includes depreciation of existing structures, property taxes, and other expenses that are unaffected by decisions about what to produce. Any expense that has already occurred previously is a fixed cost with respect to every decision that has not yet been made. Decisions should not be influenced by fixed costs.

The **contribution** of an action equals the revenue that it creates less its variable cost. This usage abbreviates the accounting phrase, "contribution toward the recovery of fixed

Table 2.1 Shop capacities (in hours per week), manufacturing times (in hours per vehicle) and contribution of each vehicle that is made.

		Manufacturing times		
Shop	Capacity	Standard	Fancy	Luxury
Engine	120	3	2	1
Body	80	1	2	3
Standard finishing	96	2		
Fancy finishing	102		3	
Luxury finishing	40			2
Contribution		$840	$1120	$1200

costs." Table 2.1 reports \$840 as the contribution of each Standard model vehicle. Thus, \$840 equals the sales price of a Standard model vehicle less the variable cost of manufacturing it. When we use the term *profit*, what we really mean is *contribution.*

Maximizing Contribution

The manager of the Recreational Vehicle plant seeks a product mix that maximizes the contribution earned per week. How many vehicles of each type should be manufactured each week? The Luxury model vehicle has the largest contribution. Each type of vehicle consumes a total of four hours of capacity in the Engine and Body shops, where congestion is likely to occur. Thus, at first glance, one might think that production should focus on the Luxury model vehicle. But we will soon see that no Luxury model vehicles should be manufactured. Eventually, we will come to understand why that is so.

2.4. FORMULATION AS A LINEAR PROGRAM

Now let's formulate the Recreational Vehicle problem as a linear program. The decision variables in this problem are the number of vehicles of each type to manufacture each week. We give them these names:

S = the number of Standard model vehicles made per week,

F = the number of Fancy model vehicles made per week, and

L = the number of Luxury model vehicles made per week.

Not all values of these decision variables are possible; first they must satisfy certain requirements or constraints. Production cannot exceed the capacity of any shop. Consider the Engine shop. The top line of Table 2.1 shows that making these numbers of vehicles consumes $3\,S + 2\,F + 1\,L$ hours of capacity of the Engine shop. The constraint

$$3\,S + 2\,F + 1\,L \leq 120$$

keeps the number of hours consumed in the Engine shop from exceeding its weekly capacity. Each shop has a capacity constraint whose data are found by reading across its line of Table 2.1. In addition, the production quantities must be nonnegative. Total contribution is given by the linear expression $(840\,S + 1120\,F + 1200\,L)$. Thus, Program 2.1 formulates the problem of manufacturing the numbers of vehicles that maximizes contribution (profit).

Program 2.1: Maximize $\{840\,S + 1120\,F + 1200\,L\}$, subject to the constraints

Engine:	$3\,S +$	$2\,F +$	$1\,L \leq$	$120,$
Body:	$1\,S +$	$2\,F +$	$3\,L \leq$	$80,$
Standard finishing:	$2\,S$		$\leq$	$96,$
Fancy finishing:		$3\,F$	$\leq$	$102,$
Luxury finishing:			$2\,L \leq$	$40,$
	$S \geq 0,$	$F \geq 0,$	$L \geq$	$0.$

To make sure that you understand Program 2.1, ask yourself what each constraint means. Ask yourself why the constraint, $S \geq 0$, is included. Ask yourself what the objective, $840\,S + 1120\,F + 1200\,L$, measures. Ask yourself why the first constraint, $3\,S + 2\,F + 1\,L \leq 120$, is written as an inequality, not as an equation.

Terminology

Every subject has its own terminology or jargon. In a linear program, the **decision variables** are the quantities that can vary. The decision variables in Program 2.1 are S, F, and L. The **objective** of a linear program is the expression that is being maximized or minimized. The objective of Program 2.1 is $\{840\, S + 1120\, F + 1200\, L\}$. The **constraints** of a linear program are the equations and inequalities that the decision variables must satisfy. Program 2.1 has eight constraints, two of which are $3\, S + 2\, F + 1\, L \leq 120$ and $L \geq 0$. All eight of these constraints happen to be inequalities. (We often abbreviate "decision variable" to **variable**; these terms are synonyms.)

A **feasible solution** to a linear program is an assignment of values to its variables that satisfies all of its constraints. Listed here are three feasible solutions to Program 2.1, with the value that each assigns to the objective:

$$\begin{array}{llll} S = 0, & F = 0, & L = 0, & \text{objective} = 0 \\ S = 0, & F = 0, & L = 20, & \text{objective} = 24{,}000 \\ S = 0, & F = 10, & L = 20, & \text{objective} = 35{,}200 \end{array}$$

Finally, a **linear program** is any optimization problem that has these properties:

- It has finitely many decision variables.
- Its objective (the quantity being maximized or minimized) is a linear expression.
- It has finitely many constraints.
- Each constraint requires a linear expression to bear one of these relations to a number: $=$, $\leq$, or $\geq$.

Program 2.1 is a linear program, and it has three decision variables. Its objective is a linear expression. It has eight constraints, each of which is a linear inequality. Program 2.1 would still be a linear program if some of its constraints were equations, if some of its data were negative, and if its objective was minimization rather than maximization.

On the other hand, Program 2.1 would *not* be a linear program if it included one or more of the following items:

- An addend $160\, S^2$ in the objective.
- An addend $2\, F\, S$ in a constraint.
- A constraint that requires S to be integer-valued.
- An inequality that is required to hold strictly, as in $L > 0$.

A linear program presumes **linearity**, which means that everything must depend linearly on the values of the decision variables. A linear program also presumes **divisibility**, which means that the decision variables can take any values, with no spaces in between. In Chapter 3, however, we will see that linear programs encompass certain types of nonlinearity.

Consistent Units of Measure?

When constructing a linear program, we must take care to measure quantities consistently. Each addend in a particular constraint must have the same unit of measure. Otherwise, we are "adding apples to oranges." Similarly, each addend in the objective must have the same unit of measure. Let us recall that the unit of measure of the product of two numbers equals the product of their units of measure.

In Program 2.1, what is the unit of measure of the objective, and is it consistent? The unit of measure of S is vehicles/week. The unit of measure of 840 is \$/vehicle. One addend in this objective is 840 S or, more precisely,

$$\begin{aligned}(840\ \$/\text{vehicle}) \times (S\ \text{vehicles/week}) &= 840\ S\ (\$/\text{vehicle}) \times (\text{vehicles/week}),\\ &= 840\ S\ (\$/\text{week}).\end{aligned}$$

Check that each addend of the objective is measured in dollars per week. Thus, the objective is consistent. It measures contribution in dollars per week.

Is the Engine shop capacity constraint consistent? One addend in this constraint is 3 S or, more precisely, (3 hours/vehicle) × (S vehicles/week) = 3 S (hours/week). Each addend in this constraint is measured in hours per week, and so is the Engine shop's capacity, which is 120 hours per week. This constraint is consistent, as are the others.

Optimal Solution

What values of the variables in Program 2.1 maximize contribution? The **greedy** solution is to make as many Luxury model vehicles as is possible because they are the most profitable (have the largest contribution). With whatever capacity that remains, make as many Fancy model vehicles as is possible because they are the next most profitable. Then use any remaining resources to make Standard model vehicles. This greedy approach sets $L = 20$, $F = 10$, and $S = 0$. It yields a weekly profit of \$35,200 because 35,200 = 20 × 1200 + 10 × 1120.

The plant manager, however, is skeptical of this naïve, greedy approach. She uses a computer code to solve Program 2.1 and finds that contribution is maximized by

$$S = 20, \qquad F = 30, \qquad L = 0$$

$$\text{objective value} = 20 \times 840 + 30 \times 1120 + 0 \times 1200 = 50{,}400.$$

In a linear program, an **optimal solution** is any feasible solution whose objective value is best. (Some linear programs have more than one optimal solution.) The value that an optimal solution assigns to the objective is called the **optimal value** of a linear program. Program 2.1 has only one optimal solution, namely, $S = 20$, $F = 30$, and $L = 0$. Its optimal value is 50,400. Each addend in the objective is measured in dollars per week, so the optimal value of Program 2.1 is \$50,400 per week.

Insight and Sensitivity Analysis

There is nothing sacrosanct about the optimal solution to a linear program. After all, the so-called optimal solution is optimal for a model, which is an approximation. The focal point lies on the insight (if any) that the linear program provides into the situation that is being modeled.

The manager of the Recreational Vehicle manufacturing facility will not be content with the optimal solution. She will wish to learn what insight it provides into the operation of the plant. Specifically, she would wish to:

- Describe this solution in a way that can be implemented when the data in the model are inaccurate, which they are.
- Learn whether this solution remains valid for a range of data that are broad enough to encompass the actual situation.
- Learn why this solution is what it is—why it is that no Luxury model vehicles should be manufactured, for instance.

These are the subjects of a **sensitivity analysis**, which is vital to the intelligent use of any decision-making model.

One of the beauties of the simplex method is that its optimal solution is accompanied by information that helps us to execute a sensitivity analysis. How? That is a main topic of Chapter 3.

As concerns linear programs, the goal of the current chapter is modest. We aimed to set the stage for solving linear programs on spreadsheets. Now that this goal has been attained, we turn our attention to spreadsheets.

2.5. A PRIMER ON EXCEL

Spreadsheet programs are strikingly similar to each other. Lotus, Quattro, and Excel can be viewed as dialects of the same language. Excel is far and above the most widely used of these spreadsheet packages, and for this reason this text focuses on Excel. Throughout, we describe the use of Excel to solve spreadsheet problems. If you are using a different spreadsheet package, you will find that nearly everything works nearly exactly as we describe it, but you will encounter slight variations.

In a few pages, this section touches on a tiny fraction of the features in Excel. Is this section too brief? That depends. If you have used the mouse and a menu-driven program (such as Microsoft Word), this section should be enough to get you started. If not, try to arrange for a knowledgeable person to be nearby when you first use Excel.

If this is your first glimpse of spreadsheets, you are in for a shock. A spreadsheet is a two-dimensional array of cells—it's nothing more than a table, a matrix! But you can place a formula behind each cell. These formulas can refer to data and function values that have been placed in other cells. And that converts a table into a potent programming language, one that has revolutionized desktop computing.

For convenient reference, this primer on Excel is organized as a series of questions and answers.

What Are Cells? How Can We Use Them?

Shown in Table 2.2 is the upper left-hand corner of a spreadsheet. A spreadsheet is nothing more than a two-dimensional array of rectangles. In Excel lingo, each rectangle is called a **cell**. Evidently, the columns are labeled by letters, the rows by numbers. When you refer to a cell, the column (letter) *must* come first; cell B4 is in the second column, fourth row.

You **select** a cell by putting the mouse pointer in that cell and then clicking it. When you select a cell, it is outlined in heavy lines, and a **fill handle** appears in the lower right-hand corner of the outline. In Table 2.2, cell D5 has been selected. Note its fill handle, which will prove to be very handy.

Excel lets you enter about a dozen different types of information into a cell, including these four—a number, a fraction, a function, and text. In Table 2.2, a number and a

Table 2.2 Excel cells that contain numbers, fractions, functions, and text.

	A	B	C	D	E
1					
2	0.3	24			
3	1/3	2.718282	mean		
4		1.414214	=mean		
5					
6					

fraction have been placed in column A, three functions have been entered in column B, and text has been entered in column C. The next few paragraphs tell how.

How Can We Put a Number in a Cell?

To enter a **number** into a cell, select that cell, type the number, then depress either the Enter key or any one of the arrow keys. To make cell A2 to look as it does, select cell A2, type 0.3, and then hit the Enter key.

How Can We Put a Fraction in a Cell?

Getting a fraction to appear in a cell is not so easy. Suppose you select cell A3, type 1/3, and then press the Enter key. What will appear in cell A3 is "3-Jan." Excel has decided that you wish to put a date in cell A3. Every correction that you subsequently enter into cell A3 will be interpreted as a date. Swell! To enter a **fraction** into cell A3 (and to get rid of the date):

- Select cell A3.
- On the Format menu, click on Cells. . . .
- In the dialog box that appears, click on Fraction.
- Type the fraction.
- Depress the Enter key or any one of the arrow keys.

How Can We Distinguish between Normal Text and Keystroke Sequences?

Throughout this text, keystroke sequences are displayed in a format that distinguishes them from normal text.

> This text displays each Excel keystroke sequence in boldface type, suppressing both:
>
> - The **Enter** keystroke that finishes the keystroke sequence.
> - Any English punctuation that is not part of the keystroke sequence.

To avoid confusion, punctuation is *omitted* from keystroke sequences, even when it leaves off the period at the end of the sentence! In Table 2.2, for instance, we selected cell A2 and then typed **=0.3**

How Can We Put a Function in a Cell?

In Excel, *every* function begins with the "=" sign. To enter a **function** into a cell, select that cell, depress the "=" key, type the function, and then depress the Enter key. Into cells B2, B3, and B4 of the spreadsheet in Table 2.2 have been entered the keystroke sequences,

=3*(2^3) **=EXP(1)** **=SQRT(2)**

Above, in boldface, are three keystroke sequences; we've omitted the **Enter** keystroke at the end of each, as well as the English punctuation.

When you enter a function into a cell, what appears there is not the function itself but the **value** that has been assigned to the function. The preceding keystroke sequences cause the appearance of:

$24 = (3)(2^3)$ in cell B2,

$2.718\ldots = e$ in cell B3,

$1.414\ldots = \sqrt{2}$ in cell B4.

In a spreadsheet, the functions themselves are *invisible*. When you enter a function into a cell, what appears there is not the function but the value that it has been assigned.

Can We Display the Functions?

Yes. If you want to display the functions in the cells, on the Tools menu, click on Options and, select the View tab and, on the Window Options list, click on Formulas, and then press the OK button on that tab. This displays the functions, but it suppresses the values that they have been assigned.

How Can We Enter Text into a Cell?

To enter **text** into a cell, type it in. If this text could be misinterpreted, begin with an *apostrophe*, which will not appear. To make cell C3 appear as it does in Table 2.2, select cell C3 and then type

'=mean

In the above, the leading apostrophe tells Excel that what follows is text, not an equation.

What Are Absolute and Relative Cell References?

An **absolute** reference to a cell includes a "$" sign before its row, before its column, or before both, as in $B4, B$4, and $B$4. A **relative** cell reference omits the $. Thus, cell B4 can be described in the four ways that are displayed below.

B4 $B4   B$4 $B$4

Inclusion of absolute and relative cell references is a clever feature of spreadsheet programs. It lets you repeat patterns. How? Read on.

What Is the SUMPRODUCT Function?

The spreadsheet that appears as Table 2.3 has numbers in columns A through E of rows 1, 2, 3, and 5.

Let's imagine that you want to do several multiplications and additions on the spreadsheet in Table 2.3. Specifically, you want to:

- Multiply each number in the first row by the corresponding number in the fifth row, add these products, and place the total in cell F1.
- Multiply each number in the second row by the corresponding number in the fifth row, add these products, and place the total in cell F2.

Table 2.3 A spreadsheet with data.

	A	B	C	D	E	F
1	0	2	0	4	0	
2	1	2	1	2	1	
3	1	0	2	0	4	
4						
5	1	-1	2	-2	3	

- Multiply each number in the third row by the corresponding number in the fifth row, add these products, and place the total in cell F3.

To do this efficiently, select cell F1, and then type

=SUMPRODUCT(A1:E1, A$5:E$5)

This function multiplies the numbers in cells A1 through E1 by the numbers in cells A5 through E5, respectively, takes the sum of these five products, and deposits that sum in the cell you have selected, which is cell F1. As a result, the number −10 appears in cell F1.

You could type similar expressions into cells F2 and F3, but that would require extra work, and it would introduce extra opportunity to make typing errors. Instead:

- Select cell F1 again. Cell F1's fill handle is the box in its lower right-hand corner.
- Put the mouse pointer on the fill handle of cell F1. The shape of that pointer, normally a Greek cross, will change to a + sign.
- While it remains a + sign, depress the mouse, slide it down across cells F2 and F3, and then release it.

When you do this, the numbers 0 and −17 will appear in cells F2 and F3. They will appear there because you just inserted into these cells the functions

=SUMPRODUCT(A2:E2, A$5:E$5)
=SUMPRODUCT(A3:E3, A$5:E$5)

The spreadsheet in Table 2.4 exhibits the results of this calculation.

In This Text, Can We Distinguish Data from Function Values?

Table 2.4 illustrates a rule that is used throughout this text and that will help you to distinguish function values from data. The rule is as follows.

> A cell that is outlined in **dotted lines** displays the value of a function. An arrow points to this cell. The function is specified at the tail of this arrow. Absolute cell references ($ signs) within the function suggest which other cells contain functions.

Table 2.4 A spreadsheet containing data and three functions.

	A	B	C	D	E	F
1	0	2	0	4	0	-10
2	1	2	1	2	1	0
3	1	0	2	0	4	17
4						↑
5	1	-1	2	-2	3	
6						
7			=SUMPRODUCT(A3:E3,A$5:E$5)			

It's evident in Table 2.4 that cell F3 exhibits the value of the function

=SUMPRODUCT(A3:E3, A$5:E$5)

Moreover, the "$" signs in this function indicate that cells F2 and F1 contain the values of the functions that are obtained by "dragging" the mouse pointer upward from cell F3.

What Does It Mean to Drag the Mouse Pointer?

To **drag** the mouse pointer is to depress the mouse button, move it while it is depressed, and then release it. To place functions in cells F2 and F3 of Table 2.4, we dragged the mouse pointer. Whenever you drag the mouse pointer, Excel tries to repeat whatever pattern you have established. It shifts the *relative* addresses by the number of rows or columns that you shift the mouse pointer.

Can We Paste Instead of Drag?

Yes. To create Table 2.4, we did not need to drag. Fixed and relative addressing has the same effect if we drag and if we copy and paste. We could have copied cell F1 onto the "clipboard" and then pasted it onto cells F2 and F3.

When we copy and paste, the relative addresses are incremented by the number of rows and/or columns that the mouse pointer has been shifted. Pasting is handy when the "target" cell is not contiguous to the one whose function is being copied.

When We Enter a Function, Must We Type the Cell References?

No. Instead of typing cell references, you can use the mouse to select blocks of cells. And clicking the mouse is more accurate.

For instance, enter the function =SUMPRODUCT(A1:E1, A5:E5) into cell F1, type as far as the left parenthesis, select cells A1 through E1, type a comma, select cells A5 through E5, and then type a right parenthesis.

By the way, you needn't capitalize "sumproduct" either; Excel does that automatically. You can also insert a space after the comma (as had been done above), which can make your function a bit easier to read.

Any Tips on Debugging?

Typically, the Excel function in a cell refers to values in other cells. It is *easy* to refer to the wrong cells, even if you use the mouse to select the cells whose values you are referring to. Excel offers two ways in which to correct this type of error. Experiment with each, and see which one works for you.

Begin with a spreadsheet that contains some functions, including functions whose arguments are values of other functions. Then select the cell that contains one of your functions. Next, on the Excel Tools menu, scroll down to the *Auditing* entry. A menu will pop up to the right of your cursor. Shift the mouse pointer to the Trace Precedents option and click. Arrows will appear; they will identify the arguments of the function in the cell that you selected. You can do this repeatedly, and you can use it to trace several levels of dependency. To get rid of the arrows, repeat the process, but click on the Remove All Arrows option. If you like this option, you may wish to use it to put the Auditing Toolbar in a convenient spot on your spreadsheet.

Again, select the cell that contains one of your functions. This function appears twice, once in the cell you clicked and once in the *Formula cell* toward the top of your spreadsheet. Move the mouse pointer to the Formula cell, just to the left of the function's "=" sign, and then click it. Each argument of your function will appear in its own color, and each cell or block of cells will be outlined in the same color. Nice! But once the pointer is on the Formula cell, you cannot move it by clicking. Instead, slide it rightward to the end of your formula and then hit the Enter key.

Sooner or later, you will want to trace dependencies. And you will find one or the other of these debugging tools to be handy.

What Is a Circular Reference?

A **circular reference** occurs when a function refers to a cell that contains a function whose value has not been fixed. Excel has no way to put the value of that function into its cell. Excel "burps" by delivering the message, "Cannot resolve circular references."

Let us illustrate. Suppose you wish to use Excel to solve the equation

$$x = 3.6 + 0.8\,x.$$

In cell A4, you might enter

=3.6 + 0.8*A4

in which case Excel 95 reports the message in Table 2.5. (Later versions of Excel have a fancier "circular reference" message, which offers to help you trace dependencies and figure out why the message is circular.)

Do not ignore a circular reference. If you do, you may get it to go away, but your spreadsheet calculations will be gibberish. This point cannot be overemphasized.

> If you ignore a circular reference, your subsequent calculations will be garbage.

This looks like very bad news. Excel cannot solve equations! At least, not directly.

Can Excel Solve Equations?

Yes! You can get around a circular reference. Modern versions of Excel contain a program called *Solver.* Solver can find a solution, if one exists, to systems of equations and inequalities. It can also find the solutions to linear programs and to other optimization problems.

Table 2.5 Circular Reference message for Excel 95.

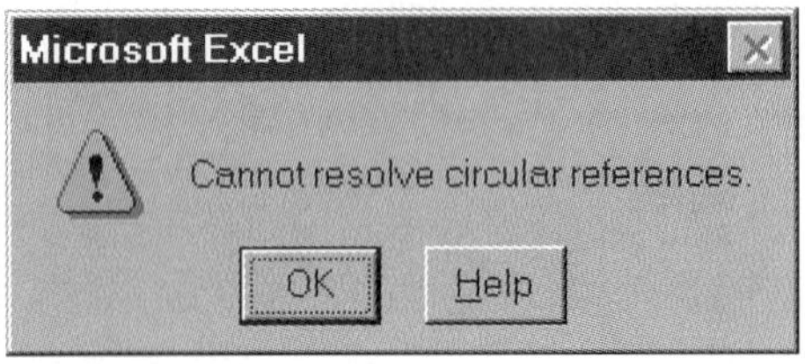

2.6. INSTALLING SOLVER

Solver will be used in nearly every chapter of this text. Before you can use this program, however, it must be installed and activated. This section tells you how to do that.

Solver is an Excel **Add-In**. You have at least one version of Solver—and probably two—available for use with this text. Both versions were written by Frontline Systems.

- The "standard" version of Solver is distributed by Microsoft. It was purchased with Excel.
- The "premium" version of Solver, whose full name is the *Premium Solver for Education*, is distributed directly by Frontline Systems; it's on the CD that accompanies this text.

If you are using a networked computer, you might be restricted to the standard version; if you are using your own personal computer, you can arrange to use either.

Which Version Should You Use?

It's a close race. For nearly all of the purposes in this text, the standard and the premium versions of Solver work equally well. They use similar algorithms for solving linear, integer, and nonlinear programs. They have the same restrictions as to problem size. (Both handle linear programs that have as many as 200 variables and as many as 100 constraints, for instance.) If you do have a choice, the premium version has the edge, for these reasons:

- As of this writing, the premium Solver has fewer bugs.
- The premium Solver includes a "genetic" algorithm that is dubbed the *Evolutionary Solver.* It finds solutions to nonlinear systems on which the standard code for nonlinear programs fails.

Is the Standard Solver Ready to Use?

Maybe. How can you tell? Open Excel. Click on its Tools menu. Is Solver listed there? If so, all is well.

If Solver is not on the Tools menu, you may only need to **activate** it. Click the Tools menu, and then select the Add-Ins menu item. If Solver is on the list of items, check it, and click on OK. This activates Solver; it places Solver on the Tools menu, where you want it.

Alas, if Solver has not been installed, you will need to find your Excel CD or disk or diskettes. Go through the Excel setup procedure and install Solver. Then activate Solver, in the manner just described.

Is the Premium Solver Easy to Install?

If you are using a networked computer, you may not be able to install the premium Solver.[1] Not to worry. The standard version is fine for our purposes.

On a personal computer, you can install and activate the premium version *after* you have installed and activated the standard version. To do so, insert the CD that accompanies

[1] Many networks prohibit users from installing Excel add-ins. Systems managers can install them. John Wiley & Sons is licensed to copy and distribute *Premium Solver for Education* for installation on networks of personal computers for use only by instructors and students who are using this text.

this text in your CD drive. Then select that drive, click on the icon marked PremSolv.exe, and follow instructions.

Can We Uninstall the Premium Solver?

If you wish, you can remove Premium Solver for Education after installing it. To do so, put the CD-ROM that accompanies this text in your disc drive and select that drive. Then double-click on the premsolv icon. On the window that pops into view, select the "Uninstall…" option, and follow the instructions.

Uninstalling the premium version removes it from memory. If you want to reinstall it, repeat the above.

Which Version of Solver Is Illustrated in This Text?

The dialog boxes that appear in this text are for the standard version of Solver, the version that comes with Excel. Table 2.6 is the only exception. It displays dialog boxes for both versions, side by side. As Table 2.6 suggests, the two dialog boxes are nearly identical.

Table 2.6 indicates that the dialog box for the premium version has one extra button, which is labeled "Premium."

Actually, *Premium Solver for Education* has two Solver dialog boxes, and you can toggle between them. Clicking on the "Premium" button in the above switches you to its sister dialog box, which has a button marked "Standard." The sister dialog box also has a window that accords you access to the Evolutionary Solver. The "Standard" dialog box in *Premium Solver for Education* behaves just like the version of Solver that Microsoft distributes, except for removal of whatever bugs have been found in that version.

Recap

Before continuing, you should have one or the other version of Solver up and running. The standard version works fine. The premium version works a little better, whether or not you use its "Premium" dialog box. The premium version also accords you access to the Evolutionary Solver, which finds solutions to some nonlinear systems that the standard version cannot.

Table 2.6 *Solver Parameters* dialog boxes for the standard version (at the left) and for the premium version (at the right).

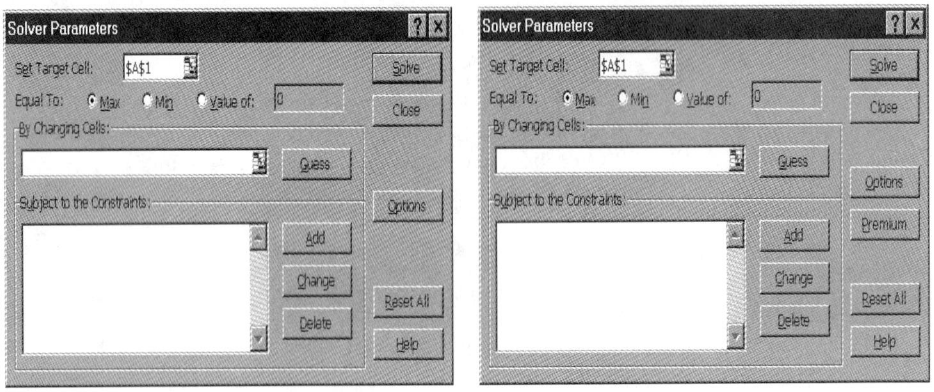

2.7. SOLVING A LINEAR PROGRAM

Solver is flexible. There are several ways in which to arrange for it to compute the optimal solution to a linear program. You are urged to begin by learning the four-step procedure that is described here. Its first two steps accomplish what Excel can do without Solver; its remaining steps use Solver to finish the job. To introduce this four-step procedure, we apply it to the Recreational Vehicle example.

Step 1: Enter the Data and Text

The first step is to open an Excel spreadsheet and enter the text and data for your linear program. For Program 2.1, it's convenient to organize these data as we have done in Table 2.7.

Remarks on Step 1:

- The letters *S*, *F*, and *L* that appear in cells B2, C2, and D2 are text (column headings), not the values of the variables.
- The values of the variables *S*, *F*, and *L* will be placed in cells B9, C9, and D9. In Solver's jargon, the cells in which the variables are placed are called **changing cells**.
- The "<=" signs in cells F3 through F7 are memory aids, nothing more.
- Cells E3 through E8 have been left blank deliberately; we will soon put a linear expression in each of these cells.

Step 2: Enter the Linear Expressions

A linear program contains several linear expressions. Its objective is a linear expression. The left-hand side of each constraint is a linear expression. On a spreadsheet, the SUMPRODUCT function makes it easy to compute linear expressions. Step 2 enters these linear expressions.

Program 2.1 contains six linear expressions: its objective and the left-hand sides of five of its constraints. Cells E3 through E8 have been reserved for these six linear expressions. Parts (a) and (b), on page 32, create all six linear expressions by specifying one SUMPRODUCT function and dragging it. Here's how.

Table 2.7 Spreadsheet after entering data and text (Step 1).

	A	B	C	D	E	F	G
1							
2		S	F	L			
3	Engine shop	3	2	1		<=	120
4	Body shop	1	2	3		<=	80
5	Standard finishing shop	2				<=	96
6	Fancy finishing shop		3			<=	102
7	Luxury finishing shop			2		<=	40
8	contribution	840	1120	1200			
9	value of variable						

(a) Select cell E3. Enter into cell E3 the function

=SUMPRODUCT(B3:D3, B$9:D$9)

(b) Select cell E3 again. Move the mouse onto cell E3's fill handle to change the pointer to a "+" sign. While the pointer remains a + sign, depress the mouse button, slide it down across cells E4 through E8, and then release it. This will deposit the functions

=SUMPRODUCT(B4:D4, B$9:D$9)
=SUMPRODUCT(B5:D5, B$9:D$9)
=SUMPRODUCT(B6:D6, B$9:D$9)
=SUMPRODUCT(B7:D7, B$9:D$9)
=SUMPRODUCT(B8:D8, B$9:D$9)

into cells E4 through E8.

(c) (*Optional, but highly recommended.*) Insert numbers into cells B9, C9, and D9, and check the arithmetic.

After you complete Step 2, your spreadsheet should look like rows 1–9 of the spreadsheet in Table 2.8. Column E of this table contains the values that have been assigned to the spreadsheet functions.

Remarks on Step 2:

- Cells B9, C9, and D9 hold the values of the decision variables.
- Cell E3 contains the amount of Engine shop capacity that is actually used.
- Cells E4 through E7 contain the amounts of the other shops' capacities that are used.
- Cell E8 measures the contribution.
- The dotted lines around cell E8 indicate that this cell contains the value of the function **=SUMPRODUCT(B8:D8;B$9:D$9)**
- The "$" signs in this function connote that cells E7 through E3 contain values of the functions that are found by dragging this function upward.

Table 2.8 Spreadsheet after entering the functions (Step 2).

	A	B	C	D	E	F	G
1							
2		S	F	L			
3	Engine shop	3	2	1	6	<=	120
4	Body shop	1	2	3	6	<=	80
5	Standard finishing shop	2			2	<=	96
6	Fancy finishing shop		3		3	<=	102
7	Luxury finishing shop			2	2	<=	40
8	contribution	840	1120	1200	3160		
9	value of variable	1	1	1			
10							
11		=SUMPRODUCT(B8:D8,B$9:D$9)					

To verify that our linear expressions are correct, we have placed numbers in the changing cells and checked the arithmetic. This is always a good idea. We emphasize:

> When building spreadsheets, insert data to check that your functions are performing the computations that you intend.

Step 3: Use Solver to Complete the Linear Program

Steps 1 and 2 have accomplished what can be done within Excel itself. Excel cannot identify any changing cells; it cannot enforce any constraints; it cannot solve a system of equations and inequalities; it cannot seek a solution that maximizes or minimizes the objective function. Solver can. In the context of Program 2.1, the following tasks remain for Solver:

- Identify the number in cell E8 as the quantity that we wish to maximize.
- Identify cells B9 through D9 as the changing cells.
- Require the numbers in cells B9 through D9 to be nonnegative.
- Require the numbers in cells E3 through E7 to be less than or equal to the numbers in cells G3 through G7, respectively.

To perform these functions is to execute Step 3. To open Solver, click on Tools, and select Solver. The *Solver Parameters* dialog box (Table 2.9 but with all cells blank) will drop onto the screen. Then execute parts (a) through (f), below.

(a) If the *Solver Parameters* dialog box obscures something that you need to see, click on its title bar and drag it out of the way.

(b) Within the *Solver Parameters* dialog box, click on the Options button. In the menu that appears, check the Assume Linear Model option. Then click the OK button. This returns you to the *Solver Parameters* dialog box. (You have just told Solver that you intend to solve a linear program.)

(c) The *Solver Parameters* dialog box should now be back on the screen. After you complete parts (d), (e), and (f) of this step, this dialog box will resemble Table 2.9.

Table 2.9 *The Solver Parameters* dialog box after completion of Step 3.

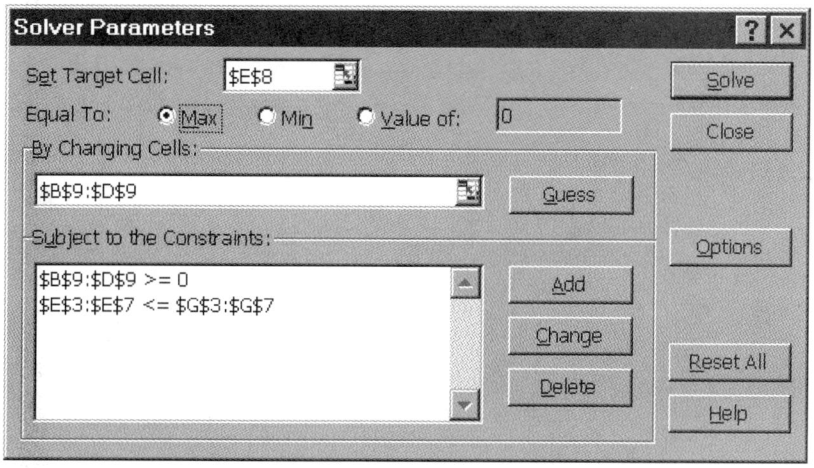

(d) In the *Solver Parameters* dialog box, click on the Set Target Cell window. Either select cell E8 or type **E8** in that window. (You need not type the "$" signs because Solver inserts them automatically.) Then click on its Max button. (You have just told Solver to maximize the value of the function in cell E8.)

(e) Click on the By Changing Cells window. Then, either select cells B9 through D9 or type **B9:D9** into that window. (You have just told Solver that the values of the decision variables are in its "changing cells.")

(f) The easy way to insert constraints is to use the Add button in the *Solver Parameters* dialog box. Click on its Add button. The *Add Constraint* dialog box will appear, and it will look like Table 2.10, with the data omitted. To require the numbers in the changing cells B9 through D9 to be nonnegative, do as follows:

- Click on the leftmost window in the *Add Constraint* dialog box. Then either select cells B9:D9 or type **B9:D9**
- Click on the small triangular button in the center window in the *Add Constraint* dialog box. Then, on the menu that pops up, select the ">=" option.
- Click on the rightmost window in the *Add Constraint* dialog box. Then type **0**

If your *Add Constraint* dialog box looks like the one in Table 2.10, you have inserted the nonnegativity constraints.

Table 2.10 Inserting the nonnegativity constraints.

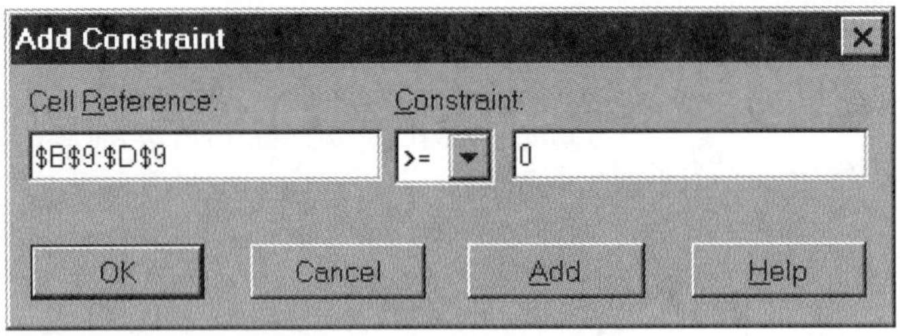

Next, to add the capacity constraints, click on the Add button in Table 2.10. A fresh *Add Constraint* dialog box will appear. To require the quantities in cells E3 through E7 to be less than or equal to the numbers in cells G3 through G7, make this *Add Constraint* dialog box look like the one in Table 2.11. To do so:

- Click on the leftmost window in the *Add Constraint* dialog box. Then either select cells E3 through E7 or type **E3:E7**
- Click on the rightmost window in the *Add Constraint* dialog box. Then either select cells G3 through G7 or type **G3:G7**
- If your *Add Constraint* dialog box looks like the one in Table 2.11, click on its OK button.

Table 2.11 Inserting the capacity constraints.

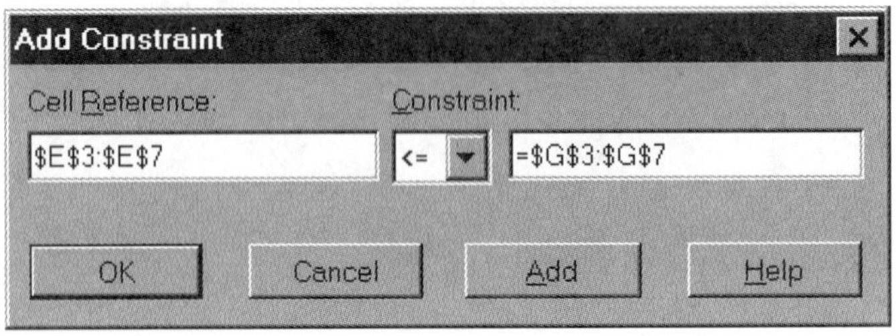

That final click caused the *Solver Parameters* dialog box to reappear. It should now look like the one in Table 2.9. If it does, you have completed Step 3.

Remarks on Step 3:

- *Solver* uses absolute addresses; it inserts the $ signs automatically.
- The *Solver Parameters* dialog box lists the constraints in alphabetical order, which may not be the sequence in which you entered them.
- You must insert the nonnegativity constraints.
- Sooner or later, you may forget part (b), which is to use the Options button to let *Solver* know that you are solving a linear model. If you forget, you will get the correct optimal solution and the correct "shadow prices," although the latter will be reported as "Lagrange multipliers."

Step 4: Solve the Linear Program

Steps 1–3 have specified the linear program. You can solve it with the push of a button. Having solved it, you can record its optimal solution.

To indicate how, we again use Program 2.1. The *Solver Parameters* dialog box should be on the screen, and it should look like Table 2.9. If this dialog box is not on the screen, click on the Excel Tools menu, and then click on Solver, which will put it back on the screen. If this dialog box does not look like Table 2.9, make it do so. To compute the optimal solution to Program 2.1 and to record it, execute parts (a) and (b):

(a) In the *Solver Parameters* dialog box, click on the Solve button. In a flash, Solver will report that it has found a solution. The solution itself will appear on your spreadsheet, which should resemble Table 2.12. As is indicated in Table 2.12, this

Table 2.12 An optimal solution to Program 2.1.

	A	B	C	D	E	F	G
1							
2		S	F	L			
3	Engine shop	3	2	1	120	<=	120
4	Body shop	1	2	3	80	<=	80
5	Standard finishing shop	2			40	<=	96
6	Fancy finishing shop		3		90	<=	102
7	Luxury finishing shop			2	0	<=	40
8	contribution	840	1120	1200	50400		
9	value of variable	20	30	0			
10							
11							
12							
13							
14							
15							
16							
17							

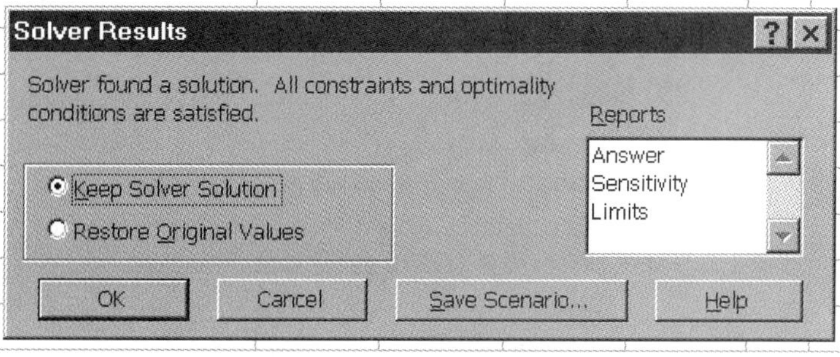

optimal solution sets $S = 20$, $F = 30$, and $L = 0$. It has an objective value of 50,400.

(b) As Table 2.12 indicates, the *Solver Results* dialog box lets you save or discard the solution, and get a solution and a Sensitivity Report. As was suggested earlier, we will find good use for the Sensitivity Report. In its *Reports* window, click on the Sensitivity entry. Then click on the OK button. This will preserve the optimal solution, and it will create a sheet entitled Sensitivity Report.

Recap

In the above four-step procedure, Step 1 enters the data, and Step 2 enters the functions. Steps 1 and 2 accomplish what Excel can do by itself. Step 3 uses Solver to record the objective, the changing cells, and the constraints. Step 4 solves the linear program and records what information you wish to keep about the optimal solution.

We learn how to use Solver by trial and error. Open a spreadsheet, and execute this four-step procedure. Print the spreadsheet that you create, which should look exactly like the one in Table 2.12. Help in printing spreadsheets can be found in the Appendix to this text.

Solving Linear Equations

In high school, you probably solved two linear equations in two unknowns and three linear equations in three unknowns—by hand. If so, you learned some important facets of linear algebra.

But, as a practical matter, solving linear equations by hand is obsolete. Solver can do it for you. How? Act as though you are solving a linear program but leave the objective blank.

2.8. OTHER USES OF SOLVER

Solver does more than solve linear programs. The *window* that is Solver's dialog box is aptly named. It accords you effortless access to these algorithms:

- An algorithm called the simplex method that solves linear programs.
- A generalization of the simplex method that solves **integer programs**, which differ from linear programs by requiring some or all of the variables to be integer-valued.
- A pair of generalizations of the simplex method that solve **nonlinear programs**, which can differ from linear programs by having nonlinear objectives and constraints.
- A "Value of" feature that seeks a solution to an equation, subject to any number of constraints.

This section describes the capabilities of Solver, and it mentions other software packages that perform similar tasks.

Solving Linear Programs

In 1947, George B. Dantzig saw that many planning problems could be posed as linear programs, and he devised a method for computing optimal solutions to them. His method is known as the **simplex method**. This method solves typical problems with blazing speed, even those that have large numbers of variables and constraints. (Chapter 18 of this text is devoted to the simplex method.)

Vast energy has been spent trying to improve on the simplex method. Only in the recent past has a class of "interior point" methods been developed that are faster than the simplex method on very large problems. For many purposes, however, the simplex method remains the best, a half century after its invention. Standard computer codes for linear programs include the simplex method as their only solver or as one of their solvers.

Solving Integer Programs

At the nub of every code that solves integer programs is an extremely simple idea that is known as "branch and bound." It's what Solver uses. It attacks an integer program by solving a sequence of linear programs. Usually, this sequence of linear programs is short, but it can be astronomically long, in which case the code fails to find a solution. (Chapter 6 of this text is devoted to integer programs.)

Solving Nonlinear Programs

For nonlinear programs, there can be no "magic bullet," no algorithm that is guaranteed to compute solutions quickly. Some nonlinear programs are easier to solve than others, however. If a nonlinear program is "convex," a well-designed algorithm should compute a solution.

But what does "convex" mean? The answer to this question will become a bit clearer in Chapter 3. Suffice to say for now that a convex nonlinear program can contain diseconomies of scale (but not economies of scale), and it can exhibit decreasing marginal return (but not increasing marginal return).

What happens if Solver (or any other algorithm) is applied to a nonconvex program? It may succeed, or it may fail in either of two ways: It can converge to a "local optimum" that is not a global optimum, and it can fail to converge at all.

Solving Equations

To find a solution to any number of linear equations, set them up as a linear program but leave the objective (Target Cell) blank. That works fine.

With the "Value of" feature of Solver, you can request a solution to a nonlinear equation, with or without constraints. If you do, Solver invokes its machinery for solving nonlinear programs. Sometimes this works, and sometimes it doesn't. As of this writing, neither Excel nor Solver contains a simple "line search" routine that solves an equation by "bisecting" an interval, repeatedly.

A Tip: Try to Start Close

When we present Solver with a system of nonlinear equations and inequalities, it accepts the values that we place in its changing cells as the first guess as to a solution. Generally speaking, Solver is more likely to find a solution to a nonlinear system if we "start close," that is, if the values in the changing cells are not too far from the actual solution.

But how can we start close? Often, a good starting point is the solution to a simpler system in which some or all of the nonlinear terms have been approximated, perhaps by zeros. By experimenting with different starting points, we may cajole Solver into solving a nonlinear system.

Limitations of Solver

The Solver code that is provided with Excel is adequate for the linear programs in this text. To solve a large linear program, you may need an upgrade or one of the commercial packages that are mentioned in the next section. To solve a large integer program, you will need an industrial-strength software package and, almost certainly, the assistance of an expert.

As concerns codes for nonlinear programs, both the "standard" and the "premium" version of Solver fall short of the state of the art. An improved code for nonlinear programs is available from Frontline Systems. This new code does not require you to "start close," and it fails less frequently than the versions provided here. For information about an upgrade, contact the vendor at **www.frontsys.com** on the world wide web.

Alternatives to Solver

Algorithms that solve linear, nonlinear, and integer programs are widely used. Building software packages and running them has become a small industry. Every few years, the journal, *ORMS Today*, updates its survey of software packages that solve linear, nonlinear, and integer programs. Its most recent survey listed more than 40 such programs.

Of the spreadsheet-based alternatives to Solver, *What's Best!* (a product of Lindo Systems Inc.) may be the most widely used. Of the non-spreadsheet-based alternatives, CPLEX, GAMS and LINDO may be among the most popular. For large problems, you will wish to use a commercial solver, and you may want to engage the help of an expert.

From our viewpoint, Solver has four virtues. It does the job, it dovetails with spreadsheets, it is widely distributed, and it is free. Solver is part of the popular spreadsheet packages, including Excel, Lotus, and Quattro. (In Quattro, it is called *Optimizer.*) If you purchase one of these packages, you get Solver with no added expense.

2.9. SOLVING A NONLINEAR SYSTEM

As just mentioned, Solver does not require linearity. To see how to use it to solve nonlinear systems, we turn our attention to:

Problem B (Solving Nonlinear Equations)

You seek values A and B that satisfy the equations

$$3A + 2B + 5\sqrt{A} = 15,$$
$$2A + 3B + 4\sqrt{B} = 20.$$

Problem B may be a bit too complicated for the methods that you learned in high school. To set this problem up for Solver, we will try to stick as close as possible to the four-step procedure for solving linear programs:

- Step 1 enters the data.
- Step 2 inserts the functions.
- Step 3 uses Solver to enforce the equations.
- Step 4 asks Solver to find a solution.

Steps 3 and 4 will be accomplished in three different ways. Why? As mentioned earlier, there is no foolproof way to solve systems of nonlinear equations, no single method that is guaranteed to work in all cases. By exploring several avenues, you might hit upon a successful route.

Table 2.13 Entering the data for Problem B.

	A	B	C	D	E	F	G	H
1	names	A	B	sqrt(A)	sqrt(B)			RHS
2		3	2	5			=	15
3		2	3		4		=	20
4	values							

Step 1: Enter the Data

We are aiming to use the SUMPRODUCT function and the SQRT() function in Excel. When entering the data, we follow the prior pattern as closely as we can. This produces Table 2.13.

In Table 2.13, row 1 contains column headings, not the values of the variables. Rows 2 and 3 contain the coefficients in the two equations. The "=" signs in column G are memory aids. Cells B4 and C4 are reserved for the values of the decision variables A and B; cells D4 and E5 for square root functions; and cells F2 and F3 for SUMPRODUCT functions.

Step 2: Enter the Functions

Table 2.14 exhibits the spreadsheet that is obtained by entering the functions =SQRT(B4) and =SQRT(C4) in cells D4 and E4 and the usual SUMPRODUCT functions in cells F2 and F3. This table also displays the values that these functions are assigned when we set $A = 2$ and $B = 4$.

As Table 2.14 indicates, dragging the function in cell D4 across cell E4 computes the square root of the numbers in cell C4. Similarly, dragging the function in cell F3 across cell F2 computes the left-hand-side value of the first equation.

Step 3: First Version

Aiming to use the optimization package within Solver, we "relax" one of the equations to a "≤" inequality and maximize its left-hand side. This produces the *Solver Parameters* dialog box in Table 2.15. In it, cells B4 and C4 are identified as changing cells, and the number in cell F2 is maximized, subject to the two constraints, one of which keeps the number in cell F2 from exceeding the number in cell H2.

Table 2.14 Entering the functions for Problem B.

	A	B	C	D	E	F	G	H
1	names	A	B	sqrt(A)	sqrt(B)			RHS
2		3	2	5		21.07	=	15
3		2	3		4	24	=	20
4	values	2	4	1.414	2			
5								
6			=SQRT(B$4)					
7								
8			=SUMPRODUCT(B3:E3,B$4:E$4)					

Table 2.15 Problem B interpreted as a maximization problem.

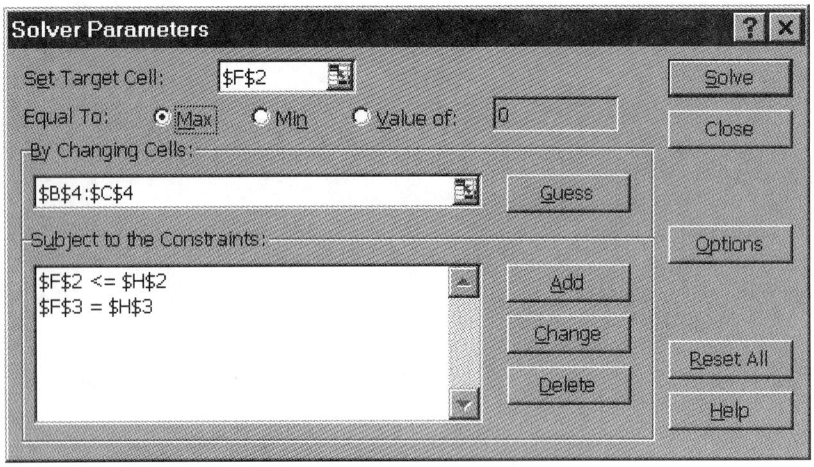

Step 3: Second Version

The second approach uses the **Value of** feature in Solver to equate the number in cell F2 to 15, subject to the single constraint, which requires the number in cell F3 to equal the number in cell H3. Implementing this strategy produces the *Solver Parameters* dialog box in Table 2.16.

Step 3: Third Version

The third approach is to use the optimization package but leave the target cell blank. This leads to the *Solver Parameters* box in Table 2.17.

Step 4: Solve the Equations

Tables 2.15, 2.16, and 2.17 present three different ways to formulate Problem B for solution by Solver. All that remains is to push the Solve button. When you do so, Solver will report that it has found a solution. Cells B4 and C4 will contain the values $A = 0.998$ and $B = 3.505$.

Table 2.16 Problem B organized to equate the value in cell F2 to 15.

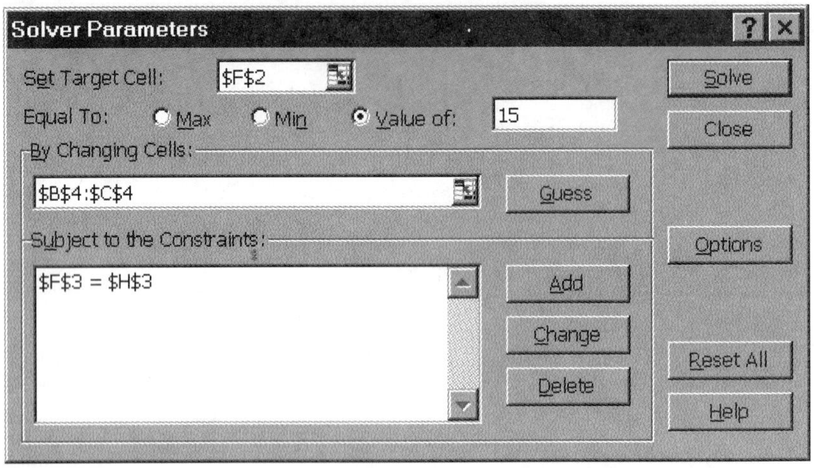

Table 2.17 Problem B organized as a maximization problem with no objective.

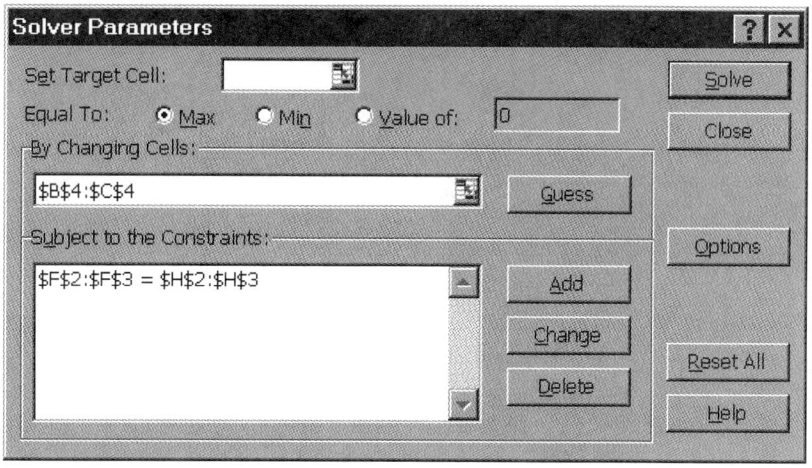

Incidentally, if you encountered difficulty, you might begin by using a tip mentioned in the prior section. Start with the solution to the linear system in which the terms entailing square roots are replaced by zeros.

As mentioned earlier, Solver *may* find solutions to nonlinear equations that have solutions. But Solver is not guaranteed to find a solution to every nonlinear equation system that has a solution. In fact, no algorithm known to humankind is guaranteed to find solutions to all systems of nonlinear equations. If Solver cannot find a solution, it tactfully reports, "Solver cannot find a solution."

A Difficulty

One difficulty with the current version of Solver is that it can "jump" out of the range for which Excel's functions are defined. For instance, in Problem B, Solver may ask Excel to compute the square root of a negative number. If this occurs, Excel grinds to a halt and reports #NUM! in the cell for which it cannot compute a value. Solver may do this *even if* we add the constraints that the values in the changing cells must be nonnegative. The Evolutionary Solver avoids this difficulty, but it's slow. You can also experiment with the Evolutionary Solver, which is on the diskette that accompanies this text.

A Maxim

This is as good a point as any to emphasize the following maxim:

> On your spreadsheet, hide no data within a spreadsheet function or in a dialog box. Instead:
>
> - Record each datum directly on the spreadsheet, and record it exactly once.
> - If your spreadsheet uses a datum more than once, enter it once and refer to the cell at which it is located, rather than entering it repeatedly.

Obeying this rule makes it easy to change the values that are assigned to data, which you will want to do when you execute a sensitivity analysis. If a datum appears only once, changing it will change all uses of it. If a datum appears in multiple places, you may overlook some of them. If a datum is hidden within a function, you may overlook it entirely.

Rarely in this text do we violate this maxim. We have violated it once, and recently. Do you see where?

2.10. OTHER USES OF EXCEL

The preceding sections contain enough information for you to tackle Chapters 3 and 4, which deal with linear programs. This section mentions the features of Excel that will prove handy in later chapters.

Charts

Excel lets you build a graph (or chart) of the output that you computed in your spreadsheet. On its *Insert* menu, you can find the *Chart* entry. Charts are very handy. The Appendix to this book contains a couple of tips that can help you to build a chart.

Functions

Excel contains an enormous variety of functions. In this text, we'll use a couple dozen of them. Nearly all of Excel's functions are entirely self-explanatory; the Appendix to this text discusses a few that are not.

The Excel functions include all of the mathematical and trigonometric functions that we will employ. To view a list of them, scroll down the Insert menu to the Function . . . entry, click on it, and then select the Math & Trig entry on the window that pops up.

The Excel functions also include most of the standard probability distributions. These will be used in the chapters on probability and in the chapters that use probability. To view a list of these probability distributions, scroll down the Insert menu to the Function . . . entry, click on it, and then select the Statistical entry on the menu that pops up.

On occasion, we will use a few of the less transparent functions, such as "IF" and "MATCH." Excel can also perform matrix arithmetic, and we will use this feature in the later chapters. We will explain what we use.

Add-ins

It's no secret that the Excel add-in called Solver will be used throughout this text. The following paragraphs mention the other add-ins that we will use.

Chapter 9 studies decision making when the future is uncertain. That chapter employs an Excel add-in called *TreePlan*, which was written by Michael Middleton. Chapter 9 explains how to use TreePlan. TreePlan itself can be found on the CD that accompanies this text. Before you can use it, however, you must install it and activate it. The Appendix to this text tells how.

Chapter 15 concerns simulation. That chapter makes extensive use of an Excel add-in called *RiskSim*. It too was written by Michael Middleton, and it is also included in the CD that accompanies this text. Chapter 15 includes a description of how to use RiskSim.

Visual Basic Macros

The CD that accompanies this text also includes three Visual Basic "macros" that were written by Kenneth Canfield, '03, a Yale College undergraduate. The Appendix describes these macros. They can be installed and activated in the same way as can TreePlan and RiskSim.

Macros can enormously enhance a spreadsheet's ability to perform particular tasks. Learning how to write macros is extremely valuable. But, within this text, we have minimized the use of macros, and we've made no attempt to teach macro programming because including it might have made this text too technical and too difficult.[2]

The Appendix and the CD

As already noted, the Appendix to this text contains instructions on how to install and activate the Excel add-ins and macros that are used here. The Appendix also explains a few of Excel's peculiarities. Scan it now, just to see what is there.

The CD that accompanies this text contains *Premium Solver for Education*, *TreePlan*, *RiskSim*, and the Visual Basic macros that we have mentioned. This CD also includes nearly all of the spreadsheets that appear within the chapters of this text. These spreadsheets can help you with the homework problems, and they can help you to understand the chapters themselves.

Advantages of Spreadsheets

Excel, Lotus, and Quattro are dialects of a single programming language, a computer code that helps us to solve problems. It may be the most widely used programming language that has yet been devised. This language has spawned an entire industry; it is powerful, and it is easy to learn.

Drawbacks

Every programming language has advantages and disadvantages. Spreadsheets do many things well, and they do some things poorly. Excel has two drawbacks that are especially important to decision science.

One is *auditibility.* When you gaze at a spreadsheet, you can see the values that the functions have been assigned, but you cannot see the functions themselves. This makes spreadsheet programs especially difficult to read and to audit.

We've tried to ameliorate this problem in our exposition. Cells that report the values of functions are enclosed in dotted lines, with arrows pointing to them from other cells where the functions are specified.

The second major drawback is that Excel affords us no way to change the content of a cell during the course of computation. We cannot *update* the number that appears in a cell. This limits the range of computation that we can execute comfortably on a spreadsheet. Evidence of this limitation can be found in Chapter 5 (Networks) and in Chapter 15 (Simulation).

Origins

Spreadsheet computation is a fabulous idea, and is credited to Dan Bricklin and Bob Frankston, the developers of *VisiCalc.* They saw how a personal computer could be programmed to automate the repetitive calculations that accountants had been forced to slog through.

Good ideas have deep roots, however, and this one goes back to the nineteenth-century English mathematician Arthur Cayley, who is best known as the originator of matrix algebra. Cayley had a second career as an accountant, and he authored a book on accounting.

---

[2] A book by Christian Albright, entitled *VBA for Modelers: Developing Decision Support Systems Using Spreadsheets* (San Francisco: Duxbury, 2001), presents the facets of Visual Basic that are particularly relevant to decision science.

Accountants had long been aware of the merit of matrix representations as a "single-entry/double-classification" alternative to double-entry bookkeeping. By the early 1960s, Yuri Ijiri was doing spreadsheet calculations on digital computers at Carnegie Mellon University. This work found its way into Ijiri's dissertation at Carnegie's Graduate School of Industrial Administration, which was widely read in book form as *Management Goals and Accounting for Control* (North Holland, 1965). In Ijiri's Ph.D. thesis, he traces these ideas to Cayley.

2.11. REVIEW

This chapter sets the stage for the use of Excel and for the further study of linear programs. You may have observed connections with high school mathematics (e.g., with linear systems) and with economics (e.g., fixed and variable cost, contribution, and the allocation of scarce resources). These connections will be strengthened in later chapters.

As concerns linear programming, the definitions in this chapter are crucial. Before turning to other chapters on linear programming, review the following terms and be certain that you understand each of them: linear expression, linear constraint, linear program, decision variable, objective, feasible solution, objective value, optimal solution, and optimal value.

The Recreational Vehicle problem is a rudimentary resource allocation problem. Program 2.1 formulates this problem as a linear program. You should understand the constraints of Program 2.1, including the reason why they are inequalities rather than equations.

From the "primer" on Excel, you should understand how to put data, text, and functions into cells. Remember that:

- Each function begins with an "=" sign.
- A leading apostrophe signals Excel that what follows is text.
- Fixed and relative addressing lets you repeat patterns, either by dragging or by using the copy and paste icons.
- The functions in Excel must refer to cells that contain data or to cells whose functions have been assigned values.

From the sections on Solver, you should master the four-step procedure for solving a linear program. Learn how to adapt it to solve equations.

When spreadsheets are described in this text, we employ two notational devices. One of them is to display a keystroke sequence in boldface type, omitting both the **Enter** keystroke at the end and any punctuation that may or may not be part of the keystroke sequence. The other device is to place dotted lines around a cell that displays the value of a function, along with a box that specifies the function.

When building your own spreadsheets, you are exhorted to follow two maxims. One of these is to insert data to check that your functions perform the calculations that you intend. The other maxim is to place all data on the spreadsheet itself and none within functions or in dialog boxes.

Solver can find solutions to systems of nonlinear equations, and it can solve nonlinear optimization problems. But neither Solver nor any other software package can be guaranteed to find a solution to every nonlinear system that has one. When you ask Solver to tackle a nonlinear system, try to "start close," to initialize Solver with values in its changing cells that are close to the solution.

2.12. HOMEWORK AND DISCUSSION PROBLEMS

1. Create a spreadsheet for the Recreational Vehicles problem, solve it, and obtain a sensitivity report. Print your spreadsheet with the optimal solution, and print your sensitivity report. Turn them in.

2. After doing Problem A, change your linear program and rerun it to answer these "what-if" questions:

(a) How do profit and production levels change if the Engine shop capacity is 121 hours rather than 120?

(b) How do profit and production levels change if the Body shop capacity is 81 hours, rather than 80, with the Engine shop capacity at its original value?

(c) How do profit and production levels change if the Body shop capacity is 81 hours and if the Engine shop capacity is 121 hours?

(d) Is there any relation between your answers to parts (a) through (c)? If so, why?

3. Do part (a) of Problem 2. Then redo part (a) with Engine shop capacity of 123 hours. Did the changes triple? Find the highest value of the Engine shop capacity for which changes occur at these rates. What causes the rates to change?

4. Vehicles exist in integer quantities. The optimal solution to Program 2.1 happens to be integer-valued, but that is not guaranteed. If the Engine shop capacity was 121 hours instead of 120, the optimal solution would have been $S = 20.5$, $F = 29.75$, and $L = 0$. Is this solution meaningful? If so, what does it mean?

5. In linear programs, we allow "$\leq$" and "$\geq$" constraints, but neither "$<$" nor "$>$" constraints. Why not? *Hint*: What happens in Program 2.1 if we require $L > 0$?

6. The optimal solution to Program 2.1 can be described in two ways. One way is to make 20 Standard model vehicles, 30 Fancy model vehicles, and 0 Luxury model vehicles. The other is to keep the Engine and Body shops busy making Standard and Fancy model vehicles and to make no Luxury model vehicles. Which description is preferable? Why?

7. (**A Woodworking Shop**) A woodworking shop makes two products, cabinets and tables. Each cabinet sells for \$3100, and each table sells for \$2100. The raw materials that are needed to produce each cabinet and table cost \$1200 and \$700, respectively. The variable labor costs needed to make each cabinet and table are \$1100 and \$900, respectively. The company's carpentry shop has a capacity of 120 hours per week. Its finishing shop has a capacity of 80 hours per week. Making each cabinet requires 20 hours of carpentry and 15 hours of finishing. Making each table requires 10 hours of carpentry and 10 hours of finishing. The company wishes to determine the product mix that maximizes profit.

(a) Create the analog for the woodworking shop of Table 2.1.

(b) Write down a linear program whose optimal solution maximizes profit.

(c) Use Solver to find this optimal solution. Explain why this solution is what it is.

(d) A way has been found to reduce the time needed to finish each cabinet from 15 hours to 12. Reformulate the linear program. Solve it again. Does the optimal solution change? If so, is there anything surprising about the way in which it changes?

8. Create a spreadsheet for Problem B that is akin to Table 2.15. Place a large number, such as 1000, in each of the changing cells. Then ask Solver to find a solution. What happens? Add constraints that require the numbers in the changing cells to be nonnegative. Try again. Any improvement?

9. A system of three equations in three variables appears below. Use Solver to find a solution. *Hint:* The Excel function LN() computes the natural logarithm of a number.

$$\begin{aligned}
3A + 2B + 1C + 5\ln(A) \qquad\qquad\qquad\qquad &= 6\\
2A + 3B + 2C \qquad + 4\ln(B) \qquad\quad &= 5\\
1A + 2B + 3C \qquad\qquad\qquad + 3\ln(C) &= 4
\end{aligned}$$

10. One table in this chapter violates the maxim, "Put all the data on the spreadsheet, none within functions and none in dialog boxes." Which table? How can it be fixed?

Part B

Using Linear Programs

Chapter 3

Analyzing Solutions of Linear Programs

3.1. PREVIEW

In Chapter 2, we saw how to use Solver to compute the optimal solution to a linear program. The optimal solution may sound like the end of an analysis, but it is only the beginning. That is because the linear program itself is a model, an approximation.

We need to ascertain whether or not the "optimal solution" to a linear program is robust and whether it remains optimal or near-optimal over ranges of data that are broad enough to encompass the situation that is being modeled. This is the subject of a sensitivity analysis. One of the beauties of linear programs is that their optimal solutions are accompanied by information that assists in a sensitivity analysis. We also need to understand what insight, if any, the optimal solution to a linear program provides into the situation that the linear program models.

3.2. WHAT CAN YOU LEARN FROM THIS CHAPTER?

This chapter focuses on two related topics, sensitivity analysis and obtaining insight from linear programs, by exploring the following questions, and others as well:

- In what ways does a linear program approximate reality?
- Can the solution to a linear program be described in a way that remains valid when its data are inaccurate? If so, how?
- What use can be made of the Sensitivity Report that Solver provides?
- How can we learn why the optimal solution takes the form that it does?
- What can we learn from geometry about the optimal solution to a linear program?
- Why is it that linear programs accommodate diseconomies of scale?
- Why is it that linear programs fail to accommodate economies of scale?

This chapter relies on and develops uses for high school math, particularly for linear equations and elementary geometry. This chapter also develops connections with economic reasoning, for instance, with "opportunity costs." Studying this chapter can tighten your grasp of high school mathematics, develop your understanding of economic reasoning, and introduce you to the issues that are listed above.

3.3. THE RECREATIONAL VEHICLE PROBLEM

To introduce sensitivity analysis, we return to the Recreational Vehicle problem, which was introduced in Chapter 2 and is, for convenient reference, recapitulated here as:

Problem A (Recreational Vehicles)

A manufacturing facility consists of five shops that are used to make three different vehicles. Table 3.1 names the vehicles and the shops. This table specifies the manufacturing

Table 3.1 Shop capacities (in hours per week), manufacturing times (in hours per vehicle), and contribution of each vehicle that is made.

		Manufacturing times		
Shop	Capacity	Standard	Fancy	Luxury
Engine	120	3	2	1
Body	80	1	2	3
Standard finishing	96	2		
Fancy finishing	102		3	
Luxury finishing	40			2
Contribution		$840	$1120	$1200

time of each vehicle in each shop, the capacity of each shop, and the contribution (profit) earned by making each vehicle. The plant manager wishes to learn what production mix maximizes the profit that she can obtain from this plant.

In Chapter 2, we formulated her profit maximization problem as a linear program. Its decision variables are:

S = the number of Standard model vehicles made per week,
F = the number of Fancy model vehicles made per week, and
L = the number of Luxury model vehicles made per week.

This linear program is:

Program 3.1: Maximize $\{840\,S + 1120\,F + 1200\,L\}$, subject to the constraints

$$\begin{array}{llllll}
\textit{Engine:} & 3\,S\,+ & 2\,F\,+ & 1\,L & \le & 120, \\
\textit{Body:} & 1\,S\,+ & 2\,F\,+ & 3\,L & \le & 80, \\
\textit{Standard finishing:} & 2\,S & & & \le & 96, \\
\textit{Fancy finishing:} & & 3\,F & & \le & 102, \\
\textit{Luxury finishing:} & & & 2\,L & \le & 40, \\
 & S \ge 0, & F \ge 0, & L \ge 0. & &
\end{array}$$

Solution on a Spreadsheet

In Chapter 2, we solved this linear program on a spreadsheet. Table 3.2 recapitulates that computation. Included in the table are a spreadsheet, a facsimile of a *Solver Parameters* dialog box, and the optimal solution that Solver obtained.

As is indicated in Table 3.2, column E contains the usual SUMPRODUCT functions. The values of the decision variables are in cells B9, C9, and D9. Solver maximizes the contribution, subject to familiar constraints. Its optimal solution sets $S = 20$, $F = 30$, and $L = 0$. It has an objective value of 50,400. Three of the cells in Table 3.2 have been shaded, and we will soon see why.

Table 3.2 A spreadsheet for Program 3.1, a facsimile of its *Solver Parameters* dialog box, and its optimal solution.

	A	B	C	D	E	F	G
1							
2	shop	S	F	L			
3	Engine	3	2	1	120	<=	120
4	Body	1	2	3	80	<=	80
5	Standard Finishing	2			40	<=	96
6	Fancy Finishing		3		90	<=	102
7	Luxury Finishing			2	0	<=	40
8	contribution	840	1120	1200	50400		
9	value of variable	20	30	0			
10							
11		=SUMPRODUCT(B8:D8,B$9:D$9)					

Maximize the number in cell E8 with B9:D9 as changing cells, subject to B9:D9 ≥ 0, E3:E8 ≤ G3:G8.

Slack and Tight Constraints

In any feasible solution to a linear program, an inequality constraint is said to be **tight** if it is satisfied as an equation and to be **slack** if it is satisfied as a strict inequality.

Program 3.1 illustrates these definitions. It has a total of eight inequality constraints, namely, the capacity constraint on each of five shops and the nonnegativity constraint on each of three variables. Column E of Table 3.2 reports the amount of each capacity that is actually used by the optimal solution. Evidently, the capacity constraints on the Engine and Body shops are tight. The other three capacity constraints are slack. The nonnegativity constraint on L is also tight. The other two nonnegativity constraints are slack.

In Table 3.2 and henceforth, an inequality constraint is **shaded** if the optimal solution causes it to be tight. It is the tight (shaded) constraints that will help us to relate the optimal solution to the situation that the linear program models.

3.4. THE PERTURBATION THEOREM

A linear program is a model; as such, it is an inexact representation of reality. Can we describe the optimal solution to a linear program in a way that can be implemented when the data are inexact? Yes. The "Perturbation Theorem" tells how.[1]

> **Perturbation Theorem.** If the data of a linear program are perturbed by small amounts, its optimal solution can change, but its tight constraints stay tight, and its slack constraints stay slack.

The Perturbation Theorem confirms our intuition. To see why, let us ask ourselves, "What is actually meant by the optimal solution to a linear program?" There are two possible answers, in the context of the Recreational Vehicle example.

- Make Standard model vehicles at the rate of 20 per week, make Fancy model vehicles at the rate of 30 per week, and make no Luxury model vehicles.
- Keep the Engine and Body shops busy making Standard and Fancy model vehicles, and make no Luxury model vehicles.

If the model is exactly correct, these answers are identical to each other. The first answer sets $S = 20$, $F = 30$, and $L = 0$, independent of the data. If the data are a little off, the first answer may not be feasible, and if it is feasible, it may not be optimal. It is the second answer that the shop manager can implement when the data are inexact. The Perturbation Theorem assures us that the second answer stays feasible and optimal for a range of data. Toward the end of this chapter, we'll develop our geometric intuition as to why the Perturbation Theorem is true.

3.5. HOW A LINEAR PROGRAM APPROXIMATES REALITY

The Recreational Vehicle example illustrates the principal ways in which linear programs approximate reality—uncertainty, aggregation, and linearization.

This model's data are **uncertain** because they cannot be measured precisely and because they can fluctuate in unpredictable ways. For instance, it is presumed that, in a

[1] In an unpleasant situation that is known as **degeneracy**, the Perturbation Theorem needs to be qualified in a way that we will discuss later in this chapter. A proof of this theorem is provided in Chapter 18.

"representative" week, 120 machine hours are available in the Engine shop; 120 equals the number of machine hours during which the Engine shop is open for business less allowances for:

- routine maintenance,
- machine breakdowns,
- shortages of vital parts,
- absence of key workers, and
- power failures, and other unforeseen events.

The actual number of machine hours available in a particular week could be larger or smaller than 120, depending on how things turned out that week. Similarly, the contribution of $840 per Standard model approximates an uncertain quantity. The actual contribution could be larger or smaller than this figure, depending on

- the current prices of raw materials,
- defects that require abnormal amounts of rework,
- market conditions that affect the sales revenues, and
- changes in inventory carrying costs.

Uncertainty in the data is one reason why models are approximate.

Models are also approximate because of **aggregation**, which refers to the lumping together of several activities in a single entity. The assembly times in the Recreational Vehicle model reflect aggregation. The Engine shop is modeled as a single entity, but it is actually a system consisting of people, tools, and machines that can produce the engines and drive trains for the three vehicles at different rates. In our simplified view, it takes three hours of Engine shop time to make each Standard model vehicle. Aggregation is useful when it avoids unimportant detail.

Linearization is the third way in which our model is approximate. The capacity constraint for the Engine shop is $2\,S + 2\,F + 1\,L \leq 120$. This constraint presumes linear interactions among the three types of vehicles that are produced there. The actual interactions are more complicated and are somewhat nonlinear. For example, this constraint accounts crudely for the setup times that are needed to change from the production of one model to another.

The Recreational Vehicle example is aggregated and simplified; it has to be. Imagine how intractable this model would become if it incorporated all of the complexities and details just mentioned. Yet, there is merit in starting with a simple and aggregated model. It will be relatively easy to build and debug. And if it is artfully built, its simplicity will cause the main insights to stand out starkly.

We would be foolish to believe the results of a simplified model without first considering how its simplifications influenced its optimal solution. Is the optimal solution to the linear program **robust**? Equivalently, would it fare well in the real world?

To answer this question, we can execute a **sensitivity analysis**: Change those data that are suspect, rerun the linear program, and measure the changes that occur in its optimal solution and in its optimal value.

One of the beauties of linear programming is that vital aspects of sensitivity analysis can be done without having to rerun the linear program! The Perturbation Theorem lets us measure the changes in the data for which the solution remains optimal. In addition, we will see that the optimal solution is accompanied by a set of "shadow prices" that let us understand why the optimal solution takes the form that it does. In the ensuing sections, we will see that these "shadow prices" can be as informative as the optimal solution itself.

3.6. A SHADOW PRICE AND ITS RANGE

To introduce the concept of a shadow price, we see how the optimal solution to Program 3.1 changes when we perturb one datum, the Engine shop capacity, by a small amount. We replace its capacity of 120 by $120 + d$, where the number d is close to zero. The Perturbation Theorem states that the optimal solution keeps the tight constraints tight, so it satisfies these three equations:

$$\begin{aligned} 3\,S + 2\,F + 1\,L &= 120 + d, \\ 1\,S + 2\,F + 3\,L &= 80, \\ L &= 0. \end{aligned}$$

Without bothering to solve the above equations, you can see that these equations are linear, so their solution varies linearly with d. For specificity, however, we have solved these equations and have gotten

$$S = 20 + 0.5\,d, \qquad F = 30 - 0.25\,d, \qquad L = 0.$$

Since this optimal solution varies linearly with d, the optimal value also varies linearly with d. By substituting the above values of S, F, and L into the objective function, we obtain

$$\begin{aligned} \text{optimal value} &= 840 \times (20 + 0.5\,d) + 1120 \times (30 - 0.25\,d) + 1200 \times 0, \\ &= 50{,}400 + 420\,d - 280\,d, \\ &= 50{,}400 + 140\,d. \end{aligned}$$

Thus, a change of d in the capacity of the Engine shop produces a change of $140\,d$ in the optimal value. This calculation illustrates the "shadow price" of a constraint, which is defined as follows.

> Each contraint's **shadow price** equals the change in the optimal value per unit change in that constraint's right-hand-side value.

We have just shown that 140 is the shadow price of the constraint on Engine shop capacity. We have calculated a shadow price in order to help you understand what they are. You do *not* need to calculate shadow prices by hand. Solver presents the shadow prices as part of its Sensitivity Report.

Unit of Measure

In Chapter 2, we observed that the unit of measure of the objective is dollars per week ($/week) and that the unit of measure of each constraint is hours per week (hours/week). If you are puzzled as to why, please review pages 21–22 of that chapter.

Each constraint's shadow price has a unit of measure, which equals the unit of measure of the objective divided by the unit of measure of the constraint. We emphasize:

> The *unit of measure* of a constraint's shadow price equals the unit of measure of the objective divided by the unit of measure of the constraint.

In Program 3.1, the objective is measured in dollars per week ($/week), and each constraint is measured in hours per week (hours/week). Each shadow price is measured in \$/hour because

$$\frac{\$/\text{week}}{\text{hours/week}} = (\$/\text{week}) \times (\text{weeks/hour}) = \$/\text{hour}$$

In particular, the shadow price for the Engine shop capacity constraint is \$140 per hour.

Breakeven Prices

The shadow prices are **breakeven** prices. The manager of the Recreational Vehicle plant should:

- Augment the Engine shop capacity somewhat if she can do so at a price below \$140 per hour.
- Not augment the Engine shop capacity if she must pay a price above \$140 per hour.
- Decrement (rent out) some Engine shop capacity if she can receive a price above \$140 per hour.
- Not decrement the Engine shop capacity at a price below \$140 per hour.

If she does add or decrease the capacity of the Engine shop, the Perturbation Theorem states that she should adjust the production levels so as to keep the tight constraints tight and the slack constraints slack. But how much capacity can she safely add or subtract at this shadow price?

The Range for Which a Shadow Price Applies

Each shadow price applies over a range of right-hand-side values. Within this range, the perturbed solution keeps the slack constraints slack, and it keeps the tight constraints tight. At each end of a range, a slack constraint becomes tight.

Let us find the range over which the shadow price of the Engine shop capacity constraint equals \$140 per hour. The right-hand-side value of this constraint, originally 120, has been replaced by $120 + d$. We have seen that keeping the tight constraints tight perturbs the production levels as follows:

$$S = 20 + 0.5\,d, \qquad F = 30 - 0.25\,d, \qquad L = 0.$$

The perturbed solution must stay feasible. The production levels must stay nonnegative, and the capacity constraint on each finishing shop must be satisfied. Thus, the amount d by which this capacity is perturbed must be close enough to zero that the perturbed solution satisfies the nonnegativity and capacity constraints

$$\begin{array}{rll} 0 \leq & S = 20 + 0.5\,d, & \text{with equality when } d = -20/0.5 = -40 \\ 0 \leq & F = 30 - 0.25\,d, & \text{with equality when } d = 30/(-0.25) = 120 \\ 96 \geq & 2\,S = 40 + d, & \text{with equality when } d = 56 \\ 102 \geq & 3\,F = 90 - 0.75\,d, & \text{with equality when } d = 12/(-0.75) = -16 \\ 40 \geq & 2\,L = 0, & \text{with inequality for all } d. \end{array}$$

The largest value d for which all of these constraints are satisfied is $d = 56$, and the smallest value of d for which all of these constraints are satisfied is $d = -16$. For $d > 56$ or

$d < -16$, this perturbed solution is not feasible, because it violates at least one of the non-negativity constraints. In this case and in general:

> Each shadow price applies to a **range** of changes in the right-hand-side value of its constraint.
>
> - The end-points of this range are called the **allowable increase** and **allowable decrease**.
> - At each end-point, a slack constraint becomes tight.

For the Engine shop capacity constraint, we have just seen that the allowable increase is 56 and that the allowable decrease is 16. In a later section, we will develop our geometric intuition as to why the shadow price is constant over this range.

3.7. SENSITIVITY TO THE RIGHT-HAND-SIDE VALUES

It is not necessary to compute, by hand, any shadow price or the range of right-hand-side values over which it applies. You can get this information from Solver. Here's how. After Solver computes the optimal solution to a linear program, it presents you with a *Solver Results* dialog box. In that box, select "Sensitivity" and then click on OK. A sheet entitled Sensitivity Report will be generated. This Sensitivity Report consists of two parts—one on the constraints, and the other on the variables. For Program 3.1, its report on the constraints looks like that in Table 3.3.

Table 3.3 reports that the shadow price for Engine shop capacity is 140 and that this price applies for increases of up to 56 hours/week and for decreases of up to 16 hours per week, exactly as we had calculated. This report shows that the shadow price for Body shop capacity constraint is 420, and it provides the range over which that price applies.

Slack Constraints

In the optimal solution, the capacity constraints on the Engine and Body shops are tight, and the capacity constraints on the finishing shops are slack. Table 3.3 reports that each of the finishing shops has 0 as its shadow price. That is no coincidence.

To see why, consider the Standard finishing shop. This shop has 96 hours of capacity, of which 40 are used and 56 are unused. Its capacity is underutilized. Added capacity in this shop has no value. Perturbing its right-hand-side value causes no change in the optimal

Table 3.3 The portion of the Sensitivity Report that concerns constraints.

Constraints

Cell	Name	Final Value	Shadow Price	Constraint R.H. Side	Allowable Increase	Allowable Decrease
$E$3	Engine shop	120	140	120	56	16
$E$4	Body shop	80	420	80	5 1/3	40
$E$5	Standard finishing shop	40	0	96	∞	56
$E$6	Fancy finishing shop	96	0	102	∞	12
$E$7	Luxury finishing shop	0	0	40	∞	40

value. Its shadow price equals zero, and this price applies to all capacity levels above 40 hours. When this shop's capacity is reduced to 40 hours, its capacity constraint, previously slack, becomes tight.

Invariably, if a constraint is slack, a small change in its right-hand-side value cannot affect the optimal value. Its shadow price must equal zero. We emphasize:

> If the optimal solution satisfies a constraint as a strict inequality, that constraint's shadow price equals zero.

Simultaneous Changes

Shadow prices and their ranges had been introduced in terms of changes in individual right-hand-side values. The Perturbation Theorem and the inherent linearity of the system being perturbed guarantee that:

> The shadow prices apply to *simultaneous* changes in the right-hand-side values if and only if these changes are small enough that the solution that keeps the tight constraints stays feasible.

The "allowable increase" and "allowable decrease" apply to changes in individual right-hand-side values. For simultaneous changes, the ranges are somewhat smaller. The fact that shadow prices apply to simultaneous changes will be very useful when we consider the "opportunity cost" of engaging in a new activity.

3.8. OPPORTUNITY COST AND MARGINAL PROFIT

Opportunity cost is fundamental to economic reasoning. Typically, in order to accomplish something new, you must reallocate resources that are productively engaged. The **opportunity cost** of doing something equals the reduction in profit that occurs if you set aside (free up) the resources that are needed to do that new thing.

When resources must be freed up, the "net" profit for doing that thing is *not* equal to its contribution. In this case and in general, the **marginal profit** for doing something equals its contribution less the opportunity cost of the resources that must be freed up to accomplish that thing.

For many of us, the concepts of opportunity cost and marginal profit are a bit nebulous. A linear program makes them crystal clear. Consider, for instance, the Luxury model vehicle. The contribution of one Luxury model vehicle is \$1200. The marginal profit for making one Luxury model vehicle is not \$1200. It is \$1200 *less* the opportunity cost of freeing up the production capacity that is needed to make one Luxury model vehicle. We will see that this opportunity cost is easy to compute on a spreadsheet.

A Spreadsheet

Table 3.4 summarizes a spreadsheet calculation of the opportunity cost and marginal profit of each vehicle. This table reproduces Table 3.2, with one additional column and two extra rows. Column H records the shadow price for each of the five constraints. We will demonstrate that row 10 computes the opportunity cost of one unit of each vehicle and hence that row 11 computes their marginal profits.

Table 3.4 Opportunity costs and marginal profits for Program 3.1.

	A	B	C	D	E	F	G	H	I
1									
2	shop	S	F	L				shadow price	
3	Engine	3	2	1	120	<=	120	140	
4	Body	1	2	3	80	<=	80	420	
5	Standard Finishing	2			40	<=	96	0	
6	Fancy Finishing		3		90	<=	102	0	
7	Luxury Finishing			2	0	<=	40	0	
8	contribution	840	1120	1200	50400				
9	value of variable	20	30	0					
10	opportunity cost	840	1120	1400 ←		=SUMPRODUCT(D3:D7,$H3:$H7)			
11	marginal profit	0	0	-200					
12		↑							
13		=B$8-B$10							

Computing the Opportunity Costs

Let us calculate the opportunity costs in row 10. We first compute the opportunity cost of one Standard model vehicle. From column B of Table 3.4, we see that making one Standard model vehicle consumes three hours of Engine shop capacity (at \$140 per hour), one hour of Body shop capacity (at \$420 per hour), and two hours of Standard finishing shop capacity (at \$0 per hour). The shadow prices apply to simultaneous variations in right-hand-side values. Thus,

$$\begin{aligned}\text{[Opportunity cost of one Standard model vehicle]} &= 3 \times 140 + 1 \times 420 + 2 \times 0,\\ &= 420 + 420 + 0,\\ &= 840.\end{aligned}$$

In a similar way, the data in columns C and D show us that

$$\begin{aligned}\text{[Opportunity cost of one Fancy model vehicle]} &= 2 \times 140 + 2 \times 420 + 3 \times 0,\\ &= 280 + 840 + 0,\\ &= 1120.\end{aligned}$$

$$\begin{aligned}\text{[Opportunity cost of one Luxury model vehicle]} &= 1 \times 140 + 3 \times 420 + 2 \times 0,\\ &= 140 + 1260 + 0,\\ &= 1400.\end{aligned}$$

It should now be clear that the function =SUMPRODUCT(D3:D7, \$D3,\$D7) in cell D10 of Table 3.4 computes the opportunity cost of one Luxury model vehicle. Dragging this function across cells C10 and B10 computes the opportunity costs of one Fancy and one Standard model vehicle.

Marginal profit equals contribution less opportunity cost. For this reason, the function =B\$8-B\$10 in cell B11 computes the marginal profit of one Standard model vehicle. Dragging this function across cells C11 and D11 computes the marginal profit of one Fancy model vehicle and of one Luxury model vehicle.

The Opportunity Cost of a Positive Variable

Table 3.4 exhibits an interesting pattern. The optimal solution sets S and F to positive values; their opportunity costs equal their contributions, and equivalently, their marginal profits equal zero. This is not a coincidence.

To understand why, imagine that the manager of the Recreational Vehicle plant offered to set aside the capacity to make one Standard model vehicle for *you*. It would still be optimal for her to make a total of 20 Standard model vehicles, 19 for the general public and 1 for you. The cost to her of setting aside the resources needed to make your Standard model vehicle must exactly offset the contribution she would have earned by selling that vehicle to the general public. In other words, its contribution equals its opportunity cost.

For the same reason, the marginal profit of one Fancy model vehicle equals zero. This illustrates a principle that holds in general.

> Suppose that the optimal solution to a linear program equates a variable to a positive value. Then:
>
> - This variable's contribution equals the opportunity cost of the resources that would need to be diverted to make one unit of the commodity that it represents.
> - Equivalently, this variable's marginal profit equals zero.

Why Is the Luxury Model Unprofitable?

The optimal solution to Program 3.1 sets $L = 0$, so it is optimal to make no Luxury model vehicles. For this reason, the marginal profit for making one Luxury model vehicle cannot be positive. In cell D11 is executed the computation:

$$\begin{aligned}[\text{Marginal profit for setting } L = 1] &= \$1200 - [\text{Opportunity cost of setting } L = 1]\\ &= \$1200 - \$1400 = -\$200.\end{aligned}$$

Thus, perturbing the optimal solution by setting $L = 1$ reduces profit by \$200. We had already known that the Luxury model vehicle is unprofitable, so it is no surprise that the marginal profit for making one of them is negative.

But the calculation of opportunity cost shows us *why* the Luxury model vehicle is unprofitable. From columns D and H of Table 3.4, we have seen that:

$$\begin{aligned}[\text{Opportunity cost of one Luxury model vehicle}] &= 1 \times 140 + 3 \times 420 + 2 \times 0,\\ &= 140 + 1260 + 0,\\ &= 1400.\end{aligned}$$

It costs a total of \$1400 to set aside the capacity needed to make one Luxury model vehicle. In the Body shop alone, making one Luxury model vehicle has an opportunity cost of \$1260 because $1260 = 3 \times 420$. The Luxury model vehicle is unprofitable because it places a heavy demand on the Body shop (three hours per vehicle), where capacity is particularly valuable (\$420 per hour, versus \$140 per hour for the Engine shop). In this case and in general:

> To learn why the optimal solution sets a variable to zero, use the shadow prices to discover *why* that variable's opportunity cost exceeds its contribution.

3.9. SENSITIVITY TO OBJECTIVE FUNCTION COEFFICIENTS

Until now, we have focused on the sensitivity of the optimal solution to data in the constraints. Let us now turn to the sensitivity of the optimal solution to the coefficients in the objective function. The Sensitivity Report in Solver provides relevant information, which is recorded in Table 3.5.

Table 3.5 The portion of the Sensitivity Report that concerns the objective.

Changing Cells

Cell	Name	Final Value	Reduced Cost	Objective Coefficient	Allowable Increase	Allowable Decrease
\$B\$9	S	20	0	840	200	280
\$C\$9	F	30	0	1120	560	100
\$D\$9	L	0	−200	1200	200	∞

A principle is at work in Table 3.5. Each "allowable increase" and "allowable decrease" specifies an end-point of an interval for which the optimal solution is unchanged. For instance, the optimal solution does not change when the contribution of a Standard model vehicle lies between \$560 (which equals \$840 − \$280) and \$1040 (which equals \$840 + \$200). In Table 3.5 and in general, the following principle holds true.

> The optimal solution is unchanged for a **range** of increases and decreases in an objective coefficient.
>
> - The end-points of this range are called the **allowable increase** and **allowable decrease**.
> - At each end-point, a tie arises as to the optimal solution. Beyond each end-point, a different solution is optimal.

In a later section, we will develop our geometric intuition as to why the above statement is correct. Each "allowable increase" and "allowable decrease" is for a change in a single coefficient. The ranges for simultaneous changes can be somewhat smaller.

The Perturbation Theorem implies much of this. To see why, we perturb the contribution of each Standard model vehicle; we let it be $840 + d$ rather than 840. The Perturbation Theorem tells us that, for values of d that are sufficiently close to zero, the optimal solution keeps the tight constraints tight. But perturbing a coefficient in the objective has no effect on the constraints. The constraints that were tight remain tight, so their solution does not change. Thus, for a range of values of d, the optimal solution to Program 3.1 remains as it was, namely, $S = 20$, $F = 30$, and $L = 0$.

The sharp-eyed reader may have observed that Table 3.5 reports "reduced costs," which have not yet been defined. Judging from their values, the phrase "reduced cost" is synonymous with "marginal profit," which is true for a maximization problem.

> In a *maximization* problem, the **reduced cost** of each variable equals its marginal profit, namely, its contribution less the opportunity cost of the resources that are needed to make one unit of the commodity that it represents.

In a maximization problem, the so-called reduced cost is actually a marginal profit. The term *reduced cost* is deeply rooted in the jargon of operations research. We are stuck with it. Later on, we will see that for a *minimization* problem, the reduced cost of each variable equals its marginal cost—its direct cost plus the opportunity cost of the resources that must be freed up to make one unit of the commodity that it represents.

3.10. COMPUTING THE SHADOW PRICES

This section presents a new way to compute the shadow prices. From a hard-nosed practical viewpoint, this section is entirely superfluous because you already know how to find the shadow prices. You can get them from the Sensitivity Report that Solver provides. Besides, in Section 3.4, we saw how to compute the shadow price of a particular constraint. By repeating that computation, you could compute the shadow price for each constraint.

In another sense, however, this section is insightful. It deepens your understanding of shadow prices and opportunity costs. We will soon see how to compute all of the shadow prices simultaneously. We have, in fact, hinted as to how to do this. We've seen that:

- If a constraint is slack, its shadow price is zero.
- If a variable is positive, its contribution equals its opportunity cost, which means that its reduced cost equals zero.

These rules will do the job.

An Equation System

To see how, let us first give names to the shadow prices in Program 3.1. It has five capacity constraints. Let us designate these shadow prices (unimaginatively) as E, B, SF, LF, and FF, respectively. (Mnemonics are being used here; E is short for the shadow price on the capacity constraint for the Engine shop, for instance.)

The constraints on the finishing shops are slack, so the first rule gives:

$$SF = 0, \qquad FF = 0, \qquad LF = 0.$$

The optimal solution equates the variables S and F to positive values. The second rule is that the reduced cost of S and F equals zero, which gives

$$0 = 840 - [3\,E + 1\,B + 2\,SF],$$
$$0 = 1120 - [2\,E + 2\,B + 3\,FF].$$

Since $SF = FF = 0$, these are two equations in two unknowns; their solution is $E = 140$ and $B = 420$. Wasn't that easy?

A Spreadsheet Computation

Let us compute the shadow prices a second time, this time on a spreadsheet. Table 3.6 contains such a spreadsheet, along with a facsimile of a Solver dialog box and the solution that Solver has computed. In Table 3.6:

- Row 9 records the optimal solution to Program 3.1.
- Column E records the familiar SUMPRODUCT functions.
- Cells H3 through H7 are changing cells. These cells contain the values of the shadow prices that Solver is being used to compute.
- The function in cell D10 computes the reduced cost of L, namely, its contribution less its opportunity cost. Sliding this function across cells C10 and B10 computes the reduced cost of F and of S.
- With a blank objective, Solver is being used to select values of the shadow prices that equate to zero the numbers in the cells H5, H6, H7, B10, and C10, which have been shaded for emphasis. (This invokes the preceding rules.)

Table 3.6 Spreadsheet computation of the shadow prices.

	A	B	C	D	E	F	G	H	I
1								shadow	
2	shop	S	F	L			RHS	price	
3	E	3	2	1	120	<=	120	140	
4	B	1	2	3	80	<=	80	420	
5	SF	2			40	<=	96	0	
6	FF		3		90	<=	102	0	
7	LF			2	0	<=	40	0	
8	contribution	840	1120	1200					
9	value of variable	20	30	0					
10	reduced cost	0	0	-200					
11									
12									
13	= D8 - SUMPRODUCT(D3:D7, $H3:$H7)								

With a blank target cell, Solver has H3:H7 as changing cells, and it enforces the constraints H5:H7 = 0 and B10:C10 = 0

- Solver has found the now-familiar solution, one that sets $E = 140$, $B = 420$, and $SF = FF = LF = 0$.

Table 3.6 is prescient. In Chapter 18, we will unearth a connection between the simplex method and Table 3.6. In Chapter 19, we will see that Table 3.6 presents the optimal solutions to two linear programs, the one under attack (Program 3.1) and its "dual."

3.11. SHADOW PRICES AND WHAT-IF QUESTIONS

The following "What-If" questions can be answered by use of the shadow prices.

An Alternate Use of Capacity?

Suppose that someone offers the manager of the Recreational Vehicle facility company \$1000 to rent five hours of its Engine shop capacity for one week. Is it profitable for her to accept this offer? If so, how profitable is it?

Table 3.3 reports that the Engine shop's shadow price equals \$140 per hour, and moreover, that this price is valid for decreases of up to 16 hours. Since five hours lies within this limit, the shadow price applies. Net profit for accepting the offer is $\$1000 - \$5 \times 140 = \$300$. Contribution increases by \$300 if she accepts this offer *and* adjusts the production quantities so as to keep the tight constraints tight.

A New Tool?

An engineer proposes a new tool that would ease the manufacture of Luxury model vehicles by reducing the time that they require in the Body shop from 3 hours to 2.6 hours. Should she buy the tool?

If the company owned this tool, the opportunity cost of one Luxury model vehicle would change, as follows:

$$
\begin{aligned}
\text{[Opportunity cost of one Luxury vehicle]} &= 1 \times 140 + 2.6 \times 420 + 2 \times 0 \\
&= 140 + 1092 + 0 \\
&= 1232.
\end{aligned}
$$

Hence, the marginal profit for perturbing the optimal solution by making one Luxury model vehicle would equal $1200 - 1232 = -32$. This tool is therefore not worthwhile, independent of the cost of acquiring it.

An Improved Tool?

The engineer has found an improved design for the same tool. The improved tool would reduce the time required in the Body shop to make each Luxury model vehicle from three hours to two hours. What then?

If the company owned the tool, the opportunity of cost of making one Luxury model vehicle would equal $1 \times 140 + 2 \times 420 + 2 \times 0 = 980$. Perturbing the optimal solution by making one Luxury model vehicle would change profit by $1200 - 980 = 220 > 0$. Luxury model vehicles would become profitable.

In order to see whether the tool is worth the price, the manager would rerun the linear program, observe that contribution increases by \$4400 per week, and judge whether that amount justifies the cost of the tool. These calculations illustrate a general principle.

> If the optimal solution equates a variable to zero, you can use the shadow prices to see how to reduce its opportunity cost below its contribution, thereby making it profitable.

A New Activity?

After a linear program has been run, we often want to see whether it would be profitable to add an activity (a decision variable) that had been omitted from the linear program. Computing the marginal profit of such an activity is known as **pricing out** that activity. Let us price out a new variable for the Recreational Vehicle example.

The marketing department proposes a new vehicle, the Nifty. Making each Nifty requires 2 hours in the Engine Shop, 2.5 hours in the Body shop, and 1.5 hours in the Standard finishing shop. Each Nifty would be sold at a price that yields a contribution of \$1300. Is the Nifty profitable?

To answer this question, we compute the opportunity cost and the marginal profit of one Nifty.

$$[\text{Opportunity cost of one Nifty}] = 2 \times 140 + 2.5 \times 420 + 1.5 \times 0 = 1330,$$
$$[\text{Marginal profit of one Nifty}] = 1300 - 1330 = -30.$$

The Nifty model is not profitable. If its opportunity cost could be reduced by more than \$30, it would become profitable.

We don't need the shadow prices to determine *whether* the Nifty model is profitable. We could find that out by adding the Nifty model to Program 3.1 and rerunning Solver. It's the shadow prices that can teach us *why* the Nifty model is unprofitable.

3.12. INSIGHT

In Chapter 1, we asserted that the primary role of modeling should be *insight*, and not merely numbers. What insight can we glean from the optimal solution to a linear program, and how? These insights arise from a sensitivity analysis. We've seen that:

- The Perturbation Theorem describes the optimal solution in a way that can be implemented when the model is inexact.

- The ranges of on the right-hand-side values and on the objective coefficients give a feel for the extent to which varying these data keeps the solution optimal.
- The shadow prices indicate why this solution takes the form that it does.

In the context of Program 3.1:

- The Perturbation Theorem says that the optimal solution is to keep the Engine and Body shops busy making Standard and Fancy model vehicles and to make no Luxury model vehicles.
- Tables 3.3 and 3.5 show that this solution remains optimal for broad ranges on each right-hand-side value and objective coefficient. It seems to be robust.
- The shadow prices show why the Luxury model vehicle is unprofitable; it places the heaviest demand on the Body shop, where capacity is particularly valuable.
- The shadow prices identify the changes that would cause the Luxury model vehicle to become profitable.

Whenever we use a model, we should ask ourselves what insight it provides into the situation that it represents. If this model is a linear program, an analysis of its Sensitivity Report can provide this insight.

3.13. MINIMIZING COST

Until now, we have focused on profit maximization. Many optimization problems cast themselves naturally as minimization problems. Why is that? One often seeks to achieve a *specific* goal at least expense. For instance:

- A commuter may wish to find the route that gets her home as quickly possible.
- An engineer may wish to complete a project as inexpensively as she can without exceeding the time allotted.

In cases like these, the value of achieving the goal is independent of the way in which it is achieved. It is natural to set aside the value of that goal and to minimize the time or cost of achieving it.

Switching the Sense of Optimization

It is easy to convert minimization and maximization problems into each other. Simply multiply each objective coefficient by -1 and reverse the sense of optimization. This works because:

$$\text{Net revenue} = \text{Income} - \text{Expenditure},$$
$$\text{Net cost} = \text{Expenditure} - \text{Income}.$$

An expense of \$5 is an income of $-$\$5, for instance.

To illustrate, we re-cast Program 3.1 as a minimization problem. To do so, we must multiply each objective coefficient by -1. This transforms a revenue of \$840 into a cost of $-$\$840, for instance. When this Program 3.1 is posed as a minimization problem, its optimal value (smallest net cost) equals $-$\$50,400. In this case and in general, switching the sense of optimization reverses the sign of the optimal value.

But you need not convert a minimization problem into a maximization problem. Solver finds the optimal solutions to minimization problems directly. Because it does so, we must interpret the Sensitivity Report that Solver provides for the case of a minimization problem.

Definition of Shadow Price and of Reduced Cost

For every linear program, Solver uses this definition of shadow price:

> In any linear program—maximization or minimization—each constraint's **shadow price** equals the change in the optimal value per unit change in that constraint's right-hand-side value.

For every linear program, Solver uses the following definition of the reduced cost.

> In any linear program—maximization or minimization—each variable's **reduced cost** has this meaning:
>
> - If the variable is positive, its reduced cost equals zero.
> - If the variable is zero, its reduced cost equals the change in the objective value if the optimal solution is perturbed by equating that variable to 1 and adjusting the values of the other variables so as to keep the tight constraints tight.

The Signs of the Reduced Costs

What can we say about the signs of the reduced costs? If a variable is positive, its reduced cost must equal zero. If a variable equals zero, we cannot improve the objective if we perturb the optimal solution by setting a variable positive. Thus:

- If the optimal solution to a *maximization* problem equates a variable to zero, that variable's reduced cost cannot be positive.
- If the optimal solution to a *minimization* problem equates a variable to zero, that variable's reduced cost cannot be negative.

Reduced Cost?

We recall that, for a maximization problem, each variable's so-called reduced cost that Solver *reports* is actually its *marginal profit*, namely, its contribution less its opportunity cost.

The situation for a minimization problem is a bit different. In a minimization problem, the coefficient of a variable in the objective is called its **direct cost** rather than its "contribution."

Being an economic concept, "opportunity cost" must have the same meaning in maximization and minimization problems. In the language of a minimization problem, the **opportunity cost** of doing something equals the increase in cost that occurs if you set aside the resources that are needed to do that thing. Similarly, in a minimization problem the **marginal cost** of doing something equals its direct cost plus its opportunity cost. We summarize:

- In a *maximization* problem, the reduced cost of each variable equals its marginal profit.
- In a *minimization* problem, the reduced cost of each variable equals its marginal cost.

This usage of English is a bit confused, but there is worse to come.

Confusion!

Throughout this text, we adhere to the definitions of shadow price and reduced cost that are stated above because these are the definitions that Solver employs. But other computer codes employ different conventions about the signs of the shadow prices and of the reduced costs. Solver (and we) may be in the minority.

LINDO@ is an excellent, widely used code that employs a different sign convention. To make LINDO output conform to our definitions, follow this recipe:

- In a maximization problem, accept the shadow prices that LINDO reports, and multiply the reduced costs that it reports by -1.
- In a minimization problem, accept the reduced costs that LINDO reports, and multiply the shadow prices that it reports by -1.

LINDO's sign convention has an advantage and a disadvantage:

- For LINDO (and not for Solver), the phrase "reduced cost" means "marginal cost."
- For Solver (and not for LINDO), the shadow prices are the optimal values of the decision variables in the "dual" linear program that we will study in Chapter 19.

Solver's sign convention meshes better with the theory that underlies linear programming and its generalizations. If you forget to check the "Assume Linear" box in Solver, it will use a code that computes solutions to nonlinear programs. It will report the correct values of the shadow prices, with the correct signs, but they will be labeled "Lagrange multipliers."

Resolving the Confusion

If you are uncertain about the sign conventions of the code that you are using to solve linear programs, test them on simple problems, such as these two:

$$\text{Maximize } \{x\}, \quad \text{subject to} \quad x + y = 1, \quad x \geq 0, \quad y \geq 0.$$
$$\text{Minimize } \{-x\}, \quad \text{subject to} \quad x + y = 1, \quad x \geq 0, \quad y \geq 0.$$

These two linear programs have the same optimal solution, which is $x = 1$ and $y = 0$. For Solver:

- In the maximization problem, the reduced cost of y equals -1 (because setting $y = 1$ reduces the optimal value by 1).
- In the minimization problem, the reduced cost of y equals $+1$ (because setting $y = 1$ increases the optimal value by 1).
- In the maximization problem, the shadow price of the constraint $x + y = 1$ equals 1 (because increasing its right-hand-side value by 1 increases the optimal value by 1).
- In the minimization problem, the shadow price of the constraint $x + y = 1$ equals -1 (because increasing its right-hand-side value by 1 decreases the optimal value by 1).

Whatever sign convention another computer code reveals on this pair of linear programs will apply to all maximization and minimization problems.

3.14. DISECONOMY OF SCALE

Prior sections have dealt almost exclusively with sensitivity analysis. This section introduces a different theme. Linear programs seem to require linearity, but we will soon see that linear programs accommodate a type of nonlinearity that is known as a "diseconomy of scale."

A function that measures total profit is said to exhibit **decreasing marginal return** if the contribution from the marginal (final) unit can decrease with the quantity that is produced but cannot increase with that quantity. Similarly, a function that measures total cost exhibits **increasing marginal cost** if the cost of the marginal (final) unit can increase with quantity but cannot decrease with quantity. Decreasing marginal return and increasing marginal cost manifest a **diseconomy of scale** because the marginal profit (or cost) can only get worse as the scale (quantity) increases. To introduce this subject, we turn our attention to:

Problem B (A Diseconomy of Scale)

The data for the Recreational Vehicle problem are exactly as in Table 3.1, with this exception: Each of the first 12 units of the Standard model vehicle has \$840 as its contribution, and any units in excess of 12 have a contribution of \$500 each. What product mix maximizes contribution?

Figure 3.1 plots total contribution versus the number of Standard model vehicles produced. This function exhibits decreasing marginal return because the return of the final item decreases with quantity.

The Standard model vehicle has become less profitable, and so we might anticipate making fewer of them. Intuitively, it's reasonable to expect to make at least 12 because the contribution of the first 12 is \$840 apiece, as it was before. Table 3.5 shows us that it will be optimal to make fewer than 20 Standard model vehicles because \$500 is below the lower bound of $\$560 = \$840 - \$280$. To see what product mix maximizes contribution, we revise Program 3.1.

Alterations to Program 3.1

We will need to change Program 3.1 in three ways. The first change is to introduce two new decision variables:

$S1$ = the number of Standard model vehicles made per week and sold at a contribution of \$840 each,

$S2$ = the number of Standard model vehicles made per week and sold at a contribution of \$500 each.

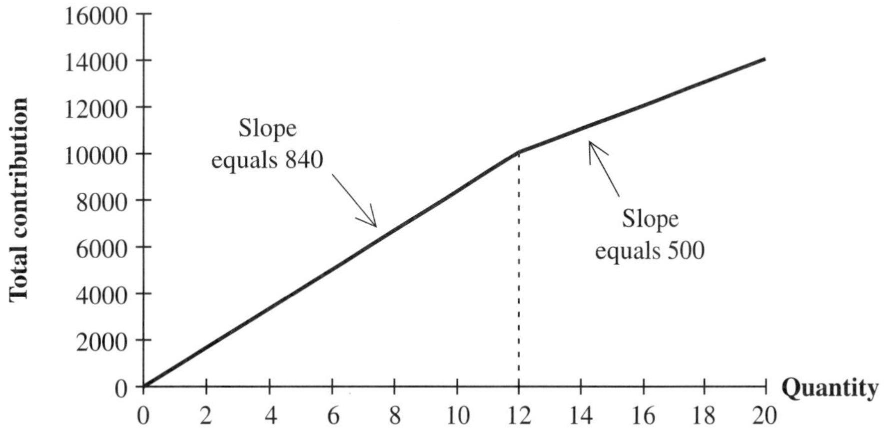

Figure 3.1 Total contribution versus quantity; an example of decreasing marginal return.

The second change appends to Program 3.1 the constraints:

$$S = S1 + S2, \qquad S1 \leq 12, \qquad S1 \geq 0, \qquad S2 \geq 0.$$

The "counting" constraint $S = S1 + S2$ keeps S equal to the total number of Standard model vehicles made each week. The constraint $S1 \leq 12$ imposes the new requirement that no more than 12 vehicles can be sold at the higher contribution of $840.

The third change alters the objective. It replaces the addend $840S$ with $(840\,S1 + 500\,S2)$. These changes produce Program 3.2.

Program 3.2: Maximize $\{0\,S + 840\,S1 + 500\,S2 + 1120\,F + 1200\,L\}$, subject to the constraints

Engine:	$3\,S +$			$2\,F +$	$1\,L \leq$	120,
Body:	$1\,S +$			$2\,F +$	$3\,L \leq$	80,
Standard finishing:	$2\,S$				$\leq$	96,
Fancy finishing:				$3\,F$	$\leq$	102,
Luxury finishing:					$2\,L \leq$	40,
Counting:	S	$-S1$	$-S2$		$=$	0,
Sales limit:		$S1$			$\leq$	12,
	$S \geq 0,$	$S1 \geq 0,$	$S2 \geq 0,$	$F \geq 0,$	$L \geq$	0.

By running Solver, we learn that the optimal solution to Program 3.2 is:

$$S1 = 12, \qquad S2 = 0, \qquad S = 12, \qquad F = 34, \qquad L = 0.$$

The product mix has changed. Now it is optimal to produce Standard model vehicles at the rate of 12 per week and Fancy model vehicles at the rate of 34 per week. The Luxury model vehicles remain unprofitable. The new optimal solution causes the capacities of the Engine and Fancy finishing shops to be fully utilized.

An Unintended Option

Program 3.2 is subtler than it may appear. The constraints in Program 3.2 allow an unintended option. Specifically, they allow $S2$ to be positive while $S1$ is less than 12. In other words, the constraints allow some Standard model vehicles to be sold at a profit of $500, while fewer than 12 are sold at a profit of $840.

This unintended option is ruled out, however, by optimization. In Program 3.2, it is feasible to sell some vehicles at the lower price and fewer than 12 at the higher price, but it is not optimal to do so.

Multiple Diseconomies of Scale

Figure 3.2 displays a different profit function that exhibits decreasing marginal return. Suppose the contribution of the Standard model vehicle is as in Figure 3.1 and the contribution of the Fancy model vehicle is as in Figure 3.2. What happens?

The analog of Program 3.2 works. In these examples and in general, decreasing marginal return introduces unintended options that are ruled out by optimization. It becomes

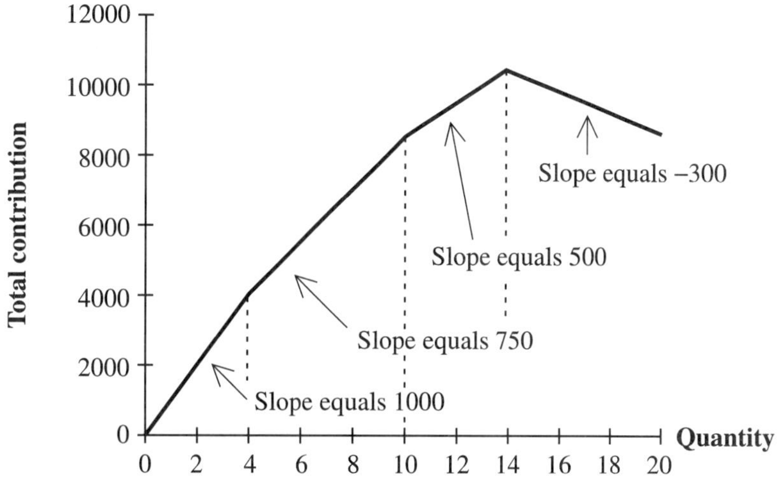

Figure 3.2 Total contribution versus quantity; a case of decreasing marginal return.

feasible to sell at a lower price without selling as many as is possible at a higher price, but it is not optimal to do so.

> A linear program readily accommodates decreasing marginal return, increasing marginal cost, and other diseconomies of scale. Its formulation introduces unintended options, but optimization rules them out.

3.15. NO ECONOMY OF SCALE!

We will soon see that linear programs fail to accommodate a type of nonlinearity that is known as an economy of scale. A function that measures total profit is said to exhibit **increasing marginal return** if the profit from the marginal (final) unit can increase with the quantity that is produced but cannot decrease with that quantity. Similarly, a function that measures total cost exhibits **decreasing marginal cost** if the cost of the marginal (final) unit can decrease with quantity but cannot increase with quantity. Increasing marginal return and decreasing marginal cost manifest an **economy of scale** because the marginal profit (or cost) can only get better as the scale (quantity) increases.

What happens when we attempt to incorporate an economy of scale within a linear program? To explore this issue, we turn our attention to:

Problem C (An Economy of Scale)

The data for the Recreational Vehicle problem are as in Table 3.1, with this exception: Each of the first 12 units of the Standard model vehicle has \$500 as its contribution, and any units in excess of 12 have a contribution of \$800 each. Which product mix maximizes contribution?

Figure 3.3 plots total contribution versus the number S of Standard model vehicles that are made. This graph exhibits increasing marginal return; as the quantity increases, so does the profit from the marginal item.

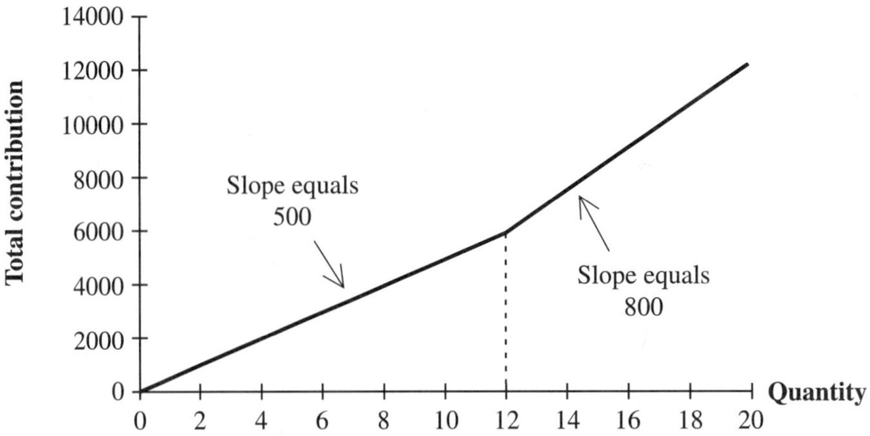

Figure 3.3 Total contribution versus quantity; an example exhibiting increasing marginal return.

Alterations to Program 3.2

Attempting to adapt Program 3.2 to this situation, we designate $S1$ as the number sold at the lower price and $S2$ as the number sold at the higher price. The constraints of Program 3.2 remain as they were, but its objective changes. The linear program becomes:

Program 3.3: Maximize $\{0\,S + 500\,S1 + 800\,S2 + 1120\,F + 1200\,L\}$, subject to the same constraints as in Program 3.2.

In Program 3.3, it is feasible to set $S2$ positive while $S1$ is below 12. In other words, it is feasible to sell at the higher price of \$800 while selling fewer than 12 at the lower price of \$500. In fact, Program 3.3 contains no incentive to sell any at the lower price. Its optimal solution is guaranteed to set $S1 = 0$. That is bad news.

Bad News

When a model exhibits an economy of scale, its formulation as a linear program introduces an unintended option, and this unintended option is selected by optimization rather than being ruled out. We emphasize:

> A linear program fails to accommodate increasing marginal return, decreasing marginal cost, and other economics of scale. Its formulation introduces unintended options, and optimization selects them.

An Integer Program

Rare is the case in which a linear program accommodates an economy of scale. Generally speaking, an economy of scale can only be handled by an "integer program." An **integer program** differs from a linear program in that one or more of its variables is restricted to integer values. Problem 22 (on page 95) sketches a method for handling the economy of scale in Figure 3.3 by introducing a variable whose values are restricted to 0 and 1. A different method is described in Chapter 6, which is devoted to integer programs.

In Solver, it's easy to require variables to be integer-valued. To do so, you need only click on the "int" button in the *Add Constraint* dialog box. The ease with which Solver lets you formulate an integer program is misleading. A large integer program presents a daunting computational burden. It is also true that in an integer program, the shadow prices lose their meaning.

3.16. LARGE CHANGES AND SHADOW PRICES

The capacity of the Engine shop is 120 hours per week. Table 3.3 reports that this shop's capacity constraint has a shadow price of \$140 per hour, with an allowable decrease of 16 and an allowable increase of 56. This price applies for capacity levels in the range between 104 hours (because $104 = 120 - 16$) and 176 hours (because $176 = 120 + 56$).

What happens to the shadow price when the Engine shop capacity goes outside the range for which the shadow price equals \$140 per hour? What do the shadow prices tell us about the effect of large variations on the right-hand-side values?

To study these questions, we use Solver to find optimal solutions to Program 3.1 for Engine shop capacities that are just below 104 and just above 176. Find the range and shadow price in each case, and repeat. The result is plotted in Figure 3.4.

In Figure 3.4, the slope decreases with quantity; this is no accident. A linear program exhibits decreasing marginal return on each capacity (right-hand-side value). On reflection, the reason is clear. When there is only a tiny amount of a particular capacity, that capacity is as profitable as it can be. As this capacity increases, other resources become more fully utilized, slack constraints become tight, and one can make less and less profitable use of the added capacity. In Figure 3.4, for instance, each breakpoint in the slope marks a point at which a slack constraint becomes tight, and added capacity becomes less profitable.

Decreasing marginal return means that, in an important sense, the current shadow prices are the most favorable. Starting with a capacity of 120, small decreases in capacity cost \$140 per unit, larger decreases cost \$165 per unit, still larger increases cost even more, and so forth. Similarly, starting with a capacity of 120, small increases earn \$140 per unit, whereas larger increases earn less. We emphasize:

> The current shadow prices are the *most favorable*; larger increases will be less profitable; and larger decreases will be more costly.

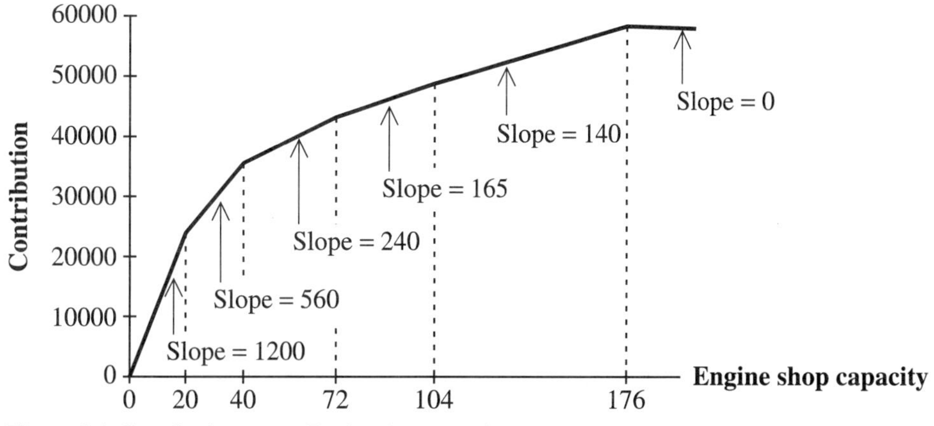

Figure 3.4 Contribution versus Engine shop capacity.

In the jargon of operations research, a plot of profit versus capacity is known as a **parametric analysis**. We have seen that, in a linear program, a parametric analysis on one or more right-hand-side values exhibits a diseconomy of scale.

3.17. LINEAR PROGRAMS AND PLANE GEOMETRY

Our attention now turns to a new topic. In this section and the next, geometry is used to provide insight into linear programs. These sections help you to:

- Visualize the set of feasible solutions to a linear program.
- Strengthen your understanding of the Perturbation Theorem.
- See what determines the "allowable increase" and "allowable decrease" in right-hand-side values and in objective coefficients.
- Learn how degeneracy can qualify the Perturbation Theorem.
- Discover the role that "extreme points" play in linear programming.

We will begin with two-dimensional (planar) geometry. To do so, we turn our attention to a linear program that has only two decision variables. This linear program is:

Program 3.4. Maximize $\{2A + 3B\}$ subject to the constraints

$$\begin{aligned} A \quad\quad &\leq 6, \\ A + \quad B &\leq 7, \\ 2B &\leq 9, \\ -A + 3B &\leq 9, \\ A \quad\quad &\geq 0, \\ B &\geq 0. \end{aligned}$$

As we shall see, the optimal solution to this linear program sets $A = 3$ and $B = 4$.

Cartesian Coordinates

Let us consider how Cartesian coordinates identify each pair (A, B) of real numbers with a point in the plane. Imagine that this page is flat. Imagine also the plane that includes this page but whose extent is infinite. And imagine that you place a dot somewhere on this page; call that dot the **origin**. Then, locate each pair (A, B) of numbers in this plane by this recipe:

- Starting at the origin, walk A units to the right.
- Then walk B units toward the top of the page.
- Place the pair (A, B) at the **point** where you ended.

Cartesian coordinates locate the pair (0, 0) at the origin. Cartesian coordinates locate the pair (3, 1) three units to the right of the origin and 1 unit above it, and so forth.

The Feasible Region

The set of all feasible solutions to a linear program is called its **feasible region**. Figure 3.5 uses Cartesian coordinates to represent this feasible region as the shaded portion of the plane. For instance, the pair (5, 1) lies within the shaded region because it is feasible to set $A = 5$ and $B = 1$.

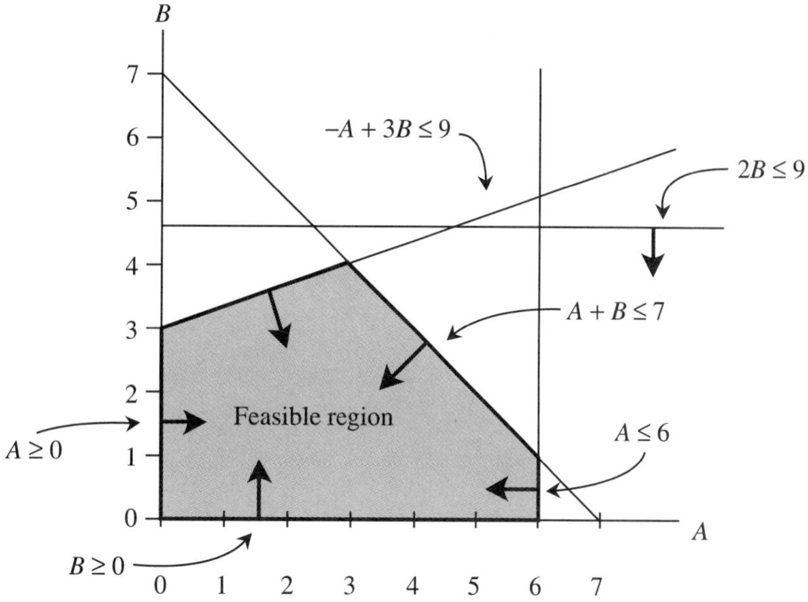

Figure 3.5 The feasible region for Program 3.4.

To indicate how Figure 3.5 was constructed, we begin with a single constraint, namely, with the constraint $A + B \leq 7$. Let us observe that:

- The points (*A*, *B*) that satisfy the constraint $A + B \leq 7$ form a region whose boundary is the line on which $A + B = 7$.
- Two points determine a line. The line $A + B = 7$ includes the points (7, 0) and (0, 7).
- The line $A + B = 7$ partitions the plane into two regions; one region satisfies the constraint $A + B \leq 7$, and the other violates it. To see which is which, we can check any point that is not on the line.
- The point (0, 0) is not on the line $A + B = 7$; since $0 + 0 < 7$, the point (0, 0) is in the region that satisfies the constraint.
- Figure 3.5 includes the line $A + B = 7$, along with a thick arrow that points from this line into the half-plane that satisfies the inequality $A + B \leq 7$, in this case, toward the origin.

Program 3.4 has six inequality constraints. Figure 3.5 contains the line on which each of these constraints holds as an equation, accompanied by a thick arrow pointing into the half-plane that satisfies this constraint as a strict inequality. The feasible region is the intersection of these half-planes, and it is shaded.

Of the six lines in Figure 3.5, one does not affect the feasible region. A constraint in a linear program is said to be **redundant** if its removal would not change the feasible region. Figure 3.5 indicates that the constraint $2B \leq 9$ is redundant.

Iso-profit Lines

Figure 3.5 provides no information about the objective of Program 3.4, which is to maximize the expression $\{2A + 3B\}$. An **iso-profit line** is a line on which profit is constant.

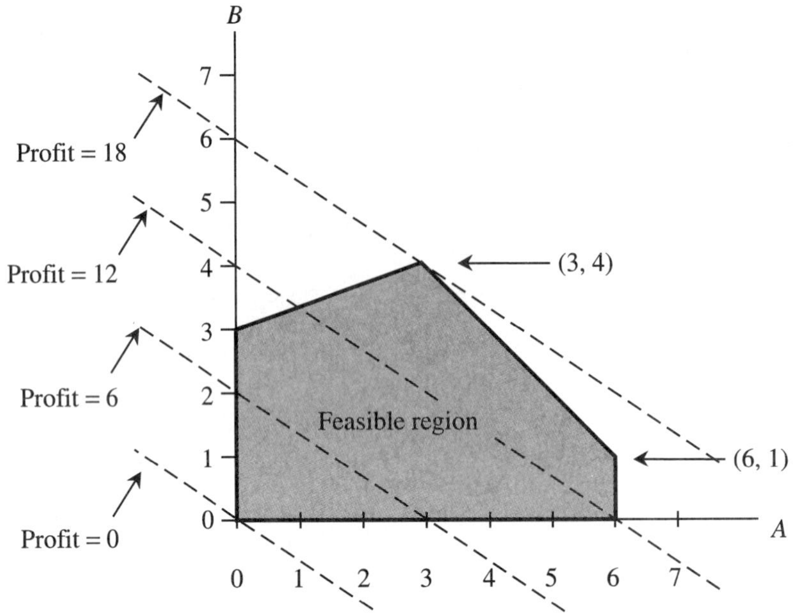

Figure 3.6 Feasible region and iso-profit lines for Program 3.4.

Figure 3.6 displays the feasible region and four iso-profit lines. Its objective, $2A + 3B$, equals 6 on the iso-profit line that contains the points (3, 0) and (0, 2). Similarly, the iso-profit line on which $2A + 3B = 12$ contains the points (6, 0) and (0, 4). In this case and in general, the iso-profit lines are parallel to each other. Notice in Figure 3.6 that the point (3, 4) has a profit of 18 and that no other feasible solution has a profit as large as 18. Thus, $A = 3$ and $B = 4$ is the unique optimal solution to Program 3.4, and 18 is its optimal value.

Objective Vector

There is another way in which to visualize the iso-profit lines. The object of Program 3.4 is to maximize the expression $(2A + 3B)$. The coefficients of A and B in this expression form the **objective vector**, (2, 3). A vector connotes motion. We think of the vector (2, 3) as moving two units to the right of the page and three units toward the top. Figure 3.7 reproduces Figure 3.6, but with the objective vector shown touching the iso-profit line $2A + 3B = 18$.

The objective vector can have its tail "rooted" anywhere in the plane. In Figure 3.7 and in general, the objective vector is *perpendicular* to the iso-profit lines. It's the *direction* in which the objective vector points that matters. In a maximization problem, we seek a feasible solution that lies farthest in the direction of the objective vector. Similarly, in a minimization problem, we seek a feasible solution that lies farthest in the direction that is opposite to the objective vector. We emphasize:

The objective vector points in the direction of increase of the objective.

Extreme Points

The feasible region in Figure 3.7 has five "corners":

(0, 0)   (0, 3)   (3, 4)   (6, 1)   (6, 0)

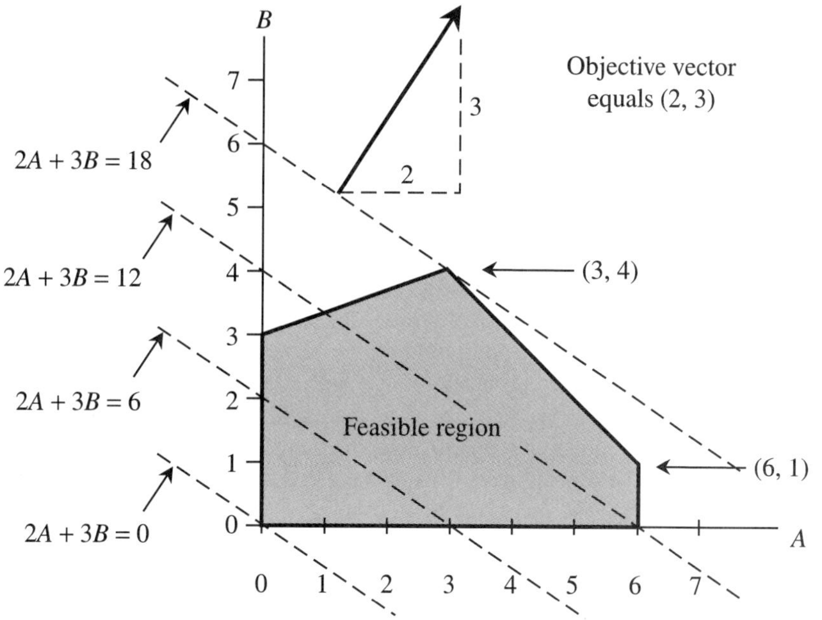

Figure 3.7 Feasible region for Program 3.4, with iso-profit lines and objective vector (2, 3).

Convince yourself that, no matter what the objective, one of these five corners is an optimal solution. Suppose, for instance, that the object of Program 3.4 was to maximize the expression $\{6A + 0.5B\}$. The objective vector and iso-profit lines would tilt clockwise, and the optimal solution would be (6, 1).

In mathematical jargon, the "corners" of the feasible region are known as **extreme points**. In Figure 3.5, each extreme point lies at the intersection of two lines. Each of these extreme points satisfies two of the six constraints as equations, and it satisfies the other four constraints as strict inequalities.

The sharp-eyed reader may have noticed that a precise definition of "extreme point" has not been provided. Since an exact definition would get us into deep water, you must rely on your geometric intuition. The final section of this chapter provides the definition. That section is starred, and it is definitely *optional.*

Multiple Optimal Solutions

More than one extreme point can be optimal. Consider this example: Maximize $\{4A + 0B\}$, subject to the constraints in Program 3.4. The feasible region is exactly the same as that in Figure 3.7, but the objective vector points to the right, and the iso-profit lines have become vertical. The extreme points (6, 1) and (6, 0) are optimal, and so is every point on the line segment that connects (6, 0) with (6, 1). In particular, the point (6, 0.5) is optimal, but it is not an extreme point. In general:

> If two points are optimal, then so is every point on the line segment that connects them.

Evidently, if a linear program has two optimal solutions, it has infinitely many optimal solutions.

The Range of an Objective Coefficient

Geometry can provide insight into the range of an objective coefficient over which the optimal solution does not change. To see how, we pose this question: For Program 3.4, what is the "allowable increase" in the contribution of A. Equivalently, by how much can the contribution of A increase before the optimal solution changes?

To study this question, we change the coefficient of A in the objective from 2 to $2 + d$ where d is a small positive number. As d increases, the objective vector $(2 + d, 3)$ and the iso-profit lines tilt clockwise. When d equals 1, the objective is to maximize $\{3\,A + 3\,B\}$, and the iso-profit lines become parallel to the line $A + B = 7$. For this value of d, there are multiple optimal solutions; the extreme points (3, 4) and (6, 1) are optimal, as is each point on the line segment connecting them. When $d > 1$, the point (6, 1) is the unique optimal solution. Thus, 1 is the allowable increase in the contribution of A.

Similarly, as d decreases, the objective vector $(2 + d, 3)$ and iso-profit lines tilt counterclockwise. When d equals -3, the objective is to maximize $\{-A + 3\,B\}$, and the iso-profit lines become parallel to the line $-A + 3\,B = 9$. For $d = -3$, the extreme points (3, 4) and (0, 3) are optimal.

The "allowable increase" and "allowable decrease" in a coefficient of the objective are the amounts by which that coefficient can be perturbed before a different extreme point becomes optimal.

The Range of a Right-Hand-Side Value

Geometry also provides insight into the range of right-hand-side values for which the optimal solution keeps the tight constraints tight. To see how, we ask ourselves what determines the "allowable increase" in the right-hand-side value of the constraint $A + B \leq 7$ of Program 3.4.

To study this question, we change the constraint from $A + B \leq 7$ to $A + B \leq 7 + d$, where d is a small positive number. Figure 3.8 shows what happens when $d = 1$. The feasible region has been enlarged by the shaded region between the lines $A + B = 7$ and $A + B = 7 + 1$. The objective vector still points in the direction (2, 3), and the iso-profit lines have not changed. The objective vector and the iso-profit lines show that the optimal solution lies at the intersection of the lines

$$-A + 3\,B = 9, \qquad A + B = 7 + d.$$

When $d = 0$, this optimal solution is the point $(A, B) = (3, 4)$. As d increases, the optimal solution shifts from (3, 4) along the dotted line in Figure 3.8. When d is large enough, the previously slack constraint $2\,B \leq 9$ becomes tight. Further increase in the dotted direction would violate this constraint. Thus, the allowable increase equals the value of d for which the perturbed optimal solution causes the constraint $2\,B \leq 9$ to become tight. The value of d for which this occurs satisfies the two equations that are displayed above, and it also satisfies $2\,B = 9$. Solving these three equations gives $B = 4.5, A = 4.5$, and $d = 2$. In brief, the allowable increase equals 2.

Geometry has helped us to understand why the Perturbation Theorem is true. As d varies, the extreme point shifts in order to keep the tight constraints tight. The allowable increase and decrease are values of d that cause previously slack constraints to become tight.

Degeneracy

It was noted earlier that the Perturbation Theorem needs to be qualified in the case of "degeneracy." The qualification is that an allowable increase or an allowable decrease can equal zero. "Degeneracy" comes in two varieties. They are easy to visualize graphically.

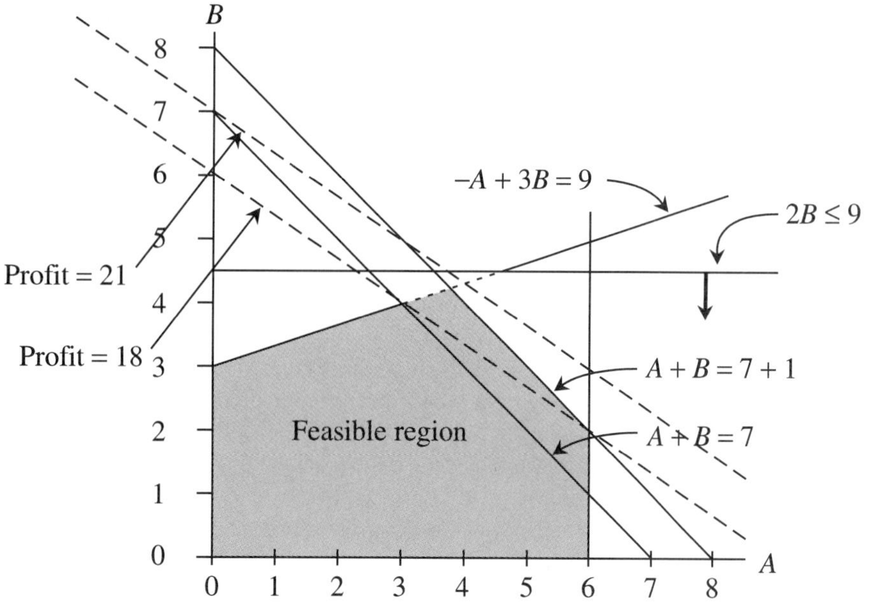

Figure 3.8 Feasible region for Program 3.4 with perturbed constraint $A + B \leq 7 + d$ for the case $d = 1$.

To exhibit one type of degeneracy, we append to Program 3.4 the constraint $B \leq 4$. What now is the "allowable increase" in the right-hand-side value of the constraint $A + B \leq 7$? Increasing this right-hand-side value has no effect on the optimal solution; this constraint's allowable increase equals zero. For the same reason, the allowable increase on the right-hand-side value of the constraint $B \leq 4$ equals zero.

To exhibit the other type of degeneracy, we change the objective of Program 3.4 to maximize $\{3A + 3B\}$. The extreme point (3, 4) remains optimal. Now, the extreme point (6, 1) is also optimal. The simplex method will pivot until it reaches extreme point (3, 4) or extreme point (6, 1). It will stop as soon as it reaches either of these extreme points. If the simplex method ends at extreme point (3, 4), it will report 0 as the "allowable decrease" on the coefficient of B and 0 as the allowable increase in the coefficient of A. Alternatively, if the simplex method stops at extreme point (6, 1), it will report 0 as the allowable decrease in the objective coefficient of A and 0 as the allowable increase on the objective coefficient of B.

These examples illustrate *degeneracy*, which limits the range of perturbation over which the solution that keeps the tight constraints tight remains optimal. When degeneracy occurs, the shadow prices retain part of their meanings as breakeven prices. If the allowable increase of a constraint equals zero, the breakeven price for added capacity cannot exceed that constraint's shadow price, but this breakeven price can be lower than the shadow price.

3.18. LINEAR PROGRAMS AND SOLID GEOMETRY

Plane geometry allows us to visualize the feasible region of a linear program that has only two variables. The Recreational Vehicle problem has three variables, which are S, F, and L. To visualize its feasible region, we need three-dimensional (or solid) geometry. Solid geometry has been familiar since birth, even if it is omitted from typical high school geometry courses; it describes the world we learned to navigate as infants.

Cartesian Coordinates

Cartesian coordinates identify each point in three-dimensional space with a triplet (*A*, *B*, *C*) of real numbers. To do so, imagine that this page is flat. Imagine the plane that includes this page but whose extent is infinite. Imagine space that includes this plane but extends above and below it, without limit. Now, place a dot somewhere on this page, and call that dot the **origin**. Then, locate each triplet (*A*, *B*, *C*) of numbers by this recipe: starting at the origin, walk *A* units toward the right of the page, walk *B* units toward the top of the page, then fly *C* units directly above the page, and locate the triplet (*A*, *B*, *C*) of numbers at the **point** that you reached. This locates the triplet (0, 0, 0) at the origin. It locates the triplet (1, 2, 3) at the point that is one unit to the right of the origin, two units toward the top of the page from the origin, and three units in front of (above) the page and so forth.

In this way, Cartesian coordinates identify every feasible solution (*S*, *F*, *L*) to Program 3.1 with a point in three-dimensional space. The feasible region (set of feasible solutions) becomes the polyhedron in Figure 3.9. For instance, the triplet (40, 0, 0) lies 40 units to the right of the origin, the triplet (0, 34, 0) lies 34 units toward the top of the page from the origin, and the triplet (0, 0, 20) lies 20 units in front of the origin.

The boundary of the feasible region in Figure 3.9 consists of seven "faces." Pause to count these faces. Each face lies in the plane that satisfies one constraint of Program 3.1 as an equation. For instance, the constraint $3\,F \leq 102$ holds as an equation on the small triangular face that has $F = 34$. Similarly, the constraint $2\,L \leq 40$ holds on the triangular face that has $L = 20$. Also, the constraint $3\,S + 2\,F + 1\,L \leq 120$ holds as an equation on the triangular face that includes these points: (20, 30, 0), (35, 0, 15), and (40, 0, 0).

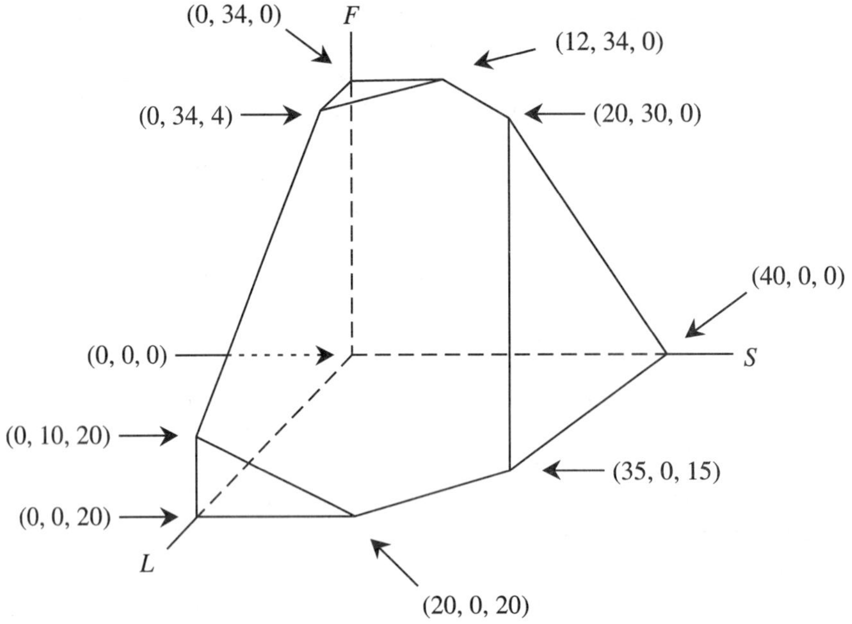

Figure 3.9 Feasible region for Program 3.1, and its extreme points.

Although Program 3.1 has eight inequality constraints, its feasible region has only seven faces. That occurs because the constraint $2\,S \le 96$ is redundant; no feasible solution has $S > 40$.

Extreme Points

The feasible region in Figure 3.9 has 10 extreme points. Each extreme point occurs at a "corner" of the feasible region. All 10 extreme points are identified in Figure 3.9. Each of these extreme points lies at the intersection of three faces, that is, in the planes on which three constraints hold as equations. For instance, the extreme point (20, 30, 0) lies in the plane $L = 0$, in the plane $3\,S + 2\,F + 1\,L = 120$, and in the plane $1\,S + 2\,F + 3\,L = 80$.

Figure 3.9 omits the objective, $\{840\,S + 1120\,F + 1200\,L\}$. This objective is constant on a plane, not on a line. These iso-profit planes are parallel to each other, and each of them is perpendicular to the objective vector (840, 1120, 1200). For this particular objective, the extreme point (20, 30, 0) will be optimal. No matter what the objective, one of the extreme points will be an optimal solution.

Extreme Points and the Simplex Method

The simplex method was devised in 1947 by George B. Dantzig, who is correctly regarded as the father of constrained optimization. A half-century later, it remains the principal tool for computing solutions to linear programs. Barring weird examples (such as Problem 17 on page 94), the simplex method "pivots" from extreme point to extreme point, improving the objective value with each pivot and stopping when it identifies an extreme point whose objective value cannot be improved upon.

A linear program can have many extreme points. Yet, for nearly every sort of linear program, the simplex method finds an optimal extreme point within $3\,m$ "pivots," where m is the number of constraints. Why the simplex method is so fast remains a bit mysterious more than 50 years after Dantzig devised it.

In Chapter 4, we will see that for certain types of linear programs, we want not merely an optimal solution but an extreme point. For these linear programs, Dantzig's simplex method has an inherent advantage over algorithms that do not pivot.

Watching Solver Pivot

When the simplex method is applied to Program 3.1, it pivots from extreme point to extreme point, improving the objective with each pivot and stopping when it encounters the extreme point (20, 30, 0). Solver allows you to watch this happen. To see it happen, do as follows:

- Set Program 3.1 up for solution by Solver.
- On the *Solver Parameters* dialog box, click on the Options button. Then on the *Solver Options* dialog box that appears:
  - Click on the Assume Linear Model button.
  - Click on the Show Iteration Results button.
  - Click on the OK button.
- When the *Solver Parameters* dialog box reappears, click on its Solve button.

Solver will begin at the extreme point (0, 0, 0). Each pivot will bring it to a new extreme point of Figure 3.9, and each extreme point will improve on the objective value of the prior extreme points. Solver will eventually reach extreme point (20, 30, 0), where it will stop. Try it, and see what happens.

Higher Dimensions

Can geometry be relevant to linear programs? Plane geometry deals with only two variables, and solid geometry with three. A linear program can easily have dozens of variables or even thousands. Plane geometry and solid geometry are easy to visualize. Luckily, results that hold for plane geometry *and* for solid geometry tend to remain valid when there are many variables. That is why geometry is relevant to linear programs. In general, the simplex method pivots from extreme point to extreme point, as we have seen.

3.19. A TAXONOMY OF LINEAR PROGRAMS

We have seen that the feasible region for Program 3.1 is bounded and that one of its extreme points is its optimal solution. The same is true of Program 3.4, but not for all linear programs. In this section, the special cases are cataloged.

A linear program is said to be **feasible** if it has at least one feasible solution. Similarly, a linear program is **infeasible** if it has no feasible solution. To construct an infeasible linear program, append to Program 3.4 the constraint $B \geq 5$. Infeasible linear programs do arise in practice. They represent situations that are so tightly constrained that they have no feasible solutions whatsoever.

A linear program is said to be **bounded** if it has at least one feasible solution and if its optimal value is finite. Similarly, a linear program is **unbounded** if it is feasible and if its optimal value is infinite. To exhibit an unbounded linear program, delete from Program 3.4 the constraints $A + B \leq 7$ and $A \leq 6$. An unbounded linear program is unlikely to occur in practice; one cannot obtain an infinite amount of anything that is desirable.

A linear program is said to have **multiple optima** if it has more than one optimal solution. To exhibit a linear program that has multiple optima, change the objective of Program 3.4 to $\{3A + 3B\}$. With this change, the points (3, 4) and (6, 1) represent optimal solutions, as does any point on the line segment that connects them. All of these points have the same objective value, which is 21.

A linear program is said to have a **bounded** feasible region if a number M exists such that, in every feasible solution, the absolute value of each decision variable does not exceed M. Programs 3.1 and 3.4 have bounded feasible regions. A linear program has an **unbounded** feasible region if no such M exists. Clearly, an unbounded linear program has an unbounded feasible region. But must a bounded linear program have a bounded feasible region? No. Consider this example: Minimize A, subject to the constraint $A \geq 0$.

Dantzig's simplex method solves *all* linear programs. The simplex method determines whether or not a linear program is feasible. If a linear program is feasible, the simplex method determines whether or not it is bounded. If a linear program is feasible and bounded, the simplex method finds an optimal solution. If there exists an optimal solution that is an extreme point, the simplex method finds one. If no optimal solution is an extreme point, the simplex method "fixes" the formulation so that an optimal solution is an extreme point, and it finds one.

3.20. GEOMETRY AND CONVEXITY*

This section[2] probes more deeply into the geometry of linear programs. In this section, we will:

- See that the feasible region of every linear program is a "convex set."
- Provide a more precise definition of an "extreme point."

[2] This section is advanced, and it is not needed for a basic understanding of linear programs. It can safely be skipped, skimmed, or postponed.

- Relate the economic concepts of increasing marginal cost and decreasing marginal return to "convex" and "concave" functions.
- Use convexity to describe the types of nonlinear programs for which Solver can be expected to find an optimal solution.

Let us begin with an idea from Euclidean geometry. It is that of the "interval" (segment of a straight line) between two points. To state this idea algebraically, we first introduce vector notation and vector arithmetic.

Vectors

Each real number x is a member of the set $\Re$ of all real numbers. Long ago, in grade school, you identified $\Re$ with the real number line.

Each pair (x_1, x_2) of real numbers is a member of the set $\Re^2$ of ordered pairs of real numbers. In Figure 3.5 (on page 73), we saw how Cartesian coordinates identify $\Re^2$ with the plane. For instance, with the pair (0, 0) located somewhere on this page, the pair (3, −2) lies three units to its right and two units toward the bottom of the page.

Similarly, each triplet (x_1, x_2, x_3) of real numbers is a member of the set $\Re^3$ of ordered triplets of real numbers. In Figure 3.9, we saw how Cartesian coordinates identify $\Re^3$ with three-dimensional space. For instance, with the triplet (0, 0, 0) placed somewhere on this page, the triplet (1, 2, 3) lies one unit to its right, two units toward the top of the page, and three units above the page.

In general, each n-tuple $(x_1, x_2, \ldots, x_n)$ of real numbers is a member of the set $\Re^n$ of all ordered n-tuples of real numbers. Relax! There will be no need to visualize $\Re^n$ because we are proceeding by analogy with plane and solid geometry. The set $\Re^n$ is called a **vector space**, and each element x of $\Re^n$ is called a **vector**.

Our initial goal is to describe the interval between two vectors in $\Re^n$. To achieve this goal, we will specify two operations—the addition of two vectors and the multiplication of a vector by a scalar.

Vector Addition

Let $x = (x_1, x_2, \ldots, x_n)$ and $y = (y_1, y_2, \ldots, y_n)$ be two vectors that have the same number n of components. The **sum**, $x + y$, of the vectors x and y is defined by

$$x + y = (x_1 + y_1, x_2 + y_2, \ldots, x_n + y_n).$$

Evidently, to take the sum of two vectors, add their components. Figure 3.10 uses Cartesian coordinates to portray the vectors (5, 1) and (−2, 3), along with their sum, which is the vector

$$(5, 1) + (-2, 3) = (5 - 2, 1 + 3) = (3, 4).$$

The gray lines in Figure 3.10 indicate that, graphically, in order to take the sum of two vectors, we can shift the "tail" of either vector to the head of the other.

Vector addition is no mystery—simply add the components. This is true of vectors in $\Re^2$, in $\Re^3$, and in higher-dimensional spaces that we cannot visualize.

Scalar Multiplication

If $x = (x_1, x_2, \ldots, x_n)$ is a vector and if c is a real number, the **scalar multiple** of x and c is defined by

$$c\,x = (c\,x_1, c\,x_2, \ldots, c\,x_n).$$

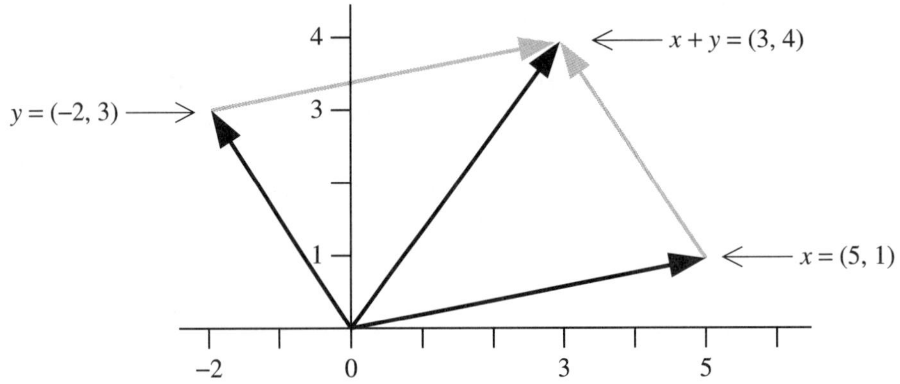

Figure 3.10 The sum $x + y$ of the vectors $x = (5, 1)$ and $y = (-2, 3)$.

Evidently, to multiple a vector x by a scalar c is to multiply each component of x by c. This scalar c can be any real number—positive, negative, or zero.

What happens when we multiply the vector x by the scalar $c = 0.75$? Each entry in x is multiplied by 0.75. This reduces the length of x without changing the direction in which it points.

What happens when we multiply the vector x by the scalar $c = -1$? Each entry in x is multiplied by -1. This reverses the direction in which x points without changing its length.

With y as a vector, we abbreviate the scalar product $(-1)\,y$ to $-y$. With x and y as two vectors that have the same number n of components, we define the **difference**, $x - y$, by

$$x - y = x + (-1)\,y = (x_1 - y_1, x_2 - y_2, \ldots, x_n - y_n).$$

Figure 3.11 plots the difference, $x - y$, of the vectors $x = (5, 1)$ and $y = (-2, 3)$. This difference is given by $x - y = (5 - (-2), 1 - 3) = (7, -2)$.

The gray line in Figure 3.11 indicates that the vector $x - y$ can be drawn as the arrow that points from y to x. This is so because $x = y + (x - y)$.

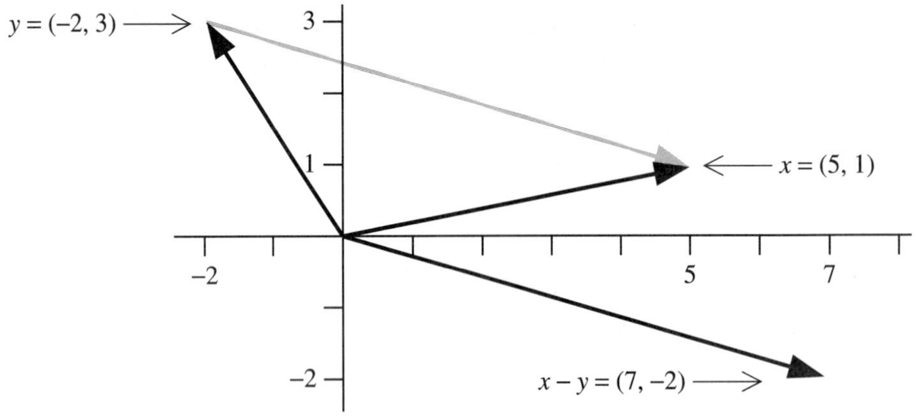

Figure 3.11 The difference $x - y$ of $x = (5, 1)$ and $y = (-2, 3)$.

Intervals

Vector arithmetic, which has just been introduced, lets us describe intervals algebraically. Let $x = (x_1, x_2, \ldots, x_n)$ and $y = (y_1, y_2, \ldots, y_n)$ be two distinct vectors that have the same number n of coordinates. The **interval** between x and y is the set of all vectors of the form

$$c\,x + (1 - c)\,y$$

as the scalar c ranges over all real numbers between 0 and 1 inclusive. This definition may look daunting, but Figure 3.12 will clarify it. In Figure 3.12, the interval between $x = (5, 1)$ and $y = (-2, 3)$ is depicted as the thick gray line segment. Note how closely Figure 3.12 resembles Figure 3.11.

To see why Figure 3.12 is correct, we reorganize the vector sum $c\,x + (1 - c)\,y$ as follows:

$$\begin{aligned} c\,x + (1 - c)\,y &= c\,x + y - c\,y \\ &= y + c\,(x - y). \end{aligned}$$

Evidently, the interval between x and y consists of each point that $y + c\,(x - y)$ that we can reach by adding to y the vector $c\,(x - y)$ as c varies from 0 to 1. Figure 3.12 depicts the vectors x, y, and $x - y$ in black lines. To obtain the vector $c\,(x - y)$, multiply $(x - y)$ by c. To obtain the sum, $y + c\,(x - y)$, place the tail of $c\,(x - y)$ at the head of y. Figure 3.12 depicts this sum for the values $c = 0, 1/4, 1/2, 3/4$, and 1. Hence, in Figure 3.12, the thick gray line displays the interval between x and y.

Intervals have been prominent in this chapter. The feasible region in Figure 3.6 (on page 74) has five edges, each of which is the interval between two extreme points. One of these edges is the interval between the extreme points (3, 4) and (6, 1). Another is the interval between (6, 1) and (6, 0).

Similarly, the feasible region in Figure 3.9 (on page 78) has 15 edges, each of which is the interval between two extreme points. One of these edges is the interval between the extreme points (20, 30, 0) and (12, 34, 0). Another edge is the interval between the extreme points (12, 34, 0) and (0, 34, 0).

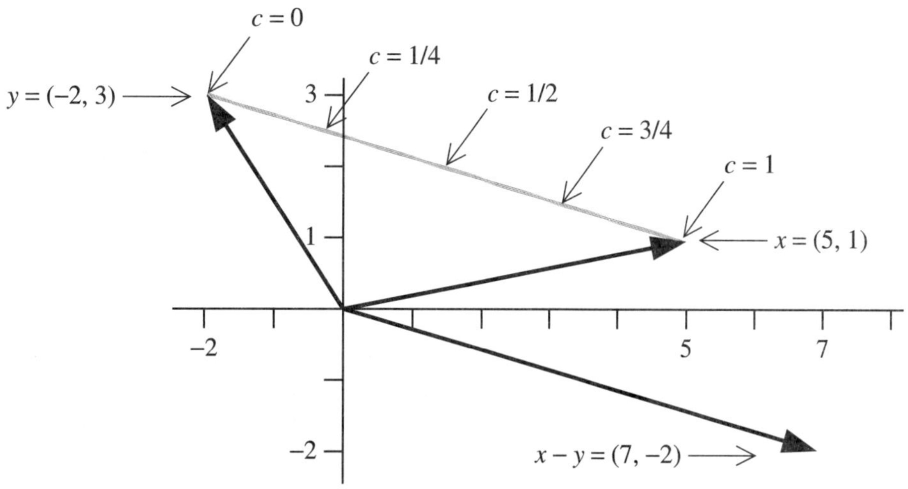

Figure 3.12 The thick gray line segment is the interval between $x = (5, 1)$ and $y = (-2, 3)$.

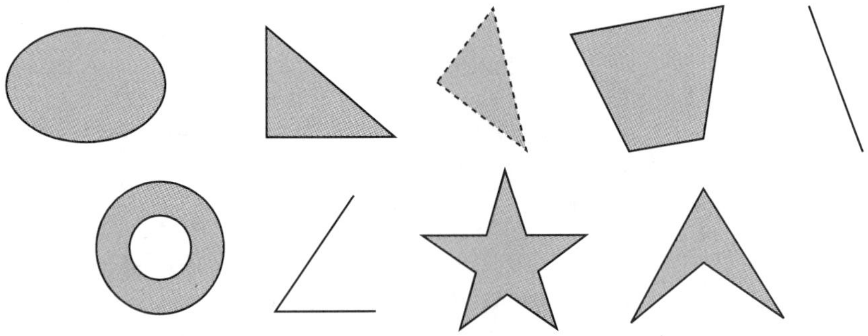

Figure 3.13 Nine sets in the plane. The top five are convex; the bottom four are not.

Convex Sets

With n as some fixed positive integer, let S denote a set of points in $\Re^n$. This set S is said to be **convex** if S contains the interval between each pair of points in S. Figure 3.13 displays nine sets in $\Re^2$ (the plane). Each of these sets is shaded. The top five are convex, and the bottom four are not convex. Can you see why?

Convex sets appeared prominently in this chapter. The feasible region in Figure 3.6 is a convex set in $\Re^2$, the plane; similarly, the feasible region in Figure 3.9 is a convex set in $\Re^3$. The fact that these feasible regions are convex is no accident. It illustrates a property that holds in general and is highlighted below.

> Every linear program has a convex set as its feasible region (set of feasible solutions).

This property is easy to verify from the following two observations:

- The set of solutions to any single linear constraint is a convex set.
- The intersection of convex sets is a convex set.

These two observations are easy enough to prove. If you are curious about that, refer to Problems 25 and 26 on page 96 for clues.

Convex Sets of Real Numbers

What do the convex sets in $\Re$ look like? The interval S between any two given numbers is a convex set. This set remains convex after either or both of its end-points are excluded. For instance, the set $T = \{z : 0 \leq z < 1\}$ is convex. The set $\Re$ is itself convex, and so is the set $S = \{z : 1/3 < z\}$. A set consisting of any single number is convex. On the other hand, no set that consists of two distinct points is convex.

Extreme Points

Earlier, we had relied on your geometric intuition of an extreme point. We can now define an extreme point algebraically. Let x and y be any two distinct points in $\Re^n$. The point z in the interval between x and y is said to be **inside** this interval if z is neither x nor y. For instance, each point inside the interval between (6, 1) and (6, 0) equals (6, c) for a number c that satisfies $0 < c < 1$.

What is special about the extreme points of a convex set S? In particular, for the feasible region in Figure 3.6, what is special about the extreme point (6, 1)? This point is in the feasible region, but it is *not* inside the interval between any two points, both of which are in the feasible region.

A point x in a convex set S is called an **extreme point** of S if x is not inside the interval between any two distinct points, both of which are in S. Please reflect on this definition. Note that for the feasible region S in Figure 3.6, both (6, 1) and (6, 0) are extreme points, but that no point (6, c) with $0 < c < 1$ is an extreme point.

Must a convex set have finitely many extreme points? No. Consider the shaded ellipse in Figure 3.13. Each point on the boundary of this ellipse is an extreme point, and there are infinitely many such points.

Must a convex set have any extreme points? Again the answer is no. The set $S = \{c : 0 < c < 1\}$ is convex, but this set has no extreme points. In fact, a convex set remains convex if we remove its extreme points from it.

A Convex Function

In economics, increasing marginal cost and decreasing marginal cost are familiar concepts. In the mathematical sciences, the same concepts are described in terms of convex functions and concave functions. We now make the connection.

Fix any positive integer n, and let S be a convex set in $\Re^n$. Consider a function f that assigns each point t in S the real number $f(t)$. This function f is said to be a **convex function** if the inequality,

$$f[c\,x + (1 - c)\,y] \le c\,f(x) + (1 - c)\,f(y),$$

holds for every point x in S, every point y in S, and every number c that is between 0 and 1. The definition may seem complicated, but a picture will make everything clear.

Let us do the case $n = 1$. A convex set S of real numbers is a familiar object. A convex function $f(t)$ is drawn in Figure 3.14. In Figure 3.14, the pairs $[x, f(x)]$ and $[y, f(y)]$ are points in the plane. The interval (line segment) that connects the points $[x, f(x)]$ and $[y, f(y)]$ is called a **chord** of the function $f(t)$. With this chord in view, we consider any point $z = c\,x + (1 - c)\,y$ where c is some number between 0 and 1. This point z is in the interval between x and y. To locate z, add to y the scalar c times $(x - y)$. The value of the function at z equals $f(z)$, and the height of the chord above z equals $[c\,f(x) + (1 - c)\,f(y)]$. The definition of a convex function requires $f(z) \le c\,f(x) + (1 - c)\,f(y)$. In brief:

> A function is convex if it lies on or below each of its chords.

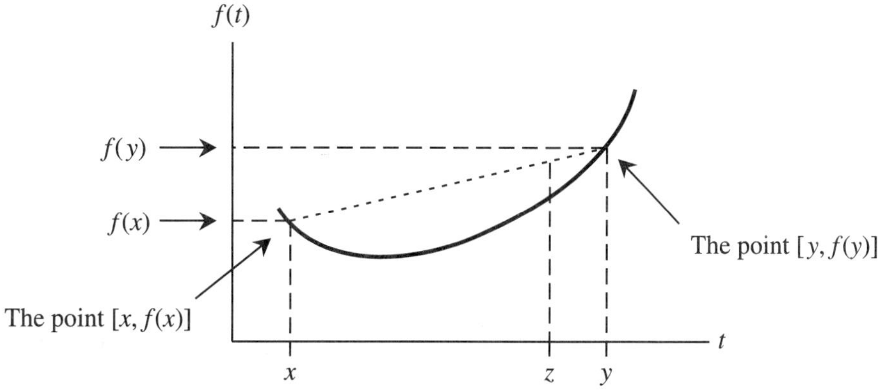

Figure 3.14 The convex function $f(t)$ and one of its chords, the latter as a dotted line.

A convex function curves upward, if at all. As a memory aide, try the jingle, "*e* to the *x* is convex." The inequality that defines a convex function need not hold strictly. A linear function is convex, for instance.

If x is a quantity and $f(x)$ is the cost of x units, then convexity of f is the same as increasing marginal cost. Similarly, if x is a quantity and $f(x)$ is the profit from x units, then convexity of f is the same as increasing marginal profit.

A Concave Function

Fix any positive integer n, and let S be a convex set in $\Re^n$. Consider a function f that assigns each point t in S the real number $f(t)$. This function f is said to be a **concave function** if the inequality,

$$f[c\,x + (1 - c)\,y] \geq c\,f(x) + (1 - c)\,f(y),$$

holds for every point x in S, every point y in S, and every number c that is between 0 and 1. The only difference between concave and convex functions is that the inequality is reversed. Evidently, a function $f(t)$ is convex if and only if $-f(t)$ is concave. For the same reason:

> A function is concave if it lies on or above each of its chords.

A concave function curves downward, if at all. The inequality that defines a concave function need not hold strictly. A linear function is both concave and convex.

If x is a quantity and $f(x)$ is the cost of x units, then the convexity of f is the same as decreasing marginal cost. Similarly, if x is a quantity and $f(x)$ is the profit from x units, then the concavity of f is the same as decreasing marginal profit.

A Convex Program

The entire apparatus of linear programming generalizes to a class of nonlinear programs, as evidenced by Solver. If you neglect to check off the "Assume Linear" option, you will get the correct optimal solution and the correct shadow prices (which will be called **Lagrange multipliers**) and the correct reduced costs (which will be called **reduced gradients**). A "convex program" can be identified as follows:

> A **convex program** takes one of these two forms:
>
> - Minimize a convex function $f(x)$ over all points x in a convex set S.
> - Maximize a concave function $f(x)$ over all points x in a convex set S.

A linear program is a convex program because its feasible region is convex and because its objective is both convex and concave.

Solver and Convex Programs

Solver contains two methods for finding solutions to nonlinear programs. Does Solver succeed? We can give a qualified answer. If the nonlinear program is convex and if it has an optimal solution, Solver can be expected to find one. Loosely speaking, this is so because :

- The "local" behavior of a nonlinear program is approximately that in which its functions are approximated by their linear derivatives.
- The reduced gradients determine either (a) that a feasible solution is a "local" optimum or (b) a "direction" that preserves feasibility and improves the objective value.
- For a convex program, a "local" optimum is guaranteed to be a "global" optimum.

Must the optimal solution to a convex program occur at an extreme point? No. Consider this example: Minimize (x^2), subject to $-1 \leq x \leq +1$.

What happens if Solver is applied to a nonlinear program that is not convex? There are no guarantees. Solver may find a global optimum. It may find a local optimum that is not a global optimum, and it may fail to converge.

Tests for Convex and Concave Functions

How can we tell whether or not a particular function is convex? This question opens a vast topic, which we shall only glimpse in this chapter.

Figure 3.14 suggests the answer for a function of *one* variable. Note that the derivative (slope) of the function $f(t)$ in Figure 3.14 is increasing, and moreover, that its second derivative is nonnegative. This connotes (correctly—we omit the proofs) that:

- A function $f(x)$ that is defined on a convex subset S of real numbers is convex if it is differentiable and if its derivative is nondecreasing on S.
- A function $f(x)$ that is defined on a convex subset S of real numbers is convex if it is twice differentiable and if its second derivative is nonnegative on S.

For instance, the function $f(x) = -x^3$ is convex on the set S of nonpositive numbers because its second derivative $f''(x)$ equals $-6x$, which is nonnegative for $x \leq 0$.

It is equally easy to check whether a function $f(t)$ is concave on a convex set S of real numbers. We know that $f(t)$ is concave if and only if $-f(t)$ is convex. Consequently:

- A function $f(x)$ that is defined on a convex subset S of real numbers is concave if it is differentiable and if its derivative is nonincreasing on S.
- A function $f(x)$ that is defined on a convex subset S of real numbers is concave if it is twice differentiable and if its second derivative is nonpositive on S.

But can we test whether a function of *several* variables is convex? To demonstrate convexity, we must show that the function lies on or below each of its chords. Without offering any evidence, we comment that this question is relatively easy to answer if the function is quadratic. An example of a quadratic function is

$$f(x_1, x_2) = 2(x_1)^2 - 2x_1x_2 + (x_2)^2$$

With x_1 fixed, this is a convex function of x_2 because its second derivative with respect to x_2 equals 2, which is nonnegative. Similarly, with x_2 fixed, this is a convex function of x_1 because its second derivative with respect to x_1 equals 4, which is nonnegative. We've shown that this function lies below the chord on which one of its variable changes and the other is fixed. But we need to show that this function lies on or below *all* of its chords. That turns out to be easy, but it entails matrix algebra that lies beyond the scope of this chapter.

3.21. REVIEW

The focal point of this chapter has been the analysis of the optimal solution to a linear program. Our review of this material is organized under several headings.

Nomenclature

This chapter has introduced terms and phrases with which you should be familiar: Perturbation Theorem, shadow price, opportunity cost, reduced cost, decreasing marginal return, increasing marginal cost, economy of scale, diseconomy of scale, feasible region, extreme point, degeneracy, infeasible linear program, and bounded linear program.

Sensitivity Analysis and Insight

Like any model, a linear program is a stylized representation of reality. A sensitivity analysis should be conducted to determine what insight the linear program provides into the situation that is being modeled.

The optimal solution to a linear program causes certain of its inequality constraints to be slack and others to be tight. The Perturbation Theorem states that the solution that keeps the tight constraints tight remains optimal if the data are somewhat inaccurate. This theorem describes the optimal solution in a way that can be implemented, and it sets the stage for a sensitivity analysis.

We've shown how to employ the information that Solver's Sensitivity Report provides in a sensitivity analysis. We've seen that the shadow prices help explain why the optimal solution is what it is—why some variables are positive and others equal zero.

Connections to Economics

Our presentation of sensitivity analysis has stressed connections to economic reasoning. The economic concepts that have been employed include the following:

- A *variable cost* is a cost that depends on (varies with) the action that is being contemplated, and a *fixed cost* is a cost that is independent of the action that is being contemplated. *Contribution* equals revenue less variable cost. When allocating resources, a profit-maximizing decision maker should maximize the contribution.
- The *opportunity cost* of doing something equals the reduction in contribution that occurs if you set aside the resources that are needed to do that thing.
- The *marginal profit* for doing something equals its contribution less its opportunity cost.
- The optimal solution to a linear program prescribes a set of *shadow prices*, one per constraint. Each constraint's shadow price is the *breakeven* price at which a profit-maximizing decision maker is indifferent to buying or selling one incremental unit of the resource whose consumption that constraint measures.
- In a profit-maximizing linear program, the shadow prices let us compute the opportunity cost of the resources needed to make one unit of each variable. To compute a variable's opportunity cost, multiply the coefficient of the variable in each constraint by that constraint's shadow price, and sum.
- In a linear program, the shadow prices satisfy (and can be calculated from) the following two rules:

  Each slack constraint has zero as its shadow price.

  Each variable that is positive has zero as its reduced cost.
- Linear programs seem to require linearity, but they readily accommodate *decreasing marginal return* in the contribution of each decision variable.
- Linear programs fail to accommodate *increasing marginal return* in the contribution of any decision variable.
- The optimal solution to a linear program exhibits *decreasing marginal return* in each right-hand-side value.

This discussion has also illustrated the economic principle of *thinking at the margin*. The value of doing something is not its contribution; rather, it is its contribution less the opportunity cost of setting aside the resources that are needed to accomplish that thing. To compute the opportunity cost of making one Nifty, we multiplied the quantity of each resource that it requires by the marginal value (shadow price) of one unit of that resource, and summed.

Connections to High School Mathematics

This chapter has relied on high school mathematics, namely, on linear equations and on geometry. We've seen how to compute the shadow prices from a system of linear equations, for instance. We've used geometry to plot the feasible region (set of feasible solutions) to a linear program. We've used geometry to indicate that, barring weird examples, the optimal solution occurs at an extreme point. We've noted that Dantzig's simplex method pivots from extreme point to extreme point, improving the objective value with each pivot and terminating when no further improvement is possible.

In this chapter, the shadow prices have been described as a property of the optimal solution. But more than that, as we will see in Chapter 18, every extreme point has a set of shadow prices, and they guide the simplex method as it pivots from extreme point to extreme point in a way that improves the objective value.

Resource Allocation

The Recreational Vehicle example is a prototypical resource allocation problem. In this example, the capacities of the shops play the role of *resources*, the vehicles play the role of *activities*, and the production quantities are the *levels* at which the activities are operated. In a resource allocation model,

- The optimal solution specifies the most efficient level of each activity.
- The shadow prices specify the value of each resource.

In a resource allocation context, the shadow prices can be as insightful as the optimal solution itself because the levels of the resources can themselves be adjusted. And the shadow prices measure the value of each adjustment.

A Starred Section

The final section of this chapter is starred because it is not needed for a basic understanding of linear programming. This starred section relates linear and nonlinear programs to convex sets, convex functions, and concave functions. Reading the starred section would deepen your geometric insight into linear and nonlinear programs. Unlike the body of this chapter, the starred section touches calculus and linear algebra.

3.22. HOMEWORK AND DISCUSSION PROBLEMS

1. For the Recreational Vehicle problem, Table 3.5 (on page 60) reports an allowable decrease on the objective coefficient of L of ∞. That is no accident. Why?

2. For the Recreational Vehicle problem, Table 3.3 (on page 56) reports a shadow price of 140 for the Engine shop constraint and an allowable decrease of 16 in this constraint's right-hand-side value. Thus, renting four hours of Engine shop for one week decreases contribution by \$560 because $560 = 4 \times 140$. *Without* rerunning the linear program, show how the optimal solution changes when the Engine shop capacity is decreased by 4. *Hint:* The Perturbation Theorem reduces this problem to solving two equations in two unknowns.

3. Perturbing the optimal solution to Program 3.1 by making one Luxury model vehicle decreases profit by \$200. Complete the following sentence and justify it: Perturbing this optimal solution by making 10 Luxury model vehicles decreases profit by at least ________ because ________.

4. Suppose, in the Recreational Vehicle problem, that the contribution of the Standard model vehicle is as plotted in Figure 3.2. Adapt Program 3.1 to this situation. Use Solver to compute its optimal solution. Describe this optimal solution in a way that can be implemented when the model's data are inexact.

5. With the Engine shop capacity fixed at 120 hours per week, use Solver to compute the optimal value to Program 1 for all values of the Body shop capacity. Plot the analog of Figure 3.4. Do you observe decreasing marginal return?

6. Consider a company that can use overtime labor at an hourly wage rate that is 50% in excess of regular time labor cost. Does this represent an economy of scale? a diseconomy of scale? Will a profit-maximizing linear program have an unintended option? If so, what will it be? Will this unintended option be selected by optimization, or will it be ruled out?

7. Consider a linear program that is feasible and bounded. Let us imagine that each right-hand-side value in this linear program was multiplied by 0.75. Complete the following sentences and justify them. (*Hint:* To educate your guess, re-solve Program 3.1 with each right-hand-side value multiplied by 0.75.)

(a) The optimal solution would be multiplied by ________, and the optimal value would be multiplied by________.

(b) On the other hand, the shadow prices would be ________.

(c) There was nothing special about the factor 0.75 because ________.

8. The shadow prices are supposed to apply to small changes in right-hand-side values. Compute the amount that the manager of the Recreational Vehicle shop could earn by renting the entire capacity of her shops at the shadow prices. Is this amount familiar? If so, why? *Hint:* It might help to review the preceding problem.

9. The sensitivity report seems to omit the shadow prices of the nonnegativity constraints. True or false:

(a) In a maximization problem, the reduced cost of each nonnegative variable x equals the shadow price of the constraint $x \geq 0$.

(b) In a minimization problem, the reduced cost of each nonnegative variable x equals the shadow price of the constraint $x \geq 0$.

10. In a linear program, a decision variable x is said to be **free** if neither the constraint $x \geq 0$ nor the constraint $x \leq 0$ is present in the linear program. A free variable is allowed to take values that are positive, negative, or zero. In the optimal solution to a maximization problem, what can you say about the reduced cost of each free variable? Why?

11. **(Subway Cars)** The Transit Authority must repair 100 subway cars per month, and it must refurnish 50 subway cars per month. Each task can be done in its own facility. Each task can also be contracted to private shops but at a higher cost. Private contracting increases the cost by \$2000 per car repaired and \$2500 per car refurnished.

The Transit Authority repairs and refurnishes subway cars in four shops. Repairing each car consumes 1/150th of the monthly capacity of its Evaluation shop, 1/60th of the capacity of its Assembly shop, none of the capacity of its Paint shop, and 1/60th of the capacity of its Machine shop. Refurnishing each car requires 1/100 of the monthly capacity if its Evaluation shop, 1/120th of the monthly capacity of its Assembly shop, 1/40th of the monthly capacity of its Paint shop, and none of the capacity of its Machine shop.

(a) Formulate the problem of minimizing the monthly expense for private contracting as a linear program. Solve it graphically or by using Solver. Describe its optimal solution in a way that can be implemented when its data are inexact.

(b) Formulate the problem of maximizing the monthly saving for repairing in the Authority's own shops as a linear program. Does this linear program have the same solution as the one in part (a)? If so, why? If not, why?

12. Write down linear programs that have each of these properties:

(a) It has no feasible solution.

(b) It is feasible and bounded, but it has an unbounded feasible region.

(c) It is feasible and unbounded.

(d) Its feasible region is bounded, and it has multiple optima.

(e) It is bounded, it has an unbounded feasible region, it has multiple optimal solutions, but only one of them is an extreme point.

13. (**A Farmer**) A 1200-acre farm includes a well that has a capacity of 2000 acre-feet of water per year. (One acre-foot is one acre covered to a depth of one foot.) This farm can be used to raise wheat, alfalfa, and beef. Wheat can be sold at \$550 per ton and beef at \$1300 per ton. Alfalfa can be bought or sold at the market price of \$220 per ton. Each ton of wheat that the farmer produces requires one acre of land, \$50 of labor, and 1.5 acre-feet of water. Each ton of alfalfa that she produces requires $1/3$ acre of land, \$40 of labor, and 0.6 acre-feet of water. Each ton of beef she produces requires 0.8 acres of land, \$50 of labor, 2 acre-feet of water, and 2.5 tons of alfalfa. She can neither buy nor sell water. She wishes to operate her farm in a way that maximizes its annual profit. Below are the data in a spreadsheet formulation, the solution that Solver has found, and a Sensitivity Report.

	A	B	C	D	E	F	G	H
1								
2	name of variable	W	AR	B	AS			RHS
3	acreage	1	1/3	0.8		800	<=	1200
4	water	1.5	0.6	2		2000	<=	2000
5	conservation		-1	2.5	1	0	=	0
6	contribution	500	-40	1250	220	700000		
7	value of variable	0	0	1000	-2500			
8								
9			=SUMPRODUCT(B6:E6,B\$7:E\$7)					

Adjustable Cells

Cell	Name	Final Value	Reduced Cost	Objective Coefficient	Allowable Increase	Allowable Decrease
\$B\$7	value of variable W	0	–25	500	25	1E+30
\$C\$7	value of variable AR	0	–30	–40	30	1E+30
\$D\$7	value of variable B	1000	0	1250	1E+30	33.3333
\$E\$7	value of variable AS	–2500	0	220	13.3333	1E+30

Constraints

Cell	Name	Final Value	Shadow Price	Constraint R.H. Side	Allowable Increase	Allowable Decrease
\$F\$3	acreage	800	0	1200	1E+30	400
\$F\$4	water	2000	350	2000	1000	2000
\$F\$5	conservation	0	220	0	1E+30	1E+30

(a) Write down the linear program. Define each variable. Give each variable's unit of measure. Explain the objective function and each constraint. Explain why the constraint $AS \geq 0$ is absent. What is the unit of measure of the objective? What is the unit of measure of each constraint?

(b) State the optimal solution in a way that she can implement when the data are inexact.

(c) As an objective function coefficient or right-hand-side value varies within its allowable range, how does she manage the farm? That is, in which activities does she engage, and which resources does she use to capacity?

(d) What would have to happen to the price of wheat in order for her to change her production mix?

(e) What would have to happen to the price of alfalfa for her to change her production mix?

Note: Parts (f) through (i) refer to the original problem and are independent of each other.

(f) The government has offered to let her deposit some acreage in the "land bank." She would be paid to produce nothing on those acres. Is she interested? Why?

(g) The farmer is considering soybeans as a new crop. The market price for soybeans is $800 per ton. Each ton of soybeans requires 2 acres of land, 1.8 acre-feet of water, and $60 of labor. Without rerunning the linear program, determine whether or not soybeans are a profitable crop.

(h) A neighbor has a 400-acre farm with a well whose capacity is 500 acre-feet per year. The neighbor wants to retire to the city and to rent his entire farm for $120,000 per year. Should she rent it? If so, what should she do with it?

(i) The variable *AS* is unconstrained in sign. Rewrite the linear program with *AS* replaced by (ASOLD − ABOUGHT), where ASOLD and ABOUGHT are nonnegative decision variables. Solve the revised linear program. Did any changes occur? Does one formulation give more accurate results than the other? If so, how and why?

14. (Pollution Control) A company makes two products in a single plant. It runs this plant for 100 hours each week. Each unit of product *A* that the company produces consumes two hours of plant capacity, earns the company a contribution of $1000, and causes, as an undesirable side effect, the emission of 4 ounces of particulates. Each unit of product *B* that the company produces consumes one hour of capacity, earns the company a contribution of $2000, and causes, as undesirable side effects, the emission of 3 ounces of particulates and 1 ounce of chemicals. The EPA (Environmental Protection Agency) requires the company to limit particulate emission to at most 240 ounces per week and chemical emission to at most 60 ounces per week. Given below are spreadsheets for this linear program and a Sensitivity Report.

	A	B	C	D	E	F
1						
2	name of variable	A	B			
3	capacity	2	1	90	<=	100
4	particulate	4	3	240	<=	240
5	chemicals		1	60	<=	60
6	contribution	1000	2000	135000		
7	value of variable	15	60			
8						
9		=SUMPRODUCT(B6:C6,B$7:C$7)				

Adjustable Cells

Cell	Name	Final Value	Reduced Cost	Objective Coefficient	Allowable Increase	Allowable Decrease
$B$7	value of variable A	15	0	1000	1 2/3	1000
$C$7	value of variable B	60	0	2000	1E+30	1250

Constraints

Cell	Name	Final Value	Shadow Price	Constraint R.H. Side	Allowable Increase	Allowable Decrease
$D$3	capacity	90	0	100	1E+30	10
$D$4	particulate	240	250	240	20	60
$D$5	chemicals	60	1250	60	20	20

(a) Write down this linear program. In this linear program, what is the unit of measure of each decision variable? of the objective function? of each shadow price?

(b) Describe its optimal solution in a way that can be implemented when its data are inexact.

(c) What is the optimal solution of this linear program? What is its optimal value?

(d) This linear program has only two decision variables, so its feasible solutions, iso-profit lines, objective vector, and optimal solution can be displayed on a plane. Do so.

(e) What is the value to the company of the EPA's relaxing the constraint on particulate emission by one ounce per week? What is the value to the company of the EPA's relaxing the constraint on chemical emissions by one ounce per week?

(f) (**An Emissions Tradeoff**) By how much should the company be willing to reduce its weekly emission of chemicals if the EPA would allow it to emit one additional ounce of particulates each week?

(g) (**An Emissions Tax**) The EPA is considering the control of emissions through taxation. Suppose that the government imposes weekly tax rates of P dollars per ounce of particulate emissions and C dollars per ounce of chemical emission. Find the smallest tax rates, P and C, that keep the company's pollutants at or below the desired levels. If these taxes were imposed, would the company prosper?

(h) (**Emissions Taxes**) Based on your answer to part (g), discuss the wisdom of controlling emissions entirely by taxation. Is it always a good idea? sometimes a good idea? never a good idea?

15. (**An Import-Export Problem**) A developing nation has shifted its economy from farming to the production of food products, yarn (woven from wool or cotton), and clothes. This nation exports the excess of its production of these goods over the amounts that it consumes. The world market prices are \$3 per unit for food, \$10 per unit for yarn, and \$25 per unit for clothes. The nation is a small factor in the world market for these commodities. The volumes it exports have negligible effect on market prices.

To produce each unit of food, the nation must import \$0.50 worth of goods (farm machinery and fertilizer), consume 0.2 units of food (e.g., fodder to feed to animals), consume 0.5 units of labor, and use 0.9 units of land. To produce each unit of yarn, the nation must import \$1.25 worth of goods, consume 1 unit of labor, and consume 1.5 units of land. To produce each unit of clothes, the nation must import \$5.00 worth of goods, consume 1 unit of yarn, and consume 4 units of labor.

The nation's population consumes 11.5 billion units of food, 0.6 billion units of yarn, and 1.2 billion units of clothes, annually. The nation possesses 65 billion units of labor and 27 billion units of land, as well as the capacity to produce clothes at the rate of 9.6 billion units per year.

The nation's economic policy calls for maximizing the net dollar value of exports for the coming year, exports being the excess of production over domestic consumption. This maximization problem has been posed as a linear program, and Solver has been used to find its optimal solution. Its spreadsheet and part of its Sensitivity Report appear next.

	A	B	C	D	E	F	G	H	I	J	K
1	variable	FP	YP	CP	imp	FE	YE	CE			
2	dollars	-0.5	-1.25	-5	1				0	=	0
3	food	0.8				-1			11.5	=	11.5
4	yarn		1	-1			-1		0.6	=	0.6
5	clothes			1				-1	1.2	=	1.2
6	labor	0.5	1	4					51.66	<=	65
7	land	0.9	1.5						27	<=	27
8	clothes cap.			1					8.775	<=	9.6
9	contribution				-1	3	10	25	126.6		
10	value	14.38	9.375	8.775	62.78	0	0	7.58			
11											
12				=SUMPRODUCT(B9:H9,B\$10:H\$10)							

Constraints

Cell	Name	Final Value	Shadow Price	Constraint R.H. Side	Allowable Increase	Allowable Decrease
$I$2	dollars	0	−1	0	1E+30	1E+30
$I$3	food	11.5	−14.6875	11.5	10.1	1.1
$I$4	yarn	0.6	−20	0.6	7.575	0.825
$I$5	clothes	1.2	−25	1.2	7.575	1E+30
$I$6	labor	51.6625	0	65	1E+30	13.3375
$I$7	land	27	12.5	27	1.2375	11.3625
$I$8	clothes cap.	8.775	0	9.6	1E+30	0.825

(a) Write down the linear program whose solution is in this spreadsheet.

(b) What is the optimal solution to this linear program? What is its optimal value? What are the units of measure of its shadow prices?

(c) Describe this optimal solution in a way that can be implemented when the data are inexact.

(d) Is there evidence in the Sensitivity Report that the country would prosper by importing one or more of the goods that it has traditionally exported? If so, what is that evidence?

16. **(An Import-Export Problem, continued)** From the preceding problem, eliminate the constraints that the variables *FE*, *YE*, and *CE* be nonnegative.

(a) Use Solver to find the optimal solution to this revised linear program. Obtain a Sensitivity Report.

(b) What changes occur in the production mix and in the value of net exports, if any?

(c) The minister of labor has vehemently insisted that no food, yarn, or clothes be imported in order to maximize employment. What merit is there to his argument?

(d) Suppose the country can increase its capacity to produce clothes at an annualized cost of $8.00 per unit. Should it do so? Without rerunning the linear program, what can you say about the number of units it should acquire at this price?

17. **(No Extreme Points)** Consider this linear program: Maximize $(A - B)$ subject to the constraints

$$A - B \le 1,$$
$$A - B \ge -1.$$

(a) Plot this linear program's feasible region. Does it have any extreme points?

(b) Does this linear program have an optimal solution? If so, name one.

(c) Use Solver to solve this linear program. What happened?

18. **(Watching Solver Pivot)** Page 79 describes a procedure for interrupting Solver at each iteration of the simplex method. Apply this procedure to Program 3.1. Record the feasible solution that each iteration produces, along with its objective value. Did Solver pivot from extreme point to extreme point? Did each pivot increase the objective value?

19. Eliminate from the Recreational Vehicle problem the Luxury model vehicles. The linear program that results has only two decision variables, so its feasible region is a portion of the plane.

(a) Write down this linear program.

(b) Display its feasible region graphically.

(c) Display its iso-profit lines and objective vector graphically.

(d) Show, graphically, that its optimal solution sets $S = 20$ and $F = 30$, so that its optimal value equals 50,400.

(e) Set aside the capacity needed to make one Luxury model vehicle. Re-solve the linear program graphically. Show that making one Luxury model vehicle decreases contribution by 200.

20. In Figure 3.8, the "allowable increase" is the value of d for which the constraint $2\,B \le 9$ becomes tight. The "allowable decrease" is the value of $-d$ for which a different constraint becomes tight. Which one? Compute the allowable decrease by hand.

21. This problem concerns the linear program: Maximize $\{A + 3\,B\}$ subject to the five constraints $A + 2\,B \ge 10$ and $2\,A - B \le -10$ and $A + B \le 12$ and $A \ge 0$ and $B \ge 0$.

(a) Graph its feasible region, and determine its optimal solution graphically.

(b) Alter its objective function so that this linear program has multiple optimal solutions, and specify three of them.

(c) Drop a constraint so that this linear program is unbounded, and specify a set of feasible solutions whose objective values approach infinity.

22. (**An Economy of Scale**) Suppose the Recreational Vehicle company can rent a tool for \$1200 per week. This tool enables them to lower the variable cost of manufacturing each Standard model vehicle, thereby increasing its contribution from \$840 to \$990.

(a) How does the availability of this tool constitute an economy of scale? At what production rate is the company indifferent between renting this tool and not?

(b) Consider the *integer program* that results when Program 3.1 is changed as follows: The addend $840\,S$ in its objective is replaced by $\{840\,S1 + 990\,S2 - 1200\,x\}$, and these six constraints are added:

$$S = S1 + S2, \qquad S1 \ge 0, \qquad S2 \ge 0,$$
$$S2 \le 40\,x, \qquad x \ge 0, \qquad x \text{ an integer.}$$

What is the role of the constraint $S2 \le 40\,x$? What is the role of the addend $1200\,x$ in the objective? Does this integer program have an unintended option? If so, is it ruled out by optimization? Does this integer program account correctly for the economy of scale? If so, why?

(c) Use Solver to find the optimal solution to the integer program in part (b). (To require that a variable be integer-valued, click on the "int" option in the Add Constraint dialog box, as shown below.) Did the optimal solution change? If so, how?

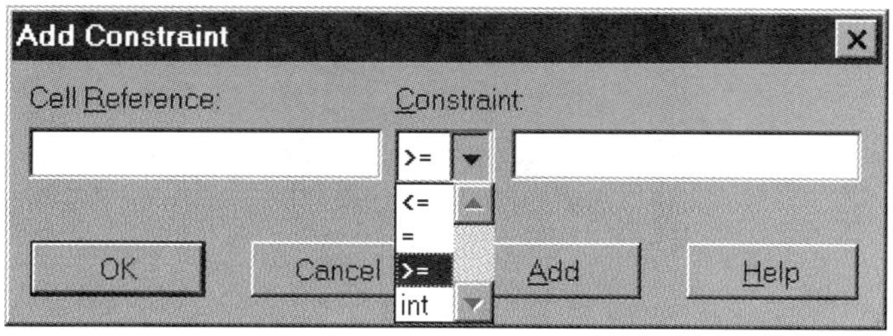

23. (**An Economy of Scale, continued**) In the preceding problem, we reformulated the profit-maximization problem as an integer program in order to accommodate an economy of scale. There are occasions in which this is unnecessary. This may have been one of them? Was it? If so, why? *Hint:* Review Table 3.5.

24. (**Edges**) In the body of this chapter, we relied on your intuitive understanding of an extreme point. In its starred section, we defined the extreme point of a convex set, but we relied on your intuitive understanding of its edges. The shaded polygon in Figure 3.6 has five edges. The polyhedron in Figure 3.9 has 15 edges. This problem asks you to develop a definition of an "edge" of a convex set. Your definition should have these properties:

- Each edge of a convex set is an interval between two extreme points, but not every interval between two extreme points is an edge.
- In Figure 3.9, the interval between extreme points (20, 0, 20) and (35, 0, 15) is an edge, but the interval between extreme points (20, 0, 20) and (20, 30, 0) is not.

Complete this definition. The interval J between two distinct extreme points of a convex set S is an **edge** of S if no point that is inside the interval J is inside any interval __________.

25. (**The Feasible Region Is Convex, Part 1**) With a_1 through a_n and a_0 as data (fixed numbers), and with x_1 through x_n as variables, let S be the set of all vectors $x = (x_1, x_2, \ldots, x_n)$ such that

$$a_1x_1 + a_2x_2 \cdots + a_nx_n \leq a_0.$$

(a) Show that S is a convex set. *Hint:* For any vectors x and y in S and any number c between 0 and 1, is it true that

$$c\,x \leq c\,a_0 \qquad \text{and that} \qquad (1 - c)\,y \leq (1 - c)a_0?$$

(b) Redo part (a) with the " $\leq$ " replaced by "=" and with the " $\leq$ " replaced by " $\geq$."

(c) What can you conclude about the set of solutions to a single linear constraint?

26. (**The Feasible Region Is Convex, Part 2**) Let S and T be convex subsets of $\Re^n$.

(a) Show that their intersection, $S \cap T$, is convex. *Hint:* Consider any points x and y in $S \cap T$. Ask yourself these questions. Is the interval between x and y in S? Is this interval in T? Is it in $S \cap T$?

(b) What can you conclude about the feasible solutions to a linear program? Why?

27. Consider the functions $f(t) = \log t$, $g(t) = 2^{3t}$, and $h(t) = \sqrt{t}$. Determine which of these functions is convex on the set S of positive numbers. Determine which is concave on the same set S.

28. True or false: In the optimal solution to a maximization problem, no variable x can have a positive reduced cost. *Hint:* Be careful.

Remark: The next problem earns its star (*) by probing beyond the scope of the discussion in this chapter.

29. *(Sketch of the **revised simplex method**) The equality-constrained version of Program 3.1 has eight variables, which are S, F, L, and the slack variable for each of its five constraints. Each extreme point of this linear program establishes a "market" for the resources according to this rule: The "agents" who engage in activities at positive levels are indifferent to buying or selling small quantities of the resources at their shadow prices and adjusting the levels of their activities accordingly.

(a) In Figure 3.9, let us begin at the extreme point (20, 0, 20), which has $S = 20$, $F = 0$, and $L = 20$. Which five variables are positive at this extreme point?

(b) What are this extreme point's shadow prices?

(c) Is there a variable, presently equal to zero, whose contribution exceeds its opportunity cost? (This could be a slack variable.)

(d) Perturb the extreme point by maximizing the variable found in part (c), keeping the other two tight constraints tight and keeping the solution feasible. Use Figure 3.9 to see what occurs.

(e) Repeat parts (a)–(d), starting at the new extreme point.

Chapter 4

A Survey of Linear Programs

4.1. PREVIEW

At the core of this chapter lies a series of problems, each of which represents an area of application of linear programming. In each instance:

- An area of application is introduced.
- An illustrative problem is posed.
- A linear program is formulated.
- This linear program is solved on a spreadsheet.
- Its optimal solution is interpreted.

Interspersed with these problems are some general observations that can help you to identify particular types of linear program and build linear programs.

4.2. WHAT CAN YOU LEARN FROM THIS CHAPTER?

There are three reasons to study this chapter. First, each problem in this chapter represents an important area of application of linear programs. Taken together, these problems can provide you with a feel for the scope of linear programming, that is, for the range of situations that can be modeled as linear programs and for the insights that linear programming can provide. When you study a problem, you may find it helpful to experiment with its spreadsheet formulation, which can be found on the CD that accompanies this text.

Second, this chapter is organized to help you master the art of formulating decision problems for solution as linear programs. Most people find that learning how to formulate a linear program is a struggle. Formulation is a difficult part of linear programming, just as "word problems" are a hard part of high school mathematics. The examples in this chapter can serve as benchmarks or templates. Interspersed with these examples are some guidelines to successful formulation. When you encounter a situation that might be formulated as a linear program, ask yourself these questions:

- Which benchmarks does the situation resemble?
- Which guidelines can help to build a linear program?

To hone your skill at formulation, pause after you read each problem. Try to build your own linear program before you read ours. If you are uncertain as to whether your linear program gives the same answer as ours, use Solver to check. Eventually, tackle some of the formulation problems at the end of this chapter.

Third, the problems in this chapter transcend linear programming. They introduce you to related topics, including network flow, integer programs, and sequential decision processes. By doing so, they help to familiarize you with other models in operations research.

This chapter poses a total of eight problems, and it uses a linear program to solve each of them. This chapter is far less "sequential" than is usual in technical writing. Sections 4.3 through 4.7 cover basic material, which you may wish to read in the normal, numerical order. But you can start anywhere. Students of economics may wish to begin with Section 4.10, which discusses models of activity analysis. You may wish to start with the Review section and then proceed to the formulation problems at this chapter's end, referring back to the text for clues and guidelines.

4.3. CRACKING AND BLENDING

With this preamble, we turn our attention to one area of application of linear programs and their generalizations. An oil refinery uses heat and pressure to **crack** complex hydrocarbon molecules into lighter constituents. These constituents are then blended into commercially useful products, such as aviation fuel, gasoline, heating oil, and petrochemicals. The object is to operate the refinery efficiently, for example, to produce the most profitable mix of products or to produce a required mix of products at least cost. In a simple model, cracking and blending take the form of linear constraints, and the optimization problem becomes a linear program.

Charnes, Cooper, and Mellon[1] pioneered the use of a linear program to model the efficient operation of an oil refinery. Their work proved to be prescient. Linear programming would soon pervade nearly every facet of the energy industry and would find other uses within the chemical processing industry. Problem A gives a flavor of the linear (and nonlinear) programs that guide the operation of oil refineries.

[1]Charnes, A.W., Cooper, W., and B. Mellon. "Blending Aviation Gasolines—A Study in Programming Interdependent Activities in an Integrated Oil Company," *Econometrica*, V. 20, pp. 135–159 (1952).

Problem A (Cracking and Blending)

A small refinery uses heat to crack crude oil into its constituents and then blends them into two products, which are Regular and Premium grades of unleaded gasoline. This refinery possesses two different cracking facilities, which are Method A and Method B. Method A converts each barrel of crude oil into one barrel of 93 octane gasoline. (A barrel is 42 gallons.) Method B converts each barrel of crude oil into one barrel of 85 octane gasoline. Cracking each barrel of crude by Method A entails a variable cost of \$4.80, compared with \$2.30 for Method B. These cracking technologies operate independently of each other. The refinery has the capacity to process 4000 barrels per month using Method A and 9000 barrels per month using Method B. Regular gasoline must have an octane rating of at least 87, whereas Premium gasoline must have an octane rating of at least 91. The octane of a blend is proportional to the fractions of its constituents. For instance, blending one barrel of 93 octane gasoline with three barrels of 85 octane gasoline produces four barrels of 87 octane gasoline because $87 = [(1)(93) + (3)(85)]/4$. The prices at which the refinery can sell this month's production are \$32.76 per barrel of Regular gasoline and \$38.22 per barrel of Premium. Currently, the refinery pays \$18.60 per barrel of crude. At current market prices, the refinery can sell all the Regular gasoline that it can produce but can sell only 3000 barrels per month of Premium. The refinery must produce at least 4500 barrels of Regular gasoline each month because it has a long-term contract for this amount. The refinery is not able to store crude or refined products. Each month, it buys the amount of crude oil that will be cracked, blended, and sold. The goal is to maximize contribution (profit). How shall the refinery be operated this month?

A Simplification

Actually, cracking separates crude oil into *several* constituents, not just one. Our model of cracking is simplistic. A realistic model of Method A cracking would specify the quantity of each constituent that is obtained by cracking one barrel of crude. These quantities would conserve mass, as well as the energy content of the crude, and the relationships would still be linear. Thus, although Problem A is simplified, it gives the flavor of realistic models of refinery operation.

A Network Flow Diagram

As a first step toward formulating Problem A as a linear program, we construct a picture of the flow of oil through the refinery. Figure 4.1 depicts the cracking and blending activities, and it names the decision variables.

Figure 4.1 is sometimes called a **network flow diagram**. In a network flow diagram, the lines with arrows are called **arcs**, and the circles are called **nodes**. In

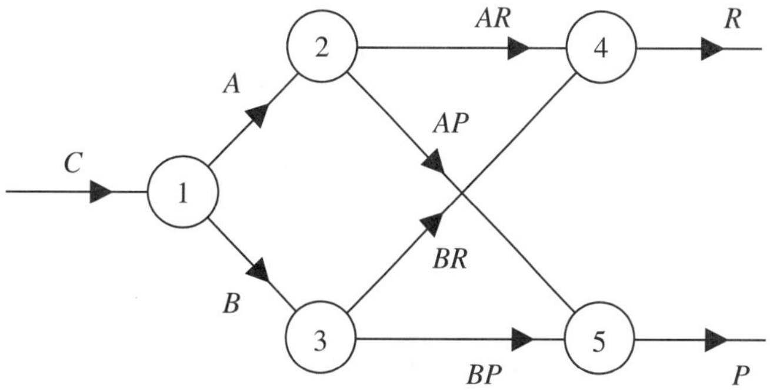

Figure 4.1 The flow of oil through a refinery.

such a diagram:

- Flow occurs along each arc in the direction of its arrow.
- Flow is conserved along each arc; what flows in the tail end of an arc comes out the head end.
- Flow is also conserved at each node. The total of the flows into a node equals the total of the flows out of it.

In Figure 4.1, the flow on each arc has been given a name. For instance, BP is the quantity that flows along the arc from node 3 to node 5. The network flow diagram in the figure identifies nine decision variables, which include these four:

C = the quantity (in barrels per month) of crude oil purchased.
A = the quantity (in barrels per month) of crude oil refined by Method A.
AR = the quantity (in barrels per month) of the output of Method A cracking that are blended into Regular gasoline.
R = the number (in barrels per month) of Regular gasoline produced.

Flow is conserved at each node. The quantity flowing into node 1 equals the quantity flowing out of that node, so

$$C = A + B.$$

This equation states that the number C of barrels of crude oil that are purchased this month equals the number A of barrels that are cracked using Method A plus the number B of barrels that are cracked using Method B. This constraint reflects the refinery's inability to inventory crude oil. The quantity flowing into node 2 equals the quantity flowing out of it, so

$$A = AR + AP.$$

This equation states that the number A of barrels of gasoline obtained from Method A cracking equals the number that is blended into Regular gasoline plus the number that is blended into Premium gasoline. Similarly, the conservation-of-flow requirements for nodes 3, 4, and 5 are

$$B = BR + BP,$$
$$AR + BR = R,$$
$$AP + BP = P.$$

Implicit in a network flow diagram is the understanding that each of the flows is nonnegative. Thus, Figure 4.1 represents the requirement that

$$C \geq 0,\ A \geq 0,\ B \geq 0,\ AR \geq 0,\ AP \geq 0,\ BR \geq 0,\ BP \geq 0,\ R \geq 0,\ P \geq 0.$$

The preceding constraints describe the flow of material through the refinery.

What Does This Network Flow Diagram Omit?

The network flow diagram omits quite a bit. The diagram in Figure 4.1 makes no mention of the following features of Problem A:

- The capacities of the two cracking facilities.
- The bounds on the amount of Regular and Premium fuel that the refinery produces.
- The octane requirement of each grade of gasoline.
- The contribution.

Let us add constraints that account for each feature. The cracking capacities are 4000 barrels per month for Method A and 9000 barrels per month using Method B, so

$$A \leq 4000, \qquad B \leq 9000.$$

The refinery must produce at least 4500 barrels of Regular gasoline this month because it has a contract for this amount. In addition, the company cannot sell more than 3000 barrels per month of Premium gasoline at current prices, so

$$R \geq 4500, \qquad P \leq 3000.$$

Regular gasoline must have an octane rating of at least 87, whereas Premium gasoline must have an octane rating of at least 91. These octane-rating requirements are represented as

$$\frac{93\,AR + 85\,BR}{R} \geq 87, \qquad \frac{93\,AP + 85\,BP}{P} \geq 91.$$

These octane-rating constraints are nonlinear. They *cannot* be part of a linear program. But note that multiplying both sides of an equality by a nonnegative number preserves the inequality. In each octane-rating constraint, the denominator is nonnegative. Multiplying both sides of the first octane-rating constraint by the number R preserves the inequality and produces the linear constraint

$$93\,AR + 85\,BR \geq 87\,R.$$

Similarly, multiplying both sides of the second octane-rating constraint by the number P produces the inequality

$$93\,AP + 85\,BP \geq 91\,P.$$

Each constraint in this refining problem has now been represented as a linear equation or as a linear inequality.

All that remains is to measure the contribution, which is given by the expression

$$\{32.76\,R + 38.22\,P - 18.6\,C - 4.80\,A - 2.30\,B\}.$$

Problem A is the problem of maximizing this expression subject to the linear constraints that have been listed above.

The Linear Program

Let us summarize. The problem of operating this refinery efficiently has been cast as a linear program, namely, as

Program 4.1: Maximize $\{32.76\,R + 38.22\,P - 18.6\,C - 4.80\,A - 2.30\,B\}$, subject to the constraints

$$\begin{aligned}
C &= A + B,\\
A &= AR + AP,\\
B &= BR + BP,\\
AR + BR &= R,\\
AP + BP &= P,\\
P &\leq 4000,\\
B &\leq 9000,\\
R &\geq 4500,\\
P &\leq 3000,\\
93\,AR + 85\,BR &\geq 87\,R,\\
93\,AP + 85\,BP &\geq 91\,P,
\end{aligned}$$

$$C \geq 0,\ A \geq 0,\ B \geq 0,\ AR \geq 0,\ AP \geq 0,\ BR \geq 0,\ BP \geq 0,\ R \geq 0,\ P \geq 0.$$

Problem A has many correct formulations as a linear program. If you did formulate it yourself, your linear program would likely differ from Program 4.1. To check whether your formulation is correct, use Solver to compute its optimal solution, and compare your results with what appears below.

"Points" and Blending Constraints

Before solving this linear program, we pause to review its blending constraints. It is natural to think of a blending constraint as a restriction on a ratio. But you may find it *easier* to write down the linearization of a blending constraint if you learn the following trick.

> To construct the linear form of a ratio constraint, count each unit of the quantity being measured as one **point**, and constrain the total number of points.

For example, 93 equals the number of "points" of octane per barrel of gasoline obtained from Method A cracking, and 93 AR equals the number of points of octane that are contributed to Regular gasoline from Method A cracking. The constraint $93\ AR + 85\ BR \geq 87\ R$ states that the total number of points of octane supplied to Regular gasoline must be at least as large as the number $87\ R$ of points that are required in R barrels of Regular gasoline.

A Spreadsheet

To build a spreadsheet for Problem A, we employ the procedure that was introduced in Chapter 2, which requires us to:

- Shift all variables to the left-hand sides of the constraints.
- Introduce a row of "changing cells" that contain the values of the decision variables.
- Insert a column of SUMPRODUCT functions.

In Program 4.1, the conservation-of-flow constraints were written as "Flow in equals flow out." In Table 4.1, they have been rewritten as "Flow in minus flow out equals zero." For instance, node 1's constraint $C = A + B$ has become $C - A - B = 0$, and the coefficients in this constraint appear in row 3. Node 2's constraint $A = AR + AP$ has become $A - AR - AP = 0$, and its coefficients are in row 4, and so forth. In Table 4.1, the values of the decision variables appear in row 15. The usual SUMPRODUCT functions appear in column K. Omitted from Table 4.1 is the Solver dialog box, which maximizes the value in cell K14 with cells B15 through J15 as changing cells, subject to constraints that keep the variables nonnegative and satisfy the equations and inequalities indicated in column L.

Table 4.1 reports the optimal solution that Solver has computed. The changing cells in row 15 report the optimal values of the decision variables. Some of the entries in column K may seem peculiar. Cell K12 contains the entry 2E-09, which is shorthand for the floating-point number 2×10^{-9}. This number is approximately zero. It and several other numbers in column K consist of **round-off error**, which occurs because Excel uses "floating-point arithmetic."

As column K of Table 4.1 indicates, this optimal solution satisfies some inequality constraints as equations, and it satisfies the rest as strict inequalities. As usual, we have shaded the tight constraints but not the slack constraints. When the optimal solution is described in terms of its tight (shaded) constraints, it is as follows:

> *This month, make full use of both cracking facilities. Mix their outputs so as to produce Regular gasoline whose octane rating is exactly 87 and Premium gasoline whose octane rating is exactly 91.*

Table 4.1 A spreadsheet for Program 4.1, which changing cells in row 15.

	A	B	C	D	E	F	G	H	I	J	K	L	M
1													
2		C	A	B	AR	AP	BR	BP	R	P			RHS
3	node 1	1	-1	-1							0	=	0
4	node 2		1		-1	-1					-3E-12	=	0
5	node 3			1			-1	-1			9E-13	=	0
6	node 4				1		1		-1		2E-11	=	0
7	node 5					1		1		-1	0	=	0
8	cap. A		1								4000	<=	4000
9	cap. B			1							9000	<=	9000
10	min. R								1		11500	>=	4500
11	max. P									1	1500	<=	3000
12	oct. R				93		85		-87		2E-09	>=	0
13	oct. P					93		85		-91	-3E-11	>=	0
14	contr.	-18.6	-4.8	-2.3					32.76	38.22	152370		
15	value	13000	4000	9000	2875	1125	8625	375	11500	1500			
16													
17						SUMPRODUCT(B14:J14,B$15:J$15)							

The Perturbation Theorem states that this solution maximizes contribution when the data are accurate and when they are somewhat inaccurate.

4.4. GUIDELINES TO FORMULATION

To formulate a linear program, you must answer these two questions:

- What are the decision variables?
- What are the constraints?

Neither answer is obvious. With Problem A in view, we now present some guidelines that can help you construct a linear program.

Identifying the Decision Variables

The following two guidelines can help you identify the decision variables in a linear program:

> Guidelines for selection of the decision variables:
>
> - A quantity is a natural candidate for a decision variable if it can vary and if a cost or revenue varies with it.
> - Try to organize the information into a diagram or table.

Program 4.1 illustrates both of these guidelines. In Problem A, the amount C of crude that is bought can vary, and a cost varies with the amount that is purchased. Similarly, the amount P of Premium gas that is sold can vary, and revenue varies with the quantity that is sold. The first guideline identifies the decision variables C, A, B, R, and P. It does not identify the variables AR, AP, BR, and BP because no cost or revenue depends on them directly. The second guideline catches those variables. In fact, the network flow diagram in Figure 4.1 identifies all nine of the decision variables.

Identifying the Constraints

Constraints come in only a few varieties. Once you learn how to recognize these "garden variety" constraints, you will be able to build linear programs that include them. Program 4.1 contains three varieties:

A **conservation constraint** reflects the fact that something is conserved. That thing might be matter, energy, airplanes, people, cash, whatever. Each node in Figure 4.1 models a conservation-of-flow constraint.

A **bounding constraint** places either a lower bound or an upper bound on some quantity. This bound may represent a **capacity**. In Program 4.1, the constraint $A \leq 4000$ is that the capacity of Method A cracking is 4000 barrels per month. This bound may represent **nonnegativity**. The constraint $P \geq 0$ reflects the fact that gasoline occurs in nonnegative quantities. The bound may represent a **minimum requirement**. The constraint $R \geq 4500$ reflects the requirement that at least 4500 barrels of Regular gasoline be produced this month because the refinery has a long-term contract for that amount.

A **ratio constraint** is the linearization of a constraint on the ratio of two linear expressions. We have seen how ratio constraints appear in blending problems. In a model of steel production, ratio constraints keep the fraction of carbon (by weight) within prescribed limits, for instance. In a model showing the busing of students to achieve racial integration, ratio constraints keep the fraction of each group of students in each school within preset limits. We recapitulate:

> When building a linear program, look for these "garden variety" constraints:
>
> - Conservation—of matter, cash, people, flow, and so forth.
> - Bounds—including nonnegativity, capacities, and minimum requirements.
> - Ratio—on the proportion of some quantity.

Later, after more examples of linear programs have come into view, we will identify a few more types of constraints.

Common Errors and Tips That Avoid Them

Let us now turn from guidelines to successful formulations to "tips" that can help avoid errors in formulation. The most rudimentary of these tips are as follows.

> Tips that avoid common errors:
>
> - To make the constraints easy to read, use mnemonics (memory aids) when you name the decision variables.
> - Remember that each decision variable is a quantity. Define it precisely, *including* its unit of measure.
> - Check for consistent units of measure. Each term in a constraint must have the same unit of measure; each term in the objective must have the same unit of measure.
> - Check that the linear program includes all of the relevant information.

We've followed the first two of these tips, using C for crude, R for Regular gasoline, and so forth. Mnemonics like these help us to read constraints, to understand Program 4.1, for instance. Mnemonics are less helpful in a Solver-based spreadsheet, however, because its changing cells contain the values of the decision variables and omit the names. In addition, we were careful to state the unit of measure of each variable, for instance, to define C as the number of barrels per month.

In Program 4.1, we neglected to check for consistent units. To do so, we need to recall that the unit of measure of the product is the product of the unit of measure. To illustrate, we examine the addend 32.76 R in the objective of Program 4.1. The unit of measure of 32.76 is dollars per barrel ($/barrel) and the unit of measure of R is barrels per month (barrels/month), so the unit of measure of 32.76 R ($/barrel) × (barrels/month) = $/month. Each addend in the objective is measured in dollars per month, so the objective is consistent, as are the constraints.

In Program 4.1, we haven't checked that it takes account of all of the relevant information. Making that check is easy if the relevant information has been assembled in a table or a figure.

Tips on Inequalities

Equality constraints represent conservation of something and are hard to get wrong. Inequality constraints are more subtle. The next tips relate to them.

> Tips on inequality constraints:
>
> - Check that the capacity constraints and the other upper and lower bounds are written as inequalities, *not* as equations.
> - Remember to include the nonnegativity constraints.

A common error is to write bounds as equations rather than as inequalities. In Program 4.1, for instance, we would have gotten the correct answer if we had required the octane rating of Regular gasoline to equal 87. But with different data we would have gotten the wrong answer. Let Solver decide which bounds to satisfy as equations.

When building a spreadsheet, we work hard at constructing the more elaborate constraints and tend to forget the easy ones. Remember to ask yourself whether you included the nonnegativity constraints.

On Being Clever

The next tip may ask you to unlearn something. In high school, for example, you may have learned that fewer variables were better. A clever maneuver might have let you formulate a problem as two equations in two unknowns rather than five equations in five unknowns. That was a good idea when you were doing the computation by hand, but now that the computer is doing the computation, when you formulate a linear program, there is good reason *not* to be clever.

Program 4.1 illustrates this point. We formulated it in terms of nine decision variables, and of these, five are superfluous. We could make do with the mixing variables, AR, AP, BR, and BP. For instance, the variable R can be eliminated because the equation $R = AR + BR$ lets us substitute $(AR + BR)$ for R wherever R appears. This substitution "simplifies" the blending constraint on Regular gasoline from

$$93\,AR + 85\,BR \geq 87\,R \quad \text{to} \quad 6\,AR \geq 2\,BR,$$

for instance. But:

- These "simplifications" introduce opportunities for error.
- Eliminating the variables R, P, C, A, and B corrupts the sensitivity analysis that Solver provides.

- The values of the variables R, P, C, A, and B are important. If you eliminate them from your linear program, you will find yourself computing them by hand.

Ugh! The general principle is to:

> Formulate a linear program in terms of its natural decision variables, which are those that are of interest to the decision maker.

In general, forget about "clever" tricks that require side calculations.

Tips on Spreadsheet Computation

The next group of tips applies to all spreadsheet computations, not just to the solutions to linear programs.

> Tips on spreadsheet computation:
>
> - Use data to check that each spreadsheet function is executing the calculation that you intend it to perform.
> - To ease revision, put each datum in a cell, preferably in only one cell. Do not hide data in spreadsheet functions.

For instance, we have used SUMPRODUCT functions to formulate linear programs. It's good practice to place simple numbers in the changing cells and to check the arithmetic by eye. Doing this helps you find errors.

You will be tempted to insert data (e.g., numerical values) directly into spreadsheet functions in order to make these functions easier to build and, possibly, easier to read. But this poses a risk. At some later time, you will almost certainly want to reexecute the computation on your spreadsheet with different data, say, to do a sensitivity analysis. That will be easy if your spreadsheet records each datum once, and *never* within a spreadsheet function.

A Tip on Computer Output

Our final tip applies to *all* computer output. Many of us fail to question the output that we obtain from computer runs. That is wrong! A computer is an idiot savant. It calculates with breathtaking speed, but it has no comprehension of what it is doing. It has no common sense. If the problem is improperly formulated, the computer is very likely to report an answer that is idiotic in some easy-to-spot way. For this reason, you are urged to:

> Examine any computer output with *intense skepticism.*

When looking at any computer output, train yourself to ask, "Does this obey or disobey the laws of common sense?" In the case of the solution to a linear program, ask yourself questions like these:

- Does this solution conserve matter?
- Does it assign the value 0 to some variable, even though it must be optimal to equate that variable to a positive value?

- Does it make sense for the tight inequality constraints to hold as equations?
- Do the signs of the shadow prices defy logic? For instance, in a maximization problem, does a constraint whose shadow prices are negative represent something that you would like to have more of?

Computers execute calculations with blinding speed and superhuman accuracy, and we could never duplicate them by hand. But computers tend to give stupid answers to incorrectly formulated problems. For this reason, you should learn to question all computer output. If there is an error in formulation, the computer is likely to give a stupid answer, and the way in which it is stupid will be a clue as to what is wrong.

Later, after more varieties of linear programs have been presented, we shall be in a position to illustrate a few more guidelines to successful formulation. Let us now resume our survey of situations that can give rise to linear programs.

4.5. SHIPPING

Our attention turns to a class of problems in which material must be shipped from place to place. This area of application of linear programming was pioneered by Tjalling C. Koopmans. During World War II, he formulated certain trans-Atlantic shipping problems as "transportation" problems, producing insights that helped propel him toward a Nobel Prize in Economics.

The Flow of Material

Linear programs and their generalizations help guide in the planning and operation of a myriad of activities that entail the flow of material. Here are some examples:

- The flow of petroleum from wells through refineries and to markets.
- The flow of manufactured goods from production facilities through warehouses to retail outlets and then to customers.
- The "flow" of commercial airplanes between cities in a way that satisfies flight schedules and provides for periodic aircraft maintenance at designated depots.
- The flow of priority mail (e.g., FEDEX) from the sender to the receiver, perhaps by air, perhaps through a hub.
- The flow of messages through a telecommunication system.
- The flow of a rental cars around a region.

In the transportation industry, it is common to solve linear programs that have thousands of variables and tens of thousands of constraints. To introduce the family of applications, we turn our attention to a simple shipping problem, one that has only a few decision variables.

Problem B (Production and Distribution)

A forest products company manufactures plywood in three plants and ships it to four depots (warehouses), from which it is sold to retailers. Each plant lies within its own timber zone, and each plant has a production capacity for the current month. Table 4.2 specifies these capacities. For instance, plant 1 has the capacity to produce as many as 2500 units of plywood this month. Each depot serves its own market, which has a demand for the current month. These demands must be satisfied exactly and are specified in Table 4.2. For instance, exactly 2000 units must be shipped this month to depot 1. Table 4.2 also specifies the unit shipping

Table 4.2 Capacities (units/month), demands (units/month) and shipping costs (dollars per unit).

	A	B	C	D	E	F
1		table of shipping costs				
2		depot 1	depot 2	depot 3	depot 4	capacity
3	plant 1	4	7	3	5	2500
4	plant 2	10	9	3	6	4000
5	plant 3	3	6	4	4	3500
6	demand	2000	3000	2500	1500	

costs. For instance, the cost of shipping each unit from plant 1 to depot 1 is $4. The cost of producing plywood has been omitted because it is the same in each plant.

The data in Table 4.2 show that the total production capacity this month is 10,000 units because 10,000 = 2500 + 4000 + 3500. Similarly, the total demand this month is 9000 units because 9000 = 2000 + 3000 + 2500 + 1500. The demands are to be satisfied exactly, so 1000 units of capacity must not be utilized. Ask yourself, "What would happen if the total demand exceeded the total production capacity?"

The Transportation Problem

Before solving Problem B, we pause to identify the type of problem that it illustrates, and to do so, we assemble its data into the network flow diagram in Figure 4.2. This diagram has three "supply" nodes (one for each plant) and four "demand" nodes (one for each depot). The arc that points into each supply node represents the production capacity in the plant that it represents, along with the fact that the capacity is an upper bound on the production level. The arc pointing out of each demand node depicts the demand at its depot, as well as the fact that the demands must be satisfied exactly. An arc points from each supply node to each demand node; this arc depicts the shipment from the indicated plant to the indicated depot. Adjacent to each of these shipping arcs is its unit cost.

In the literature, a linear program like that modeled in Figure 4.2 is known as a **transportation problem**. In a transportation problem:

- The nodes form into two groups: supply nodes and demand nodes.
- Each shipment occurs from one of the supply nodes to one of the demand nodes.

The data in a transportation problem are the amount available at each supply node, the amount required at each demand node, and the unit cost of shipping from each supply node to each demand node.

The definition of the transportation problem contains an element of ambiguity. In some formulations (including Problem B), the demands must be satisfied exactly, whereas in other formulations, the demands are lower bounds. These two variants of the transportation are genuinely different because it can be cheaper to ship more than the minimum requirements.

Transshipment and Assignment Problems

Many shipping problems fail to be transportation problems because they have transshipment points, which are nodes that can receive and send shipments. Not surprisingly, such a problem is called a **transshipment problem**.

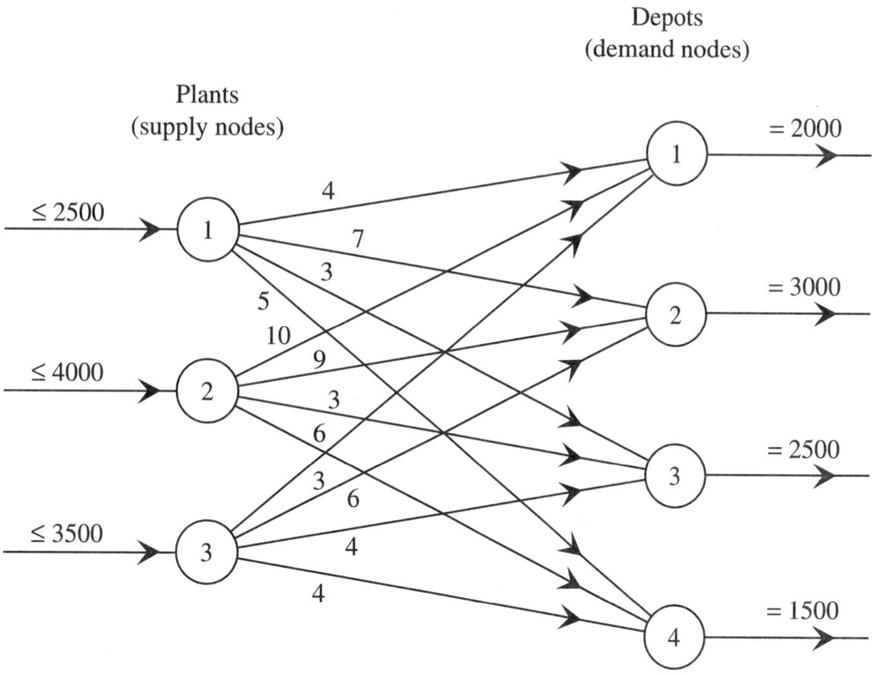

Figure 4.2 A network flow diagram for Problem B.

An **assignment problem** is the special case of the transportation problem in which exactly one unit is available at each supply node and exactly one unit is required at each demand node. The assignment problem seems a bit rudimentary. Is it too rudimentary to have any practical importance? You may think so, but read on.

A Tailored Spreadsheet

Figure 4.2 makes it clear that the key decision variables are the quantity to ship from each plant to each depot. For the cracking and blending problem, we lined the decision variable up in a row. For the transportation problem, it is easier to form the decision variables into an array that has the same shape as the shipping costs—three rows (one per plant) and four columns (one per depot). The sum across each row is the quantity shipped from that row's plant. Similarly, the sum down each column is the quantity shipped to that column's depot.

Table 4.3 presents a spreadsheet that is **tailored** to the transportation problem. The first six rows of this spreadsheet replicate Table 4.2. The rest of the spreadsheet contains changing cells and functions of the numbers in these changing cells. Specifically:

- Cells B10 through E12 are the changing cells. These cells form an array that has three rows (one per plant) and four columns (one per depot). Each changing cell contains a shipping quantity. In particular, cell C12 contains the amount to ship from plant 3 (the third row of this array) to depot 2 (its second column).
- The sum across each row of the array equals the total amount shipped from its plant. These sums appear in column F. For instance, cell F12 contains the function =SUM($B12:$E12) whose value is the amount shipped from plant 3.
- The sum down each column of the array equals the total amount shipped to its depot. These sums appear in row 13. For example, cell E13 contains the function =SUM(D$10:D$12) whose value is the amount shipped to depot 3.

Table 4.3 A tailored spreadsheet and Solver dialog box for Problem B.

	A	B	C	D	E	F	G
1		table of shipping costs					
2		depot 1	depot 2	depot 3	depot 4	capacity	
3	plant 1	4	7	3	5	2500	
4	plant 2	10	9	3	6	4000	
5	plant 3	3	6	4	4	3500	
6	demand	2000	3000	2500	1500		
7							
8	table of shipment quantities						
9		depot 1	depot 2	depot 3	depot 4	row sum	
10	plant 1	0	2500	0	0	2500	
11	plant 2	0	0	2500	500	3000	
12	plant 3	2000	500	0	1000	3500	
13	column sum	2000	3000	2500	1500		
14						=SUM($B12:$E12)	
15	cost =	41000					
16				=SUM(D$10:D$12)			
17							
18		=SUMPRODUCT(B3:E5,B10:E12)					

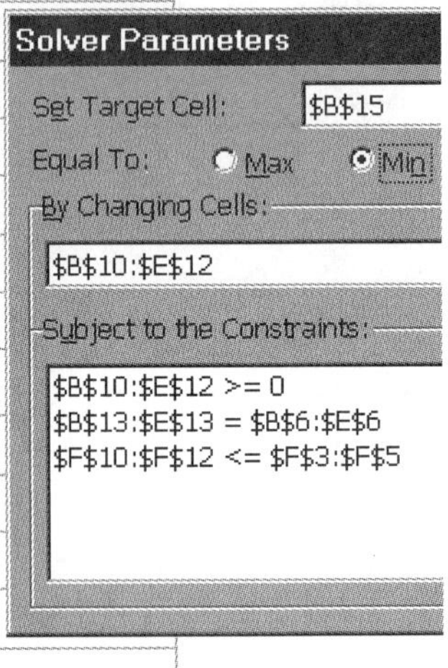

- The function =SUMPRODUCT(B3:E5,B10:E12) in cell B15 computes the total cost of the shipments.

As noted above, the function in cell F12 computes the quantity shipped from plant 2. To compute the quantity shipped from the other plants, drag this function across cells F11 and F10. Cell D13 computes the amount shipped to depot 3. To compute the quantity shipped to the other depots, drag the function in cell D13 across cells B13 through E13.

What Remains for Solver?

The spreadsheet in Table 4.3 excludes the portions of the linear program that only Solver can accomplish. Also displayed in Table 4.3 is a Solver dialog box. This dialog box:

- Minimizes the value in cell B15. (This is the total shipping cost.)
- Identifies cells B10 through E12 as the changing cells. (These cells contain the shipping quantities.)
- Keeps the values in cells B10 through E12 nonnegative. (Shipping quantities must be nonnegative.)
- Requires the values in cells B13 through E13 to equal the values in cells B6 through E6. (This requires the demand at each depot to equal the quantity shipped to it.)
- Keeps the values in cells F10 through F12 from exceeding the values in cells F3 through F5. (This keeps the shipment from each plant from exceeding its capacity.)

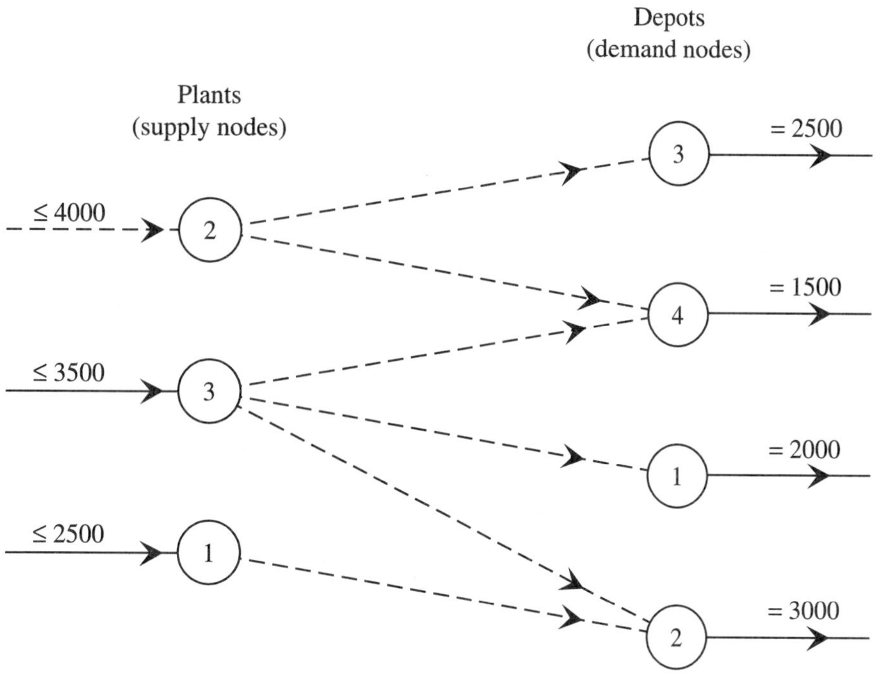

Figure 4.3 The optimal solution, with positive flows and nonbinding flows represented as dashed lines.

Included in Table 4.3 is the optimal solution that Solver has found. The tight capacity constraints have been shaded, as is usual. Evidently, plants 1 and 3 are fully utilized, and plant 2 is underutilized. Figure 4.3 presents this optimal solution graphically and in a way that indicates which constraints are slack and which are tight. In Figure 4.3:

- The shipping arcs whose flows are zero have been deleted.
- The supply and demand nodes have been re-sequenced to avoid crossing arcs. (For instance, plant 2 and depot 3 are now on top.)
- If the constraint on an arc's capacity or nonnegativity holds strictly, the arc has been re-drawn as a dashed line. (The shipping arcs are dashed because the flows on them are nonzero, and the arc pointing into plant 2 is dashed because the flow on this arc is below 4000.)

Figure 4.3 separates the positive flows into two groups. Each solid line depicts an arc whose flow equals the number adjacent to it, and each dashed line depicts an arc whose flow is unequal to its bound. Figure 4.3 lets us describe this optimal solution verbally, as follows:

Operate plants 1 and 3 at capacity, and set the production level in plant 2 just high enough to meet the total demand. Use the pattern of arcs in Figure 4.3 to ship.

As just expressed, the production and distribution plan remains optimal if the data differ somewhat from those in our linear program.

An Aside

Figure 4.3 specifies the flows on the solid arcs but not on the dashed arcs. It is very easy to compute the flows on the dashed arcs. We designate as x_{ij} the flow from supply node i to demand node j. Two dashed arcs touch the demand node for depot 2, so it is not quite clear what their flows must be. By contrast, only one dashed arc touches the supply node for plant 1, so the flow conservation constraint for this node guarantees

$$x_{12} = 2500.$$

Now only one dashed arc whose flow is unknown touches the node for depot 2, so that this node's flow conservation constraint guarantees

$$x_{32} = 3000 - x_{12} = 3000 - 2500 = 500.$$

We could next compute x_{31} or x_{23}, and so forth. Underlying this computation is a "loopless" property of extreme points.

4.6. STANDARDIZED VERSUS TAILORED SPREADSHEETS

For Problem A, we used a standardized spreadsheet, one whose decision variables are lined up in a row. By contrast, for Problem B, we employed a tailored spreadsheet because it was convenient to group the variables into an array rather than lining them up in a row.

In fact, we might have used a tailored spreadsheet for Problem A. Table 4.4 suggests how. In it, the "mixing" variables *AR*, *AP*, *BR*, and *BP* are placed in a 2×2 array, so the values of the variables *A*, *B*, *R*, *P*, and *C* become row and column sums.

To complete the spreadsheet formulation that Table 4.4 suggests, we would need to

- Drag the function in cell D3 across cell D4, and drag the function in cell B5 across cells C5 and D5.
- Identify B3:C4 as the changing cells, and insert the nonnegativity constraints on the changing cells.
- Insert the bounds on *A*, *B*, *R*, and *P*.
- Insert the blending constraints.
- Measure the contribution and maximize the contribution.

Which spreadsheet do you prefer—the standardized formulation or the one that is tailored to the specific problem? This is a matter of personal taste. People who are comfortable with spreadsheets may prefer a tailored formulation. People who are at greater ease with algebra may prefer the standard formulation. Each type of formulation has advantages, as follows.

Table 4.4 Names of the variables whose values would be in cells B3:C4 and of the functions in five other cells.

	A	B	C	D	E	F
1						
2		Regular	Premium	total		
3	Method A	AR	AP	A	←	=\$B3+\$C3
4	Method B	BR	BP	B		
5	total	R	P	C		
6		↑				
7		=B\$3+B\$4				

Advantages of the standardized formulation (e.g., Table 4.1):

- Simplicity—For small problems, it can be easier to build and to debug, and it can be easier to read.
- Sensitivity analysis—Each variable is kept in a changing cell, so the Sensitivity Report gives the range of each contribution for which the optimal solution is unchanged. In Table 4.1, for instance, *C* is kept in a changing cell, so the Sensitivity Report gives you the range of crude oil prices for which the optimal solution is as described previously. By contrast, in Table 4.4, *C* is not kept in a changing cell, and you will need to experiment to find this range.
- Flexibility—If you change the linear program, the standard formulation adapts readily. The tailored formulation may not adapt because a change can destroy the pattern that you used to arrange its variables and constraints.

Advantages of the tailored formulation (e.g., Table 4.3):

- Structure—The variables and constraints in a linear program tend to follow patterns. A specialized formulation can exploit these patterns in ways that make the constraints easy to build and easy to read.
- Size and sparsity—Typically, in a large linear program, only a small fraction of the coefficients are nonzero. The tailored formulation can group the nonzero coefficients into blocks, omitting large numbers of zeros. By contrast, the standard formulation tends to scatter the nonzero elements across many pages in ways that make it harder to build, to read, and especially to debug.
- Less algebra—Often, a tailored spreadsheet can be built without writing down any equations. You can use row sums or column sums to compute totals, for instance.

In this chapter, we employ standardized and tailored spreadsheets, depending on the setting.

4.7. THE NETWORK FLOW MODEL

The name "network flow model" describes an important class of linear programs. We interrupt our survey of linear programs to describe them.

- First, a network flow model arises frequently, either as a stand-alone linear program or as a major component of a larger linear program.
- Second, a network flow model is easy to spot, once you learn what to look for.
- Third, some network flow models can be guaranteed to have optimal solutions that are integer-valued.

The last reason is particularly handy. It shows how linear programming can be used to solve a certain type of **integer program**, which differs from a linear program by requiring that some or all of the decision variables must be integer-valued.

As noted earlier, a **directed network** consists of **nodes** and **arcs**. To visualize a directed network, we can represent each node as a circle and each arc as a line segment with an attached arrow. As Figure 4.2 suggests, each arc falls into one of two classes. Either both ends of an arc's line segment are attached to nodes, or one end of an arc's line segment is attached to a node and the other end is unattached. In a directed network, the arrow on each arc is meant to suggest a direction of "movement" or of "flow."

A **network flow model** is a linear program whose decision variables are the flows on the arcs of a directed network. Each flow occurs in the direction of its arc's arrow. These flows must satisfy two conservation laws. The quantity that flows into the "tail" end of each arc flows out the "head" end; in this way, flow is conserved along each arc. Similarly, the total of the flows on the arcs pointing into each node equals the total of the flows on the arcs pointing out of that node. In this way, flow is conserved at each node.

In addition, each arc in a network flow model has three elements of data: a lower bound, an upper bound, and a price.

- Each arc has a lower bound on its flow. This lower bound may be zero or positive.
- Each arc has an upper bound on its flow. Each arc's upper bound must be at least as large as its lower bound, and it may be as large as $+\infty$.
- In a minimization problem, each arc has a unit cost. In a maximization problem, each arc has a unit profit. The cost (or profit) of a flow on an arc equals the product of the arc's unit cost (or profit) times the amount of the flow.

Thus, a typical arc (i, j) has three elements of data: a lower bound L_{ij}, an upper bound U_{ij}, and a unit cost c_{ij}. The flow x_{ij} on arc (i, j) must lie in the interval $L_{ij} \le x_{ij} \le U_{ij}$, and this flow contributes $c_{ij}\, x_{ij}$ to the objective. An arc can have $L_{ij} = U_{ij}$, in which case its flow is fixed. An arc can have $U_{ij} = \infty$, in which case there is no upper bound on its flow. Each arc must have $L_{ij} \ge 0$, so its flow cannot be negative. An arc can have $L_{ij} = 0$, in which case its flow can be as low as zero.

Can we formulate Problem A (cracking and blending) as a network flow model? Not quite. Program 4.1 has a flow conservation constraint at each node, it has flow conservation along each arc, it has bounds on each flow, and it has a unit revenue on each flow. But its octane rating constraints keep it from being a network flow model. Program 4.1 would be a network flow model, except for the two "side constraints" that model the blending.

Can we formulate Problem B (shipping) as a network flow model? Yes. Figure 4.2 shows how.

Integer-valued Variables

In many applications, the flows in a network flow model are quantities (like airplanes, oil tankers, or people) whose values must be integers. In these applications, we have scant interest in fractional solutions, for example, in flying 0.5 airplanes from one city to another. Happily, there is a simple hypothesis under which at least one optimal solution to a network flow model is guaranteed to be integer-valued. A network flow model is said to have **integer bounds** if each arc's lower and upper bound is an integer or is plus infinity.

> **Integrality Theorem.** In a network flow model that has integer bounds, every extreme point assigns to each arc a flow whose value is an integer.

This is good news. The Integrality Theorem shows that the simplex method solves some *integer* programs! It solves each integer program that can be posed as a network flow model whose upper and lower bounds are integers. The simplex method does so because it finds an optimal solution at an extreme point.

A Warning

Solver does *not* use the simplex method unless you check off the "Assume Linear Model" box on its Options menu. If you forget to check this box, you can get an optimal solution that is fractional-valued, even if the extreme points are integer-valued. Should this occur to you, simply check off the "Assume Linear Model" box, and solve it again.

How Can You Identify a Network Flow Model?

There is an easy way to check whether a linear program has been posed as a network flow model. To do so, first write each node's conservation-of-flow equation as "Flow out minus flow in equals zero." After you do so, each decision variable can make at most five appearances. Consider a typical arc, say, an arc that points from node 2 to node 4. The flow x_{24} on this arc can appear *only* in these ways:

- In the arc's lower-bound constraint, $L_{24} \leq x_{24}$.
- In the arc's upper-bound constraint, $x_{24} \leq U_{24}$.
- In the flow conservation constraint for node 2, with a coefficient of $+1$ (because x_{24} flows out of node 2).
- In the flow conservation constraint for node 4, with a coefficient of -1 (because x_{24} flows into node 4).
- In the objective, with a coefficient of c_{24}.

If every variable appears only in these ways, it is a network flow model. If *any* variable appears in *any* other way, it is not a network flow model. For example, it is not a network flow model if one decision variable has a coefficient of $+1$ in two or more constraints, excluding upper and lower bounds.

Network Flow with Gains

The network flow model has a useful generalization that is known as **network flow with gains** or, alternatively, as **generalized network flow**. In the more general model, the amount that flows out of the "head" end of an arc is *proportional* to the amount flowing into the "tail" end of that arc rather than being equal to it.

This generalization has all of the features of the ordinary network flow model, except that each arc (i, j) has one extra datum, a **gain rate**, g_{ij}. Each unit that flows into the tail-end of an arc (i, j) is multiplied (amplified) by the factor g_{ij}. The gain rates can vary with the arc. These "gains" can be any numbers. They need not exceed 1.0; and they can be less than 1.0; and they can even be negative. Of what use are the gain rates?

- In an investment model, the quantity that flows is money, each arc represents an investment, and the arc's gain rate is the return per unit invested.
- In a model of the flow of energy through the economy, the quantity that flows is energy, each arc represents the conversion of one type of energy into another (e.g., oil into electricity), and the arc's gain rate is the conversion efficiency of its technology.

We've observed that network flow models are common, that they are easy to spot, and that they have integer-valued optimal solutions when their data are integer-valued. Let us now resume our survey of areas of applications of linear programming.

4.8. WORKFORCE PLANNING

Many organizations serve their customers for many more hours each week than any individual person can work. This is true of a police department, an emergency room, and an international airline, each of which operates around the clock. The **workforce planning**

problem is that of matching staffing levels to the company's needs. Let us consider a simple example.

Problem C (Staffing Shifts)

A priority mail company operates a distribution office around the clock. Its minimum staffing needs appear in Table 4.5. For instance, 15 employees are needed between 6 A.M. and 10 A.M. Each employee is assigned to one of six shifts, which start at 2 A.M. and every four hours thereafter. The second shift starts at 6 A.M., the third shift starts at 10 A.M., and so forth. Each shift lasts eight hours. Employees earn 20% overtime pay for each hour worked between 6 P.M. and 6 A.M. The company has asked you to set staffing levels on each shift that meet or exceed the staffing needs in each period and, subject to that requirement, minimizes the payroll cost. How can you accomplish this?

Table 4.5 Minimum staffing level in each period.

Period	2 A.M. to 6 A.M.	6 A.M. to 10 A.M.	10 A.M. to 2 P.M.	2 P.M. to 6 P.M.	6 P.M. to 10 P.M.	10 P.M. to 2 A.M.
Minimum	5	15	12	10	19	7

Evidently, the decision variables in Problem C are the number of employees to assign to each eight-hour shift. Table 4.6 organizes the data for this problem on a spreadsheet. In this spreadsheet:

- Row 3 describes the starting and ending time of each shift.
- Row 4 is reserved for Solver's changing cells, which will specify the number of employees to assign to each shift.
- Row 5 gives the cost factor for each shift. For instance, the 2 P.M. to 10 P.M. shift has 1.1 as its a cost factor because half of its hours earn the overtime bonus of 20%.
- Rows 6 through 11 describe the supply and demand of workers in each four-hour period. For instance, the employees who work the shift from 2 A.M. to 10 A.M. (column B) help

Table 4.6 A spreadsheet for Staffing Shifts (Problem C).

	A	B	C	D	E	F	G	H	I	J
1	night bonus =	0.2								
2	shift number	1	2	3	4	5	6			
3	shift starts at shift ends at	02 am 10 am	06 am 02 pm	10 am 06 pm	02 pm 10 pm	06 pm 02 am	10 pm 06 am			
4	staffing level	5	12	0	12	7	0			rqmt
5	cost factor	1.1	1	1	1.1	1.2	1.2	39.1		
6	02 am to 06 am	1					1	5	>=	5
7	06 am to 10 am	1	1					17	>=	15
8	10 am to 02 pm		1	1				12	>=	12
9	02 pm to 06 pm			1	1			12	>=	10
10	06 pm to 10 pm				1	1		19	>=	19
11	10 pm to 02 am					1	1	7	>=	7
12								↑		
13				=SUMPRODUCT(B$4:G$4,B11:G11)						

Table 4.7 A Solver dialog box for Problem C.

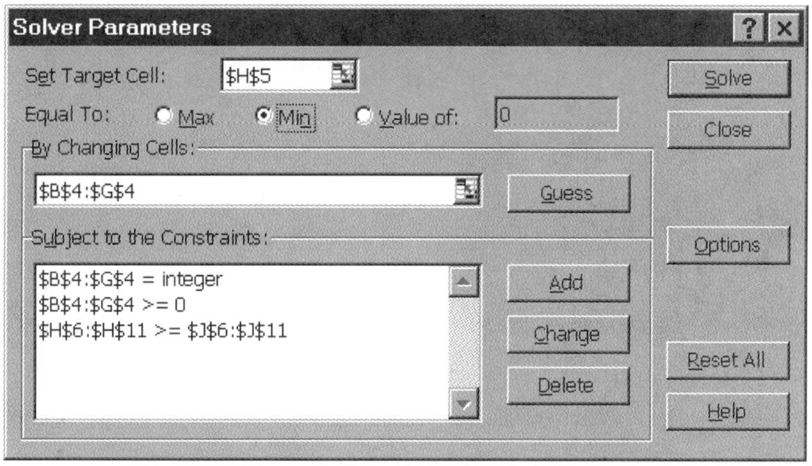

meet the need in the period from 2 A.M. to 6 A.M. (row 6) and in the period from 6 A.M. to 10 A.M. (row 7). This explains the 1 in cell B6 and the 1 in cell B7.

- Column H contains the usual SUMPRODUCT functions. For instance, cell H11 computes the number of employees who are at work between 10 P.M. and 2 A.M.
- Column J specifies the minimum staffing requirement for each period.

The object of this staffing problem is to provide adequate numbers of workers in each time period and to do so in a way that minimizes the payroll expense. Table 4.7 presents the Solver dialog box for an integer program that accomplishes this. This dialog box minimizes the number of regular-time-equivalent salaries (cell H5) subject to constraints that assign a nonnegative *integer* number of employees to each shift and that provide a staffing level in each period that is at least as large as is required.

Integer-valued Variables in Solver

Table 4.7 imposes the constraint that the values in cells B4 through G4 be integers. How do we get Solver to impose such a constraint? Typing "integer" in a window of the Add Constraint dialog box will *not* work. Instead, in Solver's Add Constraint dialog box, click on the "int" option in the small window, as is indicated in Table 4.8. When you do so, the word "integer" appears automatically in the Solver dialog box.

A Warning

Inserting the requirement that variables be integer-valued is nearly effortless. But the inclusion of integer-valued variables *severely* complicates the computation. If you require a

Table 4.8 Requiring variables to be integer-valued.

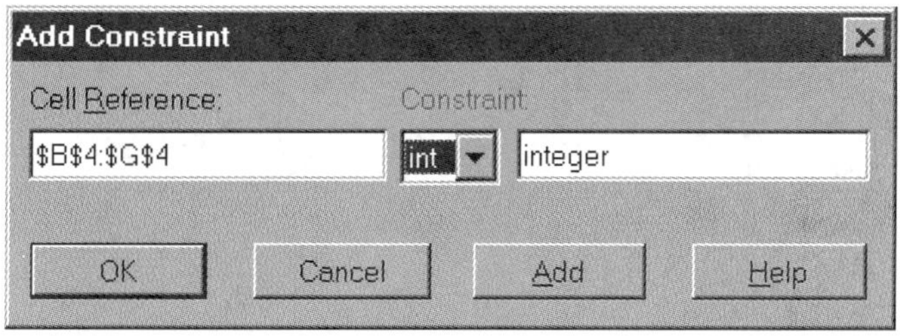

large number of variables to be integer-valued, Solver may have to solve thousands of linear programs before getting the answer. Solver may run for a very long time and may even throw up its hands in despair. Incidentally, you can interrupt Solver by hitting the **ESC** key.

Is This a Linear Program?

Let us forget the requirement that the variables be integer-valued. With these constraints eliminated, let us run Solver. What happens? It reports integers in cells B4 through G4. Specifically,

$$x_1 = 5, x_2 = 12, x_3 = 0, x_4 = 12, x_5 = 7, x_6 = 0,$$

where, for instance, x_1 equals the number of people who are assigned to shift 1, which is from 2 A.M. to 10 A.M. Was it a happy accident that these staffing levels are integers? Or is Solver guaranteed to provide an integer-valued solution?

If we formulated Problem C as a network flow problem with integer bounds, the Integrality Theorem would guarantee an integer-valued optimal solution. Did we? That's not quite clear. Rows 6 through 11 seem to model "conservation-of-flow" constraints. Each variable appears with a coefficient of $+1$ in *two* of these constraints. If this were a network flow model, each variable could be made to appear in at most two of those constraints, once with a coefficient of $+1$ and once with a coefficient of -1.

But there are six such constraints, and six is an even number. Let us multiply the constraints for the *even-numbered* periods by -1. In Table 4.6, we reverse all signs in rows 7, 9, and 11, and we switch their constraints from "$\leq$" to "$\geq$." This has no effect on the feasible region (set of feasible solutions). Now, each variable appears exactly twice in these "conservation-of-flow" constraints, once with a coefficient of $+1$ and once with a coefficient of -1. Thus, Problem C is a network flow problem with integer bounds, thinly disguised. The Integrality Theorem does guarantee that the simplex method finds an optimal solution that is integer-valued. The requirement that the variables be integer-valued is unnecessary. It's a linear program, not an integer program.

4.9. MAXIMIZING THE FLOW

One often wishes to maximize the total quantity that can flow from one place to another. A **max-flow problem** is a network flow model that has two distinguished nodes, which are called the **source** and the **sink**. The object of a max-flow problem is to send the maximum flow from the source node to the sink node, while satisfying the capacity constraint on each arc and the flow conservation constraint at each intermediary node. Let us examine a simple instance of this.

Problem D (An Electrical Emergency)

Because of an electrical breakdown, a city is in need of electrical power. There is ample reserve power at a nearby hydroelectric station. Figure 4.4 depicts the portion of the power grid that connects the hydroelectric station to the city. Each power line is represented by an arc or by a pair of arcs. The hydroelectric station is connected by a power line (arc) to node 1 of this figure. This city is connected by a power line (arc) to node 8. Each node depicts the junction of at least three power lines. Adjacent to each intermediary arc is the spare capacity of its power line in the arc's direction of flow. For instance, the power line between nodes 3 and 6 has eight units of spare capacity in the 3-to-6 direction and three units of spare capacity in the 6-to-3 direction. Transmission losses along the power lines are negligible, so flow is conserved along each arc. Flow is also conserved at each node of the power grid. How much added power can be supplied to this city? How can it be routed to the city?

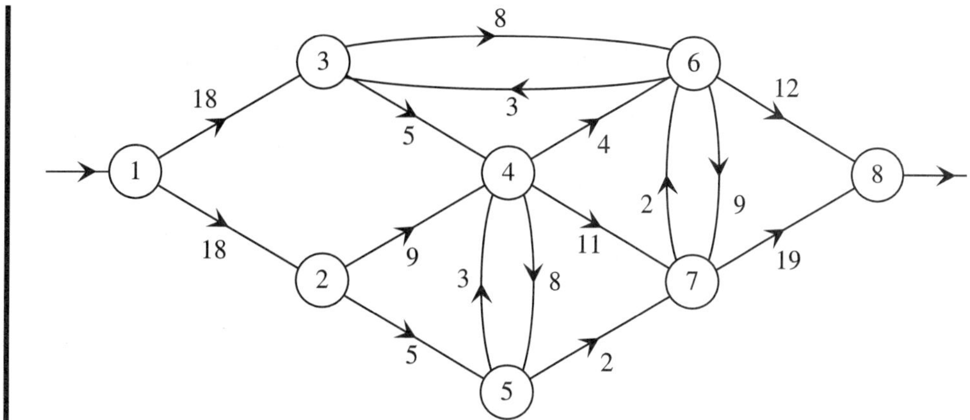

Figure 4.4 A power grid, with the residual capacity on each arc.

Figure 4.4 is a directed network, and all but two of its arcs have upper bounds on their flows. Flow is conserved along each arc and at each node. The object is to maximize the flow out of node 8. The flow out of node 8 (to the city) equals the flow into node 1 (from the hydroelectric station).

Table 4.9 presents a spreadsheet for this linear program, along with its optimal solution. In Table 4.9:

- Row 1 identifies the arcs in Figure 4.4.
- Row 2 contains the changing cells, and each changing cell contains the flow that Solver has assigned to its arc.
- Row 3 specifies the capacity of each arc.
- Rows 4 through 11 describe the flow conservation constraints for the nodes. Each flow conservation constraint has been written as "Flow out minus flow in equals zero."
- Column T contains the usual SUMPRODUCT functions.

Table 4.9 A spreadsheet for max flow (Problem D), with the changing cells in row 2.

	A	B	C	D	E	F	G	H	I	J	K	L	M	N	O	P	Q	R	S	T	U	V	W
1	arc	S1	12	13	24	25	34	36	45	46	47	54	57	63	67	68	76	78	8D				
2	flow	25	14	11	9	5	3	8	0	4	11	3	2	0	-0	12	0	13	25				shadow
3	cap.		18	18	9	5	5	8	8	4	11	3	2	3	9	12	2	19					price
4	node 1	-1	1	1																0	=	0	-1
5	node 2		-1		1	1														0	=	0	-1
6	node 3			-1			1	1						-1						0	=	0	-1
7	node 4				-1		-1		1	1	1	-1								0	=	0	-1
8	node 5					-1			-1			1	1							0	=	0	-1
9	node 6							-1		-1				1	1	1	-1			0	=	0	0
10	node 7										-1		-1		-1		1	1		0	=	0	0
11	node 8															-1		-1	1	0	=	0	0
12																							
13									=SUMPRODUCT(B11:S11,B$2:S$2)														

Table 4.10 Solver dialog box for Problem D.

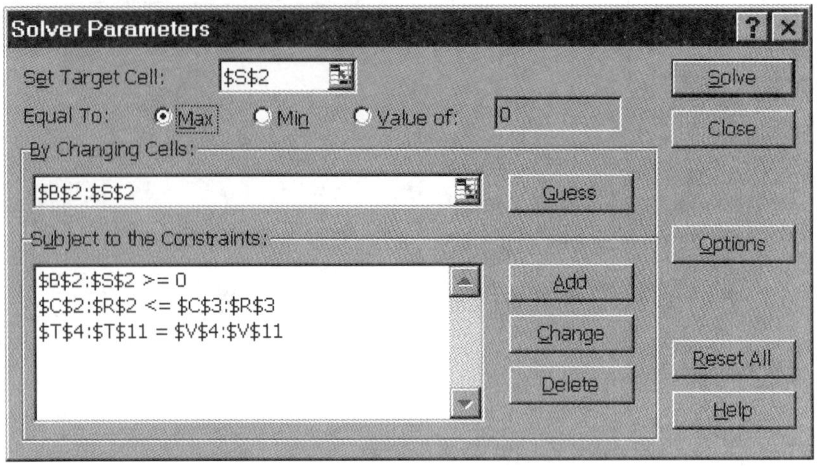

To illustrate, we examine column Q of Table 4.9. Cell Q1 records the fact that this column models arc (7, 6), which points out of node 7 and into node 6. The entry in cell Q2 is the flow that Solver has assigned to this arc, namely, 0. The entry in cell Q3 is the capacity of arc (7, 6), which equals 2. Cell Q10 contains +1 because this arc points out of node 7. Cell Q9 contains −1 because this arc points into node 6.

The Solver dialog box for this spreadsheet appears as Table 4.10. Solver identifies the flows as the changing cells. Its constraints keep these flows nonnegative, and they keep each flow from exceeding the capacity of its arc. Its constraints also require that the flow out of each node minus the flow into that node equals zero. Subject to these constraints, Solver maximizes the flow into the city (the value in cell S2).

Table 4.9 reports the optimal solution that Solver has found for this linear program. Exactly 25 units of flow can be sent from the hydroelectric plant to the city, and the flow variables in Table 4.9 show how to get it there. Also recorded in this table are the shadow prices for the nodes. These shadow prices are linked to a lovely theorem. To introduce this theorem, we must define the terms *cut* and *cut capacity*, terms whose meanings are easy to grasp graphically. In a max-flow problem:

- A **cut** is any set S of nodes that includes the source but not the sink.
- Each cut S **deletes** the set $A(S)$ of arcs that consists of each arc (i, j) having i in S and j not in S.
- Each cut S has a **capacity** $c(S)$ that equals the sum of the capacities of the arcs in $A(S)$.

There are many cuts. In Figure 4.4, one cut consists of the set $S = \{1, 3\}$ of nodes. Notice in Figure 4.4 that the cut $S = \{1, 3\}$ deletes the set $A(S)$ that consists of arcs (1, 2), (3, 4), and (3, 6). In this figure, the cut $S = \{1, 3\}$ has $c(S) = 31$ because 31 equals the sum of the capacities of arcs (1, 2), (3, 4), and (3, 6).

For each cut S, the deleted arc-set $A(S)$ makes it impossible to ship anything from the source to the sink. The capacity $c(S)$ of each cut S places an upper bound on the maximum flow from the source to the sink.

We know that the maximum flow in Figure 4.4 is 25 units. Is there a cut whose capacity equals 25? Yes. In 1956, L.R. Ford and D.R. Fulkerson caused a sensation with the publication of their Max-Flow Min-Cut theorem.

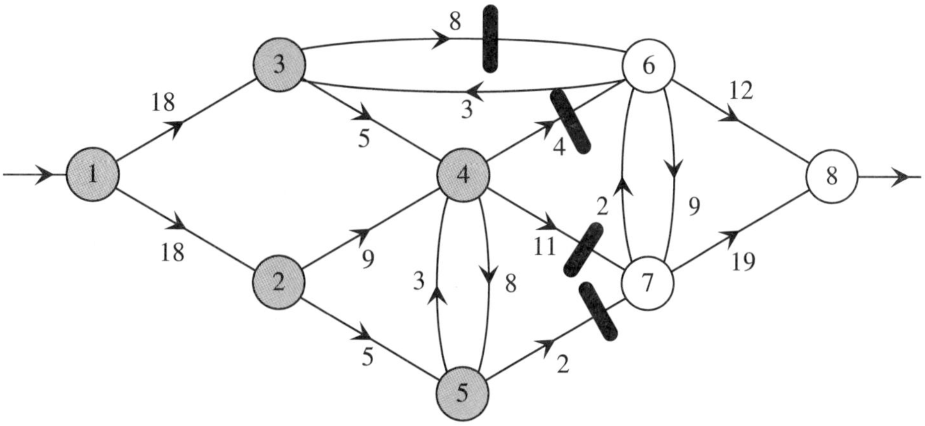

Figure 4.5 The cut $S = \{1, 2, 3, 4, 5\}$ consisting of each node whose shadow price equals that of the source.

> **Max-Flow Min-Cut Theorem.** The maximum flow equals the smallest of the capacities of the cuts.

We find a cut whose capacity is smallest from the shadow prices of the nodes. Notice in Table 4.9 that there are only two prices: The source has one price, which equals -1; and the sink has another price, which equals 0. Let us examine the cut S that consists of each node whose shadow price equals that of the source. The shadow prices in Table 4.9 show us that this cut is given by $S = \{1, 2, 3, 4, 5\}$. In Figure 4.5, the nodes in S have been shaded, and a thick line has been drawn through each arc in $A(S)$. It's evident from Figure 4.5 that $c(S) = 8 + 4 + 1 + 2 = 25$.

Figure 4.5 is no accident. For *every* max-flow problem, the simplex method terminates as follows: The set S of nodes whose shadow prices equal that of the source form a cut whose capacity $c(S)$ equals the maximum flow.

There is nothing "linear" about a cut; a node is either in the cut or is not. Nonetheless, the simplex method finds the cut whose capacity is smallest. For a potent demonstration of the power of linear programming, that is hard to beat.

4.10. ACTIVITY ANALYSIS

An **activity analysis** is described in terms of **goods** and **technologies**. Each technology transforms one bundle of goods into another. The **inputs** to a technology are the goods that it consumes, whereas the **outputs** of a technology are the goods that it produces. Each technology can be operated at any nonnegative **level**. The decision variables in an activity analysis are the levels at which to operate the technologies.

The simple activity analysis that we are building has constant returns to scale. Thus, tripling both the level of a technology triples both the consumption of each of its inputs and the production of each of its outputs.

Activity analysis has played a central role in economics. For several decades, Wassily Leontief built activity analyses of the American economy, obtaining important insights from them. For this work, he was awarded the Nobel prize in economics in 1973. As we shall see in Chapter 19, activity analysis also plays a key role in "general equilibrium," which is the cornerstone of microeconomic theory.

A Sector

An activity analysis can model the production capabilities of a plant, a single firm, a sector of the economy, or an entire economy. Let us imagine an activity analysis of the sector of the economy that produces iron. In one such model, the technologies are mining, transportation (of ore to smelters), and smelting. The inputs to smelting are iron ore, energy, and capital, whereas the outputs are iron ingots, slag, and heat.

In this model, both the goods and the technologies are aggregated. The good "iron ore" describes iron from several different mines, each with its own mineral content. The technology of "smelting" includes several smelters, each of which has its own operating characteristics. Building an activity analysis is an exercise in engineering—in making useful approximations.

Balance of Trade

An activity analysis can be a bit subtle. Later in this section, we develop a general model of activity analysis. Let us begin with a simple example, one in which each technology has only one output.

Problem E (Olde England)

Olde England engaged in three major technologies: the production of food, yarn (spun from wool or flax), and cloth (woven from yarn). Table 4.11 describes the inputs to each technology, the domestic consumption of each good, and various capacities. Specifically:

- The first three columns of Table 4.11 specify the inputs to each technology. For instance, the production of each unit of food consumed 0.2 units of food (such as grains fed to animals), 0.10 units of labor, 0.15 units of land, and 0.50 £ (pounds) of imported goods.
- The fourth column of Table 4.11 specifies the annualized domestic consumption of food, yarn, and cloth, for example, consumption of 0.06 billion units of yarn.
- The final column specifies the annualized capacities, which include the capacity to spin 2.5 billion units of yarn as well as 0.85 billion units of land.

The amounts of food, yarn, and cloth that Olde England produced in excess of domestic consumption were sold on the world market at these prices: 3 £ (pounds) per unit for food, 12 £ per unit for yarn, and 25 £ per unit for cloth. If Olde England produced less than its domestic consumption of food, yarn, or cloth, any deficit was purchased on the

Table 4.11 The inputs per unit output of each technology, annual levels of domestic consumption, and annual capacity.

	Technology: Production of				
Good	Food	Yarn	Cloth	Consumption (units/year)	Capacity (units/year)
Food (units)	0.2			2.3 billion	2.8 billion
Yarn (units)		1.3		0.06 billion	2.5 billion
Cloth (units)			1	0.20 billion	2.1 billion
Labor (units)	0.10	0.25	0.60		1.9 billion
Land (units)	0.15	0.20			0.85 billion
Imports (£)	0.50	1.25	5.00		

world market at the same prices. Olde England wished to maximize the value of its net exports in pounds (£), less the cost of any imports, subject to the requirement that it serve the needs of its population. Which technologies should Olde England have operated and at what levels?

A Linear Program for Olde England

To formulate Olde England's problem as a linear program, we employ two different types of decision variable. One type specifies the level of each technology, and the other specifies the "net" amount of each good that Olde England exports. Let us designate:

F = Olde England's food production (in billions of units per year),
Y = Olde England's yarn production (in billions of units per year),
C = Olde England's cloth production (in billions of units per year),
FE = Olde England's net export of food (in billions of units per year),
YE = Olde England's net export of yarn (in billions of units per year), and
CE = Olde England's net export of cloth (in billions of units per year).

The production quantities must be nonnegative. By contrast, the net exports can be negative. **Net** export means the amount exported *less* the amount imported. If, for instance, $FE = -0.6$, then Olde England imports 0.6 billions of units of food per year.

Olde England's net balance of trade equals the value of its net exports less the cost of the imported goods that it requires for production. Its net balance of trade is given by the expression

$$\{3\,FE + 12\,YE + 25\,CE - 0.50\,F - 1.25\,Y - 5.00\,C\}.$$

We can use one set of variables for net exports because the same market price applies to purchases and sales. For instance, the addend $3\,FE$ accounts correctly for the contribution of net food exports when FE is positive and when FE is negative.

The top three rows of Table 4.11 account for the production and uses of the three goods that Olde England trades on the world market. Each of the these rows gives rise to an equation of the form:

$$(\text{production}) = \begin{pmatrix}\text{consumption by}\\ \text{technologies}\end{pmatrix} + \begin{pmatrix}\text{consumption by}\\ \text{population}\end{pmatrix} + (\text{net export}).$$

Thus, reading across the first three rows of Table 4.11 allows us to build equations for food, yarn, and cloth, which are

$$F = 0.2\,F + 1.3\,Y + 2.3 + FE,$$
$$Y = 1.0\,C + 0.06 + YE,$$
$$C = 0.20 + CE.$$

To interpret the first of these equations, we note that the quantity F of food produced equals the sum of:

$0.2\,F + 1.3\,Y$ (the amount of food consumed by the technologies),
2.3 (the amount of food consumed by the population),
FE (the net export amount of food).

In these equations, it is essential that net export quantities can be negative. If total food consumption exceeds food production, then FE is negative and $-FE$ equals the amount of food that is imported.

The capacities in Table 4.11 give rise to these five constraints:

$$F \leq 2.8,$$
$$Y \leq 2.5,$$
$$C \leq 2.1,$$
$$0.10\, F + 0.25\, Y + 0.6\, C \leq 1.9,$$
$$0.15\, F + 0.20\, Y \leq 0.85.$$

The entire linear program has now been displayed, except for the nonnegativity constraints on the variables F, Y, and C.

A Spreadsheet for Olde England

Table 4.12 organizes this linear program on a spreadsheet. Row 1 labels its variables, Row 2 contains its changing cells, Row 3 contains the coefficients of its objective, Rows 4, 5, and 6 contain the data in its the three equality constraints, with the variables shifted to their left-hand sides, and Rows 7 through 11 contain the data for its five capacity constraints. Column H contains the usual SUMPRODUCT functions.

Table 4.13 displays the Solver dialog box for this spreadsheet. This dialog box maximizes the quantity in cell H3, with B2 through G2 as the changing cells, subject to constraints that keep the quantities in cells B2 through D2 nonnegative and impose the equalities and inequalities indicated by column I of Table 4.12. Table 4.13 omits nonnegativity constraints on the values in cells E2, F2, and G2. Why is that?

Table 4.12 displays the optimal solution that Solver has found. As is usual, the inequality constraints that hold as equations have been shaded. Evidently, Olde England should have made full use of its labor, land, and capacity to spin yarn. The nation should have been a net exporter of yarn and cloth and an importer of food.

The Technology Matrix

Olde England served to introduce activity analysis. Our model of Olde England exploited the fact that its technologies have one output apiece, enabling us to specify (in Table 4.11) the inputs needed to produce one unit of the output of each technology. If a technology has two or more outputs, we must describe it in a different way. A general model of an activity

Table 4.12 A spreadsheet for Olde England.

	A	B	C	D	E	F	G	H	I	J
1		F	Y	C	FE	YE	CE			
2	values	2.333	2.5	1.736	-3.68	0.704	1.536			
3	contribution	-0.5	-1.25	-5	3	12	25	22.827		
4	food prod.	0.8	-1.3		-1			2.3	=	2.3
5	yarn prod.		1	-1		-1		0.06	=	0.06
6	cloth prod.			1			-1	0.2	=	0.2
7	food cap.	1						2.3333	<=	2.8
8	yarn cap.		1					2.5	<=	2.5
9	cloth cap.			1				1.7361	<=	2.1
10	labor cap.	0.1	0.25	0.6				1.9	<=	1.9
11	land cap.	0.15	0.2					0.85	<=	0.85
12								↑		
13			=SUMPRODUCT(B$2:G$2,B11:G11)							

Table 4.13 Solver dialog box for Olde England.

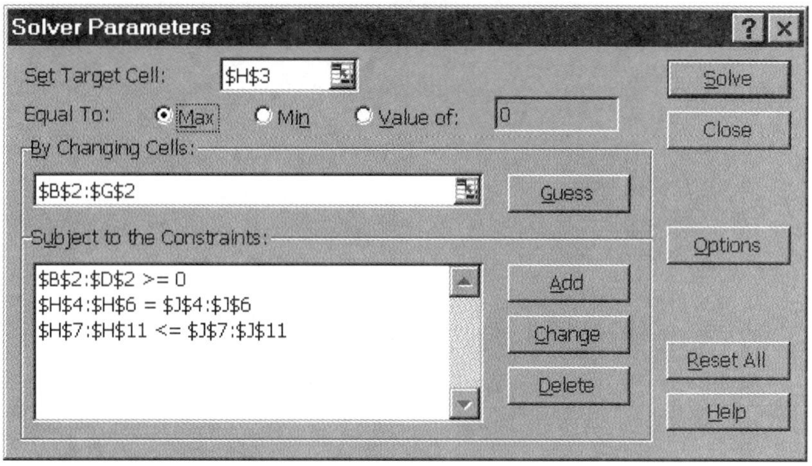

analysis entails a "technology matrix." To describe this matrix, we reserve the integers m and n for the meanings that are specified below:

$$m = \text{the number of goods,}$$
$$n = \text{the number of technologies.}$$

It is convenient to number the goods *1* through m and the technologies *1* through n. In an activity analysis, the **technology matrix** A has m rows (one for each good) and n columns (one for each technology); the entry A_{ij} at the intersection of its ith row and its jth column is given by

$$A_{ij} = \text{the } \textit{net} \text{ production of good } i \text{ that is obtained by setting the level of technology } j \text{ equal to } 1.$$

As is usual, a net quantity can be positive, negative, or zero. If A_{ij} is positive, then good i is an output of technology j. If A_{ij} is negative, then good i is an input to technology j. If A_{ij} is zero, then good i is neither an input nor an output of technology j.

The decision variables in an activity analysis are the levels at which the technologies are to be operated. There are n decision variables, one per technology, and we denote them by

$$x_j = \text{the level at which technology } j \text{ is operated,} \quad \text{for } j = 1, \ldots, n.$$

Technology levels must be nonnegative, so these decision variables must satisfy the constraints

$$x_j \geq 0, \quad \text{for } j = 1, \ldots, n.$$

The technology matrix is an elegant tool for describing an activity analysis, but it has a peculiar feature. When we use a technology matrix, the net production of a good can be *negative*. To illustrate, we suppose (as is the case in Olde England) that labor is an input to each technology. The net production of labor will be negative if any activity levels are positive. Net production of -6 units of labor is a consumption of $+6$ units of labor.

Activity Analysis of a Plant

To illustrate the technology matrix and negative net quantities, we construct an activity analysis of a plant that produces some goods and consumes others. Let us consider:

Problem F (Activity Analysis of a Plant)

In a company that engages in centralized control, plant 7 has been directed to set its production levels so as to meet or exceed its net demand of each of its goods. The plant manager seeks the least expensive way to accomplish this goal. This plant's goods are labeled 1 through m, its technologies are labeled 1 through n, and its data are the numbers, A_{ij}, c_j, u_j, and b_i whose definitions are:

$$
\begin{aligned}
A_{ij} &= \text{the net production of good } i \text{ per unit of technology } j, && \text{each } i \text{ and } j,\\
c_j &= \text{the unit cost of operating technology } j, && \text{each } j,\\
u_j &= \text{the capacity (maximum level) of technology } j, && \text{each } j,\\
b_i &= \text{the minimum } \textit{net} \text{ demand of good } i, && \text{each } i.
\end{aligned}
$$

Program 4.2, formulates the plant's decision problem as a linear program. In this program, the first m constraints keep the net production of each good at least as large as the net demand. Its remaining constraints keep the levels of the technologies within bounds.

Program 4.2: Minimize $\{c_1 x_1 + c_2 x_2 + \cdots + c_n x_n\}$, subject to the constraints

$$
\begin{aligned}
A_{11} x_1 + {} & A_{12} x_2 + \cdots + A_{1n} x_n \geq b_1,\\
A_{21} x_1 + {} & A_{22} x_2 + \cdots + A_{2n} x_n \geq b_2,\\
& \vdots\\
A_{m1} x_1 + {} & A_{m2} x_2 + \cdots + A_{mn} x_n \geq b_m,\\
0 \leq x_1 \leq u_1,\ & 0 \leq x_2 \leq u_2, \ldots, 0 \leq x_n \leq u_n.
\end{aligned}
$$

To illustrate Program 4.2, we turn from a general presentation of the plant's data to the specific numbers in Table 4.14. This plant engages in three technologies and produces or consumes five goods. Except for the costs, each datum in Table 4.14 has been taken from Olde England.

Evidently, goods 1, 2, and 3 are outputs of technologies 1, 2, and 3, respectively. Good 4 is an input to each technology, and good 5 is an input to technologies 1 and 2. The data in Table 4.14 particularize Program 4.2 to:

Program 4.3: Minimize $\{5.5x_1 + 12.5x_2 + 25x_3\}$, subject to the constraints

$$
\begin{aligned}
0.8x_1 - 1.3x_2 & \geq 2.3,\\
1x_2 - 1x_3 & \geq 0.06,\\
1x_3 & \geq 0.20,\\
-0.1x_1 - 0.25x_2 - 0.6x_3 & \geq -1.9,\\
-0.15x_1 - 0.20x_2 & \geq -0.85,\\
0 \leq x_1 \leq 2.8,\ 0 \leq x_2 \leq 2.5,\ & 0 \leq x_3 \leq 2.1.
\end{aligned}
$$

Table 4.14 Plant 7's technology matrix and other data.

	Technology j				
Good i	1	2	3	Good i	b_i
1	0.80	−1.30		1	2.3
2		1.00	−1.00	2	0.06
3			1.00	3	0.20
4	−0.10	−0.25	−0.60	4	−1.9
5	−0.15	−0.20		5	−0.85
u_j	2.8	2.5	2.1		
c_j	5.50	12.0	25.00		

To interpret Program 4.3, we suppose that good 4 is labor. Evidently, labor is an input to each technology. The net demand for labor is -1.9 units, and equivalently, the net supply of labor is $+1.9$ units. A negative net demand is equivalent to a positive net supply.

Negative Net Quantities

The tricky part of an activity analysis is perhaps to interpret net quantities that are negative. Negative net production is positive consumption, negative net demand is positive supply, and negative net export is positive import.

Multiplying an inequality constraint by -1 reverses the sense of each quantity and the inequality as well. Consider a constraint that requires net production to be at least as large as net demand. When multiplied by -1, this constraint keeps net consumption from exceeding net supply.

4.11. DYNAMIC NETWORK FLOW

A network flow model is said to be **dynamic** if its flows occur at different moments in time and if time matters. In a dynamic network flow model, each arc describes a flow (movement) that begins at a certain time and place that ends at a later time and, perhaps, at a different place.

To illustrate, we model the assignment of aircraft to flights. Let us imagine that KLM (the Royal Dutch Airline) uses a dynamic network flow model to assign aircraft to its daily schedule of passenger flights. Each scheduled departure is responsible for an arc whose flow is *fixed* (its upper bound and its lower bound equal one). One such arc depicts the flight from Berlin to Amsterdam that departs from Berlin at 0900 GMT (Greenwich mean time) and reaches Amsterdam 110 minutes later. One unit of flow *must* occur on that arc. Similarly, one unit of flow must occur on each arc that depicts a scheduled flight.

The remaining arcs in KLM's model account for what happens to each aircraft *after* it arrives at its destination. That aircraft can be held at its destination for reassignment to some later flight. Alternatively, KLM can **deadhead** that aircraft (fly it without passengers) to an airport at which it is needed. Each possible reuse of an aircraft after it arrives at its destination is responsible for an arc. The decision variables are the flows on these "reuse" arcs.

A realistic model of the assignment of commercial aircraft to flights will have gigantic numbers of variables and constraints. The airlines use these models because they are profitable.

Problem G (Cargo Routing)

A pint-sized cargo-routing company owns three ships and transports cargo between three ports. The ports are labeled A, B, and C. At this moment, the beginning of day 1, one vessel is at each port. The company needs to return one vessel to each port at the beginning of day 12. For the intervening period, the company has accepted six contracts. You have been asked (belatedly) whether these contracts can be fulfilled and, if so, how to fulfill them with the smallest possible number of days of deadheading. The spreadsheet in Table 4.15 contains the relevant data. The array at its left lists the travel times from port to port; the array at its center lists the six shipping contracts that the company has accepted; and the array at its right contains no new information, consisting of derived information that will soon be described.

Our formulation of the cargo-routing problem as a linear program entails a large spreadsheet. Table 4.15 contains its top 12 rows. The CD that accompanies this book includes the entire spreadsheet. You may find it convenient to open that spreadsheet and explore it as you read this section.

Table 4.15 Travel times and scheduled trips for Problem G.

	A	B	C	D	E	F	G	H	I	J	K	L	M
1	travel times					scheduled trips					scheduled trips		
2	(in days)					trip	leave	on	for		from	to port	arrive
3		A	B	C		nmbr	port	day	port		port #	#	on day
4	A	0	2	3		1	A	1	B		1	2	3
5	B	2	0	1.5		2	B	1	C		2	3	2.5
6	C	3	1.5	0		3	C	3	B		3	2	4.5
7						4	A	4	B		1	2	6
8	return date =			12		5	A	6	C		1	3	9
9						6	C	9	A		3	1	12
10													
11					=MATCH(G9,A$4:A$6,0)								
12					=MATCH(I9,B$3:D$3,0)					=H9+OFFSET($A$3,K9,L9)			

When Trips End

Our first task is to compute the time at which each trip ends. In Table 4.15, this is accomplished by the MATCH functions in columns K and L and by the OFFSET functions in column M. Evidently, columns K and L assign the integers 1, 2, and 3 to ports A, B, and C, and column M computes the time at which each vessel appears at its destination. In particular:

- The function =MATCH(G9,A$4:A$6,0) in cell K9 has the value 3 because the entry in cell G9 is the third entry in the array A$4:A$6.
- The function =MATCH(I9,$B3:$D3,0) in cell L9 has the value 1 because the entry in cell I9 is the first entry in the array $B3:$D3.
- The function =H9+OFFSET($A$3,K9,L9) in cell M9 has the value $12 = 9 + 3$ because the entry that is three rows below cell $A$3 and one row to its right equals 3.

Dragging these functions upward computes the proper entries in cells K4 through M8. You may be able to guess how the MATCH and OFFSET functions work from the above. You could also look them up in the Appendix.

Ship Reassignment

In the cargo-routing problem, what are the decision variables? The decision variables cannot be the scheduled trips because these trips are fixed. The crucial issues are what to do with the ships that are now at Ports A, B, and C and what to do with each ship after it completes each scheduled trip. The decision variables determine how each vessel can *next* be used and at what cost. Table 4.16 presents the relevant data. Specifically:

- Each row of Table 4.16 describes a port and a time at which a vessel becomes available at that port.
- Each column of Table 4.16 describes a port and a time at which a vessel is required at that port.
- At the intersection of each row and column lies the deadhead time for the reassignment of the ship in its row to the use in its column.

Table 4.16 Potential reassignments, with the cost of each.

	A	B	C	D	E	F	G	H	I	J	K	L
13	prior	arrival data										
14	trip	port #	date	table of deadhead costs								
15	a	1	1	0	100	100	0	0	3	0	2	3
16	b	2	1	100	0	1.5	2	2	1.5	2	0	1.5
17	c	3	1	100	100	0	3	3	0	3	1.5	0
18	1	2	3	100	100	100	100	2	1.5	2	0	1.5
19	2	3	2.5	100	100	0	100	3	0	3	1.5	0
20	3	2	4.5	100	100	100	100	100	1.5	2	0	1.5
21	4	2	6	100	100	100	100	100	1.5	2	0	1.5
22	5	3	9	100	100	100	100	100	0	3	1.5	0
23	6	1	12	100	100	100	100	100	100	0	100	100
24	next	departure data		1	1	3	4	6	9	12	12	12
25	trip	from port #		1	2	3	1	1	3	1	2	3
26		on trip #		1	2	3	4	5	6	a	b	c

The top three rows of Table 4.16 describe the location of each vessel at the start of day 1. The remaining six rows of Table 4.16 identify the port that each trip reaches and the time at which the ship arrives. Similarly, the first six columns of the table describe the six scheduled trips, and the remaining three columns describe the need to return one ship to each port at the start of day 12. In particular, row 19 describes a vessel that will become available on day 2.5 at port 3. Similarly, column H describes a trip that must depart on day 6 from port 1. And cell H19 contains the number 3 because it takes three days to deadhead from port 3 to port 1 and because $2.5 + 3 \leq 6$, so the ship arrives in time. If this ship arrived after the scheduled departure, we would record 100 as the deadhead time. (Any large number would do as well.)

Table 4.16 has been built from the information in Table 4.15. For instance, columns B and C of Table 4.16 are copied from columns L and M of Table 4.15. Rows 24, 25, and 26 of Table 4.16 are copied, respectively, from columns H, K, and F of Table 4.15. The IF function was used to determine whether a "reuse" arrives too late and, if so, to equate its deadhead cost to 100.

An Assignment Problem

In Table 4.16, the deadhead times form the 9×9 array in cells D15 through L23. Each row has one unit of a supply, each column has one unit of demand, and the numbers in the array are the shipping costs. The cargo-routing problem has now been reduced to an assignment problem. Did you see that coming?

Table 4.17 displays the decision variables in this assignment problem, along with functions that compute their row and column sums, as well as the total shipping cost. These decision variables form the 9×9 array in cells D30 through L38. The functions in column M compute the row sums; the functions in row 40 compute the column sums; and the function in cell E42 computes the cost of this assignment by multiplying each shipping quantity in Table 4.17 by its cost, which appears in Table 4.16.

Table 4.17 Cargo routing (Problem G) as an assignment problem.

	B	C	D	E	F	G	H	I	J	K	L	M
28		prior										row
29		trip #				table of changing cells						sum
30		a	1	0	0	-0	0	0	0	0	0	1
31		b	0	1	0	0	0	0	0	0	0	1
32		c	0	0	-0	1	0	0	0	0	0	1
33		1	0	0	0	0	1	0	0	0	0	1
34		2	0	0	1	0	0	0	0	0	0	1
35		3	0	0	0	0	0	0	0	1	0	1
36		4	0	0	0	0	0	1	0	0	0	1
37		5	0	0	0	0	0	0	0	0	1	1
38		6	0	0	0	0	0	0	1	0	0	1
39	next trip #		1	2	3	4	5	6	a	b	c	
40	column sum		1	1	1	1	1	1	1	1	1	
41												
42		cost =		6.5		=SUM(G30:G38)				=SUM(D38:L38)		
43												
44				=SUMPRODUCT(D30:L38,D15:L23)								

Table 4.18 presents the Solver dialog box for this linear program. This LP minimizes the value in cell E42, with cells D30 through L38 as changing cells, subject to constraints that keep the values in the changing cells nonnegative, that keep each column sum equal to 1, and that keep each row sum equal to 1.

Table 4.17 exhibits the optimal solution that Solver has found to this linear program. This table reports an optimal value of 6.5, so the schedule can be met, and 6.5 days of deadheading are required. The simplex method has found an optimal solution that is integer-valued, as is guaranteed by the Integrality Theorem. From Table 4.17, we see that this optimal

Table 4.18 Solver dialog box for Problem G.

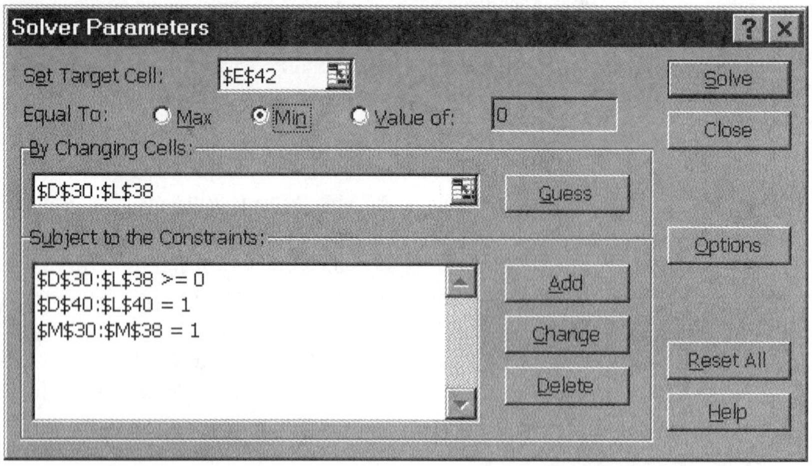

solution is to route the three vessels as follows:

$$a \to \text{trip \#1} \to \text{trip \#5} \to c,$$
$$b \to \text{trip \#2} \to \text{trip \#3} \to b,$$
$$c \to \text{trip \#4} \to \text{trip \#6} \to a,$$

where, for instance, lower case "*a*" describes a ship that begins or ends at port A.

If you forgot to check the "Assume Linear" button before solving this linear program, Solver would not use the simplex method. It might not find an integer-valued solution, and it might run and run and run. Fortunately, you can interrupt Solver by hitting the ESC key.

In building the spreadsheet for Problem G, we have practiced what we preach. Each datum appears only once, and no data are buried within Excel functions, thereby easing sensitivity analysis. For instance, changing one travel time in Table 4.15 would alter a large number of deadhead costs.

Realistic Routing Problems

Our tiny tanker routing problem turned into a 9 × 9 assignment problem, which involves 81 decision variables. A major airline has many flights per day. The problem of assigning aircraft to one week's flights has a gigantic number of decision variables. And there are "side constraints" that:

- allow for substitution of a larger plane for a smaller one,
- require each plane to be cycled into the depot for periodic overhaul after a prescribed number of flying hours, and
- keep the number of planes sent each day to the depot from exceeding the capacity of the depot.

These side constraints make the problem an integer program rather than a network flow model. These integer programs are enormous, but successful algorithms have been devised for them. These algorithms exploit the fact that a network flow problem lies at the root of this integer program.

4.12. A MULTIPERIOD PRODUCTION PLAN

Many products have seasonal demand; examples include toys, sporting equipment, bikes, and air conditioners. For many retail stores, the shopping period between Thanksgiving and Christmas accounts for over half of the annual sales revenue.

Level and Chase Strategies

How should a firm supply a product whose demand is seasonal? To **level** is to produce at a roughly constant rate throughout the year and to build inventory in the hope that the demand will materialize. To **chase** is to match the production rate to the demand rate. For a number of reasons, neither strategy is ideal.

- Leveling entails added investment in inventory (unsold merchandise) that builds up in anticipation of the periods of peak demand.
- Leveling risks producing the wrong products.
- Chasing requires extra production capacity because capacity is fully utilized only during periods of peak demand.

- Chasing requires either that workers be left idle during slack periods or, alternatively, that the firm incurs the added expenses of hiring, training, and firing.

To produce a rational production plan for a seasonal product, we must assess the trade-off and strike a balance. Such a plan is usually set on an **aggregated** basis; for example, the firm sets the level of capacity to produce skis without committing itself to the exact amounts of each type and size of ski. An **aggregate plan** is a profile of aggregated capacity over the season or calendar year that strikes a balance between the various costs and uncertainties.

Problem H (Tire Fabrication)

It is now October 1. A tire manufacturer has contracted to deliver snow tires to Sears in these quantities: 1000 units (each unit being eight dozen) in October, 2600 units in November, 2000 units in December, 3000 units in January, 2500 units in February, and 1000 units in March. The snow tires that are required in a given month can be produced during that month and shipped directly to Sears. Snow tires can also be produced in earlier months and held in inventory until they are needed. The company has no snow tires on hand now, and it wants to have none in inventory on April 1.

The company employs 90 trained tire fabricators now, and it needs to employ exactly 100 trained tire fabricators on April 1. Each trained employee can make 18 units of tires per month without working overtime. The normal work day is eight hours long, and trained employees can work up to two hours of overtime each day. While working overtime, trained employees make tires at the normal rate. Trainees make only eight units during their training period, which takes one month. Trainees are not allowed to work overtime. Workers can be hired and fired effective on the first day of each month.

The costs are as follows. Wage costs total \$2600 per employee per month, with an increase of 50% for overtime hours. It costs \$7000 to fire a worker; \$3000 to hire and train a worker, not counting that person's lost productivity during training; and \$40 to hold each unit of tires in inventory each month. This inventory carrying cost is applied to each unit of tires that is on hand on the first day of the month. You are to find the employment and production schedule that satisfies the Sears contract at least cost.

This tire fabrication problem has many decision variables. There are several months, and for each month, we must keep track of:

- the number of trained workers,
- the number of workers hired,
- the number fired,
- the number of units produced,
- the amount of overtime labor used, and
- the number of units in inventory.

For this problem, a tailored spreadsheet emerges naturally because it is convenient to place the decision variables in a two-dimensional array. We will assign a row to each type of variable and a column to each month. To label the columns, we number the months of October through April with the integers 1 through 7, respectively.

Which Variables Go into Changing Cells?

The idea is to put enough of the decision variables into changing cells that the others can be described as functions of them. We have a choice. The variables that we elect to place in changing cells are, for $k = 1, \ldots, 7$,

$$\begin{aligned} H_k &= \text{the number of workers hired on the first day of month } k, \\ F_k &= \text{the number of workers fired on the first day of month } k, \\ R_k &= \text{the number of units made with regular-time labor during month } k, \\ O_k &= \text{the number of units made with overtime labor during month } k. \end{aligned}$$

These decision variables are enough, enabling us to keep track of the number of workers employed each month, the regular-time and overtime production capacity in each month, and the inventory on hand at the beginning of each month.

The Workforce

These decision variables are now used to keep track of the workforce. Let us designate:

$$\begin{aligned} W_k &= \text{the number of workers who are employed on the first day of month } k \text{ but} \\ &\quad\ \text{before any hiring or firing occurs,} \\ T_k &= \text{number of trained workers during month } k. \end{aligned}$$

One group of equations keeps track the number W_k of workers from month to month. Specifically,

$$\begin{aligned} W_1 &= 90, \\ W_{k+1} &= W_k + H_k - F_k, \qquad \text{for } k = 1, 2, \ldots, 6, \\ W_7 - F_7 &= 100. \end{aligned}$$

The first of these equations states the fact that 90 workers are on hand at the beginning of October. With $k = 1$, the second constraint states that the number of workers on hand at the beginning of November equals the number on hand at the beginning of October plus the number hired on October 1, less the number fired at that time. The final equation states the requirement that 100 trained workers are needed on April 1.

Training takes one month. The W_k workers who are employed at the start of the month have been employed for at least one month, and hence are trained. Thus, the number T_k of trained workers during month k is given by

$$T_k = W_k - F_k, \qquad \text{for } k = 1, \ldots, 7.$$

Production Capacity

Next, we describe the production capacities. Trained employees can produce tires at the rate of 18 units per month, and trainees can produce tires at the rate of 8 units per month. Thus, the regular-time production capacity during month k equals $18T_k + 8H_k$. The number R_k of units made with regular-time labor must satisfy the constraint

$$R_k \le 18T_k + 8H_k, \qquad \text{for } k = 1, 2, \ldots, 7.$$

Only trained workers can work overtime. Each trained worker can spend one-fourth of his or her time working overtime, so the overtime production capacity during month k equals $(18/4)T_k$. Thus, the number O_k of units made with overtime labor must satisfy the constraint

$$O_k \le (18/4)T_k, \qquad \text{for } k = 1, 2, \ldots, 7.$$

Table 4.19 Demands by month, October being month 1.

Month k	1	2	3	4	5	6	7
Demand D_k	1000	2600	2000	3000	2500	1500	0

The Inventory Position

We must also keep track of the inventory position at the start of each month. Let us designate:

I_k = the number of units of tires that are in inventory at the start of month k.

To ease the task of accounting for I_k, we label the demands in months 1 through 7 as D_1 through D_7, as shown in Table 4.19.

The equations that keep track of the inventory position are

$$I_1 = 0,$$
$$I_{k+1} = I_k + R_k + O_k - D_k, \qquad \text{for } k = 1, 2, \ldots, 6,$$
$$I_7 = 0.$$

In this group of equations, the first states the fact that beginning inventory equals zero, and the last states the requirement that ending inventory must equal zero. The remaining equations require the inventory at the beginning of next month to equal the inventory at the beginning of this month plus the total of regular and overtime production during this month, less the demand during this month.

Each start-of-month inventory position I_k must be nonnegative; if I_k was negative, the contract with Sears would be violated because fewer units had been fabricated during months 1 through $k - 1$ than the demand during those months. To avoid "backorders," we must include the constraints that I_k be nonnegative for each k.

A Spreadsheet for Tire Fabrication

Before accounting for the costs, we assemble the preceding information into a spreadsheet and a Solver dialog box. This spreadsheet appears as Table 4.20. Its first 11 rows contain data; for instance, row 11 specifies the demand for each period. Its changing cells are in rows 12 through 15, and these changing cells specify values for the variables H_k, F_k, R_k, and O_k for each k. The functions in rows 16 through 20 specify other decision variables in terms of the values in the changing cells. In particular,

- The function =C16+C12−C13 in cell D16 equates the number W_2 of workers at the start of period 2 to $W_1 + H_1 - F_1$.
- The function =C17+C14+C15−C11 in cell D17 equates the inventory I_2 at the start of period 2 to $I_1 + R_1 + O_1 - D_1$.
- The function =C16−C13 in cell C18 equates the number T_1 of trained workers during period 1 to $W_1 - F_1$.
- The function =$E5*C18+$E6*C12 in cell C19 equates the regular-time production capacity during period 1 to $18T_1 + 8H_1$.
- The function =$E7*$E5*C18 in cell C20 equates the overtime production capacity during period 1 to $(0.25)(18)T_1$.

In Table 4.20, dragging rightward repeats the patterns established in cells D16, D17, C18, C19, and C20.

Table 4.20 Spreadsheet with changing cells C12 through I15 and with functions in rows 16 through 20.

	A	B	C	D	E	F	G	H	I
1			initial conditions			final conditions			
2			I1 =	0		I7 =	0		
3			W1 =	90		T7 =	100		
4									
5	production rate for trained workers =				18				
6	production rate for trainees =				8				
7	overtime limit: fraction of mo. =				0.25				
8									
9	quantity	symbol	Oct	Nov	Dec	Jan	Feb	Mar	Apr
10	month	k	1	2	3	4	5	6	7
11	demand	Dk	1000	2600	2000	3000	2500	1500	0
12	nbr. hired	Hk							
13	nbr. fired	Fk							
14	nbr. prod. reg. time	Rk							
15	nbr. prod. ovr. time	Ok							
16	nbr. wkrs at start	Wk				← =C16+C12-C13			
17	inventory at start	Ik				← =C17+C14+C15-C11			
18	nbr. trained wkrs	Tk			← =C16-C13				
19	reg. time capacity	--			← =\$E5*C18+\$E6*C12				
20	overtime capacity	--			← =\$E7*\$E5*C18				

What Table 4.20 Omits

Table 4.20 does not account for the cost, nor does it account for the following constraints:

- The values in cells C12 through I18 must be nonnegative.
- Each period's regular-time production (in row 14) must not exceed its regular-time capacity (in row 19).
- Each period's overtime production (in row 15) must not exceed its overtime capacity (in row 20).
- The initial conditions, $W_1 = 90$ and $I_1 = 0$.
- The final conditions, $I_7 = 0$ and $T_7 = 100$.

The Solver dialog box in Table 4.21 identifies cells C12 through I15 as the changing cells, and it enforces the above constraints, including the initial and final conditions. It also identifies the quantity in cell D28 as the number that we wish to minimize.

Accounting for the Cost

It remains to account for the cost in cell D28 of this spreadsheet. Over the seven-month period, the total number H of workers hired and the total number F of workers fired are given by

$$H = H_1 + H_2 + \cdots + H_7,$$
$$F = F_1 + F_2 + \cdots + F_7.$$

Table 4.21 Solver dialog box for Problem H.

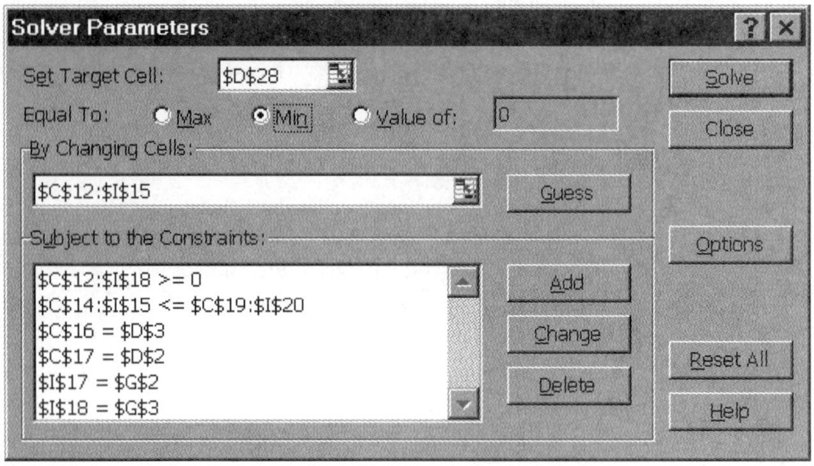

The total number of employees who receive salary during October equals W_2. Similarly, the total number W of worker-months for which salary must be paid from October 1 through March 30 is

$$W = W_2 + W_3 + \cdots + W_7.$$

The number OT of employee-months for which overtime wages are incurred equals the number of units of overtime production divided by 18:

$$OT = (O_1 + O_2 + \cdots + O_7)/18.$$

The total number I of units of inventory for which carrying costs must be paid is given by

$$I = I_1 + I_2 + \cdots + I_7.$$

In Table 4.22, the functions in cells C23 through E27 compute H, F, W, OT and I from the above equations. The function in cell D28 computes the total cost, which is given by the expression

$$3000\,H + 7000\,F + 2600\,W + (2600)(1.5)\,OT + 40\,I.$$

An Aggregate Plan

Table 4.22 also records the optimal solution that Solver has computed. The workforce must increase from 90 to 100 over the production interval. This goal is accomplished by hiring 14.08 workers on October 1 and firing 4.08 workers on March 1. As is usual, the tight inequality constraints are shaded. These shaded constraints show that regular-time capacity is fully utilized during October through February, that overtime capacity is fully utilized during November through February, and that no inventory is on hand on March 1. Once the workforce has been determined, the aggregate plan *levels* during October through February and matches demand (*chases*) during March.

An Integer-Valued Solution

This "solution" is nonsense, however. We must hire and fire integer numbers of workers. Technically, the optimization problem is an integer program, not a linear program. We could introduce the constraint that the number of workers hired and fired in each period be integer-valued and solve the resulting integer program.

Table 4.22 Optimal solution of the tire fabrication problem.

	A	B	C	D	E	F	G	H	I
9	quantity	symbol	Oct	Nov	Dec	Jan	Feb	Mar	Apr
10	month	k	1	2	3	4	5	6	7
11	demand	Dk	1000	2600	2000	3000	2500	1500	0
12	nbr. hired	Hk	14.08	0	0	0	0	0	0
13	nbr. fired	Fk	0	0	0	0	0	4.08	0
14	nbr. prod. reg. time	Rk	1733	1873.47	1873	1873	1873	1500	0
15	nbr. prod. ovr. time	Ok	0	468.367	468.4	468	468.4	0	0
16	nbr. wkrs at start	Wk	90	104.082	104.1	104	104.1	104	100
17	inventory at start	Ik	0	732.653	474.5	816	158.2	-0	-0
18	nbr. trained wkrs	Tk	90	104.082	104.1	104	104.1	100	100
19	reg. time capacity	--	1733	1873.47	1873	1873	1873	1800	1800
20	overtime capacity	--	405	468.367	468.4	468	468.4	450	450
21									
22		symbol	nmbr	unit cost					
23	number hired	H	14.08	3000					
24	number fired	F	4.082	7000					
25	number paid	W	620.4	2600					
26	nmbr paid overtime	OT	104.1	3900		overtime factor =			1.5
27	nmbr units held	I	2182	40					
28	total cost			2177061					
29				↑					
30				=SUMPRODUCT(C23:C27,D23:D27)					

An alternative to integer programming is to *round up*—that is, to increase H_1 from 14.08 to 15 and see what happens to the optimal solution to the linear program. When we do that, the optimal solution increases F_6 to 5, and the optimal value increases from \$2,437,061 to \$2,437,600. The increase—\$539—is an insignificant fraction of the total. Evidently, the increase in employment costs is almost entirely offset by substituting regular-time labor for overtime labor and by reducing the number of units in inventory.

4.13. A SEQUENTIAL DECISION PROCESS

The tire fabrication problem is now used to illustrate a scheme for describing decision problems that evolve over time. This scheme is very flexible, and is therefore well worth learning.

Roughly speaking, a **sequential decision process** is a decision problem that evolves from state to state in accord with these rules:

- At particular moments in time, the decision maker observes the "state" of the system and then selects the "action" to take.
- The decision maker then waits for the occurrence of a "transition" to a new state. This new state can depend on the current state and action, but not on prior states and actions.
- Similarly, the cost or revenue that is incurred between the moment the current state is observed and the moment at which the next state is observed can depend solely on the current state and action.

In a sequential decision process, the **state** of the system is a summary of its prior history that suffices to evaluate current and future actions. For the tire fabrication problem, decisions can be taken at the beginning of each month. The state of its system consists of the triplet (k, W_k, I_k), where (as before):

- It is now the start of month k.
- At this moment, before any hiring or firing occurs, exactly W_k trained tire fabricators are employed.
- At this moment, exactly I_k units of tires are in inventory.

To evaluate current actions, the triplet (k, W_k, I_k) suffices; no other information need be known. We need not know how long these I_k units have been in inventory nor when these W_k employees were hired.

In the jargon of operations research, the components of a state are called **state variables**. For the tire fabrication example, the state variables are k (the beginning of the kth month), W_k (the number of workers on hand at the beginning of month k), and I_k (the inventory on hand at the beginning of month k).

The **actions** at a particular state are the decisions one can then take. For the tire fabrication problem, the actions for state (k, W_k, I_k) are:

- The number H_k of workers to hire.
- The number F_k of workers to fire.
- The number R_k of units of tires to make with regular-time labor,
- The number O_k of units of tires to make with overtime labor.

The state constrains the actions. We cannot fire more workers than we employ. and the number of units that we produce with regular-time labor cannot exceed our regular-time production capacity. Nor can we produce more units with overtime labor than our overtime capacity. In other words,

$$\begin{aligned} F_k &\le W_k, \\ R_k &\le 18(W_k - F_k) + 8H_k, \\ O_k &\le (18/4)(W_k - F_k). \end{aligned}$$

In a sequential decision process, the **law of motion** from state to state can depend only on the current state and action, not on any prior states or actions. For the tire fabrication problem, the future state $(k + 1, W_{k+1}, I_{k+1})$ is given in terms of the current state (k, W_k, I_k) and the set $\{H_k, F_k, R_k, O_k\}$ of actions by

$$\begin{aligned} W_{k+1} &= W_k + H_k - F_k, \\ I_{k+1} &= I_k + R_k + O_k - D_k. \end{aligned}$$

Similarly, in a sequential decision process, the cost that accrues between transitions can depend solely on the current state and action. For the tire fabrication problem, this cost is given by the expression

$$3000\, H_k + 7000\, F_k + 2600\, (W_k - F_k) + (2600)(1.5)O_k + 40I_k.$$

Finally, a sequential decision process may have constraints on its **initial conditions** and/or on its **final conditions**. For the tire fabrication problem, the initial conditions are

$$I_1 = 0 \quad \text{and} \quad W_1 = 90,$$

and the final conditions are

$$I_7 = 0 \quad \text{and} \quad W_7 - F_7 = 100.$$

A sequential decision process is itself a template, a guideline to formulation. The key idea is that of a state.

> **Property of states.** In a sequential decision process, each transition occurs from state to state. Each state is a summary of what came before and contains enough information to describe the income or cost that occurs prior to transition and the law of motion that governs transition.

The term *state* is used consistently here. In a Markov chain (Chapter 13), transitions occur from state to state, and each state is a summary of what came before that suffices to describe the law of motion (probability distribution) that governs transition to the next state. In a simulation (Chapter 15), the state contains the realization of enough random variables that the law of motion can be specified.

Prior familiarity with sequential decision processes would have helped you to formulate the tire fabrication problem as a linear program. It would have been clear what the states must be, what the actions must be, how the state constrains the actions, and how transition occurs from state to state. In a tailored spreadsheet, the actions will be the decision variables.

The family of techniques used to solve sequential decision processes is known as **dynamic programming**. In this text, dynamic programming is used to solve shortest-path problems (Chapter 5) and Markov decision problems (Chapter 9). Nearly every sequential decision process can be formulated as a linear program. For this reason, linear programming is an important technique of dynamic programming.

4.14. REVIEW

We now recapitulate the problems in this chapter and the subjects that each problem introduced. Interspersed with this review are general observations about network flow, integer programs, and the importance of linear programming.

Cracking and Blending

The Recreational Vehicle problem in Chapter 3 introduced you to models of technology-based production systems. Problem A (Cracking and Blending) was another example of the same sort, with blending constraints added in. In this type of model, technology changes the form of the material. The principal data are the technological coefficients. These data are often linear, or nearly so, which makes these models particularly amenable to linear programming.

Technology-based production systems abound in manufacturing and the chemical processing, forest products and steel industries. In some instances, we seek the most efficient use of the available resources, and in other cases, we wish to achieve given levels of outputs as efficiently as possible.

Production and Distribution

Problem B (Production and Distribution) introduced you to models of systems in which material must be moved efficiently from each source to each destination. The transportation industry is rife with applications of this type. Often, the quantities that are shipped must be integers, for example, whole airplanes or whole ships. Integer-valued solutions, as we have seen, can occur naturally. They may require explicit introduction of constraints that keep the variables integer-valued.

Problems A and B illustrated the nomenclature of *network flow.* In a network flow model, the flows occur on the *arcs*, and flow is conserved at each *node.* Network flow models occur frequently, either as stand-alone linear programs or as components of larger models. We've described four special classes of network flow problems—*transportation*, *transshipment*, *assignment*, and *max-flow* problems. We have also mentioned one generalization, which is network flow with *gains*. The *Integrality Theorem* demonstrates that a network flow problem has an optimal solution that is integer-valued if it has integral data.

The literature on networks is vast, with much of it focusing on rapid solution of particular types of network flow models. Efficient special-purpose algorithms are important to the gigantic network flow problems that arise in practice.

Problem B also introduced the "tailored" spreadsheet. You may prefer a tailored spreadsheet under all circumstances. It is certainly preferable when the decision variables array themselves naturally into rectangles.

From Problems A and B, we abstracted some guidelines and tips for successful formulation of linear programs. The most important of these tips may be the one that applies to all computer output, not just to linear programs: Be suspicious of all computer output. If you formulate a problem incorrectly, the computer is likely to present you with a solution that defies common sense and offers clues as to what is wrong.

This illustrates a particular advantage of spreadsheet modeling—it can be a theater for **interactive learning**. You have an idea, and you try it out on a spreadsheet. When you get a silly answer, you have a clue as to what's wrong. You then refine your idea and get another silly answer—which leads to yet another try. Eventually, aha!

Staffing Shifts

Problem C (Staffing Shifts) introduced you to the scheduling model. In this type of model, items (people, aircraft, or whatever) must be assigned to specific tasks, and typically, some or all of the variables must be integer-valued. A linear program becomes an *integer program* when one or more of its variables is required to be integer-valued. The staffing problem initially appeared to be an integer program, but the Integrality Theorem guarantees that the simplex method will solve it.

Electrical Emergency

Problem D (Electrical Emergency) introduced the *max-flow* problem and the related *min-cut* problem. The Max-Flow Min-Cut Theorem demonstrates that the maximum value of the flow equals the smallest of the capacities of the cuts. We saw how the shadow prices for the max flow determine a cut whose capacity is smallest.

Olde England

Problem E (Olde England) and Problem F (Activity Analysis of a Plant) introduced *activity analysis*, which is an elegant model of resource allocation. Each *technology* converts *inputs* into *outputs*. In an activity analysis, the decision variables are the *levels* at which to operate the technologies. An activity analysis forces us to grapple with "net" quantities whose values can be negative. In Chapter 19, we will see that an activity analysis is the natural way to describe the production side of an economy that is in general equilibrium.

Cargo Routing

Problem G (Cargo Routing) illustrated a type of dynamic routing problem that is common in the transportation industry. From a mathematical viewpoint, this problem boils down to an assignment problem. Realistic routing problems become enormous assignment problems, with side constraints.

Tire Fabrication

Problem H (Tire Fabrication) introduced a multiperiod planning problem. This problem had many variables and constraints, but they formed a pattern. The same type of variables and constraints linked each period to the next. Exploiting this pattern enabled us to build the constraints with a "drag."

This example introduced models that may be used in the face of a seasonal demand. *Leveling* requires extra inventory and risks manufacturing the wrong items, whereas *chasing* sales requires extra capacity. An *aggregate plan* strikes a balance between the two.

Problem H also introduced us to the language of a *sequential decision process*, which describes a decision problem that unfolds over time. The key idea here is that a *state* is a summary of the prior history that suffices to evaluate current actions.

Integer Programs

Several of the problems in this chapter turned out to be integer programs. Such programs can be devilishly difficult to solve, for no algorithm known to humankind can solve all large integer programs problems quickly. When you attack an integer program, it is good practice to begin with the *relaxation* of the integer program that omits the integrality constraints, for the following reasons.

- The relaxation is a linear program, so the simplex method finds an optimal solution quickly.
- If the relaxation happens to be a network flow model with integral data, the Integrality Theorem guarantees that the simplex method finds an integer-valued optimal solution to it.
- The relaxation eliminates constraints, which can only improve the optimal value. Thus, the relaxation provides a *bound* on the optimal value, which can only worsen when the integrality constraints are reimposed.
- You may get lucky: The optimal solution to the relaxation may happen to be integer-valued. If not, you may be able to "round off" the relaxation's optimal solution in a way that produces a feasible integer-valued solution whose objective value is only slightly worse. (We did this in Problem H.)
- If you don't get lucky, you can insert some or all of the integrality constraints and hope for the best.

Over the decades, spectacular progress has been made in the computation of optimal solutions to the integer programs that arise in practice. Chapter 6 introduces branch and bound, which is vital to solving large-scale integer programs that arise in practice. Chapter 6 also shows how to use integer-valued variables to model decision-making problems.

Guidelines to Successful Formulation

Previously, we observed that the constraints in a linear program come in three "garden" varieties—conservation, bound, and ratio. There is one other: A **counting** or **measuring** constraint counts (measures) a quantity of interest. This quantity may be the net output of a set of production activities, or it may be the cost of a set of activities.

The most incisive guideline to formulation may be to regard the examples in this chapter as "templates" (patterns) and to ask yourself which of these templates a particular situation resembles. Is it a resource allocation problem? a network flow? a blending problem? an activity analysis? a sequential decision process? a combination?

When you encounter a sequential decision process, remember that its structure helps you to build the linear program and you should specify the states and actions. In a tailored spreadsheet, the actions will be the decision variables. Some constraints will specify the ways in which the states and actions constrain each other; others will specify the law of motion from state to state. And don't forget the constraints that impose the initial and final conditions.

The Importance of Linear Programming

Linear programs are truly *ubiquitous*. This chapter has provided evidence of the range of problems that can be formulated as linear programs; the problems at the end of this chapter extend this range somewhat. In Chapter 19, we will use a linear program and its "dual" program to formulate other classes of decision problems as linear programs.

The simplex method is extremely *fast*, solving practical linear programs with blinding speed. It is so fast in fact that it is a "subroutine" in algorithms that solve nonlinear programs. For instance, the "branch-and-bound" method solves an integer program by repeated solution of linear programs.

Both the model of a linear program and the simplex method *generalize*. We've seen that linear programs easily encompass diseconomies of scale. In a starred section of Chapter 3, we observed that Solver computes optimal solutions to convex nonlinear programs. A deft variant of the simplex method solves certain nonlinear programs. The same algorithm solves problems having multiple decision makers who compete with each other. A related variant computes a general equilibrium of an economy.

A linear program provides *mathematical insight* at which we have only hinted. Underlying linear programming is a theory of "duality." When we seek the optimal solution to a particular linear program, we end up with optimal solutions to a pair of linear programs—the one under attack and its "dual" program. Duality may sound arcane, but it sheds deep insight into linear algebra, analysis, economics, and computer science. (Incidentally, the min-cut problem is the dual of the max-flow problem.)

For these reasons—ubiquity, speed of solution, generality, and mathematical insight—linear programming may mark the most significant development in applicable mathematics in the last half century. It is a subject that richly rewards its students.

The current chapter does not close our discussion of linear programs. Chapter 5, for example, uses linear programs to solve different sorts of network optimization problems than those treated here. In addition, Chapter 18 presents Dantzig's simplex method; it solves linear programs quickly, and adaptations of it solve nonlinear programs. Finally, Chapter 19 presents duality, which enlarges the scope of linear programming, and uses a linear program to construct a general equilibrium.

4.15. HOMEWORK AND DISCUSSION PROBLEMS

1. **(Cracking and Blending,** p. 103) Create the spreadsheet in Table 4.1, use Solver to compute the optimal solution and issue a sensitivity report, and then answer these questions:

(a) By how much would the price of crude oil have to decrease for it to become optimal to change the set of tight constraints?

(b) By how much would the price of Premium have to increase for it to be optimal to change the set of tight constraints?

(c) Was it wise to accept a long-term contract to sell Regular gasoline at market prices?

(d) Would it be profitable for the refinery to sell Premium in excess of 3000 barrels per month at an intermediate price—between \$32.76 and \$38.22 per barrel?

2. (**Cracking and Blending,** p. 99) A blending constraint can be expressed as either a ratio or a linear expression. For instance, in Program 4.1 the blending constraint for Regular gasoline was expressed in two ways, as follows. Which representation is more accurate, and why?

$$\frac{93AR + 85BR}{R} \geq 87, \qquad 93AR + 85BR \geq 87R.$$

3. (**Cracking and Blending,** p. 99) Suppose that the refinery's price on Regular gasoline was optimistic. This month, the first 5000 gallons of Regular gasoline sell at \$32.76 per gallon. Regular gasoline in excess of this amount sells at a discount of \$4.00 per gallon, except that Regular gasoline in excess of 10,000 gallons per month sells at a discount of \$7.00 per gallon. Re-compute the optimal solution, and interpret it.

4. (**Cracking and Blending,** p. 99) Suppose that each cracking method has 20% waste, which can be disposed without cost. Hence, each barrel of crude oil that is cracked by Method A yields 0.8 barrels of 93 octane gasoline. Alter Program 4.1 to account for this, and re-compute its optimal solution. What changes? What does not change?

5. (**Cracking and Blending,** p. 99) Suppose that the marketing manager reviews the output and comments, "This solution calls for me to sell about eight gallons of Regular gasoline for every gallon of Premium. By pressing hard, I can sell four gallons of Regular for every gallon of Premium, but not more." Alter the linear program to account for this, and re-solve it. Describe the optimal solution numerically and verbally.

6. (**Cracking and Blending,** p. 99) Complete the tailored spreadsheet that Table 4.4 begins. Solve its linear program.

7. A new type of steel consists of iron plus controlled amounts of carbon, chrome, manganese, and silicon. The table that follows describes the cost and composition (other than iron) of 10 prospective components of this steel. This table also prescribes minimum and maximum concentrations of each material.

(a) Formulate a linear program whose optimal solution finds the cheapest blend of these materials into steel that has the required properties. Solve this linear program.

(b) How robust is its optimal solution? Try to pick a material that is not present in the optimal blend whose substitution would not appreciably increase the cost of the steel. Adapt your linear program to determine which material this new material would substitute for and that amount by which its substitution would increase cost.

Material	Cost (\$/lb)	Carbon (%)	Chrome (%)	Manganese (%)	Silicon (%)
1	0.03	4	0	0.9	2.3
2	0.065	0	10	4.5	15
3	0.07	0	0	0	45
4	0.06	0	0	0	42
5	0.10	0	0	60	18
6	0.13	0	20	9	30
7	0.12	0	8	33	25
8	0.08	15	0	0	30
9	0.02	0.4	0	0.9	0
10	0.02	0.1	0	0.3	0
Minimum (%)		3.0	0.3	1.35	2.7
Maximum (%)		3.5	0.45	1.65	3.0

8. A city produces 7000 tons of trash per day, some of which is burned and the rest is transported to landfill. Although burning is cheaper, it produces particulates and sulfur dioxide. The state requires the town to produce not more than 8000 units of particulates and 16,000 units of sulfur dioxide per day.

Incinerator	Capacity (tons/day)	Emissions (units/ton burned)	
		Particulates	Sulfur dioxide
A	3200	2.5	2
B	1800	1.1	3.5
C	1000	1.5	3

(a) Formulate a linear program that minimizes the cost of disposing of the town's trash while meeting the state's pollution control requirements.

(b) Solve the linear program that you formulated in part (a). Describe its optimal solution in a way that remains valid when its data are perturbed somewhat.

(c) Hauling costs the town \$50/ton more than burning. Incinerator A can be fitted with a smokestack device that reduces its sulfur dioxide emissions by 10% at an added operating cost of \$3.50 per ton burned. Reformulate the linear program to include this device. Resolve it, and determine whether or not this device is worthwhile.

9. **(Production and Distribution,** p. 107) Suppose that the production cost in plant 2 is cheaper than that in plants 1 and 3 by \$4.00 per unit. What effect does this differential have on the linear program? on its optimal solution?

Remark: In a network flow model, one or more of the shipping costs can be negative.

10. **(Production and Distribution,** p. 107) Directly from Figure 4.3:

(a) Compute the optimal solution to Problem B.

(b) Compute the allowable increase and the allowable decrease in the capacity of plant 2.

11. **(Production and Distribution,** p. 107) For the example in Table 4.3, the constraints for plants 1 and 3 have negative shadow prices, which seems to suggest that shipping more can cost less. Can that be so? If so, why? Be specific.

12. **(Production and Distribution,** p. 107) Formulate and solve the variant of the production and distribution problem in which planning occurs for two consecutive months. The data for the two months are identical. Finished goods (plywood) can be stored at any depot for one month, at a cost of \$0.50 per unit.

13. **(Production and Distribution,** p. 107) The optimal solution in Table 4.3 happens to have integer-valued flows. Does this surprise you? Why?

14. Invent a simple transportation problem whose shipping costs are nonnegative but for which shipping more costs less. *Hint:* You will need only two supply nodes and only two demand nodes. Arrange that increasing the supplies and demands lets you ship more "directly."

15. A Recreational Vehicle (RV) company has four plants that are located in different midwestern cities. Production costs are the same in each plant. The company ships from these plants to distributors in each of six regions. The following table states the plant capacities (in vehicles per month), the amount required by each distributor (also in vehicles per month), and the unit shipping costs (in dollars per vehicle). The company pays the shipping costs. Profit margins are slim. Company policy precludes shipping on routes whose costs exceed \$500 because excess shipping costs raise its cost above the price it receives from its distributors.

Plant	Production Capacity	Distributors					
		1	2	3	4	5	6
A	300	120	720	600	320	920	480
B	450	300	480	280	400	880	640
C	400	400	680	320	200	540	560
D	500	480	920	680	160	480	160
Requirement		250	300	200	275	300	325

(a) Formulate a linear program that determines the least-cost shipping plan that supplies the distributors with their monthly requirements and is acceptable to the company. *Hint:* An easy way in which to do part (a) is to increase each shipping cost that exceeds \$500 by a factor, such as 100, that is large enough to be prohibitive.

(b) Solve this linear program.

(c) Re-solve this linear program for the case in which shipping on any route is allowed.

(d) Does the company's policy of avoiding expensive routes make economic sense? Explain why or why not.

(e) For the linear program that you solved in part (c), draw the analog of Figure 4.3.

16. Do a diet problem as a field exercise. On the web, look up the minimum nutritional dinner requirements for people in your cohort. Go to a local fast-food place or campus eatery and ask for the nutritional content of each of its items. Find a least-cost diet that satisfies or exceeds the nutritional requirements. If you wish, add palatability constraints and use binary variables. For the really ambitious: minimize the total cost of two item-disjoint diets, each of which satisfies or exceeds the nutritional requirements.

Remarks: The diet problem spans the modern history of computation. During World War II, the distinguished economist, George Stigler, sought to formulate field rations that provided minimum daily requirements of about 10 substances. One such problem was solved (more or less) with Herculean effort by parallel computation on a roomfull of mechanical calculators. In an amusing reminiscence, George Dantzig recounted the enormous difficulties he encountered in the 1950s when he attempted to wrest a solution to his own personal diet problem from an early computer at The Rand Corporation. Today, solving the diet problem is a breeze. It was Robert Bosch, a professor at Oberlin, who thought of doing the diet problem as a field exercise.

17. In a hospital's surgery unit, a prescribed minimum number of medical residents must be on duty each day: at least 8 on Monday, 7 on Tuesday, 6 Wednesday, 6 on Thursday, 5 on Friday, 3 on Saturday, and 3 on Sunday. Each medical resident works one shift each week. This shift lasts 48 hours and begins on midnight of a particular day. How many residents are required to cover this schedule? Is this an integer program or a network flow problem? If it is not a network flow problem, what property of a network flow model does it violate?

18. Police officers work for eight consecutive hours. They are paid 20% above their normal rate for hours worked between 22:00 (10 P.M.) and 06:00 (6 A.M.). The police chief wishes to schedule the force to provide at least the following minimum number of officers.

Period	02:00–06:00	06:00–10:00	10:00–14:00	14:00–18:00	18:00–22:00	22:00–02:00
Minimum	12	20	18	24	29	18

(a) Write a linear program whose optimal solution realizes his goal. Why is it not necessary to require the variables to be integer-valued?

(b) Solve this linear program.

(c) Does the optimal solution that you obtained in part (b) minimize the total number of officers that are needed to provide the minimum coverage in each period?

(d) Reformulate your linear program, if necessary, so that Solver's Sensitivity Report determines whether the 20% overtime bonus is relevant, that is, whether or not you get the same optimal solution for any bonus.

19. A town has 39 police officers, each of whom works five days a week. Each officer has the same days off week after week, and the days off may be consecutive or nonconsecutive. The town needs to have at least 30 officers on duty on Monday, 24 on Tuesday, 25 on Wednesday, 22 on Thursday, 20 on Friday, 37 on Saturday, and 35 on Sunday.

(a) Among the plans that meet this requirement, find one that minimizes the number of officers whose days off are not consecutive.

(b) Suppose that not more than nine officers can have the same pair of days off. Does this make a difference? If so, what difference does it make?

20. (**Electrical Emergency,** p. 118) In the solution to a max-flow problem, let D denote the set of arcs whose flows are at their upper bounds. Designate as $k(D)$ the sum of the capacities of the arcs in D. Argue that $k(D)$ must be at least as large as the capacity of the minimum cut. Argue that $k(D)$ can exceed the capacity of the minimum cut. *Hint:* Review Table 4.6.

21. (**An Electrical Emergency,** p. 118) Suppose that there is a 3% transmission loss on each power line, except on the (short) lines between hydroelectric station and node 1 and between node 8 and the city. Suppose also that power is currently flowing on each line from the lower-numbered to higher-numbered node.

(a) Alter the spreadsheet in Figure 4.4 to account for these transmission losses.

(b) True or false: The spreadsheet in part (a) describes a generalized network flow model in which each arc's gain rate is equal to 1.0 or to 0.97.

(c) After thinking carefully about part (b), solve your linear program.

(d) True or false: Maximizing the flow into node 1 is equivalent to maximizing the flow out of node 8. (This was true when the gains equaled 1.0.)

22. (**Olde England,** p. 122) The defense minister of Olde England wants the country to stockpile 0.5 units of food. How expensive would this be? Can this question be answered without rerunning the linear program? If so, why?

23. (**Olde England,** p. 122) The prime minister of Olde England wants the country to import no food. What changes would occur if her view prevailed?

24. (**Olde England,** p. 122) To develop its economy, Olde England could invest to expand its capacity to spin yarn at an annualized cost of 2.00 £ per unit and to increase its arable land at an annualized cost of 2.50 £ per unit. At these prices, what levels of investment maximize the net value of exports? What changes occur in the mix of activities?

25. (**Olde England,** p. 122) For Olde England, we can replace each free (unconstrained) variable by the difference of two nonnegative variables, for example, FE by $(FEXP - FIMP)$.

(a) Would these changes have any effect on the optimal solution? If so, what?

(b) Would these changes have any effect on the sensitivity analysis information that Solver provides. If so, what?

(c) In general, is it a good idea or a bad idea to replace free variables by the difference of two nonnegative variables? Why?

26. (**Plant 7,** p. 125) Formulate Program 4.3 on a spreadsheet and apply Solver. Interpret the output that Solver provides. *Hint:* Was it possible for Olde England to be self-sufficient in food, yarn, and clothes?

27. (**Cargo Routing,** p. 127) In Problem G, one ship has been required to be at each port on day 12. Remove this requirement; let the ships end anywhere. How does the linear program change? What happens to its spreadsheet? to its optimal solution?

28. (**Cargo Routing,** p. 127) Prove or disprove that there must be more than one optimal solution to Problem G if the optimal routing plan it reports in Table 4.17 allows two ships to be at the same port at the same time.

29. (**Cargo Routing,** p. 127) Parts (a)–(c) describe variations on this problem. For each part, state what changes you would make in Table 4.15, describe the types of changes that would occur in the spreadsheet, make the changes, and report the new optimal solution.

(a) One ship must be at each port on day 13 rather than on day 12.

(b) Trip 3 leaves port C on day 4 rather than on day 3.

(c) Trip 3 leaves port C on day 3 but goes to port A rather than to port B.

(d) At day 12, two ships must be at port A, one at port B.

30. (**Tire Fabrication,** p. 132) Inserting the constraint $H_1 \geq 15$ increased the optimal value to \$2,437,600. Solve the tire fabrication problem as an integer program. Did you get the same optimal value?

Are there any advantages to substituting the constraint $H_1 \geq 15$ for the integer program? If so, what are they?

31. (**Tire Fabrication,** p. 132) Suppose that it takes two months, not one, to train each tire fabricator. As before, while in training each employee can make tires at the rate of 8 units per month. After training, an employee can make tires at the rate of 18 units per month.

(a) What change is needed in the function that defines the number T_k of trained workers during period k? What change occurs in the regular-time production capacity for period k?

(b) Re-solve the linear program. To what extent does this optimal solution level? chase?

(c) Re-solve the linear program in which H_1 is rounded up to the nearest integer. What happens?

32. (**Tire Fabrication,** p. 132) Suppose Sears is willing to accept backorders but at a penalty of $400 per unit backordered per month.

(a) What changes would be needed to solve this variant of the tire fabrication problem? *Hint:* For each month k, you might introduce a decision variable B_k that is required to satisfy $B_k \geq 0$ and $B_k \geq -I_K$.

(b) On a spreadsheet, solve this variant of the tire fabrication problem.

(c) Can you tell from a sensitivity analysis on the original linear program whether or not the variant would decrease cost? If so, how?

33. A Rent-a-Car Company has a training program for its agents. Training takes one month. Trainees do not meet the public during their month of training. Eighty % of each month's trainees complete the program successfully, and the remaining 20% are fired at the end of their month of training. During training, each trainee requires 20 hours of supervision by a trained employee. Each trained agent can meet the public for as many as 160 hours per month, less any time spent supervising trainees. Ten % of the trained agents leave for other jobs at the end of each month. Trainees receive half pay during their month of training. There are 70 trained agents on hand on March 1. The requirement for trained agents for the next seven months is presented in the following table.

Month	March	April	May	June	July	August	Sept.
Requirement	60	65	85	95	110	95	80

(a) Create a linear program that satisfies these requirements as cheaply as possible. In part (a), allow the number of agents hired each month to be any number; do not require it to be an integer. Solve this linear program.

(b) Round off the solution you obtained in part (a) in a way that satisfies all needs and hires integer numbers of trainees each month. By how many person-months did the payroll cost increase?

(c) Impose the constraint that the number of trainees in each month must be an integer. Solve it. How close was your guess in part (b) to the least-cost integer solution?

34. As part of an airline's schedule, a particular airplane flies the following "four-leg rotation" over and over: JFK → LAX → O'Hare → Dallas → JFK. The following considerations affect the amount of fuel in the aircraft at takeoff:

- To protect against bad weather congestion, the aircraft must plan on landing with a reserve of at least 3000 gallons of fuel in its tanks.
- The fuel tanks can hold at most 18,000 gallons of fuel.
- To protect its landing gear, the aircraft can land with at most 8000 gallons of fuel aboard.

Fuel consumption varies with the leg, and fuel prices vary from airport to airport. The airline wishes to fuel its planes in a way that will minimize its total fuel bill. **Tankered** fuel—fuel in excess of the requirement for the leg—increases takeoff weight, which increases the rate of fuel consumption during flight, so a fraction of the tankered fuel is consumed during flight. The relevant data appear below. These data indicate that the aircraft must take off on the JFK → LAX leg with at least $10.5 + (3)/(1 - 0.22)$ thousand gallons of fuel, for instance. The airline wishes to minimize the cost of the fuel needed to fly this "rotation" repeatedly, with the same fuel in the airplane when it lands at JFK at the end of each rotation.

Leg	JFK → LAX	LAX → O'Hare	O'Hare → Dallas	Dallas → JFK
Fuel consumed (thousands of gallons)	10.5	7.4	4.2	6.3
Fraction of tankered fuel consumed	0.22	0.12	0.08	0.11
Price of fuel at port of origin ($/gallon)	1.07	0.85	0.98	0.84

(a) What constitutes a state? an action?

(b) Build a linear program that solves this problem. *Hint:* This problem resembles an "aggregate plan" with this complication; the amount of fuel at the beginning of a cycle must equal the amount of fuel at the end.

(c) Solve the linear program that you build in part (b). *Hint:* If you use Excel to write the "wraparound constraint" that equates the starting fuel level to the ending fuel level, Excel will report that you have created a "circular reference." To avoid that, create the "wraparound" constraint within Solver.

(d) State its linear program's optimal solution in a way that would remain valid if the data were slightly different.

35. For planning purposes, an appliance company aggregates its products into "units," each of which represents 1000 appliances of a typical mix. The monthly demand for appliances (measured in units) follows the cyclical pattern that is shown below. A large peak occurs in the summer (months 5–8) and a smaller peak in the winter holiday season.

Month	1	2	3	4	5	6	7	8	9	10	11	12
Units	300	350	425	550	700	800	725	650	400	250	400	600

Each employee can produce appliances at the rate of 0.6 units per month while working regular time. Each employee can work as many as two hours of overtime for every eight hours of regular time, during which time he or she produces appliances at the normal rate. The company possesses enough machinery to make appliances at the rate of 540 units per month without overtime. Employees earn at the rate of $3750 per month of regular-time employment. They earn 50% more than their hourly rate while working overtime. Hiring costs $1000 per employee, whereas firing costs $4500 per employee. Each unit that is in inventory at the start of the month costs $6000. The company begins month 1 with 600 units on hand. For parts (a)–(c), allow any end-of-year inventory.

(a) What are the state variables? What are the actions?

(b) Write a linear program that solves this aggregate planning problem. Write it in such a way that a sensitivity analysis would reveal the benefit of increasing the manufacturing capacity.

(c) Solve your linear program. To what extent, if any, does it level or chase? What is the value of one added unit of manufacturing capacity? By how much is cost reduced if the company declines to meet one unit of demand in March? in June?

(d) Suppose that the monthly demand period repeats, year after year. How many units should the company have on hand at the beginning of each year? To what extent, if any, should it level or chase? *Hint:* Rerun your linear program with a "wraparound" constraint.

36. **(A Hierarchy of Goals)** Your company has a local advertising budget of $65,000. The following table lists its goals by priority and estimates the number of persons in each group that are reached by each type of ad. This table also specifies the cost of each type of ad. Use one or more linear programs to find the largest value of k such that you can achieve goals 1 through k.

Priority and Goal	Type of Ad		
	Sports	News	Plot
1. Reach at least 70,000 young women	3000	6000	7000
2. Reach at least 50,000 young men	9000	3000	2000
3. Reach at least 60,000 elders	4000	2000	8000
Cost ($ per TV ad)	8000	4000	6000

Remark: The preceding problem represents a brief foray into a large and important area that is known as *multicriterion* decision making. An alternative to the hierarchical approach is to explore the trade-off or efficient frontier between various criteria.

37. (**A Regression**) On a slow day, a realtor wondered whether the prices of the houses in her town that had recently sold might follow a simple linear model, at least approximately. She guessed that the sales price of a home might depend linearly on the size of the home (in square feet), on the size of the lot (in front feet), and on the quality (scored from 0 to 10), as follows:

$$\begin{pmatrix}\text{sales price}\\ \text{of home } k\end{pmatrix} = \begin{pmatrix}\text{no. of square ft}\\ \text{of home } k\end{pmatrix} A + \begin{pmatrix}\text{no. of ft of frontage}\\ \text{of home } k\end{pmatrix} B + \begin{pmatrix}\text{quality of}\\ \text{home } k\end{pmatrix} C + D + E_k,$$

where E_k is the "error" in the predicted price. She wondered what values of the numbers A, B, C, and D provided the best fit to the last 10 sales, where the "best fit" is one that minimizes the sum of the absolute values of the errors. Data for the last 10 sales appear in the table that follows this problem.

(a) Formulate this problem as a linear program, and solve it. *Hint:* One of many ways in which to tackle this problem is to minimize $(x_1 + \cdots + x_{10})$ subject to the constraints $x_k \geq E_k$ and $x_k \geq -E_k$ for each k.

(b) Of these 10 houses, is there any single home that accounts for more than 25% of the total error? If so, which one?

(c) Of the four factors, is there one that is responsible for less than 1% of the total error? If so, which one?

House k	Price ($)	Size (sq ft)	Frontage (ft)	Quality (0–10)
1	144,000	2100	60	5
2	84,000	1200	60	4
3	175,000	2600	100	6
4	120,000	1650	65	8
5	200,000	2850	140	10
6	160,000	2300	110	7
7	94,000	1250	60	5
8	143,000	2050	105	9
9	209,000	3100	150	7
10	124,000	1700	70	6

Remark: A classical approach to *regression* problems like this one is to minimize the sum of the squares of the errors. The classical approach can be formulated as a nonlinear program, and it can also be solved explicitly, using calculus.

Chapter 5

Networks

5.1. PREVIEW

In Chapter 4, directed networks were used to model a class of optimization problems whose decision variables were the flows on the arcs. In the current chapter, the focus is on the network itself, particularly on its "paths" and their lengths.

To introduce this subject, we'll need to learn some nomenclature. Fortunately, pictures make everything clear. To begin, we recall that each node in a directed network can be pictured as a circle with an identifying symbol (such as i) inside, and each directed arc (i, j) can be represented as a line segment connecting nodes i and j with an arrow pointing from node i to node j. Figure 5.1 depicts a network that has five nodes and seven directed arcs. Its nodes are numbered 1 through 5. This network contains directed arc (1, 2), but it does not contain directed arc (2, 1).

In Chapter 4, it was convenient to allow one end of a directed arc to be attached to a node, and to let the other end be free. Here, where the focus is on the network, we require both ends of each directed arc to be attached to nodes. In this chapter, the phrase "directed arc" is often abbreviated "arc" because we shall not consider networks whose arcs lack directions (arrows).

The motivating idea for a path is that directed arc (i, j) connotes direct *movement* from node i to node j, a path being a sequence of such movements. Mathematically, a **path** is defined to be a sequence $(i_1, i_2, \ldots, i_n)$ of at least two nodes with the property that (i_k, i_{k+1}) is a directed arc for $k = 1, 2, \ldots, n - 1$. The network in Figure 5.1 contains many paths, including (1, 2), (1, 2, 3), and (2, 3, 4). This network does not contain path (4, 3, 2) because (3, 2) is not one of its arcs.

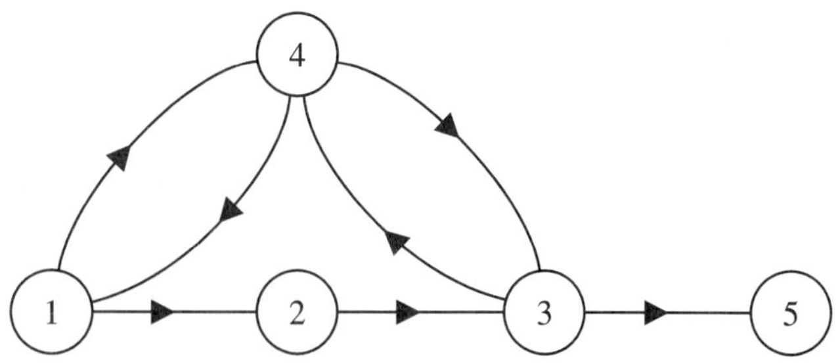

Figure 5.1 A directed network.

Path $(i_1, i_2, \ldots, i_n)$ is said to be a path **from** node i_1 **to** node i_n. The network in Figure 5.1 contains many paths from node 1 to node 4, including path (1, 4), path (1, 2, 3, 4), and path (1, 4, 1, 4).

A path $(i_1, i_2, \ldots, i_n)$ is called a **cycle** if i_1 equals i_n. The network in Figure 5.1 contains many cycles, including cycles (1, 4, 1) and (1, 2, 3, 4, 1).

A directed network is said to be **cyclic** if it contains at least one cycle. The network in Figure 5.1 is cyclic. A directed network is said to be **acyclic** if it contains no cycle. The network in Figure 5.1 becomes acyclic if arcs (4, 1) and (3, 4) are deleted.

In the path $(i_1, i_2, \ldots, i_n)$, nodes i_2 through i_{n-1} are said to be **intermediary** nodes. For example, path (1, 4) has no intermediary nodes. In path (1, 4, 1, 4), nodes 1 and 4 are intermediary. A cycle is said to be **simple** if its intermediary nodes are distinct from each other and from its "from" node.

Optimization problems arise when each arc (i, j) is assigned a **length** $c(i, j)$. Typically, the **length** of a path is taken to be the sum of the lengths of its arcs. Typical optimization problems are to find the shortest path from one node to another, or the longest path from one node to another.

A network can fail to have longest paths. In Figure 5.1, for instance, suppose that $c(3, 4) = c(4, 3) = 5$. In this case, cycle (3, 4, 3) has 10 as its length. Because this cycle can be repeated any number of times, the network can have no longest path from any node to any other.

Arc lengths *can* be negative, and networks can also fail to have shortest paths. In Figure 5.1, for example, suppose that $c(3, 4) = 2$ and $c(4, 3) = -5$. The length of the cycle (3, 4, 3) equals -3. No shortest path exists from any node to any node.

5.2. WHAT CAN YOU LEARN FROM THIS CHAPTER?

This chapter consists of two segments, one of which is basic and the other advanced. The basic segment focuses on shortest-path problems and on longest-path problems.

In the basic segment, you will see that the natural way in which to solve these problems is to begin by taking a step backward. Suppose, for instance, that you seek the shortest path from node b to node c of a network. You will see that an efficient way to solve this problem is to "embed" it in a family of shortest-path problems—for instance, to find the shortest path from each node to node c.

The basic segment describes two settings in which shortest-path and longest-path problems arise. One of these settings is a method for solving optimization problems that's known as dynamic programming. Often, a shortest-path problem is the end result of using dynamic programming to formulate an optimization problem for solution. The other setting is a widely used model of project management, which leads to the computation of longest paths in an acyclic network.

The basic segment also includes two methods for solving shortest-path and longest-path problems. One of these methods, a recursion, succeeds when the network is acyclic. The other method, a linear program, succeeds whenever the problem is well-posed—that is, whenever a shortest path or longest path exists.

The advanced segment is more technical and so is considerably more difficult. It presents a method for solving a shortest-path problem in a network that can be cyclic but whose arc lengths are nonnegative. Ideas introduced for that problem are adapted to solve the assignment and max-flow problems, both of which were illustrated in Chapter 4.

5.3. A SHORTEST-PATH PROBLEM

The simplest network optimization problem is to find, in an acyclic network, the shortest path from one node to another. To introduce this shortest-path problem, we focus on:

Problem A (Pete, the Cyclist)

Pete cycles to work each morning. He has explored several routes and recorded the travel times that appear in Figure 5.2. In this figure, each node depicts the junction of two or more roads Pete might travel. Each directed arc represents a road he might take, the direction in which he might cycle it, its travel time, and the nodes it connects. For instance, it takes 12 minutes to get from junction 2 to junction 5 via the road that connects them. Pete wants to know which route gets him to the office as quickly as possible and how long it takes.

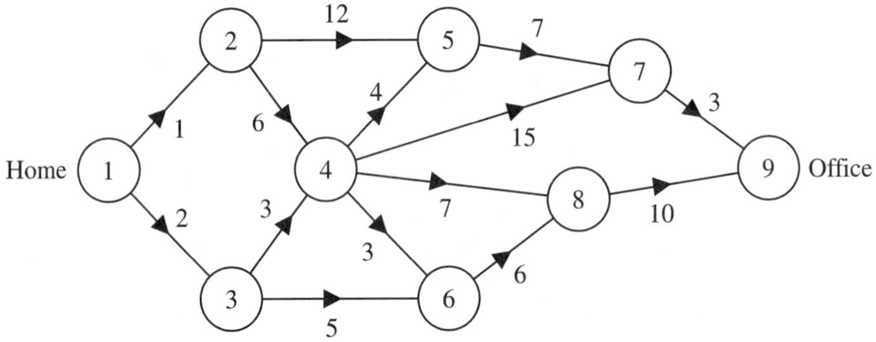

Figure 5.2 Travel times between junctions.

Pete's shortest-path problem is simple enough to solve by the "eyeball" method. The shortest path from node 1 (Pete's home) to node 9 (his office) is (1, 3, 4, 5, 7, 9), and its travel time (length) is 19 minutes. The second shortest path is (1, 2, 4, 5, 7, 9), and, incidentally, it is 2 minutes longer.

Embedding

Having solved Pete's problem by inspection, we use it to introduce a methodology that works for all shortest-path problems, however large their networks may be. The idea is to **embed** the problem we wish to solve in a family of related problems, link their solutions, and solve them all, coherently.

We'll embed Pete's problem in a family of shortest-path problems, one per node. With $f(9) = 0$, let us define, for each node i other than node 9,

$$f(i) = \text{length of the shortest path from node } i \text{ to node } 9.$$

Linking Solutions

Aiming to link the solutions to these shortest-path problems, we interpret

$$c(i, j) + f(j), \qquad \text{for each arc } (i, j),$$

as the length of the shortest path from node i to node 9 whose first arc is arc (i, j). In other words, $c(i, j) + f(j)$ equals the time it takes Pete to get from node i to the office if he first traverses arc (i, j) and then goes from node j to the office as quickly as possible.

The sum $c(i, j) + f(j)$ equals the length of *a* path from node i to node 9, and $f(i)$ is the length of the shortest path from node i to node 9, which guarantees

$$f(i) \leq c(i, j) + f(j), \qquad \text{for each arc } (i, j). \tag{5.1}$$

But the fastest path from node i to node 9 traverses *some* arc (i, j) first and then proceeds as quickly as possible from node j to node 9. So some j satisfies inequality (5.1) as an equality, and

$$f(i) = \min_j \{c(i, j) + f(j)\}, \qquad \text{for } i \neq 9. \tag{5.2}$$

In Equation (5.2), minimization is taken over those j such that (i, j) is an arc; if (i, j) is not an arc, $c(i, j)$ has no meaning. Equation (5.2) relates the solutions to the shortest path problems. The value it specifies for $f(i)$ depends on the values of $f(j)$ for those j such that (i, j) is an arc.

A Recursion

The nodes in Figure 5.2 have been numbered so that each arc (i, j) has $i < j$. This means that $f(i)$ can be computed from Equation (5.2) as soon as $f(j)$ has been computed for each j that exceeds i. In other words, Equation (5.2) can be solved recursively, in decreasing i, first for $i = 8$, then for $i = 7, \ldots,$ ending with $i = 1$.

To illustrate, let us suppose that $f(5)$ through $f(8)$ have been determined. From Equation (5.2), we can then calculate $f(4)$ because

$$f(4) = \min \begin{cases} c(4, 5) + f(5) \\ c(4, 6) + f(6) \\ c(4, 7) + f(7) \\ c(4, 8) + f(8) \end{cases} = \min \begin{cases} 4 + 10 \\ 3 + 16 \\ 15 + 3 \\ 7 + 10 \end{cases} = 14.$$

Now that $f(4)$ is also known, we can compute $f(3)$, then $f(2)$, and finally $f(1)$ from Equation (5.2), as in:

$$f(3) = \min \begin{cases} c(3, 4) + f(4) \\ c(3, 6) + f(6) \end{cases} = \min \begin{cases} 3 + 14 \\ 5 + 16 \end{cases} = 17,$$

$$f(2) = \min \begin{cases} c(2, 4) + f(4) \\ c(2, 5) + f(5) \end{cases} = \min \begin{cases} 6 + 14 \\ 12 + 10 \end{cases} = 20,$$

$$f(1) = \min \begin{cases} c(1, 2) + f(2) \\ c(1, 3) + f(3) \end{cases} = \min \begin{cases} 1 + 20 \\ 2 + 17 \end{cases} = 19.$$

This recursion is efficient; it requires one calculation per arc in the network.

A Linear Program

There is another way to solve Equation (5.2), which may be a bit less efficient computationally but is easy to execute on a spreadsheet. It has other advantages, too. This method is by solving a linear program. The data in this linear program are the arc lengths, the $c(i, j)$'s. Its decision variables are $f(1)$ through $f(8)$. This linear program is:

Program 5.1: Maximize $\{f(1) + f(2) + \cdots + f(8)\}$, subject to the constraints

$$f(i) - f(j) \le c(i, j), \qquad \text{for each arc } (i, j),$$
$$f(9) = 0.$$

We shall see that the optimal values of the decision variables in Program 5.1 solve Equation (5.2). Let us first interpret its constraints. To recognize the constraint $f(i) - f(j) \le c(i, j)$, we shift $f(j)$ to its right-hand side and get

$$f(i) \le c(i, j) + f(j), \qquad \text{for each arc } (i, j), \tag{5.3}$$

which is identical to expression (5.1).

The objective of Program 5.1 may seem odd: We seek the shortest paths, but the linear program *maximizes* something. Why? Expression (5.3) holds for each arc (i, j), and that guarantees that each feasible solution to Program 5.1 has

$$f(i) \le \min_j\{c(i, j) + f(j)\}, \qquad \text{for } i \ne 9. \tag{5.4}$$

A feasible solution can satisfy

$$f(i) < \min_j\{c(i, j) + f(j)\}, \tag{5.5}$$

for one or more of the values of i. But a feasible solution that satisfies inequality (5.5) cannot be optimal because increasing $f(i)$ preserves feasibility and improves the objective value. Thus, the optimal solution to Program 5.1 cannot satisfy inequality (5.5). In other words, optimization forces each inequality in expression (5.4) to hold as an equation. In this way, maximization forces a solution to Equation (5.2).

A Spreadsheet

The spreadsheet in Table 5.1 formulates Program 5.1 for solution by Solver. Its changing cells are cells B3 through I3, which contain the values of $f(1)$ through $f(8)$. Cell *J*3 is not a changing cell; it contains the datum 0, which enforces the constraint $f(9) = 0$. Column K contains the usual SUMPRODUCT functions, and the constraints K4:K17 <= M4:M17 enforce the inequality constraints in Program 5.1. Solver has maximized the expression in cell K18. Table 5.1 reports the optimal solution that Solver has found. Cell B3 shows that $f(1) = 19$, which is no surprise.

In Table 5.1, each inequality constraint that holds as an equation has been shaded, and each shaded inequality satisfies

$$f(i) - f(j) = c(i, j). \quad \text{Equivalently,} \quad f(i) = c(i, j) + f(j).$$

In other words, each shaded constraint prescribes the first arc on a shortest path from node i to node 9. Figure 5.3 preserves the shaded arcs and records each node's f-value adjacent to it.

We had sought the shortest path from Pete's home to his office. Figure 5.3 provides more information than this, showing how to get to the office as quickly as possible from every road junction (node) that Pete might encounter, not just from home.

Trees

Figure 5.3 illustrates a useful definition, that of a "tree." In Figure 5.3, node 9 plays a distinguished role. For each node i other than node 9, the network in Figure 5.3 contains exactly one path from node i to node 9. Its path from node 2 to node 9 is (2, 4, 5, 7, 9), for instance.

Table 5.1 Optimal solution to the shortest-path problem.

	A	B	C	D	E	F	G	H	I	J	K	L	M
2	name	f(1)	f(2)	f(3)	f(4)	f(5)	f(6)	f(7)	f(8)	f(9)			arc
3	value	19	20	17	14	10	16	3	10	0			length
4	(1,2)	1	-1								-1	<=	1
5	(1,3)	1		-1							2	<=	2
6	(2,4)		1		-1						6	<=	6
7	(2,5)		1			-1					10	<=	12
8	(3,4)			1	-1						3	<=	3
9	(3,6)			1			-1				1	<=	5
10	(4,5)				1	-1					4	<=	4
11	(4,6)				1		-1				-2	<=	3
12	(4,7)				1			-1			11	<=	15
13	(4,8)				1				-1		4	<=	7
14	(5,7)					1		-1			7	<=	7
15	(6,8)						1		-1		6	<=	6
16	(7,9)							1		-1	3	<=	3
17	(8,9)								1	-1	10	<=	10
18		1	1	1	1	1	1	1	1	1	109		

Solver maximized the value in cell K18,
with B3:I3 as changing cells,
subject to K4:K17 <= M4:M17

=SUMPRODUCT(B$3:J$3,B18:J18)

A directed network is said to be a **tree** if it has a distinguished node, called its **root**, and if one of the following conditions is met:

- Either for each node *i* other than the root, the network contains exactly one path from node *i* to the root.
- Or for each node *j* other than the root, the network contains exactly one path from the root to node *j*.

Figure 5.3 illustrates the first type of tree, that of paths to node 9. We'll soon illustrate the second type of tree, that of paths from node 1 to all of the others.

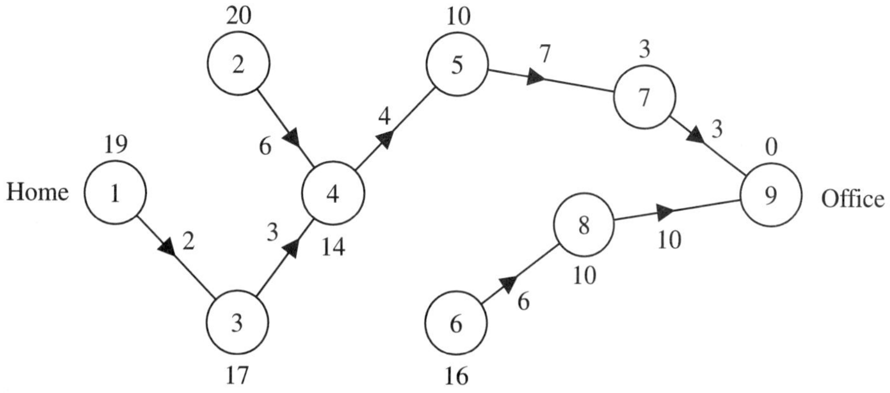

Figure 5.3 Shortest paths to node 9 and their lengths.

Recap

Let us review what has been accomplished. Pete's shortest-path problem has been embedded in a family of shortest-path problems, one per starting node. Equation (5.2) links their solutions. This equation holds because, for each node i other than node 9, the shortest path from node i to the office takes some arc (i, j) and proceeds as quickly as possible from node j to the office. Since each arc (i, j) has $i < j$, Equation (5.2) can be solved recursively, in decreasing i. This shortest-path problem has also been solved as a linear program. Either way, the solution takes the form of a tree of shortest paths to node 9. (With different data, there could have been ties, and so multiple trees of shortest paths.)

The recursion works when the network is acyclic. The linear program works more generally and does not require the network to be acyclic. It works when some arc lengths are negative, provided the network has no cycle whose length is negative. If a network has a cycle whose length is negative, it has no shortest path from any node in that cycle to any other node in that cycle. Thus, the linear program constructs a tree of shortest paths whenever one exists.

A Different Embedding and a Different Tree

We embedded Pete's problem in a family of shortest-path problems, one per node, but that could have been done in a different way. Specifically, with $g(1) = 0$, we could have defined $g(j)$ for each node j other than node 1 by

$$g(j) = \text{length of the shortest path from node 1 to node } j.$$

These optimization problems are linked by the analog of Equation (5.2), which is

$$g(j) = \min_{i}\{g(i) + c(i, j)\}, \qquad \text{for } j \neq 1. \tag{5.6}$$

Equation (5.6) can be solved recursively, in increasing j, and can also be solved by linear programming. Either way, its solution specifies the tree of shortest paths from node 1 to the others that appear in Figure 5.4.

Figure 5.4 reports that the shortest path from node 1 to node 9 is (1, 3, 4, 5, 7, 9) and that its length equals 19.

Network Flow Formulations

The network flow model in Chapter 4 can also be used to compute shortest paths. To find a shortest path from node 1 to node 9, force one unit of flow into node 1, force one unit of

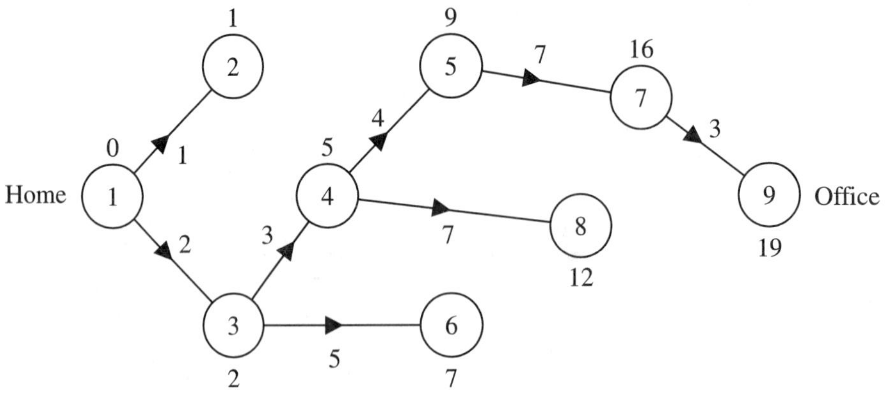

Figure 5.4 The tree of shortest paths from node 1 to the others.

flow out of node 9, require flow to be conserved at each node, and minimize the cost of the flow.

Similarly, to compute the tree of shortest paths from node 1 to the other nodes in Figure 5.4, force eight units of flow into node 1, remove one unit of flow from nodes 2 through 9, require flow to be conserved at each node, and minimize the cost of the flow.

Longest-Path Problems

We will soon have reason to study a longest-path problem in an acyclic network. It, too, can be solved in two ways. For the network in Figure 5.2, we could set $F(9) = 0$ and, for each node i other than node 9, define $F(i)$ by

$$F(i) = \text{the length of the longest path from node } i \text{ to node 9.}$$

These $F(i)$'s are the solution to

$$F(i) = \max_j \{c(i, j) + F(j)\}, \qquad \text{for } i \neq 9. \tag{5.7}$$

We can solve Equation (5.7) recursively, in decreasing i, or by linear programming. (This linear program is a minimization problem, with "≥" constraints.) Either solution method produces a tree of longest paths to node 9.

Alternatively, with $G(1) = 0$, we can define $G(j)$ for each node i other than node 1 by

$$G(j) = \text{the length of the longest path from node 1 to node } j.$$

These $G(j)$'s satisfy

$$G(j) = \max_i \{G(i) + c(i, j)\}, \qquad \text{for } j \neq 1. \tag{5.8}$$

Equation (5.8) can be solved recursively or by linear programming. Either way, its solution specifies the tree of longest paths from node 1 to the others.

5.4. DYNAMIC PROGRAMMING

Section 5.3 illustrates a methodology that solves a broad range of optimization problems. This technique, dynamic programming, consists of several steps. Each step entails some jargon, which the shortest-path problem will be used to illustrate. The first step is to cast the decision problem as a sequential decision process.

What Is a Sequential Decision Process?

As was the case in Chapter 4, a **sequential decision process** entails a *cycle of events* in which a decision maker:

- observes the state,
- selects an action,
- receives a payoff or incurs a cost, as a result,
- waits for transition to occur to some other state, and
- observes that state,

whereupon the cycle is renewed. In many sequential decision processes, this cycle of events terminates after finitely many transitions; in others, it can continue indefinitely.

What Is an Action? What Is a State?

The cycle of events introduced two terms: action and state. An **action** is a choice that can be made when a state is observed, and a **state** is a summary of the prior history of the process that suffices to determine:

- Which actions the decision maker can elect.
- The payoff or cost for picking each action; this payoff (or cost) is allowed to depend on the state that the decision maker observes and on the action that the decision maker elects, but neither on prior states nor prior actions.
- The law of motion that governs transition to the next state; it, too, can depend on the state that the decision maker observes and the action that the decision maker elects.

Thus, in a sequential decision process, the actions available to the decision maker can depend on the current state but not on prior states and prior actions. Similarly, the payoffs or costs and the law of motion can depend on the current state and action, but not on prior states or actions.

Illustration

To illustrate, we interpret a shortest-path problem as a sequential decision process and use it to highlight the role played by states. To do so, we regard:

- Each node i from which at least one arc emanates as a state.
- Each arc (i, j) as an action.

The decision maker observes state (node) i, and having done so, selects one of the arcs (i, j) whose "from" node is state i. Selecting arc (i, j) costs $c(i, j)$ and causes transition to state (node) j, whereupon the cycle of events is renewed. A node from which no arcs emanate is *not* a state; when such a node is reached, decision making stops.

The actions that the decision maker can select can depend on the current state (node)—the one that the decision maker observes—but not on prior states and not on prior actions. Similarly, the cost of selecting an action can depend solely on the current state and action, as can the state to which transition will occur.

Embedding

Viewing an optimization problem as a sequential decision process **embeds** the problem of interest in a family of optimization problems, one for each state. The problem for state i imagines that the decision maker begins by observing state i. Typically, a decision maker is interested in only one of these problems, namely, the one whose starting state is where she is at.

In Problem A, for instance, Pete was interested in the shortest path from his home (state 1) to the office. Before we solved this problem, we took a step backward. We embedded it in a family of optimization problems, one per starting state; we sought the shortest path from each state i to the office.

What Is Dynamic Programming?

The methodology of **dynamic programming** consists of these steps:

- Embed the problem of interest in a sequential decision process, which is a family of optimization problems, one per starting state.

- Link the solutions to these optimization problems by a so-called functional equation.
- Find an efficient way to solve them all.

That's how we've solved the shortest-path and longest-path problems in prior sections.

Richard Bellman coined the terms *dynamic programming* and *functional equation* and showed how dynamic programming solves an extremely broad array of optimization problems.

What Is a Functional Equation?

But what is a functional equation? Although a precise answer to that question lies beyond our reach, we have illustrated them. Equations (5.2), (5.6), (5.7), and (5.8) are functional equations. In each case, the **functional equation** has these properties:

- It is actually a system of equations, one per state.
- On the left-hand side of its equation for a particular state is the optimal value of the sequential decision process that is associated with that state.
- On the right-hand side of its equation for a particular state can be the optimal values of the optimization problems for various states.
- To the right of each "=" sign can be an operator, such as "max" or "min."

In each of the functional equations that has been displayed, we are adding values, though that isn't necessary. What is necessary is a type of **monotonicity**; increasing function values on the right-hand side of an equation cannot decrease the function value on the left-hand side. For instance, Equation (5.2) would still be a functional equation if it was

$$f(i) = \min_j\{\max[c(i, j), f(j)]\}, \qquad \text{for } i \neq 9, \tag{5.9}$$

because increasing $f(j)$ cannot decrease $f(i)$.

For a sequential decision process that marches forward in time and has a fixed end-point, the functional equation is easy to solve through a procedure known as **backwards recursion**; start at the end and work back toward the beginning. We solved Equation (5.2) in that way. Often, the solution to a functional equation can be found by linear programming. We have solved Equation (5.2) in that way too.

What's New?

In one sense, nothing is new, for all we have done is to introduce a lot of jargon that isn't needed to solve shortest-path and longest-path problems.

In another sense, a great deal is new. The methodology of dynamic programming is handy and is well worth learning. The key idea is to embed the problem of interest in a class of optimization problems, to link their solutions via a functional equation, and to solve them all efficiently.

The hard part of dynamic programming is to identify the states. Remember that a state is a summary of what's come before that suffices to evaluate the current actions.

An Example

Many optimization problems turn out to be shortest-path or longest-path problems after being formulated as sequential decision processes. Here is one.

Problem B (Manufacturing Turbines)

Your company manufactures turbines. It is the beginning of the month, and you have one turbine on hand now. Orders for this month and the following three are for 2, 1, 1, and 0 turbines, respectively. You need to have one turbine on hand at the beginning of the fifth month. So a total of four turbines must be produced over the next four months. Orders for a particular month can be filled with that month's production, or from inventory. Each month, your company can manufacture 0, 1, or 2 turbines at a variable cost of \$0, \$7000, and \$10,000, respectively. The cost of having each turbine in inventory at the start of the month is \$2000. Find the least expensive production schedule.

To formulate this problem as a sequential decision process, we must identify the states and actions. It's pretty clear that the actions are the production quantities and that the state must keep track of two pieces of information—the month and the inventory position (the number of turbines on hand) at the start of the month. Not all production quantities are feasible for each state. In the first month, for instance, you must produce at least one turbine because you have one on hand and you will need to ship two.

Let us label the five months alphabetically, a through e. A state consists of a letter and a number; the letter identifies the month, and the number is the inventory position at the start of the month. For instance, state $(c, 2)$ means that two turbines are on hand at the start of month c. We begin at state $(a, 1)$ because one turbine is on hand at the start of the first month. We must end at state $(e, 1)$ because one turbine must be on hand at the start of the fifth month.

A Network Formulation

Figure 5.5 models the turbine production problem as a shortest-path problem in an acyclic network. To avoid clutter, the production quantities have been omitted from the figure. If, for instance, you produce two turbines in month a, transition occurs from state $(a, 1)$ to state $(b, 1)$, at a cost of $10 + 2$, this being the sum of the cost of producing two turbines in a single month and the cost of having one turbine on hand at the start of that month.

You seek the shortest path from state $(a, 1)$ to state $(e, 1)$. From Figure 5.5, it's easy to compute the tree of shortest paths to node $(e, 1)$ and to see that the least-cost production plan is to produce two units in the first month, two in the third month, and none in the other months.

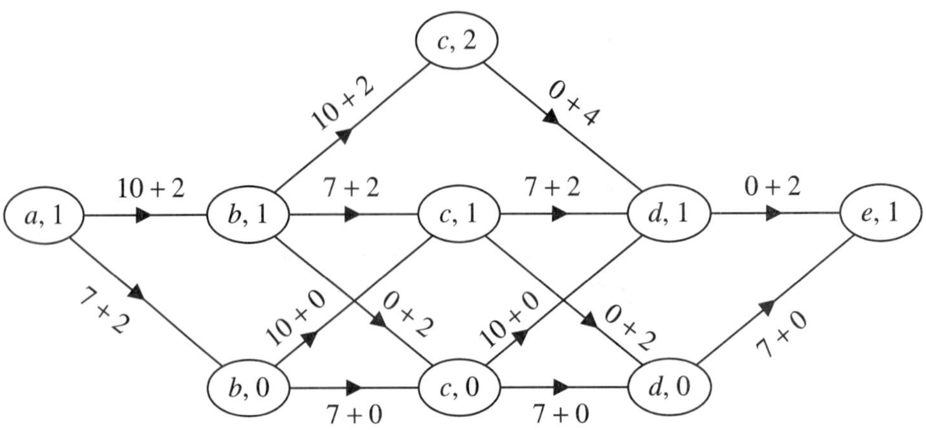

Figure 5.5 A network formulation of Problem B.

Deterministic Dynamic Programs

An optimization problem is said to be **deterministic** if chance plays no role in it. What happens when a deterministic optimization problem is solved by dynamic programming? As a general rule:

- Its solution takes the form of a shortest-path or a longest-path problem.
- This shortest-path or longest-path problem inherits the structure of the optimization problem that it models.

This helps explain why shortest-path and longest-path problems arise. They can result from formulating an optimization problem for solution by dynamic programming. When they do, they inherit the structure of the underlying problem, and that structure can be used to accelerate the computation of shortest paths.

Similarly, an optimization problem is said to be **stochastic** if chance plays a role in its description. Later, in Chapters 8 and 9, we will see how dynamic programming solves stochastic optimization problems.

5.5. PROJECT MANAGEMENT WITH CPM AND PERT

The critical path method (CPM) and the program evaluation and review technique (PERT) are quantitative techniques for managing projects. Both techniques model a project as consisting of several **activities**, each of which has two attributes. One of these attributes is the activity's duration, that is, the length of time needed to accomplish it. The other attribute is the activity's **predecessors**, which are those activities that must be completed before it can begin.

Constructing a home illustrates this model. It entails several activities—laying a foundation, framing, roofing, plumbing, exterior painting, and so forth. Each activity has a time duration, and each activity can have one or more predecessors. For instance, roofing cannot begin before framing is complete.

A Bit of History

CPM and PERT were developed in the 1950s. At that time, CPM assumed that each activity had a fixed duration time, and it aimed to minimize the length of time needed to complete the project. PERT was designed to deal with the case of uncertain duration times. Its goal was to maximize the probability that the project was completed on time. At present, nearly a half century later, each of these techniques has absorbed the best ideas of the other and they have nearly coalesced.

CPM and PERT are routinely used to manage large-scale construction projects, such as rebuilding the World Trade Center after a bomb exploded in its parking garage in February of 1993. These techniques are also widely used to manage development projects, particularly those that entail technological innovation. The pioneering application of PERT was in the development of the Polaris submarine, the world's first nuclear-powered ship, which entailed considerable technological innovation.

A Project

This section introduces project management for the case of deterministic activity durations. Uncertain durations are discussed briefly at the end of this section. When the activities have fixed duration times, the project management problem takes the form of a longest-path problem. To see how this occurs, we focus on:

Problem C (Kathy's Renovation Project)

Kathy is renovating her home. Her project requires 10 different activities, which she has labeled *a* through *j*. Table 5.2 lists these activities, specifies the time needed to accomplish each of them, measured in weeks, and lists its predecessors. For instance, activity *e* requires five weeks, and work on activity *e* cannot begin until activities *c* and *d* have been completed. Kathy wants to complete this project as quickly as possible. She wonders how long it will take. She also wonders which activities cannot be delayed without extending the project's completion time.

Table 5.2 Activities, their completion times (in weeks), and their predecessors.

Activity	*a*	*b*	*c*	*d*	*e*	*f*	*g*	*h*	*i*	*j*
Duration	3	6	8	4	5	1	9	2	7	10
Predecessors	—	—	*a*	*a*, *b*	*c*, *d*	*b*, *c*	*b*	*g*	*f*, *h*	*e*, *f*

A Novel Network

Although it seems natural to model this project by representing each activity as a directed arc, it is simpler to represent each activity as a node and to represent each precedence relation as a directed arc. The network model of this project will be a bit different from those seen previously. Its *nodes* have lengths, the length of each node being the duration of the activity that the node represents. Its arcs do not have lengths; effectively, their lengths are zero.

Figure 5.6 models Kathy's renovation project in this way. Its network has one node per activity, and it has two other nodes, *S* and *F*, which depict the start of the project and its finish. Each activity's node has two labels. One of these labels is the activity that the node represents. The other is that activity's duration. This network has one arc for each precedence relation. For instance, Figure 5.6 has an arc pointing from node *a* to node *d* because Table 5.2 states that activity *a* must be completed before activity *d* can begin.

Path Lengths

In this model, the **length** of a path from node *p* to node *q* is defined to be the sum of the lengths of its *intermediary* nodes (i.e., those nodes other than *p* or *q*). This rule assigns to the path (*b*, *d*, *e*, *j*) the length of 9 = 4 + 5 because the existence of this path means that at

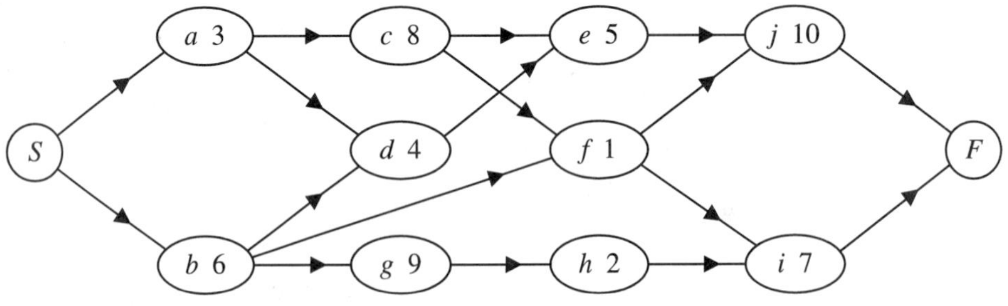

Figure 5.6 A network representation of Kathy's renovation project.

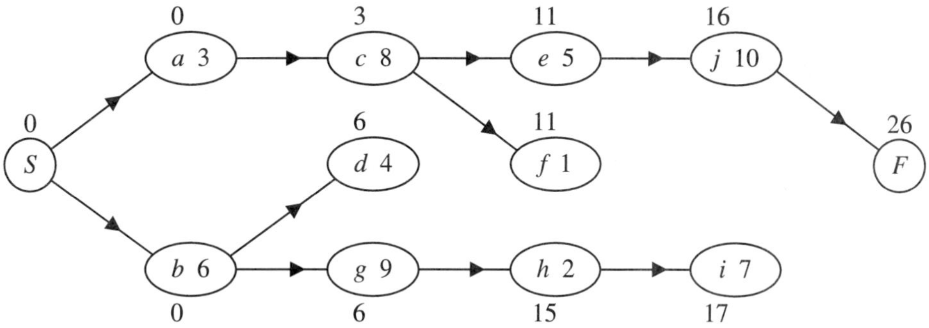

Figure 5.7 Tree of longest paths from node *S* to the others and earliest start times.

least nine weeks must elapse between the moment at which activity *b* ends and the moment at which activity *j* commences. The time required to complete the project equals the length of the *longest* path from node *S* to node *F*.

To analyze Kathy's renovation problem, we will solve two longest-path problems. We'll find the tree of longest paths from node *S* to the other nodes as well as the tree of longest paths from each node to node *F*.

Earliest Start Times

The earliest time at which an activity can start equals the length of the longest path from node *S* to that activity's node. We can find these earliest start times recursively, working from left to right in Figure 5.6. Figure 5.7 reports the earliest start times and the activities that determine them.

Adjacent to each node in Figure 5.7 is the length of the longest path from node *S* to that node. This is the earliest time at which the node's activity can begin. Evidently, it takes 26 weeks to reach node *F*, that is, to complete this project. Figure 5.7 reports that 11 weeks must elapse before work on activity *e* can commence. Its path (*S*, *a*, *c*, *e*) has 11 as its length because $11 = 3 + 8$. To check that this is the longest path from node *S* to node *e*, note in Figure 5.6 that the other two paths from node *S* to node *e* have lengths of 7 and 10.

Critical Path and Critical Activity

In the critical path method, a **critical path** is a sequence of activities that cannot be delayed without increasing the project completion time. For Kathy's renovation project, Figure 5.7 shows that there is exactly is one critical path, which is the sequence (*a*, *c*, *e*, *j*) of activities. The duration times of the activities in this critical path total 26 weeks because $26 = 3 + 8 + 5 + 10$.

An activity is said to be **critical** if it is on a critical path. In Problem C, activities *a*, *c*, *e*, and *j* are critical. The "critical" activities are those that cannot be delayed without increasing the time at which the project is completed.

Latest Completion Times

The length of the longest path from each node to node *F* determines the latest completion time of that node's activity. Thus, we can find the latest completion times by solving a family of longest-path problems. Figure 5.8 reports their solutions. In it, each node has been labeled

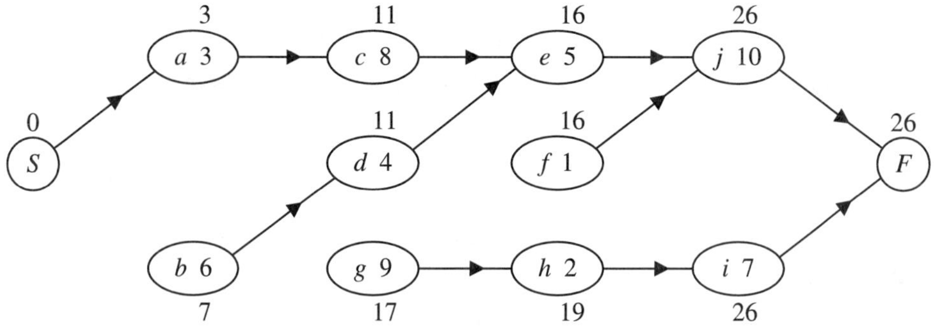

Figure 5.8 Tree of longest paths to node *F* and latest completion times.

with 26 less the longest path length from it to node F. This is the latest time at which the node's activity can be completed without delaying completion of the project.

Table 5.3 consolidates the information found in Figures 5.7 and 5.8. This table specifies each activity's earliest start time, latest completion time, duration, and "maximum delay," the last of which is defined by

$$\begin{bmatrix}\text{earliest start}\\ \text{time}\end{bmatrix} + \begin{bmatrix}\text{duration}\\ \text{time}\end{bmatrix} + \begin{bmatrix}\text{maximum}\\ \text{delay}\end{bmatrix} = \begin{bmatrix}\text{latest completion}\\ \text{time}\end{bmatrix}.$$

Thus, an activity's maximum delay equals the time by which it can be delayed without increasing the project completion time, assuming that no other activities are delayed.

Table 5.3 shows, for instance, that the start of activity *f* can be delayed four weeks without increasing the project duration time. By contrast, activity *b* or *d* can be delayed by only one week. If activity *b* is delayed by one week, activity *d* must not be delayed at all; it becomes critical.

Problem C had only one critical path. Paths can have the same length, and several paths can be critical. In general, an activity is critical if and only if its maximum delay equals zero. Each critical activity is a node in at least one critical path.

"Crashing" and the Cost-Time Tradeoff

A project can entail a large investment that typically can begin to earn income as soon as the project is completed, but not before. For instance, late completion of a construction project can be very costly, and early completion can be very profitable. In brief, time is money.

It is often possible to invest in ways that shorten the duration times of the activities, for instance, by using overtime labor. When overtime is used, a pair of resource allocation problems emerges. They are to:

- Allocate a fixed "overtime" budget B to the activities so as to minimize the project completion time.
- Determine the "overtime" budget B that maximizes net profit from the project.

Table 5.3 Activities, earliest start times, latest completion times, and maximum delays.

Activity	*a*	*b*	*c*	*d*	*e*	*f*	*g*	*h*	*i*	*j*
Latest completion time	3	7	11	11	16	16	17	19	26	26
Earliest start time	0	0	3	6	11	11	6	15	17	16
Duration	3	6	8	4	5	1	9	2	7	10
Maximum delay	0	1	0	1	0	4	2	2	2	0

A project network is acyclic, and its tree of earliest start times is easy to compute recursively. Solving these problems by linear programming might seem like overkill, but when a budget is introduced, linear programs come to the fore, as do nonlinear programs.

Linear and Nonlinear Programs

Suppose that the duration time of each activity decreases linearly with the overtime budget allocated to it. Assume for instance, that the duration $D(e)$ of activity e is given in terms of the overtime budget $B(e)$ allocated to activity e by

$$D(e) = 5 - 0.1B(e), \qquad \text{for } 0 \leq B(e) \leq 10. \tag{5.10}$$

Thus, the duration of activity e is reduced from five weeks to four by allocating 10 units of money (each unit being \$100, say) to that activity. When duration times vary in this way, the budget allocation problem remains much like Program 5.1, with a constraint that allocates the overtime budget to the activities. Similarly, if the net profit from the project varies linearly with its completion time, the profit-maximization problem becomes a linear program in the same way.

If the duration time of each activity exhibits decreasing marginal return with the overtime budgets allocated to it, the budget allocation problem becomes a nonlinear program whose solution Solver can easily find. And when the value of the project exhibits decreasing marginal return, the profit-maximization problem becomes a nonlinear program whose solution Solver can easily find.

A Critique of CPM

A principal merit of the critical path method is that it requires us to think systematically about a project—to parse it into its constituent activities, to estimate the duration time of each activity, and to identify the ways in which the activities interact with each other. Building the model helps us to understand the project.

Computing the earliest start times and the latest completion times determines which activities require surveillance and, possibly, which deserve to be speeded up.

No model of a project can be guaranteed to be accurate because over time, unforeseen events can occur. When they do, we update the network, re-compute its critical path, and re-determine which activities need close monitoring.

The critical path method captures the most salient ways in which the activities in a project relate to each other. It is based on precedence, the fact that certain activities cannot begin until others are finished.

One drawback of CPM is that activities can also interfere with each other through *nonconcurrence*; a pair of activities can be performed in either sequence but not at the same time. For instance, you can paint the walls before you sand the floors, you can paint the walls after you sand the floors, but you should not paint the walls while you sand the floors. To incorporate nonconcurrence, we would need to switch from a linear programming formulation to an integer program.

If a project entails research, some of its activities may model *alternative* ways in which to accomplish an objective. Each such activity may have a cost, a duration time, and a probability of success. In this setting, we do not seek success in all activities but only in enough activities to achieve the desired technological breakthroughs. A "critical path" fails to capture the essence of the issue, which is to achieve the new technology with minimal time and expense. Quite a different model is needed.

Finally, the duration times of the activities can be *uncertain*, in which case the uncertainty must be modeled in a way that reflects the project manager's goals. He might wish to maximize the probability that the project is completed within a predetermined length of time. Alternatively, he might seek to minimize the expectation of the time needed to complete the project.

Uncertain Duration Times*

This subsection[1] builds a model that has uncertain duration times. Pedagogically, this subsection is out of place, drawing in the language and concepts of probability that are introduced in Chapters 6 through 10. This subsection is placed here, however, because its subject is project management.

How can we handle the case of uncertain activity durations? There is no single answer to that question. Sketched below is one heuristic (approximation) that was used with success in many applications of PERT. This heuristic consists of two steps.

Step 1: For each activity, estimate these three elements of data:

- Its most optimistic (smallest possible) duration, A.
- Its most pessimistic (largest possible) duration, B.
- Its most likely duration, M.

Model the duration time of this activity by a random variable X whose mean and standard deviation are given in terms of A, B, and M by

$$E(X) = \frac{A + 4M + B}{6}, \qquad StDev(X) = \frac{B - A}{6}.$$

Comments: It seems appropriate to equate $E(X)$ to a weighted average of A, M, and B and to place most of the weight on M, as is done above. The formula given for $StDev(X)$ is more problematic. The variable X takes values between A and B. Equating the standard deviation of X to $(B - A)/6$ equates the range of X to *six* standard deviations. This may significantly understate the uncertainty in the activity's duration time.

Traditionally, the random variable X is modeled as a Beta distribution whose mean and variance are given above and whose range is the interval between A and B. The density of this Beta distribution equals zero at A, it rises to a maximum, and then it decreases, taking the value zero at B, but its maximum is not guaranteed to occur at M.

Step 2: Working solely with the most likely duration times of the activities, solve the critical path problem. Then, for the actual situation (in which the duration time of each activity is uncertain), assume that:

- The critical path for the most-likely-duration-times case is unique, and it remains critical when the duration times are perturbed.
- The duration times of the activities in the critical path are mutually independent.
- There are a large number of activities on the critical path, and each is of short duration relative to the completion time of the entire project.

Comments: The first two of these assumptions are especially questionable. With uncertain duration times, a noncritical path can easily become critical. In addition, the duration times of different activities can be affected by a single cause (e.g., a labor shortage or a strike in the shipping industry), which can make them heavily dependent, not independent.

---

[1] This starred subsection treats a special topic that can be omitted with no loss of continuity. It draws on information in Chapters 6–10.

Together, these assumptions let us use the Central Limit Theorem (Chapter 11) to approximate the project's duration time by a normal distribution, one whose mean equals the sum of the means of the critical activities and whose variance equals the sum of the variances of the critical activities. Thus, whatever need be known about the project duration time can be computed, including the probability that it will be completed within the allotted interval of time.

This heuristic approximates the effect of uncertain task durations. Although using this method requires a leap of faith, intelligent use of it has proven to be effective.

5.6. ALGORITHMS FOR NETWORKS

Networks have long fascinated mathematicians, computer scientists, and operations researchers. Many problems are easy to pose in the context of a network; some turn out to be easy to solve, but others are truly daunting.

The next three sections glimpse the literature on algorithms for networks. These sections present network algorithms that:

- Find a tree of shortest paths from a particular node to all others in a network whose arc lengths are nonnegative.
- Solve the assignment problem.
- Solve the max-flow problem.

Each of the next three sections describes an efficient method for solving a problem that *can* be posed as a linear program. Thus, these sections do not enlarge the scope of solvable problems. For a hard-nosed practical introduction, they are unnecessary. That's one reason why they are starred.

These starred sections are included because they introduce simple ideas that have many uses. In fact, we will see that similar ideas are used to attack each of these problems.

The algorithms in the starred sections are interesting for a second reason. As a rule, Dantzig's simplex method has proved to be very hard to beat. Enormous effort, mostly fruitless, has been expended searching for algorithms that run faster than the simplex method. Network optimization problems are one arena in which this effort has borne fruit. For network optimization problems, the algorithms in the next three sections outperform the simplex method.

5.7. DIJKSTRA'S METHOD FOR SHORTEST PATHS*[2]

In 1959, E. Dijkstra published a simple method for finding a tree of shortest paths from one node of a network to the others. His method does not require the network to be acyclic, but it does require the arc lengths to be nonnegative.

As is usual, the length of directed arc (i, j) is denoted $c(i, j)$. Let i be the node from which we seek the tree of shortest paths. With $f(i) = 0$ we designate, for each node j other than i,

$$f(j) = \text{the length of the shortest path from node } i \text{ to node } j.$$

The shortest path from node i to node j has final arc (k, j) for some k, so that

$$f(j) = \min_k \{f(k) + c(k, j)\} \qquad \text{for each } j \neq i. \tag{5.11}$$

[2]This is the first of three starred sections that can be skipped with no loss of continuity.

If k achieves the minimum in Equation (5.11), then $f(j) = f(k) + c(k, j)$, and, since each arc length $c(k, j)$ is assumed to be nonnegative, it must be that $f(k) \leq f(j)$. From this observation, it's easy to see, as Dijkstra did, that we can calculate shortest-path lengths by increasing value of $f(k)$, as follows.

Dijkstra's Method:

1. (*Initialization*) Color node i blue. Set $v(i) = 0$ and, for each node j other than node i, set $v(j) = +\infty$.
2. (*Termination*) Stop if no nodes are blue.
3. Otherwise, among the blue nodes, select a node k for which $v(k)$ is smallest. Color node k gray. Then, for each arc (k, j) for which node j is not gray, color node j blue and replace $v(j)$ by the smaller of itself and $v(k) + c(k, j)$. Return to Step 2.

The first three of the following comments on Dijkstra's method can be cobbled into an inductive proof that ends with $v(k) = f(k)$ for each k.

- Each node j other than node i is colored blue as soon as a path from node i to node j is found.
- While node j is blue, $v(j)$ equals the length of the shortest path from node i to node j, each of whose intermediary nodes is gray.
- Each blue node k becomes gray when Dijkstra's method determines that $v(k) = f(k)$.
- To make Dijkstra's method easy to read, we've shown how it computes the f-values but not the tree itself. To record the tree, use a pointer $p(j)$ for each node j other than i and update its value as follows: In Step 3, set $p(j) \leftarrow k$ whenever arc (k, j) reduces $v(j)$. These pointers let us "backtrack" from node j to identify a shortest-path node i to node j.
- Dijkstra's method continues to work if the comparisons in Step 3 are executed for each arc (k, j), even if node j is gray. No such comparison could reduce $v(j)$, which is why they aren't needed.

An Example

To illustrate Dijkstra's method, we will apply it to the shortest-path problem in:

Problem D (A Tree of Shortest Paths)

Rows 2 through 8 of the spreadsheet in Table 5.4 specify the arc lengths of a six-node network. For this network, find the tree of shortest paths from node 3 to the others.

Rows 2 through 8 of Table 5.4 label the nodes 1 through 6 and report the length of each arc that leads from one node to another. Cell D6 reports that $c(4, 2) = 15$, for instance. The diagonal entries in Table 5.4 have been left blank because, being nonnegative, they can affect no shortest paths.

A Spreadsheet

Table 5.4 also shows the progress of Dijkstra's method after the first and second execution of Step 3. The first execution of Step 3 colors node 3 gray. Then, for each node j other than node 3, it colors node j blue and sets $v(j) = v(3) + c(3, j) = 0 + c(3, j)$. This is accomplished by dragging the function in cell H11 across row 11.

Table 5.4 Arc lengths for Problem D and v-values after the first and second execution of Step 3 of Dijkstra's method.

	A	B	C	D	E	F	G	H	I	J
2			1	2	3	4	5	6		
3		1		5	5	4	10	7		
4		2	12		10	12	10	2		
5		3	19	3		1	16	5		
6		4	16	15	18		7	3		
7		5	2	18	5	8		6		
8		6	7	8	4	5	3			
9										
10			v(1)	v(2)	v(3)	v(4)	v(5)	v(6)		=H$5
11	Execution #1		19	3	0	1	16	5		
12	Execution #2		17	3	0	1	8	4		=MIN(H$11,$F$11+H$6)

Of the blue nodes in row 11, node 4 has the smallest label. Hence, the second execution of Step 3 colors node 4 gray and, for each j, replaces $v(j)$ by the smaller of itself and $v(4) + c(4, j)$. Dragging the function in cell H12 across row 12 accomplishes this. For instance, $v(1)$ gets reduced from $19 = c(3, 1)$ to $17 = v(4) + c(4, 1) = 1 + 16$. Of the blue nodes in row 12, node 2 has the smallest label. The next iteration will therefore color node 2 gray and, for each j, will replace $v(j)$ by the smaller of itself and $v(2) + c(2, j)$, and so forth.

Snap Shots

The pictures in Figure 5.9 record the progress of Dijkstra's method after the first, third, and sixth execution of Step 3. Each picture is a tree whose root is node 3. In each tree:

- Each node's v-value is adjacent to it.
- For each node j other than node 3, the label $v(j)$ equals the length of the tree's path from node 3 to node j.
- Each gray node j has $v(j) = f(j)$.

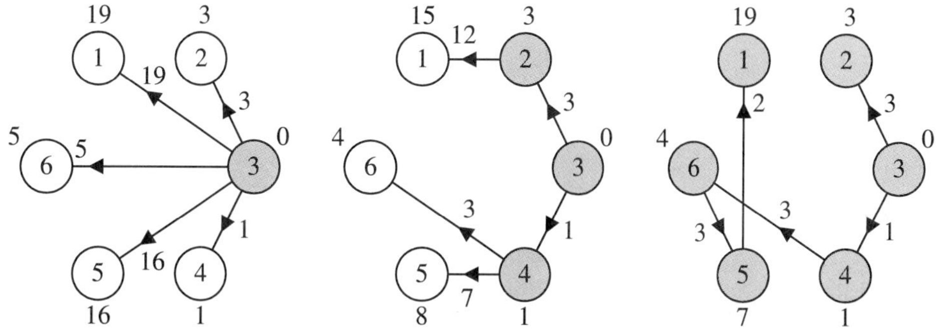

Figure 5.9 Trees and v-values after the first, third, and sixth execution of Step 3 of Dijkstra's method.

- In Figure 5.9, the "blue" nodes are unshaded, which saves on typesetting expense. Each "blue" node j has $v(j)$ equal to the length of the shortest path from node 3 to node j, all of whose intermediary nodes are gray.

In the central tree, node 6 has the smallest label among the "blue" (unshaded) nodes. At the next execution of Step 3, node 6 will be colored gray, and each label $v(j)$ will be replaced by the smaller of itself and $v(6) + c(6, j)$.

Subtrees

Figure 5.9 illustrates a useful definition—that of a "subtree." A portion of the network is called a **subtree** if it is a tree for the nodes that *it* includes but not necessarily for the nodes in the entire network.

In each diagram, the network consisting of the gray nodes and the arcs connecting them is a subtree. In the central diagram, this subtree consists of nodes 2, 3, and 4 and of arcs (3, 2) and (3, 4). This network is a tree whose root is node 3, but it omits some nodes of the original network.

In Dijkstra's method, each execution of Step 3 *enlarges* the subtree of shortest paths by adding one node and one arc to it. For instance, the fourth execution of Step 3 enlarges the central subtree in Figure 5.9 by adding node 6 and arc (4, 6) to it.

Running Time

Let n denote the number of nodes in the network. The running time for a naïve implementation of Dijkstra's method is proportional to n^2. The reason is as follows. Step 3 requires computation of $v(k) + c(k, j)$ for approximately half of the arcs, roughly $n^2/2$. Also, each execution of Step 3 requires comparison of the v-values of the blue nodes, for a total of roughly $n^2/2$ comparisons.

Large, Sparse Structured Networks

In practice, the networks to which Dijkstra's method is applied are large, sparse, and structured. In this context, **large** means that the number n of nodes is enormous (e.g., in the hundreds of thousands); **sparse** means that the number of arcs is a tiny fraction of n^2; and **structured** means that the arcs form a pattern that is inherited by their tree of shortest paths. Typically, when a deterministic optimization problem is formulated as a dynamic program, the network inherits the structure of the optimization problem.

If the network is sparse, the most time-consuming part of Dijkstra's method is to determine the blue node k whose label is lowest. This step can be speeded up by keeping the blue nodes' v-values in a data structure called a heap. This step can be avoided entirely for networks whose ratio M/m is modest, where M is the length of the longest arc and m is the length of the smallest arc. To see how a system of roughly M/m "buckets" suffices, see Problem 14, page 192.

Dijkstra's method illustrates the concept of **reaching**, which consists of "reaching" out from a node whose f-value is known to update the v-values for other nodes. If the network has structure during computation, reaching lets us "prune" it of arcs that cannot be part of the shortest-path tree. For large structured networks, the fastest shortest-path methods may be those that incorporate reaching, pruning, and buckets.[3]

[3] For an account of structured shortest-path problems, the reader may refer to Chapter 2 of *Dynamic Programming: Models and Applications*, (Englewood Cliffs, NJ: Prentice Hall, 1982), by E. V. Denardo.

5.8. THE HUNGARIAN METHOD FOR THE ASSIGNMENT PROBLEM*

The assignment problem, the simplest of the network flow problems, was introduced in Chapter 4. In an assignment problem, we reserve the symbol m for the number of "source" nodes, which equals the number of "demand" nodes. Specifically, the **assignment problem** has:

- A given number m of source nodes, with a supply of one unit at each source node.
- The same number m of demand nodes, with a requirement of one unit at each demand node.
- A cost $c(i, j)$ of shipping one unit from each source node i to each demand node j.
- The constraint that each shipping quantity must equal zero or one.
- The goal of minimizing the total shipping cost.

The decision variables in the assignment problem are the shipping quantities, and we denote:

$$x(i, j) = \text{the quantity shipped from source node } i \text{ to demand node } j.$$

For the assignment problem, a **shipping plan** is a set of values of the shipping quantities such that the total quantity shipped out of each source node i equals 1, the total quantity shipped into each demand node i equals 1, and each shipping quantity equals zero or 1. The assignment problem is to find a shipping plan whose total shipping cost is smallest.

Technically, the assignment problem is an integer program because the shipping quantities must be integers. The simplex method solves the assignment problem nonetheless because the Integrality Theorem guarantees an integer-valued solution. For assignment problems with very large numbers of nodes, however, the simplex method can perform poorly. Its worst-case running time can grow in proportion to 2^m, which is very slow indeed.

In 1955, H. W. Kuhn published an ingenious method for solving the assignment problem in which the worst-case running time grows only as m^4, which is a vast improvement for large values of m. Kuhn generously dubbed his method the **Hungarian method** to acknowledge his debt to the work done by the Hungarian mathematicians, J. Egerváry and D. König, long before the era of linear programming.

An Example

A general description of the Hungarian method would overwhelm us with notation. Instead, we'll introduce the Hungarian method in the context of:

Problem E (Assignment)

Solve the 5×5 assignment problem whose shipping costs are given by the spreadsheet in Table 5.5.

Column B of the table indicates that the source nodes are labeled 1 through 5; row 2 indicates that its demand nodes are labeled 6 through 10, and cells C3 through G7 specify the shipping costs. For instance, cell G6 indicates that $c(4, 10) = 7$, namely, that the cost of shipping one unit from source node 4 to demand node 10 equals 7. Row 8 of this table records the least of the shipping costs into each demand node.

Changing the Shipping Costs

Table 5.5 notes that 2 is the least of the shipping costs into node 6. What happens if the number 2 is subtracted from each shipping cost into node 6? Every feasible solution to the

Table 5.5 Spreadsheet for a 5 × 5 assignment problem.

	A	B	C	D	E	F	G
2			6	7	8	9	10
3		1	2	5	5	4	10
4		2	12	12	10	12	10
5		3	11	3	13	1	16
6		4	16	15	18	16	7
7		5	18	18	5	8	6
8	col. min.		2	3	5	1	6

assignment problem must ship one unit into node 6, so this change reduces the cost of *every* shipping plan by two units. The relative desirability of different shipping plans is unaffected. Each optimal shipping plan remains optimal. Each nonoptimal shipping plan remains nonoptimal, and one of the shipping costs becomes zero.

Similarly, if the number 10 was subtracted from each shipping cost out of node 2, every shipping plan's cost would be reduced by 10, the optimal shipping plans would stay optimal, and two of the shipping costs would be reduced to zero.

The Hungarian method consists of a deft sequence of subtractions from the rows and columns of the cost matrix, aiming to bring an optimal solution into view. The "inner loop" of the Hungarian method changes these costs in a way that enlarges a subtree of zero-cost arcs. It does so in search of a path along which shipment can be increased at zero cost.

A "Hot Start"

The Hungarian method can be initialized with any nonnegative cost matrix, for instance, with the costs in Table 5.5. But it is easy to "hot start" the Hungarian method with a cost matrix, each of whose elements is nonnegative, with at least one zero in each row and at least one zero in each column. To accomplish this:

1. First, subtract the smallest shipping cost into each demand node from all shipping costs into that node.
2. Then, for the shipping costs that result from Step 1, subtract the smallest shipping cost out of each source node from all shipping costs out of that node.

Table 5.5 records the least shipping cost into each demand node. Step 1 subtracts 2 from each shipping cost in column C, it subtracts 3 from each shipping cost in column D, and so forth. Executing Step 1 transforms Table 5.5 into Table 5.6. The costs in Table 5.6 are nonnegative, and at least one cost in each column equals zero.

Table 5.6 records the smallest shipping cost out of each source node. Step 2 subtracts 4 from each shipping cost out of node 2, and it subtracts 1 from each shipping cost out of row 4. Executing Step 2 transforms Table 5.6 into Table 5.7.

As noted above, the cost matrices in Tables 5.5 through 5.7 have the *same* optimal shipping plans. In Table 5.7, all costs are nonnegative, each row contains at least one zero, and each column contains at least one zero.

A Partial Shipping Plan

It is not possible to ship one unit from each source node to each demand node using only the zero-cost arcs in Table 5.7. Four units can be shipped but not five. One way (there is

Table 5.6 Shipping costs after execution of Step 1.

	A	B	C	D	E	F	G	H	I
9									
10			6	7	8	9	10		row min.
11		1	0	2	0	3	4		0
12		2	10	9	5	11	4		4
13		3	9	0	8	0	10		0
14		4	14	12	13	15	1		1
15		5	16	15	0	7	0		0

Table 5.7 Shipping costs after execution of Step 2.

	A	B	C	D	E	F	G
17			6	7	8	9	10
18		1	0	2	0	3	4
19		2	6	5	1	7	0
20		3	9	0	8	0	10
21		4	13	11	12	14	0
22		5	16	15	0	7	0

another) to ship four units using only the zero-cost arcs is to set

$$x(1, 6) = x(2, 10) = x(3, 7) = x(5, 8) = 1, \tag{5.12}$$

with the other $x(i, j)$'s (shipping quantities) equal to zero. This shipping pattern satisfies the requirements at demand nodes 6, 7, 8, and 10 but not at node 9. It ships one unit out of source nodes 1, 2, 3, and 5. Thus, this shipping pattern fails to ship anything out of source node 4 or into demand node 9.

In this example and in general, the Hungarian method is initialized with a matrix (array) of nonnegative shipping costs and with a **partial shipping plan** that:

- ships only on arcs whose current shipping costs equal zero,
- ships at most one unit from each source node,
- ships at most one unit to each demand node,
- ships quantities that equal 0 or 1,
- ships no units from at least one source node and so ships no units to at least one demand node.

The spreadsheet in Figure 5.10 initializes the Hungarian method with the costs in Table 5.7 and with the partial shipping plan given by Equation (5.12). This shipping plan is recorded by shading each cell that corresponds to a positive shipment. For instance, cell C25 is shaded because this shipping plan sets $x(1, 6) = 1$. Figure 5.10 contains other information that will be soon explained.

Thinking at the Margin

Economics is not the only discipline in which it pays to think at the margin. Often, in a network flow problem, we are interested in *perturbing* the flow that currently exists. In the

	A	B	C	D	E	F	G	H
23								
24			6	7	8	9	10	
25		1	0	2	0	3	4	
26		2	6	5	1	7	0	-1
27		3	9	0	8	0	10	
28		4	13	11	12	14	0	-1
29		5	16	15	0	7	0	
30							1	

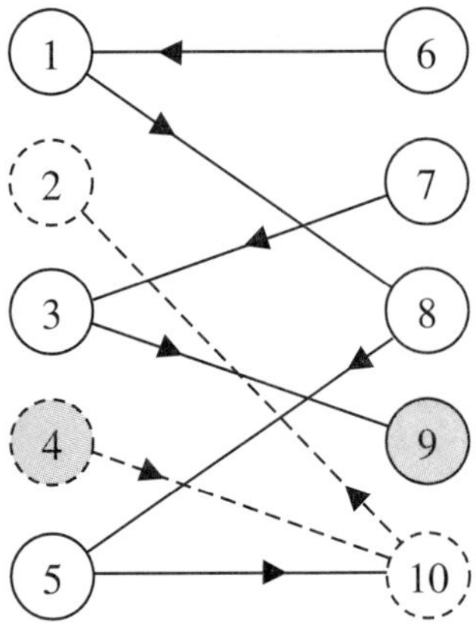

Figure 5.10 Partial shipping plan, costs, incremental network, and zero-cost subtree.

assignment problem, the flows cannot exceed 1. If a flow equals 0, the only way to perturb it is to increase it; if a flow equals 1, the only way to perturb it is to decrease it. Decreasing a positive flow on an arc (i, j) amounts to causing a flow to occur on this arc but in the *reverse* direction.

The Incremental Network

In general, the "incremental" network records the directions in which flows can be perturbed. The Hungarian method seeks to perturb the flow using only the zero-cost arcs. For it, each current cost matrix and partial shipping plan are now said to prescribe an **incremental network** whose arcs are determined as follows.

From the original network:

- Delete each arc (i, j) whose current shipping cost $\bar{c}(i, j)$ is positive.
- Delete each arc (i, j) whose shipping quantity $x(i, j)$ equals 1, but replace this arc with the **reverse** arc (j, i).

Here and hereafter, $\bar{c}(i, j)$ denotes the *current* cost of shipping on arc (i, j), not the original cost of shipping on this arc. And $x(i, j)$ equals the amount shipped on arc (i, j) in the *current* partial shipping plan.

Figure 5.10 presents the incremental network for the costs and partial shipping plan in its spreadsheet. This network preserves the four zero-cost arcs on which no flow occurs, namely, arcs (1, 8), (3, 9), (4, 10), and (5, 10). It reverses the four arcs on which flows occur; the reverse arcs are (6, 1), (7, 3), (8, 5), and (10, 2). Shipping one unit on a reverse arc, such as arc (6, 1), corresponds to decreasing the flow on arc (1, 6) from one to zero.

Largest Zero-cost Subtree

As noted earlier, the partial shipping program in Equation (5.12) fails to ship anything from source node 4 to demand node 9, which is why nodes 4 and 9 are lightly shaded in the incremental network. The Hungarian method builds an incremental network that contains a path from node 4 to node 9. It accomplishes this in several iterations, each of which:

- Identifies, in the incremental network, a subtree of arcs whose root is node 4 which includes as many nodes as is possible.
- Notes whether this subtree includes node 9 and, if not, changes the shipping costs in a way that enlarges this subtree.

For the incremental network in Figure 5.10, the *dashed* arcs and nodes comprise the largest subtree of zero-cost arcs whose root is node 4. This subtree contains arcs (4, 10) and (10, 2), which are drawn as dashed lines. This subtree includes nodes 4, 10, and 2, whose circles are drawn in dashed lines.

The spreadsheet in Figure 5.10 records the same subtree. In this spreadsheet, the cells that are outlined in dashed lines correspond to the arcs and nodes of this subtree. Cells G26 and G28 are outlined because the subtree contains reverse arc (10, 2) and forward arc (4, 10). Cells B26, B28, and G24 are outlined because this subtree includes nodes 2, 4, and 10.

Changing the Costs

H. W. Kuhn saw how to change the costs so as to expand this subtree. His insight is highlighted in the "Subtree Expander" routine that appears next. In it, $\bar{c}(p, q)$ denotes the *current* value of the shipping cost from source node p to demand node q, not the original value of this shipping cost.

Subtree Expander: Among all arcs (p, q) for which source node p is in the zero-cost subtree and demand node q is not in the zero-cost subtree, pick a pair (i, j) for which current shipping cost $\bar{c}(i, j)$ is smallest. Set $z = \bar{c}(i, j)$. Then:

- For each source node p in the subtree, subtract z from each current shipping cost out of node p.
- Then, for each demand node q in the subtree, add z to each current shipping cost into node q.

For the spreadsheet in Figure 5.10, the Subtree Expander picks $(i, j) = (2, 8)$ and sets $z = \bar{c}(2, 8) = 1$. Invariably, z is a positive number. (If z was zero, then node j would have been included in the subtree.) For the spreadsheet in Figure 5.10, the Subtree Expander subtracts 1 from the costs in rows 26 and 28, and then it adds 1 to the costs in column G.

The spreadsheet in Figure 5.10 is coded in a way that records these changes. If a source node is outlined, z is to be subtracted from its row of costs. If a demand node is outlined, z is to be added to its column of costs. Column H indicates that 1 is to be subtracted from the costs in rows 26 and 28, and row 30 indicates that 1 is to be added to the costs in column G.

An Enlarged Zero-cost Subtree

These changes in cost are reminiscent of the "hot start" procedure. Like the earlier changes in shipping costs, they have no effect on the set of optimal solutions. In Figure 5.10 and in general, these changes have the following properties:

- For each arc (p, q) in the prior subtree, the current shipping cost on arc (p, q) remains equal to zero (because z was subtracted from this cost and was then added to it).
- For each arc (p, q) whose shipping quantity equals 1, the current shipping cost remains unchanged and so is equal to zero.
- The current shipping cost on arc (i, j) has been reduced to zero.
- All current shipping costs remain nonnegative.

Executing the Subtree Expander produces the cost matrix in Figure 5.11, which also records the new incremental network and (in dashed lines) the new zero-cost subtree.

	A	B	C	D	E	F	G	H
31								
32			6	7	8	9	10	
33		1	0	2	0	3	5	
34		2	5	4	0	6	0	-4
35		3	9	0	8	0	11	
36		4	12	10	11	13	0	-4
37		5	16	15	0	7	1	-4
38					4		4	

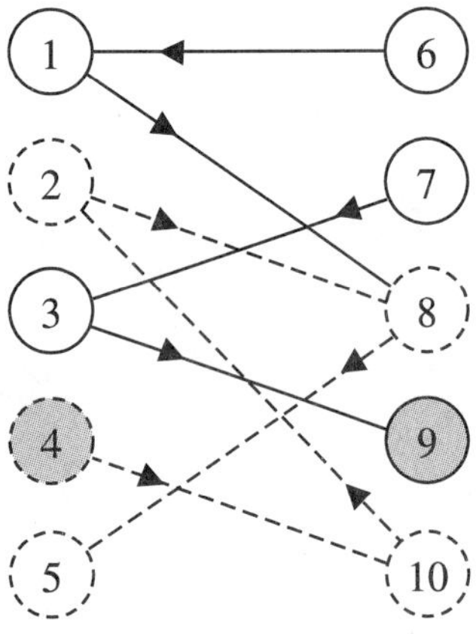

Figure 5.11 Shipping plan, current shipping costs, updated incremental network, and subtree.

The shipping cost $c(2, 8)$ has been reduced to zero, but shipping cost $c(5, 10)$ has become positive. Thus, arc (2, 8) joins the incremental network, and arc (5, 10) departs. Please check in Figure 5.10 that the arc (5, 10) was not dashed; it was not part of the prior zero-cost subtree.

The largest subtree of zero-cost arcs whose root is node 4 has expanded. Arcs (2, 8) and (8, 5) have been added to it, and arcs (4, 10) and (10, 2) have remained in it. This subtree includes nodes 2, 4, 5, 8, and 10 but not node 9.

A Still Larger Subtree

The second application of the Subtree Expander proceeds exactly as does the first. In Figure 5.11, the smallest cost $\bar{c}(i, j)$ with i in the subtree and j not in the subtree is $\bar{c}(2, 7) = 4$. Subtracting 4 from the indicated rows and then adding 4 to the indicated columns produces Figure 5.12.

Arc (2, 7) has joined the incremental network, and arc (1, 8) has departed. The subtree of zero-cost arcs whose root is node 4 has expanded to include arcs (2, 7), (7, 3), and (3, 9). This subtree includes node 9. Success!

	A	B	C	D	E	F	G
39							
40			6	7	8	9	10
41		1	0	2	4	3	9
42		2	1	0	0	2	0
43		3	9	0	12	0	15
44		4	8	6	11	9	0
45		5	12	11	0	3	1

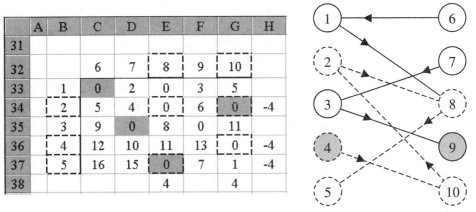

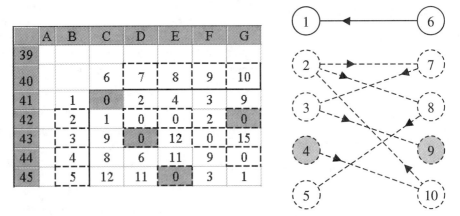

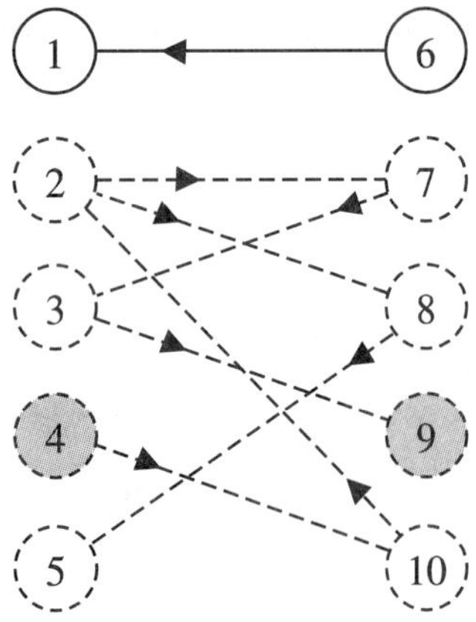

Figure 5.12 Revised shipping costs, incremental network, and subtree.

Table 5.8 An optimal solution; flows on shaded arcs equal 1.

	A	B	C	D	E	F	G
46							
47			6	7	8	9	10
48		1	**0**	2	4	3	9
49		2	1	**0**	0	2	0
50		3	9	0	12	**0**	15
51		4	8	6	11	9	**0**
52		5	12	11	**0**	3	1

Improving the Partial Flow

In this subtree, we backtrack from node 9 to node 4 to identify the path (4, 10, 2, 7, 3, 9) in the incremental network from node 4 to node 9. Each arc in this path has zero as its cost. The effect of perturbing the flow by shipping one unit via the arcs on this path is to set

$$x(4, 10) = x(2, 7) = x(3, 9) = 1 \quad \text{and} \quad x(2, 10) = x(3, 7) = 0.$$

(The flows on the reversed arcs have been decreased from one to zero.) This perturbation increases the total flow by one unit, and it produces the pattern of flows shown in Table 5.8.

In Table 5.8, one unit flows on each shaded arc. The shipping plan in the table satisfies the constraints of the assignment problem because one unit flows out of each source node and one unit flows into each demand node. To see that this shipping plan is optimal, we observe that:

- Each shipment occurs on an arc whose current shipping cost equals zero.
- The current cost of each arc is nonnegative.
- These costs were obtained by adding and subtracting constants from the rows and columns of the original cost matrix, which preserves the set of optimal solutions to the assignment problem.
- Consequently, the shipping plan is optimal.

The Hungarian method has completed its work.

General Discussion

In general, the Hungarian method proceeds exactly as we have described it, with one slight emendation. We initialized the Hungarian method with a partial shipping plan that satisfied the requirement at all but one of the demand nodes.

In general, a partial shipping plan can satisfy all but r of the requirements with r greater than 1. In this case, the incremental network contains a set of r *disjoint subtrees*, each of which is rooted at a source node from which nothing is being shipped and each consists solely of zero-cost arcs. At each iteration, the Subtree Expander enlarges at least one of these subtrees without shrinking any of the others. Eventually, these disjoint subtrees include a demand node to which nothing had been shipped. At that point, backtracking identifies the path on which shipping can be augmented at zero cost, and r decreases by 1.

We've described subtrees visually. An algorithm much like Dijkstra's method "identifies" them, as shown in the following:

Subtree Identifier:

1. (*Initialization*) Color each node i from which nothing is being shipped blue.
2. (*Termination*) Stop if no nodes are blue.
3. Else, pick any blue node k and color it gray. For each arc (k, j) in the incremental network for which node j has not been colored, color node j blue and set $p(j) \leftarrow k$. Return to Step 2.

The effect of coloring node j blue in Step 3 is to add arc (k, j) to the subtree(s). Recording the pointer $p(j) \leftarrow k$ lets us backtrack from node j to a node from which nothing is being shipped. The Subtree Identifier stops when no more nodes can be added to the subtree(s).

If the Subtree Identifier colors a node j to which nothing is being shipped, the pointers let us identify a path in the incremental network along which flow can be increased by 1. If not, the next iteration of the Subtree Expander will enlarge the subtree(s).

It's easy to organize the Hungarian method so that each iteration of the Subtree Expander requires a number of computer operations that is proportional to m^2, at worst. The number of iterations of the Subtree Expander is proportional to m^2, at worst. Thus, the Hungarian method solves the assignment problem with a total number of computer operations that is proportional to m^4, at worst.

The Transportation Problem

The assignment problem is the simplest of the network flow problems that we encountered in Chapter 4. The transportation problem is a bit more complex. The Hungarian method generalizes to solve the transportation problem without requiring any new idea.

For the transportation problem, a "partial shipment" ships no more from any source node than its supply, it ships no more to each demand node than its requirement, and it ships only on zero-cost arcs. Each cost matrix that is nonnegative and each partial shipment plan prescribe an "incremental network" whose arcs are determined as follows:

- Omit each arc (i, j) whose current cost $\bar{c}(i, j)$ is positive.
- If $\bar{c}(i, j) = 0$ and $x(i, j) = 0$, include arc (i, j).
- If $\bar{c}(i, j) = 0$ and $x(i, j) > 0$, include arcs (i, j) and (j, i).

With this slight change, the Hungarian method solves the transportation problem.

The Primal Dual Method for Network Flow

In general, a network flow model can include "transshipment" nodes, namely, those having arcs that point in and arcs that point out. The full generalization of the Hungarian method is known by the esoteric name, the **primal dual method** for network flow. This method solves network flow problems with a number of computer operations that is proportional to $n^3 K$, at worst, where n equals the number of nodes in the network and where K equals the sum of the fixed flows into the nodes. The primal dual method gets its name from the theory of duality in optimization, to which the final chapter of this text is devoted. Yet, as Kuhn noted, the key to it predates linear programming.

5.9. METHODS FOR MAX FLOW*

The max-flow problem first appeared in Chapter 4. In the present chapter, we require each arc to be connected to two nodes. For that reason, we must describe the max-flow problem slightly differently. Here, the **max-flow** problem is a network flow problem in which the network has:

- Two distinguished nodes, one called the source and the other the sink.
- A wraparound arc, from the sink to the source, whose capacity is infinite.
- A finite upper bound on the flow on each of the other arcs, which is called that arc's capacity.
- The constraint that each arc's flow lie between 0 and its capacity.
- The constraint that flow be conserved at each node.
- The goal of maximizing the flow on the wraparound arc.

The flow on the wraparound arc is the net flow into the source; it is also the net flow out of the sink.

A Bit of History

In 1954 at The RAND Corporation, Ted Harris and General F. S. Rose proposed the max-flow problem in the context of interdiction of a rail network. Almost immediately, L. R. Ford and D. R. Fulkerson discovered the Max-Flow Min-Cut Theorem and invented a "labeling" method to prove it by finding both the maximum-flow and the minimum-capacity cut. It was soon discerned that their Max-Flow Min-Cut Theorem was related to earlier theorems of Egerváry and König and of K. Menger (see Problems 20 and 21, page 193).

The labeling method of Ford and Fulkerson is elegantly simple. Each iteration conserves flow at each node, keeps each arc's flow between zero and that arc's capacity, and increases the flow on the wraparound arc. The running time of their labeling method increases, of course, with the size of the network. But its running time also depends on the capacities of the arcs, in awkward ways. A series of papers refined their labeling method, ameliorating its various difficulties.

Then, in 1974, the Russian mathematician A. V. Karzanov published a novel "preflow" method that does not conserve flow at the nodes, except at the end. Preflow methods begin by "pushing" the largest possible flow from the source node into the network. They end with as much as possible of this flow at the sink, the remainder back at the source. Preflow methods seem to be the quickest of the many algorithms that have been devised for the max-flow problem.

An Example

This section introduces a labeling method and a preflow method, both of which are described by reference to the max-flow problem in:

Problem F (Max Flow)

Figure 5.13 depicts a network whose source is node a and whose sink is node g. Its wraparound arc is shown in two pieces, one pointing into the source and the other pointing out of the sink. Adjacent to each of the other arcs in this network is a pair of numbers. The left-hand member of this pair is the flow on the arc, currently zero. The right-hand member

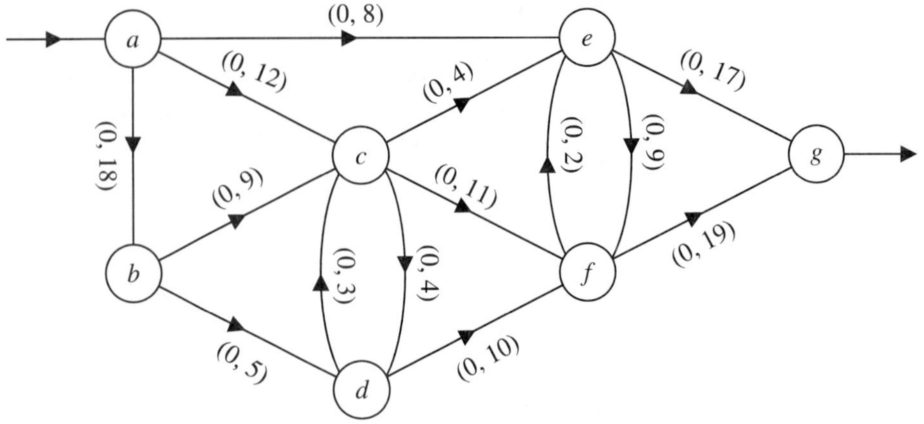

Figure 5.13 Arcs, capacities, and flows (currently zero) for a max-flow problem in which a is the source node and g is the sink.

of each pair is the arc's capacity. Arc (a, c) has 12 as its capacity, for instance. A flow is said to be **feasible** if it conserves flow at each node and keeps each arc's flow between zero and its capacity. The **max-flow** problem is to find the feasible flow that maximizes the flow on the wraparound arc.

In a max-flow problem, our concern lies with paths from the source to the sink. Nodes other than the source and the sink are now called **intermediary** nodes. In Figure 5.13, nodes b, c, d, e, and f are intermediary.

Drawing the Incremental Network

As was the case in the Hungarian method, our focus is on perturbing the flow. Now, however, the flows can be perturbed in slightly more complex ways. For instance, a flow that is positive but below its upper bound can be increased as well as decreased.

To depict a network, each arc (i, j) has been drawn as a line between nodes i and j with an arrow pointing from node i to node j. We now wish to depict the incremental network in the same diagram as the network itself. To do so, we:

- Add a gray arrow in the "reverse" direction when the arc's flow is positive and hence can be decreased.
- Omit the black arrow when the arc's flow is at its capacity and hence cannot be increased.

Figure 5.14 illustrates this scheme for an arc (i, j) whose capacity equals 8. If its flow $x(i, j)$ is below its capacity, the black arrow is included. If its flow $x(i, j)$ is above zero, the gray reverse arrow is included.

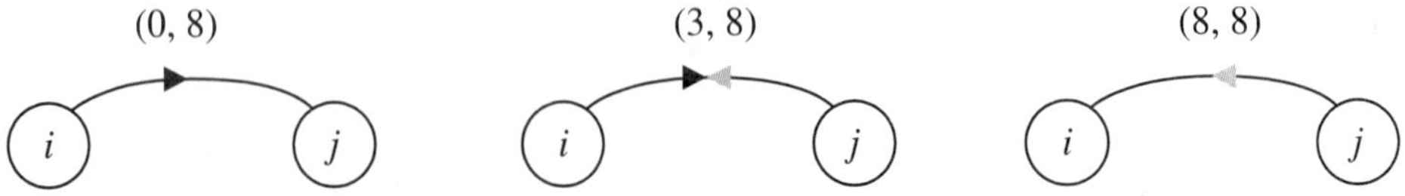

Figure 5.14 Depicting arc (i, j) whose flow equals zero (at the left), equals its capacity (at the right), and lies strictly between (at the center).

The wraparound arc is special; we want the flow on it to increase but not to decrease. Thus, in the incremental network, the gray reverse arrow is omitted from the wraparound arc, independent of the flow on it.

Labeling

The **labeling method** of Ford and Fulkerson is now known as an Augmenting Path Method. It can be initialized with any nonnegative values of the flows that preserve flow at each node and satisfy each arc's capacity constraint. It proceeds as follows:

Augmenting Path Method:

1. In the incremental network, initialize the Subtree Identifier with the source colored blue and the other nodes uncolored. Execute it. Stop if it does not color the sink.
2. Otherwise, backtrack from the sink to identify a source-to-sink path in the incremental network along which flow can be increased. Increase the flow by the largest amount that keeps each arc's flow between 0 and its capacity. Return to Step 1.

The Subtree Identifier contains an element of ambiguity, and this ambiguity is inherited by the Augmenting Path Method. The Subtree Identifier lets us pick any blue node in its Step 3 and color it gray. To simplify the discussion—and only for that reason—we elect to resolve this ambiguity *lexicographically*, by selecting the blue node that comes earliest in the alphabet and coloring it gray.

First Flow Augmentation

To illustrate this Augmenting Path Method, we apply it to the network in Figure 5.13, initialized with each flow equal to zero. In this case, Figure 5.13 also depicts the incremental network (because all flows equal zero). In Step 1, the Subtree Identifier (on page 178) first colors node a gray and colors nodes b, c, and e blue. Of the blue nodes, b comes earliest in the alphabet. Thus, the Subtree Identifier next colors node b gray and then colors node d blue. It then colors node c gray and node f blue. Eventually, the Subtree Identifier colors all nodes gray; it constructs the tree in Figure 5.15.

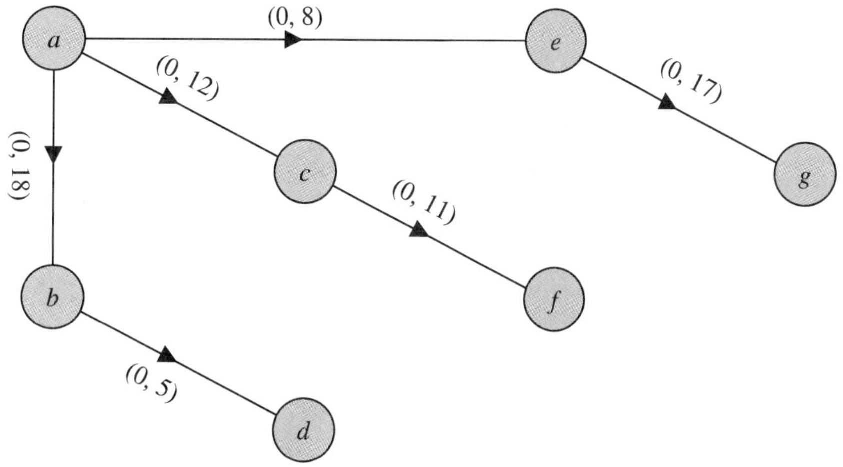

Figure 5.15 The initial tree.

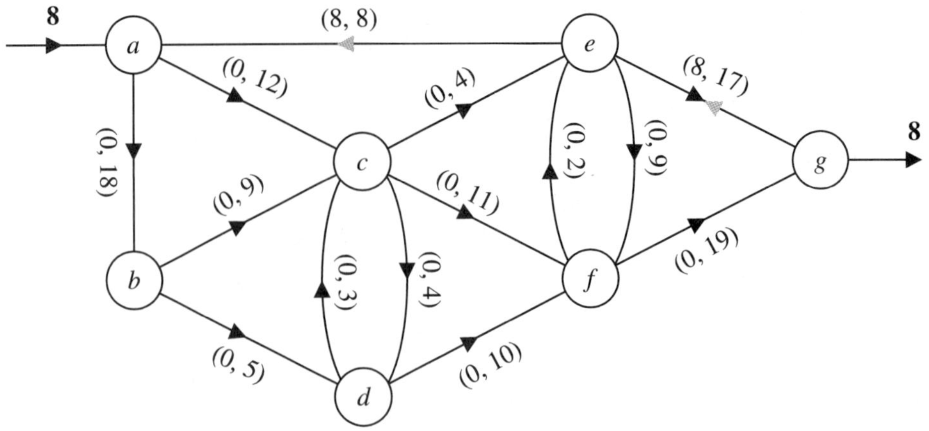

Figure 5.16 Flows and incremental network after the first flow augmentation.

The Subtree Identifier has colored the sink (node g) gray. Backtracking from node g identifies the path (a, e, g) along which flow can be augmented. The largest value of flow on this path that keeps the arcs's flows within their bounds equals 8. So Step 2 of the Augmenting Path Method causes eight units to flow on this path, which produces the flows and incremental network in Figure 5.16. Note that arc (a, e) is a reverse arc and that arc (e, g) has become "bidirectional."

Second Flow Augmentation

For the flows in Figure 5.16, Step 1 of the Augmenting Path Method constructs the incremental network's largest subtree rooted at node a. Figure 5.17 displays this subtree.

This tree includes the sink (node g). Once again, backtracking from the sink identifies the path (a, c, e, g) along which flow can be augmented, this time by four units. Doing so results in the flows and incremental network seen in Figure 5.18.

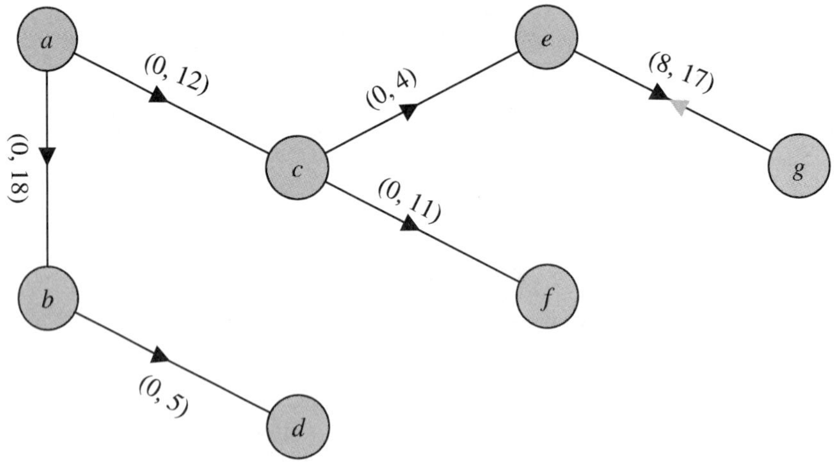

Figure 5.17 The second tree.

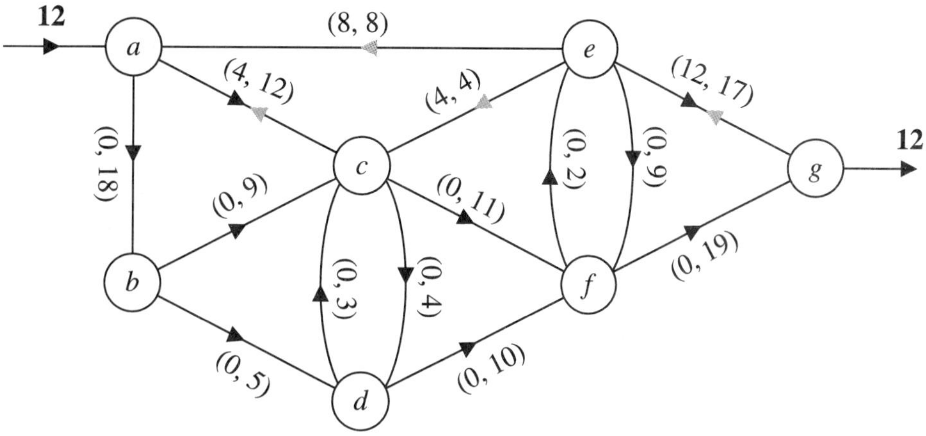

Figure 5.18 Flows and incremental network after two flow augmentations.

Termination

After several more flow augmentations, the flows in Figure 5.19 are obtained. The Subtree Identifier stops with nodes a, b, and c colored gray. It can color no other nodes because each arc pointing from a gray node to an uncolored node has its flow equal to its capacity (is reversed). A thick black line "severs" each such arc.

Figure 5.19 shows how 32 units can flow from the source node to the sink. Notice that the capacities of its severed arcs total 32 because $32 = 8 + 4 + 11 + 4 + 5$.

The Max-Flow Min-Cut Theorem

This reminds us of the Max-Flow Min-Cut Theorem. Let us recall from Chapter 4 that in a max-flow problem:

- A **cut** is any set S of nodes that includes the source but not the sink.
- Each cut S **deletes** the set $A(S)$ of arcs that consists of each arc (i, j) having i in S and j not in S.
- Each cut S has a **capacity** $c(S)$ that equals the sum of the capacities of the arcs in $A(S)$.

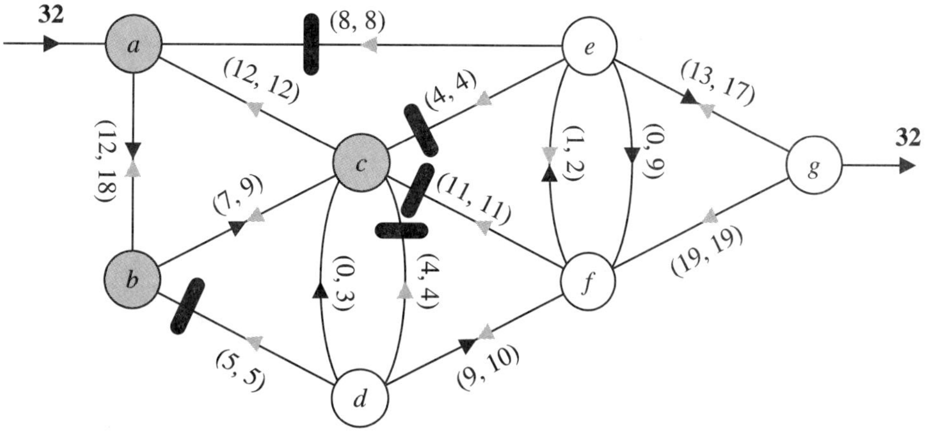

Figure 5.19 Only the shaded nodes can be reached from the source in the incremental network.

In Figure 5.14, nodes a, b, and c form a cut whose capacity equals 32. The flow augmentation method of Ford and Fulkerson has found a cut whose capacity is smallest, as is demonstrated below.

Theorem 1 (The Max-Flow Min-Cut Theorem). When flow augmentation terminates, its gray nodes are a cut whose capacity is smallest, and this cut's capacity equals the maximum value of the flow.

Proof[4]**.** First, we consider any cut T and a feasible solution to the max-flow problem that maximizes the quantity z that flows on the wraparound arc. Flow is conserved at each node. Summing the conservation-of-flow equation over each node in T shows that:

$$\sum_{(j,i):j\notin T, i\in T} x(j,i) = \sum_{(i,j):i\in T, j\notin T} x(i,j). \tag{5.13}$$

The sum on the left includes the flow z on the wraparound arc. Each addend on the right is no greater than the capacity of its arc, so their sum is not greater than $c(T)$. As a consequence, $z \leq c(T)$. This inequality holds for every cut T, so

$$z \leq \min\{c(T) : T \text{ is a cut}\}. \tag{5.14}$$

Now, let's examine flows for which flow augmentation terminates, with z^* denoting its flow on the wraparound arc. Since this flow is feasible, we have

$$z^* \leq z. \tag{5.15}$$

Also, let T^* denote the set of gray nodes at termination of the flow augmentation method. The set T^* includes the source but not the sink, so T^* is a cut. Furthermore, T^* cannot be enlarged. Equation (5.13) holds with T^* substituted for T. Its left-hand side contains the wraparound arc. All of the other arcs (j, i) on its left-hand side have $x(j, i) = 0$ because, otherwise, j would be added to T^*. Each arc (i, j) on its right-hand side has $x(i, j)$ equal to arc (i, j)'s capacity because, otherwise, j would be added to T^*. Thus, equation (5.13) gives

$$z^* = c(T^*) \tag{5.16}$$

Since T^* is a cut, expressions (5.14), (5.15), and (5.16) combine into

$$c(T^*) = z^* \leq z \leq \min\{c(T) : T \text{ is a cut}\} \leq c(T^*).$$

In the above, each inequality must hold as an equation. Thus, the maximum value z of the flow equals the minimum value of the cut capacity, and both are equal to the capacity $c(T^*)$ of the cut with which flow augmentation terminates. The theorem has been proved. ◆

Finite Termination?

Intuitively, it seems clear that flow augmentation must terminate finitely, but that is not quite correct. We consider three cases, only two of which guarantee finite termination.

- First, suppose that each arc's capacity is a positive integer. In this case, each flow augmentation increases the total flow from the source to the sink by at least one unit, and finite termination is guaranteed.

[4] The proof is technical and can be skipped with no loss of continuity.

- Next, let's consider the case in which each arc's capacity is a rational number (the ratio of integers). In this case, each flow augmentation must increase the total flow by at least $1/K$, where K equals the product of the denominators of the capacities. That, too, assures finite termination.
- Now, consider the most general situation in which the capacities need not be rational numbers. In this case, flow augmentation may fail to converge in finitely many flow augmentations. It can even converge, as the number of iterations becomes infinite, to a flow that is not maximum.

Flow augmentation has the virtue of simplicity, allowing each nonterminal iteration to increase the flow on the wraparound arc. But flow augmentation also has drawbacks. The flows on other arcs can go up and then down. When the capacities are irrational, the flow on an arc can go up and down, infinitely often.

Irrational Capacities?

In a sense, no one cares about irrational capacities because they are impractical. But the misbehavior of flow augmentation when the capacities are irrational suggests that it can require enormous numbers of iterations for capacities that are integer-valued but large. Prior to 1974, many papers were written on schemes that circumvented the difficulties in the original labeling method.

In 1974, Karzanov changed the landscape by publishing a "preflow" method that avoids these difficulties entirely. In general, preflow methods require a number of computer operations that is proportional to n^3, where n is the number of nodes in the network. This bound is independent of the sizes of the capacities.

Karzanov's idea was to begin by "pushing" as much flow as possible across each arc that points away from the source and then, while possible, push that flow toward the sink. When it becomes impossible to push flow closer to the sink, push any that remains back toward the source. His scheme is called a preflow method because it violates the flow conservation constraints at the nodes, except at the end.

Preflow Terminology

There are several preflow methods. The one that we will describe here "pushes" flow "down" from the "hilltop." To set the stage, we introduce some nomenclature. In the max-flow problem, each arc has a capacity but no cost. Path lengths are now defined with respect to the incremental network. In the incremental network, we define the **length** of each path as the number of arcs that it contains. Thus, a shortest path from one node to another is a path with the fewest arcs. The incremental network determines the **height** of each node by these rules:

- The sink has height zero.
- The source has height n, where n is the number of nodes in the network.
- Each intermediary node's height equals the length of the incremental network's shortest path from it to the sink, if such a path exists. If not, the node's height equals the sum of n and the length of the incremental network's shortest path from it to the source.

The **excess** at each node equals the amount that has flowed into that node but has not flowed out of it. If an intermediary node has excess and has height below n, flow can be (and

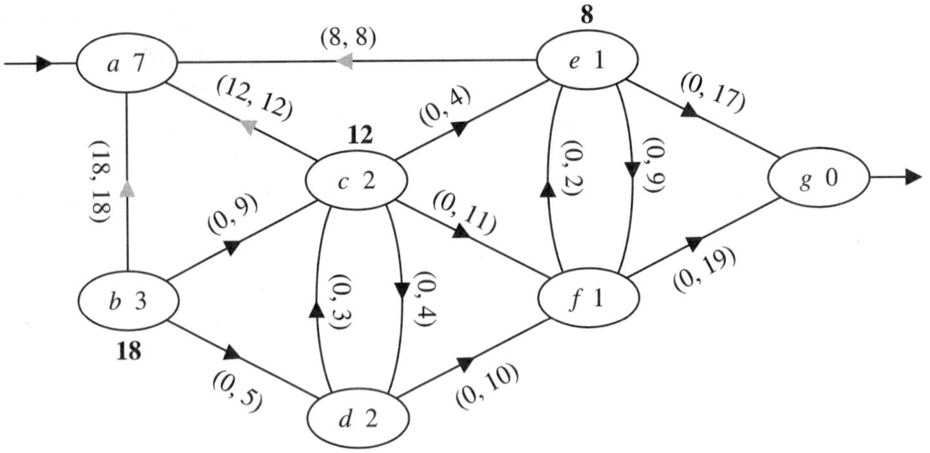

Figure 5.20 Pushing from hilltop, initialized.

is) pushed from it across an arc to a node whose height is lower. When that is no longer possible, any excess that remains at intermediary nodes will be pushed back to the source.

The Initial Preflow

Figure 5.20 indicates how this method is initialized. The flow on each arc whose "from" node is the source is set equal to that arc's capacity, and that amount becomes the excess at the arc's "to" node. For instance, arc (a, b) is assigned a flow of 18, and an excess of 18 appears adjacent to node b. With these flow values, each node's height in the incremental network appears inside its circle, which has been stretched to make room. For instance, the height of node b equals 3 because the shortest path from node b to node g (the sink) contains three arcs.

With this initialization, the preflow method that we are developing is stated below. We recall that the intermediary nodes are those other than the source and sink.

Pushing from the Hilltop:

1. As long as some intermediary node whose height is below n has excess, select the highest such node, and push as much as is possible across an arc from it to a node whose height is lower.
2. When no intermediary node whose height is below n has excess, select the highest node, and push as much as is possible across an arc from it to a node whose height is lower.
3. Stop when no intermediary node has any excess.

The First Pushes

In Figure 5.20, node b is the highest node that has excess. From node b, this method pushes nine units across arc (b, c) and pushes five units across arc (b, d). This reduces the excess at node b by a total of 14 units, and it creates excesses at nodes c and d of 21 and 5 units, respectively. Figure 5.21 results from these two pushes.

In Figure 5.21, the height of node b increased to 8 because the incremental network contains no path from node b to the sink, but it does contain a path of length 1 from node b to the source.

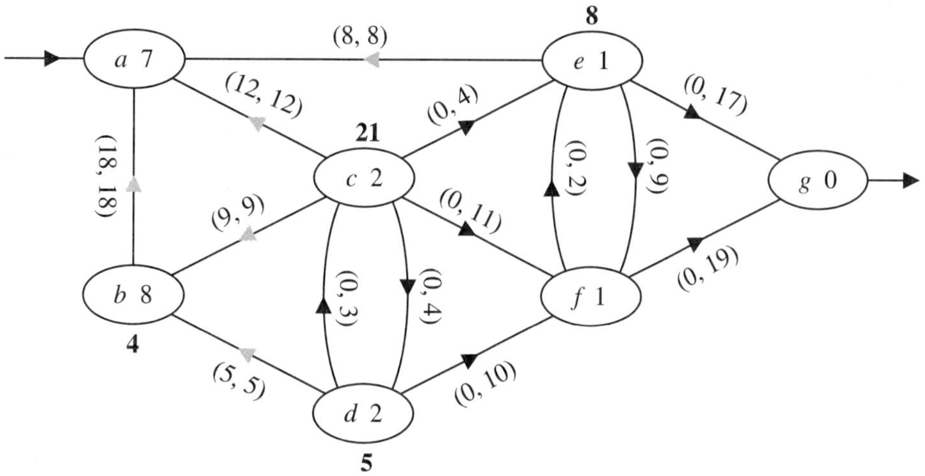

Figure 5.21 Pushing from hilltop, after two pushes from node *b*.

The Next Pushes

In Figure 5.21, nodes *c* and *d* have the highest height among those whose height is below *n*. The next push selects either of these nodes and shifts as much excess as is possible from it across an arc to a node whose height is lower. Pushing from node *d* and then pushing twice from node *c* results in Figure 5.22.

In Figure 5.22, the excess at node *d* has been shifted to node *f*, and the excess at node *c* has been reduced from 21 to 6 by pushing four units on arc (c, e) and 11 units on arc (c, f). The height of node *c* has increased from 2 to 3. The next push will shift four units from node *c* to node d, which will increase the height of node *c* from 3 to 8. After several more pushes, the excess at nodes *d*, *e*, and *f* will have been pushed to the sink, which will result in Figure 5.23.

In Figure 5.23, exactly 32 units have reached the sink. Nodes *b* and *c* have excesses, but the incremental network allows none of it to be shifted toward the sink. For the next pushes,

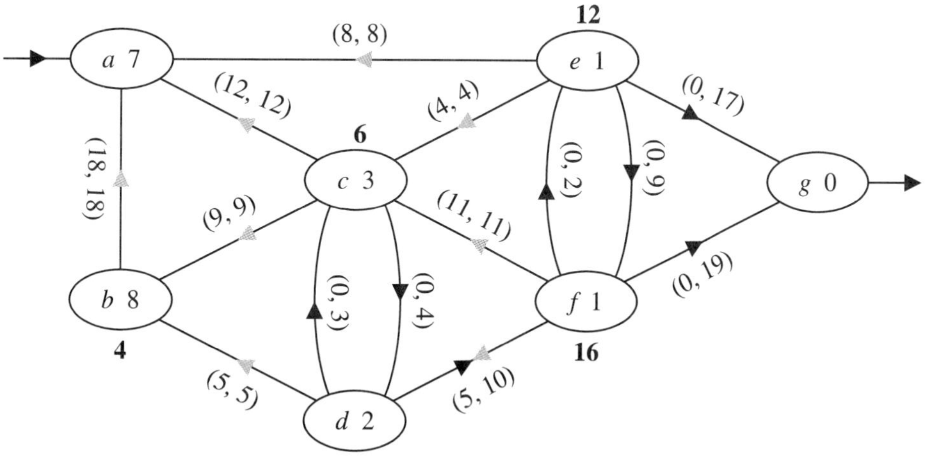

Figure 5.22 Pushing from hilltop, after three further pushes, one from node *d* and then two from node *c*.

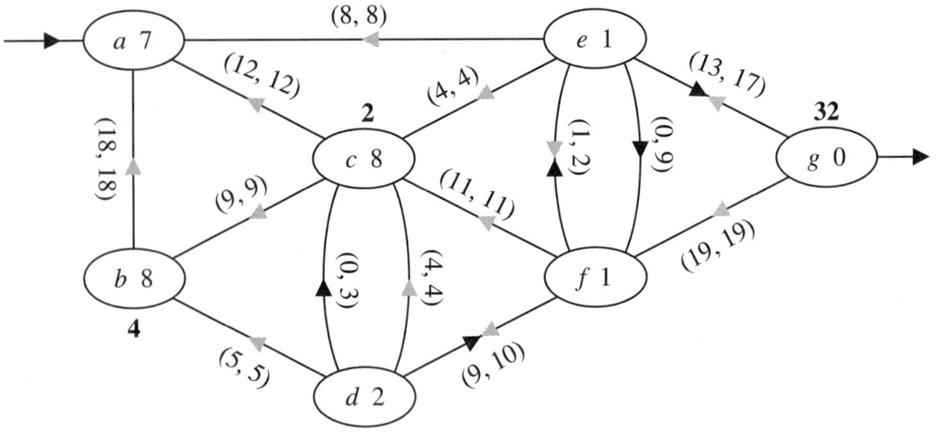

Figure 5.23 Pushing from hilltop, interrupted as soon as no intermediary node whose height is below n has any excess.

nodes b and c have the highest heights, and their excesses are shifted back to the source, producing the flow and incremental network in Figure 5.24.

The flows in Figure 5.24 cause the algorithm to stop because no intermediary node has any excess. In Figure 5.24, 32 units have reached the sink, 6 units have been returned to the source, and the flows indicate how these 32 units got to the sink. The incremental network can shift no more flow to the sink because the only arcs on which flow can be shipped are to nodes whose heights exceed n.

Running Time

The preflow method we've presented has a lovely monotonicity property. Figures 5.20 through 5.24 illustrate the fact that, at each iteration, the heights of the nodes can only increase; they cannot decrease. Each node's height is an integer between 0 and $2n$, so the total number of iterations at which at least one height increases is not greater than $2n^2$. It's not hard to show that the number of iterations at which no node's height increases is at most n^3, and moreover, that the total number of computer operations needed to execute it is not more than n^3.

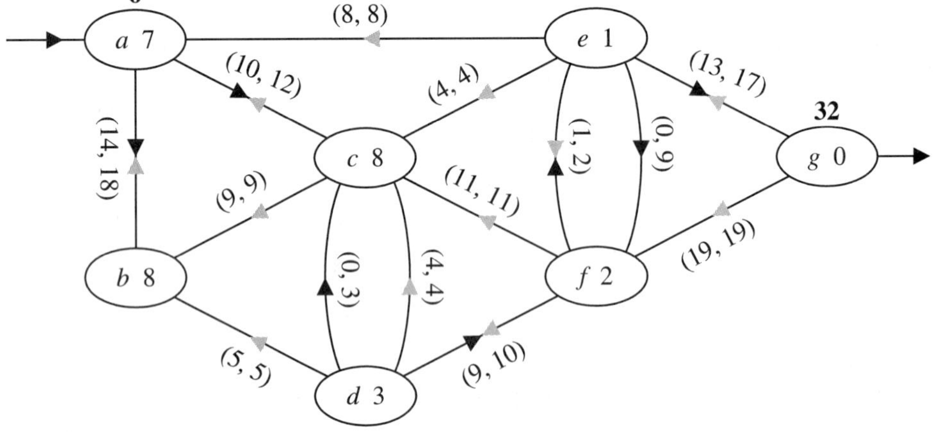

Figure 5.24 Pushing from hilltop, at termination.

5.10. REVIEW

This chapter has been all about trees of paths. Sometimes our interest focuses on a tree of paths from one node to all others and sometimes on a tree of paths to one node from all others. Sometimes these have been trees of longest paths, more often of shortest paths.

For acyclic networks, we have seen how to "grow" these trees recursively. We've also seen how to find them from the solution to a linear program. The linear program is a bit odd in that it uses maximization to compute a tree of shortest paths, and uses minimization to find a tree of longest paths.

The starred sections are, once again, all about growing trees. Dijkstra's method grows a tree of shortest paths from one node to all others. With equal ease, it can grow a tree of shortest paths to one node from all others. The Hungarian method solves the assignment problem by growing trees of zero-length paths from source nodes to demand nodes. The labeling method solves the max-flow problem in the same way.

This chapter has focused on network optimization problems that are easy to state and easy to solve. To balance the account, we close with a pair of problems that are easy to state and hard to solve.

The Traveling Salesperson

In a directed network, interpret the length of each arc as its travel time. The "traveling salesperson" starts at her or his home node. This salesperson must visit each of the nodes exactly once and end by returning home. The traveling salesperson wishes to do this as quickly as is possible. In a directed network, a **tour** is a cycle that visits each node *exactly* once. The **traveling salesperson** problem is that of finding a tour whose length is smallest.

The traveling salesperson problem is beguilingly easy to state but torturously difficult to solve. No known algorithm for the problem works well in the worst case. Effective heuristics (methods that often work but cannot be guaranteed) have been devised, and the better ones make deft use of ideas developed in this chapter. Chapter 6 formulates the traveling salesperson problem as an integer program and suggests a method for solving it.

The traveling salesperson problem appears in many disguises. One of them is the problem of finding the shortest of the *simple* cycles in a network whose paths can have negative lengths.

The Four-Color Problem

In many network optimization problems, the arcs lack direction, in which case the word "graph" is used. A **graph** consists of two objects, a finite set of nodes, and a finite set of edges, each **edge** consisting of a set $\{i, j\}$ of two distinct nodes. A graph is said to be **planar** if it can be drawn on a plane with no crossing edges. The **four-color** problem is as follows: Each node of a planar graph is to be colored, and no two nodes that share an edge can have the same color. How many colors are needed?

Again, the problem is easy to state but hard to solve. It's easy to show that the answer is at least four and not more than five. For decades, nearly everyone thought that four colors were enough, but no one could prove it. Finally, a proof was provided that four colors are enough. The proof turns out to be ungainly. It uses mathematics to reduce the number of cases that need be tested to a finite number that is enormous but within the computational power of digital computers.

5.11. HOMEWORK AND DISCUSSION PROBLEMS

1. (**A Network in Excel?**) On an Excel spreadsheet, think of each cell as a node. And if a function in cell i makes reference to information in cell j, interpret (i, j) as a directed arc. What sort of network gives rise to a "circular reference" message?

2. For the network in Figure 5.2, with $G(1) = 0$, Equation (5.8) specifies the length of the longest path from node 1 to node j for each j. On a spreadsheet, solve this equation recursively, in ascending j.

3. For the network in Figure 5.2, construct the linear program that computes, for each node j, the length $G(j)$ of the longest path from node 1 to node j, with $G(1) = 0$.

4. A 5×7 array of integers appears below this problem. Suppose that each movement in this array can occur rightward by one column or downward by one row. You wish to move from the upper left corner of this array to its lower right corner by a sequence of such movements that minimizes the sum of the integers encountered.

(a) Set this up as a shortest-path problem through a directed acyclic network, and write its functional equation. *Hint:* Each state (node) is an ordered pair (p, q) in which p is the row and q is the column.

(b) Solve this functional equation on a spreadsheet. On the same spreadsheet, use IF statements to record the optimizers, for example, "R" for shift to the right, "D" for shift down.

0	4	3	6	4	4	5
7	8	6	8	9	3	1
2	3	1	8	7	8	6
6	2	9	3	0	2	3
5	6	2	3	1	5	0

5. In the array of numbers that appears below this problem, positive numbers are arc lengths, and blanks denote missing arcs; the network has no arcs whose "from" node is node 1, for instance. Find the tree of shortest paths to node 1. Is there a longest path from node 3 to node 1?

	1	2	3	4	5	6
1						
2	2					
3	16	11		2	5	3
4		1	2			
5		8		5		
6			3		2	

6. **(A Non-Additive Functional Equation)** The terrain between Pete's home and office is hilly. The length of each arc in Figure 5.2 measures the slope of the hill that Pete must climb if he travels its road in the indicated direction. For each node, Pete wishes to select the path to the office that minimizes the steepest of the hills he must climb.

(a) Define $f(i)$ appropriately, write a functional equation for the $f(i)$'s, and solve it. Do you still get a tree of shortest paths?

(b) Write a linear program whose optimal solution is the $f(i)$'s. Solve your linear program. Did you get the same tree?

7. For the project whose data follow, determine the minimum completion time, the earliest start time of each activity, the latest completion time of each activity, and the critical activities.

Activity	a	b	c	d	e	f	g	h	i	j
Duration	4	5	8	6	4	3	7	4	3	2
Predecessors	—	a	a	c	b, d	c	c	f, b	b, g	g

8. **(A Linear Program for CPM)** For the project network whose data are in Table 5.2, write a linear program whose optimal solution gives the tree of earliest start times, and use Solver to solve it. Then write another linear program whose optimal solution gives the tree of latest completion times, and use Solver to solve it.

9. **(CPM with Resource Allocation)** Consider a project whose data are given by Table 5.2. Assume, however, that each activity's duration time can be reduced by up to 30% by "crashing." Up to this 30% limit, investing K dollars in any activity decreases that activity's duration by $0.002\,K$ weeks. Recall that this project can be completed in 26 weeks with no investment in crashing. Suppose that the value of completing this project X weeks early equals \$5000 X.

(a) Write a linear program that schedules the tasks so as to complete the project in the most profitable way. *Hints:* The objective of your linear program may measure net benefit. Its constraints may resemble those in the variant of Program 5.1 that solves a longest-path problem. One or more of its constraints may measure *X*.

(b) Solve your linear program. Which tasks are critical? Why is the budget allocated as it is?

10. **(CPM with a Labor Pool)** Consider a project whose data are given by Table 5.2 but with a somewhat different interpretation. A pool of labor is available, and at the beginning, the entire pool of labor rests at the start node. The time it takes to complete each activity equals the "duration" given in Table 5.2 divided by the fraction of the labor pool that is allocated to it. If, for instance, 20% of the labor pool is assigned to activity a, this activity takes $3/0.2 = 15$ weeks to complete. Labor can migrate to each activity from its predecessors, for example, from node S to nodes a and b, from node a to nodes c and d, and so forth.

(a) Write the (network flow) constraints that describe the flow of labor through the project's activities.

(b) Write a nonlinear program that allocates the labor so as to minimize the project completion time. *Hint:* This program is in the spirit of Program 5.1, except that it's a minimization problem and that $3/y(a)$ plays the role of the cost, $y(a)$ being the fraction of the labor pool that is allocated to activity a.

(c) Solve your nonlinear program. Is there anything noteworthy about its solution?

Remark: Proof that the solution to the preceding problem has the form you identified in part (c) can be found in a 1994 paper by E. V. Denardo, A. J. Hoffman, T. MacKensie, and W. R. Pulleyblank, "A Nonlinear Allocation Problem," *IBM J. Res. Dev.*, 36 (1994), pp. 301–306.

11. **(Locating Bus Stops)** A long street consists of 150 blocks of equal length. A bus runs "uptown" on this street. A fixed number n of bus stops is to be located on this street so as to minimize the total distance walked by the population of bus users. Each person who takes an uptown trip walks from his or her origin to the nearest bus stop, gets on the bus, rides it to the bus stop nearest his or her destination, gets off, and walks the rest of the way. It is known that, during the course of the day:

Exactly $B(j)$ people begin trips at block j, for $j = 1, 2, \ldots, 150$.
Exactly $C(j)$ people complete trips at block j, for $j = 1, 2, \ldots, 150$.

Necessarily,

$$\sum_{i=1}^{j} B(i) \geq \sum_{i=1}^{j} C(i), \qquad \text{for } j = 1, 2, \ldots, 150,$$

with equality holding when $j = 150$. This issue is where to locate these n bus stops.

(a) Suppose that bus stops are located at block p and at block $q > p$, but not in between. Show that the total distance walked to and from blocks p through q is given by $W(p, q)$, where

$$W(p, q) = \sum_{i=p}^{q} [B(i) + C(i)] \min(i - p, q - i).$$

(b) If the first stop is located at block p and if the last stop is located at block q, interpret

$$A(p) = \sum_{i=1}^{p} [B(i) + C(i)](p - i),$$

$$B(q) = \sum_{i=q}^{150} [B(i) + C(i)]\,(i - q).$$

(c) Suppose that the n bus stops are located at blocks $p_1, p_2, \ldots, p_n$ with $1 \leq p_1 < p_2 < \cdots < p_n \leq 150$. Interpret the sum

$$A(p_1) + W(p_1, p_2) + \cdots + w(p_{n-1}, p_n) + B(p_n).$$

(d) For this bus stop problem, what information should a state incorporate? In what way, if any, is its solution related to the equations that appear below?

$$f(1, p) = A(p),$$
$$f(k, p) = \min\{f(k - 1, q) + A(q, p) : 1 \leq q < p\},$$
$$F(150) = \min\{f(n, q) + B(q) : n \leq q \leq 150\}.$$

12. **(Bus Stops, continued)** Suppose each block takes five minutes to walk, suppose that the bus travels at a rate of two blocks per minute, and suppose that it takes the bus an additional three minutes to decelerate to a stop, unload passengers, load passengers, and accelerate back to full speed. Write a functional equation whose solution selects the number of stops and locates them so as to minimize the total travel time of the population of bus users.

13. **(Dijkstra's Method)** Table 5.4 reports the v-values and permanent nodes at the first two executions of Step 2 of Dijkstra's method. On a spreadsheet, repeat these steps and execute the next four. *Remark:* When Dijkstra's method is applied to a large sparse network, finding the temporary node k whose label $v(k)$ is smallest can require a large fraction of the computer time. The next problem uses "buckets" [see part(d)] to avoid this.

14. **(Dijkstra's Method with Buckets of Width 3)** In Table 5.4, only arc (3, 4) has 1 as its length. The remaining arcs (k, j) have $c(k, j) \geq 2$. For this network, consider the application of Dijkstra's method to find the tree of shortest paths from node 3 to the other nodes.

(a) Is it true that no execution of Step 3 can reduce any label $v(j)$ for which $v(j) < v(k) + 3$? If so, why?

(b) Is it true that each node j whose label $v(j)$ is below $v(k) + 3$ has $v(k) = f(k)$? If so, why?

(c) Is it necessary to select, in Step 3, the temporary node k whose label is smallest? Or will any node whose label is within 3 of the minimum suffice?

(d) Suppose the temporary nodes are grouped in buckets, where the nth **bucket** contains the nodes j whose labels satisfy $3(n-1) \leq v(j) < 3n$. At each iteration of Step 3, the lowest-numbered nonempty bucket is found, and the comparison in Step 3 is executed for each arc (k, j) with k in this bucket and with j temporary, shifting node j to the appropriate bucket if $v(j)$ is reduced. Does this work? If you recycle buckets when they are emptied, will a total of six buckets suffice? Why?

15. **(Hungarian Method for the Assignment Problem)** For the cost matrix shown in Table 5.5, initiate the Hungarian method with the partial shipping plan that sets $x(1, 6) = x(2, 10) = x(3, 9) = x(5, 8) = 1$. Execute it, using a spreadsheet to produce analogs of Figures 5.10–5.12.

16. **(A Hot Start for the Assignment Problem)** There are many heuristics for "hot starting" the assignment and transportation problems. **W. R. Vogel** thought to pay attention to the difference between the cheapest and second cheapest cost in each row and column. For the data in Table 5.5, record the difference between the smallest cost in each row and the second smallest cost in that row. Do the same for the columns. Subtract to create a zero in a row or column whose difference is largest. Update the differences and repeat. Do so until you achieve a cost matrix with a zero in each row and column. For the matrix with which you end, initialize the Hungarian method.

17. **(Hungarian Method for the Transportation Problem)** The spreadsheet following this problem specifies a transportation problem with four source nodes and six demand nodes. Its supply nodes are labeled 1 through 4, and its demand nodes 5–10. Its supplies are in column H, its requirements are in row 7, and its shipping costs are in cells B3 through G6. The problem is to find the least-cost shipping plan that satisfied the demands. The sum of the supplies equals 17, and so does the sum of the demands. (To "hot start" the Hungarian method, we've arranged for at least one zero in each row and column of the shipping costs.)

	A	B	C	D	E	F	G	H
1								
2		5	6	7	8	9	10	supply
3	1	2	0	3	0	3	2	3
4	2	0	1	6	0	0	6	4
5	3	0	6	0	3	4	0	2
6	4	4	1	4	0	3	4	8
7	requirement	3	3	6	2	1	2	

(a) Find a partial shipping plan that exhausts the supplies at source nodes 1, 2, and 3, and ships only on zero-cost arcs. Which demand(s), are incompletely satisfied?

(b) Draw the incremental network, and identify the largest subtree rooted at node 4. *Hint:* The incremental network includes arc (i, j) if $c(i, j)$ equals zero *or* if $c(j, i)$ equals zero and $x(j, i)$ is positive.

(c) On a spreadsheet, change the costs in a way that enlarges this subtree.

(d) Repeat step (c) until your subtree includes a demand node that is incompletely satisfied. Backtrack to identify the path from supply node 4 to this demand node. Ship as much as is possible on this path.

18. (**Flow Augmentation**) Starting with Figure 5.18, continue the flow augmentation procedure to completion. Draw the analogs of Figures 5.17 and 5.18 for each iteration.

19. (**Pushing from the Hilltop**) Starting with Figure 5.22, continue "pushing from the hilltop" until no excess remains at any node whose height is below n. Draw the analog of Figure 5.22 for each iteration.

Remark: The next problem concerns a theorem that was proved well before Ford and Fulkerson discovered the Max-Flow Min-Cut Theorem. This theorem entails two definitions. A **bipartite** network is a network whose nodes partition themselves into two sets, S and T, so that each arc (i, j) has i in S and j in T. (The transportation network is bipartite.) Arc (i, j) is said to **touch** nodes i and j. Consider this:

> Theorem of Egerváry and König. *In a bipartite network, the maximum number of arcs no two of which touch the same node equals the minimum number of nodes in a set* A *such that each arc in the network touches at least one node in* A.

20. Use the Max-Flow Min-Cut Theorem to prove the theorem of Egerváry and König. *Hint:* Add a source node s, a sink node t, an arc (s, i) for each i in S and an arc (j, t) for each j in T, with the capacity on each new arc being l, and the capacities on the original arcs being huge.

Remark: The next problem concerns a more general theorem, also proved well before Ford and Fulkerson discovered the Max-Flow Min-Cut Theorem. Consider this:

> Theorem of Menger. *In a directed network, let* S *and* T *be disjoint sets of nodes. Then the maximum number of node-disjoint paths from nodes in* S *to nodes in* T *equals the minimum number of nodes in a set* A *such that each path from a node in* S *to a node in* T *touches at least one node in* A.

21. Use the Max-Flow Min-Cut Theorem to prove Menger's Theorem. *Hint:* It's the same hint as for the preceding problem.

22. (**Shortest Cycle**) In a directed network, each arc (i, j) has length $c(i, j)$. Take the length of each path as the sum of the lengths of its arcs. Assume that no cycle has negative length. Guess at a linear program whose solution identifies the shortest cycle. *Hint:* The variables are the flows on the arcs, the flow into each node equals flow out of it, and the sum of the flows equals 1.

23. (**The Traveling Salesperson**) Suppose you had an efficient algorithm for finding the shortest simple cycle in a directed network whose arcs can have negative lengths. (No such algorithm has been found.) Show how this algorithm would solve the traveling salesperson problem. *Hint:* Subtract a large number M from the cost $c(i, j)$ of each arc (i, j).

Chapter 6

Integer Programs

6.1. INTRODUCTION

Chapters 3, 4, and 5 presented a variety of examples of linear programs, along with a few "integer programs." An **integer program** differs from a linear program by requiring that one or more of the decision variables be integer-valued.

This chapter surveys the use of integer-valued variables to solve optimization problems. One obvious use of integer-valued variables is to measure decision variables (such as people and airplanes) that exist in integer amounts. Less obvious uses of integer-valued variables are to model nonlinearities and logical requirements, such as penalties for violating constraints.

Does the restriction that some or all of the decision variables be integer-valued make the optimization problem easier to solve or harder? More difficult by far! Solver finds the optimal solution to an integer program by solving a sequence of linear programs, and this sequence can run into the dozens, the thousands, or more. Because integer programs can be difficult to solve, this chapter also surveys the situations in which they seem to be needed but can be avoided.

6.2. WHAT CAN YOU LEARN FROM THIS CHAPTER?

This chapter acquaints you with the roles of integer-valued variables in optimization problems. From its main sections, you can learn:

- How to use integer-valued variables to model nonlinearities and logical requirements.
- How Solver finds optimal solutions to integer programs.
- When the need to solve an integer program can be circumvented.

- How integer programming attacks a famous (and notoriously difficult) problem that's known as the traveling salesperson problem.

Collectively, these sections indicate what integer-valued variables can accomplish and how to make effective use of them.

6.3. SOLVING AN INTEGER PROGRAM

We begin with a sketch of the principal method for solving an integer program: the branch-and-bound method. This section shows how it works, why it can be time consuming, and what it can and cannot accomplish.

A Relaxation

A **relaxation** of an optimization problem is the result of weakening or eliminating one or more of its constraints. Many integer programs constrain variables to take the value 0 or 1, as in:

$$0 \le x \le 1 \quad \text{and} \quad x \text{ an integer.}$$

A typical relaxation is to eliminate the constraint that x be an integer, thereby allowing x to take any value between 0 and 1, as well as the values 0 and 1 themselves. In this instance and in general:

> A relaxation can only *expand* the set of feasible solutions: Every solution that was feasible for the original problem remains feasible for its relaxation, but not every solution that is feasible for the relaxation need be feasible for the original problem.

What effect does a relaxation have on the optimal value, that is, on the objective value of the optimal solution? Relaxing a maximization problem can only increase its optimal value, whereas relaxing a minimizing problem can only decrease its optimal value.

Typically, a relaxation replaces a difficult problem with an easier one. Suppose that an optimal solution to a relaxation happens to satisfy the constraints of the original problem. In this case, the two problems have the same optimal value, and this optimal solution to the relaxation is also optimal for the original problem. That's a happy outcome.

The LP Relaxation

In the case of an integer program (IP), the relaxation that omits only the requirements that variables be integer-valued is called the **LP relaxation** of this integer program. Being a linear program, this relaxation is easy to solve. If the solution to the LP relaxation happens to satisfy the missing integrality constraints, we have gotten lucky because we have found an optimal solution to the integer program itself.

Typically, the optimal solution to the LP relaxation violates one or more integrality constraints. In this case, an optimal solution to the integer program can be found by solving a series of linear programs. We will use Figure 6.1 to indicate how.

A Bound

The integer program under attack in Figure 6.1 is a minimization problem, and its integer-valued variables include x, y, and z. Each box in Figure 6.1 describes a linear program.

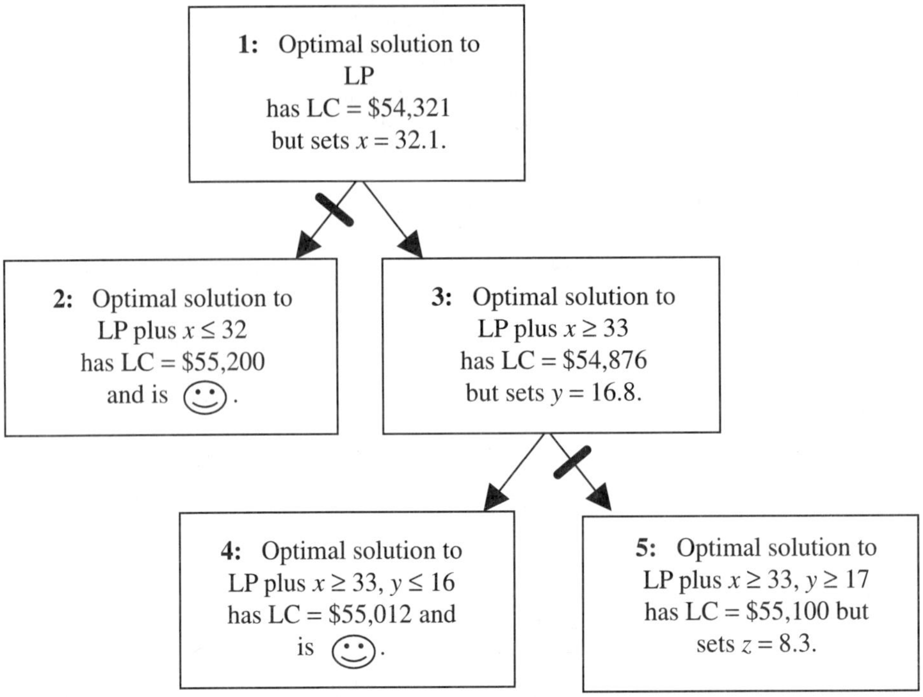

Figure 6.1 The LP relaxation of a cost-minimizing IP, with branching and pruning.

The topmost box in the figure, which is labeled box 1, records information about the optimal solution to the LP relaxation of the integer program. This box's least cost (LC for short) equals $54,321, and the variable x (which is supposed to be integer-valued) takes the value 32.1 in the optimal solution to the LP relaxation. (It is quite possible that other variables, such as y or z, violate constraints that they be integer-valued, but x has been singled out in box 1.)

From box 1 alone, we have learned something useful, namely, that the least cost of the integer program must be at least as large as $54,321, for this is the least cost of its LP relaxation.

Branching

What to do? A simple-minded strategy is to **branch** on the variable $x = 32.1$ by replacing box 1 with a pair of linear programs. One of the programs is the LP relaxation with the added constraint $x \leq 32$, while the other is the LP relaxation with the added constraint $x \geq 33$. Boxes 2 and 3 report the results of solving these two linear programs. In both cases, the LC cannot be less than $54,321. Do you see why? The optimal solution to the original integer program must be feasible for one of these boxes. Again, do you see why?

In box 2, we got lucky. This box's least cost is $55,200, and its smiley face means that the optimal solution to this box's linear program assigns an integer value to each variable that is supposed to be integer-valued. A feasible solution to the integer program has been found. And we now know that the optimal value (least cost) of the integer program lies between $54,321 and $55,200.

If the LC in box 3 had been $55,200 or greater, we would have found an optimal solution to the integer program. Do you see why?

Branching a Second Time

Box 3 reports a least cost that is below \$55,200. Moreover, the variable y (that is supposed to be integer-valued) takes the value 16.8. A second branch is called for. We replace the LP in box 3 with a pair of linear programs. One of these linear programs differs from that in box 3 by appending to it the constraint $y \leq 16$. The other differs by appending the constraint $y \geq 17$.

Box 4 reports on the optimal solution to the LP relaxation plus the constraints $x \geq 33$ and $y \leq 16$. This LP's least cost equals \$55,012, and the smiley face in box 4 indicates that each variable that was supposed to be integer-valued is integer-valued. We got lucky again.

Box 5 reports on the optimal solution to the LP relaxation plus the constraints $x \geq 33$ and $y \geq 17$. That LP's least cost equals \$55,100, and its optimal solution assigns z (which is supposed to be integer-valued) the value 8.3.

Pruning

If we have at hand an LP whose optimal solution satisfies the integrality constraints of the original IP, we can **prune** (eliminate) each linear program (box) whose objective value is equal or worse. For our minimization problem, box 4 lets us prune boxes 2 and 5. The thick gray lines in Figure 6.1 record the pruning of these two boxes. No variables remain to branch upon, and the feasible solution to box 4 is the optimal solution to the original integer program.

Branch and Bound

The scheme we have just sketched, known in the literature as **branch and bound** and (less commonly but more aptly) as **branch and prune**, solves a series of LPs. It begins with the LP relaxation that suppresses all integrality constraints. Each branch identifies a variable whose value should be an integer but is not. Each branch replaces a solved linear program by two new linear programs, each of which appends a constraint on the identified variable. One new LP constrains this variable to be no greater than the next lower integer; the other requires it to be no smaller than the next larger integer.

If branch and bound finds a feasible solution to the integer program, it prunes (eliminates) all LPs whose objective values are not better than this feasible solution's objective. Branch and bound stops when nothing remains to branch upon. The feasible solution with which it stops is the optimal solution to the IP.

A Bad Idea?

Branch and bound sounds like a very bad idea. Indeed, its worst-case performance is catastrophic. If an integer program has 200 integer-valued variables, it may be necessary to solve an astronomical number of linear programs before the first feasible solution to the integer program is encountered.

In practice, however, deft implementations of branch and bound—those that make clever choices as to which variables to branch upon—solve large-scale integer programs within reasonable lengths of time. Branch and bound cannot be guaranteed to be fast, but it often works.

A Requirement?

Does branch and bound require the relaxation of the original integer program in which the integrality constraints are dropped to be a linear program? Technically, no; the relaxation *could* be a nonlinear program. If it is a nonlinear program, it had better be easy to solve (convex). As a practical matter, branch and bound works best when the integer program differs from a linear program *only* by requiring certain variables to be integer-valued.

What happens if the integer program includes the product $y\,z$ of two decision variables? That's bad news; the product $y\,z$ is neither convex nor concave. What happens if the integer program includes the constraint $y > 0$? That, too, is bad news; strict inequality constraints can't be part of linear or nonlinear programs.

Sensitivity Analysis?

The optimal solution to a linear program is accompanied by a set of shadow prices that form the basis for a sensitivity analysis. These shadow prices help us to understand why the solution takes the form that it does. Each shadow price describes the marginal benefit of a small change in a right-hand-side value. For larger changes, the shadow prices provide an upper bound on the benefit, which exhibits decreasing marginal return in the size of the change.

In an integer program, there are no shadow prices, and it is not the case that adding resources produces decreasing marginal benefit. Indeed, in an integer program, increasing a right-hand-side value by 2 may be *more* than twice as beneficial as increasing it by 1. When Solver computes an optimal solution to an integer program, it does not report shadow prices. There are none.

When you need information about the sensitivity of the optimal value of an integer program to a right-hand-side value, you have no choice but to solve the integer program repeatedly, once with each right-hand-side value that is of interest to you.

6.4. WHEN AN INTEGER PROGRAM IS NOT NEEDED

Branch and bound is the principal tool for solving integer programs. It may solve an integer program quickly, but it can run for a very long time without even finding a feasible solution to the integer program.

Fortunately, not all of the situations that seem to require an integer program actually do require the full force of branch and bound. This section catalogs optimization problems that appear to be integer programs but whose solutions can be found by simpler methods. Some items in this catalog will be familiar because they made their initial appearances in prior chapters.

Decreasing Marginal Return

Initially, linear programming seemed to require all relationships to be linear, but that is not so. It was noted in Chapter 3 that linear programming encompasses the case of decreasing marginal return.

If the marginal benefit of an activity decreases with its level, we need only break the range of that activity into intervals of (possibly approximate) linearity and allow the linear program to operate in whatever intervals it chooses. When this linear program is solved, a more profitable interval will be exhausted before a less profitable interval is used at all. Precisely the same method works in the case of increasing marginal cost; a less expensive interval will be used to capacity before a more costly interval is used at all.

This trick introduces an unintended option, which is to use a less profitable interval without making full use of a more profitable interval. In the case of increasing marginal cost or decreasing marginal return, optimization rules out the unintended option.

By contrast, the same trick *fails* in the case of increasing marginal return because the optimizer makes full use of the more profitable interval before making any use of the less profitable one.

Network Flow

The decision variables in an optimization problem may measure quantities (such as trucks or people) that exist in integer amounts. When they do, the natural formulation of the optimization problem is as an integer program, not a linear program.

But if this integer program is a network flow model whose fixed flows (bounds) are integer-valued, the Integrality Theorem in Chapter 4 guarantees that the simplex method finds an optimal solution that equates each decision variable to a value that is an integer. For these network flow problems, an integer program is not required.

Getting Lucky

In the case of an integer program (IP), a convenient relaxation is the linear program (LP) that results from deleting only the constraints that decision variables be integer-valued. Being a linear program, this relaxation is easy to solve. If the solution to this relaxation happens to satisfy the integrality constraints, we got lucky because we found a solution to the integer program.

In the case of an integer program that is a network flow problem with integral data, the Integrality Theorem guarantees that you will get lucky when you solve this relaxation by the simplex method.

If the integer program is a network flow problem with a few side constraints, we may get lucky. If we don't get lucky, we should branch only on those variables that appear in the side constraints, counting on the simplex method to provide an integer-valued solution to the network flow constraints whenever it can.

Rounding Off

As noted earlier, the optimal value of the LP relaxation provides a bound on the optimal value of the IP itself. Moreover, it is often possible to round off the optimal solution to the LP relaxation in a way that satisfies the integrality constraints without degrading the objective value too much. This can be accomplished manually. It can also be accomplished by adding a few constraints to the LP relaxation, much as branch and bound would do.

Suppose, for instance, that the optimal solution to the LP relaxation of an IP sends 32.1 trucks to Chicago and has a total cost of $54,321. An obvious alternative is to send 32 or 33 trucks. Suppose, for instance, that sending 32 trucks to Chicago satisfies the IP's integrality constraints and increases cost to $55,200. The optimal solution to the IP may be less than $55,200, but it is not lower than $54,321. Rounding off may be close enough.

On the other hand, suppose that the "optimal" solution to the LP relaxation calls for opening 0.5 warehouses in St. Louis. It may not be possible to "round" this solution off to 0 or 1; if it is possible, it may degrade performance enormously.

An Economy of Scale

As mentioned earlier, the trick that accommodates diseconomies of scale fails to accommodate economies of scale. But there is a way in which we can get lucky. To see how, we return to:

Problem A (Recreational Vehicles)

Let us reconsider the Recreational Vehicle example, Problem A in Chapter 3, with one change. Now, the contribution of the first 12 Standard model vehicles is $700 apiece, and the contribution of any in excess of 12 is $1000 apiece. Does this economy of scale change the optimal solution?

The Recreational Vehicle problem had been solved with a contribution of $840 for each Standard model vehicle. Part of Solver's Sensitivity Report tells how sensitive the optimal solution is to the objective coefficients. That information was reported as Table 3.5 of Chapter 3, and it is reproduced here.

The Allowable Increase and Allowable Decrease in Table 6.1 prescribe a range of values of the contribution of Standard model vehicles for which the optimal solution is unchanged. This range consists of all values between 560 (which equals 840 − 280) and 1040 (which equals 840 + 200). Thus, the optimal solution remains unchanged if the contribution earned by the Standard model vehicles lies within its allowable range, that is, between 560 and 1040.

But more can be said. This range also applies to the contribution of each Standard model vehicle, *individually.* It applies, for instance, when the first 12 Standard model vehicles have a contribution of 700 and the remainder have a contribution of 1000. It does so because 700 and 1000 lie within the allowable range. For this economy of scale, the optimal solution continues to set $S = 20$, $F = 30$, and $L = 0$. The optimal value changes, but the optimal solution does not.

Table 6.1 The portion of the Sensitivity Report that concerns the objective.

Changing Cells

Cell	Name	Final Value	Reduced Cost	Objective Coefficient	Allowable Increase	Allowable Decrease
$B$9	S	20	0	840	200	280
$C$9	F	30	0	1120	560	100
$D$9	L	0	–200	1200	200	∞

6.5. BINARY VARIABLES AND THEIR USES

Our attention now turns to the ways in which integer-valued variables arise in the formulation of optimization problems for solution.

A particularly handy integer-valued variable is aptly called a **binary variable**; it can take only two values, which are 0 and 1. A binary variable often represents a "yes–no" decision, 1 being short for "yes," 0 for "no." Binary variables often called **zero-one** variables; these terms are synonyms. Through a series of examples, we will see how binary variables can model logical requirements and nonlinearities.

An Either–Or Requirement

Let us begin with an example of an **either–or** requirement, namely, a stipulation that at least one of two constraints must be satisfied.

Example 6.1. Suppose that the variables x and y in an optimization problem lie between 0 and 88, and suppose that at least one of these variables must equal zero.

To implement the requirement in Example 6.1, we introduce the binary variable f and the constraints

$$f \text{ is binary}, \quad x \leq 88f, \quad y \leq 88(1 - f).$$

Equating f to 0 forces $x = 0$ and allows $y \le 88$. Similarly, equating f to 1 forces $y = 0$ and allows $x \le 88$. In this way, the yes–no variable f forces at least one of the variables x and y to equal zero, while letting the other fall anywhere within its normal range.

An If–Then Requirement

The next example illustrates an **if–then** requirement, which imposes a constraint if a condition is satisfied.

Example 6.2. As in the prior example, suppose that the variables x and y in an optimization problem lie between 0 and 88. Now suppose that y must equal zero if x is positive, but y can take any value in its range if x is zero.

To fulfill the requirement of Example 6.2, we use the binary variable f and the constraints

$$f \text{ is binary}, \quad x \le 88(1 - f), \quad y \le 88f.$$

If $x = 0$, the middle constraint allows $f = 1$, which lets y take any value within its normal range. On the other hand, if $x > 0$, the middle constraint requires $f = 0$, which forces y to equal zero, exactly as desired.

Counting Successes or Failures

In a maximization problem, we sometimes wish to declare a "success" if a decision variable lies on one side of a particular number and a "failure" if it lies on the other side. An example follows.

Example 6.3. The smokestack of a coal-burning electrical generation plant releases several chemicals that harm the environment. This example concerns the plant's rate p of emission of sulfur dioxide, measured in kilos per day.

- No matter what fuel the plant purchases and no matter what technology is employed, not more than 88 kilos per day of sulfur dioxide will be emitted. The plant's rate p of sulfur dioxide emission can be controlled by purchasing low-sulfur coal and by investing in cleaner technology.
- The government has set a bound b on p, and this bound is well below 88 kilos per day.
- The government imposes a pollution penalty of \$3000 for each day in which the company emits more than b kilos of sulfur dioxide.

Is this penalty sufficient to induce the company to comply?

A complete answer to this question would require more information than is provided in Example 6.3, namely, the cost of operating the plant with a rate p of sulfur dioxide emission. We provide part of this answer by using a binary variable f to ascertain whether the penalty need be paid, as in:

$$f \text{ is binary}, \quad p \le 88 + (b - 88)f.$$

If $f = 0$, the above inequality reduces to $p \le 88$, which allows the variable p to take any value up to its normal limit. On the other hand, if $f = 1$, the same inequality becomes $p \le b$, which keeps the pollution level from exceeding the government's bound. Thus, including the addend $(1 - f)(3000)$ in the per-day cost of operating the plant imposes the penalty when the company is not in compliance, that is, when p is not constrained to lie below b.

This scheme is a bit subtle but in ways that are familiar. Like the model of a diseconomy of scale, this scheme creates an unintended option, which is to equate f to zero (and incur the pollution fee) even if the emissions bound is met. That option is ruled out by optimization. Like Examples 6.1 and 6.2, this example allows p to fall anywhere in its normal range when f takes the value zero.

An Unwieldy If–Then Constraint

Not every logical constraint can be satisfied with yes–no variables in an otherwise linear program. An example follows.

Example 6.4. As before, we suppose that the variables x and y in a linear program lie between 0 and 88. If x is zero, y must equal 0. If x is positive, then y can take any value.

The natural way to implement this "if–then" requirement is through the binary variable f and the constraints

$$f \text{ is binary}, \quad f + x > 0, \quad y \leq 88\,(1 - f).$$

If x is positive, the constraint $f + x > 0$ allows f to equal zero, in which case the remaining constraint becomes $y \leq 88$, which allows y to take any value within its range. Alternatively, if x is zero, the constraint $f + x > 0$ implies $f = 1$, which guarantees $y = 0$, as desired.

But the constraint $f + x > 0$ cannot be part of an integer program, for the same reason that its relaxation in which the integrality requirement on f is suppressed cannot be part of a linear program. Constraints can be equations, "$\leq$" inequalities, and "$\geq$" inequalities, but they cannot be strict inequalities. Branch and bound does not dovetail with "$>$" constraints because its linear programs can lack optimal solutions.

The if–then requirement in Example 6.4 cannot be enforced by any integer program. It can be approximated by "tightening" the constraint $f + x > 0$ to $f + x \geq 0.001$, which enforces $y = 0$ when $x < 0.001$.

Set Covering Constraints

In some situations, an organization wishes to place facilities at enough locations that each customer has reasonable access to at least one facility.

Example 6.5. Let there be n customers and m sites at which facilities can be located. Number the customers 1 through n, and number the sites 1 through m. A facility at a particular site can serve certain customers but not others. The array A of data determine which is which:

$$A(i, j) = \begin{cases} 1 & \text{if customer } j \text{ can be served by a facility at site } i \\ 0 & \text{otherwise} \end{cases}$$

It costs $c(i)$ to locate a facility at site i, for $i = 1, 2, \ldots, m$. We wish to minimize the cost of locating facilities in such a way that each customer can be served by at least one facility.

Example 6.5 leads us naturally to yes–no variables $y(i)$ through $y(m)$, where $y(i)$ equals 1 if a facility is located at site i and equals zero otherwise. The facility location problem becomes this integer program:

Minimize $\sum_{i=1}^{m} c(i)\,y(i)$ subject to

$$\sum_{i=1}^{m} y(i)A(i,j) \geq 1, \qquad \text{for } j = 1, 2, \ldots, n,$$

$$y(i) \text{ is binary}, \qquad \text{for } i = 1, 2, \ldots, m.$$

The jth of the above inequalities imposes the requirement that customer j be served by at least one facility. Collectively, the constraints of this integer program are called **set covering** constraints; they locate the facilities so that each customer is "covered" by at least one facility.

A Fixed Charge

The previous examples used binary variables to model logical requirements. We now present a pair of examples in which binary variables model nonlinearities.

In a linear program, cost can vary linearly with quantity. The next example uses a binary variable to model a discontinuous cost function, one that jumps upward as the quantity becomes positive.

Example 6.6. Let us suppose that the cost of ordering widgets includes a **fixed charge** of \$10, which occurs if any widgets are purchased, and a per-unit cost of \$1.40 per widget. Let us also suppose that not more than 88 widgets can be ordered.

To model the cost of ordering w widgets, we introduce the binary variable f and the continuous-valued variable w, with these interpretations:

$$f = \begin{cases} 1 & \text{if any widgets are ordered} \\ 0 & \text{if no widgets are ordered} \end{cases}$$

$$w = \text{the number of widgets that are ordered.}$$

The constraints include

$$f \text{ is binary}, \quad w \geq 0, \quad w \leq 88f.$$

If $f = 1$, the inequality $w \leq 88f$ allows w to take any value within its normal range. Alternatively, if $f = 0$, then $w = 0$. In a cost-minimization problem, the objective includes $10\,f + 1.40\,w$, which measures the cost of producing widgets.

The resulting integer program includes an unintended option that is ruled out by optimization. The unintended option is to set $f = 1$ and $w = 0$. This option cannot be optimal because setting $f = 0$ and $w = 0$ costs \$10 less.

Piecewise Linear Continuous Function

Binary variables are now used to specify a function that is piecewise linear and continuous within a finite range. This function can have any number of pieces. The idea is to introduce one binary variable and one continuous variable per interval of linearity. The sum of the binary variables is required to equal 1, so exactly one of them equals 1. An interval's continuous variable is constrained to equal zero unless its binary variable equals 1. To introduce this technique, we employ an example that exhibits decreasing marginal cost.

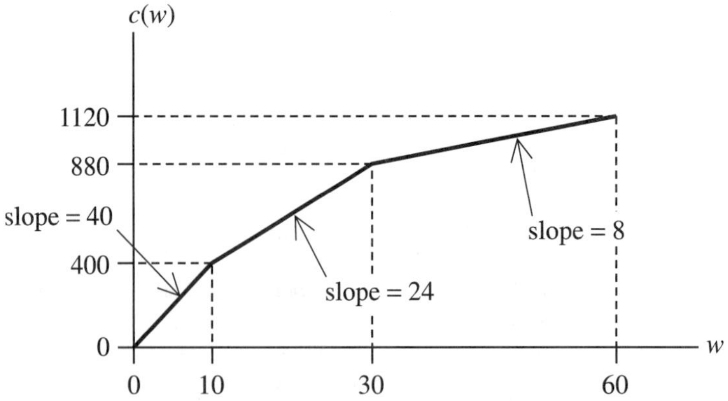

Figure 6.2 The cost $c(w)$ of producing w widgets.

Example 6.7. Suppose that the cost $c(w)$ of producing w widgets is as depicted in Figure 6.2. Evidently, the first 10 widgets cost \$40 apiece to produce, the next 20 widgets cost \$24 apiece to produce, and the final 30 cost \$8 apiece to produce.

In Figure 6.2, the range between 0 and 60 breaks naturally into three intervals of linearity. Let us label the intervals from left to right; interval 1 lies between 0 and 10, interval 2 between 10 and 30, and interval 3 between 30 and 60. Let us introduce three binary variables, one per interval, with

$$f_i \text{ is binary,} \quad \text{for } i = 1, 2, 3,$$
$$f_1 + f_2 + f_3 = 1.$$

These constraints require one of the f_i's to equal 1, and the rest to equal 0. The idea is that f_i equals 1 when w lies in the ith interval. To implement this idea, we introduce three continuous variables, one per interval, with

$$0 \le z_i \le f_i, \quad \text{for } i = 1, 2, 3.$$

Evidently, z_i cannot be positive unless f_i equals 1. The idea is that z_i should measure the fraction of the way that w lies into the ith interval, measured from its left-hand end-point. For instance, z_2 equals $1/5$ when w is one-fifth of the way into the second interval, measuring from its left-hand end-point. This is accomplished by setting

$$w = 0f_1 + z_1(10 - 0) + 10f_2 + z_2(30 - 10) + 30f_3 + z_3(60 - 30).$$

For instance, the values $f_2 = 1$ and $z_2 = 1/5$ give $w = 10 + (1/5)(30 - 10) = 10 + 4 = 14$, which is one-fifth of the way into the second interval, measuring from its left-hand end-point.

Similarly, if $f_i = 1$, the cost $c(w)$ equals the cost at the left-hand end-point of the ith interval plus the fraction z_i of the change in cost at the right-hand end-point. For the function $c(w)$ plotted in Figure 6.2, we get

$$c(w) = 0f_1 + z_1(400 - 0) + 400f_2 + z_2(880 - 400) + 880f_3 + z_3(1120 - 880).$$

For instance, the values $f_2 = 1$ and $z_2 = 1/5$ give $c(w) = 400 + (1/5)(880 - 400) = 400 + 96 = 496$.

The technique that has just been illustrated applies to *any* piecewise linear function on any interval whose width is finite. This function need not exhibit decreasing marginal cost.

Recap

We've seen how binary variables model a variety of logical requirements, setup costs, and piecewise linear functions. In several of the preceding examples, a binary variable has been used to control the value of a continuous variable that is required to lie in a *finite* interval. We often used the number 88 as the width of the interval. In some of these examples, a binary variable introduced an unintended option that was ruled out by optimization.

6.6. HOW SOLVER ENFORCES INTEGRALITY REQUIREMENTS

We have not yet indicated how to formulate an integer program for solution by Solver. That's easy. Suppose, for instance, that cells B6 through F6 contain decision variables whose values you wish Solver to restrict to the integers 0 and 1. To have it do so, follow these steps:

1. In the Solver Parameters dialog box, click on the Add button.
2. In the Add Constraint dialog box that pops up, click in the Cell Reference window, and then select cells B6 through F6.
3. Next, click on the button in the central window of the Add Constraint dialog box. It will then present the options in the left-hand side of Table 6.2. Scroll down to the "bin" option (short for binary) and click.
4. After that click, the Add Constraint dialog box will appear as it does on the right-hand side of Table 6.2. Click on the OK button, which returns you to the Solver Parameters dialog box.

Alternatively, if you wish to restrict cells B6 through F6 to integer values, proceed exactly as above but scroll down to the "int" option and click.

6.7. THE TRAVELING SALESPERSON

When the "traveling salesperson" problem is first presented, it may seem a bit whimsical, but it arises frequently and in several guises. Its data are the number n of cities and the travel times from city to city. We number these cities 1 through n. For each city i and for each city j other than i, we denote as $c(i, j)$ the number of units of time that it takes to travel from city i to city j.

Tours and Their Travel Times

A salesperson is now at city 1. A **tour** consists of visiting cities 2 through n *exactly* once and then returning from the last city visited to city 1. The travel time of a tour is defined to be the sum of its travel times from city to city.

Table 6.2 Using the Add Constraint dialog box to restrict the values of cells B6 through F6 to the integers 0 and 1.

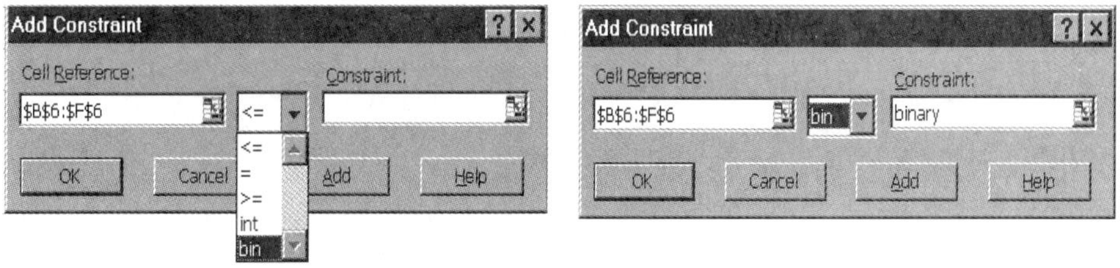

Suppose, for instance, that $n = 5$; one tour is described by the sequence (1, 3, 5, 2, 4, 1). This sequence begins with city 1, it visits each of the other four cities exactly once, and it ends by returning to city 1. This tour's travel time is given by

$$c(1, 3) + c(3, 5) + c(5, 2) + c(2, 4) + c(4, 1).$$

The **traveling salesperson** problem is that of finding a tour whose travel time is smallest. The $c(i, j)$'s have been called travel times, but these data can take any values, including negative ones.

This section sketches one approach to solving the traveling salesperson problem. We'll see how to formulate it as an integer program. Problem 22 sketches a different formulation as an integer program.

A Related Assignment Problem

Both formulations build on an assignment problem that appears below. Its data are the number n of cities and the travel times, the $c(i, j)$'s. Its decision variables are the $x(i, j)$'s, where assigning $x(i, j)$ the value 1 corresponds to visiting city j immediately after visiting city i. This assignment problem is:

Minimize $\sum_{i=1}^{n} \sum_{j=1}^{n} c(i, j) x(i, j)$, subject to the constraints

$$\sum_{k=1}^{n} x(k, i) = 1, \qquad \text{for } i = 1, 2, \ldots, n,$$

$$\sum_{j=1}^{n} x(i, j) = 1, \qquad \text{for } i = 1, 2, \ldots, n,$$

$$x(i, j) \geq 0, \qquad \text{for all } i \text{ and } j.$$

Each tour prescribes a feasible solution to this assignment problem. Each tour puts $x(k, i) = 1$ if city k is visited immediately before city i, and it puts $x(i, j) = 1$ if city j is visited immediately after city i.

This assignment problem omits the requirement that the $x(i, j)$'s be integers, but that presents no problem for the simplex method. The Integrality Theorem guarantees an optimal solution that is integer-valued.

Technically, this assignment problem is incorrect because it includes the variable $x(i, i)$ for each i. That's easy to fix. Either remove each variable $x(i, i)$, or leave these variables in and equate $c(i, i)$ to a prohibitively large number.

What Is Missing?

Unfortunately, the constraints of this assignment problem do not require that the pattern of travel be a single tour. They allow a **subtour**, namely, a set of fewer than n cities that are visited in sequence. For example, with $n = 5$, the constraints of the assignment problem are satisfied by assigning each variable the value zero, except for

$$x(2, 5) = x(5, 3) = x(3, 2) = 1 \quad \text{and} \quad x(1, 4) = x(4, 1) = 1,$$

which describe the subtours (2, 5, 3, 2) and (1, 4, 1). Collectively, these subtours visit each city exactly once, so they satisfy the constraints of the assignment problem.

Subtour Elimination Constraints

In theory, at least, it's easy to rid ourselves of subtours. Of the three variables in the subtour (2, 5, 3, 2), at most two can be positive in a tour, so we can add to the assignment problem the constraint

$$x(2, 5) + x(5, 3) + x(3, 2) \leq 2,$$

which renders the subtour (2, 5, 3, 2) infeasible. In general, each subtour visits a number k of cities, where k lies below n. This subtour can be precluded by a **subtour elimination constraint** that requires at most $k - 1$ of its cities to be visited.

Is This Practical?

From a practical viewpoint, subtour elimination constraints seem problematic. One difficulty is that they are legion; roughly 2^n constraints would be needed to eliminate all subtours. A 20-city traveling salesperson problem would require roughly 1 million subtour elimination constraints, for instance.

Another difficulty with adding subtour elimination constraints is that they produce an LP that is no longer an assignment problem, so we are no longer guaranteed an integer-valued optimal solution it. The integrality constraints may need to be reimposed.

Surprisingly, it is often possible to solve a traveling salesperson problem by enforcing a small number of subtour elimination constraints. It can be effective to "branch" as follows: Identify a subtour in the optimal solution to a linear program, and then solve the linear program that differs from it by adding the constraint that eliminates this particular subtour.

A Job Sequencing Problem

To illustrate this branching strategy, we turn our attention to the problem of finding a least-cost sequence in which to perform a group of jobs on a particular machine. At first blush, it may not look like a traveling salesperson problem, but it is.

Problem B (Job Sequencing)

Five different jobs must be done on a single machine. The **setup time** that is needed to reset the machine to perform a particular job depends on what job preceded it. Table 6.3 specifies these setup times in minutes. This table shows, for instance, that doing job A first entails a 3-minute setup and doing job D immediately after job A requires a 17-minute setup. (The entries of 100 minutes indicate prohibited sequences, such as doing a job after itself.) The goal is to sequence the jobs so as to minimize the sum of the setup times.

Table 6.3 Setup time (in minutes) to switch from the job in a given row to the job in a given column.

	A	B	C	D	E	F	G
2		A	B	C	D	E	finish
3	start	3	5	11	13	7	100
4	A	100	11	21	17	15	0
5	B	9	100	17	25	19	0
6	C	17	19	100	21	7	0
7	D	11	13	27	100	11	0
8	E	21	15	13	29	100	0

This job sequencing problem is a six-city traveling salesperson problem; the machine starts empty, it "visits" each of five jobs exactly once, and then it returns to the empty state.

Problem 24 invites you to set this up as a 6×6 assignment problem and solve it. You will find that its optimal solution includes the subtour (C, E, C), in which jobs C and E follow each other. If you augment the assignment problem with the constraint that eliminates this particular subtour, you will find that its optimal solution prescribes a tour that solves the job sequencing problem.

6.8. REVIEW

In this chapter, we have seen how to use binary variables to model fixed costs, piecewise linear functions, and several different sorts of logical requirements. We cannot model all logical requirements. When a binary variable controls the value of a decision variable (or, equivalently, of a linear expression), its natural range must be finite, for instance. A binary variable can introduce an unintended option provided that it is ruled out by optimization.

The principal solution method for integer programs is known as branch and bound. It attacks the integer program by solving a sequence of linear programs. This sequence may be short, but it can run into the tens of thousands, or higher. The basic idea of branch and bound is to *divide and conquer.* If a linear program's optimal solution equates a variable that should be integer-valued to a value k that is not an integer, that linear program is replaced by a pair. One of the pair requires the variable to be at least as large as the smallest integer that exceeds k, and the other requires the variable to be not greater than the largest integer that lies below k. Whenever a feasible solution to the integer program is found, eliminate all linear programs whose optimal values exceed its objective value.

A section of this chapter describes several situations in which integer programs seem to be necessary but are not. That section shows how sensitivity analysis can circumvent the need to deal with an economy of scale, for instance.

Another section of this chapter probes the traveling salesperson problem. We've seen how to formulate it as an integer program that has a huge number of constraints, one per subtour. We've also seen how to attack it by sequentially inserting subtour elimination constraints as they are needed.

To obtain a feel for the settings in which integer programming is useful, you might scan the titles of the homework and discussion problems at the end of this chapter. In some of these problems, a schedule must be set. In others, a set of objects must be partitioned.

6.9. HOMEWORK AND DISCUSSION PROBLEMS

1. A feasible solution to an integer program has an objective value of 23.5; an optimal solution to the linear program relaxation in which the integrality requirements are dropped is 21.6.

(a) Is this integer program a maximization problem or a minimization problem?

(b) What can you say about its optimal value?

2. Suppose, in Figure 6.1, that box 3 had LC = \$55,300 rather than \$54,876. Would it be necessary to branch on the variable y? Why?

3. Suppose, in Figure 6.1, that box 5 had LC = \$54,900 rather than \$55,100. How close to optimum would the solution in box 4 be? How would branch and bound proceed? What would constitute good luck?

4. Figure 6.1 illustrates branch and bound for a minimization problem. Draw an analog of Figure 6.1 for a maximization problem, and discuss the analogous branch-and-bound scheme for your figure.

5. **(Which Variables to Branch Upon)** Consider a network flow problem that has 1003 nodes, 10,800 arcs, integer-valued bounds, constraints that all flows be integer-valued, and the "side" constraint that at most two of the variables $x(1, 2)$, $x(2, 3)$, $x(3, 4)$, $x(4, 5)$, and $x(5, 6)$ can be positive.

(a) Suppose, in addition, that the constraints keep the flows on arcs (1, 2) and (2, 3) and (3, 4) and (4, 5) and (5, 6) from exceeding 1. Show how to express the above "side" constraint within an integer program.

(b) For the situation in part (a), which variables in the LP relaxation are good candidates for branching? Which are not? Why?

(c) Redo part (a) for the case in which the flows on arcs (1, 2) and (2, 3) and (3, 4) and (4, 5) and (5, 6) cannot exceed 88 rather than 1. *Hint:* Introduce five binary variables, one for each of these arcs.

(d) For the situation in part (c), which variables in the LP relaxation are good candidates for branching? Which are not? Why?

6. Consider the problem of maximizing $\{6\,x + 14\,y\}$, subject to four constraints, which are $2\,x + y \leq 25$ and $x + 2\,y \leq 12$ and $y \geq 0$, and that x must equal one of the values 0, 1, 4, and 6.

(a) Formulate this as an integer program. (You may need four yes–no variables.)

(b) Use Solver to find its optimal solution.

7. Consider the problem of maximizing $\{c(x) + 14\,y\}$ subject to the constraints $2\,x + y \leq 25$, $x + 2\,y \leq 12$, $x \geq 0$, and $y \geq 0$, where the function $c(x)$ of x is piecewise linear, with $c(0) = 0$ and with slopes that are as follows:

End-points	0 and 2	2 and 3	3 and 4	4 and ∞
Slope	8	−6	11	8

(a) Formulate this problem as an integer program. *Hint:* Is there a maximum value on x?

(b) Use Solver to find its optimal solution.

8. **(Infeasibilities)** Our description of branch and bound was slightly incomplete because we did not account for the possibility of infeasibility.

(a) Suppose that the LP relaxation of an integer program has no feasible solution. What then?

(b) Suppose that one of the boxes encountered by branch and bound describes a linear program that has no feasible solution. What then?

(c) Suppose that the integer program has no feasible solution but that its LP relaxation does have feasible solutions. What then?

9. In which examples and problems within this chapter did binary variables introduce unintended options that were ruled out by optimization?

10. In which examples and problems within this chapter was it assumed that a continuous decision variable has finite upper bound? Why?

11. **(Sensitivity Analysis and Multiple Economies of Scale)** This problem refers to the variant of the Recreational Vehicle problem in which the contribution of the first 12 Standard model vehicles is \$750 apiece and the contribution of any in excess of 12 is \$1000 apiece. Parts (a) and (b) of this problem are independent of each other.

(a) In addition to the economy of scale on Standard model vehicles, the contribution of the first four Luxury model vehicles is \$1100 apiece and the contribution of any in excess of 4 is \$1400 apiece. Without introducing integer-valued variables, argue that the optimal solution remains $S = 20$, $F = 30$, and $L = 0$. *Hint:* You might wish to rerun the original LP with the contribution of S as 750 and check its sensitivity analysis.

(b) In addition to the economy of scale on Standard model vehicles, the contribution of the first 16 Fancy model vehicles is \$1000 apiece, and the contribution of any in excess of 16 is \$1500 apiece. Without introducing integer-valued variables, argue that the optimal solution remains $S = 20$, $F = 30$, and $L = 0$.

12. **(Scheduling Maintenance)** Over the next eight weeks, a company must perform periodic maintenance on five items of heavy equipment, which are labeled A through E. Periodic maintenance of each item takes three weeks, but the company has some flexibility as to the week in which each item's maintenance can commence. The following table indicates, for instance, that maintenance of item A can commence as early as the first week or as late as the fourth week, that item A requires 8 units of labor in the first week it is in periodic maintenance, 11 hours in the its second week in maintenance, and 6 in its third week.

			Units of Labor by Week		
Item	Earliest Start	Latest Start	1	2	3
A	1	4	8	11	6
B	1	3	6	4	9
C	2	5	13	3	2
D	2	6	2	5	12
E	3	5	15	17	5

(a) Suppose that the company wishes to minimize the largest number of units of labor required in any week. Formulate this problem as an integer program. (You may need one yes–no variable for each week in which maintenance on each item can commence.)

(b) Use Solver to compute the optimal solution to the integer program in part (a).

(c) Suppose that the company wishes to maximize the smallest number of units of labor required in any week. Formulate that problem as an integer program.

(d) Use Solver to compute the optimal solution to the integer program in part (c). Is it different from that in part (b)?

13. (Assigning Departure Gates to Flights) As the schedule setter for an airline, you must schedule exactly one early-morning departure from Pittsburgh to each of four cities. Due to competition, the contribution earned by each flight depends on its departure time, as indicated below. For instance, the most profitable departure time for O'Hare is at 7:30 A.M. Your airline has permission to schedule these four departures at any time between 7 A.M. and 8 A.M, but you have only two departure gates, and you cannot schedule more than two departures at any time.

Contribution per flight, in thousands of dollars

Time	Laguardia	O'Hare	Logan	National
7:00 A.M.	8.2	7.0	5.6	9.5
7:30 A.M.	7.8	8.2	4.4	8.8
8:00 A.M.	6.9	7.8	3.1	7.0

(a) Formulate the problem of maximizing contribution as an integer program.

(b) Solve the integer program you formulated in part (a).

(c) Another airline wishes to rent one departure gate for the 7:00 A.M. time. What is the smallest rent that would be profitable for you to charge?

14. (An Equitable Partition) After a few weeks of marital bliss, Alison and Bill discussed the household chores. Alison, who majored in Industrial Engineering, noticed that each chore has an economy of scale, one that suggests that each chore should be done exclusively by Bill or herself. She estimated the number of hours per week that each of them takes to perform each chore. These data appear in the following table. They show, for instance, that Alison gets the cooking done more quickly than Bill. Alison and Bill agreed that each chore should be done in its entirety by one of them and that they should also divide up the chores so that the maximum amount of time that either spends doing chores is minimized.

	A	B	C	D	E	F
2	**activity**	**cook**	**clean up**	**shop**	**laundery**	**gardening**
3	Alison's time	7.8	5.0	3.6	3.5	6.8
4	Bill's time	10.1	4.0	4.7	4.2	5.5

(a) On a spreadsheet, formulate their chore-allocation problem as an integer program.

(b) Solve the integer program you formulated in part (a). Who does which chores? Who spends more time doing chores?

(c) Suppose that Alison and Bill wish to minimize the *total* time that they spend doing chores. Would an integer program be necessary? Who would do each chore? Who would spend more time doing chores?

15. (Locating Depots) A trucking company has contracted to serve five major customers. It must lease depots from which to serve these customers. Listed in the following table are the capacity of each depot (number of trucks), the leasing cost of each depot (dollars per month), the demand of each customer (truckloads per month), and the shipping cost from each depot to each customer (dollars per truckload). The company wishes to lease depots and ship in a way that will minimize its monthly expense.

Shipping Cost from Depot to Customer							
1	2	3	4	5	Depot	Capacity of Depot	Leasing Cost of Depot
4.0	7.6	6.6	7.6	8.4	A	350	800
6.6	6.0	6.0	8.0	7.0	B	200	300
8.0	8.4	3.6	7.0	7.6	C	450	1100
9.0	9.2	5.2	5.2	6.7	D	250	440
150	200	230	110	80		customer demand	

(a) On a spreadsheet, formulate the company's optimization problem as an integer program.

(b) Solve the integer program you formulated in part (a). Which depots get leased? What is the shipping pattern?

16. (Partitioning a Region) A region is broken up into seven districts, which are labeled 1 through 7. The company wishes to assign representatives to districts so as to maximize the number of potential customers that can be visited. The company can assign a sales representative to a single district or to a pair of adjacent districts. It does not split districts between sales representatives. The following table shows, for instance, that district 1 is adjacent to districts 2 and 3 and that district 1 has 18 potential customers.

District	1	2	3	4	5	6	7
Adjacent to	2, 3	1, 3, 4, 5	1, 2, 4	2, 3, 5, 6, 7	2, 4, 6	4, 5, 7	4, 6
Customers	18	15	20	11	30	8	36

(a) Suppose that the company has three sales representatives. Formulate an integer program that assigns each representative to a pair of districts so as to maximize the total number of potential customers who can be visited. (This may require as many as 11 yes–no variables.)

(b) Use Solver to solve the integer program that you formulated in part (a).

(c) Imagine that you are one of these sales representatives. If you were assigned district 1, you would prefer district 3 as your second district because that pairing would allow you to visit a total of 38 potential customers. Make similar statements for districts 2 through 7.

(d) For the values $k = 1, 2,$ and 3, use part (c) to show how to assign k sales representatives to districts so as to maximize the number of potential customers that are visited. What feature of the solution to part (c) are you exploiting?

17. (Continuation of Problem 16) Suppose that the company has four sales representatives and wishes to assign them as equitably as possible without splitting districts.

(a) Formulate an integer program that assigns a single district or two adjacent districts to each sales representative and does so in a way that minimizes the excess of the largest number of potential customers seen by a sales representative over the smallest.

(b) Use Solver to solve the integer program you formulated in part (a).

18. (**Utilizing Restricted Endowment**) In addition to its general scholarship fund, a university's alumni have given it endowment for M restricted scholarships, each of which can be used to support one entering student of a particular sort, such as a person with a Huguenot surname, a woman of Russian Orthodox faith, and so on. The admissions officer has accepted N applicants who qualify for financial aid. Many of them meet the restrictions of none of these restricted scholarships, some meet the restrictions for one, and some qualify for more than one. The datum $Q(n, m)$ equals 1 if the nth of these N applicants qualifies for restricted scholarship m, and this datum equals 0 otherwise. Each applicant can be awarded at most one restricted scholarship. The university wishes to award as many of these restricted scholarships as is possible.

(a) Formulate this allocation problem as an integer program.

(b) Speculate on the chance of getting lucky by solving the relaxation of the integer program you formulated in part (a) in which the requirements that the allocations be integers is suppressed.

19. For Example 6.7, three binary variables were used. Are two binary variables sufficient? If so, how?

20. (**Redistricting a State**) A small state is comprised of 10 counties. In the most recent reapportionment of the House of Representatives, this state has been allocated three seats (Congressional Districts). By longstanding agreement between the parties, no county can be split between two Congressional Districts. Each Congressional District must represent 520,000 and 630,000 persons. The governor wishes to assign each county to a Congressional District in a way that maximizes the number of seats in which Democrats are at least 52% of the population. Listed below are the total population of each county and the number of Democrats in it, both in thousands.

County	1	2	3	4	5	6	7	8	9	10
Population	226	218	76	75	234	255	168	155	47	274
Democrats	128	129	44	25	122	96	128	114	31	112

(a) In an allocation of counties to districts, let the decision variable $T(i)$ denote the total population of district i, and let $D(i)$ denote the number of Democrats in district i. Interpret the yes–no variable $f(i)$ that satisfies the constraints

$$f(i) \text{ is binary}, \quad D(i) \geq 0.52\, T(i) + 630 \times 0.52 \times [f(i) - 1],$$

and explain why the above inequality allows $D(i)$ as low as zero and $T(i)$ as high as 630.

(b) Formulate the governor's optimization problem as an integer program that has not more than 23 yes–no variables.

(c) Use Solver to find an optimal solution. (Solver may ask your permission to exceed its normal time limit. Allow it to run for a few extra minutes; you can always shut it off by hitting the ESC key.) Interpret the solution that Solver provides.

21. (**Capital Budgeting**) A real estate developer can elect to invest in any of m projects. Each project has a natural starting time and a natural completion time. All project completion times lie within the next 60 months. If project i is undertaken, it will return $A(i, j)$ dollars during month j, for $j = 1, 2, \ldots, 60$. Some of these returns will be negative; for a typical project, the returns in the early months are negative, and the payoff occurs at the project completion time. She can invest net revenue in the bank at an interest rate of 0.5% per month. At the start of the period, she has no money in her bank account. She has obtained a line of credit of \$5 million from the bank at an interest rate of 1.5% per month. This means that her outstanding debt to the bank can be any amount up to \$5 million at any time during the next 60 months, and she must pay 1.5% per month on the amount she has actually borrowed. Interest is paid at the end of the month and is compounded monthly. She wishes to undertake a group of projects that maximizes the amount in her bank account at the end of the 60-month period. Build an integer program that accomplishes her goal.

22. (**Traveling Salesperson**) For the traveling salesperson problem in Section 6.7, we designate $x(i, j)$ as a binary variable and require

$$\sum_{j=1}^{n} x(i, j) = 1, \quad \text{for } i = 1, 2, \ldots, n,$$

$$\sum_{k=1}^{n} x(k, i) = 1, \quad \text{for } i = 1, 2, \ldots, n,$$

As was indicated in Section 6.6, these constraints do not, by themselves, eliminate subtours. Let us introduce the continuous decision variables $u(2)$ through $u(n)$ and the constraints

$$u(i) - u(j) + n\,x(i, j) \le n - 1, \quad \text{for each pair } (i, j) \text{ with } i \ne j,\ i \ge 2, \text{ and } j \ge 2.$$

Note: Being continuous decision variables, $u(1)$ through $u(n)$ can take any values that satisfy the above constraints.

(a) Consider a subtour that does not include city 1. Is it precluded by the above constraints? If so, why? *Hint:* Sum the above over each pair (i, j) for which $x(i, j) = 1$ in this subtour, and note that the u-values cancel out.

(b) Consider a subtour that includes city 1. Is it precluded by the above constraints? *Hint:* If there is a subtour that includes city 1, must there be a subtour that excludes city 1?

(c) Consider a tour that includes city 1. Is it precluded by the above constraints? *Hint:* Try setting $u(j) = k$ if city j is the kth to be visited on this tour.

(d) Use the constraints listed above to formulate the traveling salesperson problem as an integer program.

Remark: Before working the next problem, you might wish to review the idea of a "state" and the characteristics of a "network flow problem," as they were discussed in Chapter 4.

23. (**Assigning Aircraft to Flights**) A small airline possesses a fleet that consists of ten large airplanes and four small ones. This airline flies between cities A, B, and C. This airline has morning and afternoon flights between each pair of cities. Its flight schedule is summarized in the following table. For instance, the entry from city A to city B in the morning reads "2L, 1S," which means that the demand is for two large planes and one small plane. Each plane that arrives at a city in a morning flight can leave that city on the same day's afternoon flight but not before. Each plane that arrives at a city on an afternoon flight can leave that city on a morning leg on the following morning but not before. Large planes can be substituted for small planes. The variable cost of operating large and small planes are \$15,000 and \$10,000, respectively. These costs apply to deadheading as well as to regular flights. To avoid complicated schedules, the airline insists that the same pattern of flights occur on successive days.

Morning flights

	To		
From	A	B	C
A	—	2L, 1S	1S
B	1L, 2S	—	1L, 1 S
C	1L	1L, 2S	—

Afternoon flights

	To		
From	A	B	C
A	—	1L, 1S	1L
B	1L, 3S	—	2L
C	1L	1L, 1S	—

(a) What are the state variables at the start of each period? What would they be if the airline allowed the pattern of flights to repeat every other day rather than every day?

(b) Which of the following constraints might your linear program include? Exclude?

$$\text{L1AB} \ge 2, \quad \text{L1AB} + \text{S1AB} \ge 3, \quad \text{L1AB} = 2, \quad \text{L1AB} + \text{S1AB} = 3.$$

(c) Formulate this problem as an integer program. If the integrality constraints were ignored, would it be a network flow problem? Why or why not? *Hint:* This linear program is a natural for a tailored spreadsheet; its decision variables form themselves naturally into four groups, with repeating patterns of coefficients.

(d) Solve the *relaxation* in which the integrality constraints are deleted. Did you get a fractional solution? If so, try to round if off to an integer-valued solution that has the same objective value. If that fails, impose enough integrality constraints to solve it.

(e) Route each of the 14 aircraft in a way that implements the optimal solution that you found in part (d).

(f) This company can lease an extra small plane for $6000 per day. Should it do so? Why?

24. (**Sequencing Jobs on a Machine**) This concerns Problem B in Section 6.6. Its goal is to sequence five jobs on a single machine so as to minimize the sum of the setup times. As was noted in Section 6.6, this is a six-city traveling salesperson problem.

(a) On a spreadsheet, set up and solve the 6×6 assignment problem in which each activity must be followed by another.

(b) Does the optimal solution that you obtained in part (a) sequence the jobs? Alternatively, it might sequence job Q immediately after job P and sequence job P immediately after job Q. If it does the latter, add to your linear program the subtour elimination constraint, $x(P, Q) + x(Q, P) \leq 1$.

(c) After including this subtour elimination constraint, re-solve your linear program. What happens? Does its optimal solution sequence the jobs? If it did not, how might you proceed?

Part C

Probability for Decision Making

Chapter 7

Introduction to Probability Models

7.1. PREVIEW

Throughout history, uncertainty has influenced human affairs and has often led to efforts to provide protection from its ill effects. Ancient Egyptian records, for example, describe the steps that the Egyptians took to protect their crops from the uncertainties caused by the spring floods. Similarly, during the Italian Renaissance, merchants learned to insure themselves against the uncertainties of inter-city commerce. In more recent times, Benjamin Franklin organized the first American insurance company in 1752 to protect individuals from losses due to fire damage to their homes.

Uncertainty pervades modern life. The price of a corporation's stock one year from now is uncertain, as is the peak-load demand for electricity. The extent to which you will enjoy school is uncertain, as is the benefit that you will glean from it.

Probability models shed light on the uncertainty in each of these situations, historic and current. It must be admitted, however, that probability models developed in a different context—in the context of games of chance.

Some take a fatalistic attitude toward uncertainty and firmly believe that "Whatever will be, will be." Others struggle to deal with it and so make rational decisions. This chapter introduces the science of probability, which describes and measures uncertainty; it is a prelude to rational decision making in the face of uncertainty.

7.2. WHAT CAN YOU LEARN FROM THIS CHAPTER?

This chapter has only two goals: to familiarize you with the language of probability, which has a considerable vocabulary; and to show you how to use this language to compute various probabilities. Many beginners find the language a bit alien and the computations a bit daunting. To ease access, we present each concept in the context of a decision that needs to be taken.

7.3. A MODEL OF UNCERTAINTY

When we describe a model of uncertainty, the terms *outcomes* and *probabilities* have specific meanings. The **outcomes** are the things that might happen, the ways in which the uncertainty might be resolved. Outcomes are required to be **mutually exclusive** and **collectively exhaustive**. "Mutually exclusive" means that only one of the outcomes can occur, whereas "collectively exhaustive" means that nothing other than an outcome can occur.

The **probability** of an outcome is the likelihood that it will occur when the uncertainty is resolved. The probabilities are required to be nonnegative and to sum to 1. Specifically, the probability of each outcome is a number between 0 and 1, and the sum of the probabilities of the outcomes equals 1. A **probability model** of an uncertain situation is a list of its outcomes, accompanied by the probability of each outcome.

One Balanced Coin

Let us build a probability model of one toss of a balanced coin. In this (somewhat simplified) model, the outcomes are "heads" and "tails." The coin cannot end on an edge, and it cannot roll out of sight. Since the outcomes are mutually exclusive and collectively exhaustive, exactly one of them occurs when the uncertainty is resolved.

This coin is balanced, so that the probability that it comes up heads equals the probability that it comes up tails. The sum of these probabilities equals 1, so each of them equals $1/2$. In short, the probability that the coin will come up heads is $1/2$, and the probability that it will come up tails equals $1/2$.

Repeated Tosses

To help motivate probabilities, we imagine that we flip this balanced coin 1000 times. What will happen? It will come up heads roughly 500 times ($1/2$ of the total), and it will come up tails the remainder of the times. The fraction of the time that it will come up heads is roughly equal to the probability that it comes up heads on any single flip.

Figure 7.1 simulates 100 tosses of a fair coin. It plots the fraction of times that heads come up versus the total number of tosses. In this particular experiment, a total of 46 heads

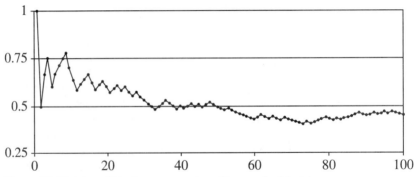

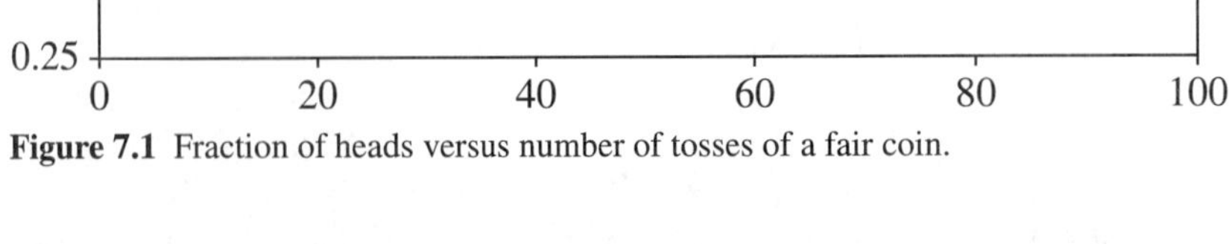

Figure 7.1 Fraction of heads versus number of tosses of a fair coin.

Table 7.1 Probability model for one throw of a balanced white die.

Outcome	Probability
$W = 1$	1/6
$W = 2$	1/6
$W = 3$	1/6
$W = 4$	1/6
$W = 5$	1/6
$W = 6$	1/6

occurred in the first 100 tosses. If this experiment were to be repeated, the graph would look different, and a different number of heads would be likely to occur in the first 100 tosses. As the length of the experiment increased, the fraction of heads would settle down to 0.5. Figure 7.1 was not created by tossing a coin; instead, Excel was used to simulate coin tossing. Problem 25 on page 242, indicates how to accomplish that.

One Balanced Die

Imagine that we are about to throw one perfectly balanced white die on a table. This die has six faces, which are numbered 1 through 6. Since this die is perfectly balanced, it is equally likely to come to rest with each face pointing up. Thus, each face has a probability of 1/6 of being "up" when the die comes to rest. The phrase "$W = 5$" means that this white die comes to rest with the face marked 5 pointing up. Our probability model of this throw is given in Table 7.1.

Let us recall that no model is an exact representation of reality; no die is perfectly balanced. The dice that we play with at home are slightly unbalanced because the pips are dug into their faces. The face with one pip dug into it is opposite to the face with six pips dug into it. This shifts the die's center of mass die away from the 6 face and toward the 1 face, making 6 the most likely outcome and 1 the least likely outcome.

In statistics, a probability model is called an **experiment**. In a probability model (or an experiment), the set of all possible outcomes is called the **sample space** and is often denoted by the Greek letter Ω (Omega).

7.4. EVENTS AND THEIR PROBABILITIES

In a probability model, the word "event" has a specific meaning: an **event** is a set (or collection) of outcomes. The **probability** of an event is defined as the sum of the probabilities of the outcomes in it. Suppose we toss our balanced white die. It comes up "odd" if it shows 1, 3, or 5, and the probability of that event is given by

$$\begin{aligned} P(W \text{ is odd}) &= P\{W \text{ equals 1 or 3 or 5}\} \\ &= P(W = 1) + P(W = 3) + P(W = 5) \\ &= 1/6 + 1/6 + 1/6 \\ &= 3/6 = 1/2. \end{aligned}$$

In short, the event "W is odd" occurs with probability of 1/2.

Events enrich the vocabulary of probability, enabling us to describe complicated situations in simple language. To illustrate, we turn our attention to an example.

Problem A (Two Balanced Dice)

Imagine that you toss two balanced dice, one of which is white and the other is red.

- What are the outcomes, and what is the probability of each outcome?
- What is the probability of the event $W = 1$, that is, that the white die comes up a 1?
- What is the probability that the sum S of the pips (dots) on both dice is 3 or less?
- Given $S \leq 3$, what is the probability that $W = 1$?

This section answers the first three questions and then introduces some general ideas that they illustrate. The fourth question will lead us to "conditional events" and "conditional probability," which are the subjects of a separate section.

Two Balanced Dice

For Problem A, we first describe the outcomes and their probabilities. When the two dice are tossed together, there are 36 different outcomes, for there are six possibilities for each of two dice, and $36 = 6 \times 6$. Each die is balanced, and neither die influences the other, so each of these 36 outcomes is equally likely to occur. Table 7.2 lists these outcomes and records the probability of each outcome, which equals $1/36$. In Table 7.2 and throughout, a sequence of three dots (an ellipsis) means "follow the pattern that has been established."

If the white and red dice were not balanced, the list of outcomes in Table 7.2 would not change, but their probabilities would no longer be equal.

For this example, the event $W = 1$ contains six outcomes, which are listed on the left side of Table 7.3. Similarly, the event $S \leq 3$ contains the three outcomes that are listed on the right side of Table 7.3.

In Problem A, each outcome has probability of $1/36$, so Table 7.3 indicates that

$$P(W = 1) = 6 \times 1/36 = 7/36,$$
$$P(S \leq 3) = 3 \times 1/36 = 1/12.$$

Of the four questions posed by Problem A, the first three have been answered.

Table 7.2 Probability model of the throw of two balanced dice, one white, the other red.

Outcome	Probability
$W = 1, R = 1$	1/36
$W = 1, R = 2$	1/36
$W = 1, R = 3$	1/36
$W = 1, R = 4$	1/36
$W = 1, R = 5$	1/36
$W = 1, R = 6$	1/36
$W = 2, R = 1$	1/36
$W = 2, R = 2$	1/36
$\vdots$ $\vdots$	$\vdots$
$W = 6, R = 4$	1/36
$W = 6, R = 5$	1/36
$W = 6, R = 6$	1/36

Table 7.3 The event $W = 1$ and the event $S \leq 3$.

Outcomes in the Event $W = 1$		Outcomes in the Event $S \leq 3$
$W = 1, R = 1$	$W = 1, R = 4$	$W = 1, R = 1$
$W = 1, R = 2$	$W = 1, R = 5$	$W = 1, R = 2$
$W = 1, R = 3$	$W = 1, R = 6$	$W = 2, R = 1$

Terminology

The fourth question in Problem A deals with conditional probability. Before tackling it, we will introduce some terminology and ideas that the first three questions serve to illustrate.

Throughout this chapter, the letters A and B stand for two events in the same probability model. In probability models, the adverb "not" and the conjunctions "and" and "or" have specialized meanings that are described in Table 7.4.

To illustrate the definitions in Table 7.4, we interpret A as the event $W = 1$ and B as the event $S \leq 3$. The event $\overline{W = 1}$ contains the 30 outcomes in which W is not equal to one. Each of these 30 outcomes occurs with probability of $1/36$, so

$$P(\overline{W = 1}) = 30/36.$$

For our example, the event $W = 1$ and $S \leq 3$ contains the outcomes that are listed in both events in Table 7.3. There are two such outcomes. Consequently,

$$\begin{aligned} P(W = 1 \text{ and } S \leq 3) &= P(W = 1, R = 1) + P(W = 1, R = 2) \\ &= 2 \times 1/36 = 1/18. \end{aligned}$$

Similarly, the event $W = 1$ or $S \leq 3$ includes the seven outcomes that are listed at least once in Table 7.3. Thus,

$$P(W = 1 \text{ or } S \leq 3) = 7 \times 1/36 = 7/36.$$

Evidently, the phrases "A and B" and "B and A" describe the same event. The event A and B is sometimes written as A, B. The same event, A and B, is also written as $A \cap B$; here, "$\cap$" is the **intersection** symbol from set theory. The event A and B can be written in any of these six ways:

$$A \text{ and } B \quad B \text{ and } A \quad A, B \quad B, A \quad A \cap B \quad B \cap A$$

Occasionally, we speak of $P(A \text{ and } B)$ as the **joint probability** of the events A and B.

Similarly, the phrases "A or B" and "B or A" describe the same event. The event A or B is sometimes written as $A \cup B$; here, "$\cup$" is the **union** symbol from set theory. The event A or B can be written in these four ways:

$$A \text{ or } B \quad B \text{ or } A \quad A \cup B \quad B \cup A.$$

In set theory, the event $\overline{A}$ is called the **complement** of A and is sometimes written A^c instead of $\overline{A}$.

Table 7.4 Descriptions of three events.

Event	Outcomes in the Event
$\overline{A}$, read as "not A"	Each outcome that is not in A
A and B, read as "A and B"	Each outcome that is both in A and in B
A or B, read as "A or B"	Each outcome that is in A or in B or in both

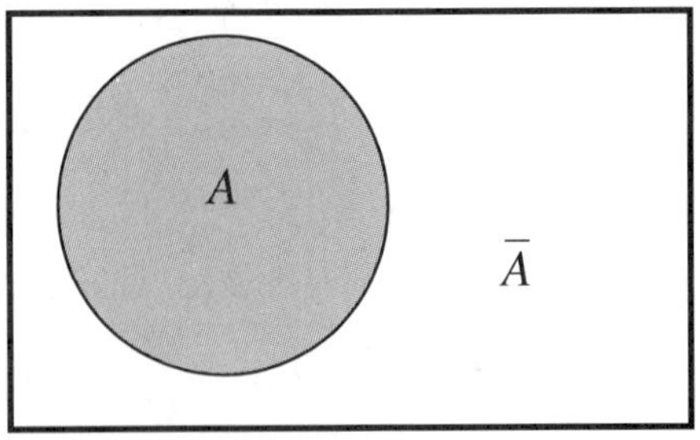

Figure 7.2 A Venn diagram in which the event A is depicted as the shaded circle and $\bar{A}$ is the unshaded portion of the rectangle.

Venn Diagrams

Simple pictures help us to visualize the events $\bar{A}$, $A \cup B$, and $A \cap B$ and to compute their probabilities. These pictures are called **Venn diagrams**. Figure 7.2 is a Venn diagram in which the set Ω of all outcomes is represented as the area enclosed by the rectangle, the event A is represented as the round shaded area, and the event $\bar{A}$ is represented as the unshaded portion of the rectangle, the part that excludes event A.

With Figure 7.2 in view, we assert that

$$P(A) + P(\bar{A}) = 1. \tag{7.1}$$

Why? The set $\bar{A}$ consists of those outcomes that are not in A, and the sum of the probabilities of all of the outcomes equals 1.

Equation (7.1) provides a second way in which to compute $P(A)$. It is to enumerate the outcomes in $\bar{A}$, compute $P(\bar{A})$, and set $P(A) = 1 - P(\bar{A})$. This formula works well when the outcomes in $\bar{A}$ are easier to enumerate than the outcomes in A.

Figure 7.3 is also a Venn diagram. In this Figure, the event A is depicted as a shaded circle, and the event B is depicted as a different shaded circle. The intersection of these circles depicts the event $A \cap B$. The shaded region depicts the event $A \cup B$.

The naïve way in which to compute $P(A \cup B)$ is to enumerate the outcomes that are in either or both of the events A and B and to sum their probabilities. There are other ways in which to compute $P(A \cup B)$, and they can be easier to implement.

A Formula for *P*(*A* or *B*)

The Venn diagram in Figure 7.3 suggests ways in which to compute $P(A \cup B)$ without enumerating unions. We assert that

$$P(A \cup B) = P(A) + P(B) - P(A \cap B). \tag{7.2}$$

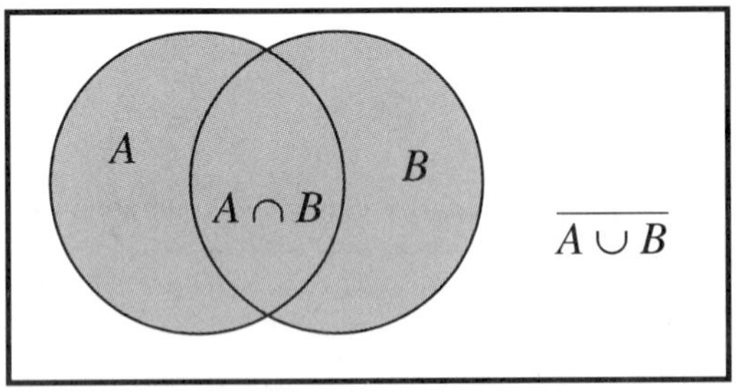

Figure 7.3 A Venn diagram. The event $A \cup B$ is shaded, and the event $\overline{A \cup B}$ is the unshaded portion of the rectangle.

Why? The shaded region in Figure 7.3 consists of the outcomes in $P(A \cup B)$. The expression $P(A) + P(B)$ double-counts the probability of each outcome in the event $A \cap B$. Hence, subtracting $P(A \cap B)$ corrects the count.

To illustrate Equation (7.2), we return to Problem A. We had used Table 7.3 to enumerate the seven outcomes in the event $W = 1$ or $S \le 3$, thereby learning that $P(W = 1 \text{ or } S \le 3) = (7)(1/36) = 7/36$. It is easy to see that $P(W = 1 \text{ and } S \le 3) = 2/36$. For this reason, Equation (7.2) gives

$$\begin{aligned} P(W = 1 \text{ or } S \le 3) &= P(W = 1) + P(S \le 3) - P(W = 1 \text{ and } S \le 3) \\ &= 6/36 + 3/36 - 2/36 = 7/36. \end{aligned}$$

Disjoint Events

Two events are said to be **disjoint** if they have no outcomes in common. If events A and B are disjoint, then $A \cap B = \emptyset$, where $\emptyset$ denotes the empty set. The probability of any event is the sum of the probabilities of the outcomes in it. As there are no outcomes in the empty set, $P(\emptyset) = 0$. Note from Equation (7.2) that

$$P(A \cup B) = P(A) + P(B) \quad \text{if} \quad A \cap B = \emptyset,$$

so the probabilities of disjoint events add.

Other Formulas for *P*(*A* or *B*)

To discover another formula for $P(A \cup B)$, we reexamine the shaded region in Figure 7.3. The circle at the left depicts the event A. The remainder of the shaded region is crescent-shaped; it depicts the event $\overline{A} \cap B$ in which A does not occur and B does occur. It is evident from Figure 7.3 that the events A and $\overline{A} \cap B$ are disjoint. Their union is the shaded region in Figure 7.3. The probabilities of disjoint events add, so the probability of the shaded region is given by

$$P(A \cup B) = P(A) + P(\overline{A} \cap B). \tag{7.3}$$

The same Venn diagram provides a third formula for $P(A \cup B)$, and it too is handy. Note in Figure 7.3 that the outcomes that are *not* in the event $A \cup B$ are those that are not in A and are not in B. In other words,

$$\overline{A \cup B} = \overline{A} \cap \overline{B}.$$

We recall from Equation (7.1) that the probability of an event and the probability of its complement sum to 1. We have just seen that the complement of the event $A \cup B$ is the event $\overline{A} \cap \overline{B}$, so

$$P(A \cup B) = 1 - P(\overline{A} \cap \overline{B}). \tag{7.4}$$

Sometimes it is easiest to compute $P(A \text{ or } B)$ by enumeration, sometimes from Equation (7.2), sometimes from Equation (7.3), and sometimes from Equation (7.4).

Computing *P*(*A* or *B* or *C*)

Equations (7.2), (7.3), and (7.4) generalize to the union of any number of events. Let us consider the union of three events. The shaded region in Figure 7.4 depicts the event $A \cup B \cup C$.

To adapt Equation (7.3) to the union of three events, we note that the shaded region in Figure 7.4 consists of seven pieces. Let us observe that:

- The bottom-most piece depicts the event $\overline{A} \cap \overline{B} \cap C$ in which events A and B do not occur but C does.

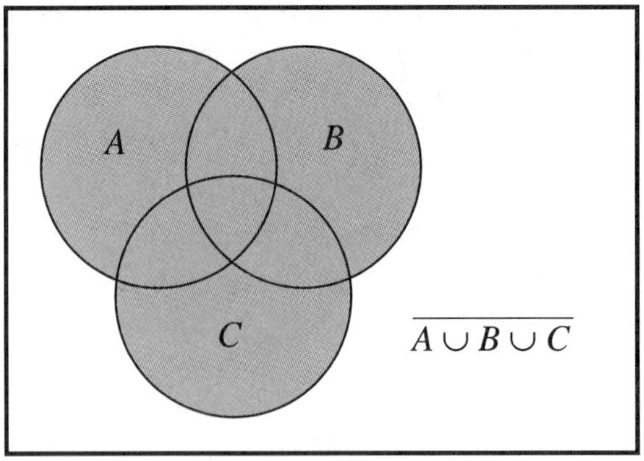

Figure 7.4 Venn diagram for the event $A \cup B \cup C$.

- When taken together, the other six pieces depict the event $A \cup B$, and Equation (7.3) gives $P(A \cup B) = P(A) + P(\overline{A} \cap B)$.

The events $A \cup B$ and $\overline{A} \cap \overline{B} \cap C$ are disjoint, so their probabilities add. Thus,

$$P(A \cup B \cup C) = P(A) + P(\overline{A} \cap B) + P(\overline{A} \cap \overline{B} \cap C). \tag{7.5}$$

Equation (7.5) adapts Equation (7.3) to the union of three events. Can you guess its generalization to the union of four events?

To adapt Equation (7.4) to the union of three events, we observe that the unshaded region in Figure 7.4 contains the outcomes that are not in the event $A \cup B \cup C$. Note from Figure 7.4 that

$$\overline{A \cup B \cup C} = \overline{A} \cap \overline{B} \cap \overline{C}$$

because the set of outcomes that are not in $A \cup B \cup C$ equals the intersection of the outcomes that are not in A, not in B and not in C. The probability of an event and its complement sum to 1, so

$$P(A \cup B \cup C) = 1 - P(\overline{A} \cap \overline{B} \cap \overline{C}). \tag{7.6}$$

Equation (7.6) adapts Equation (7.4) to three events.

To compute the probability of the union of two or more events, we must avoid "double counting." This can be tricky. Equations (7.2) through (7.6) provide ways to compute these probabilities. In each case, a Venn diagram has helped.

7.5. CONDITIONAL EVENTS AND CONDITIONAL PROBABILITY

Of the questions posed by Problem A, one has not yet been answered. We have not yet learned how to compute the probability that $W = 1$ given $S \leq 3$. To do this, we introduce the idea of a condition. When we speak of an event B as a **condition**, we mean that one of the outcomes in B has occurred, but we do not know which of the outcomes in B has occurred, and this changes the probabilities, as follows:

- The probability of each outcome that is not in B equals zero.
- The probability of each outcome that is in B is divided by $P(B)$, so that the probabilities of the outcomes in B sum to 1.0.

Effectively, the outcomes that are not in B are discarded, and the outcomes that are in B have their probabilities re-scaled so that they sum to 1.

In probability models, the vertical line has special meaning. The phrase "$A|B$" is read as "A given B." The phrase "$A|B$" describes the **conditional event** that A occurs given the condition that B occurs. For instance, $W = 1|S \leq 3$ describes the event in which the white die is a 1 given the condition that the sum S is 3 or less. The **conditional probability** $P(A|B)$ is defined by the equation

$$P(A|B) = \frac{P(A \text{ and } B)}{P(B)}. \tag{7.7}$$

Equation (7.7) reflects the fact that the event $A|B$ ignores the outcomes that are in $A \cap \overline{B}$ and re-scales the outcomes in $A \cap B$ by dividing their probabilities by $P(B)$.

Intuition

To develop our intuition for Equation (7.7), we return to Problem A and compute some conditional probabilities. Let us recall from earlier calculations (and tables) that

$$P(W = 1) = 6/36 \quad P(W = 1 \text{ and } S \leq 3) = 2/36 \quad P(S \leq 3) = 3/36. \tag{7.8}$$

To compute the probability that $W = 1$ given the condition $S \leq 3$, we use Equations (7.7) and (7.8) to get

$$P(W = 1|S \leq 3) = \frac{P(W = 1 \text{ and } S \leq 3)}{P(S \leq 3)} = \frac{2/36}{3/36} = 2/3.$$

Does the equation $P(W = 1|S \leq 3) = 2/3$ make sense? We recall from Table 7.3 that the event $S \leq 3$ consists of three outcomes. These three outcomes are equally likely, and two of them have $W = 1$. So it is natural that $P(W = 1|S \leq 3) = 2/3$.

Similarly, to compute the probability that $S \leq 3$ given the condition $W = 1$, we use Equations (7.7) and (7.8) to get

$$P(S \leq 3|W = 1) = \frac{P(S \leq 3 \text{ and } W = 1)}{P(W = 1)} = \frac{2/36}{6/36} = 1/3.$$

Is it reasonable that $1/3 = P(S \leq 3|W = 1)$? We recall from Table 7.3 that the event $W = 1$ consists of six outcomes, and of these, exactly two have $S \leq 3$. All six outcomes are equally likely, so it makes sense that $P(S \leq 3|W = 1) = 2/6 = 1/3$.

These calculations have indicated that the definition of conditional probability is consistent with the meaning of the word "condition."

A Handy Formula

An important formula is obtained when we clear the denominator in Equation (7.7). Specifically, when we multiply both sides of Equation (7.7) by $P(B)$, we obtain

$$P(A|B)P(B) = P(A \text{ and } B).$$

This equation, like any other, must hold with the roles of A and B reversed, so $P(B|A)P(A) = P(B \text{ and } A)$. The phrases "$A$ and B" and "B and A" describe the same event, so we have verified both parts of the formula

$$P(A \text{ and } B) = P(A|B)P(B) = P(B|A)P(A). \tag{7.9}$$

Equation (7.9) presents a fact that is extremely important. Please memorize the following:

> The joint probability of two events equals the probability of either event times the conditional probability of the other event.

Independent Events

The events A and B are now said to be **independent** of each other if $P(A|B) = P(A)$. Thus, the events A and B are independent if knowing that B will occur provides no information about the probability that A will occur. This definition seems to be asymmetric, but it is not. A bit of algebraic manipulation (that we omit) verifies that independent events have $P(B) = P(B|A)$. The notion of independent events is also important enough to memorize:

> Two events are independent of each other if and only if knowing that one of them has occurred does not affect the probability of the other.

Similarly, the events A and B are said to be **dependent** if $P(A|B) \neq P(A)$. This also looks asymmetric, but a bit of algebraic manipulation would verify that dependent events have $P(B) \neq P(B|A)$. Thus, two events are dependent if knowing that one of them has occurred affects the probability of the other.

To see whether the events $W = 1$ and $S \leq 3$ are independent or dependent, we recall from previous calculations that

$$P(W = 1|S \leq 3) = 2/3 \quad \text{and} \quad P(W = 1) = 1/6.$$

Since $P(W = 1|S \leq 3) \neq P(W = 1)$, the events $W = 1$ and $S \leq 3$ are dependent. Is it reasonable that the events $W = 1$ and $S \leq 3$ are dependent? Of course. Knowing that $S \leq 3$ increases the likelihood that $W = 1$.

Are the events $W = 1$ and $R = 5$ independent or dependent? The definition of conditional probability gives

$$P(W = 1|R = 5) = \frac{P(W = 1, R = 5)}{P(R = 5)} = \frac{1/36}{6/36} = 1/6.$$

We also know that $P(W = 1) = 1/6$. Thus, knowing that $R = 5$ does not change the probability that $W = 1$. These two events are independent.

Is it reasonable that the event $W = 1$ and the event $R = 5$ are independent? Yes! For these events to be dependent, the dice would need to be "watching" each other.

Tests for Independent Events

Let A and B be independent events. Intuitively, do you feel that $P(A|\overline{B}) = P(A)$? If so, you are right. If two events are independent, knowing that one of them has *not* occurred has no effect on the probability of the other. Table 7.5 presents eight tests for independence. If any one of these tests holds, they all hold, and the events are independent. If one these tests fails, they all fail, and the events are dependent.

Table 7.5 provides eight different tests for the independence of two events, but it does not exhaust the possibilities. It is easy to check that the events A and B are independent if and only if

$$P(A \text{ and } B) = P(A)\,P(B).$$

Table 7.5 Eight tests for independence of the events A and B.

$P(A) = P(A\|B)$	$P(B) = P(B\|A)$
$P(A) = P(A\|\overline{B})$	$P(B) = P(B\|\overline{A})$
$P(\overline{A}) = P(\overline{A}\|B)$	$P(\overline{B}) = P(\overline{B}\|A)$
$P(\overline{A}) = P(\overline{A}\|\overline{B})$	$P(\overline{B}) = P(\overline{B}\|\overline{A})$

With this test for independence in view, we ask ourselves: Do the events $W = 1$ and $R = 5$ continue to be independent if the dice are unbalanced? Suppose $P(W = 1) = 1/5$ and $P(R = 5) = 1/7$. If the events $W = 1$ and $R = 5$ were dependent, the joint probability $P(W = 1 \text{ and } R = 5)$ could not equal $(1/5)(1/7)$. For these two events to be dependent, the two dice would have to collude. Inanimate objects do not collude.

7.6. OBJECTIVE AND SUBJECTIVE PROBABILITY

What is the root cause of the uncertainty in our models? Is it randomness that is inherent in nature? Is it a lack of information? Or is it some combination of the two?

Suppose you throw a fair die on a roulette table. The uncertainty in the outcome is due to inherent randomness rather than to lack of information. Hardly anyone thinks that he or she can throw a die so precisely as to affect the chance of its coming to rest with any particular face up. Gambling is one situation in which uncertainty is due purely to inherent randomness. Can you think of another?

Objective uncertainty is due to inherent randomness, whereas **subjective uncertainty** is due to a lack of information. The uncertainties that we face in life tend to be mixtures of the two.

Suppose the weather forecaster announces that the chance of rain tomorrow is 30%. What is meant by this? Is it that the random events that have not yet occurred will, with probability of 0.3, congeal into rain tomorrow? Or is it that in 30% of the days in which weather patterns similar to today's have occurred, rain fell on the next day? More likely, it's the latter, a subjective probability.

This chapter began with a discussion of flips of a coin and rolls of a die. These examples of objective uncertainty provide clear-cut illustrations of outcomes, events, conditional probability, and so forth. They let us focus on the fundamental ideas without the added complication of subjective uncertainty.

A subjective probability is an *opinion*. When the uncertainties are somewhat subjective or entirely subjective, different individuals can have different information. Hence, they can have different assessments of the probabilities. When the uncertainty is subjective, it is proper to speak of *your* probability of an event, not *the* probability of an event. Furthermore, you can use information to revise your probabilities.

A subjective probability can express an opinion about an event that will occur or that has occurred. It can also express an opinion about a fact.

7.7. A MEDICAL TEST

Let us now probe a setting that blends objective and subjective elements. A medical test for a condition takes a sample of a body fluid (such as blood), measures its constituents, and classifies them into one of two categories. One category is called a **positive** outcome (+ for short); this category suggests the presence of the condition. The other category is called a **negative** outcome (− for short), and it suggests the absence of the condition.

Problem B (A Blood Test)

Imagine that you are the president of the National League. Pressure has mounted for you to require your baseball players to submit to tests for the use of intravenous drugs. You estimate that 2% of the baseball players in your league are intravenous drug users. Since drug tests are far from your realm of expertise, you consult with a medical doctor who is an authority on the subject. She informs you that no test does a perfect job of diagnosis, but that there is a well-studied blood test for intravenous drug use. This test has been administered to large numbers of people known to be intravenous drug users as well as to large numbers of people known not to be intravenous drug users. Of the people who are intravenous drug users, 90% test positive; of those who are not intravenous drug users, 99% test negative. You ask yourself, "Is this test accurate enough to administer to the athletes in the National League?"

This problem has only three pieces of data, each of which is a probability. Each probability will be used to specify an equation. The equation

$$P(ID) = 0.02$$

expresses your understanding that a randomly selected ballplayer will be an intravenous drug user (abbreviated ID) with probability 0.02. The blood test is described by two conditional probabilities. The equation

$$P(+|ID) = 0.9$$

records the fact that a positive test outcome occurs with probability of 0.9 when the blood test is administered to an intravenous drug user. The equation

$$Pr(-|\overline{ID}) = 0.99$$

records the fact that a negative test outcome occurs with a probability of 0.99 when the blood test is administered to an individual who is not an intravenous drug user.

Let us emphasize that the events $+|ID$ and $ID|+$ are very different from each other. In a conditional event, what follows the vertical line has occurred. In particular:

$+|ID$ denotes the event in which a person has a positive test outcome given that this person is an intravenous drug user.

$ID|+$ denotes the event in which a person is an intravenous drug user given that this person has a positive test outcome.

Subjective Probability?

Is the statement $P(ID) = 0.02$ objective or subjective? It depends. On one hand, if you could be certain that exactly 2% of the players in your league were intravenous drug users and if this statement was made about a member of your league who has not yet been chosen at random, the statement $P(ID) = 0.02$ would be objective. If you thought that approximately 2% of the players in your league were intravenous drug users, this statement would be somewhat subjective. And if you make this statement about a particular player, say, Joe, it is your opinion about a fact; Joe either is an intravenous drug user or he is not.

Sensitivity and Specificity

In medical parlance, the probability $P(+|ID)$ that the test gives a positive result when applied to a person with the condition is known as the **sensitivity** of the test. A perfectly sensitive test for a condition gives a positive result to everyone who has the condition.

Similarly, the probability $P(-|\overline{ID})$ that the test gives a negative result when applied to a person who does not have the condition is called the **specificity** of the test. A perfectly specific test for a condition gives a negative result to everyone who does not have the condition.

A **perfect test** for intravenous drug users would have $P(+|ID) = 1$ and $P(-|\overline{ID}) = 1$. Perfect tests do not exist in medicine. Actual tests misclassify a few people who have the condition and a few people who do not have it.

The number $P(-|ID)$ is the probability that a person who is an intravenous drug user tests negative. Probabilities sum to one; a person who is an intravenous drug user tests positive or negative, so

$$P(+|ID) + P(-|ID) = 1.$$

We know that $P(+|ID) = 0.9$, so

$$P(-|ID) = 1 - P(+|ID) = 1 - 0.9 = 0.1.$$

The equation $P(-|ID) = 0.1$ states that 1/10th of the ballplayers who are intravenous drug users will test negative; these are "false negatives." A similar calculation gives

$$P(+|\overline{ID}) = 1 - P(-|\overline{ID}) = 1 - 0.99 = 0.01,$$

so this test misclassifies 1/100 th of the ballplayers who are not intravenous drug users; these are "false positives."

Testing Ballplayers

In your role as president of the National League, you are likely to be particularly concerned about the possibility of false positives—about players who test positive but are not intravenous drug users. You want to learn what a positive test outcome means. What is the probability $P(ID|+)$ that a person who tests positive is an intravenous drug user? And what is the probability $P(\overline{ID}|+)$ that a person who tests positive is not an intravenous drug user?

After mulling things over, you hit upon the idea of computing some joint probabilities. From the probabilities that appear above, you calculate:

$$\begin{aligned} P(+, ID) &= P(+|ID)P(ID) &&= 0.9 \times 0.02 = 0.018, \\ P(-, ID) &= P(-|ID)P(ID) &&= 0.1 \times 0.02 = 0.002, \\ P(+, \overline{ID}) &= P(+|\overline{ID})P(\overline{ID}) &&= .01 \times (1 - 0.02) = 0.0098, \\ P(-, \overline{ID}) &= P(-|\overline{ID})P(\overline{ID}) &&= .99 \times (1 - 0.02) = 0.9702. \end{aligned}$$

The Joint Probability Table

You assemble these probabilities into Table 7.6, which is aptly called a **joint probability table**. The numbers inside this table are the probabilities of the joint events that they depict, for example, $0.018 = P(+, ID)$. The probabilities to the right are the row sums, and the probabilities at the bottom are the column sums.

Imagine that you administer this test to a typical athlete. What is the probability of a positive test result? The probability of any event equals the sum of the probabilities of the outcomes in it. To compute $P(+)$, you sum the probabilities of the outcomes in the first column of Table 7.6, getting

$$P(+) = P(+, ID) + P(+, \overline{ID}) = 0.018 + 0.0098 = 0.0278.$$

Table 7.6 Joint probabilities for National League athletes.

	+	−	
ID	0.018	0.002	0.02
$\overline{ID}$	0.0098	0.9702	0.98
	0.0278	0.9722	

Thus, the probability is 0.0278 that a typical National League athlete will test positive. By using the formula for conditional probability, you compute

$$P(ID|+) = \frac{P(ID,+)}{P(+)} = \frac{0.018}{0.0278} = 0.6475$$

$$P(\overline{ID}|+) = \frac{P(\overline{ID},+)}{P(+)} = \frac{0.0098}{0.0278} = 0.3525.$$

What do these conditional probabilities tell you? Given a positive test outcome, the probability that an athlete is an intravenous drug user is roughly $2/3$ Of the individuals who tested positive, roughly $2/3$ would be intravenous drug users. But of the individuals who test positive, roughly $1/3$ would *not* be intravenous drug users. This test is dangerous. This test might be worth doing if athletes who tested positive were required to take further tests and without any publicity.

Testing for a Rare Event

The result we obtained in Problem B is typical of tests for rare events, for example, a test for colon cancer that is administered to a 50-year-old male or a test for the HIV antibody that is administered to a pregnant woman. In tests for rare events, a positive outcome on a single test can be inconclusive.

7.8. BAYES' RULE

The equation $P(ID) = 0.02$ records your probability that a randomly selected athlete is an intravenous drug user. If that athlete tests positive, the subjective probability $P(ID|+)$ that he is an intravenous drug user is updated using the formula for conditional probability. The process of using information to update subjective probabilities is known as **Bayes' rule** (see Figure 7.5).

A sense of time is implicit in Figure 7.5. Initially, before event B is observed, you have an estimate $P(A)$ of the probability of the event A. After event B is observed, you have a revised estimate $P(A|B)$ of the probability of the event A. In this context, $P(A)$ is called the **prior** probability, and $P(A|B)$ is called the **posterior** probability.

In Bayes' rule, an easy way to compute $P(A|B)$ is through a joint probability table, such as Table 7.6. That table implements this calculation:

$$P(A|B) = \frac{P(A \cap B)}{P(B)} = \frac{P(B \cap A)}{P(B \cap A) + P(B \cap \bar{A})},$$

$$= \frac{P(B|A)P(A)}{P(B|A)P(A) + P(B|\bar{A})P(\bar{A})}.$$

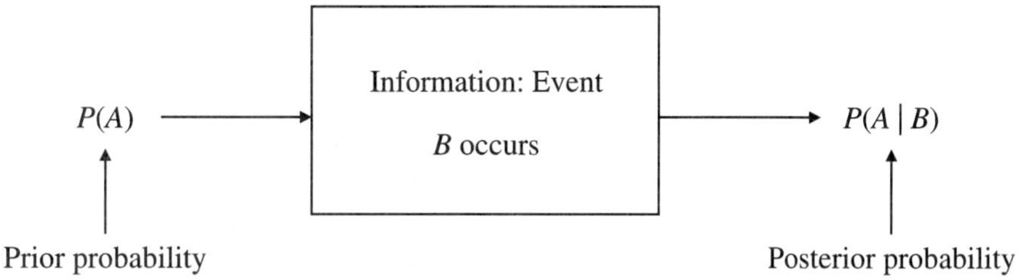

Figure 7.5 Bayes' rule updates the probability of the event A, given the information in the event B.

This formula for $P(A|B)$ may look formidable, but it contains nothing new. Its numerator is an entry in the joint probability table, and its denominator is a column sum in that table. To check that this is so, substitute $+$ for B and ID for A.

We can use Bayes' rule repeatedly to process two or more pieces of information, provided that the events they describe are independent. When we use Bayes' rule repeatedly, the posterior probability given the earlier information becomes the prior probability for the later information.

7.9. A PROBABILITY TREE

The drug testing example is now used to illustrate a useful diagram that is called a probability tree. A **probability tree** consists of two elements: nodes and arcs. Each node is drawn as a circle, perhaps with a number or letter inside, and depicts a point at which an element of uncertainty is resolved. Each arc is drawn as a line segment and corresponds to a resolution of the uncertainty in the node to its left. Probability trees are also called **chance trees**.

We read a probability tree from left to right. When we are at a node, everything to the left of it is given, equivalently, is known to have occurred. When the uncertainty at that node is resolved, we move to the right, along one of the arcs that emerge rightward from the node. The arcs that emerge to the right of a node represent the events that resolve this uncertainty. These are *conditional* events, the condition being that we are at the node. These conditional events are mutually exclusive and collectively exhaustive. Their probabilities sum to 1.

The Data Tree for the Drug Testing Problem

The data for the drug testing problem are assembled in the probability tree in Figure 7.6. We read this tree, like any probability tree, from left to right. Node 1 resolves the uncertainty as to whether or not an athlete is an intravenous drug user. Two arcs emerge to the right of node 1 and represent the events ID and $\overline{ID}$. Note that these events are mutually exclusive and collectively exhaustive. Their probabilities sum to 1 and are recorded on the tree.

Examine node 2 of Figure 7.6. When we are "at" node 2, we know everything in the path to its left. So we know that the athlete is an intravenous drug user. Node 2 resolves the uncertainty as to whether this athlete tests positive or negative. The upper arc depicts the event $+|ID$, and the lower arc depicts the event $-|ID$. The events $+|ID$ and $-|ID$ are mutually exclusive and collectively exhaustive. Their probabilities sum to 1 and are recorded on their arcs.

Node 3 is much like node 2 except that when we are "at" node 3, the athlete is known not to be an intravenous drug user.

Listed at the right of Figure 7.6 is the event that each end-point represents, along with the probability of this event. The topmost end-point describes the event ID and $+$. We recall

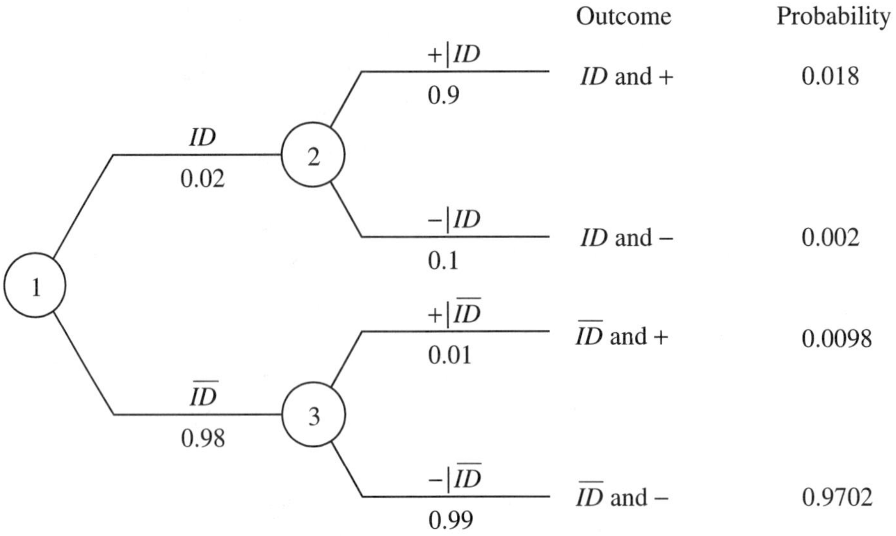

Figure 7.6 Data tree for blood test on National League athletes.

from Equation (7.9) that $P(ID, +) = P(ID)P(+|ID)$. In this case and in general:

> The probability of each end-point in a probability tree equals the product of the probabilities in the path that leads to that end-point.

The tree in Figure 7.6 is logically correct, but it is not chronological. As president of the National League, you do not know whether individual athletes are intravenous drug users.

The Chronological Tree

The chronological tree is the **flipped** version of the tree in Figure 7.6, with the test result first and the conditions reversed. The tree in Figure 7.7 is chronological. In this figure, node a resolves the uncertainty about the outcome of a test on a randomly selected athlete. Being at node b means that the test outcome is +. Node b resolves the uncertainty about whether a person who tests positive is an intravenous drug user.

How can we assign probabilities to the flipped tree? The end-points of the flipped tree depict the same outcomes as in the original tree, with the same probabilities. (Recall that "*A* and *B*" and "*B* and *A*" are the same event.) This specifies the probability of each outcome in Figure 7.7. The probability $P(+)$ equals the sum of the probabilities of the two outcomes in that event.

$$P(+) = P(+ \text{ and } ID) + P(+ \text{ and } \overline{ID}) = 0.018 + 0.0098 = 0.0278.$$

To compute $P(ID|+)$, we use the conditional probability formula, getting

$$P(ID|+) = \frac{P(ID, +)}{P(+)} = \frac{0.018}{0.0278} = 0.6475.$$

The other probabilities in Figure 7.7 are computed in similar ways.

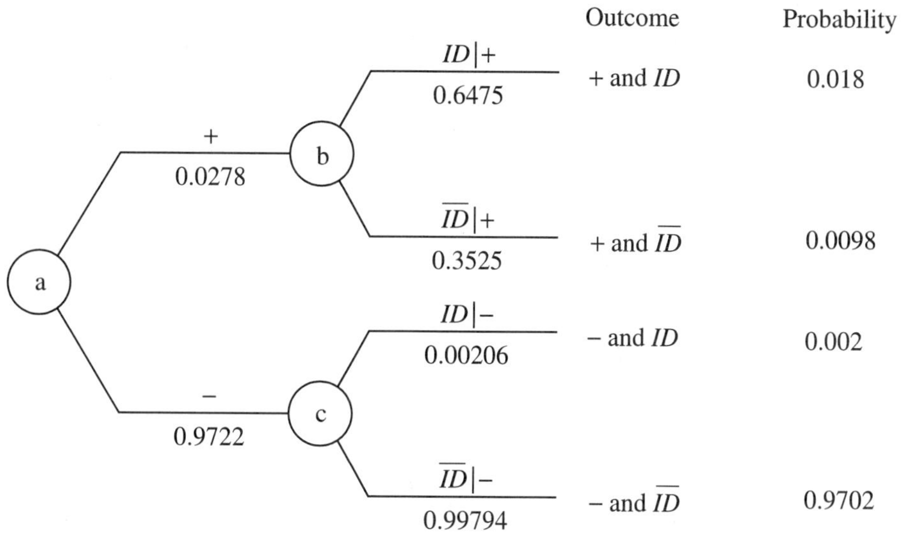

Figure 7.7 Chronological (flipped) tree for blood test on National League athletes.

The Value of a Probability Tree

The probability trees in Figures 7.6 and 7.7 contain nothing new. The information in them can be found from the joint probability table. What good are they?

The mathematical description of a conditional event and a conditional probability can be a bit obscure. In a probability tree, the condition is natural. It is that you know the information to the left of the node. The arcs that emerge rightward from a node resolve an element of uncertainty. These arcs depict conditional events that are mutually exclusive and collectively exhaustive. Their conditional probabilities sum to 1.

In addition, a probability tree is a clear visual representation of an uncertain situation. It is easy to understand. Figures 7.6 and 7.7 describe the drug testing problem so clearly that a person who is not schooled in probability can grasp it. Probability trees will appear throughout this text because they represent uncertain quantities so clearly.

7.10. A SPREADSHEET

The calculations entailed by the drug testing problem are easy to execute on a spreadsheet. Table 7.7 shows how. The data for Problem B appear in cells B4, D4, and C5; the other numbers in this spreadsheet are functions of these data. Column F displays these functions. Dragging the function in cell C10 computes the joint probabilities, and dragging the function in cell C17 computes the posterior probabilities.

Conditional probability is a tricky concept. It's easy to make mistakes—getting the condition backwards, for instance. Before you work with a spreadsheet, be sure you have the idea of conditional probability under your belt. That said, spreadsheets are very handy. Table 7.7 automates the arithmetic and facilitates a sensitivity analysis. To see what happens if 3% of the athletes are intravenous drug users, you need only change the number in cell D4 from 0.02 to 0.03, for instance.

Table 7.7 Spreadsheet for the drug testing problem.

	A	B	C	D	E	F
2		**Conditional probabilities**				
3	**event**	**P(+ | event)**	**P(- | event)**	**P(event)**		= 1 - B4
4	ID	0.9	0.1	0.02		
5	not ID	0.01	0.99	0.98		= 1 - D4
6						= 1 - C5
7						
8		**Joint probability table**				
9	**event**	**P(+ , event)**	**P(- , event)**	**P(event)**		= C4 * $D10
10	ID	0.018	0.002	0.02		= $D4
11	not ID	0.0098	0.9702	0.98		
12		0.0278	0.9722			=C$10+C$11
13						
14						
15		**Posterior probabilities**				
16	**event**	**P(event | +)**	**P(event | -)**			
17	ID	0.6475	0.0021			= C10/C$12
18	not ID	0.3525	0.9979			

7.11. THE PROBABILITY TREE AS A COMPUTATIONAL TOOL

In an earlier section, we used a probability tree to describe a model of uncertainty. We have "flipped" a probability tree in a way that processes information. A probability tree can also be used to compute the probability of a complicated event. To see how, we pose the question in:

Problem C (A Bridge Hand)

A deck of 52 playing cards contains four aces—the Ace of Spades, Hearts, Diamonds, and Clubs, which we denote as *AS*, *AH*, *AD*, and *AC*, respectively. You shuffle the deck thoroughly and then deal yourself a "hand" that consists of 13 cards. What is the probability that your hand contains at least one ace?

We need to compute the probability of the event in which your hand contains an ace. This event, $AS \cup AH \cup AD \cup AC$, is the union of four simpler events. We will use the probability tree in Figure 7.8 to compute its probability. Node 1 resolves the uncertainty as to whether you have the Ace of Spades. There are 13 positions in your hand and 52 cards in the deck, so the probability that you have the Ace of Spades equals 13/52. Node 2 in this tree resolves the uncertainty as to whether you have the Ace of Hearts, given that you do not have the Ace of Spades. The Ace of Spades is not in your hand; the probability that you have the Ace of Hearts given that you do not have the Ace of Spades equals 13/51 because you have 13 cards and there are 51 cards unaccounted for, the Ace of Spades being somewhere else. Node 3 resolves the uncertainty as to whether you have the Ace of Diamonds, given that you have neither the Ace of Spades nor the Ace of Hearts. And so forth.

Each end-point in Figure 7.8 depicts an event. All such events are mutually exclusive and collectively exhaustive, and their probabilities sum to 1. The probability of reaching each end-point equals the product of the probabilities in the path that leads to it. In the top four

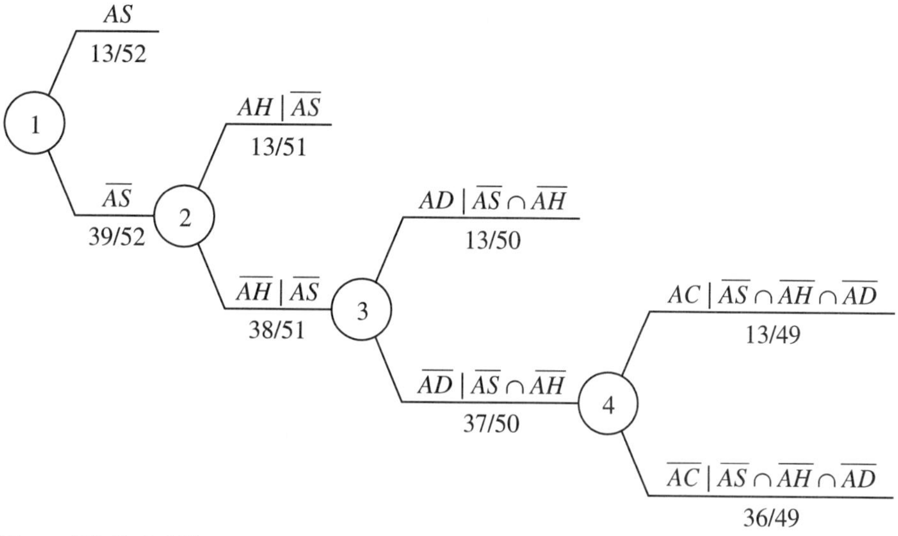

Figure 7.8 Probability tree to compute P (at least one ace).

end-points, you have at least one ace; in the bottom end-point, you do not. So Figure 7.8 demonstrates that

$$P(\text{at least one ace}) = 13/52 + 39/52 \cdot 13/51 + 39/52 \cdot 38/51 \cdot 13/50 + 39/52 \cdot 38/51 \cdot 37/50 \cdot 13/49 = 0.696, \quad (7.10)$$

$$P(\text{no ace}) = 39/52 \cdot 38/51 \cdot 37/50 \cdot 36/49 = 0.304, \quad (7.11)$$

where both probabilities are rounded to three significant digits. Equation (7.10) specifies the probability that you have an ace. Actually, so does Equation (7.11) because

$$P(\text{at least 1 ace}) = 1 - P(\text{no ace}) = 1 - 39/52 \cdot 38/51 \cdot 37/50 \cdot 36/49 = 1 - 0.304 = 0.696. \quad (7.12)$$

The expressions in Equations (7.10) and (7.12) may look different, but they are equal. In fact, both expressions are familiar. In Equation (7.10), we generalized Equation (7.5) to four events. In Equation (7.12), we generalized Equation (7.6) to four events. This suggests, correctly, that we could have used a Venn diagram to solve Problem C. We used a probability tree because the tree makes clear what condition holds at each node and what the probabilities are.

7.12. A PROBABILITY TREE TO IMPLEMENT BAYES' RULE

In this section, a probability tree is used to implement Bayes' rule. Let us consider:

Problem D (A Thief)

A theft has occurred, and either Matt or Wendy is the thief. You know that they did not collaborate. You feel that Matt is somewhat more likely to be the thief and that he is the thief with probability of 0.6. You estimate that each of them has a 90% chance of passing a lie detector test if he or she is innocent and that each of them has a 20% chance of passing a lie detector test if he or she is guilty. You feel that the test results on the two individuals are independent of each other. Suppose Matt fails his lie detector test and that Wendy passes hers. Given this information, what is the probability that Matt is the thief?

Let M designate the event in which Matt is the thief, and let W designate the event in which Wendy is the thief. The events M and W are mutually exclusive (Matt and Wendy did

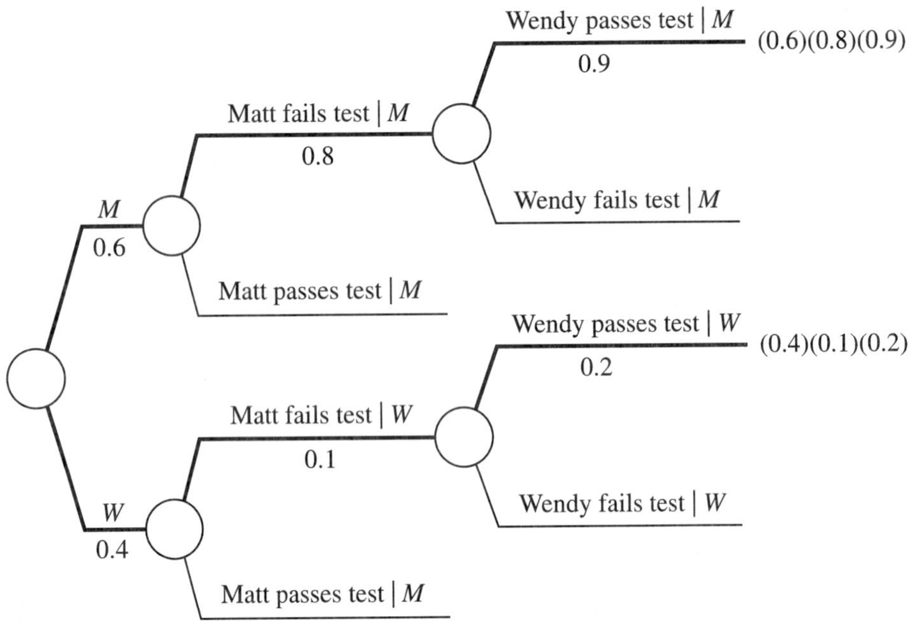

Figure 7.9 Probability tree to process (in thick lines) the event B in which Matt fails his lie detector test and Wendy passes hers.

not collaborate) and are collectively exhaustive (one of them is the thief). Prior to application of the lie detector tests, your subjective probabilities are $P(M) = 0.6$ and $P(W) = 0.4$.

Let B designate the event in which Matt fails his lie detector test and Wendy passes hers. We need to compute the posterior probability $P(M|B)$. The event B forms the dark lines in the tree in Figure 7.9.

With B as the event in which Matt fails his test and Wendy passes hers, it is evident from Figure 7.9 that

$$\begin{aligned} P(M \text{ and } B) &= (0.6)(0.8)(0.9) = 0.432, \\ P(W \text{ and } B) &= (0.4)(0.1)(0.2) = 0.008, \\ P(B) &= 0.432 + 0.008 = 0.440, \\ P(M|B) &= \frac{P(M \text{ and } B)}{P(B)} = \frac{0.432}{0.440} = 0.982. \end{aligned}$$

Thus, the posterior probability that Matt is the thief equals 0.982.

The probability $P(M|B)$ can also be calculated from a joint probability table or from a spreadsheet. The advantage of a probability tree is that it clarifies the conditional probabilities.

7.13. REVIEW

Probability is the language of uncertainty, and like any language, this one has a substantial vocabulary. To use probability, you must become familiar with its vocabulary, which includes these terms: probability model, outcome, mutually exclusive, collectively exhaustive, event, union, intersection, complement, disjoint events, conditional event, conditional probability, independent events, dependent events, joint probability table, Bayes' rule, prior probability, information, posterior probability, probability tree, and flipped tree.

We've seen that probability can model objective uncertainty, which is due to randomness that is inherent in nature, and subjective uncertainty, which is due to a lack of information.

In the case of subjective uncertainty, different observers can have different assessments of the probabilities.

This chapter introduced two tools for computing the probability of complicated events: the Venn diagram and the probability tree. Within this chapter, conditional probabilities have been computed in three ways—via a joint probability table, a probability tree, and Bayes' rule.

Probability trees will be used in subsequent chapters. A probability tree has these advantages:

- It provides a clear visual picture of the uncertainty that is being modeled.
- It clarifies conditional probability; the condition that holds at each node consists of the information in the path to its left.
- It parses a complicated event into simpler constituents.

The calculations in this chapter have been simple enough to execute by hand. We have also seen how to perform them on a spreadsheet. A spreadsheet lets the computer do the numerical work, and it facilitates sensitivity analysis. A spreadsheet proves to be especially handy in cases where a "drag" can be used, for instance, in the repeated application of Bayes' rule.

Many people find "elementary" probability to be elusive. The probability of a complicated event can be hard to specify, and conditional events can be especially puzzling. That is one reason why we have provided several ways in which to compute things. To develop your intuition for elementary probability, you might tackle some of the problems that accompany this chapter.

7.14. HOMEWORK AND DISCUSSION PROBLEMS

1. A white die and a red die are tossed. Both dice are perfectly balanced. Let w and r denote the numbers appearing of the top face of the white and red die, respectively.

(a) What is the probability that $w = r$?

(b) What is the probability that $w < r$?

(c) What is the probability that $w = 3|w < r$?

(d) Are the events $w = 3$ and $w < r$ dependent or independent?

(e) Are the events $w = 3$ and $w \neq r$ dependent or independent?

2. Urn A contains 4 white balls and 8 black balls. Urn B contains 9 white balls and 3 black balls. A die will be tossed. If the die comes up a 1 or 2, a ball is selected from Urn A. If the die comes up 3–6, a ball is selected from Urn B.

(a) Draw a probability tree that depicts this experiment.

(b) Find the probability that a white ball will be drawn.

(c) If a white ball is drawn, find the probability that it was taken from Urn A.

3. In a city, 87% of the households have TVs, 45% have stereos, and 38% have both. What is the probability that a randomly selected household has at least one of these appliances?

4. **(Independence)** Suppose that $P(A) = P(A|B)$. Demonstrate that $P(B) = P(B|A)$.

5. **(Independence)** Suppose that $P(A) = P(A|B)$. Demonstrate that $P(A) = P(A|\overline{B})$. *Hint:* Start with $P(A) = P(A \cap B) + P(A \cap \overline{B})$, and use Equation (7.9) twice.

6. On a slow day, an automobile dealer classified sales and leases for the past year and counted the number in each of six categories. Assume that these data typify the dealer's business.

Type of Car	Buy-Cash	Buy-Finance	Lease
New	22	15	34
Used	51	18	0

(a) What is the probability that the next car to be sold or leased is new?

(b) What is the probability that the next car to be sold is new and is financed?

(c) Excluding leases, do you think the type of car sold is dependent of the method of payment?

7. (Venn Diagram)

(a) Copy Figure 7.4 and shade the event $(A \cup B) \cap C$.

(b) Copy Figure 7.4 and shade the event $A \cup (B \cap C)$.

(c) Copy Figure 7.4 and shade the event $(A \cup \overline{B}) \cap C$.

8. (The Analog of Equation (7.4) for the Union of Three Events) Parts (a)–(c) relate to Figure 7.4, which depicts three events.

(a) Copy Figure 7.4. Inside each of its seven shaded regions, write the number of times the outcomes in this region are counted by $P(A) + P(B) + P(C)$.

(b) Copy Figure 7.4. Inside each of its shaded regions, write the number of times the outcomes in each region are counted by $P(A \cap B) + P(A \cap C) + P(B \cap C)$.

(c) Justify the following formula:

$$P(A \cup B \cup C) = P(A) + P(B) + P(C) - P(A \cap B)$$
$$- P(A \cap C) - P(B \cap C) + P(A \cap B \cap C).$$

(d) Guess a generalization of the above to the union of four events, but do not prove it. *Remark:* A relatively simple proof would proceed by induction, for example, by using part (c) with C replaced by $C \cup D$.

(e) Generalize Equation (7.6) to $P(A \cup B \cup C \cup D)$.

(f) Which formula is simpler, your answer to part (d) or to part (e)?

9. (Evidence?) In a widely publicized court proceeding, a man was tried for the murder of his wife, who had been stabbed to death. The man was a known wife-beater. The man's lawyer asserted that the fact that the defendant was a known wife beater was not relevant because only 2% of the men who beat their wives subsequently murder them. Relate this assertion to conditional probability. By the way, in about two-thirds of the convictions of husbands for the murder of their wives, the husbands had a history of wife beating.

10. (Independent Events) The dice in Problem A (two balanced dice) are balanced. Devise a pair of unbalanced dice for which the events $W = 1$ and $S \leq 3$ are independent. *Hint:* There are many ways to do this; an easy way is to let R be balanced and to let W be balanced, except that $P(W = 1)$ exceeds $1/6$.

11. (Pairwise Independent Events) A fair coin is tossed twice. Event A is that the first toss is a heads. Event B is that the second toss is a heads. Event C is that exactly one of the two tosses is a heads.

(a) Prove or disprove: The events A and B are independent of each other; the events A and C are independent of each other; and the events B and C are independent of each other.

(b) Prove or disprove: All three events are independent of each other.

12. (A Probability Tree) Poker is played with a normal 52-card deck. In Poker, the ace can count as "high" or "low," but the other cards are in their natural order, 2, 3, . . . , 10, J, Q, K. A five-card poker hand is called a "straight" if the cards are contiguous. The "lowest" straight is A-2-3-4-5, and the "highest" is 10-J-Q-K-A. K-A-2-3-4 is not a straight because the Ace must be either high *or* low. You have been dealt a five-card poker hand. Use a probability tree to compute:

(a) The probability that this hand is the straight 6-7-8-9-10.

(b) The probability that this hand is any straight.

13. (Craps) Two balanced dice are rolled. What is the probability that the sum of their pips equals 7? *Hint:* A probability tree might help.

14. (**Birthdays**) In a class of K students, no one was born on February 29, and no one is related to anyone else. Denote as $P(K)$ the probability that these K students have K different birthdays.

(a) Is $P(K + 1) = P(K)(365 - K)/365$ and, if so, why?

(b) Find the smallest class size K for which the probability that two or more students have the same birthday exceeds 0.5. (*Hint:* On a spreadsheet, drag to compute K and $P(K)$ in ascending K.)

(c) In part (a), did you use the fact that no one is related to anyone else?

(d) Adapt the formula in part (a) and redo part (b) for the case of a grammar school in which twins, if any, are put in the same class. Suppose that each pregnancy produces twins (who have the same birthday) with probability of $1/84$. *Hint:* List the K children alphabetically.

15. (**A Famous Game Show**) Having done well in several rounds of a game show, you are confronted with three boxes labeled A, B, and C. The Master of Ceremonies (MC) tells you that a check for \$50,000 is in one of the boxes, and that it is equally likely to be in any one of them. You are told that after you select a box, the MC, who knows which box contains the prize, will open a box that you did not select and that does not contain the prize. You will then be allowed to change your mind.

(a) You choose box A. Before the MC opens a box, what is the probability that the prize is in box A?

(b) You choose box A, and then the MC opens one of the other boxes, which is empty. What now is the probability that the prize is in box A? What is the probability that you will win if you switch from A to the unopened box? Justify your answers by using a probability tree that consists of one node and two branches.

(c) Adapt your analysis to the case of four boxes.

16. (**A Test Market**) Before deciding to launch a new toy, a manufacturer assembles a focus group of children and evaluates their reactions to the toy. The following table summarizes its experience with toys that have subsequently been produced. Assume that these data are typical.

Market Response	Focus Group's Reaction: Favorable	Neutral	Unfavorable
A hit	30	15	4
A flop	10	15	28

(a) Consider a new toy that is brought before a focus group and is produced. What is the probability that this toy will be a hit?

(b) Suppose, in addition, that the focus group reacted favorably. What now is the probability that this toy will be a hit?

(c) The company is ecstatic about the prospects for this particular toy. Before seeking the focus group's reaction, the company's expert feels that the toy has an 80% chance of being a hit. Suppose the focus group reacted favorably. What then is the probability that the toy will be a hit?

17. (**Serial Bayes**) To solve Problem D (A Thief) serially, by repeated use of Bayes' rule (Figure 7.5), do parts (a) and (b).

(a) Find the probability that Matt is guilty given the information that he fails his lie detector test.

(b) With the answer you calculated in part (a) as the prior probability, compute the probability that Matt is guilty given the information that Wendy passes her lie detector test.

(c) Compare your answer with the one computed in the text.

(d) Is it correct to process pieces of information that are dependent serially? Justify your answer, perhaps by reference to a probability tree.

18. (**A Confirming Test**) Suppose, in Problem B (A Blood Test), that there were a second test for IV drug use, one whose sensitivity is 0.95 and whose specificity is 0.99. The second test is 10 times as expensive as the first. The outcomes of the two tests are independent events.

(a) If an athlete tests positive on both tests, what is the probability that this ball player is an intravenous drug user?

(b) Why might the president of the National League administer the more accurate test only to the athletes who tested positive on the less accurate test?

19. (**Polish Paternity**) A court-appointed statistician has determined that 65% of the males who are sued for paternity in Poland are guilty. Mr. Z is being sued in a Polish court by a lady who claims that he is the father of her baby. The baby's blood type is A positive. Each person inherits the blood type of one of that person's parents. The baby's mother does not have A positive blood, so the father does. Ten percent of the males have A positive blood.

(a) True or false: The probability that Mr. Z's blood is type A positive is 0.1.

(b) True or false: The probability that Mr. Z's blood type is A positive is greater than 0.65.

(c) True or false: If Mr. Z is not the father, the probability that his blood type is A positive is 0.1.

(d) Now suppose that Mr. Z's blood is tested and found to be type A positive. What is the probability that he is the father?

(e) For this part, suppose that Mr. Z's blood is tested and found to be type O. What is the probability that he is the father?

20. (**Hemophilia**) Hemophilia is a disease that only men get. To explain why women do not get the disease, we review a bit of genetics.

- Women have two X chromosomes. Men have one X chromosome and one Y chromosome.
- Hemophilia is due to a rare and recessive defect (allele) in the X chromosome.
- A female fetus both of whose X chromosomes have the allele dies in the womb.
- A girl, one of whose X chromosomes has the allele, does not get the disease because hemophilia is "recessive"; this girl is called a "carrier."
- A boy whose X chromosome has the hemophiliac allele (defect) gets the disease.
- Each fetus (unborn child) inherits one chromosome from each parent. Each fetus is equally likely to inherit each of its father's chromosomes and is equally likely to inherit each of its mother's chromosomes.
- Thus, a son inherits his father's Y chromosome and one of his mother's X chromosomes. Similarly, a daughter inherits her father's X chromosome and one of her mother's X chromosomes.

Claire's father is not a hemophiliac, but Claire's brother is. Thus, her brother inherited a defective X chromosome from their mother. Claire's mother has one defective X chromosome.

(a) What is the probability that Claire is a carrier? Is this probability objective or subjective?

(b) Claire gives birth to her first child, a boy, Adam, who is not a hemophiliac. What now is the probability that Claire is a carrier?

(c) Claire gives birth to her second child, a boy, Brent, and he too is not a hemophiliac. What now is the probability that Claire is a carrier?

(d) Claire give birth to her third child, another boy, Drew, and he too is not a hemophiliac. What now?

21. (**Is It Catching?**) A new virus has been identified. It has been transmitted by shared blood, for example, by a transfusion of blood taken from a person who has the virus. This virus has also been transmitted by having sexual intercourse with a person who has it. Whether or not it can be transmitted by social contact is unknown. "Social contact" includes touching, being together, and kissing, but not sex. Medical experts estimate that:

- There is a 50–50 chance that this virus can be transmitted by social contact.
- If the virus can be transmitted by social contact, there is one chance in five that it will be transmitted within a family by social contact within one year, given that one member of the family has it at the beginning of the year.
- Transmissions of this virus within different families are independent of each other.

Let n denote the number of families in which one member is known to be infected with the virus at this moment. Let B_n denote the event in which no transmissions occur within any of these n families within one year. Let C denote the event in which this virus can be transmitted by social contact.

(a) Does the statement $P(C) = 0.5$ describe objective or subjective probability? Why?

(b) Without doing any calculation, guess the smallest value of n such that $P(C|B_n)$ does not exceed 1/100,000. (There is no wrong answer to part (b).)

(c) Compute $P(B_1|C)$. Compute $P(B_2|C)$. Compute $P(B_n|C)$.

(d) Compute $P(B_n|\overline{C})$.

(e) Compute $P(C|B_n)$.

(f) Find the smallest n such that $P(C|B_n) \leq 1/100{,}000$. *Remark:* Part (f) can be done in at least three ways—algebraically, with a drag, or by using Solver.

(g) At the beginning of the year, one member of each of 100 families had the virus. During the course of the year, no transmissions of this virus occurred in any of these families. Qualitatively, what does this say about C?

22. (**High-risk Drivers**) An automobile insurance company's customers fall into two classes. Class S (short for safe) customers, who are 80% of the total, have a probability of 1/12th of filing at least one claim in a given year. Class R (short for risky) customers, who are 20% of the total, have a probability of 1/6th of filing at least one claim in a given year.

(a) A customer files a claim during a certain year. What is the probability that this is a class R customer?

(b) A customer filed a claim in a certain year, no claim in the next year, and one claim in the year after that. What is the probability that this is a class R customer?

(c) The state insurance commissioner allows companies to charge a higher premium to a customer if the probability is at least 0.7 that, over a long period of time, this customer files claims more frequently then does the company's median customer. Is the company authorized to charge a higher premium for next year to a customer who files one claim in the current year? to a customer who files two claims in a three-year period?

(d) A customer has filed claims in each of the first n years he had a policy. Find the smallest value of n such that the company is authorized to charge a higher premium in year $n + 1$.

(e) Can this company profit significantly by differentiating its customers on the basis of their filing patterns?

23. (**False Positives**) To estimate the prevalence of HIV, the State of New York implemented a policy of testing women who gave birth. The test that was used is known as ELISA; it is actually a test for the HIV antibody. Assume that ELISA has a sensitivity of 0.998 and a specificity of 0.998. (Both numbers are high; in particular, ELISA gives more false positives than these numbers indicate.) An article in the *American Journal of Public Health* reports on the administration of the ELISA test to 108,562 women in rural upstate New York who gave birth between November of 1987 and March of 1990. Of these women, 82 tested positive. From this observation, the authors conclude that the prevalence of HIV among rural women in New York state is approximately 0.076% because (82)/(108,562) = 0.00076.

(a) What error did the authors commit?

(b) Approximately how many false positives would this test produce?

(c) What do you suppose was going on?

24. (**Beer**) Each of five males ranked the four most popular American beers from best to worst, with no ties allowed. Suppose that the rankings were mutually independent and completely random, for example, that each ranking is equally likely.

(a) What is the probability that a particular beer received no first-place votes?

(b) What is the probability that at least one beer received no first-place votes?

(c) What is the probability that one of the beers received a majority of the first-place votes? *Hint:* Have the males cast their votes in sequence, and use probability trees. For example, compute the probability

$p(1,1,0,0)$ that males #1 and #2 rank different beers first, and, subsequently, compute the probability $p(2,1,0,0)$ that two of the first three males rank the same beer first.

Remark: The following table indicates how Figure 7.1 was built. This table shows how to use Excel to simulate three flips of a fair coin rather than 100. In each row of this table:

- Column A counts the total number n of coin tosses.
- The Excel function =RAND() in column B returns a randomly generated number that is equally likely to lie anywhere between 0 and 1.
- The IF function in column C interprets numbers below 0.5 as heads (coded 1) and numbers that are 0.5 or greater as tails (coded 0).
- Column D counts the total number of heads in the first n tosses.
- Column E computes the average number of heads in the first n tosses.

	A	B	C	D	E
1					
2	n		1 means Heads	sum	average
3	1	0.3626524	1	1	1
4	2	0.820633	0	1	0.5
5	3	0.1952094	1	2	0.666667
6	↑	↑	↑	↑	↑
7					
8	=A4+1	=RAND()	=IF(B5<0.5,1,0)	=D4+C5	=D5/E5

25. (Simulating Tosses of a Fair Coin)

(a) Use an Excel spreadsheet to simulate 600 flips of a fair coin. Plot an analog of Figure 7.1. *Notes:* The remark that precedes this problem may help. The Appendix may also help; it describes the functions that appear above, and it gives tips on creating charts.

(b) Adapt part (a) to simulate 100 flips of an unbalanced coin whose probability of coming up heads equals 0.55.

26. (Lead Poisoning) A budget of $100,000 has been allocated for testing children in a large city for lead poisoning. You have two tests for lead poisoning. A blood test costs $10 for each administration; its sensitivity and specificity equal 1.0. A urine test costs $2 for each administration; its sensitivity equals 0.8, and its specificity equals 0.9 . Hence, 80% of the children who have lead poisoning get positive test results on the urine test, and 10% of the children who do not have lead poisoning get positive test results on this test. You estimate that 3% of the inner-city children have lead poisoning. Your goal is to allocate the available resources to identify as many of them as possible.

(a) Suppose you administer only the blood test. Approximately how many cases of lead poisoning will it identify?

(b) Suppose you administer the urine test and follow up with the blood test on those children who test positive on the urine test. To approximately how many children can you administer this procedure? Approximately how many cases of lead poisoning will it identify?

(c) Which procedure is preferable? What fraction of the children would need to have lead poisoning for the two procedures to identify approximately the same number of cases?

27. (Orange or Crimson?) At dusk, while crossing the street of a major city, Jon was struck by a hit-and-run vehicle. An eyewitness reported that the vehicle that struck Jon was a taxi, that it was orange, and that it was going the wrong way on a one-way street. The city has two taxi companies. The Crimson company owns 85% of the taxis, each of which is painted crimson. The Orange company owns the remaining 15%, each of which is painted orange. Jon sued the Orange company. In court, the eyewitness's testimony was determined to be totally credible except as to the color of the cab. Crimson and orange cabs look somewhat similar at dusk, and the probability that the eyewitness mistook either for the other is 0.2. In cases like this, the judge must find against the defendant if the "preponderance of the evidence" is that the defendant is the guilty party. This particular judge, Judy, has had a course in decision science, and she interprets preponderance of the evidence to mean a probability of at least 0.7 of guilt.

(a) How does Judge Judy find? Why?

Now, suppose that before Jon brought the lawsuit, he had a chat with his wife, Sue, a smart lawyer. She went to the police department to inquire about reckless driving convictions. Sue learned that, during the past year the Crimson company had 188 reckless driving citations and the Orange company had had 12.

(b) Given that a randomly selected taxi is driven recklessly, what is the probability that it is owned by the Orange company? by the Crimson company?

(c) In Judge Judy's court, who did Sue sue? Did she win? Why?

Chapter 8

Discrete Random Variables

8.1. PREVIEW

In Chapter 7, a probability model was described as a list of outcomes, with a probability for each outcome. A **random variable** is a particular type of probability model, one that assigns a numerical value to each outcome.

Typically, a capital letter (such as X) is used to describe a random variable, and the corresponding lower-case letter (such as x) is used to describe a value that the random variable might take. The probability tree in Figure 8.1 represents a random variable X that can take n different numerical values, which are labeled x_1 through x_n. Figure 8.1 indicates that the random variable X takes the value x_1 with probability p_1, that it takes the value x_2 with probability p_2, and so forth. The outcomes of X are mutually exclusive and collectively exhaustive, so $p_1 + p_2 + \cdots + p_n = 1$.

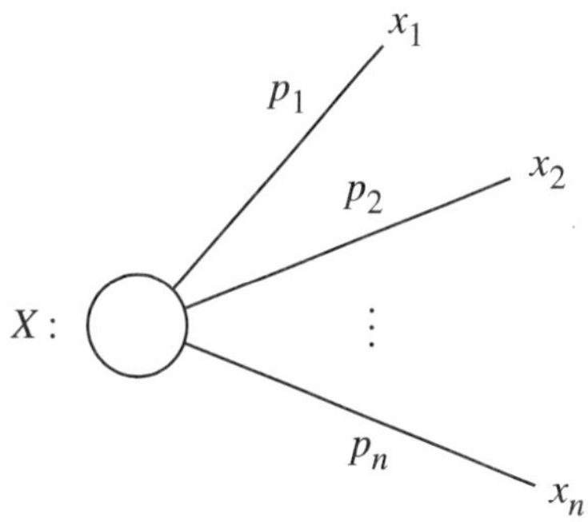

Figure 8.1 The random variable X as a probability tree.

What distinguishes a random variable from an ordinary probability model? A random variable assigns a number to each outcome. In Figure 8.1, for instance, x_1 through x_n are numbers. The term *random variable* is firmly ensconced in the literature. That's a pity, for *uncertain quantity* would have been more descriptive. A random variable is a *quantity* whose value is *uncertain*.

Random variables are so important that two chapters of this book are devoted to them. The present chapter studies the case of a **discrete** random variable X, which means that X takes isolated values, with spaces between them. In a later chapter, our attention will turn to "continuous" random variables, whose values can vary continuously, with no spaces in between.

Why are continuous random variables discussed in a separate chapter? A continuous random variable cannot be described as a list of values, with the probability of each. For it, Figure 8.1 will be meaningless. To handle continuous random variables, we will need to build a different probability model.

Table 8.1 describes the same random variable X that appears in Figure 8.1. Table 8.1 lists the values that X can take and specifies the probability that X takes each value. For instance, the equation $p_1 = P(X = x_1)$ states that X takes the value x_1 with probability of p_1.

The **probability distribution** of a discrete random variable is a list of the values that it might take, along with the probability of each. Table 8.1 presents the probability distribution of X, as does Figure 8.1. The probability distribution of a discrete random variable is often referred to as its **probability mass function**; these terms are synonyms.

8.2. WHAT CAN YOU LEARN FROM THIS CHAPTER?

This chapter introduces the basic properties of discrete random variables, namely:

- The "expectation," as a measure of the center of its probability distribution.
- The "variance," as a measure of the spread in its probability distribution.

Table 8.1 The random variable X.

Value of X	Probability
x_1	$p_1 = P(X = x_1)$
x_2	$p_2 = P(X = x_2)$
$\vdots$	$\vdots$
x_n	$p_n = P(X = x_n)$

- The sum of two or more random variables.
- The expectation and variance of the sum of random variables.
- The normal approximation to the sum of a large enough number of independent random variables.

When a random variable measures profit, expectation is a measure of desirability (high is good), and variance is a measure of undesirability (high is bad). For operational decisions (with small stakes), we advocate using the expectation as a figure of merit. When the stakes are larger, it can be prudent to explore the tradeoff between expectation and variance.

Later in this chapter, our attention turns from general properties of discrete random variables to a few of the discrete random variables that appear most frequently in applications, including the "binomial" and the "Poisson." We will see that the binomial distribution is natural for counting and that the Poisson distribution is the natural model for arrivals at a service facility.

Spreadsheets are used extensively in this chapter. They will help us compute expectations and variances, explore the shapes of probability distributions, and execute calculations that would be cumbersome without them.

8.3. HOW RANDOM VARIABLES ARISE

Very frequently, a random variable arises indirectly from a more detailed probability model. To illustrate, consider the total number N of heads in three flips of a fair coin. The probability model of three flips has eight different outcomes, these outcomes are equally likely, and each outcome occurs with probability of $1/8$. The triplet (H, H, T) describes the outcome in which the first flip is heads (H), the second flip is heads (H), and the third flip is tails (T). Table 8.2 lists each outcome, its probability, and the value that it assigns to the number N of heads.

In Table 8.2, the event $N = 1$ includes three outcomes, and this event occurs with a probability of $3/8$. Table 8.3 records the probability distribution of N, but this table omits the outcomes of the coin-toss experiment that lead to each value of N.

Examples of random variables that summarize more detailed probability models are as follows.

- Suppose a college offers admission to 1200 applicants. The number K of applicants who will accept their offers of admission is a random variable.

Table 8.2 Three flips of a fair coin, with the total number N of heads.

Outcome	Probability	Value of N
(T, T, T)	$1/8$	0
(H, T, T)	$1/8$	1
(T, H, T)	$1/8$	1
(T, T, H)	$1/8$	1
(H, H, T)	$1/8$	2
(H, T, H)	$1/8$	2
(T, H, H)	$1/8$	2
(H, H, H)	$1/8$	3

Table 8.3 Probability distribution of the number N of heads in three flips of a fair coin.

Value of N	$P(N = n)$
0	1/8
1	3/8
2	3/8
3	1/8

- The number C of individuals who will place telephone calls within the next hour is a random variable.
- The total sales revenue R that a supermarket will receive tomorrow is a random variable.

A great deal can be said about these three random variables. Later in this chapter, we will see that, typically, the number K of acceptances has the "binomial" distribution, the number C of callers has the "Poisson" distribution, and the total sales revenue R has the "normal" distribution.

8.4. THE EXPECTATION, A MEASURE OF THE CENTER

A random variable has two types of summary measure. One type measures the "center" of its probability distribution, and the other measures the "spread" in its probability distribution.

To introduce these measures, we place a simple random variable in view. Suppose that you are offered an investment Y in which you lose \$1000 with probability of 1/4, win \$4000 with probability of 3/8, and win \$6000 with probability of 3/8. Figure 8.2 records the probability distribution of the profit Y that you will receive if you accept this gamble.

The **expectation** of a random variable is found by multiplying each value that the random variable can take by the probability that this value occurs and summing the result. The expectation of the random variable X is denoted $E(X)$. For the random variable X in Table 8.1 and in Figure 8.1,

$$\begin{aligned} E(X) &= x_1P(X = x_1) + x_2P(X = x_2) + \cdots + x_nP(X = x_n) \\ &= x_1p_1 + x_2p_2 + \cdots + x_np_n. \end{aligned}$$

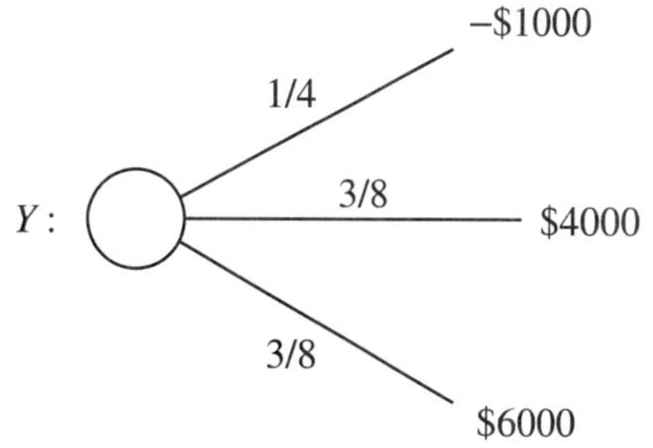

Figure 8.2 The profit Y from an investment.

The expectation $E(Y)$ of the random variable Y in Figure 8.2 is given by

$$\begin{aligned} E(Y) &= (-\$1000)(1/4) + (\$4000)(3/8) + (\$6000)(3/8), \\ &= -\$250 + \$1500 + \$2250, \\ &= \$3500. \end{aligned}$$

The expectation need not be a value that the random variable can assume. We just saw that the random variable Y has $E(Y) = \$3500$; the random variable Y takes the values -1000, $+4000$, and $+6000$, but not 3500.

The expectation of a random variable has several synonyms, notably, **expected value**, **mean**, **expected monetary value** (in the case of an uncertain monetary payoff), and **expected monetary cost** (in the case of an uncertain monetary payout). The mean of a random variable is often denoted by the English letter m or the Greek letter μ (pronounced mu).

The Median

The expectation is not the only sensible measure of the "center" of a probability distribution. The **median** of the random variable X is a number t such that $P(X \leq t)$ and $P(X \geq t)$ are closest to each other. The median of Y equals 4000 because no other number t has $P(Y \leq t)$ closer to $P(Y \geq t)$.

The median is particularly appropriate when we are dealing with population statistics. Consider, for instance, the annual income W earned by a randomly selected American family. The median of W is a number t such that roughly (or exactly) half the families earn t or less and roughly (or exactly) half earn t or more. In this sense, the median income is the income of the "typical" family.

In this text, the expectation plays a more important role than the median. Some virtues of the expectation can be presented now. Others will become apparent as the chapter proceeds.

Repeated Trials

One way to motivate the expectation is by "repeated trials." Imagine that you could make 100 different investments, each independent of the others and each having the same probability distribution as does Y. If you did so, you would lose \$1000 roughly 25 times, win \$4000 roughly 37.5 times, and win \$6000 roughly 37.5 times, for an average winning of \$3500. In this sense, the expectation is the average that you would win per investment.

Operational Decisions

We recall from Chapter 1 that an *operational* situation is one whose outcomes do not threaten the person's or the organization's way of life. By contrast, a *strategic* situation is one that could imperil the individual or organization's way of life.

A subject called utility theory (which lies a bit beyond the scope of this text) offers justification for the use of expectation of profit as a figure of merit when making operational decisions. It is reasonable for an organization or an individual to have a utility function on wealth whose "proportional risk aversion" is constant or decreasing. If this condition is satisfied, maximizing the expectation on each of sufficiently many operational decisions is optimal.

But suppose that *you* face an investment opportunity whose payoff is the random variable Y. Would you accept it? If you accept that investment, you will earn an expected profit of \$3500, and there is a probability of $3/4$ that you will earn at least \$4000, but there is a probability of $1/4$ that you will lose \$1000. If you cannot afford to lose \$1000, the investment opportunity is strategic, and the gamble is not worthwhile.

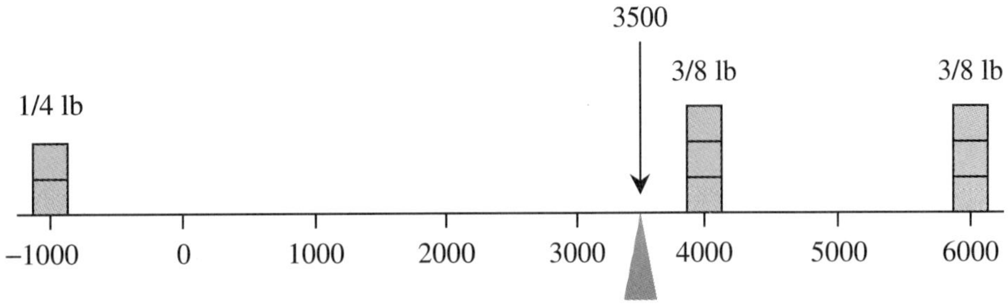

Figure 8.3 Interpretation of $E(Y) = 3500$ as a fulcrum of Y, each weight (shaded square) being $1/8$ lb (pound).

A Physical Interpretation

The expectation also has a physical interpretation—as the fulcrum or balance point. We mark a rod with each value that the random variable can take and place a weight on each mark that is proportional to the probability that the random variable takes its value. The expectation of the random variable is the fulcrum. Figure 8.3 illustrates this for the random variable Y. A weight of $1/4$ pound has been placed at -1000, a weight of $3/8$ pound at 4000, and a weight of $3/8$ pound at 6000. The fulcrum (balance point) will be at 3500, which equals the expectation of Y.

Later, we will discover other reasons explaining why the expectation is such an important measure of the center of a probability distribution.

8.5. THE VARIANCE, A MEASURE OF SPREAD

The expectation of a random variable does not tell us whether the values that it can take are close to its mean or are far away. We will soon introduce and compare several measures of the "spread" of a random variable, of the extent to which its values differ from its mean. To illustrate the spread, suppose you could choose between two investments, one whose profit Y is given in Figure 8.2, and another whose payoff Z is given in Table 8.4.

Let us compute the expectation of Z:

$$E(Z) = (\$3000)(0.5) + (\$4000)(0.5) = \$3500.$$

The random variables Y and Z have the same expectation. The values of Z differ from its mean by \$500. The values of Y differ from its mean by as much as \$4500. The distribution of Y is more "spread out" than that of Z.

If you could choose between Y and Z, would you be indifferent, or would you prefer one to the other? Most people would choose Z. Although Z forgoes the possibility of winning \$6000, it also avoids the possibility of losing \$1000. Indeed, Z guarantees a payoff of

Table 8.4 The profit Z you could receive from an investment.

Value of Z	Probability
\$3000	0.5
\$4000	0.5

at least \$3000. The random variable Z seems to entail less "risk"; its outcomes are bunched more tightly around its mean in ways that will soon be measured.

The Range

A particularly simple measure of the spread of a random variable is its **range**, which equals the largest value that the random variable can take less the smallest value. The range of X is written as Range(X). The random variables Y and Z that are described above have

$$\text{Range}(Y) = \$6000 - (-\$1000) = \$7000,$$
$$\text{Range}(Z) = \$4000 - \$3000 = \$1000.$$

By this measure, the random variable Z has one-seventh the spread of Y.

A Function of a Random Variable

The range of a random variable is determined exclusively by its largest and smallest values. The probabilities have no effect on the range, for which reason the range is a coarse measure of risk. Before presenting less coarse measures of risk, we pause to introduce a technical subject, namely, a function of a random variable. You are no doubt familiar with elementary functions, for example, with the function $g(x) = (x - 3)^2$ or the function $h(x) = |x - 3|$. But what is meant by a function of a random variable?

To answer this question, we return to the random variable X in Figure 8.1. The random variable X takes the value x_1 with probability p_1, it takes the value x_2 with probability p_2, and so forth. The **function** $g(X)$ of X is the random variable that takes the value $g(x_1)$ with probability p_1, that takes the value $g(x_2)$ with probability p_2, and so forth. Figure 8.4 presents the probability tree of $g(X)$.

Figures 8.1 and 8.4 highlight the difference between the random variables X and $g(X)$. The probabilities are the same, but the outcomes change. For each j, the random variable $g(X)$ replaces the outcome x_j by $g(x_j)$.

To compute the expectation of a random variable, multiply each value that it can take by the probability that it takes that value, and form the sum. Figure 8.4 makes it clear that the expectation $E[g(X)]$ of the random variable $g(X)$ is given by:

$$\begin{aligned} E[g(X)] &= g(x_1)P(X = x_1) + g(x_2)P(X = x_2) + \cdots + g(x_n)P(X = x_n) \\ &= g(x_1)p_1 + g(x_2)p_2 + \cdots + g(x_n)p_n. \end{aligned}$$

With $\mu = E(X)$, we will soon use the functions $|X - \mu|$ and $(X - \mu)^2$ of the random variable X to measure its spread.

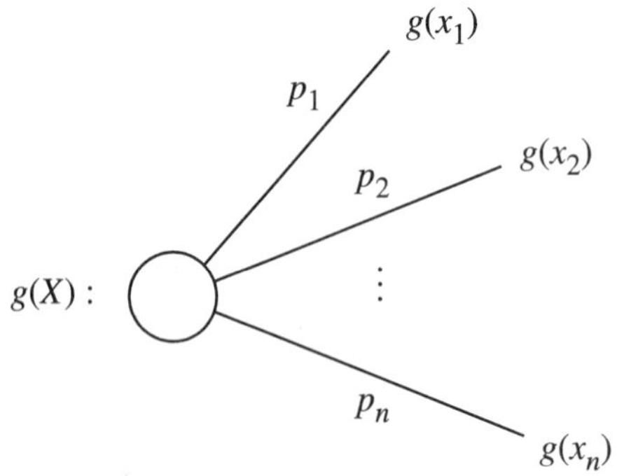

Figure 8.4 The random variable $g(X)$.

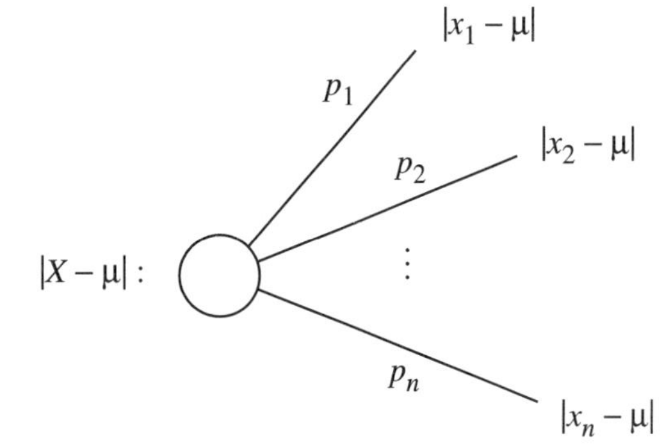

Figure 8.5 The random variable $|X - \mu|$.

The Deviation from the Mean

With $E(X) = \mu$, the random variable $|X - \mu|$ measures the absolute value of the deviation of X from its mean. To describe $|X - \mu|$, we specialize Figure 8.4 to the function $g(X) = |X - \mu|$ and obtain Figure 8.5.

Thus, the random variable $|X - \mu|$ takes the value $|x_1 - \mu|$ with probability p_1, the value $|x_2 - \mu|$ with probability p_2, and so forth. The expectation of $|X - \mu|$ is called the **mean absolute deviation** of X and is abbreviated MAD(X). To compute this expectation, we multiply each outcome by its probability and take the sum. It is evident from Figure 8.5 that:

$$\text{MAD}(X) = E(|X - \mu|) = |x_1 - \mu|p_1 + |x_2 - \mu|p_2 + \cdots + |x_n - \mu|p_n.$$

To illustrate the mean absolute deviation, we calculate MAD(Y) and MAD(Z) for the investment opportunities Y and Z whose probability distributions appear in Figure 8.2 and Table 8.4. Since $E(Y) = E(Z) = \$3500$, we get

$$\begin{aligned}\text{MAD}(Y) = E(|Y - \mu|) &= |-\$1000 - \$3500| \times 1/4 + |\$4000 - \$3500| \times 3/8 \\ &\quad + |\$6000 - \$3500| \times 3/8 \\ &= \$2250. \\ \text{MAD}(Z) = E(|Z - \mu|) &= |\$3000 - \$3500| \times 1/2 + |\$4000 - \$3500| \times 1/2 \\ &= \$500.\end{aligned}$$

Evidently, MAD(Z) is well below MAD(Y). By this measure, Z has considerably less spread than does Y.

The Squared Deviation from the Mean

With $\mu = E(X)$, the most useful measure of the spread of a random variable X is the expectation not of $|X - \mu|$ but of $(X - \mu)^2$, namely, the expectation of the *square* of the difference between X and its mean. Why this is so will become clear later in the chapter. Substituting $(X - \mu)^2$ for $g(X)$ in Figure 8.4 shows us that $(X - \mu)^2$ is the random variable in Figure 8.6.

The expectation of $(X - \mu)^2$ is called the **variance** of X and is denoted Var(X). It is clear from Figure 8.6 that:

$$\text{Var}(X) = E[(X - \mu)^2] = (x_1 - \mu)^2p_1 + (x_2 - \mu)^2p_2 + \cdots + (x_n - \mu)^2p_n \quad (8.1)$$

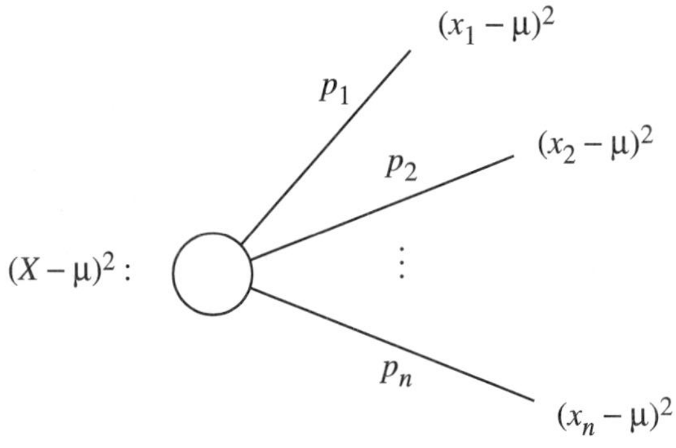

Figure 8.6 The random variable $(X - \mu)^2$.

For example, the random variable Y in Figure 8.2 has $E(Y) = \$3500$, so its variance is given by

$$\begin{aligned}\text{Var}(Y) &= (-\$1000 - \$3500)^2 \times 1/4 + (\$4000 - \$3500)^2 \times 3/8 \\ &\quad + (-\$6000 - \$3500)^2 \times 3/8, \\ &= 7{,}500{,}000\ \$ \times \$.\end{aligned}$$

This calculation provides you three reasons to be skeptical of the variance.

- First, the variance weighs the probability of each outcome by the square of the difference between that outcome and the mean, thereby putting heavy and possibly disproportionate weight on those outcomes that are far from the mean.
- Second, the unit of measure of the variance is the square of the unit of measure of the variable itself, $ \$ \times \$ $ in the case of Y. That is a weird unit of measure.
- Third, the numerical value of the variance can be enormous.

Nonetheless, the variance will turn out to be the most important measure of spread of a distribution.

The Standard Deviation

To measure spread in the same unit as the variable itself, we take the square root of the variance and obtain the **standard deviation**. The standard deviation of X is abbreviated StDev(X) and is given by

$$\text{StDev}(X) = \sqrt{\text{Var}(X)} = \sqrt{E[(X - \mu)^2]}, \quad \text{with } \mu = E(X).$$

For example, the random variable Y whose variance is given above has

$$\text{StDev}(Y) = \sqrt{7{,}500{,}000 \times \$ \times \$} = \$2739.$$

The standard deviation of a random variable has the same unit of measure as the random variable itself and has a value commensurate with the spread in the random variable's distribution.

Throughout the literature on random variables, the lower-case Greek letter σ (pronounced sigma) denotes the standard deviation of a random variable, and σ^2 (pronounced sigma squared) denotes the variance. We often abbreviate Var(X) to σ^2 and StDev(X) to σ. In brief, we often take

$$\mu = E(X), \quad \sigma^2 = \text{Var}(X), \quad \sigma = \text{StDev}(X).$$

Later in this chapter, we will present some of the reasons why the mean and standard deviation are important.

Table 8.5 Spreadsheet computation of the mean, variance, and standard deviation of Y.

	A	B	C	D	E
3	**probability**	**value of Y**	**Y - E(Y)**		
4	1/4	-1,000	-4,500	◄	= B4 - B$7
5	3/8	4,000	500		
6	3/8	6,000	2,500		
7	E(Y) =	3,500	◄		= SUMPRODUCT(A4:A6, B4:B6)
8	Var(Y) =	7,500,000	◄		= SUMPRODUCT(A4:A6, C4:C6, C4:C6)
9	StDev(Y) =	2,738.61	◄		= SQRT(B8)
10	E(Y^2) =	19,750,000	◄		= SUMPRODUCT(A4:A6, B4:B6, B4:B6)
11	Var(Y) =	7,500,000	◄		= B10 - B7 * B7

A Spreadsheet

It is easy to compute the mean, variance, and standard deviation on a spreadsheet. Table 8.5 accomplishes this for the random variable Y. In Table 8.5:

- Cells A4 through B6 specify the probability distribution of Y.
- The SUMPRODUCT function in cell B7 computes $E(Y)$.
- The functions in cells C4 through C6 specify the probability distribution of the random variable $Y - E(Y)$.
- The function in cell B8 computes Var(Y). This function multiplies the entries in cells A4 through A6 by the *square* of the corresponding entries in column C and takes the sum.

A Second Formula for the Variance

Rows 10 and 11 of Table 8.5 use a second formula to compute the variance. To explain it, we substitute X^2 for $g(X)$ in Figure 8.4 and compute its expectation, getting

$$E(X^2) = (x_1)^2 p_1 + (x_2)^2 p_2 + \cdots + (x_n)^2 p_n.$$

As row 11 suggests, the variance of X can be calculated from the formula,

$$\text{Var}(X) = E(X^2) - \mu^2, \quad \text{with} \quad E(X) = \mu. \tag{8.2}$$

Equation (8.2) states that the variance of a random variable equals the expectation of its square less the square of its expectation.

Equation (8.2) is easy to verify. For the reader who may be interested in a proof, we write the function $(x - \mu)^2$ of x as $(x^2 - 2\mu x + \mu^2)$, and we look at the three terms, getting:

$$\begin{aligned} \text{Var}(X) = \sum_{i=1}^{n} (x_i - \mu)^2 p_i &= \sum_{i=1}^{n} [(x_i)^2 - 2\mu x_i + \mu^2] p_i \\ &= \sum_{i=1}^{n} (x_i)^2 p_i - 2\mu \sum_{i=1}^{n} x_i p_i + \mu^2 \sum_{i=1}^{n} p_i \\ &= E(X^2) - 2\mu\mu + \mu^2 \\ &= E(X^2) - \mu^2. \end{aligned}$$

Now that the expectation and variance have been described, we turn our attention to the reasons they are important.

8.6. THE SUM OF RANDOM VARIABLES

In many situations, the random variable in which we are interested is actually the sum of other random variables. The amount of energy that the customers of a utility company will use tomorrow equals the sum of the amounts used by its individual customers. The total profit from an investment portfolio for the coming year equals the sum of the profits on the individual investments. And so forth.

But what exactly is meant by the sum of random variables? To explore this subject, we pose the question in:

Problem A (Three Investment Opportunities)

There are three projects, which are labeled 1, 2, and 3. You may invest in any project or in any combination of these projects for a period of one year. The profit (return less investment) that you would earn from each project depends on the state of the economy at the end of the year, and that is uncertain. There are three possible states of the economy, which are labeled *a*, *b*, and *c*. Thus, the profit X_1 of project 1 is uncertain, and it is determined by the state of the economy at the end of the year. The same is true of the profit X_2 for project 2 and the profit X_3 for project 3. Table 8.6 specifies the profit for each project given each state of the economy, along with the probability that this state will occur. (Table 8.6 includes a column, labeled *T*, whose meaning will soon be explained.) What happens if you invest in each of these projects? in all three?

The Sum

Before answering the question posed in Problem A, we interpret the random variable *T* that appears in Table 8.6. Under *each* outcome, the value that is assigned to *T* equals the sum of the values of X_1, X_2, and X_3. That is the meaning of the equation

$$T = X_1 + X_2 + X_3.$$

Suppose, for instance, that outcome *a* occurs. In this case, project 2 yields a profit of 4, but projects 1 and 3 incur losses, and the total profit *T* on all three projects equals -1 because $-1 = -2 + 4 - 3$. Similarly, if outcome *c* occurs, projects 1 and 3 will be profitable, but project 2 will yield a loss, and the total income *T* will equal 3 because $3 = 3 - 5 + 5$.

The random variable *T* describes your profit if you invest in all three projects. Figure 8.7 presents a probability tree for *T*. This tree recognizes the fact that, under each outcome, *T* equals the sum of the values of the random variables X_1, X_2, and X_3.

An interesting feature of this example is that the total profit is less variable than the profit on any of the individual projects. Figure 8.7 indicates that the range of *T* equals 4, while the ranges of X_1, X_2, and X_3 are 5, 9, and 8, respectively. To an extent, the risks in the individual projects are offsetting each other.

Table 8.6 States of the economy, their probabilities, the profits of projects 1, 2, and 3, and the total profit from all three.

Outcome	Probability	X_1	X_2	X_3	T
a	0.2	−2	4	−3	−1
b	0.5	0	2	−1	1
c	0.3	3	−5	5	3

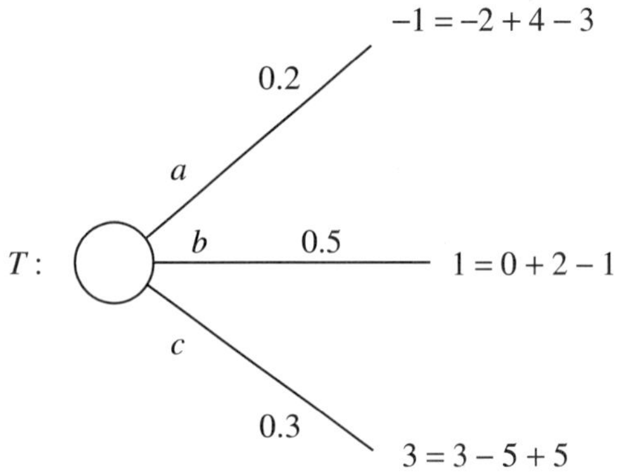

Figure 8.7 The random variable $T = X_1 + X_2 + X_3$ as a probability tree.

The Expectation of the Sum

Let us compare the expectation of the profit on each project with the expectation of the total profit. We compute:

$$\begin{aligned} E(X_1) &= (0.2) \times (-2) + (0.5) \times (\ \ 0) + (0.3) \times (\ \ 3) = 0.5 \\ E(X_2) &= (0.2) \times (\ \ 4) + (0.5) \times (\ \ 2) + (0.3) \times (-5) = 0.3 \\ E(X_3) &= (0.2) \times (-3) + (0.5) \times (-1) + (0.3) \times (\ \ 5) = 0.4 \\ E(T) &= (0.2) \times (-1) + (0.5) \times (\ \ 1) + (0.3) \times (\ \ 3) = 1.2 \end{aligned}$$

The expectation $E(T)$ of the total profit equals 1.2. The sum, $0.5 + 0.3 + 0.4$, of the expectations of the individual profits also equals 1.2. It is no accident that the expectation of the sum equals the sum of the expectations. The reason this is so is evident in the above calculations; each addend in the equation for $E(T)$ is the sum of the addends above it, for example, $(0.2) \times (-1) = (0.2) \times (-2 + 4 - 3)$ because $-1 = -2 + 4 - 3$.

Expectations Add

This property of expectations holds in general. For any random variables X_1 through X_n, we have

$$E(X_1 + X_2 + \cdots + X_n) = E(X_1) + E(X_2) + \cdots + E(X_n).$$

We have just discovered a key reason for using the expectation as a measure of the center of a probability distribution. Expectations add! You are urged to memorize the following.

> The expectation of the sum of random variables equals the sum of their expectations.

For example, the expectation of the total profit on a portfolio of investments equals the sum of the expected profits on the individual investments.

Do the Variances Add?

Does the variance of the sum equal the sum of the variances? For the example in Table 8.6, routine calculations, which we omit, produce:

$$\text{Var}(X_1) = 3.25, \quad \text{Var}(X_2) = 12.61, \quad \text{Var}(X_3) = 9.64, \quad \text{Var}(T) = 1.96.$$

The variance of the sum need not equal the sum of the variances; the variance of the sum may be smaller or larger than the sum of the variances. In this instance, the variance of the sum happens to be smaller than any of the individual variances.

The question posed in Problem A is whether you should invest in any or all three of these projects. We can offer a qualified answer: If you accept expected profit as a measure of desirability and the variance in profit as a measure of undesirability, you would prefer to invest in all three than in any one of them.

8.7. INDEPENDENT RANDOM VARIABLES

We will soon describe a situation in which the variance of the sum does equal the sum of the variances. In preparation, we recall from Chapter 7 that the events A and B are *independent* of each other if knowing that one has occurred has no effect on the probability of the other. In other words, the events A and B are independent if $P(A|B) = P(A)$.

There is a similar notion of independence among random variables. Two random variables are said to be "independent" of each other if knowing the value taken by one of them provides no information about the other. Specifically, the random variables X and Y are now said to be **independent** of each other if, for every value x that X can assume and for every value y that Y can assume,

$$P(X = x) = P(X = x \mid Y = y). \qquad \textit{test for independence}$$

This equation states that knowing the value of Y provides no information about the probability distribution of X.

This test for independence appears to be asymmetric, but it is not. As was the case in Chapter 7, it's easy to check that the random variables X and Y are independent if

$$P(Y = y) = P(Y = y|X = x)$$

for each value x that X can take and for each value y that Y can take. Thus, if two random variables are independent, knowing the value of either provides no information about the probability distribution of the other.

It is also easy to check that the random variables X and Y are independent if

$$P(X = x, Y = y) = P(X = x)P(Y = y)$$

for each x and y. In other words, the random variables X and Y are independent if the joint probability distribution of X and Y equals the product of their individual probability distributions.

The random variables X and Y are said to be **dependent** if they are not independent of each other. Thus, if two random variables are dependent, knowing the value of one provides information about the probability distribution of the other.

For the investment example in Table 8.6, are the random variables X_1, X_2, and X_3 independent or dependent? They are dependent. If you know the value taken by any one of these random variables, you also know the values taken by the other two. If, for instance $X_1 = -2$, then $X_2 = 4$ and $X_3 = -3$. It is certainly not the case that $P(X_3 = -3) = P(X_3 = -3|X_1 = -2)$. These random variables also "feel" dependent; a high profit on project 2 or on project 3 guarantees a low profit on the other, for instance.

Several random variables are said to be **mutually independent** if information about values taken by any group of them has no effect on the probability distribution of any others. Variances add if the random variables are mutually independent. We emphasize:

> The variance of a sum of mutually independent random variables equals the sum of their variances.

In mathematical terms, *if* the random variables X_1 through X_n are mutually independent, then

$$\text{Var}(X_1 + X_2 + \cdots + X_n) = \text{Var}(X_1) + \text{Var}(X_2) + \cdots + \text{Var}(X_n).$$

This equation articulates an advantage of the variance over the mean absolute deviation. The variances of independent random variables add. The mean absolute deviations of independent random variables need not add.

8.8. THE COVARIANCE

We have seen that the variance of a sum of independent random variables equals the sum of their variables. What about the variance of the sum of *dependent* random variables? This section introduces a formula that works.

Let X and Y denote two random variables, which may or may not be independent of each other, with $\mu_X = E(X)$ and with $\mu_Y = E(Y)$. The **covariance** of X and Y is defined to be the expectation of the random variable $(X - \mu_X)(Y - \mu_Y)$. The covariance of X and Y designated Cov(X, Y). Thus:

$$\text{Cov}(X, Y) = E[(X - \mu_X)(Y - \mu_Y)].$$

If X and Y are independent, it is clear intuitively (and is not hard to verify) that Cov(X, Y) = 0.

Note that Cov(X, Y) and Cov(Y, X) equal each other because both are the expectation of the same quantity, which is $(X - \mu_X)(Y - \mu_Y)$. The covariance is not entirely unfamiliar because $\text{Cov}(X, X) = E[(X - \mu_X)(X - \mu_X)] = \text{Var}(X)$.

The Variance of the Sum

If X and Y are dependent, Cov(X, Y) can have any sign. This covariance can be positive or negative; it can also be zero. A famous formula for the variance of the sum of two random variables states:

$$\text{Var}(X + Y) = \text{Var}(X) + \text{Var}(Y) + 2\,\text{Cov}(X, Y). \tag{8.3}$$

For the reader who may be interested, we sketch a proof of Equation (8.3). Let us write

$$\begin{aligned}[X + Y - (\mu_X + \mu_Y)]^2 &= [(X - \mu_X) + (Y - \mu_Y)]^2 \\ &= (X - \mu_X)^2 + (Y - \mu_Y)^2 + 2(X - \mu_X)(Y - \mu_Y).\end{aligned}$$

In the above, we take the expectation, recall that the expectation of the sum equals the sum of the expectations, and see that Equation (8.3) results.

Equation (8.3) generalizes to compute the variance of the sum of any number of random variables. We highlight:

> The variance of the sum of several random variables equals the sum of their variances plus twice the sum of their covariances.

For example,

$$\begin{aligned}\text{Var}(X + Y + Z) = \text{Var}(X) &+ \text{Var}(Y) + \text{Var}(Z) + 2\,\text{Cov}(X, Y) \\ &+ 2\,\text{Cov}(X, Z) + 2\,\text{Cov}(Y, Z).\end{aligned} \tag{8.4}$$

A Sample Calculation

Covariances are easy to calculate on a spreadsheet. To see how, we return to the random variables X_1, X_2, and X_3 in Table 8.6. Their covariances are computed in Table 8.7.

Table 8.7 Computing means, variances, and covariances on a spreadsheet.

	A	B	C	D	E	F	G	H
3	**Outcome**	**Probability**	**X1**	**X2**	**X3**	**X1 - m1**	**X2 - m2**	**X3 - m3**
4	a	0.2	-2	4	-3	-2.5	3.7	-3.4
5	b	0.5	0	2	-1	-0.5	1.7	-1.4
6	c	0.3	3	-5	5	2.5	-5.3	4.6
7								
8			**m1**	**m2**	**m3**			
9			0.5	0.3	0.4		=C6-C$9	
10								
11			= SUMPRODUCT($B4:$B6, C4:C6)					
12		**Table of covariances**						
13			X1	X2	X3			
14		X1	3.25	-6.25	5.5			
15		X2	-6.25	12.61	-11.02			
16		X3	5.5	-11.02	9.64			
17								
18			= SUMPRODUCT($B$4:$B$6, $H$4:$H$6, F4:F6)					

In this table:

- Cells B4 through E6 record the probability distributions of X_1, X_2, and X_3.
- The function in cell C9 computes $m_1 = E(X_1)$. Dragging the function in cell C9 across cells D9 and E9 computes $m_2 = E(X_2)$ and $m_3 = E(X_3)$.
- Dragging the function in cell F6 across cells F4 through H6 specifies the values that each outcome assigns to $(X_1 - m_1)$, to $(X_2 - m_2)$, and to $(X_3 - m_3)$.
- The function in cell C16 executes the calculation

$$\text{Cov}(X_3, X_1) = (0.2)(-3.4)(-2.5) + (0.5)(-1.4)(-0.5) + (0.3)(4.6)(2.5) = 5.5.$$

- Dragging the function in cell C16 across cells D16 and E16 calculates

$$\text{Cov}(X_3, X_2) = (0.2)(-3.4)(3.7) + (0.5)(-1.4)(1.7) + (0.3)(4.6)(-5.3) = -11.02,$$
$$\text{Cov}(X_3, X_3) = (0.2)(-3.4)(-3.4) + (0.5)(-1.4)(-1.4) + (0.3)(-5.3)(-5.3) = 9.64.$$

- Similar functions compute the other two rows of the array of covariances.

The array of covariances in Table 8.7 is symmetric. This occurs because $\text{Cov}(X, Y)$ and $\text{Cov}(Y, X)$ are the expectation of the same quantity, $(X - \mu_X)(Y - \mu_Y)$. The diagonal terms in this array are the variances. That occurs because $\text{Cov}(X, X)$ equals the expectation of $(X - \mu_X)(X - \mu_X) = (X - \mu_X)^2$.

As a check on the calculations in Table 8.7, we use it to compute the variance of T. Since $T = X_1 + X_2 + X_3$, Equation (8.4) gives

$$\begin{aligned}\text{Var}(T) &= \text{Var}(X_1 + X_2 + X_3),\\ &= \text{Var}(X_1) + \text{Var}(X_2) + \text{Var}(X_3) + 2\text{Cov}(X_1, X_2) + 2\text{Cov}(X_1, X_3)\\ &\quad + 2\text{Cov}(X_2, X_3),\\ &= 3.25 + 12.61 + 9.64 + (2)(-6.25 + 5.5 - 11.02) = 1.96.\end{aligned}$$

Previously, we had reported that $\text{Var}(T) = 1.96$, so the above is no surprise.

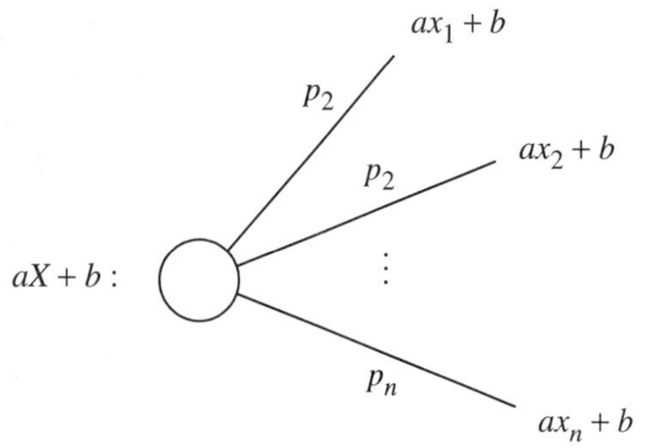

Figure 8.8 The function $aX + b$ of X.

8.9. A LINEAR FUNCTION OF A RANDOM VARIABLE

In preparation for the following sections, we study a linear function of a random variable. With a and b as constants, this section focuses on the linear function $(aX + b)$ of the random variable X. Let us recall from Figure 8.1 that X takes the values x_1 through x_n with probabilities p_1 through p_n, respectively. The probability tree in Figure 8.8 reports that the function $g(X) = aX + b$ of X takes the values $ax_1 + b$ through $ax_n + b$ with probabilities p_1 through p_n, respectively.

It is easy to see from Figure 8.8 that adding b to each outcome adds b to the expectation. It's equally clear from the figure that multiplying each outcome by the number a multiplies the expectation by a. In brief, Figure 8.8 indicates that

$$E(aX + b) = aE(X) + b. \tag{8.5}$$

A bit of algebra (that we omit) would verify that

$$\text{Var}(aX + b) = a^2\text{Var}(X), \quad \text{StDev}(aX + b) = |a|\,\text{StDev}(X). \tag{8.6}$$

Equation (8.6) shows that adding b to each outcome has no effect on the variance. This occurs because adding b to each outcome adds b to the mean, and these b's cancel out when we take the squared deviation from the mean. Equation (8.6) also shows that multiplying each outcome by a multiplies the variance by a^2, and hence multiplies the standard deviation by the absolute value of a.

To complete a discussion of linear functions, we state the analog for the covariance of Equation (8.6). It is:

$$\text{Cov}(aX + b, cY + d) = ac\text{Cov}(X, Y). \tag{8.7}$$

In particular, adding a constant to each outcome has no effect on the covariance, just as it has no effect on the variance. Equation (8.7) specializes to Equation (8.6) when $cY + d$ is replaced by $aX + b$.

8.10. MEAN-VARIANCE ANALYSIS

An individual or a company that invests capital chooses among several investment opportunities. Each investment opportunity has an uncertain profit (return less investment), and the profits from different investment opportunities can be dependent, as has been illustrated by Problem A. A **portfolio** is a set of investments, and a goal of investment is to find an attractive portfolio.

An investor seeks a portfolio whose profit is a large value but has little spread (risk). These can be conflicting objectives. Investment opportunities that offer large profits tend to be riskier, and the risks in different investments can reinforce or offset each other. If we accept the expectation as a measure of profit and the variance as a measure of risk, we seek a portfolio whose profit has a large expectation and a low variance.

A **mean-variance analysis** explores the tradeoff between the expectation and the variance of the profit that an individual or organization can earn. A portfolio is said to be **efficient** if it achieves the smallest variance in its profit over all portfolios whose mean profit are at least as large as its mean profit.

Harry Markowitz pioneered mean-variance analysis of investment portfolios. He observed that an efficient portfolio can be found by solving an optimization problem that is known as a "quadratic program." He observed that *diversification*—investing in many assets—can reduce risk (variance) at any desired level of mean return. For this work, Markowitz shared a Nobel Prize in economics. To introduce mean-variance analysis, we turn our attention to:

Problem B (A Portfolio)

Suppose that the random variables X_1, X_2, and X_3 in Table 8.6 describe the profit you would earn by investing in one share of companies 1, 2, and 3, respectively, for a period of one year. What are the mean and variance of the profit that you would obtain by investing in a portfolio that consists of s_1 shares of the first company, s_2 shares of the second company, and s_3 shares of the third company?

Your total profit R for the year is a random variable, and it is given by the equation

$$R = s_1X_1 + s_2X_2 + s_3X_3 \tag{8.8}$$

because your profit from investing in s_k shares of company k equals s_kX_k and because your total profit equals the sum of the profits on your shares in the three companies.

Expected Total Profit

Problem B asks you to compute the mean and variance of R. The expectation of the sum equals the sum of the expectations, so Equation (8.8) gives

$$E(R) = E(s_1X_1) + E(s_2X_2) + E(s_3X_3).$$

To check that $E(s_kX_k) = s_kE(X_k)$, note that s_k is a number, and use Equation (8.5). Thus,

$$E(R) = s_1E(X_1) + s_2E(X_2) + s_3E(X_3).$$

In Table 8.7, we saw that $E(X_1) = 0.5$, that $E(X_2) = 0.3$, and that $E(X_3) = 0.4$. Hence,

$$E(R) = 0.5s_1 + 0.3s_2 + 0.4s_3. \tag{8.9}$$

At an intuitive level, Equation (8.9) is clear. It states that your expected total profit sums the product of the expected profit per share of each stock times the number of shares of that stock that you own.

Variance of the Total Profit; the Covariance Method

We'll develop two ways to compute the variance of the total profit. Since $R = s_1X_1 + s_2X_2 + s_3X_3$, we know that the variance of R equals the sum of the variances plus twice the sum of the covariances:

$$\begin{aligned}\text{Var}(R) = {} & \text{Var}(s_1X_1) + \text{Var}(s_2X_2) + \text{Var}(s_3X_3) + 2\text{Cov}(s_1X_1, s_2X_2) \\ & + 2\text{Cov}(s_1X_1, s_3X_3) + 2\text{Cov}(s_2X_2, s_3X_3).\end{aligned}$$

Table 8.8 The probability distribution of R.

Outcome	Probability	X_1	X_2	X_3	$R = s_1X_1 + s_2X_2 + s_3X_3$
a	0.2	−2	4	−3	$-2s_1 + 4s_2 - 3s_3$
b	0.5	0	2	−1	$2s_2 - 1s_3$
c	0.3	3	−5	5	$3s_1 - 5s_2 + 5s_3$

Since s_i and s_j are numbers, Equations (8.6) and (8.7) give

$$\text{Var}(s_iX_i) = (s_i)^2\text{Var}(X_i) \quad \text{and} \quad \text{Cov}(s_iX_i, s_jX_j) = s_is_j\,\text{Cov}(X_i, X_j).$$

Substituting and then using the numbers in Table 8.7 provides:

$$\begin{aligned}\text{Var}(R) &= (s_1)^2\text{Var}(X_1) + (s_2)^2\text{Var}(X_2) + (s_3)^2\text{Var}(X_3)\\ &\quad + 2\,s_1s_2\text{Cov}(X_1, X_2) + 2\,s_1s_3\text{Cov}(X_1, X_3) + 2\,s_2s_3\text{Cov}(X_2, X_3),\\ &= 3.25(s_1)^2 + 12.61(s_2)^2 + 9.64(s_3)^2\\ &\quad + (2)(-6.25)s_1s_2 + (2)(5.5)s_1s_3 + (2)(-11.02)s_2s_3. \end{aligned} \tag{8.10}$$

Once the spreadsheet in Table 8.7 has been built, it is easy to compute the expectation of the total profit from Equation (8.9) and the variance of the total profit from Equation (8.10).

Variance of the Total Profit; the Scenario Method

For our portfolio problem, the variance of R can be computed directly, without using covariances. Table 8.8 suggests how. It records the probability distributions of the random variables X_1, X_2, X_3, and R. Each outcome in Table 8.8 is called a scenario. For instance, the top line of Table 8.8 shows that when outcome (scenario) a occurs, the total profit R takes the value $(-2s_1 + 4s_2 - 3s_3)$. The other two lines specify R for the other two scenarios.

Table 8.8 lets us compute the expectation of R and R^2. To do so, we multiply the probability of each outcome by the value of R (or R^2) it prescribes and sum, getting

$$\begin{aligned}E(R) &= (0.2)(-2s_1 + 4s_2 - 3s_3) + (0.5)(2s_2 - 1s_3)\\ &\quad + (0.3)(3s_1 - 5s_2 + 5s_3),\\ E(R^2) &= (0.2)(-2s_1 + 4s_2 - 3s_3)^2 + (0.5)(2s_2 - 1s_3)^2\\ &\quad + (0.3)(3s_1 - 5s_2 + 5s_3)^2.\end{aligned}$$

Let us recall from Equation (8.2) that the variance of a random variable equals the expectation of its square less the square of its expectation. Specifically:

$$\text{Var}(R) = E(R^2) - [E(R)]^2.$$

The preceding three equations specify Var(R) without computing any covariances.

A Spreadsheet

Table 8.9 uses a spreadsheet to implement the scenario method for computing $E(R)$ and Var(R). In Table 8.9:

- Cells B4 through E6 specify the probability distribution of X_1, X_2, and X_3.
- Cells C7, D7, and E7 are reserved for the values of s_1, s_2, and s_3. Table 8.9 records the values $s_1 = 10$, $s_2 = 20$, and $s_3 = 30$. To perform the same calculation for a different portfolio, change these numbers.

Table 8.9 Computing the mean and standard deviation of R via the scenario method.

	A	B	C	D	E	F	G
3	**Outcome**	**Prob.**	**X1**	**X2**	**X3**	**R**	
4	a	0.2	-2	4	-3	-30	
5	b	0.5	0	2	-1	10	
6	c	0.3	3	-5	5	80	
7	number of shares		10	20	30	↑	
8			= SUMPRODUCT(C6:E6, C$7:E$7)				
9							
10	E(R) =	23	←	= SUMPRODUCT(B4:B6, F4:F6)			
11	E(R^2) =	2150	←	= SUMPRODUCT(B4:B6, F4:F6. F4:F6)			
12	Var(R) =	1621	←	= C11 - C10 * C10			
13	StDev(R) =	40.26	←	= SQRT(C12)			

- The functions in cell F6 compute the value of R under outcome c. Dragging this function upward computes the value of R under the other two outcomes.
- The SUMPRODUCT functions in cells B10 through B13 compute $E(R)$ and StDev(R) in familiar ways.

Table 8.9 sets the stage for computation of an efficient portfolio. To do so, we use Solver to minimize the value in cell B12 (the variance) with cells C7, D7, and E7 as the changing cells, subject to certain constraints. One constraint allocates the budget, and another keeps the quantity in cell B10 (the mean profit) at least as large as the desired threshold.

Efficient portfolios can be computed by the covariance method (Table 8.7) or from the scenario method (Table 8.9). For this example, the scenario method is easier.

8.11. THE SUM OF i.i.d. RANDOM VARIABLES

This section develops some important reasons why the expectation is the most important measure of central tendency and why the variance is the most important measure of spread. Here, we study the sum of independent random variables that have the same probability distribution.

A sequence $X_1, X_2, \ldots$ of random variables is said to be **independent and identically distributed** (abbreviated **i.i.d.**) if each of them has the same probability distribution and if knowing the values taken by any group of them provides no information about the others.

The Sum

Let $X_1, X_2, \ldots$ be independent and identically distributed (i.i.d.) random variables, with common mean $\mu = E(X_k)$ and common variance $\sigma^2 = \text{Var}(X_k)$. The symbol S_n denotes the sum of X_1 through X_n, so that

$$S_n = X_1 + X_2 + \cdots + X_n.$$

We will verify that these i.i.d. random variables have

$$E(S_n) = n\mu \quad \text{and} \quad \text{Var}(S_n) = n\sigma^2. \tag{8.11}$$

Evidently, S_n is the sum of n random variables. Each of these n random variables has μ as its expectation, and the expectation of the sum equals the sum of the expectations, so

$E(S_n) = n\mu$. Each of these n random variables has σ^2 as its variance. These random variables are independent, so the variance of their sum equals the sum of the variances; thus, $\text{Var}(S_n) = n\sigma^2$. This verifies Equation (8.11).

Since the standard deviation is the square root of the variance, Equation (8.11) shows that i.i.d. random variables have

$$\text{StDev}(S_n) = \sqrt{n}\sigma. \tag{8.12}$$

Expression (8.12) is remarkable. For some values of k, the random variable X_k exceeds its mean. For other values of k, the random variable X_k lies below its mean. To an extent, these events offset each other. Equation (8.12) indicates that the extent to which they do not offset each other grows *exactly* as $\sqrt{n}$.

The expectation of S_n grows linearly in n, but its standard deviation grows only as the square root of n. This manifests an economy of scale that has important implications for retailers, insurers, and others.

The Central Limit Theorem

More can be said of S_n than is evident in Equations (8.11) and (8.12). We must postpone until Chapter 11 a description of the "normal" distribution, whose "bell-shaped" curve may be familiar to you. The normal distribution gets its name from a famous theorem, as follows.

> **Central Limit Theorem:** For sufficiently large values of n, the sum S_n of n i.i.d. random variables is well-approximated by the normal distribution having the same mean and standard deviation as does S_n.

The Central Limit Theorem is a key reason for taking the expectation as a measure of the center and the variance as a measure of the spread. When a large number of independent random variables are summed, only their means and variances affect the sum. Except for the mean and variance, detail about the individual random variables *vanishes* when many of them are summed. The Central Limit Theorem will be illustrated later in this chapter and will be described more completely in Chapter 11.

8.12. BERNOULLI TRIALS

We now shift our attention from general properties of random variables to particular types of random variable that occur frequently in applications. These random variables have their origin in "Bernoulli trials," a phrase whose meaning is suggested by a sequence of coin flips, each having probability p of getting heads. Specifically, **Bernoulli trials** describes a sequence of experiments that have these properties:

- Each experiment has only two outcomes.
- Each experiment has the same probability distribution over its two outcomes.
- Each experiment is independent of the others.

Each of these experiments may be the toss of a coin, the application of a medical test to a member of a population, or anything else with these properties. When discussing Bernoulli trials, we often refer to the two outcomes as "success" and "failure."

A random variable X is said to have the **Bernoulli** distribution with parameter p if X takes the value 1 with probability p and if X takes the value 0 with probability $1 - p$. The probability tree in Figure 8.9 depicts this random variable.

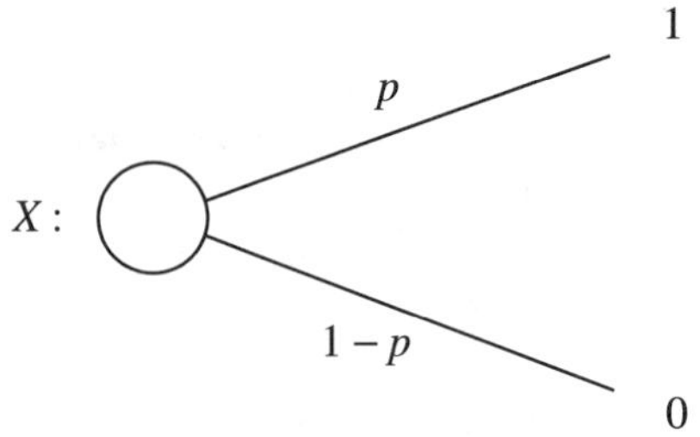

Figure 8.9 A random variable X having the Bernoulli distribution with parameter p.

The Bernoulli distribution describes the simplest discrete random variable that has any uncertainty in it. The Bernoulli distribution is the basis for *counting*; it counts a success as 1, a failure as 0. A success occurs with probability p, a failure with probability $1 - p$.

Let us compute the expectation and variance of this Bernoulli random variable X. It is clear from Figure 8.9 that $E(X) = p$. An easy way to compute $\text{Var}(X)$ is to observe in Figure 8.9 that $X = X^2$, and hence that $E(X^2) = p$. Thus, Equation (8.2) shows that $\text{Var}(X) = E(X^2) - [E(X)]^2 = p - p^2 = p(1 - p)$. In brief:

$$E(X) = p \quad \text{and} \quad \text{Var}(X) = p(1-p) \quad \text{when } X \text{ is Bernoulli.} \tag{8.13}$$

8.13. THE BINOMIAL DISTRIBUTION

The random variable S_n is said to have the **binomial distribution** with parameters n and p if S_n equals the number of successes that will occur in n independent Bernoulli trials, each having probability p of success. To express this random variable S_n mathematically, we write

$$S_n = X_1 + X_2 + \cdots + X_n,$$

where X_1 through X_n denote random variables that are mutually independent, each of which has the Bernoulli distribution with the same parameter p. The random variable S_n can take any value between 0 and n. For instance, the event in which S_n equals 0 occurs if X_1 through X_n equal 0, and $P(S_n = 0) = (1 - p)^n$. Similarly, $P(S_n = n) = p^n$.

Since S_n is the sum of n i.i.d. random variables, each of which has mean p and variance $p(1 - p)$, we get

$$E(S_n) = np \quad \text{and} \quad \text{Var}(S_n) = np(1 - p) \quad \text{when } S_n \text{ is binomial.} \tag{8.14}$$

Solely to complete the record, we record the probability distribution of S_n. The values that S_n might take are the integers 0 through n, and for each integer k between 0 and n, inclusive,

$$P(S_n = k) = \binom{n}{k} p^k (1 - p)^{n-k} \quad \text{with} \binom{n}{k} = \frac{n!}{k!(n - k)!}, \tag{8.15}$$

where $n!$ (pronounced n factorial) is defined by $n! = n \times (n - 1) \times \cdots \times 1$ for $n \geq 1$, with $0! = 1$.

The formula in Equation (8.15) is unwieldy. Imagine computing Equation (8.14) by hand. Hand computation is not necessary because an Excel function specifies the binomial probability distribution. Specifically:

- For a random variable Y that has the binomial distribution with parameters n and p, the Excel function =BINOMDIST(k, n, p, 0) reports $P(Y = k)$.
- For the same random variable Y, the Excel function =BINOMDIST(k, n, p, 1) reports $P(Y \leq k)$.

Table 8.10 The binomial distribution with $n = 7$ and $p = 0.2$.

	A	B	C	D	E	F	G
2	n =	7					
3	p =	0.2					
4							
5	**k**	**P(Sn = k)**	**P(Sn <= k)**		= BINOMDIST(A6, B$2, B$3, 0)		
6	0	0.20972	0.20972		= BINOMDIST(A6, B$2, B$3, 1)		
7	1	0.36700	0.57672				
8	2	0.27525	0.85197				
9	3	0.11469	0.96666				
10	4	0.02867	0.99533				
11	5	0.00430	0.99963				
12	6	0.00036	0.99999				
13	7	0.00001	1.00000				

The Binomial Distribution with $n = 7$ and $p = 0.2$

Table 8.10 uses these Excel functions to compute the binomial distribution for a binomial random variable S_n with parameters $n = 7$ and $p = 0.2$. This table reports $P(S_n = k)$ and $P(S_n \leq k)$ for k between 0 and 7, inclusive.

The random variable S_n in Table 8.10 has $E(S_n) = np = (7)(0.2) = 1.4$. Table 8.10 indicates that 1 is the most likely value of S_n, that $P(S_n = 5)$ is tiny, and that $P(S_n \geq 6)$ is smaller still. Excel's "chart" capability has been used to plot the probability distribution of S_n that appears as Figure 8.10.

Figure 8.10 indicates that this probability distribution is asymmetric. This figure indicates that that 1 is the most probable value and that values in excess of 5 are unlikely to occur.

The Binomial Distribution with $n = 100$ and $p = 0.07$

Let us now examine the probability distribution of a random variable S_n whose distribution is binomial, with parameters $n = 100$ and $p = 0.07$. This random variable represents the number of successes in 100 Bernoulli trials, each having a probability 0.07 of success. This

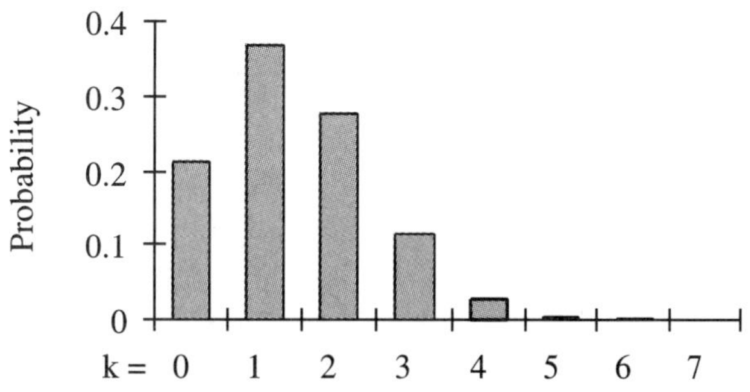

Figure 8.10 The binomial distribution with parameters $n = 7$ and $p = 0.2$.

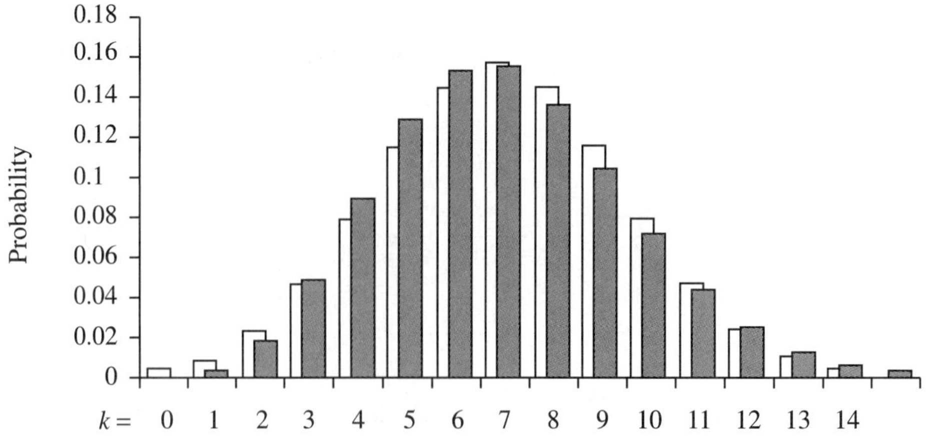

Figure 8.11 Binomial probability distribution with parameters $n = 100$ and $p = 0.07$ (shaded) and normal approximation to it (unshaded).

random variable S_n can take the values 0 through 100, with expectation $E(S_n) = np = (100)(0.07) = 7$. The shaded portion of Figure 8.11 plots its probability distribution.

Although this random variable S_n can take any value k up to 100, its distribution is nearly symmetric and "bell-shaped," and the probability that S_n exceeds 14 is tiny.

The Normal Approximation to the Binomial

The unshaded probability distribution in Figure 8.11 is the "normal" approximation to the binomial with parameters $n = 100$ and $p = 0.07$. Figure 8.11 shows that this normal distribution is a decent fit. This figure illustrates the following rule:

> The normal distribution provides a decent fit to the binomial if the expected number np of successes is at least 7 and if the expected number $n(1 - p)$ of failures is at least 7.

In other words, when $np \geq 7$ and $n(1 - p) \geq 7$, the binomial distribution sums enough Bernoulli random variables that the Central Limit Theorem has taken effect.

Counting Acceptances

The binomial distribution counts successes. To illustrate its use, we turn our attention to a problem faced by college admissions officers.

Problem C (College Admissions)

A college has room for 300 new students. By agreement with its competitors, the college's admissions officer must mail all offers of admission on the same date. Each admitted applicant accepts her or his offer with probability of 0.6. Different applicants make their decisions independently. The college wishes to make as many offers as is possible but is willing to run only 1 chance in 20 that more than 300 students accept their offers. Its director of admissions has asked you how many offers of admission should be made.

Table 8.11 Bracketing the number n of offers to make.

	A	B	C	D	E	F	G
1							
2	k =	300		min =	400		
3	p =	0.6		incr =	20		
4		=E2	= B6 + $E3				
5		▼	▼				
6	n =	400	420	440	460	480	500
7	P(Sn <= k) =	1.000	1.000	1.000	0.991	0.878	0.517
8		▲					
9		= BINOMDIST($B2, B6, $B3, 1)					

You designate as n the number of offers of admissions to make. The number S_n of students who will accept their offers is binomial with parameters n and 0.6. The college wants you to pick the largest value of n for which $P(S_n > 300)$ does not exceed $1/20$. Equivalently, it seeks the largest value of n for which $P(S_n \leq 300)$ is at least $1 - 1/20 = 19/20 = 0.95$.

Aiming for an overestimate, you calculate the value of n for which the expected number of acceptances equals 300. Setting $np = 300$ gives $n = 300/p = (300)/(0.6) = 500$. Evidently, 500 offers is somewhat too large because $P(S_{500} \leq 300)$ is about 0.5.

Using Excel, you can easily home in on the correct value of n by a process called *bracketing*. Table 8.11 indicates how. In Table 8.11:

- Cell E2 records the minimum value (of 400), and cell E3 records the increment (of 20).
- A drag across row 6 computes the values 420, 440, 460, 480, and 500.
- A drag across row 7 computes $P(S_n \leq 300)$ for the values of n in row 6.

From Table 8.11, you learn that 460 offers is too few because $P(S_{460} \leq 300)$ equals 0.991 and that 480 is too many because $P(S_{480} \leq 300)$ is 0.878. This "brackets" the number n of offers; n lies between 460 and 480. To shrink the bracket, you can change the "minimum" in cell E2 to 460 and the "increment" in cell E3 to 4. After repeated shrinking, you determine that 471 is the largest value of n that keeps $P(S_n \leq 300)$ above 0.95. Offers of admission should be made to 471 applicants.

8.14. THE CUMULATIVE DISTRIBUTION FUNCTION

For some purposes, we wish to know the probability that a random variable falls within a certain range. This was the case in Problem C, for instance, where we wanted to know the probability $P(X \leq 300)$. In this type of situation, the "cumulative distribution" is particularly useful.

The **cumulative distribution function** (abbreviated **CDF**) of the discrete random variable X is the function $F(t)$ of t that is defined by $F(t) = P(X \leq t)$ for each value of t. Evidently, at each value x_k that X attains, the function $F(t)$ jumps upward by the amount $P(X = x_k)$, and $F(t)$ is flat between successive values that X attains. Table 8.10 specifies the CDF of the random variable Y at each point where it increases. Figure 8.12 plots this CDF.

The CDF jumps upward at each value that the discrete random variable takes. For instance, $P(Y \leq 0) = 0.2097$ and $P(Y \leq t) = 0$ for all $t < 0$.

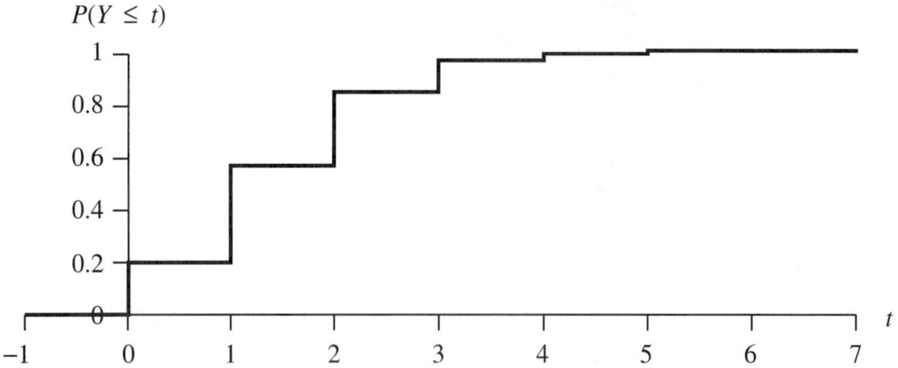

Figure 8.12 The CDF of a random variable Y whose distribution is binomial with parameters $n = 7$ and $p = 0.2$.

The Excel functions include several discrete probability distributions. For each, Excel reports both the probability distribution and its CDF. To distinguish between the two, Excel employs this convention:

- To get Excel to report the probability distribution itself, use "0" as the final argument.
- To get Excel to report the cumulative distribution function, use "1" as the final argument.

The functions =BINOMDIST(k, n, p, 0) and =BINOMDIST(k, n, p, 1) illustrate this convention, the latter being the CDF.

8.15. THE GEOMETRIC DISTRIBUTION

The random variable N is said to have the **geometric distribution** with parameter p if N equals the number of independent Bernoulli trials that are needed to obtain the first success, where each Bernoulli trial has probability p of success. Figure 8.13 presents the probability distribution of N.

The event $N = 1$ occurs if the first trial is a success, and this occurs with probability p, which explains the first row of Figure 8.13. The event $N = 2$ occurs if the first trial is a failure and the second trial is a success; that occurs with probability of $(1 - p)p$, which explains the second row of Figure 8.13. In general, the event $N = k + 1$ occurs if the first k trials are failures and the next trial is a success, which occurs with probability $(1 - p)^k p$. The dots at

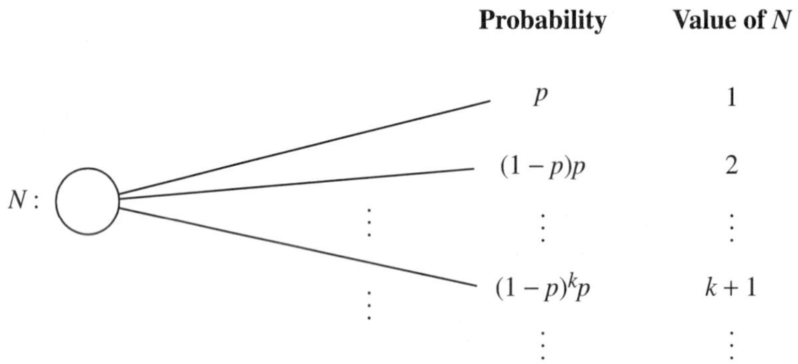

Figure 8.13 A geometric random variable N.

the bottom of Figure 8.13 recognize the fact that there is no largest value of k. The geometric random variable N has *infinitely* many outcomes because the first success could occur after any number k of failures.

A Puzzle

To encompass the geometric distribution, we must allow the probability model to have as many outcomes as there are positive integers. This seems innocuous, but it opens a Pandora's box, presenting us with a mathematical and philosophical puzzle. We shall identify the puzzle and then pass on. Suppose you flip a fair coin until the first heads appears. You will need to flip this coin a number N times that has the geometric distribution with parameter $p = 0.5$. It is conceivable that you will never get a head; the sequence of flips could exist exclusively of tails. The sequence $(T, T, T, \ldots)$ that consists exclusively of tails occurs with probability of 0, but it can occur. Let us agree to stay clear of the mathematics of events that can occur but do occur with probability of zero!

A bit of computation (that we omit) shows that the mean and variance of N are given by:

$$E(N) = \frac{1}{p} \quad \text{and} \quad \text{Var}(N) = \frac{1-p}{p^2} \quad \text{when } N \text{ is geometric.} \tag{8.16}$$

No Memory

The geometric distribution has no "memory" in a sense that we now identify. Let us compute the conditional probability that the first success occurs on trial $n + k$ given that the first n trials were failures. The formula for conditional probability gives:

$$P(N = n + k|N > n) = \frac{P(N = n + k, N > n)}{P(N > n)} = \frac{P(N = n + k)}{P(N > n)}.$$

We compute the numerator and denominator of this ratio. We get $P(N = n + k) = (1 - p)^{n+k-1}p$ because the event $N = n + k$ occurs if we toss $n + k - 1$ tails, followed by a head. We get $P(N > n) = (1 - p)^n$ because the event $N > n$ occurs if our first n tosses are tails. Substituting gives

$$P(N = n + k|N > n) = \frac{(1-p)^{n+k-1}p}{(1-p)^n} = (1-p)^{k-1}p = P(N = k).$$

Evidently, the probability that we must wait k *more* tosses for the first head is independent of the number n of tails that have occurred so far. This is the **memoryless property** of the geometric distribution. The coin that we are flipping has no memory; its probability of coming up heads on the next flip is independent of whatever happened on prior flips.

8.16. THE POISSON DISTRIBUTION

Of all of the discrete probability distributions, the most useful may be the Poisson, which is the natural model for arrivals at service facilities.

The Poisson

The Poisson distribution has only one parameter, and the symbol λ (pronounced lambda) is often reserved for this parameter. A random variable whose distribution is Poisson can take any value that is a nonnegative integer. Its probability law may seem to be a bit intricate.

Specifically, a random variable N is said to have the **Poisson** distribution with parameter λ if N can take the values 0, 1, 2, . . . and if

$$P(N = k) = e^{-\lambda}\frac{\lambda^k}{k!}, \quad \text{for} \quad k = 0, 1, 2, \ldots, \tag{8.17}$$

where e is **Euler's constant**, whose value is approximately 2.71828. There is no upper bound on the value k that a Poisson random variable can take. A bit of algebra (that we omit) verifies that

$$E(N) = \lambda \quad \text{and} \quad \text{Var}(N) = \lambda \quad \text{when } N \text{ is Poisson.} \tag{8.18}$$

Thus, the parameter λ of the Poisson distribution equals its expectation and its variance.

The Poisson Distribution in Excel

Equation (8.17) does not lend itself easily to hand computation. Again, imagine computing $P(N = 15)$ from Equation (8.15) by hand. Hand computation is not necessary because an Excel function specifies the Poisson distribution. Specifically:

- For a random variable Y that has the Poisson distribution with parameter λ, the Excel function =POISSON(k, λ, 0) reports $P(Y = k)$.
- For the same random variable Y, the Excel function =POISSON(k, λ, 1) reports $P(Y \leq k)$.

Figure 8.14 presents two Poisson distributions, one with parameter (mean) λ of 2, and the other with parameter λ of 7. Excel's charting capability was used to produce these distributions. When $\lambda = 2$, the probability distribution of N is asymmetric. By contrast, when $\lambda = 7$, the probability distribution is nearly symmetric and bell-shaped.

Figure 8.14 suggests (correctly) that as λ increases, the distribution of the Poisson becomes more and more symmetric and bell-shaped. Later in this section, we will see why. Our first task, however, is to indicate why the Poisson is the natural model for arrivals at service facilities. To accomplish it, we introduce some vocabulary.

The Superposition of Point Processes

The terms *superposition* and *point process* are best introduced by an example. In Figure 8.15, time increases linearly, to the right. Each of the five dashed lines depicts the call initiation process on a particular telephone. Each "×" on a particular dashed line marks the time at which a call is initiated on that line's telephone. Each dashed line in the figure is called a **point process**, and each "×" is said to be an **arrival**.

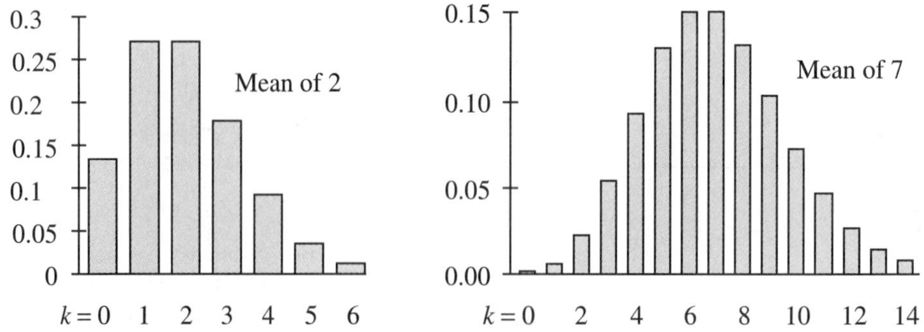

Figure 8.14 The Poisson distribution with parameter (mean) λ of 2 and of 7.

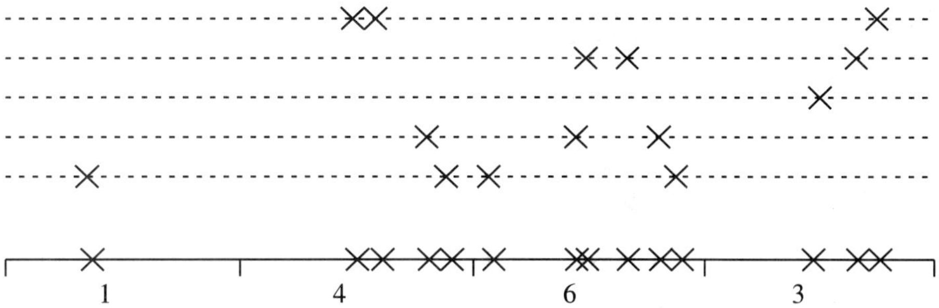

Figure 8.15 Five point processes and their superposition.

The solid line is the **superposition** of the dashed lines. Each "×" on the solid line marks a time at which a call is initiated by one of the telephones whose arrival processes have been superimposed (aggregated). The superposition is itself a point process.

In Figure 8.15, time has been divided into four intervals, and the aggregate number of arrivals in each interval has been recorded (e.g., 1 arrival in the first interval, 4 in the second, 6 in the third interval, and 3 in the fourth).

Information about the aggregated (superimposed) arrival process is important to the operator of the facility that serves these arrivals. We will soon see that the aggregated process can have a simple structure. In preparation, we introduce a definition. A point process is said to be a **Poisson** process if these two conditions are met:

- The number of arrivals that will occur in any interval of time has the Poisson distribution.
- The numbers of arrivals in nonoverlapping intervals of time are independent of each other.

We will soon see why arrivals at a service facility form a Poisson process. The solid line in Figure 8.15 is characteristic of a Poisson process—periods of heavy activity punctuated by periods of inactivity.

The Superposition Theorem

The Superposition Theorem describes the superposition of a large number of point processes. To introduce it, we pursue the telephone example. A central office serves thousands of telephones. At each telephone, calls are placed intermittently, with some uncertainty as to the time between calls. Calls at different telephones are placed independently of each other, though not necessarily independently of the calendar time. No single telephone accounts for a significant fraction of the calls that the central office experiences. For these reasons, the call initiation process at a central office satisfies the hypothesis of the following theorem.

Superposition Theorem. Consider a facility that serves many customers. Suppose that:

- Each customer uses the facility intermittently, with some uncertainty in the time between successive arrivals of that customer.
- Each customer decides when to use the facility independently of the others.
- No single customer accounts for a significant fraction of the arrivals.

Then the superposition of their arrival processes is well-approximated by a Poisson process.

The gist of the Superposition Theorem is that the "superposition" of many low-intensity independent arrival processes is a Poisson process. Except for their mean inter-arrival times, details about the individual processes have no effect on the superposition.

In a Poisson process, the number of arrivals in every period is Poisson. The number of arrivals in nonoverlapping 15 minute periods is Poisson, but they need not have the same means (parameters). The mean arrival rate can vary with the time of day, for instance.

To illustrate the Superposition Theorem, we pose three rhetorical questions:

- Does the number N of people who will initiate telephone calls in Copenhagen within a given 10-minute period have the Poisson distribution? Yes. There are many potential callers in Copenhagen, each of whom initiates calls intermittently and independently of the others, and no single customer accounts for a significant fraction of the traffic.
- Does the number N of customers who will arrive at a fish store within a given 30-minute period have the Poisson distribution? Yes (ho, ho), and for the same reasons.
- Does the number N of birds who arrive at a bird feeder within a given 5-minute period have the Poisson distribution? No. Birds do not make arrival decisions independently. "Birds of a feather flock together."

In brief, if a facility serves many customers who time their arrivals independently, the aggregate arrival process will be Poisson.

The Superposition of Poisson Processes

Let us suppose that we superimpose two Poisson processes. What happens? To develop an intuition for this issue, let's consider the arrivals at a hospital's emergency room. This emergency room serves a large population. Each member of this population appears intermittently and independently of the others. No single patient is responsible for a significant fraction of the arrivals. The Superposition Theorem guarantees that each of the following is a Poisson process:

- The arrival process of adult female patients at an emergency room.
- The arrival process of adult male patients at this emergency room.
- The arrival process of juvenile patients at this emergency room.
- The arrival process of all patients at this emergency room.

Each of these arrival processes is Poisson. The first three processes are independent of each other. (The only thing that would make them dependent is the simultaneous appearance of multiple family members, each requiring treatment, but that is rare enough to ignore.) The fourth process is the superposition of the first three. This illustrates a point that we highlight.

The superposition of independent Poisson processes is a Poisson process.

For instance, the superposition (aggregate) of the arrival processes at the emergency room of adult male patients and adult female patients is a Poisson process.

The Sum of Poisson Random Variables

Let N_1 and N_2 be independent random variables, let N_1 be Poisson with parameter (mean) of λ_1, and let N_2 be Poisson with parameter of λ_2. What can we say about their sum, $N_1 + N_2$? The fact that the expectation of the sum equals the sum of the expectations guarantees $E(N_1 + N_2) = E(N_1) + E(N_2) = \lambda_1 + \lambda_2$, the last because the parameter of a Poisson random variable equals its expectation.

But more can be said. This sum, $N_1 + N_2$, has the Poisson distribution, a fact that we could verify algebraically. The algebra is unnecessary, however, because the result that is highlighted above demonstrates that the sum is Poisson. Do you see why? We emphasize:

> **Sum of Poisson random variables:** The sum of independent Poisson random variables is Poisson. The parameter (mean) of the sum is the sum of the parameters (means).

The Normal Approximation to the Poisson

Let us now consider the sum of S_n of n independent Poisson random variables each of which has 1 as its parameter (mean). We have just learned that S_n has the Poisson distribution, with $E(S_n) = n$. The Central Limit Theorem guarantees that the sum S_n is approximately normal when n is large enough. Figure 8.14 suggests (correctly) that $\lambda = 7$ is large enough for the normal to provide a decent fit.

A Random Split of a Poisson Process

To introduce the idea of a random split, we use an example. In an emergency room, a "triage" nurse determines the category of treatment that each patient requires. Let us number the categories 1 through C. Let category 1 be routine (nonemergency) care, category 2 a cardiac case, category 3 a surgical case, and so forth. A certain fraction of the patients will be assigned each category. We denote as p_k the fraction of the arrivals that will be assigned to category k, for $k = 1, 2, \ldots, C$. Thus, a randomly selected arrival has a category X whose probability law is given in Figure 8.16.

In a **random split** of an arrival process, the same probability law assigns a category to each arrival, and these assignments are independent of each other. Independence is reasonable in the case of triage; the fact that a given arrival requires cardiac care does not affect the disease category of the next arrival. By assigning a category to each arrival, this random split separates one arrival process into several. The properties of this split are highlighted below.

> A random split of a Poisson process into C categories results in C independent Poisson processes, one per category.

This result seems innocuous, but it has surprising implications. To bring one of these implications into view, we suppose that the probability law in Figure 8.16 has been used to

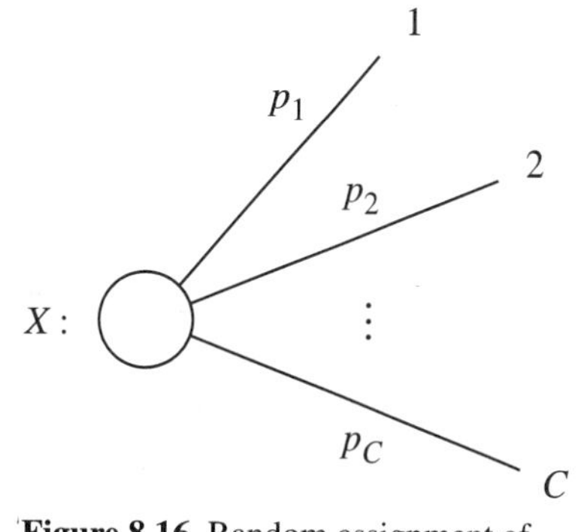

Figure 8.16 Random assignment of category.

assign categories to arrivals in a Poisson process. Denote as N the total number of arrivals during a particular period. Since the arrival process is Poisson, we know that the probability distribution of N is Poisson. Of these N arrivals, denote as N_k the number that get assigned to category k, for each k. The equation

$$N = N_1 + N_2 + \cdots + N_C \tag{8.19}$$

holds because each of these N arrivals is assigned to exactly one category.

The result that is highlighted here states that the random variables N_1 through N_C are mutually independent and that each of them has the Poisson distribution. Since each arrival is assigned to category k with probability p_k, we also know that $E(N_k) = p_k E(N)$. The random variables N_1 through N_C are *not* independent of N; knowing N determines the value of their sum, for instance. But the random variables N_1 through N_C are independent of *each other*.

Recap

The Poisson distribution has a single parameter, λ, which equals its mean and its variance. In a Poisson process, the number of arrivals in each interval of time has the Poisson distribution, and the numbers of arrivals in nonoverlapping intervals of time are independent of each other. We've identified these facts about the Poisson.

- The Superposition Theorem shows that a Poisson process is the natural model for the number of arriving customers at a service facility.
- The superposition of Poisson processes is a Poisson process.
- The sum of independent Poisson random variables is Poisson.
- The random split of a Poisson process into C categories produces C Poisson processes that are independent of each other.
- The random split of a Poisson random variable into C categories produces C Poisson random variables that are independent of each other.

These results connote—correctly—that the Poisson will play a prominent role in later chapters that describe the operation of systems.

8.17. REVIEW

A discrete random variable is a probability model that assigns a number to each outcome. An easy way in which to describe a discrete random variable is through its probability tree, as in Figure 8.1. A different way in which to describe a random variable is through its cumulative distribution function (CDF), as in Figure 8.15. The CDF is handy when we want to compute the probability that a random variable X falls into a given interval. Excel reports both the probability distribution and the CDF of the random variables that appear among its functions.

Two important measures of a random variable are its expectation (or mean) and its variance. The expectation measures the center of the probability distribution, whereas the variance measures the spread in the probability distribution. Listed below are the properties of means and variances.

- The expectation of the sum of two or more random variables equals the sum of their expectations, whether or not they are independent.
- The variance of the sum of random variables equals the sum of their variances plus twice the sum of their covariances.
- If random variables are mutually independent, their covariances equal zero, and the variance of their sum equals the sum of their variances.

- The unit of measure of the variance is the square of the unit of measure of the random variable.
- The standard deviation equals the square root of its variance. The standard deviation has the same unit of measure as the random variable itself.
- The sum of sufficiently many independent random variables has the bell-shaped distribution of the "normal" random variable. This normal distribution (which is studied in Chapter 11) preserves no information about the random variables that have been summed other than their means and variances.

A probability tree is a handy way in which to visualize a discrete random variable. A probability tree also lets us visualize a function of a random variable. Just as X takes the value x_k with probability p_k, the function $g(X)$ takes the value $g(x_k)$ with probability p_k. With μ denoting $E(X)$, the variance of X can be computed in either of two ways. The variance of X equals the expectation of the function $(X - \mu)^2$, and it also equals the expectation of the function $(X^2 - \mu^2)$.

We have indicated that, for operational decisions entailing monetary payoffs, maximizing expected profit is sensible. We have also suggested that when the payoffs are substantial, it can be reasonable to examine the tradeoff between the expectation and variance of the profit.

This chapter introduces four discrete probability distributions, namely, the Bernoulli, the binomial, the geometric, and the Poisson. Let us recall that:

- The binomial distribution is the natural model for counting; the binomial with parameters n and p counts the number of successes in n independent Bernoulli trials, each having probability p of success.
- The geometric counts the number of Bernoulli trials needed to obtain the first success. The geometric has a "memoryless" property; the number of trials that remain until the first success is independent of the number of failures that have occurred so far.
- The Poisson is the natural model for arrivals at a service facility; the superposition of a large number of independent low-intensity arrival processes is Poisson.

In particular, the Poisson plays the same role for the superposition that the normal plays for the sum. In the superposition, details about the individual arrival processes become unimportant, except to the extent that they affect the mean arrival rate.

Table 8.12 summarizes the probability distributions of the discrete random variables that we have studied.

Spreadsheets have appeared throughout this chapter. They have been used to compute means, variances, and covariances, as well as to explore the shapes of the binomial and Poisson distributions. Spreadsheets have been used to compute probabilities that would be hard to calculate by hand. Spreadsheets have also helped us to identify conditions under which the normal distribution is a good approximation to the binomial and the Poisson.

Table 8.12 Four discrete random variables.

Random variable X	Parameter(s)	Values of k	$P(X = k)$	$E(X)$	Var(X)
Bernoulli	p	$\{0, 1\}$	$p^k(1-p)^{1-k}$	p	$p(1-p)$
Binomial	n, p	$\{0, 1, \ldots, n\}$	$\binom{n}{k}p^k(1-p)^{n-k}$	np	$np(1-p)$
Geometric	p	$\{1, 2, \ldots\}$	$(1-p)^{k-1}p$	$\frac{1}{p}$	$\frac{1-p}{p^2}$
Poisson	λ	$\{0, 1, \ldots\}$	$e^{-\lambda}\frac{\lambda^k}{k!}$	λ	λ

8.18. HOMEWORK AND DISCUSSION PROBLEMS

1. List the reasons why the expectation and the variance are used as measures of the center and the spread of a probability distribution.

2. Let the random variable Y in Figure 8.2 and Z in Table 8.4 be independent of each other.

(a) Draw a probability tree that for the random variable $T = Y + Z$.

(b) Represent the probability tree in part (a) on a spreadsheet. On that spreadsheet, compute the expectation of T, the expectation of T^2, and the variance of T.

(c) Repeat parts (a) and (b) for the random variable $U = Y - Z$.

(d) Did the means and variances that you obtained in parts (b) and (c) look familiar? If so, why?

3. The random variable Z in Table 8.4 takes the values 3000 and 4000 with probability of 0.5. Compute MAD(Z) and StDev(Z). State in words why MAD(Z) and StDev(Z) take the values that they do.

4. As an alternative to the expectation, one could employ the "average value" of a random variable, namely, the sum of the values, divided by their number. By this measure, the average value of the random variable Y in Figure 8.2 would be $(-\$1000 + \$4000 + \$6000)/3 = \$3000$. Intuitively, why is the expectation more descriptive than the average?

5. The variance is a measure $M(X)$ of the spread of the distribution of the random variable X. The variance is an "additive" measure in the sense that independent random variables X and Y have $M(X + Y) = M(X) + M(Y)$. True or false: Of the measures of spread that were discussed in Section 8.4, only the variance is additive in this sense.

6. The random variable Y in Figure 8.2 takes the values -1000, 4000, and 6000 with probabilities $1/4$, $3/8$, and $3/8$, respectively.

(a) Plot the cumulative distribution function $F(t) = P(Y \leq t)$ of t.

(b) Shade the region consisting of each point (x, y) in which $x \geq 0$, and $F(x) \leq y \leq 1$. Compute the area of the shaded region. Does it equal $E(Y)$?

7. Suppose that X is equally likely to take the values 1, 2, 3, 4, 5, and 6. Specify the probability distribution of X. Compute $E(X)$, $\text{Var}(X)$, and $\text{StDev}(X)$.

8. Suppose that X and Y are independent random variables, each of which is equally likely to take the values 1, 2, 3, 4, 5, and 6. Specify the probability distribution of the random variable $T = X + Y$.

9. How many times must you flip a coin in order for the sum of the number of heads to be approximately normal?

10. Let the random variable Y denote an amount that your firm will earn today in U.S. dollars. Suppose that you wish to convert Y to Japanese yen. Your bank converts U.S. dollars to Japanese yen at the rate of c yen per dollar, and it imposes a fixed charge of d yen for each transaction. Interpret $cY - d$. What is its unit of measure? What is the unit of measure of its variance? Interpret the equation $\text{Var}(cY - d) = c^2\,\text{Var}(Y)$ in terms of dollars and yen.

11. The random variable W takes the values 1.5, 3, 6, and 7 with probabilities 0.2, 0.25, 0.4, and 0.15, respectively.

(a) Compute the expectation of W and W^2. Compute the variance of W and its standard deviation.

(b) Define the (cumulative distribution) function $F(t)$ of t by $F(t) = P(W \leq t)$. Plot the function $F(t)$ versus t.

(c) Shade the region consisting of each point (x, y) in which $x \geq 0$, and $F(x) \leq y \leq 1$. Compute the area of the shaded region. Does it equal $E(W)$?

12. **(The Expectation as an Area)** For the random variable X in Figure 8.1, suppose that $0 \leq x_1 < x_2 < \cdots < x_n$, and set $F(t) = P(X \leq t)$ for each t. Show $E(X)$ equals the area of the region that contains each pair (x, y) in which $x \geq 0$ and $F(x) \leq y \leq 1$. *Hint:* Break the area into n rectangles whose areas are $p_k x_k$ for $k = 1, 2, \ldots, n$.

13. **(Efficient Portfolios)** You have a budget of \$10,000 to invest for one year in the common stock (shares) of three companies. Today's market price for each company's common stock is \$10 per share. Table 8.6 specifies the profit per share of each stock under each state of the economy at the year's end, as well as the probability that each state will occur. Table 8.9 computes the expectation and variance of the profit on a portfolio that includes s_1, s_2, and s_3 of shares of stock in company 1, 2, and 3, respectively.

(a) Suppose you wish to guarantee an expected profit of at least 4% on your \$10,000 portfolio, equivalently, that $0.5s_1 + 0.3s_2 + 0.4s_3 \geq 0.04 \times 10{,}000$. Use Solver to find the portfolio that minimizes the variance subject to constraints that keep the mean return at or above the desired threshold, enforce your budget constraint, and keep the purchase quantities nonnegative.

(b) Redo part (a) several times, each time for an expected profit threshold that is at least some fixed percent x of your \$10,000 investment. Plot the **efficient frontier**, which describes the tradeoff between the mean profit and its standard deviation. Describe the way in which the portfolio varies as the mean profit level increases.

(c) Redo parts (a) and (b) for the case in which **short sales** are allowed. A short sale occurs if you borrow stock, sell it now, and buy it back later. Selling shares in company k short for a period of one year amounts to removing the constraint that S_k be nonnegative. For the case in which short sales are allowed, why do the efficient portfolios take the form that they do?

14. Let X be binomial with parameters $n = 10$ and $p = 0.5$. Compute:

(a) $P(X = 5)$.

(b) $P(3 \leq X \leq 6)$.

(c) $P(X \leq 3)$.

15. Let Y be binomial with parameters n and p. Specify $E(Y^2)$. *Hint*: This should require no computation.

16. A consultant to your election reports having done a survey that indicates that between 56% and 60% of the electorate prefers you to your opponent. This news seems a bit too good to be true. Before paying the consultant's fee, you ask a friend to sample 10 randomly selected voters and confirm the consultant's finding.

(a) Your friend calls 10 voters and learns that 3 prefer you and 7 prefer your opponent. If the consultant is right, how likely is a result this far below his estimate?

(b) You ask your friend to call 10 more voters. Your friend does so and learns that 4 of them prefer you and 6 prefer your opponent. If the consultant is right, how likely are results this far below his estimate? Should you question him about his survey?

17. **(In Control)** A process for making circuit boards is either "in control" or "out of control." When the process is in control, each circuit board that it makes is defective with probability 0.02. When it is out of control, each circuit board that it makes is defective with probability of 0.10. Historically, this process has been in control 80% of the time. Inspection of circuit boards is imperfect. If a circuit board is defective, inspection identifies it as such with probability of $\frac{3}{4}$. If a circuit board is not defective, inspection identifies it as nondefective with probability of 1.0.

(a) While the process is in control, 100 circuit boards are inspected. Describe the probability distribution of the number X of defective circuit boards that will be detected.

(a) Repeat part (a) for a process that is out of control.

(b) Suppose that a batch of 100 circuit boards was inspected and that five have been found to be defective. What is the probability that the process is out of control?

(c) Designate as k the number of circuit boards that are found to be defective by testing a batch of 100. Calculate the probability that the process is out of control as a function of k. Find the smallest value of k for which this probability is at least 0.95.

18. Ten percent of the samples of air contain a rare molecule. Designate as N the number of samples that must be taken until one is found that contains this molecule.

(a) What can you say about the probability distribution of N? What is its expectation? What is its standard deviation?

(b) How many samples must you take in order to assure a probability of 0.9 that at least one of them contains this molecule?

19. **(Betting Systems)** For an evening of entertainment, Jack and Jill have joined an American roulette table, which works as follows. A ball is spun around the wheel. It is equally likely to come to rest at any of 38 slots that are numbered 1 through 36, "O," and "OO." (The "O" and "OO" slots are neither even nor odd.) Jack's habit is to bet \$5.00 either on "even" or on "odd" at each roll. Either way, Jack wins \$5.00 with probability of $18/38$ and loses \$5.00 with probability of $20/38$. Jill's habit is to bet \$5.00 on a "number" at each roll. No matter which number she chooses, Jill wins 35 times the amount of her bet with probability of $1/38$ and loses the amount that she bet with probability of $37/38$. Over the course of the evening, each of them will place 100 bets. The outcomes of different spins of the roulette wheel are independent.

(a) Let the random variable W equal Jack's net profit for the evening. Compute the expectation, variance, and standard deviation of W.

(b) Do the same for Jill.

(c) True or false: Jack's net profit W is given by $W = 10X - 500$ where X has the binomial distribution with parameters $n = 100$ and $p = 18/38$. Use an Excel spreadsheet to compute the probability that W is positive.

(d) Do the analog of part (c) for Jill. Which of them is more likely to earn a profit?

Note: The **positive part** of the number x is the function $(x)^+ = \max\{x, 0\}$. Thus, if x is a positive number, $(x)^+ = x$, and if x is a negative number, $(x)^+ = 0$. The positive part plays an important role in inventory problems, as is indicated below.

20. **(Lost Sales and Unsold Units)** You sell an item, retail. You have q units of this item on hand at the start of today. No further units will appear until tomorrow. The demand D that will occur today for this item is random. To the extent that you cannot satisfy the demand, you lose the sales. (Potential customers go elsewhere; they do not wait for you to restock the item.) Which of the following statements are true? Why?

(a) $x = (x)^+ - (-x)^+$.

(b) $(D - q)^+$ equals the number of lost sales opportunities that will occur today.

(c) $(q - D)^+$ equals the number of unsold units you will have on hand at the end of the day.

(d) $(q - D)^+ = q - D + (D - q)^+$.

(e) $E[(q - D)^+] = q - E(D) + E[(D - q)^+]$.

(f) The expected number of unsold units is determined by the expected demand, q, and the expected number of lost sales.

(g) $(q - D)^2 = [(q - D)^+]^2 + [(D - q)^+]^2$.

21. You are the admissions officer for a residential college. In recent years, 55% of the applicants who were offered admission accepted their offers and came; you have no reason to believe that the accept rate will be any different this year. Dorm space is a problem; you have room for 300 freshmen, and you want to make enough offers that there is only 1 chance in 10 that the number of arriving freshmen will exceed the college's dorm capacity.

(a) If you make n offers of admission, the "yield" (number of people who become freshmen) is a random variable S_n whose distribution is ________ (fill in the blank).

(b) Use an Excel function to determine how large n should be.

Note: Parts (c), (d), and (e) refer to the value of n that you computed in part (b).

(c) Designate as Y the number of freshmen who appear for which your college has no dorm rooms. Thus, Y equals 0 if $S_n \le 300$, Y equals 1 if $S_n = 301$, and so forth. On a spreadsheet, specify $P(Y = k)$ for the values $k = 0, 1, 2, \ldots, 30$. (The probability that Y exceeds 30 is small enough to ignore.) On the same spreadsheet, compute $E(Y)$ and, $E(Y^2)$, $\text{Var}(Y)$ and $\text{StDev}(Y)$.

(d) Specify the probability distribution of Y given the condition that more than 300 students accepted offers of admission. Compute the mean, variance, and standard deviation of this random variable.

(e) Designate as Z the number of unfilled dorm rooms. Note that $Z = 0$ if S_n is at least 300, that $Z = 1$ if $S_n = 299$, and so forth. Compute $E(Z)$. *Hint:* Does the preceding problem guarantee $E(Z) = 300 - (n)(0.55) + E(Y)$?

22. (**Insurance Reserve**) A company insures each of n individual policyholders. The company reimburses its policyholders for losses on their policies. Losses on different policies are independent of each other. Each policy sustains a loss of \$100,000 with probability of 0.02 and sustains no loss with probability of 0.98. The total amount T_n that the company will pay out to policyholders during the year is a random variable.

(a) In the case of automobile liability insurance, is it reasonable that the losses on different policies be independent?

(b) In the case of flood or hurricane insurance, is it reasonable that the losses on different policies be independent?

(c) True or false: $T_n = (\$100{,}000)\ S_n$, where S_n has the binomial distribution with parameters n and $p = 0.02$.

(d) Express $E(T_n)$ and $\text{StDev}(T_n)$ as functions of n.

(e) To protect its policyholders against a bad year, the company sets aside a reserve that equals four times the standard deviation of its total payout for the year. On a spreadsheet, show how the reserve per policy varies with the number n of policyholders. Is there an economy of scale here?

23. (**Exam Strategy**) A multiple-choice test consists of 50 questions and has 5 choices for each answer. Your score on this test equals the number of correct answers less one-fourth the number of incorrect answers.

(a) Imagine that you make a pure guess on one question. Let X denote the change in your score due to this guess. What is the probability distribution of X? What is its mean? its standard deviation?

(b) Is there a sense in which this scoring system is fair?

(c) Redo part (a) for the case in which you can eliminate two of the choices as being definitely wrong.

Note: For parts (c) through (f), make these assumptions: On each of 22 questions, you can eliminate two of the five choices as being definitely wrong. Among the three choices that remain, you guess. Let N denote the number of correct guesses, and let Y denote the net increase (possibly negative) in your score that is due to guessing.

(d) What is the expectation of Y? What is its standard deviation?

(e) The number N of correct guesses is ________ (fill in the blank).

(f) True or false: $Y = (5N - 22)/4$.

(g) Compute the probability that guessing *de*creases your score.

(h) Suppose that on each of n questions, you eliminate two answers and guess among the remaining three. For values of n between 1 and 22, compute the probability that this strategy decreases your score. *Hint:* Part (e) gives a clue.

24. (**Lunch**) To reduce its enormous drag while flying at supersonic speed, the Concorde was designed to be light and compact. It accommodates 125 passengers and 135 lunches. After reaching cruising altitude, each passenger is offered a choice of two entrées, le Plat du Belge (red cabbage with sausage and mashed potatoes, a true delight) or French toast (bread dipped in a batter of reconstituted milk and eggs, fried, refrigerated, reheated, and doused with Lyle's golden syrup). Passengers choose entrées independently, and 60% select French toast. Before takeoff, the airline must stock the plane with some number b of servings of le Plat du Belge and $f = 135 - b$ servings of French toast. Today's flight is full. Let B and F denote the number of passengers on today's flight who will select le Plat du Belge and French toast, respectively.

(a) True or false: B and F are independent random variables.

(b) What is the probability distribution of B?

(c) Use a spreadsheet to find the number b that maximizes the probability that each customer receives his or her first-choice entrée. Compute this probability. *Hint:* you want to maximize $P(B \le b, F \le f)$.

(d) By how much would the probability that all customers can be satisfied increase if room could be made for one more lunch?

25. Suppose that arrivals satisfy the hypothesis of the Superposition Theorem. The expectation of the number of arrivals between noon and 1 P.M. is 26, and the expectation of the number of arrivals between 1 P.M. and 2 P.M. is 13. What can be said about the number of arrivals between noon and 2 P.M.? Why?

26. Customer arrivals at a shop are Poisson with a steady rate of six per hour.

(a) What is the probability that six or fewer customers will arrive within a one-hour period?

(b) What is the probability that three or fewer customers will arrive within a one-hour period?

(c) What is the probability that one or fewer customers will arrive within a 10-minute period?

(d) For what interval of time, in minutes, is it equally likely that no customers will arrive and that at least one customer will arrive?

27. Consider a Poisson process with a steady rate of 60 arrivals per hour. Consider a random split of these arrivals in accord with Figure 8.16, with $p_1 = 0.2$ and $p_2 = 0.3$. Let N denote the number of arrivals in a 10-minute period. Specify:

(a) $P(N_1 = 3)$ and $P(N_2 = 2)$.

(b) $P(N_1 = 3 | N_2 = 2)$ and $P(N_1 = 3, N_2 = 2)$.

(c) $P(N_1 = 3 | N = 5)$ and $P(N_1 = 3, N = 5)$.

28. (Spatial Distribution and the Poisson) We have argued that when points are distributed at random on the time line, the number falling into each interval is Poisson. A similar argument applies to the random distribution of points in *space*.

In a famous study of 537 flying bombs dropped on London during World War II, the south of London was divided into 576 regions of equal area (0.25 square kilometers), and the number of bomb hits on each region was counted. If bombs were dropped randomly in space, the number N of bombs hitting a particular region would have the Poisson distribution with parameter $\lambda = 537/576$. The expectation of the number of regions receiving k hits would equal $(576)\ P(N = k)$.

The actual bomb hits are tabulated below; of the eight regions that were hit by at least four bombs, seven regions were hit exactly four times, and one region was hit seven times.

	A	B	C	D	E	F
2	**# of hits**	**0**	**1**	**2**	**3**	**>= 4**
3	# of regions	229	211	93	35	8

(a) Compute the expectation of the number of regions that would receive the indicated number of hits if the spatial Poisson model was correct. How does the fit look, visually?

(b) Use the CHITEST function in Excel to measure the closeness of the fit. How does the fit look, mathematically?

29. (A Service Level) A retailer restocks each item that she sells after the close of each business day. The daily demands for these items are independent of each other. The daily demand for each item is Poisson, and she knows the mean daily demand for each item. Each night, after the store is closed for business, she restores her inventory position on each item to a level just high enough to keep the probability of satisfying the entire next day's demand for that item at least 19/20. Her **safety stock** on each item is the excess of its start-of-day inventory position over the mean demand for the day. On a spreadsheet, compute her safety stock for items whose mean demands are 5, 10, 20, 40, and 80 units. Plot the safety stock versus the square root of the demand. Does this safety stock exhibit an economy of scale?

30. (Field Exercise) Select a service facility (bank, library, eatery, or the like) and a period during which its mean arrival rate is fairly stable. Treat each group of people (e.g., family) who arrive together as one "arrival."

(a) Count the number of arrivals in a 10-minute period. Then choose an interval length such that the mean number of arrivals per interval is at least 1.5.

(b) Observe your system for roughly 60 intervals, and record the number of arrivals during each of these intervals. (If your intervals are 2 minutes long, it will take 120 minutes to record 60 intervals.)

(c) Tabulate your observations on a spreadsheet in a way that is suggested by cells B3 through C9, below.

	A	B	C	D
2	**number n per period**		**empirical**	**expected**
3		0	6	7.226
4		1	20	15.295
5		2	14	16.187
6		3	8	11.421
7		4	6	6.044
8		5	5	2.558
9	at least	6	1	1.269
10	number of intervals		60	60.000
11	number of arrivals		127	
12	mean		2.117	
13	chi-squared statistic		0.498	

(d) Use functions to compute the analogs of your data of the numbers in cells C10, C11, and C12.

(e) Create the analog for your data of column D. (In the above table, the value of the function =C10*POISSON(B3,C$12,0) appears in cell D3.)

(f) Make an Excel chart that compares your empirical observations with the expected numbers. You should *not* expect the fit to be exact. Why? Visually, does it look as though the Poisson is a credible approximation to your data?

(g) Use the Excel statistical function, CHITEST, to compute the χ^2 (pronounced chi squared) probability that random fluctuations would produce a statistic at least as large as the one that you observed. In the preceding table, the function =CHITEST(C3:C9,D3:D9) is in cell 13.

31. Let X be a discrete random variable with expectation μ and variance σ^2.

(a) Fix any positive number t. Show that

$$\sigma^2 \geq t^2 P(|X - \mu| \geq t).$$

Hint: When calculating σ^2, approximate $(x_k - \mu)^2 p_k$ by zero if $|x_k - \mu|$ is below t and by $t^2 p_k$ if not.

(b) Prove **Chebychev's inequality**, which is that

$$P(|X - \mu| \geq t) \leq \frac{\sigma^2}{t^2}.$$

32. (**The Weak Law of Large Numbers**) Let S_n denote the sum of n i.i.d. random variables, each having mean μ and variance σ^2. The random variable $A_n = S_n/n$ denotes the average of these random variables.

(a) Express the mean and variance of A_n in terms of μ and σ^2.

(b) Use Chebychev's inequality to show that

$$P(|A_n - \mu| > t) \leq \frac{\sigma^2}{n(t - \mu)^2}.$$

(c) The weak law of large numbers is that for every positive number ε

$$P(|A_n - \mu| > \varepsilon) \to 0 \quad \text{as} \quad n \to \infty.$$

Prove the weak law of large numbers.

Remarks: The weak law of large numbers justifies our interpretation of μ as the long-run average of repeated trials. The **strong law of large numbers** is a far stronger statement of the same sort. The strong law states that

$$1 = P(\mu = \lim_{n\to\infty} A_n).$$

In other words, the strong law is that the probability of the event in which A_n fails to converge to μ equals 0.

33. (**Another Formula for the Covariance**) Demonstrate that $\text{Cov}(X, Y) = E(X \cdot Y) - \mu_X \mu_Y$.

Remark: Probabilities and expectations can often be computed by using the idea of a **regeneration point**, which is a situation in which the past does not matter. The next two problems are instances of this.

34. Let the random variable N have the geometric distribution with parameter p.

(a) Argue that N and $\hat{N}$ have the same probability distributions, with $\hat{N}$ specified in the probability tree that appears below.

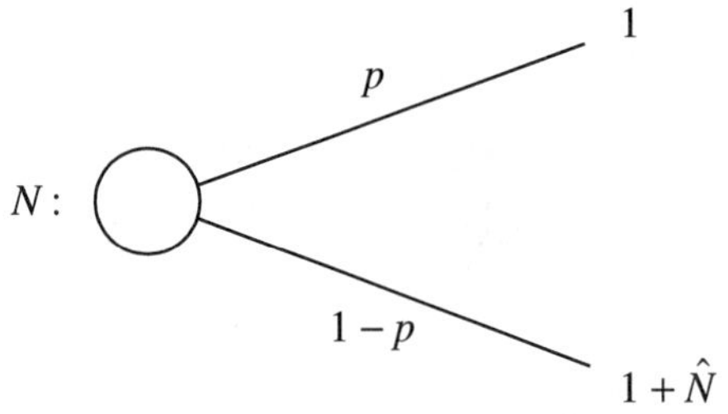

(b) From the preceding tree, compute the expectation of N, the expectation of N^2, and the variance of N.

35. (**Making Your Point**) The game of craps is played with two balanced dice. Your first throw of these was a "5," which means that the total of the numbers of pips on the two dice was 5. Your "point" is a 5. You will make your point (win) if you throw another 5 before throwing a 7. You will lose if you throw a 7 before throwing another 5. The probability of throwing a 5 on a single throw equals $4/36$, and the probability of throwing a 7 on a single roll is $6/36$. Let X equal 1 if you make your point, and let X equal 0 if you do not. As a consequence, $E(X)$ equals the probability that you will make your point.

(a) Argue that X and $\hat{X}$ have the same probability distributions, with $\hat{X}$ specified in the probability tree that appears below.

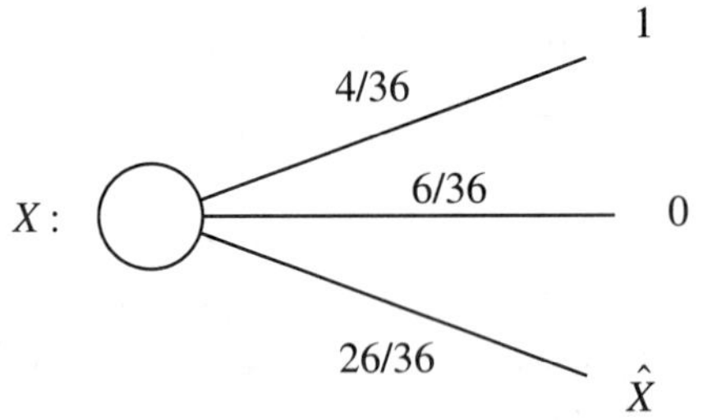

(b) Compute $E(X)$, the probability that you make your point.

(c) Suppose your first roll was a 6. You will make your point (win) if you throw another 6 before throwing a 7. What is the probability of making your point?

Chapter 9

Decision Trees and Generalizations

9.1. PREVIEW

Uncertainty about the future lies at the heart of many decision-making situations. A company that decides to purchase a piece of equipment has imprecise knowledge of its useful lifetime, reliability, and maintenance costs. A store that buys merchandise has inexact knowledge of what its customers will wish to buy. An airline purchases aircraft and sets flight schedules without knowing what demands will occur. A medical patient selects a course of treatment without knowing its outcome.

Although the future is unknown, it may obey a simple probability law. For instance, in Chapter 8 we learned that a natural model for arrival processes is the Poisson distribution. Similarly, the uncertainties in the course of a medical treatment can be expressed as probabilities, for example, the probability that chemotherapy will eradicate a particular type of cancer in a certain class of patient. In this context, the probabilities may be objective or subjective, and they can model inherent randomness or our lack of information.

This chapter introduces the subject of rational decision making under uncertainty. The simplest model of decision making under an uncertain future is the so-called decision tree. A decision tree embellishes a probability tree by having a second type of node. In addition to chance nodes, at which elements of uncertainty are resolved, this tree has one or more "choice nodes," at which decisions are made. In a decision tree, the probabilities can be objective or subjective, but they are assumed to be known. For a decision tree to be practical, the probabilities must be known, at least approximately, and the number of nodes must not be too large.

9.2. WHAT CAN YOU LEARN FROM THIS CHAPTER?

The decision tree is the simplest decision-making model in the face of an uncertain future. In such a model, a plan of action must account for all contingencies (chance outcomes) that can arise. A decision tree represents the uncertainty and the choices graphically. This makes it easy to visualize the contingency plans, which are called strategies.

In this chapter, you will see how to draw a decision tree and how to "roll it back" to find a strategy whose expected net cost is lowest and, equivalently, whose expected net profit is highest. You will also see how to use TreePlan (an Excel add-in that comes with this text) to execute the rollback automatically.

Decision trees bring into view two ways in which one strategy (or random variable) can dominate another. You will see that no decision maker who prefers more income to less would pick a strategy dominated in either of these ways.

In Chapter 7, "information" was interpreted as learning a chance outcome, and Bayes' rule measured the effect of information on probabilities. Here, you will see how to measure the economic value (increase in expected profit) of the same type of information.

Toward the end of this chapter, the methodology for analyzing decision trees is adapted to tree-like structures in which choice nodes can be visited only once. It will also be generalized to a case in which choice nodes can be revisited. These generalizations will introduce you to Markov decision processes.

9.3. AN EXAMPLE: MEDFLY ERADICATION

To introduce decision trees, we pose the rhetorical question in:

Problem A (Medfly Eradication)

You are the Florida state official who is responsible for controlling crop damage. A localized infestation of medflies has just been detected, threatening the citrus region, where it would destroy a crop worth billions of dollars annually. The governor has just telephoned you. In exactly one day, you are to present him your plan for ridding Florida of this pest. He has made it clear that he will entertain no plan that does not eradicate the medfly within two months.

In your efforts to rid Florida of the medfly infestation, you have three tools at your disposal—burning, spraying, and the release of sterile males. The characteristics of these options are as follows:

- Burning the infested area will kill the medflies. But burning is expensive, it destroys all flora and fauna, and it may destroy some private property.
- You can spray the infested area with malathion to get rid of the medflies. If you spray, you will learn in one month whether or not the spray has eradicated the pest.
- If, after one month, you learn that the spray failed, you can spray a second time. You will learn whether the second spraying eradicates the pest one month after you undertake it.

Table 9.1 Actions for medfly eradication, with the cost and probability of success of each.

Action	Cost (millions of dollars)	Probability of Success
Burn now	60	1
Burn in one month	84	1
Burn in two months	117.6	1
Spray now	3	1/3
Spray in one month	4.2	1/3
Incubate sterile males and release if needed	6	1/2

- You can raise a crop of sterile males; this takes one month. If the initial spray fails, you can decide to release the sterile males. Since a female medfly mates but once, if she mates with a sterile male, she will have no offspring. You will learn whether the sterile males eradicate the medflies one month after releasing them. Destroying sterile males costs nothing.

The cost and probability of success of your various options are presented in Table 9.1. These costs increase by 40% per month because the infested area will expand at that rate.

Choice Nodes and Chance Nodes

To reveal your strategies and their effects, we shall draw a decision tree, which differs from the probability trees in Chapters 7 and 8 by having a second type of node, one at which a decision is made. A **decision tree** consists of nodes and arcs, in which

- A decision tree is drawn so that time moves from left to right. When you are "at" a node in a decision tree, you know everything to its left, as in a probability tree.
- Each **chance node** (still) describes a situation in which an element of uncertainty is resolved. Each way in which this uncertainty can be resolved is represented by an arc that leads rightward from its chance node, either to another node or to an end-point. The probability on each such arc is a conditional probability, the condition being that one is at the chance node to its left. These conditional probabilities sum to 1, as they do in a probability tree.
- Each **choice node** describes a situation in which an action must be selected. Each action is represented as an arc that leads rightward from its choice node, either to another node or to an end-point of the tree. The cost of each action is annotated on its arc.

In a decision tree, choice nodes are drawn as squares, chance nodes as circles. At each chance node, an element of uncertainty is resolved; at each choice node, the decision maker must select one of the available actions. Each action entails a cost, which may be negative.

A Decision Tree for Medfly Eradication

Figure 9.1 presents a decision tree for your medfly eradication problem. Its choice nodes have been assigned numbers, and its chance nodes have been assigned letters. You begin at the left, at choice node 1. To its right are three arcs, each of which represents an action that you

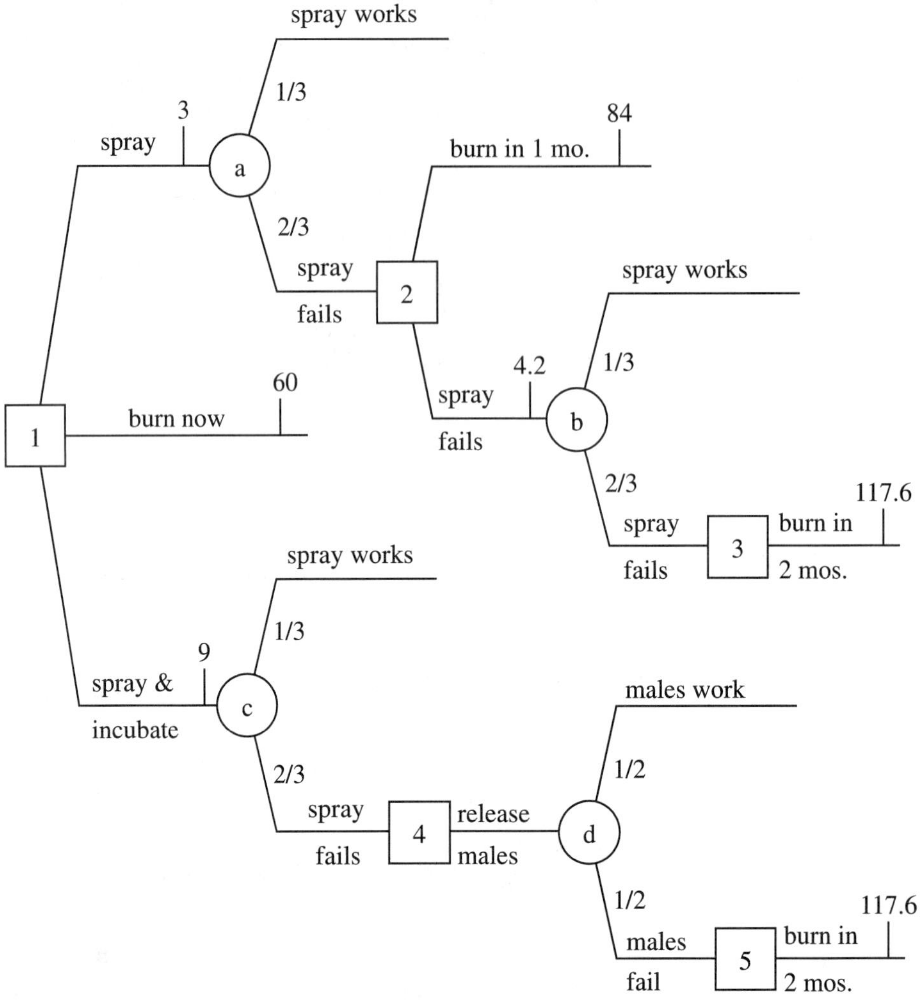

Figure 9.1 A decision tree for medfly eradication.

could take now. These actions are to burn, to spray, and to spray while incubating a crop of sterile males. The cost of each action is annotated on its arc. The "spray" action leads to chance node a, where the uncertainty about whether the spray works is resolved. Conditional probabilities appear on the arcs to the right of each chance node, exactly as in a probability tree. At "choice" nodes 3 and 5, only one action is available to you; two months have elapsed, and you must burn.

Some actions have been omitted from Figure 9.1. At choice node 1, you could decide to wait for one month and then burn. At choice node 4, you could burn rather than releasing the sterile males. The tree would be correct if you included these actions, but you need not include them. Do you see why?

In Figure 9.1, each action entails a cost, which is annotated on its arc. In general, costs can be positive or negative. Each cost in a decision tree must have the same unit of measure, of course. In Figure 9.1, the costs are measured in millions of dollars.

The amounts of money involved in this decision are substantial, but they are a tiny fraction of the annual budget of the State of Florida. We shall act as though this situation is operational, and we shall minimize the expectation of total cost. Later on, we will suggest that this situation has strategic overtones.

Minimizing the expectation of cost is not the only rational criterion, of course. A different criterion, called **min-max regret**, is to minimize the largest cost that could occur. In Figure 9.1, the min-max regret strategy is to burn now; this costs \$60 million, but it is the only way to eradicate the medfly without risking the loss of more than \$60 million.

Strategies

In a decision tree, a **strategy** is a plan of action that accounts for every possible chance outcome or contingency. In Figure 9.1, there are exactly four strategies:

1. Spray now. If the spray works, do nothing more. If the spray fails, burn in one month.
2. Spray now. If the spray works, do nothing more. If the spray fails, spray again in one month. If the second spraying works, do nothing more. If the second spraying fails, burn in two months.
3. Burn now.
4. Spray now and simultaneously incubate sterile males. If the spray works, destroy the males and do nothing more. If the spray fails, release the males in one month. If the males work, do nothing more. If the males fail, burn in two months.

We can compute the expectation of the cost of each strategy and then choose a strategy whose expectation of cost is smallest. Since there are only four strategies, the volume of computation is not onerous, and it is made simpler if we recall that the expectation of the sum equals the sum of the expectations. Specifically:

$$E(\text{cost of strategy 1}) = 3 + (2/3) \times (84) = 59$$
$$E(\text{cost of strategy 2}) = 3 + (2/3) \times (4.2) + (2/3) \times (2/3) \times (117.6) = 58.067$$
$$E(\text{cost of strategy 3}) = 60$$
$$E(\text{cost of strategy 4}) = 9 + (2/3) \times (1/2) \times (117.6) = 48.2.$$

Let us justify the expression for the expected cost of strategy 1. It costs \$3 million now. The expectation of its cost one month from now equals $(2/3) \times (84) + (1/3) \times (0)$, and the expectation of the sum equals the sum of the expectations. Similar calculations justify the expressions for the expected costs of the other strategies.

A Recursion

In a larger tree, the strategies can be too numerous to enumerate. Fortunately, there is a simple recursion that finds the best strategy without enumerating all of them. This recursion consists of two steps.

Recursion to find the best strategy:

1. (*Evaluating end-points*) Adjacent to each end-point of the tree, accumulate the net total cost that accrues if this end-point is reached.
2. (*Rolling back*) Then, moving from right to left:
   - **(a)** At each chance node, compute the expectation of cost by multiplying the probability of each arc to its right by the expected cost of reaching the node or end-point to which it leads, and summing.
   - **(b)** At each choice node, pick an action whose expectation of cost is lowest.

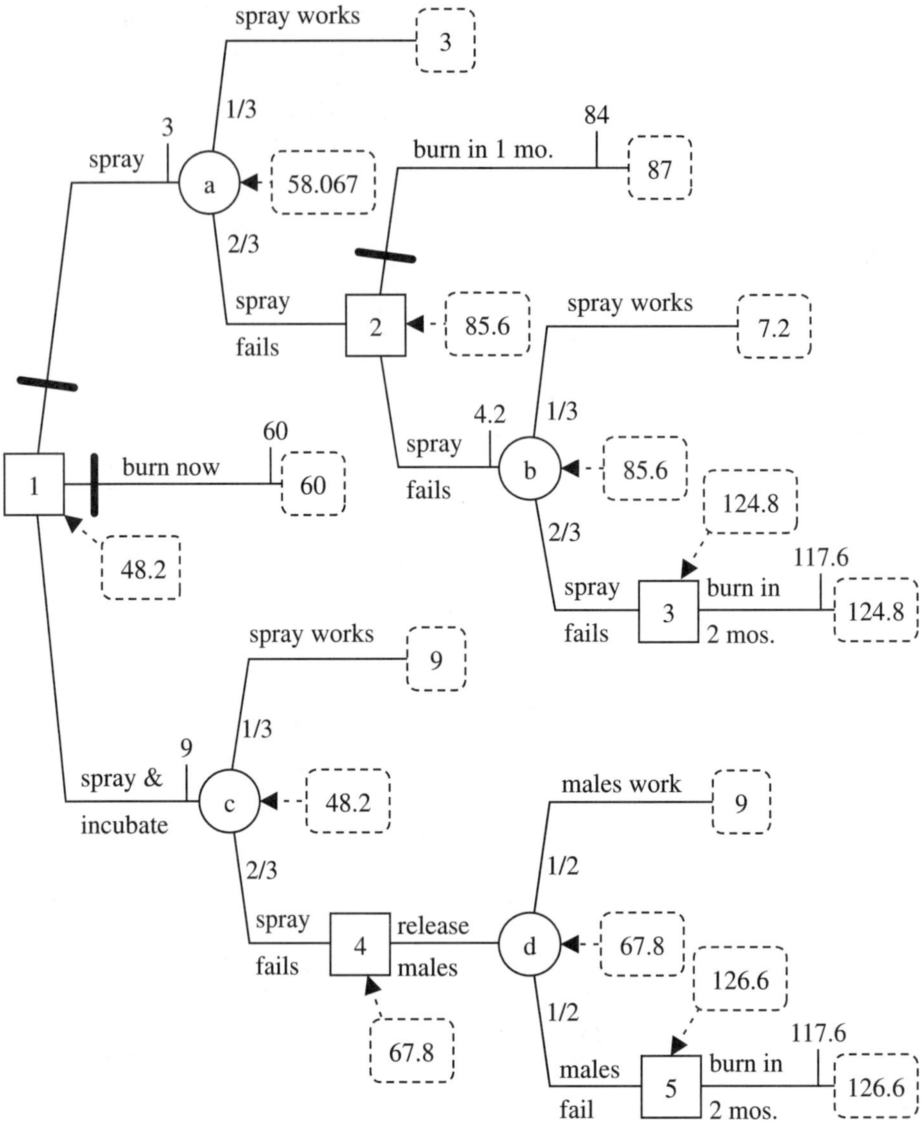

Figure 9.2 Rollback of the medfly decision tree.

The medfly example is now used to illustrate this recursion. In Figure 9.2, the cost of each of its end-points is recorded to the right of that end-point, inside a dashed box. The topmost end-point occurs if you spray once and it works; if you attain this end-point, you will have spent \$3 million. The next-to-top end-point occurs if you spray, if it fails, and if you then burn; if you reach that end-point, you will have spent 87 = 84 + 3 million dollars. And so forth.

The idea of rolling back is to evaluate each node *after* having evaluated all nodes to its right. Figure 9.2 illustrates the rollback in which each node's cost appears in a dotted box, to its right. Let us see how these costs are computed. Having evaluated the end-points, you can evaluate choice nodes 3 and 5, each of which happens to have only a single choice. Choice node 3 is assigned the cost of its sole choice, which is 124.8. Choice node 5 is assigned cost of 126.6. Having evaluated these choice nodes, you can now evaluate chance nodes b and d:

$$\text{Expected cost for chance node } b = (1/3) \times 7.2 + (2/3) \times 124.8 = 85.6$$

$$\text{Expected cost for chance node } d = (1/2) \times 9 + (1/2) \times 126.6 = 67.8.$$

After evaluating chance node b, you can evaluate choice node 2. Its actions have expected costs of 87 and 85.6. You choose the smaller. In Figure 9.2, a heavy line "cuts" the higher-cost option. Similar calculations evaluate choice node 4 and chance node c.

Finally, after evaluating chance nodes a and c, you can evaluate choice node 1. It has three actions. Again, you pick the cheapest, which is to spray while incubating sterile males. Again, the actions that do not achieve minimum cost are blocked (cut) by heavy lines.

In Figure 9.2 and in general, the best strategy is the contingency plan that is not cut off, and the expected cost of the best strategy is annotated at the leftmost node. Figure 9.2 makes it clear that the best strategy is to incubate while spraying, stop if the spray works, release the sterile males if the spray fails, stop if the males work, and burn in two months if they fail. This is strategy 4, and its expected cost is \$48.2 million, exactly as we observed earlier.

The rollback assigns a cost to each chance node and to each choice node. What meaning do these costs have? The answer is as follows:

> Rolling back assigns to each node the least expected total cost given that this node is reached.

The rollback assigns to node 2 the cost of 85.6; this is the least expected total cost given that node 2 is reached. Later in this chapter, we describe a slightly different rollback, one that assigns to each node the least expected *future* cost given that this node is reached.

Dominated Strategies

A strategy is said to be **dominated** if there is another strategy that costs no more under every chance outcome and costs less under at least one chance outcome. Because no dominated strategy can minimize expected cost nothing is gained by including a dominated strategy in your decision tree.

In Figure 9.1, for instance, we omitted certain actions. We could have done nothing for one month and then sprayed; we could have sprayed a second time even if the first spray worked; and so forth. The strategies that we omitted from Figure 9.1 are dominated.

There is nothing wrong with including dominated strategies in your decision tree. If you are in doubt about whether including an arc leads only to domination, leave it in. When you roll back the tree, you will block each dominated strategy, as its expected cost cannot be smallest.

Sensitivity Analysis

For Problem A, Figure 9.2 exhibits the strategy whose expected cost is lowest, and Figure 9.2 shows that the expected cost of this strategy is \$48.2 million. Figure 9.2 also reports that the second-best strategy has an expected cost of \$58.067 million, which is nearly \$10 million higher.

The best strategy in Figure 9.2 looks good, but some of the data in the tree may be subjective and problematical. If so, you should do a sensitivity analysis on these data to see if this strategy is *robust*, that is, if it remains best or near-best for ranges of data that are broad enough to encompass the situation that actually prevails.

A Breakeven Calculation

Let us now do a sensitivity analysis, one that illustrates a useful concept. You took as $1/2$ the probability that sterile males would destroy what remains of the medfly infestation if the spray failed. Suppose that you are uncertain about this probability. A **breakeven analysis** replaces this datum by a variable and finds the range on that variable for which the best strategy is unchanged. Let us take p as the probability that the sterile males wipe out what remains of the medfly infestation if the spray fails. Since p affects only strategy 4, this breakeven analysis is rather simple. You need to find the range of p for which

$$9 + (2/3)(1 - p)117.6 \leq 58.067.$$

A little algebra shows that the preceding inequality holds for $p \geq 0.374$. Thus, strategy 4 is best whenever $p \geq 0.374$, where p is the probability that the sterile males destroy what remains of the medfly infestation after spraying.

Tests for robustness need not insist on optimality. Typically, you would wish to execute a sensitivity analysis on several elements of data, individually and collectively. Typically, you are looking for a strategy that performs well (best or nearly best) over a range of data.

The preceding breakeven analysis was relatively simple; p entered only the cost of one of the strategies, and p entered linearly. Other breakeven analyses can be more complex. Consider, for instance, the possibility that the infestation is spreading at a rate $(1 + r)$ per month, where r may differ from 0.4. You wish to find the range on r for which strategy 4 has the least expected cost. Doing this by hand could be laborious because the expected cost of two of the strategies is quadratic in r.

Breakeven computations need not be done by hand. Later in the chapter, we will see how to use a spreadsheet to draw decision trees and roll them back automatically. Spreadsheets make quick work of such questions as "What if the infestation spreads at a rate that is somewhat below 40% per month?"

Briefing the Governor

How should you handle tomorrow's meeting with the governor? By all means, draw the decision tree and bring it. In this case and in general, a decision tree may be worth thousands of words. A decision tree has these merits:

- It lays the model out visually, for each participant in the decision to understand and critique.
- It parses a complicated issue into its constituents, models each constituent, and shows how to piece together an overall strategy.
- It displays the strategies.

Your decision tree displays the ways in which the pest can be eliminated within the desired time period. It specifies the cost of each action and the probability that it will succeed. It parses the medfly eradication problem into components, and it shows how they interact. It presents the governor with the courses of action that he might follow, and it measures their effects.

After describing your tree to the governor, explain that on an expected-cost basis, the least unattractive strategy is to spray immediately, follow that up, if necessary, with a flock of sterile males, and burn as a last resort. Note that this choice is best by about \$10 million. Discuss the extent to which it is robust. Emphasize that this strategy entails less spraying than the second-best strategy, that it minimizes the probability of burning, but that it runs the risk of having to burn a large region.

After presenting your tree, attempt to guide the discussion toward information that it fails to capture, such as environmental damage. Spraying with malathion kills insects indiscriminately; it also disrupts the ecosystem. Burning also kills indiscriminately but more extensively. As concerns environmental damage, delaying action is the worst alternative because the infestation will spread during the delay. If the governor has particular concerns, help him to incorporate them into a rational analysis.

Medfly Postscript

The medfly example was written with the experience of former Governor Edmond G. (Jerry) Brown, Jr., of California in mind. Long ago, in June of 1980, the Mediterranean fruit fly appeared in Santa Clara County, a wealthy residential district south of San Francisco, but not far from the San Joaquin and Imperial valleys, which are major agricultural areas, with crop output of about $15 billion in 1980.

Adult medflies burrow into fleshy fruit to lay eggs. Their larvae feed on the pulp, turning it to mush. Adult medflies can be destroyed with pesticides, including Malathion. Pesticides are ineffective on medfly larvae, which are protected by the skins of their host fruit. One way to kill the larvae is to burn the trees that host them. Another is to harvest the fruit and then fumigate it with toxic gas.

Between June of 1980 and the end of that year, there was much debate but little action. Throughout this period, Governor Brown was vehemently opposed to aerial spraying with Malathion. By January of 1981, the area of infestation had expanded from 30 square miles to 150 square miles. At that time, a four-pronged attack was announced:

- Strip the fruit trees in the infected areas and carry the fruit away for burial in special landfill areas.
- Spray the fruit trees with Malathion from the ground.
- Release sterile males.
- Prohibit citizens from carrying fruit out of a larger 500 square mile region.

No part of this program proved to be effective. During the cool weather of January, the medfly is relatively dormant and so is less susceptible to spraying. Hauling fruit seemed to spread the medfly to new sites. A huge quantity of sterile males was required. Some of them were imported, and one imported batch proved to be virile; they had not had their radiation doses. As a result, by July of 1981, the medfly had spread to a 620 square mile region.

On July 8, 1981, Governor Brown again rejected aerial spraying of Malathion as an unacceptable health risk. On July 10, at the request of the great agricultural State of Texas, federal officials announced a quarantine, requiring fumigation of all produce leaving California. Two days later, on July 12, Governor Brown lamented that a "political" quarantine forced him to reverse himself and order aerial spraying. In early August 1981, the medfly appeared in the San Joaquin Valley. At that point, the Japanese embargoed all California produce, $118 million that year. The pest was reined in by September 1982 but continued to appear sporadically throughout the state. By September 1981, political savants had foretold the end for Jerry ("Moonbeam") Brown.

It is often argued that rational models are of scant value in the public sector where each action has a range of consequences and where each consequence benefits some constituencies and harms others. That is not our view. It is especially true in the public sector that a systems analysis can bring an issue into focus.

In particular, a decision tree might have brought Governor Brown to a rational analysis. First, it would be evident from a quantitative model that delaying action enlarges the region of infestation (at a rate of 40% per month, as it turned out), which exasperates whatever remedy is implemented. Second, spraying was the only strategy that offered a reasonable prospect for ridding Santa Clara County of the medfly, short of setting the county ablaze. And spraying a relatively small area early is far better from every viewpoint (economic, ecological, and political) than spraying a huge area a year later.

9.4. MINIMIZATION AND MAXIMIZATION

For medfly eradication, each action entails a cost, and it is natural to formulate the problem as the minimization of cost. In a minimization problem, the expenses have positive signs, any revenues have negative signs, and one minimizes the expectation of net cost.

In other settings, it can be natural to formulate an optimization problem as a maximization problem. In a maximization problem, the revenues have positive signs, the expenses have negative signs, and one maximizes the expectation of net profit. To roll back a maximization problem, at each choice node, pick an action whose expected profit is largest. In a profit-maximization problem, a strategy is **dominated** if there is some other strategy whose profit is at least as large under every chance outcome and is larger under at least one chance outcome.

It goes without saying, perhaps, that maximizing net profit is equivalent to minimizing net cost. To convert either problem to the other, multiply each element of expense and revenue by -1.

9.5. TREEPLAN

Any of several Excel "add-in" packages let you draw a decision tree, roll it back automatically, and, best of all, engage in a sensitivity analysis that determines whether a strategy is robust. Of these packages, TreePlan is one of the most user-friendly. TreePlan was written by Michael Middleton, and it is included in the CD that accompanies this text.

Before you can use TreePlan, it must be installed and activated. If these steps have been accomplished, TreePlan will appear on your list of Excel tools. If TreePlan does not appear among your tools, you may refer to the Appendix, which tells how to copy TreePlan into your Excel library folder and to activate it.

A Guide

A step-by-step introduction to TreePlan will soon be provided. Before doing so, we present this brief guide:

- Before using TreePlan, open a spreadsheet and put the problem's data in its upper left-hand corner. Then select the cell in which you wish the upper left-hand corner of the tree to appear.
- Type **Ctrl+T** to call up TreePlan.
- To add an arc to your tree, put the cursor on the node to which it is to be attached, then type **Ctrl+T**
- To delete an arc, put the cursor on its rightmost node, then type **Ctrl+T**
- To switch from maximization to minimization, move the cursor away from the tree, then type **Ctrl+T** and, on the menu that pops up, click on the Options . . . button.

Table 9.2 Data for medfly eradication.

	A	B	C	D	E
1	**Data for the medfly decision tree**				
2					
3	action	cost	p(works)	cost in 1 mo.	cost in 2 mos.
4	burn now	60	1/3	84	117.6
5	spray now	3	1/2	4.2	5.88
6	incubate	6			
7	inflation factor	1.4		= B5 * B$7	= D5 * B$7

Then, on the *TreePlan . . . Options* menu that pops up, click on Minimize Costs, and click on OK.

- To prepare for sensitivity analysis, enter no data on your tree; instead, insert references to the cells where these data reside.
- TreePlan puts formulas in certain cells; leave those cells alone!

TreePlan is user-friendly. With this guide in view, you may wish to familiarize yourself with it by trial and error. Alternatively, you may prefer to follow the more methodical procedure that appears below.

A Decision Tree for Medfly Eradication

To introduce TreePlan, we now describe a step-by-step procedure that uses it to set up a decision tree for the medfly problem.

Step 1: Open an Excel worksheet and enter the data for the medfly decision tree in the upper left-hand corner of a worksheet, as in Table 9.2. *Remark:* placing the data in the upper-left-hand corner facilitates sensitivity analysis. Not all of the numbers in Table 9.2 are data, however; cells D4 through E5 contain functions. Do you see why?

Step 2: Select the cell at the upper left-hand corner of the area in which you want your tree to appear (e.g., cell A8). Then either type **Ctrl+T** or select TreePlan on the Excel Tools menu. (Hitting **Ctrl+T** is the quick way to call up TreePlan, which you will need to do repeatedly.) Either way, the menu that appears in Table 9.3 will pop up. Click on the New Tree button. Your spreadsheet should now resemble Figure 9.3.

Step 3: The initial node of the tree for medfly eradication has three branches—burn now, spray now, and spray and incubate. To add a third branch to your tree, select its initial node

Table 9.3 The TreePlan . . . New menu.

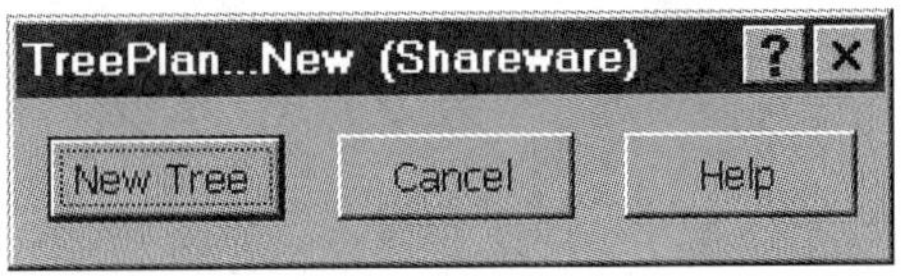

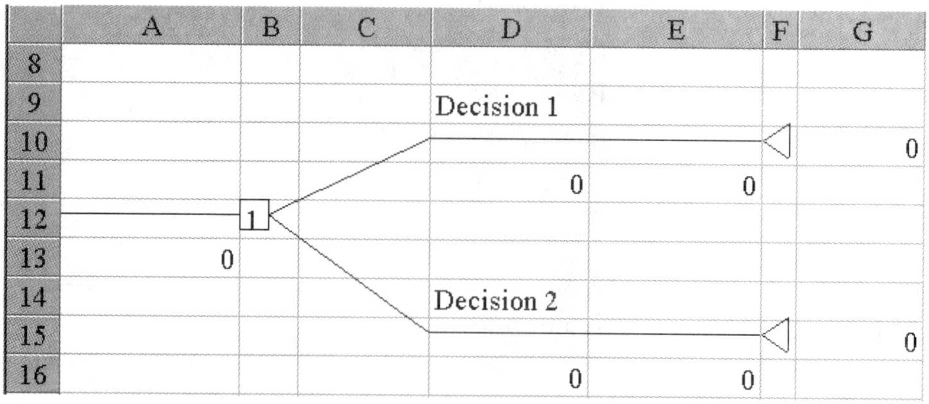

Figure 9.3 The start of a decision tree.

(cell B12), then hit **Ctrl+T** and note that the menu in Table 9.4 appears. Click on its Add branch button and then click on OK. This produces the decision tree in Figure 9.4.

Step 4: In the cell labeled Decision 1 (cell D9 in Figure 9.4), type the name of the first decision. In the cell below Decision 1 that contains a zero (cell D11 in Figure 9.4), enter a pointer to the cost of this decision. For instance, type **burn now** in cell D9 and **=B4** in cell D11. Do the same for the other two actions. This should change your tree into the form shown in Figure 9.5. *Remark:* do *not* enter data into cells (such as cell E11) into which TreePlan has entered functions.

Step 5: TreePlan has assigned a value of 60 to the initial node of the above tree. Evidently, TreePlan is maximizing expected profit. You want it to minimize expected cost. To accomplish that, move the cursor away from the tree, hit **Ctrl+T**, and then, on the *TreePlan . . . Select* menu that appears, click on the *Options . . .* button. Then, on the *TreePlan . . . Options* menu that pops up, click on Minimize Costs and then click on OK. The value of the initial node will switch to 3, indicating that TreePlan is now minimizing expected cost.

Table 9.4 The TreePlan . . . Decision menu.

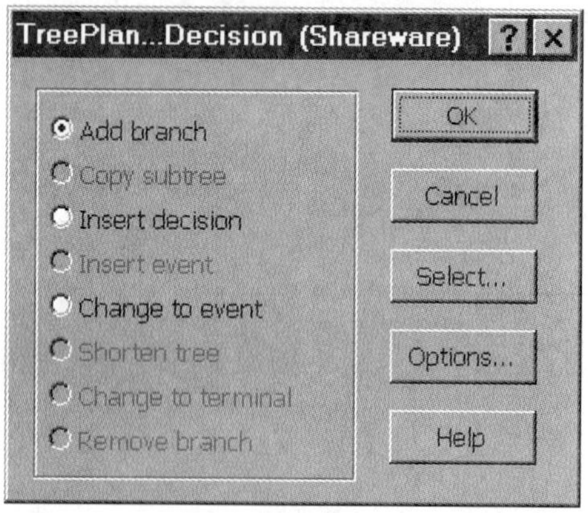

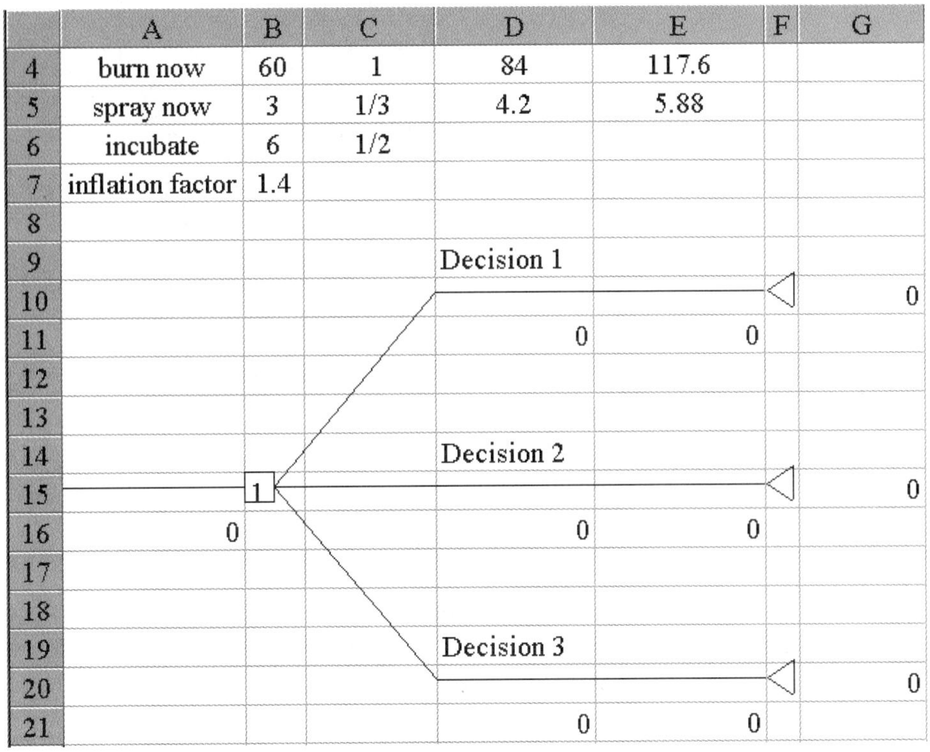

Figure 9.4 A third decision.

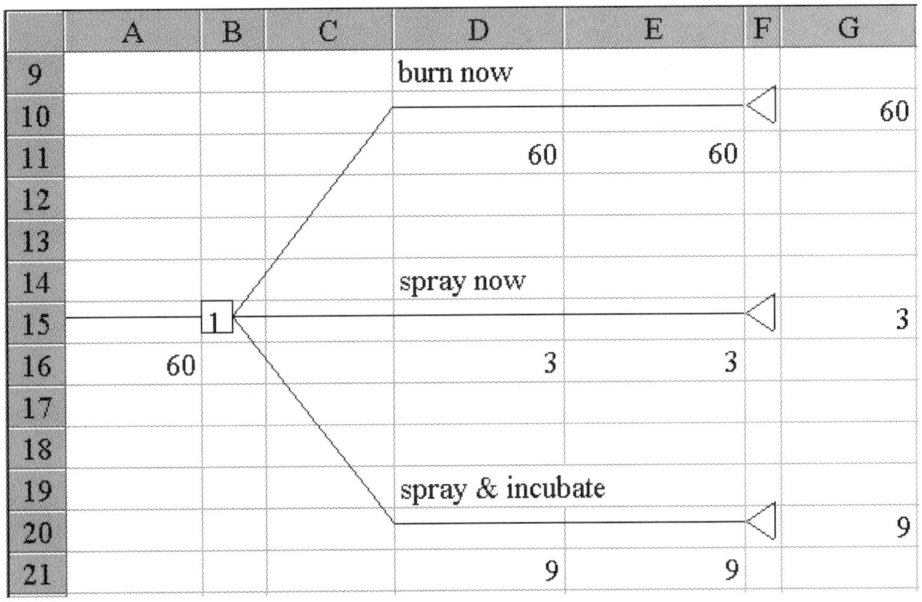

Figure 9.5 Three choice nodes.

Table 9.5 Creating branches.

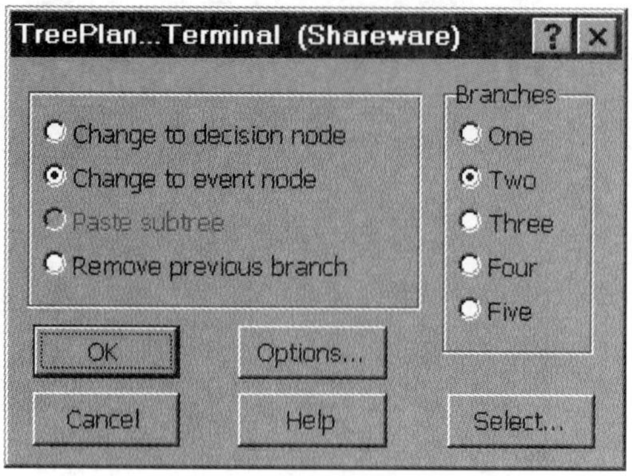

Step 6: Each end-point of the tree is a triangle. To replace an end-point with a choice or chance node, select its triangle and proceed as follows.

To insert a chance node after the "spray now" action, select the triangle at the right-hand end of that action's branch (cell F15), type **Ctrl+T**, and note that the *TreePlan . . . Terminal* menu in Table 9.5 will pop up. To change the triangle to an event node having two branches, fill it out as shown in Table 9.5 and then click on OK.

On each branch that emerges rightward from the event node, enter a name, a probability (above the name), and, for some branches, a cost (below the name). After you do so, your tree will resemble Figure 9.6.

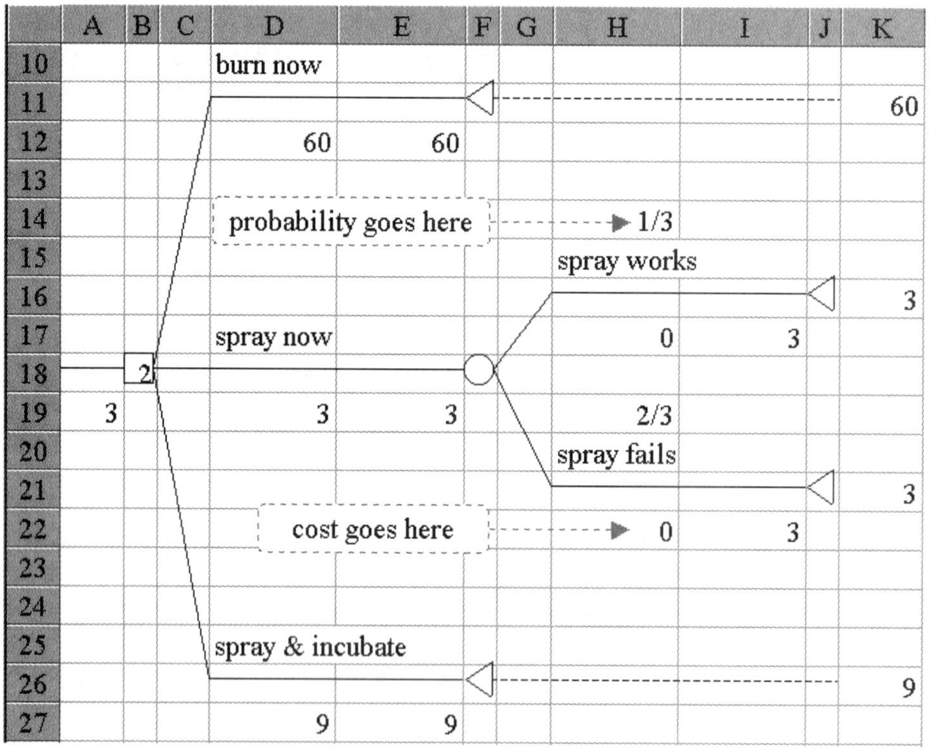

Figure 9.6 A partial decision tree.

To complete the tree, you must add more chance nodes and choice nodes. To do that, repeat the preceding steps. For instance, to insert a choice node that follows failure of the spray, select cell J21, type **Ctrl+T** and then use the "TreePlan . . . Terminal" menu that pops up.

Eventually, you may want to print your tree. There is a *wrong* way to do that. If you click on File, Print, All, a few hundred blank sheets may print because TreePlan uses the "bottom" of the worksheet as a scratch pad, and Excel prints blank sheets. Instead, follow the instructions in:

Step 7: To print your decision tree within Excel, follow these steps:

a. Select the portion of the worksheet that you wish to print.

b. If you want to omit the gridlines from your printout, on the Excel Tools menu, click on Options, select the View tab, and remove the check from the box marked "Gridlines." *Note:* To put the gridlines back, re-check this box after you print your tree.

c. If you wish to print the entire tree on one or two sheets, on the Excel File tab, click on Page Setup, check the "fit to" box, and select the number of pages "wide" and "tall."

d. Finally, to print your tree, on the Excel File tab, click on Print, and check Selection.

A good way to familiarize yourself with TreePlan is to build a spreadsheet for the medfly problem. Admittedly, building this spreadsheet takes a bit of work. The work pays off when you use Excel to execute a sensitivity analysis.

9.6. MODELING AN INTANGIBLE

A key parameter in a decision problem can be unknown or "intangible." The fact that we do not know a parameter may not preclude a rational analysis. We can do a breakeven analysis on its value. To illustrate, we pose a rhetorical question in:

Problem B (R&D Management)

You manage a plant for a high-technology company. Production of a new product must begin nine months from now. An investment in technology will be required, and you have some alternatives as to what to do. Adapting current technology to this product is guaranteed to work, will take six months, and will cost \$10 million. At a cost of \$15 million, you can adapt current technology in half the time. You can also try to develop a new technology, whose variable production costs will be somewhat smaller, and which may have other uses as well. The new technology requires success on both of two innovations, A and B. Neither innovation is guaranteed to succeed, but the successes of innovations A and B are independent of each other. You can undertake innovations A and B in either sequence, or simultaneously. Table 9.6 displays the cost, the time required, and the probability of success of each innovation. These costs are within the operational range for your company, which is content to minimize the expectation of net cost. How do you analyze this problem?

Problem B confronts you with two goals. Your company wants you to minimize the expectation of the investment that is needed to produce the new product within nine months. Your company is also interested in developing a new technology. The issue is to bring the tension between these goals into focus.

Table 9.6 Developing a new product.

Technology	Time Needed (mos.)	*P*(works)	Cost (millions of $)
Current	6	1	10
Current, with overtime	3	1	15
Innovation A	3	4/5	3.5
Innovation B	3	2/3	2.5

It is left to you, the reader, to build a decision tree for this R&D management problem and to find a way to do a breakeven analysis on the value of acquiring the new technology.

9.7. INFORMATION AND ITS VALUE

In Chapter 7, information was used to update probabilities. This information consisted of learning one or more chance outcomes. For instance, we saw how learning the outcome of a medical test updates the probability that a patient has a disease.

In a probabilistic setting, **information** consists of learning a chance outcome. In a decision problem, information has economic value because knowledge of which chance outcome occurs affects one's best strategy. In general, the **value of information** is defined by the equation:

$$\text{Value of Information} = E(\text{net revenue of best strategy given that information}) \\ - E(\text{net revenue of best strategy without that information}).$$

To illustrate the value of information, we pose the rhetorical question in:

Problem C (Drilling for Oil)

Your company has leased an offshore parcel that may contain oil. You must decide whether to drill for oil or to abandon the lease. Your geologist estimates 0.4 as the probability that the parcel contains oil. Drilling will cost $400,000. Drilling will find the oil, if it is there. If you drill and find oil, you will recover $1.5 million. If you drill and find no oil, you will recover nothing, of course. What strategy maximizes expected net profit? What is the value of learning whether or not the parcel contains oil before you decide to drill?

The information in Problem C is whether or not the parcel contains oil. We will compute the value of this information. Before we do so, you are urged to pause and estimate the amount of money that this information is worth. Then, after we have computed its value, you will be able to see how accurate your estimate was.

A Decision Tree with No Information

To calculate the value of this information, we compute the value of the lease twice, once with the information and once without it. Figure 9.7 describes the decision problem without the information. Evidently, it pays to drill. Drilling has expected net revenue of $(0.4)(1.1) + (0.6)(-0.4) = 0.44 - 0.24 = 0.2$ million dollars.

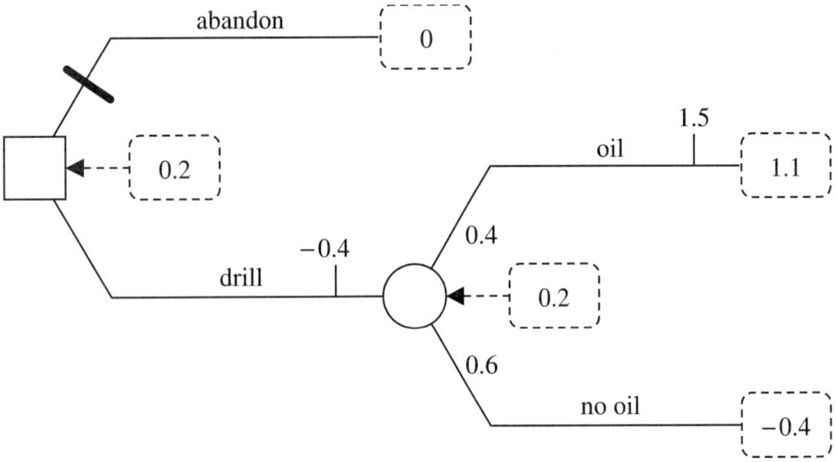

Figure 9.7 The expected value of the lease, with no information.

A "Flipped" Decision Tree

Learning whether or not the parcel contains oil does *not* affect the probability that there is oil. The sole advantage of learning this information is to let you make choices after learning it. This information "flips" (reverses) the chance and choice nodes in Figure 9.7 and produces the tree in Figure 9.8.

Figure 9.8 presents the decision tree in which you learn whether or not the parcel contains oil *before* deciding whether to drill. The chance node at which this information is learned has been moved to the left. The probability that the parcel has oil has not changed;

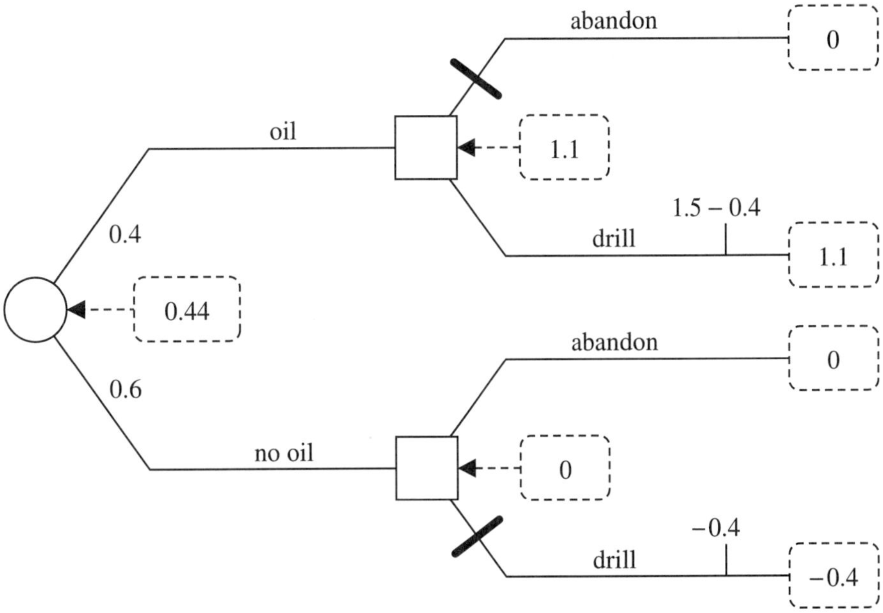

Figure 9.8 The expected value of the lease, given the information as to whether or not the parcel has oil.

it still equals 0.4. The computation in Figure 9.8 shows that the value of the lease given this information equals 0.44 million dollars.

EVPI

The information we have evaluated is whether or not the parcel contains oil. The **expected value of perfect information (EVPI)** equals the expected profit given this information less the expected profit given no information. From Figures 9.8 and 9.7, we have learned that

$$\text{EVPI(whether the parcel contains oil)} = \$440{,}000 - \$200{,}000 = \$240{,}000.$$

The value of information is a *breakeven* value; it is the most that you would pay to learn the outcome. If a clairvoyant could tell you, unequivocally, whether your parcel contained oil, the most you would pay for that information would be $240,000. In this instance, was the EVPI smaller than your estimate?

Problem C (continued)

Let us suppose that your geologist can offer a seismic test that, like any test, is imperfect. This test costs $50,000, its outcomes are + and −, and its sensitivity and specificity are 75%, so $0.75 = P(+|\text{oil}) = P(-|\text{no oil})$. What is the value of learning the test outcome?

When computing the value of a test, by convention we assume that it is free. Thus, the value of the test is a breakeven value, namely, the largest fee that we would pay to learn the test outcome. In the jargon of decision analysis, the value of a test is known as its **expected value of sample information**, abbreviated **EVSI**. An imperfect test cannot be worth more than a perfect test. In this case, EVSI cannot exceed $240,000. What is your guess as to the EVSI for this test?

The Value of Imperfect Information

The straightforward way in which to compute EVSI is to roll back the decision tree in which the outcome of the test is revealed before any decisions need be made, thereby computing the expected value given the sample information. From this value, subtract $200,000, which is the expected value given no information.

We will pursue that route and then take a shortcut. To compute the probabilities, $P(+)$ and $P(-)$, of positive and negative test outcomes, we resurrect an idea from Chapter 7, namely, the joint probability table. To create Table 9.7, we recall that the joint probability of two events equals the probability of either times the conditional probability of the other. Thus, $P(\text{oil},+) = P(\text{oil})P(+|\text{oil}) = (0.4) \times (0.75) = 0.3$.

The decision tree in Figure 9.9 describes the problem in which you know the test outcome before deciding whether or not to drill. The probabilities in this tree are taken from Table 9.7; for example, $P(+) = 0.45 = 9/20$ and $P(\text{oil}|+) = P(\text{oil}, +)/P(+) = 0.3/0.45 = 2/3$.

To compute the expected value of the lease given knowledge of the test outcome, we can roll back the tree in Figure 9.9 and obtain a value of $270,000. Let us recall from Figure 9.7

Table 9.7 Joint probabilities for test information.

	+	−	
Oil	0.3	0.1	0.4
No oil	0.15	0.45	0.6
	0.45	0.55	

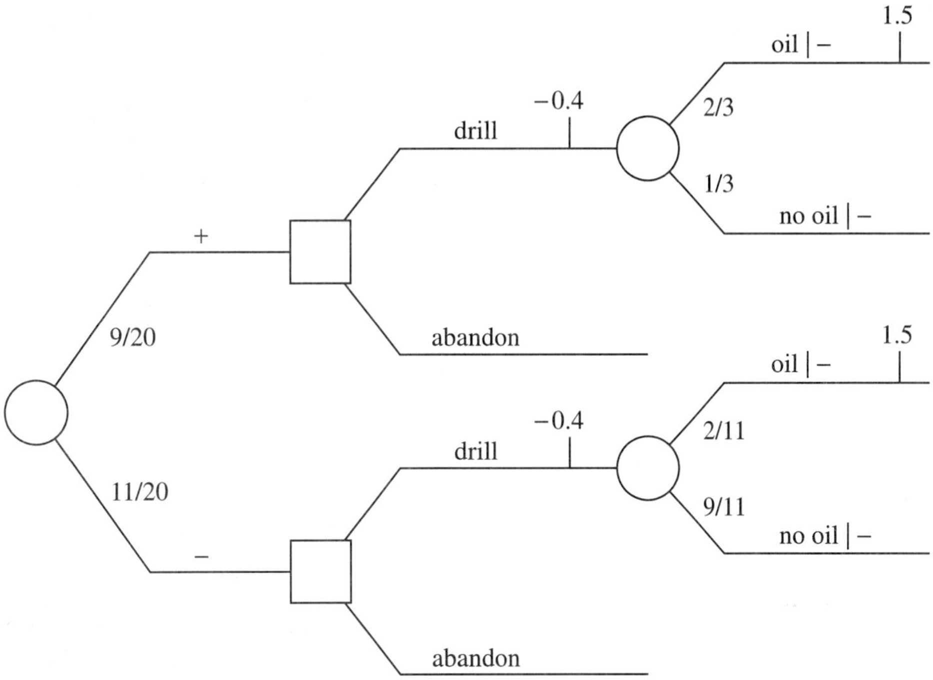

Figure 9.9 Expected value of the lease given knowledge of the test information.

that with no knowledge of the test outcome, the lease is worth \$200,000. By taking the difference, we see that

$$E(\text{value of test information}) = \$270{,}000 - \$200{,}000 = \$70{,}000.$$

In brief, this imperfect test has EVSI of \$70,000.

The value of information measures its worth; if it isn't worth much, it isn't important. We have seen that the value of information about the seismic test is \$70,000. Since this test costs \$50,000, doing it improves your expected profit by \$20,000 = \$70,000 − \$50,000. If the cost of this test exceeded \$70,000, you would be better off not doing it.

The sensitivity and specificity of this test are $3/4$, but its value (of \$70,000) is far less than $3/4$th of the value (of \$240,000) of perfect information. Do you see why that occurs?

A Shortcut

Rolling back the decision tree in Figure 9.9 is not necessary; in fact, there is no need to draw the tree at all. To see why, we ask ourselves what we can hope to learn from this tree. It has four strategies. which are labeled A through D:

Strategy A: Test. Drill if $+$, abandon if $-$.

Strategy B: Test. Drill if $-$, abandon if $+$.

Strategy C: Test. Drill regardless of the test outcome.

Strategy D: Test. Abandon regardless of the test outcome.

Strategies C and D make no use of the test information; they were evaluated in Figure 9.7. Strategy B is **perverse** in the sense that it goes against the test outcome; why make poor use of the test outcome? Only Strategy A accords you the opportunity to improve on the net revenue you can get with no information. And the net profit for Strategy A can be found directly from Table 9.7. Specifically,

$$E(\text{net profit for Strategy A}) = (1.5)P(\text{oil}, +) + (-0.4)P(+)$$
$$= (1.5)(0.3) - (0.4)(0.45) = 0.27.$$

This simple calculation demonstrates that the value of the lease given knowledge of the test outcome is \$270,000.

Pruning Perverse Actions

We have just illustrated a principle that lets you prune a decision tree. Let us consider a test for the event A. Typically, its sensitivity and specificity are well above 0.5, so that the sum, $P(+|A) + P(-|\bar{A})$, of its sensitivity and specificity exceeds 1.0. If you suspect that such a test has $P(A|+) > P(A)$ and $P(\bar{A}|-) > P(\bar{A})$, you are right. (See Problem 7 on page 316 for details.) You have no motive to bet against such a test. We emphasize:

> Having decided to undertake a test whose sensitivity and specificity sum to more than 1.0, you may omit the arcs that bet "against" the test outcome.

In Figure 9.9, for instance, if the test outcome is "+," you can omit the arc "abandon"; similarly, if the test outcome is "−," you can omit the arc "drill." When you prune this tree of these arcs, all that remains is Strategy A. Its expected net cost can be found directly from its joint probabilities, as we have done.

Pruning a decision tree in this way eases computation. Failure to prune does not make the tree incorrect. If you are in doubt as to whether or not an arc can be pruned, leave it in. No harm is done except, possibly, for making you perform extra labor.

Value of Information in a Minimization Problem

The value of information has been introduced in the context of profit maximization. In a cost-minimization problem, the value of the information is the decrease in cost that the information makes possible, so that

$$\text{Value of Information} = E(\text{net cost of best strategy without that information})$$
$$- E(\text{net cost of best strategy with that information}).$$

In either case, the value of information measures the importance of the information. If the value of a piece of information is low, that information is not worth learning.

9.8. COMPARING RANDOM VARIABLES*

Two separate reasons have been found to prune a decision tree. Each of these reasons will now be shown to correspond to a way in which one random variable (or strategy) dominates another. When x and y are numbers, we have known since grade school what it means for x to be less than y. The comparable question about random variables has not yet been broached. It appears below:

> *When X and Y are random variables, what does it mean for the random variable X to be less than the random variable Y?*

---

*This section can be skipped with no loss of continuity.

Table 9.8 Random variables X and Y for the same probability model.

Outcome	Probability	Value of X	Value of Y
o_1	p_1	x_1	y_1
o_2	p_2	x_2	y_2
$\vdots$	$\vdots$	$\vdots$	$\vdots$
o_n	p_n	x_n	y_n

We will provide two different answers to this question, both of which are reasonable. Each answer motivates a type of pruning that has already occurred.

An Example

To begin, we examine two random variables, X and Y, that share the same underlying probability model. Table 9.8 lists the outcomes in this probability model, and it records the value that each outcome assigns to X and Y.

In Chapter 8, we wrote $X = Y$ if each outcome assigned the same value to X and to Y. In other words, the random variable X **equals** Y (written $X = Y$) if

$$x_i = y_i \quad \text{for each value of } i.$$

We now say the random variable X is **greater than or equal** to Y (written $X \geq Y$) if

$$x_i \geq y_i \quad \text{for each value of } i.$$

The statement that X is greater than or equal to Y requires *each* outcome to assign X a value that is at least as large as the value that it assigns to Y. Finally, we say that the random variable X is **greater than** Y (written $X > Y$) if

$$\left.\begin{array}{ll} x_i \geq y_i & \text{for each value of } i \\ x_k > y_k & \text{for at least one value } k \end{array}\right\}. \tag{9.1}$$

In other words, X is greater than Y if each outcome assigns to X a value that is at least as large as the value it assigns to Y and if at least one outcome assigns to X a larger value than it assigns to Y.

A Basis for Preference

In Chapter 8, we argued that the expectation of profit is a reasonable measure of desirability (more is better) and that the variance of profit is a reasonable measure of undesirability (less is better).

By comparing random variables, we have found a more fundamental way in which one random variable is better than another. In a profit-maximizing setting, it is reasonable to prefer random income X to Y if $X > Y$. In a cost-minimizing setting, it is reasonable to prefer random cost X to Y if $Y > X$.

Omitted Arcs

This type of preference has arisen. To see how, we reconsider Problem A (Medfly Eradication). Listed below are two strategies, one of which had been omitted from Figure 9.1.

Strategy 1: Spray now. If the spray works, do nothing more. If the spray fails, burn in one month.

Table 9.9 The cost X_1 of Strategy 1 and the cost X_5 of Strategy 5.

Event	Probability	Value of X_1	Value of X_5
Spray works	1/3	3	3
Spray fails	2/3	$87 = 3 + 84$	$120.6 = 3 + 117.6$

Strategy 5. Spray now. If the spray works, do nothing more. If the spray fails, pause, then burn in two months.

Let the random variable X_1 denote the cost of Strategy 1, and let X_5 denote the cost of Strategy 5. The values of X_1 and X_5 depend on whether or not the spray works. Table 9.9 tells how.

If the spray works, the two strategies have the same cost. If the spray fails, Strategy 1 costs less. Evidently, $X_1 < X_5$. Any decision maker who prefers less cost to more prefers Strategy 1 to Strategy 5.

In a cost-minimization situation, let strategies A and B have random costs X_A and X_B, respectively. Strategy A is now said to **dominate** strategy *B* if $X_A < X_B$. In Figure 9.1, each arc that we omitted led only to dominated strategies; that is why these arcs were omitted.

Comparing Probability Distributions

There is a second answer to the question of whether *X* is larger than *Y*. It, too, leads to domination and tree-pruning. Let us now compare the probability distributions of different random variables. In general, the random variable *X* is **equal in distribution** to *Y* (written $X \sim Y$) if

$$P(X = t) = P(Y = t) \quad \text{for every value of } t.$$

If the random variables *X* and *Y* are independent and identically distributed, they satisfy $X \sim Y$ (because they have the same distribution), but they violate $X = Y$ unless they have only one outcome (no uncertainty).

We now say that the random variable *X* is **greater than or equal in distribution** to *Y* (written $X \succeq Y$) if

$$P(X \geq t) \geq P(Y \geq t) \quad \text{for every value of } t.$$

To interpret the statement $X \succeq Y$, think of the random variables *X* and *Y* as the amounts of money that you would receive if you used two different strategies. Then, the expression $X \succeq Y$ holds if, for every amount *t* of money, the probability that *X* gives you at least *t* is at least as large as the probability that *Y* gives you *t*.

Finally, we say that the random variable *X* is **greater in distribution** than *Y* (written $X \succ Y$) if

$$\left.\begin{array}{ll} P(X \geq t) \geq P(Y \geq t) & \text{for every number } t \\ P(X \geq t) > P(Y \geq t) & \text{for at least one number } t \end{array}\right\}. \tag{9.2}$$

Thus, the expression $X \succ Y$ holds if $X \succeq Y$ and if, for at least one value of *t*, the probability that *X* is at least as large as *t* exceeds the probability that *Y* is at least as large as *t*.

In a profit-maximization problem, it is reasonable to prefer *X* to *Y* if $X \succ Y$; in a cost-minimization problem, it is reasonable to prefer *X* if $Y \succ X$.

Table 9.10 Net profit X_A for betting with and against the test outcome and net profit X_B for betting against the test outcome.

Event	Probability	Value of X_A	Value of X_B
Oil, +	0.30	1.05 = 1.5 − 0.4 − 0.05	−0.05
Oil, −	0.10	−0.05	1.05 = 1.5 − 0.4 − 0.05
No oil, +	0.15	−0.45 = −0.4 − 0.05	−0.05
No oil, −	0.45	−0.05	−0.45 = −0.4 − 0.05

Perverse Actions

To see how this weaker sort of preference occurs, we review Problem C (Drilling for Oil) and compare the strategies that make constructive and perverse use of the test information. These two strategies are listed below.

Strategy A: Test. Drill if +, abandon if −.

Strategy B: Test. Drill if −, abandon if +.

Let the random variable X_A denote the net profit for Strategy A. Similarly, let the random variable X_B denote the net profit from Strategy B. The joint probabilities in Table 9.7 determine the values of X_A and X_B. Table 9.10 tells how.

Note in Table 9.10 that neither $X_A > X_B$ nor $X_B > X_A$. Each strategy gives a larger payoff under two of the outcomes and a smaller payoff under the other two. Table 9.10 reports that X_A and X_B take three values, which are 1.05, −0.05, and −0.45. Figure 9.10 presents the probability distributions of X_A and X_B.

Evidently, X_A places lower weight (probability) on the smallest value and larger probability on the largest value. Table 9.11 reports $P(X_A \geq t)$ and $P(X_B \geq t)$ for the same three values of t.

Table 9.11 demonstrates that for any given amount t of money, Strategy A gives you as high or higher a probability of earning t as does B. Moreover, for values of t that are between −0.05 and 1.05, Strategy A gives you a higher probability of earning t. In brief, $X_A \succ X_B$. Thus, in this weaker sense, an individual who prefers more money to less prefers Strategy A to Strategy B.

In a profit-maximization problem, let strategies A and B have random profits X_A and X_B, respectively. Earlier, we had said that Strategy A *dominates* Strategy B if $X_A > X_B$. We now say that Strategy A **stochastically dominates** Strategy B if $X_A \succ X_B$. A strategy that makes

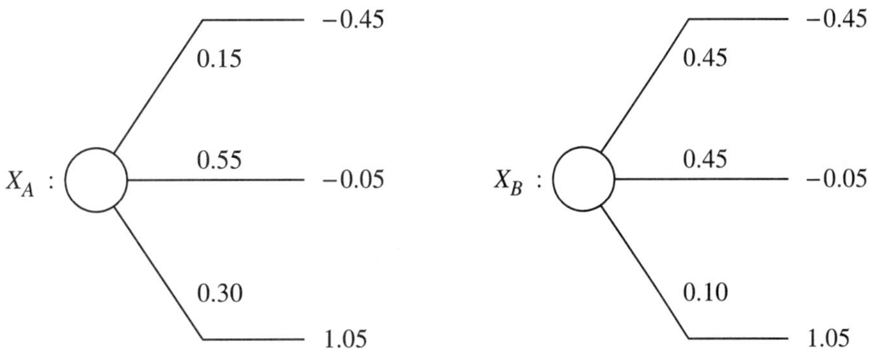

Figure 9.10 The probability distribution of X_A and X_B.

Table 9.11 The probability that X_A and X_B are at least as large as each value t that either takes with positive probability.

Value t	$P(X_A \geq t)$	$P(X_B \geq t)$
−0.45	1.0	1.0
−0.05	0.85	0.55
1.05	0.30	0.10

perverse use of information is dominated by the strategy that differs from it by making constructive use of the same information.

Recap

The random variables X and Y satisfy $X > Y$ if every chance outcome assigns X a value that is at least as high as the value it assigns to Y and if at least one chance outcome assigns X a higher value. If two random variables satisfy $X > Y$, one of them dominates the other. This type of dominance justifies omission of "silly" arcs, arcs that can only increase cost.

The random variables X and Y satisfy $X \succ Y$ if the probability distribution of X is "above" that of Y. If two random variables satisfy $X \succ Y$, then one of them stochastically dominates the other. That type of dominance justifies pruning arcs that bet against the test outcome.

There is a still weaker form of dominance, which represents a preference for avoiding risk (randomness). We mention it solely to complete the record. Let the random variables X and Y measure cost. The random income X is said to exhibit **second-order stochastic dominance** over Y if $Y \sim (X + Z)$ where the random variable Z has $E(Z|X = t) \geq 0$ for every number t.

9.9. A SEQUENTIAL DECISION PROCESS

When we rolled back a decision tree, we did something that may have seemed a bit unnatural. We associated with each node the least value of the *total* expected cost, given that this node would be reached. A more natural approach would be to associate with each node the least value of the expected current and future cost. This more natural approach will prove to have advantages.

The more natural approach takes the perspective of a "sequential decision process." You may have encountered sequential decision processes in Chapters 4 and 5, but those chapters aren't prerequisite to this one. In a **sequential decision process**, the primitive concepts are "states, "actions," and "transitions," which have these properties:

- Transitions occur from state to state.
- At each moment in time at which a state is occupied, an action must be selected.
- The law of motion that governs the transition from state to state can depend on the state that is currently occupied and on the action that is selected, but not on prior states and not on prior actions.
- Selecting an action can incur a cost (possibly negative); this cost can depend on the current state and action, but not on prior states and not on prior actions.

A decision tree *is* a sequential decision process. The choice nodes are the states. The arcs that point rightward from the choice nodes are the actions. The law of motion (transition probabilities) depends on the current state (node) and action (arc), but not on prior states or actions. The cost depends solely on the current action (arc).

The Role of States

Each state in a sequential decision process contains enough of the prior history of the process to evaluate the current and future actions. Formulating an optimization problem as a sequential decision process can require ingenuity. The key is to identify the states. It helps to understand that each state contains enough information about the past that the rest is irrelevant.

The methodology for analyzing sequential decision processes is described in general terms and is then applied to medfly eradication. The general description may look perplexing, but the application will make everything clear.

Embedding

For each state i in a sequential decision process, we now define the quantity $f(i)$ by

$$f(i) = \text{the least expected present and future cost given that state } i \text{ is observed now.}$$

Typically, our interest lies in $f(i)$ for a particular state, not for every state. The problem of computing $f(i)$ has just been **embedded** in a family of optimization problems, one per state. Embedding appears to be a step backwards. Rather than attacking the problem of interest, embedding introduces a family of optimization problems, one per state.

The f-value for each state equals the optimal value of the sequential decision process in which decision making starts at that state. The f-values for the various states are linked by a so-called **functional equation**, which is actually a system of equations, one per state. For a sequential decision process that marches forward in time and has a fixed end-point, the functional equation is easy to solve by a procedure that is known as **backwards recursion**; start at the end and work back toward the beginning.

An Illustration

This methodology will now be applied to medfly eradication, and it will look very familiar. What are the states? It's clear from Figure 9.1 that the choice nodes are the states. Figure 9.1 numbers the states (nodes) 1 through 5.

For each state i, the number $f(i)$ equals the least expected current and future cost given that state i is now observed. The functional equation links the $f(i)$'s for the various states. Since transitions occur to higher-numbered states, we will be able to solve the functional equation in decreasing i, starting with state 5 and ending with state 1.

It is evident from Figure 9.1 that state 5 has only one action, which is to burn at a cost of 117.6:

$$f(5) = 117.6.$$

State 4 also has only one action, which is to release the sterile males. As Figure 9.1 indicates, the sterile males have already been paid for, but they have a probability (1/2) of failing, in which case you reach state 5:

$$f(4) = (1/2)(0) + (1/2)f(5) = (1/2)(117.6) = 56.8.$$

State 3 also has only one action, which is to burn at a cost of 117.6:

$$f(3) = 117.6.$$

At state 2, we must choose between burning and spraying a second time. Figure 9.1 indicates that burning costs 84. Similarly, spraying a second time costs 4.2 now and leads to state 3 with probability 2/3, so

$$f(2) = \min\begin{cases} 84 \\ 4.2 + (1/3)(0) + (2/3)f(3). \end{cases}$$

We have already computed $f(3)$. By substituting the value of $f(3)$ in the above, we get

$$f(2) = \min\begin{cases} 84 \\ 4.2 + (2/3)(117.6) \end{cases} = \min\begin{cases} 84 \\ 82.6 \end{cases} = 82.6.$$

Finally, at state 1, there are three actions. Each line of the equation that follows accounts for one of these actions.

$$f(1) = \min\begin{cases} 3 + (2/3)f(2) \\ 60 \\ 9 + (2/3)f(4). \end{cases}$$

By substituting the f-values for states 2 and 4, we obtain

$$f(1) = \min\begin{cases} 3 + (2/3)(82.6) \\ 60 \\ 9 + (2/3)(56.8) \end{cases} = \min\begin{cases} 58.0667 \\ 60 \\ 48.2 \end{cases} = 48.2$$

This computation demonstrates (once again) that each strategy that gets rid of the medfly has an expected cost of at least \$48.2 million and that one strategy has an expected cost of \$48.2 million. The strategy whose expected cost equals \$48.2 million selects the action that attains each state's f-value. At state 1, it selects the bottom-most of the three actions, which is to spray while incubating sterile males. At state 2, it selects the bottom action, which is to spray a second time.

What's New?

In one sense, very little is new. The recursion for the f-values differs only slightly from the rollback that appeared earlier in the chapter. The f-value assigned to each state equals its least *future* cost. The number assigned to that state in the prior rollback equals the least *total* cost given that that state will be occupied. These numbers differ by the cost of reaching the state. To illustrate, consider state 4. Its f-value equals 56.8. In Figure 9.2, node 4 was assigned the value 67.8 . The cost of reaching node 4 equals 9. And 67.8 equals 9 + 56.8.

On the other hand, the current recursion can be more efficient. To see why, imagine a state that can be reached by several paths. In a decision tree, paths are not allowed to "merge." A decision tree must include one node per path to this state, and it executes one calculation per node. That's extra work.

9.10. EXAMPLE OF A SEQUENTIAL DECISION PROCESS

Let us now consider a problem that can be formulated in two different ways. One is a decision tree, and the other is more efficient. This problem is:

Problem D (The Chess Match)

Chess is a Russian pastime, and the Russians are good at it. The game is played between two players, and each game ends with a victory or a draw. Either one player wins the game and the other loses, or the game is a draw. An American and a Russian are to play a three-game chess match, whose rules are as follows:

- The player who wins the most games wins the match.
- If, however, the players are tied at the end of three games (with one win apiece or with three draws), they continue playing until someone wins a game, and that person wins the match.

The Russian is the stronger chess player, and he plays consistently. The American has two tactics, bold and timid. If she plays boldly, she wins with probability $p = 0.46$ and loses with probability $1 - p = 0.54$. If she plays timidly, she draws with probability $q = 0.9$ and loses with probability $1 - q = 0.1$. With either tactic, the outcomes of different games are independent of each other. The American has asked you to advise her as to what strategy she should use.

Let's imagine that your first thought is to draw a decision tree. At each end-point of this tree, the American either wins the tournament or loses it. You plan to assign each end-point of this tree a payoff of 1 if the American wins the match and a payoff of 0 if the American loses it. Consequently, maximizing the expectation of the American's payoff maximizes the probability that she will win the match.

This line of attack will succeed. If you pursue it, you will discover that the decision has many nodes, and moreover, that its nodes record unnecessary detail. If, for instance, the American loses the first game, she faces precisely the same situation at two different nodes—the one in which she played boldly and lost and the one in which she played timidly and lost. What matters is that two games remain and that she is one win behind; how she got to this situation is irrelevant.

States

It occurs to you that thinking in terms of states can eliminate this redundancy. At any point in the match, a state must be a sufficient summary of what went before. The state must keep track of two quantities—the number of games that have been played and the American's "score," that is, the number of games she has won less the number she has lost. In other words, each state is a pair (k, s) in which

$k =$ the number of games that have been played, and

$s =$ the American's current score.

Aiming for a functional equation, you define $f(k, s)$ by

$f(k, s) =$ the probability that the American will win the match given that state (k, s) is observed now and that she chooses her best strategy.

You have just embedded the decision problem of interest in a family of optimization problems, one per starting state.

A Functional Equation

The functional equation specifies $f(k, s)$ for each state. This functional equation is actually a system of equations, one per state. To construct the functional equation, we first consider a state (k, s) in which $k = 3$. Three games have been played (because $k = 3$). If the American's score s is positive, she wins. If her score s is negative, she loses. If her score

s equals zero, the best she can do is to play boldly, so

$$f(3, s) = \begin{cases} 1 & \text{if } s > 0 \\ 0.46 & \text{if } s = 0. \\ 0 & \text{if } s < 0 \end{cases} \tag{9.3}$$

Let us now examine a state (k, s) with $k \leq 2$. At least one more game must be played. In the current game, the American can choose bold play or timid play. If she chooses bold play, her score increases by 1 with probability 0.46, and it decreases by 1 with probability 0.54. If she chooses timid play, her score stays equal to s with probability 0.9 and decreases by 1 with probability 0.1. With either strategy, the number k of games she has played increases by 1. Bold and timid play account for the top and bottom lines in

$$f(k, s) = \max \begin{cases} (0.46)f(k+1, s+1) + (0.54)f(k+1, s-1) \\ (0.9)f(k+1, s) + (0.1)f(k+1, s-1) \end{cases} \quad \text{for } k \leq 2. \tag{9.4}$$

A Spreadsheet

Equations (9.3) and (9.4) are easy to solve on a spreadsheet. Equation (9.4) is a recursion, and once it has been set up for a particular state, a drag adapts it to the other states.

Table 9.12 computes $f(k, s)$ for each state (k, s) that can arise in a three-game match. Rows 9 through 12 of the table correspond to having played 0, 1, 2, and 3 games. Columns B through H correspond to scores between -3 and $+3$. Row 12 executes the computation in Equation (9.3). The function in cell E9 executes the computation in Equation (9.4) for state (0, 0). Dragging this function over the cells in rows 8 and 9 computes $f(k, s)$ for each state that can be reached from state (0, 0).

Table 9.12 records the American's probability $f(k, s)$ of winning the match for each state (k, s) that she might reach. The match begins at state (0, 0) because no games have been played and no one has won any games. Evidently, she can win this match with probability 0.527, even though she is the less skilled player. Had you seen that coming? She would have been even better off with a two-game match, by the way. Do you see why?

Table 9.12 Probability of winning and optimal strategy for the chess match; bold play in the unshaded states, timid play in the shaded states.

	A	B	C	D	E	F	G	H	I	J
2			lose	tie	win					
3		bold	0.54		0.46		= MAX($C$3*E9 + $E$3*G9,			
4		timid	0.1	0.9			$C$4*E9 + $D$4*F9)			
5										
6	score s =		-3	-2	-1	0	1	2	3	
7	number k of games played	0				0.527				
8		1			0.212	0.549	0.897			
9		2		0	0.212	0.46	0.946	1		
10		3	0	0	0	0.46	1	1	1	
11										
12						= E3				

To record the American's optimal Strategy, we need to see which alternative achieves the maximum in each state. Table 9.12 omits this calculation. We have done it, and we have shaded the cells at which she plays timidly. The American's optimal strategy is to play boldly when her score is not positive and to play timidly when her score is positive.

9.11. A MARKOV DECISION PROCESS*

In Problem A (Medfly Eradication), each choice node can be reached by only one path. In Problem D (The Chess Match), each choice node can be reached by several paths, but each is visited at most once during the course of the match. A Markov decision process differs from these examples in only one way. In it, states can be revisited. This section introduces Markov decision processes and provides an example.

The Data

Let us consider a sequential decision process whose objective is to minimize the expectation of net cost. Being a sequential decision process, its data are allowed to depend solely on the state that is observed and the action that is selected. Let the symbol i stand for a state, and let the symbol k stand for an action. When the object is minimization, the data in a **Markov decision process** are:

C_i^k = the immediate cost (possibly negative) if state i is observed and action k is selected,
P_{ij}^k = the probability of transition to state j given that state i is observed and action k is selected.

Consider, for instance, Program 9.1 (Medfly Eradication). For state 1, there are three actions, which we designate as s (for spray), b (for burn), and $s\&i$ for (spray and incubate). It is evident from Figure 9.1 that

$$C_1^s = 3, \quad C_1^b = 60, \quad C_1^{s\&i} = 9, \quad P_{12}^s = \tfrac{2}{3}, \quad P_{14}^{s\&i} = \tfrac{2}{3},$$

and that all other transition probabilities from state 1 equal zero.

A Functional Equation

Aiming for a functional equation, we define the quantities $f(i, k)$ and $f(i)$ as follows:

$f(i, k)$ = the least value of the expected present and future cost given that state i is observed now and action k is selected.

$f(i)$ = the least value of the expected present and future cost given that state i is observed now.

To specify $f(i, k)$, note that observing state i and selecting action k incur cost C_i^k and cause transition to state j with probability P_{ij}^k. Also note that the least cost you can incur if state j is observed equals $f(j)$. The expectation of the sum of the current and future costs equals the sum of their expectations, so

$$f(i, k) = C_i^k + \sum_j P_{ij}^k f(j), \quad \text{for each state } i \text{ and action } k.$$

* This section and the next can be omitted with no loss of continuity.

When in state i, you pick the action k that minimizes expected present and future cost. Thus,

$$f(i) = \min_k\{f(i,k)\}, \quad \text{for each state } i.$$

Combining the preceding two equations produces

$$f(i) = \min_k\left\{C_i^k + \sum_j P_{ij}^k f(j)\right\} \quad \text{for each state } i. \tag{9.5}$$

The system of equations in (9.5) forms the so-called **functional equation** for this Markov decision process. System (9.5) links the f-values for the various states.

Equation (9.5) holds for decision trees. It holds for Markov decision problems in which decision making is guaranteed to stop eventually. It even holds for a class of Markov decision models in which decision making might continue indefinitely, provided the costs are bounded and termination can be guaranteed to occur with probability of at least $\alpha > 0$ after some fixed number N of transitions.

A Linear Program

The solution to a functional equation can often be found recursively, as was done for medfly eradication and for the chess match. In nearly every case, the solution to a functional equation can also be found by solving a linear program. In Program 9.1, each state i is assigned a decision variable, $g(i)$. For decision trees and for the Markov decision problems that satisfy the above property, there is only one optimal solution to Program 9.1, and it has $g(i) = f(i)$ for each state i.

Program 9.1: Maximize$\{\sum_i g(i)\}$, subject to the constraints

$$g(i) \le C_i^k + \sum_j P_{ij}^k g(j), \quad \text{for each state } i \text{ and action } k.$$

It may strike you as odd that the minimization problem has been formulated as a linear program whose objective is to maximize something. Intuitively, here's what is going on: For each state i, maximization forces $g(i)$ upward so that an equality is achieved for at least one k. In this way, maximization forces a solution to Equation (9.5). The same trick was used in Chapter 5 where a shortest-path problem was formulated as a linear program whose objective is maximization.

Profit Maximization

The Markov decision model has been introduced in the context of a cost minimization. To adapt this model to profit maximization, make these changes:

- C_i^k becomes the immediate reward (possibly negative) for observing state i and selecting action k.
- $f(i)$ becomes the greatest expected present and future profit given that the sequential decision process begins with the observation of state i.
- In Equation (9.5), "min" becomes "max."
- In Program 9.1, "Maximize" becomes "Minimize," and the sense of each constraint switches from $\le$ to $\ge$.

The next section illustrates a profit-maximizing Markov decision process.

9.12. MAIL ORDER CATALOG SALES*

Many families receive stacks and stacks of unsolicited mail order catalogs. Why? The suggestion of an answer may be found in:

Problem E (Mail Order Catalogs)

It costs a mail order company \$1.80 to print and mail each catalog. The company earns an average of \$25 on each order that it receives in response to a catalog mailing. From its records, the company has ascertained that:

- If a catalog is mailed to a target customer who has received no prior catalogs from the company, the probability that this person places an order is 0.05.
- If a catalog is mailed to a customer who placed an order k catalogs ago and not since then, the probability $p(k)$ that this person will place an order is as given in Table 9.13.

Is it profitable to mail a single catalog to a potential customer? What strategy maximizes the expected profit per customer?

The company's expected net profit for mailing exactly one catalog to a potential customer equals $-\$1.80 + (0.05)(\$25) = -\$0.55$. Hence, it is not profitable to mail a single catalog to a potential customer. But it may be profitable to send a catalog to a potential customer in the hope of repeat orders.

States

Can we model Problem E as a decision tree? No. We would find that it has no natural "end-point" because a customer can order again and again, without limit.

Problem E can be formulated as a Markov decision process. To do so, we observe that the current and future income depend solely on the number k of catalogs that have been sent to an individual since he or she last placed an order. This number k need not exceed six because Table 9.13 shows that a person who placed his or her last order six catalogs ago will place no more orders. Thus, Problem E can be formulated as a Markov decision process that has these seven states:

- By state 0 is meant a target customer to whom no catalog has ever been sent.
- For the integers k between 1 and 6, state k denotes a customer who placed his or her most recent order k catalogs ago.

Table 9.13 Probability $p(k)$ of an order versus the number of mailings since the most recent order.

k	1	2	3	4	5	≥6
$p(k)$	0.4	0.2	0.1	0.03	0.01	0

*This section concerns a specialized topic that can be omitted with no loss of continuity.

A Functional Equation

Let $f(k)$ denote the largest expected current and future profit that the company can earn from a customer whose state is k. Note that

$$f(6) = 0,$$

because the data in Table 9.13 indicate that a customer whose most recent order was placed six catalogs earlier will place no further orders and so should be sent no further catalogs.

Consider a customer who placed an order k catalogs ago but not since, where the integer k is between 1 and 5, inclusive. The company can elect to send no further catalogs, in which case it earns no further income, which accounts for the top line in the equation

$$f(k) = \max\begin{cases} 0 \\ -1.80 + p(k)[25 + f(1)] + [1 - p(k)]f(k + 1) \end{cases} \quad \text{for } k = 1, \ldots, 5.$$

The bottom line of this equation accounts for the cost of sending a catalog to the customer, for the expected profit of $p(k)$ 25 from that catalog, and for the expectation of the future profit. Finally,

$$f(0) = \max\begin{cases} 0 \\ -1.80 + (0.05)[25 + f(1)] \end{cases}$$

because the company can choose to send no catalog or at least one. The seven equations that are displayed above constitute a functional equation for catalog sales.

A Linear Program

For each state k, the optimal strategy selects an action whose expected current and future income equals $f(k)$. Computation of the numbers $f(0)$ through $f(6)$ and the optimal strategy can be posed as the linear program:

Table 9.14 A linear program for Problem E.

	A	B	C	D	E	F	G	H	I	J	K	L
2		catalog cost =		1.8								
3		profit per sale =		25					= - $D$2 + B6 * $D$3			
4												
5	k	p(k)	0	1	2	3	4	5	6			
6	0	0.05	1	-0.05						-0.55	>=	-0.55
7	1	0.40		0.60	-0.60					8.2	>=	8.2
8	2	0.20		-0.20	1	-0.80				3.2	>=	3.2
9	3	0.10		-0.10		1	-0.90			0.7	>=	0.7
10	4	0.03		-0.03			1	-0.97		-0.73	>=	-1.05
11	5	0.01		-0.01				1	-0.99	-0.24	>=	-1.55
12	6	0.00		0.00					1	0	>=	-1.8
13	obj. coef. =		1	1	1	1	1	1	1	38.52		
14		f(k) =	0.66	24.2	10.54	3.12	0	0	0			

Solver minimized the value in cell J13 with C14:I14 as changing cells, subject to C14:I14 >= 0 and J6:J12 >= L6:L12

= SUMPRODUCT(C13:I13, C$14:I$14)

Minimize $\{f(0) + f(1) + \cdots + f(6)\}$, subject to the constraints

$$f(0) \geq -1.80 + (0.05)[25 + f(1)]$$

$$f(k) \geq -1.80 + p(k)[25 + f(1)] + [1 - p(k)]f(k+1) \quad \text{for } k = 1, \ldots, 5,$$

$$f(k) \geq 0, \quad \text{for } k = 0, 1, \ldots, 6.$$

The above adapts Program 9.1 to a maximization problem. The following spreadsheet solves it.

The optimal strategy is found from the tight constraints in Table 9.14, and these constraints have been shaded. Evidently, the optimal strategy is to mail one catalog to a potential customer and, if that person places an order, to continue to send catalogs until no order has been placed for four consecutive catalogs. This strategy earns the company an expected profit of \$0.66 per candidate customer.

This introductory text presents only one of the many applications of Markov decision processes. Another application is to the pricing of financial derivatives.

9.13. REVIEW

This chapter has introduced a substantial vocabulary. You should now be facile with these terms: decision tree, strategy, pruning, rollback, breakeven computation, value of information, tree-flipping, sequential decision process, and functional equation. If you have read the starred sections, you should also be familiar with the terms domination, stochastic domination, and Markov decision process.

The decision tree is the simplest model of decision-making under uncertainty. A decision tree elaborates on the probability tree by including choice nodes as well as chance nodes. The choices entail costs or profits. The chance nodes have conditional probabilities of transition to other nodes. For a decision tree, a strategy is a plan that specifies an action for each possible contingency—for each choice node that can be reached.

Throughout this chapter, we have taken the expectation as the measure of cost or profit. Thus, the best strategy is one whose expectation of net cost is lowest and, equivalently, whose expectation of net profit is largest. A simple recursion computes the best strategy. In this recursion, all costs (or profits) are accumulated at the end-points of the decision tree, and then a "rollback" is executed. At each chance node, the rollback computes the expected cost (or profit). At each choice node, the rollback picks the action whose cost (or profit) is best. In Chapter 10, we shall see that precisely the same "rollback" adapts to more general performance criteria than the expectation of profit or of cost.

An Excel add-in called TreePlan lets you build a decision tree on a spreadsheet. TreePlan rolls it back automatically. Building a decision tree on a spreadsheet can require effort, but the spreadsheet greatly eases sensitivity analysis, which determines whether or not a strategy is robust. One facet of sensitivity analysis is a "breakeven" calculation on the value of an intangible.

The rollback assigns to each node the expected *total* net profit (or net cost) given that this node is visited. A slightly different recursion assigns to each node the expected *future* net profit (or net cost), given that this node is now visited. The second recursion allows for multiple paths to the same node. It can be more efficient. When a decision problem has a pattern, the second recursion can be particularly easy to execute on a spreadsheet, without constructing a decision tree.

A pair of starred sections generalizes decision trees in a way that allows nodes (states) to be revisited. The generalization is a Markov decision process whose optimal strategy can be found by solving a linear program.

9.14. HOMEWORK AND DISCUSSION PROBLEMS

1. (Medflies) In this problem, the spray works with probability of $1/3$. Find the range on this probability for which Strategy 4 has the least expected cost.

2. (Medflies) In this problem, costs escalate at the rate of 40% per month. This figure may be inaccurate. Find a range of rates of escalation for which Strategy 4 has the lowest expected cost.

3. (Medflies) In this problem, suppose that the governor is willing to entertain plans that eradicate the medfly within three months, not two. How does that affect your decision tree? your recommendation?

4. (Medflies) In this problem, let the random variable X denote the cost of the strategy that sprays and incubates sterile males, sprays a second time if the first spray fails, and burns if the second spray fails. Identify a strategy whose cost Y satisfies $Y < X$ and demonstrate that it does.

5. (R&D Management) For this problem:

(a) Use a decision tree to find the strategy that minimizes the expected cost of developing the necessary production capacity within nine months.

(b) Find the range of values of the new technology for which it is best to develop the new technology. If you develop it, what strategy should you use? *Hint:* To each end-point in which new technology is obtained, include a value v of obtaining it.

(c) Suppose there is added value to getting the new technology in six months rather than nine. Might this affect your optimal strategy? If so, how?

6. (The Chess Match) For this problem, p equals the probability that she wins the game if she plays boldly, and q equals the probability that she draws the game if she plays timidly. The spreadsheet in Table 9.12 was done with $p = 0.46$ and $q = 0.9$.

(a) With q fixed at 0.9, find the smallest value of p for which her probability of winning the match is at least 0.5?

(b) Repeat part (a) with $q = 0.95$.

(c) Is her best strategy robust? Why?

(d) With $p = 0.46$ and $q = 0.9$, how does she maximize the probability of winning a five-game tournament, and what is that probability?

7. (Test information) Let event A have probability $P(A)$ that lies strictly between 0 and 1. A test as to whether or not A occurs has sensitivity $s = P(+|A)$ and specificity $\sigma = P(-|\bar{A})$.

(a) Show that

$$P(A|+) = \frac{s\,P(A)}{s\,P(A) + (1-\sigma)P(\bar{A})} = \frac{s}{s + (1 - s - \sigma)P(\bar{A})}\,P(A).$$

(b) Use part (a) to demonstrate that $P(A|+)$ exceeds $P(A)$ if and only if $s + \sigma > 1$.

(c) Adapt parts (a) and (b) to show that $P(\bar{A}|-)$ exceeds $P(\bar{A})$ if and only if $s + \sigma > 1$.

(d) How would you use this test if $s + \sigma$ is below 1? Would you call the test perverse?

8. (Farmer John) The demand for perishable farm products tends to be "inelastic," and this has a perverse effect. In a foul-weather year, harvests are low, but the price is bid up because the demand is unabated and consumers are bidding for a smaller crop. In a bountiful year, harvests are large, but demand does not increase, and the abundant supply forces prices down. Farmers tend to come out ahead in a foul-weather year and behind in a bountiful year. To quantify this statement, let's consider a three-state model: normal weather, fair weather, and foul weather. The weather and its effect on Farmer John's crop and prices are as follows:

State of Nature	Probability	Market Price ($/unit)	Farmer John's Crop (units/year)
Fair	1/3	7	120,000
Normal	1/3	10	100,000
Foul	1/3	15	80,000

Omitted from the above are John's cash outlay for seed, fertilizer, mortgage interest, rental of machinery, and so forth. This outlay amounts to $850,000 per year, independent of the weather.

(a) Describe the probability distribution of John's net revenue. Compute its expectation.

(b) At the beginning of the year, John can purchase a "forward contract" whose terms are as follows: John agrees to sell 100,000 units of grain at harvest time for which he will receive a total of $1 million. *Note:* if John contracts to deliver 100,000 units of grain, he must deliver that amount. Compute the expected profit for the person or company that sells this contract to John. Is it large or small?

(c) Suppose John buys this contract. Draw John's probability tree given that he has done so. Has John reduced his risk? Explain your answer. Some farmers *do* buy forward contracts of this sort; what do you suppose happens to them in bad-weather years?

(d) Suppose John could take a contract to deliver any number z of units of grain at harvest time for $10z$ dollars. Find the value of z that minimizes the range of his net profit. For this contract amount, what is the expectation of his profit?

(e) Repeat part (d), but select z to minimize the variance of his net profit.

9. **(Treasure Hunt)** A Spanish ship carrying $25 million worth of gold bullion, at current prices, sank in the seventeenth century, and your company may have discovered the region in which it sank. The region in question is a 10 square mile area off the coast of Florida. Your expert estimates a probability of 0.3 that the gold is actually there. To locate it, you must dive. Each dive has a 10% chance of finding the treasure, if it is there. Different dives are independent. Each dive costs $0.2 million. Should the gold be found, the cost of recovering it will be minuscule. For gains and losses below $25 million, your company is risk-neutral; it acts to maximize its expected net revenue. *Remark:* You could find the optimal strategy through a decision tree, but the work would be heroic.

(a) Specify formulas for:

$$\begin{aligned}
a(n) &= P(n \text{ dives are unsuccessful, gold is there}),\\
b(n) &= P(n \text{ dives are unsuccessful, gold is not there}),\\
c(n) &= P(n \text{ dives are unsuccessful}),\\
d(n) &= P(\text{gold is there} \mid n \text{ unsuccessful dives}),\\
e(n) &= P(\text{find gold on the } n\text{th dive}).
\end{aligned}$$

Hint: Express $e(n)$ in terms of $d(n-1)$.

(b) Specify a formula for:

$$f(n) = \text{the expectation of the net profit on the } n\text{th dive given that the first } n-1 \text{ dives have been unsuccessful.}$$

Hint: The quantity $f(n)$ excludes the (sunk) cost of the first $n-1$ dives. Might it equal $25e(n) - 0.2$, in millions?

(c) Open an Excel spreadsheet and compute the solutions to parts (a) and (b) for the values $n = 0, 1, 2, \ldots, 20$. What strategy maximizes the expectation of net profit?

(d) On your spreadsheet, compute the expected total profit for the strategy you specified in part (c). *Hint:* sum $d(n-1)f(n)$ over certain values of n.

10. **(Pascal's Triangle on a Spreadsheet)** It can be handy to think of a "state" as a regeneration point. Let us consider a sequence of n Bernoulli trials, each of which is either a "success" or a "failure." The notation (S, S, F) is shorthand for the sequence of three trials in which the first two are successes and the last is a failure. Designate

$$f(k, n) = \text{the number of different sequences of } n \text{ trials in which } k \text{ successes occur.}$$

(a) Show that $f(0, n) = 1$ and that $f(n, n) = 1$.

(b) Show that $f(k, n) = f(k-1, n) + f(k, n-1)$ for $0 < k < n$. *Hint:* The first trial is either a success or a failure.

(c) On a spreadsheet, compute $f(k, n)$ for each pair (k, n) that satisfies $0 \le k \le n \le 10$. Do these numbers look familiar?

11. **(Domination and Stochastic Domination)** Suppose that random variables X and Y satisfy $X > Y$. Show that $X \succ Y$. (In other words, show that domination implies stochastic domination.)

12. **(Medflies)** Cast this problem in the format of Program 9.1 and use Solver to compute its optimal solution. Does Solver identify the same optimal value and optimal strategy that you got from your tree?

13. **(Howie Slams)** Howie Slams, the tennis great, has analyzed his service. He has two serves, the "smash" and the "spin." Against today's opponent, he wins only 55% of the points on which his spin is in bounds and 75% of the points on which his smash is in bounds. The probability that his spin is in bounds is 0.9. Howie's smash is a bit inconsistent; it varies from day to day, and it can vary over the course of a match against a single opponent. The probability p that Howie's smash is in bounds can be as low as 0.4 and as high as 0.85. Howie wants to serve in way that maximizes the probability that he wins the point. (The server gets two chances—if the first serve is out of bounds, the server gets to try again.) He has asked you for advice.

(a) Draw a decision tree that lists all four of his strategies and evaluates the probability that each strategy wins the point.

(b) On a spreadsheet, compute the probability each strategy wins the point for various values of p.

(c) Do a breakeven analysis, and make a recommendation.

14. **(Fish Ladders)** Environmentalists have just won a major lawsuit. An electric utility must construct a fish ladder that allows salmon to swim upstream beyond the company's main hydroelectric facility. Three designs for the fish ladder have been proposed. Their costs, construction times, and probability of success are as follows:

Type of Ladder	Cost (millions)	Time (years)	*P*(succeeds)
A	80	3	1
B	70	2	0.8
C	60	1	0.7

The utility must also pay the state a fee until it can demonstrate that the salmon are successfully circumventing its dam. The trial judge has set the fee, tentatively, at $10 million per year. Design A is proven. If Design A is built, the company will need to pay the fee until construction is complete. Designs B and C are novel. It will not be known whether they succeed until one year after they are completed, and the company will need to pay the fee for that year. If it is learned that Design B has failed, Design B can be converted to Design A in one year at a cost of $30 million, at which time the company would be relieved of the fee. A fish ladder using Design C has just been completed elsewhere. In one year, it will be known whether Design C succeeds there. If Design C succeeds there, it will succeed here, and the company will be relieved of the fee as soon as it is constructed. Similarly, if Design C fails there, it will fail here. Design C cannot be converted to either of the other designs. There is room for only one ladder. The utility wishes to build a successful ladder at minimum expected total expense.

(a) Could there be merit in doing nothing for one year?

(b) Represent the utility company's decision problem as a decision tree. At each end-point of this tree, list three elements of data—the construction cost of reaching that end-point, the number of years that fees are paid, and the number of years until a successful ladder is built.

(c) State one dominated strategy, and name a strategy that dominates it.

(d) With a fee of $10 million per year, find the best strategy.

The fish have their own perspective on the fish ladder. They wish to minimize the expectation of the time until they get a working ladder. In certain cases, the utility company will continue to pay the fee for a year after a working ladder is built, but, of course, that is of no concern to the fish.

(e) What strategy is best from the viewpoint of the fish?

(f) At a hearing on the fee, the environmentalists have argued that the judge should set the fee high enough to motivate the company to do what is best for the fish. Is there economic merit to this argument?

15. (**Bold Play**) Your current asset position is $3000. You desperately need to increase it quickly to $10,000. The only alternative available to you is to bet on a roulette wheel. At each spin of the wheel you win an amount of money equal to your bet with probability $p = 18/38$ and lose the amount that you bet with probability $20/38$. This particular wheel accepts "high rollers" only, namely, those who bet a number of "chips" that are worth $1000 apiece. Your goal is to maximize the probability of reaching an asset position of $10,000.

(a) Identify the states in this Markov decision problem.

(b) Write a functional equation whose solution gives the probability of reaching $10,000 given each state i.

(c) Use Solver to compute the probability that you attain your goal of $10,000 . For each state i, specify the optimal amount(s) to bet. (*Hint:* Adapt Program 9.1 to a maximization problem.) Is "bold play" one of the optimal strategies?

(d) Redo part (c) with $p = 18/37$.

(e) Redo part (c) if you needed only $9000 rather than $10,000.

16. (**A Subcontractor**) Art's machine shop may bid for a contract to produce 100,000 units of a certain part. Art is hesitating between placing a bid of $8.25 per unit and a bid of $8.50 per unit. He guesses that a bid of $8.25 has a 70% chance of being accepted and that a bid of $8.50 has a 50% chance of acceptance. Art has two production processes available for making this part. Process A costs $7.75 per unit. Process B is novel, and its cost is uncertain; Art estimates its cost as equally likely to be $7.00, $8.00, and $9.00. There is time before Art must submit a bid for a test run on Process B. This test run would cost $10,000 and outcome would be "Low" or "High." A "Low" test outcome would suggest a low unit cost for Process B but would not guarantee it. Based on past experience, Art estimates the conditional probabilities as follows:

Condition: Cost of Process B Equals	*P*(Low)	*P*(High)
$7.00	0.8	0.2
$8.00	0.5	0.5
$9.00	0.1	0.9

(a) First, omit the possibility of a test. Formulate Art's problem as a decision tree and roll it back to find the strategy that maximizes Art's net profit.

(b) Redo part (a), but this time include the possibility of a test before Art submits a bid. What changes?

(c) Redo part (b), but allow Art to do the test either before he submits a bid or after it is accepted, but not both. What changes?

17. (**Mail Order Catalog Sales**) For Problem E, suppose that catalogs must be recycled, that it costs $0.25 to recycle each catalog, and that this cost must be borne by the mail order companies. How does this recycling cost affect the functional equation? its solution?

18. (**A Bonus Round for Jeopardy**) Having been the winner on a competitive quiz show you get to play in the bonus round. Dollar amounts have been written on the backs of cards A, B, and C. The bonus is the largest of the amounts on these cards. You will win the bonus if you can guess the card on which it is written. You have no information about the amounts on the cards, except that they are positive. Pat, the quiz master, assures you that the prizes have been placed randomly on the cards. His assistant, Vanna, will turn over card A, then card B, and then card C. She will pause after turning over each card to offer you the opportunity to announce, "That one." If the card she just turned over when you announce "That one" has the largest amount, you win that amount as a bonus. Otherwise, you earn no bonus.

(a) If you announce "That one" after she turns over card A, what is the probability that you will win?

(b) Is there a way in which you win with probability of $1/2$? *Hint:* Let Vanna turn over two cards before taking any action. Then what?

Chapter 10

Utility Theory and Decision Analysis

10.1. PREVIEW

Chapter 9 focused on decision trees, which were rolled back to maximize the expectation of net profit, and equivalently, to minimize the expectation of net cost. It was asserted that maximizing expected profit is reasonable for operational decisions. The present chapter also focuses on decision trees, but here, the perspective is more fundamental. We aim to specify the performance criterion that is appropriate when a particular individual contemplates a particular decision.

John Von Neumann and Oskar Morgenstern are rightly credited for developing **utility theory**, which shows how an individual who accepts certain postulates of rational behavior reaches a decision that is consistent with his or her beliefs. Utility theory applies to *all* decision-making situations that one person might face while acting alone. It applies when the consequences of one's actions include large monetary gains and losses; it applies when the consequences are not monetary; and it applies to operational and to strategic situations.

As used in this chapter, the phrase **decision analysis** describes the application of utility theory to a particular decision. Decision analysis measures an individual's preferences as to the events and outcomes in a decision tree. Decision analysis determines whether the conclusions reached by utility theory are robust and whether the postulates of rational behavior are appropriate for that situation.

10.2. WHAT CAN YOU LEARN FROM THIS CHAPTER?

This chapter aims to teach you what utility theory is, how to use it, and what its limitations are. It introduces you to the postulates of rational behavior. It interprets these postulates. It also indicates how a decision maker who accepts these postulates can reach a decision that is consistent with his or her preferences. Finally, this chapter critiques these postulates and discusses how a decision maker's preferences can violate them.

To simplify the discussion, we restrict ourselves to trees that have a finite number of end-points. The postulates of rational behavior will imply that:

- Among these end-points, there will be at least one that the decision maker most prefers and at least one that the decision maker least prefers.
- For each end-point e, the decision maker is indifferent to substitution of a "lottery" that gives the worst end-point with probability $w(e)$ and the best end-point with probability $b(e) = 1 - w(e)$.
- After substitution of this lottery for each end-point, the decision maker's most preferred strategy is found by rolling back the tree to minimize the probability of getting the worst end-point and, equivalently, to maximize the probability of getting the best end-point.

Different decision makers can have different indifference probabilities and still be rational. You will see how to elicit a particular decision maker's indifference probabilities.

These postulates of "rationality" will seem plausible, but clear-thinking decision makers can act in ways that violate them. This chapter surveys the main critiques of these postulates, as well as the other limitations of utility theory.

10.3. LARGE STAKES

To introduce the issues with which decision analysis must grapple, we turn our attention to a decision whose outcomes entail substantial amounts of money.

Problem A (The Quiz Show)

You are a contestant on a TV quiz show for which you have trained diligently. You have accumulated winnings of \$48,000. At this, the climactic moment in the show, the quizmaster offers you the notorious Double-Or-Nothing question. If you decline this question, you keep your current winnings. If you accept this question, you will receive \$96,000 if you answer it correctly, and \$0 if you answer it incorrectly. Based on your study of previous questions in its category, you estimate your probability of getting it right as 3/4. What do you do?

On an expected-profit basis, you should accept this question. If you do accept it, your expected earnings will be $\$72{,}000 = (3/4) \times \$96{,}000 + (1/4) \times \$0$. That is 50% more than the \$48,000 that you would get by declining it. But would *you* accept it? Many of us would not. Let us consider why.

Risk Aversion

"Risk aversion" is one reason why a person might decline this question. An individual is said to be **risk averse** if, for any random payoff, he or she would prefer to receive the expectation of the payoff to the payoff itself. Many decision makers are averse to risk. Those of us who are significantly risk averse would prefer \$48,000 for certain to a gamble that has an expectation of \$72,000, but has a significant chance of earning \$0. For these people, the expectation of net profit is not the whole story when the stakes are large.

Regret and Elation

"Regret" is a more subtle reason why a person might decline this question. An individual is said to feel **regret** if the plan that he or she employs leads to an undesirable outcome that would have been avoided by a different plan. If you accept the gamble and lose, you do not merely end up with \$0. You have also given up \$48,000. If you gamble and lose, you may regret having done so when you mull over the many ways in which \$48,000 would have improved your life—or when your "significant other" regales you with the ways in which \$48,000 would have improved the common welfare. Regret can be a factor in rational decision making.

Similarly, an individual is said to feel **elation** if the plan that he or she employs leads to a desirable outcome that could not have been attained by another plan. If you accept the gamble and win, you may be a hero among your macho office-workers.

Utility theory envisions the decision problem as a choice between probability trees. Regret and elation challenge this paradigm. They suggest that a thoughtful decision maker might also concern herself or himself with the outcomes that could have occurred if a different choice had been made.

10.4. THE AMNIOCENTESIS DECISION

The bulk of this chapter concerns a decision whose main consequences are medical rather than monetary. This decision is whether or not a pregnant woman should undergo a medical procedure that is known as amniocentesis.

Many pregnant women do undergo amniocentesis; some do not. Needless to say, perhaps, different women can reach different decisions, each in accord with the postulates of rational behavior. This decision will be used to introduce the postulates of rational behavior, to interpret them, and to study their implications.

Amniocentesis

Every pregnancy, if carried to term, could result in the birth of a child with a chromosomal abnormality that is known as Down's syndrome. A Down's syndrome child may be severely retarded, require continuous care, and die at a young age. A test known as **amniocentesis** provides a pregnant woman with information about whether her fetus has Down's syndrome. A needle is thrust through the wall of her uterus, a sample of her amniotic fluid is taken, and the sample is analyzed. The outcomes of this analysis are "+" and "−." Like any test, amniocentesis is imperfect. Its sensitivity is 99% and its specificity is 99.5%, which means

$$P(+|D) = 0.99 \quad \text{and} \quad P(-|U) = 0.995,$$

where D denotes the event in which the fetus has Down's syndrome and where U denotes the event in which the fetus is *u*naffected, that is, does not have Down's syndrome. Being an invasive procedure, amniocentesis entails a risk. It punctures the uterus, which increases the probability of a miscarriage (spontaneous abortion).

Although we focus on amniocentesis, the approach that we will describe applies directly to medical tests and procedures that share these properties:

- False positives—positive test outcomes on individuals who do not have the condition.
- False negatives—negative test outcomes on individuals who do have the condition.
- Undesirable side effects.

Many medical tests and procedures share these properties of amniocentesis. And, as mentioned earlier, the general approach applies broadly, not merely to medical decisions.

Predictive Value

The **positive predictive value** of a test equals the probability that the person has the condition given a positive test outcome. The **negative predictive value** of a test equals the probability that the person does not have the condition given a negative test outcome. Bayes' rule relates the predictive value of a test to its sensitivity, its specificity, and the prevalence of the condition in the population.

For amniocentesis, the positive predictive value is the probability $P(D|+)$ that a woman's fetus has Down's syndrome given a positive test outcome, and the negative predictive value is the probability $P(U|-)$ that her fetus is unaffected by Down's syndrome given a negative test outcome. The prevalence of the condition is the probability $P(D)$ that a pregnant woman's fetus has Down's syndrome.

The probability $P(D)$ increases with the woman's age, and it varies with other risk factors. The probability that a pregnant woman will have a miscarriage also increases with her age. For these reasons, the risks and benefits of amniocentesis vary from woman to woman. To focus on the decision faced by a specific individual, we consider:

Problem B (Ms. C)

Ms. C is a 36-year-old industrial engineer who is married to a 38-year-old research psychologist. They have one child, a 6-year-old son who is healthy. Ms. C has had no prior miscarriages. She has been pregnant for 16 weeks. For her and for women in her risk group, the relevant probabilities are:

0.005 = the probability that her fetus has Down's syndrome.

0.032 = the probability that her fetus will spontaneously abort if she does not have amniocentesis.

0.037 = the probability that her fetus will spontaneously abort if she does have amniocentesis.

Should Ms. C undergo amniocentesis? Who should decide?

Neither question has a single correct answer. Some women would make this decision themselves, and other women would involve their religious leader, their doctor, or their husband. We shall see how a woman who accepts the postulates of rational behavior would grapple with this issue and reach her decision. Some of these women would elect amniocentesis; others would reject it.

Predictive Value of Amniocentesis

To calculate the predictive value of the amniocentesis test, we first use a spreadsheet to compute the joint probabilities. Table 10.1 executes a procedure that is familiar from Chapter 7. The function =B5*B6 in cell C11 computes the joint probability

$$P(D, +) = P(D)P(+|D) = 0.005 \times 0.99 = 0.00495,$$

for instance.

Suppose Ms. C undergoes amniocentesis. The numbers in Table 10.1 let us compute the predictive value of the test. If Ms. C gets a negative test outcome, the probability $P(U|-)$

Table 10.1 Joint probabilities for the amniocentesis test on Ms. C.

	A	B	C	D	E
2	**event**	**probability**			
3	miscarriage | no amnio.	0.032			
4	miscarriage | amnio.	0.037			
5	D	0.005			
6	+ | D	0.99			
7	– | U	0.995			
8	D	0.005			
9					
10			+	–	
11		D	0.004950	0.000050	0.005
12		U	0.004975	0.990025	0.995
13			0.009925	0.990075	

that her fetus does not have Down's syndrome is given by

$$P(U|-) = \frac{P(U, -)}{P(-)} = \frac{0.990025}{0.990075} = 0.99995.$$

This test has excellent negative predictive value. If Ms. C has the test and learns that the test outcome is negative, there is roughly 1 chance in 20,000 that her fetus has Down's syndrome.

Suppose Ms. C gets a positive test outcome; the probability $P(D|+)$ that her fetus has Down's syndrome is given by

$$P(D|+) = \frac{P(D, +)}{P(+)} = \frac{0.00495}{0.009925} = 0.498.$$

The positive predictive value of this test is modest, at best. If Ms. C elects amniocentesis and gets a positive test outcome, there is a 50% chance that she is carrying an *unaffected* fetus, one that does not have Down's syndrome.

Ms. C and her spouse face a tradeoff. Undergoing amniocentesis, intending to abort her pregnancy if the test outcome is positive, lessens the risk of giving birth to a Down's syndrome child, but it increases the risk of aborting an unaffected fetus.

This chapter draws upon a classic study by Pauker and Pauker.[1] Over an extended period, they used decision analysis to help hundreds of couples evaluate this tradeoff, that is, how to decide whether or not to undergo amniocentesis. A substantial majority of these couples did elect amniocentesis, even though the test's positive predictive value is far from perfect. Since their study was published, a second, confirming test has entered medical practice; that test improves—modestly—upon the positive predictive value of the amniocentesis test.

[1] Susan P. Pauker and Stephen G. Pauker, "The Amniocentesis Decision: Ten Years of Decision Analytic Experience," *Birth Defects: Original Article Series* 23 (1987), pp. 151–159.

10.5. EXECUTING A DECISION ANALYSIS

The amniocentesis decision will be used to introduce the methodology of decision analysis. In a nutshell, this methodology consists of identifying the decision maker or decision makers and eliciting from them the information on which a rational decision can be made.

The Decision Maker

The first step in a decision analysis is to identify the decision maker(s) and to determine the factors that impact on their decision. We assume that:

- Ms. C and her husband feel that the amniocentesis decision is hers to make.
- She fears giving birth to a child with Down's syndrome. The child could require full-time care. In that eventuality, one parent would need to abandon his career, the family would be financially strained, and the marriage itself might be threatened. Given the choice, she would abort a Down's syndrome fetus rather than give birth to a child who had Down's syndrome.

These are *assumptions*. Other couples could view this situation differently, of course.

Ranking Outcomes

A so-called rational decision maker is a person whose preferences satisfy four conditions or axioms. The first of these axioms is:

> ***Axiom 1 (Complete and Transitive Ordering).*** The decision maker is able to rank the outcomes from best to worst, with ties allowed.

Each "outcome" in Axiom 1 consists of one or more end-points in the decision tree. Ms. C ranks the outcomes from first (her most preferred end result) to fourth (her least preferred end result) as follows:

Outcome 1: Give birth to an unaffected child.

Outcome 2: Abort a Down's syndrome fetus.

Outcome 3: Abort an unaffected fetus.

Outcome 4: Give birth to a Down's syndrome child.

Thus, Ms. C's most favored outcome is to give birth to an unaffected child, and her least favored outcome is to give birth to a Down's syndrome child. If a fetus is to be aborted, she would prefer that it be a Down's syndrome fetus.

Let it be emphasized that this is Ms. C's preference ranking. Some women rank every abortion below any birth. It would be irrational for those women ever to undergo amniocentesis.

In this context, even the list of "outcomes" can vary with the decision maker. By grouping the end-points as she has, Ms. C has expressed herself as equally loath to abort an unaffected fetus under two different circumstances—through miscarriage and through planned abortion after getting a positive test outcome on amniocentesis. If Ms. C were not indifferent, she would have more than four outcomes to grapple with.

When counseling patients, Pauker and Pauker were able to group the two abortion outcomes together. They did not find it necessary to distinguish between the abortion of an unaffected fetus and of a Down's syndrome fetus.

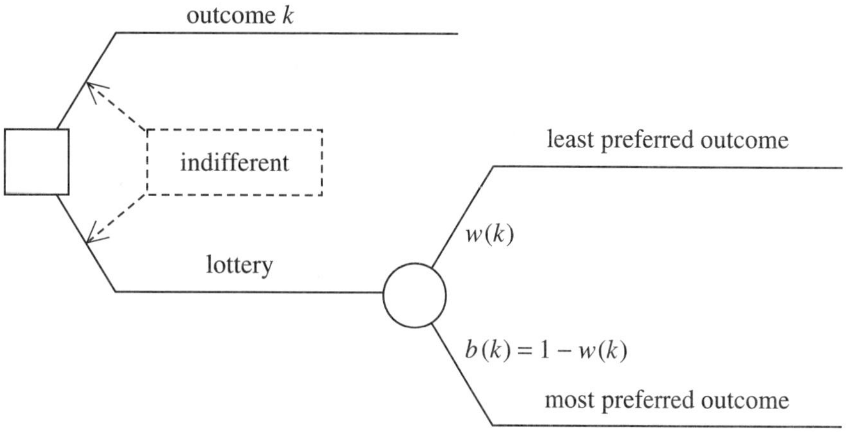

Figure 10.1 Indifference between outcome k and a lottery.

Indifference Probabilities

In the jargon of utility theory, the word **lottery** describes any probability tree each of whose end-points identifies an outcome. A lottery can concentrate its mass on one of the outcomes, and it can spread its mass between any group of outcomes.

Axiom 1 ranked outcomes from best to worse. The next axiom establishes an indifference between a single outcome and a lottery over two others. It is:

> ***Axiom 2 (Continuity).*** Suppose the decision maker prefers outcome A to outcome B and prefers outcome B to outcome C. Then there is a probability p that is strictly between *0* and *1* such that the decision maker is indifferent between receiving outcome B and receiving a lottery that obtains outcome C with probability p and obtains outcome A with probability $1 - p$.

One implication of Axiom 2 is that the outcome that Ms. C ranks kth has **indifference probability** $w(k)$ such that she is indifferent between receiving outcome k and receiving a lottery that results in her least preferred (worst) outcome with probability $w(k)$ and in her most preferred (best) outcome with probability $b(k) = 1 - w(k)$. See Figure 10.1.

Since outcome 1 is most preferred, it is evident in Figure 10.1 that $w(1) = 0$. Since outcome 4 is least preferred, it is equally evident that $w(4) = 1$. Axiom 2 states that Ms. C has indifference probabilities $w(2)$ and $w(3)$.

We assume that $w(2) = 0.05$ and that $w(3) = 0.30$. Later in the chapter, we will see how Ms. C can estimate her indifference probabilities. The statement that $w(2)$ equals 0.05 means that Ms. C is indifferent between aborting a Down's syndrome fetus (ranked second) and a lottery that gives her a Down's syndrome baby (ranked *w*orst) with probability $w(2) = 0.05$ and an unaffected baby (ranked *b*est) with probability $b(2) = 1 - w(2) = 0.95$.

The outcomes that pertain to Ms. C's decision have now been ranked, and the strength of her preference between these outcomes has been weighed. Table 10.2 summarizes this information.

Substitution

A lottery is any probability tree, each of whose end-points prescribes one of the outcomes that the decision maker has ranked. In a lottery's probability tree, a **subtree** consists either of an end-point or of a node and all arcs to its right. Each subtree describes a lottery.

Table 10.2 Ms. C's ranking of outcomes and her indifference probability $w(k)$ for each outcome, k.

		Indifference probabilities	
Outcome k	Interpretation	$w(k)$	$b(k)$
1	Give birth to unaffected baby	0.00	1.00
2	Abort Down's syndrome fetus	0.05	0.95
3	Abort unaffected fetus	0.30	0.70
4	Give birth to Down's syndrome baby	1.00	0.00

The next axiom states that if a decision maker is indifferent between two lotteries, he or she is indifferent to substitution of either for the other in a more complicated lottery. Specifically:

> ***Axiom 3 (Substitution).*** Suppose a decision maker is indifferent between receiving lotteries L' and L''. Then in any probability tree that includes lottery L' as a subtree, this decision maker is indifferent to substitution of L'' for L'.

In particular, Axiom 3 states that if a decision maker is indifferent between receiving an outcome and a lottery, then that person is indifferent to substitution of the lottery for the outcome. Figure 10.2 shows how substitutions for the intermediate outcomes reduce each lottery L to an equivalent lottery over the outcomes that Ms. C ranks best and worst.

Ms. C is indifferent between lotteries L and L'. Lottery L' has only two outcomes. It is evident in Figure 10.2 that lottery L' obtains outcome 4 (her least preferred) with probability $W(L)$ and that it obtains outcome 1 (her most preferred) with probability $B(L)$, where

$$W(L) = P(L = 2)w(2) + P(L = 3)w(3) + P(L = 4), \tag{10.1}$$

$$B(L) = P(L = 1) + P(L = 2)b(2) + P(L = 3)b(3). \tag{10.2}$$

Lottery L

Probability	Outcome
$P(L=1)$	1
$P(L=2)$	2
$P(L=3)$	3
$P(L=4)$	4

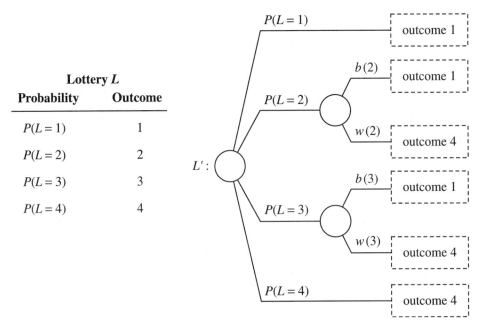

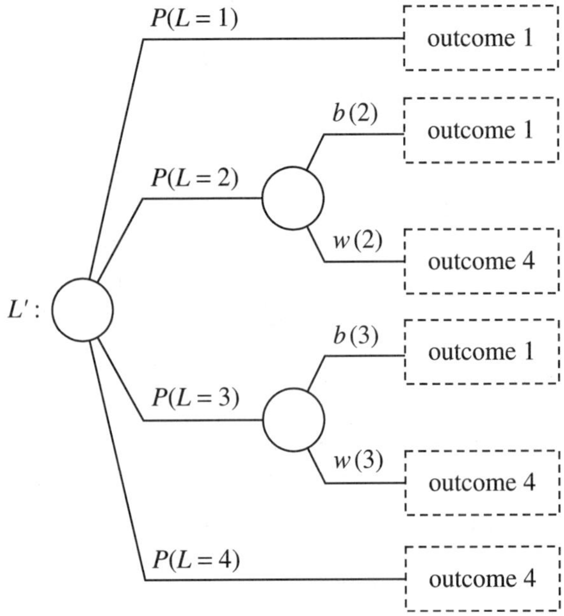

Figure 10.2 A lottery L and the lottery L' obtained from L by substituting lotteries for intermediate outcomes.

In this way, the indifference probabilities boil *every* lottery L down to the equivalent two-outcome lottery that gets the decision maker's least preferred (worst) outcome with probability $W(L)$ and most preferred (best) outcome with probability $B(L)$.

Preference

Axioms 1–3 make no assumption about which lotteries a decision maker prefers. The final axiom does this.

> ***Axiom 4 (Monotonicity).*** Suppose the decision maker prefers outcome A to outcome B. Then, between two lotteries that have A and B as their only outcomes, the decision maker prefers the lottery with the higher probability of obtaining outcome A.

Axiom 4 states that a decision maker prefers a higher probability on a more desirable outcome. Doesn't this sound familiar? It was the basis for both types of pruning in Chapter 9, for domination and for stochastic domination.

Rationality

A decision maker is said to be **rational** with respect to a decision tree if that person can group its end-points into outcomes, with preferences over outcomes that satisfy Axioms 1 through 4. We assume that Ms. C is rational with respect to the decision tree that she faces as she contemplates amniocentesis. Consequently, she most prefers the lottery L for which $W(L)$ is smallest, and equivalently, the lottery for which $B(L)$ is largest.

Since $w(1) = 0$ and $w(4) = 1$, Equation (10.1) shows that, when comparing lotteries, we can assign to each outcome k the *cost* $w(k)$ and minimize expected cost. Similarly, since $b(1) = 1$ and $b(4) = 0$, Equation (10.2) shows that we can assign to each outcome k the *profit* $b(k) = 1 - w(k)$ and maximize expected profit.

10.6. A DECISION TREE

The decision tree in Figure 10.3 models Ms. C's three principal strategies: (1) She could do nothing; (2) she could undergo amniocentesis and make constructive use of the test information; and (3) she could have an abortion without a test. Each end-point in which outcome k occurs has been assigned the cost $w(k)$. For instance, a cost of $w(2) = 0.05$ has been assigned to the four end-points in which she aborts a Down's syndrome fetus. (For this tree, numerical computation is eased if we minimize expected cost rather than maximizing expected profit.) The conditional probabilities in Figure 10.3 have been calculated from the joint probability table in the usual way, for example, $P(D|+) = P(D,+)/P(+)$. This tree has been rolled back to find her least-cost strategy.

Although this tree models only three strategies, it is *bushy* in the sense that it has lots of arcs. It would be even bushier if we had included some strategies that are dominated (e.g., to undergo amniocentesis and make perverse use of the test information).

Ms. C's Optimal Strategy

The computation in Figure 10.3 shows that Ms. C's best strategy is to undergo amniocentesis, elect an abortion if the test outcome is positive, and reject an abortion if the test outcome is negative. The expected cost of this strategy is 0.01278. As noted earlier, expected cost is an indifference probability. Ms. C is indifferent between this strategy and receiving a lottery that gets her worst choice (a Down's syndrome baby) with probability 0.01278 and gets her best choice (an unaffected baby) with probability of $0.98722 = 1 - 0.01278$.

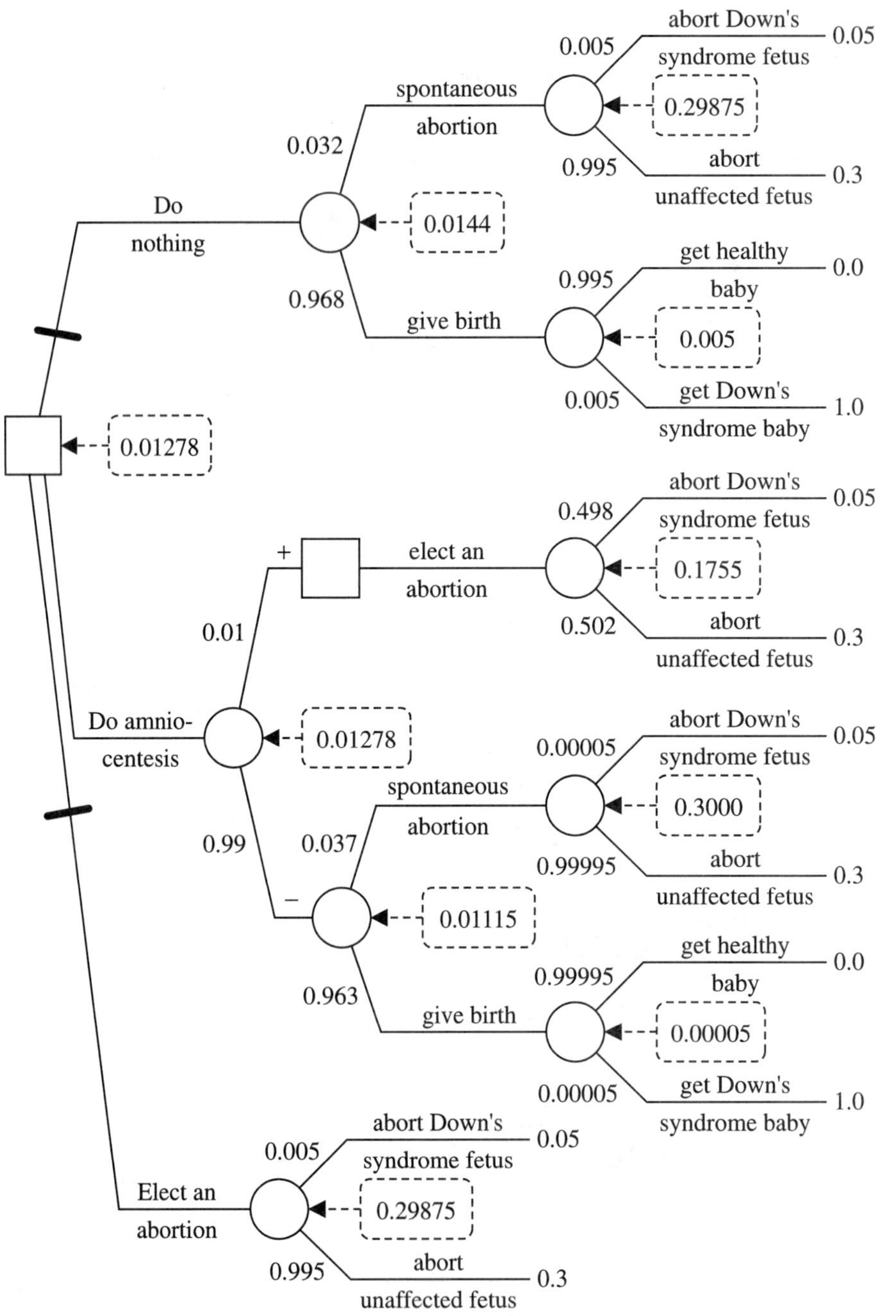

Figure 10.3 Ms. C's decision tree.

Reasons for Skepticism

It seems reasonable for Ms. C to accept the postulates of rational behavior, but there is ample room for skepticism about our analysis of her decision tree. Specifically:

- Ms. C's indifference probabilities $w(2)$ and $w(3)$ may be difficult to specify with precision and confidence.

- The best two strategies in Figure 10.3 have costs of 0.01278 and 0.0144, which are rather close to each other.

This raises a question. Is the conclusion that she undergo amniocentesis robust? We will soon see whether it is.

Let us first level a more mundane critique of her decision tree. Drawing Figure 10.3 and rolling it back requires unnecessary work. There are only three strategies. We will soon see that, on a spreadsheet, it is easier to compute each strategy's probability distribution over the outcomes directly rather than from Figure 10.3.

Risk Profiles

The **risk profile** of a strategy is its probability distribution over the outcomes. A risk profile omits the value placed on each outcome. A risk profile can be important when the values of the outcomes are subjective.

The amniocentesis decision will be used to illustrate risk profiles and their uses. Let us label Ms. C's principal strategies mnemonically, as follows:

Strategy NP: Have a *N*ormal *P*regnancy (no amniocentesis).

Strategy AB: Do an *Ab*ortion.

Strategy AM: Do *Am*niocentesis. Abort the pregnancy if the test outcome is positive. Do not abort if the test outcome is negative.

The risk profile for each of these strategies is easily found from the joint probability table. Although you may find it unnecessary, we present as Figure 10.4 a probability tree for Strategy AM, which is the most complex of the three.

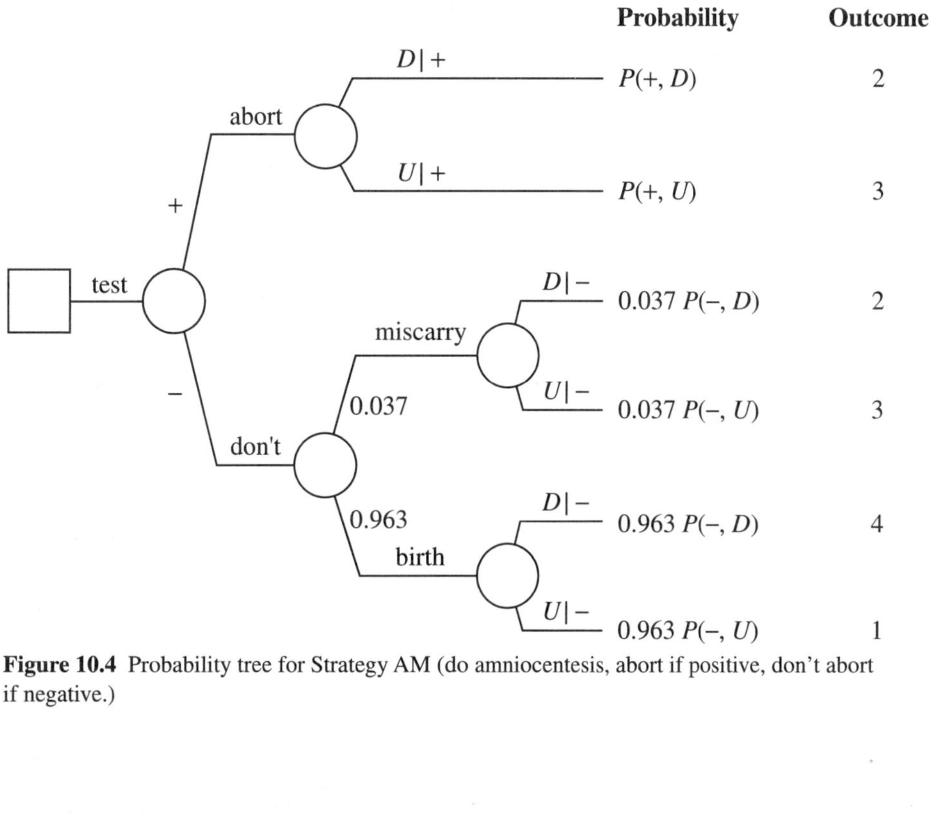

Figure 10.4 Probability tree for Strategy AM (do amniocentesis, abort if positive, don't abort if negative.)

Table 10.3 Risk profiles for the amniocentesis decision.

	A	B	C	D	E
2	**event**	**probability**			
3	miscarriage | no amnio.	0.032			
4	miscarriage | amnio.	0.037			
5	D	0.005			
6	+ | D	0.99			
7	– | U	0.995			
8			+	–	
9		D	0.004950	0.000050	0.005
10		U	0.004975	0.990025	0.995
11			0.009925	0.990075	
12					
13	**outcome**	**NP**	**AB**	**AM**	**w(k)**
14	#1: healthy baby	0.96316		0.9533941	0
15	#2: abort Down's fetus	0.00016	0.005	0.0049519	0.05
16	#3: abort healthy fetus	0.03184	0.995	0.0416059	0.3
17	#4: Down's syndrome baby	0.00484		0.0000481	1
18	E(cost) =	0.014400	0.298750	0.012778	

Figure 10.4 shows, for instance, that strategy AM obtains outcome 2 with probability of $P(+, D) + 0.037P(-, D)$. Rows 16–19 of the spreadsheet in Table 10.3 compute the risk profile for the strategies NP, AB, and AM. For instance, cell D17 contains the function =C11+B4*D11 that computes the probability that strategy AM obtains outcome 2. Table 10.3 omits the functions themselves.

Ms. C's principal choices are Strategy NP (normal pregnancy) and Strategy AM (do amniocentesis and make constructive use of the test outcome). The risk profiles in columns B and D of Table 10.3 measure the tradeoff between these strategies. Evidently, Strategy AM decreases the probability of getting a Down's syndrome baby from 0.0048 (one chance in 200) to 0.000048 (one chance in 20,000), but it increases the probability of aborting a healthy fetus from 0.0318 (three chances in 100) to 0.0416 (four chances in 100).

A Breakeven Computation

Column E of Table 10.3 records the indifference probabilities $w(1)$ through $w(4)$. Row 20 computes the expected cost of each strategy. Row 20 obtains the same expected costs as in Figure 10.3, and it does so without drawing the tree.

In this case and in general, each strategy's cost varies *linearly* in the numbers $w(1)$ through $w(4)$. Since $w(1) = 0$ and $w(4) = 1$, only the values $w(2)$ and $w(3)$ can be problematic. Let us observe that:

- Ms. C's principal strategies are NP and AM.
- Each strategy's cost varies linearly with $w(2)$ and $w(3)$, so the values of $w(2)$ and $w(3)$ for which she is indifferent between strategies NP and AM form a line.
- Two points determine a line, and Solver can be used to find two points on this line.

One point on this line is $w(2) = 0$ and $w(3) = 0.491$. Another point is $w(2) = 1$ and $w(3) = 0$. To find the first of these points, we inserted 0 in cell E15 and used Solver to

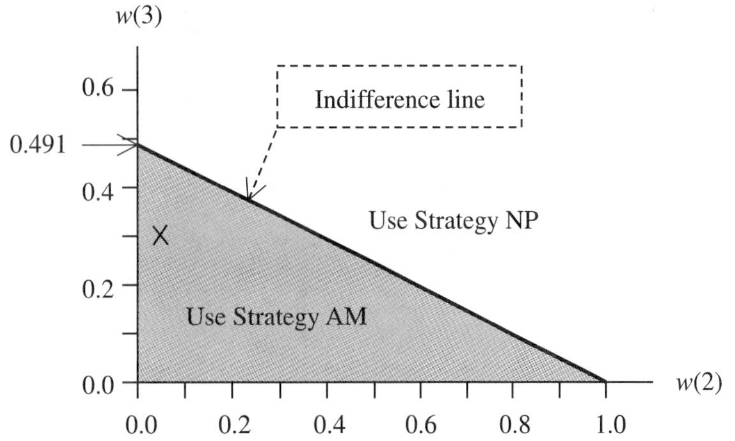

Figure 10.5 Regions for which Ms. C prefers Strategy AM (do amniocentesis) and Strategy NP (normal pregnancy).

select the value in cell E16 that equates to zero the difference between the values in cells D18 and B18. To find the other, we repeated the procedure with 1 in cell E15. Figure 10.5 plots the breakeven line.

The "×" in Figure 10.5 lies at the point (0.05, 0.30), which corresponds to Ms. C's estimates $w(2) = 0.05$ and $w(3) = 0.30$ of her indifference probabilities. The point (0.05, 0.30) lies comfortably within the region for which Strategy AM is best. Evidently, for Ms. C, the decision to undergo amniocentesis is robust.

Two uses have been made of the risk profiles in Table 10.3. These risk profiles quantify the tradeoff between different strategies. They also enable breakeven calculations that determine the extent to which the optimal strategy is sensitive (or insensitive) to the decision maker's evaluation of the outcomes.

10.7. ASSESSING SUBJECTIVE PROBABILITIES

An important issue has not yet been addressed. How can Ms. C estimate her indifference probabilities, $w(2)$ and $w(3)$? These probabilities are subjective. A reasonable way in which to estimate any subjective probability is to place it in a decision-making context and try to find the value for which the decision maker is indifferent between two alternatives.

To illustrate this estimation procedure, we consider how to estimate Ms. C's indifference probability $w(3)$ for the outcome that she ranks third, which is to abort an unaffected fetus. Figure 10.6 asks her to choose between this outcome and a lottery that gets her least favored outcome with probability p and her most favored outcome with probability $1 - p$.

Bracketing

A process that is aptly called **bracketing** can help her estimate $w(3)$. Bracketing pursues the following line of questions:

Q. Ms. C, in Figure 10.6, which choice do you prefer when $p = 1$?

A. That's easy. When $p = 1$, I abort, as giving birth realizes my least preferred outcome, a Down's syndrome child.

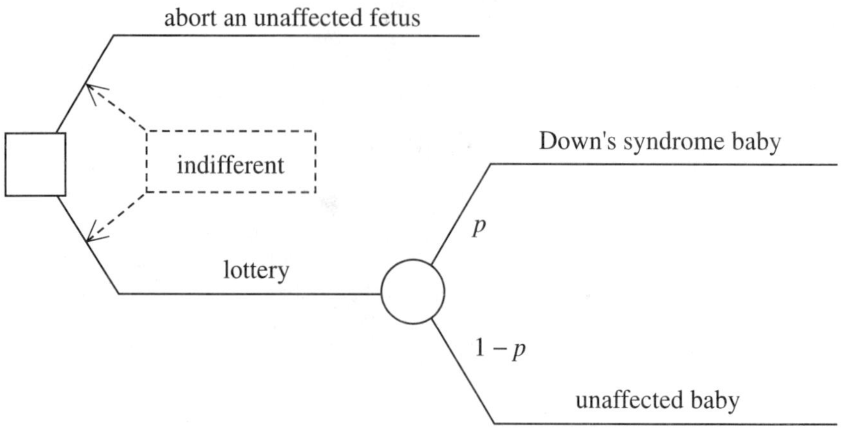

Figure 10.6 Ms. C's breakeven analysis for $p = w(3)$.

Q. Ms. C, which choice do you prefer when $p = 0$?

A. Again, that's easy. When $p = 0$, I give birth, as that gets my most favored choice, an unaffected baby.

Q. Ms. C, what about the halfway point, which is $p = 1/2$?

A. Well, that's a painful decision. But with even odds on getting a Down's syndrome child, I'd prefer to abort.

At this point, her indifference probability $w(3)$ has been "bracketed"; it lies between 0 and 1/2. Repeating the question with $p = 1/4$ halves the range of ambiguity in $w(3)$.

Bracketing is inexact; the subjective probabilities that one obtains from it should therefore be viewed as estimates. Ms. C's indifference probabilities are approximately 1/20 and 3/10. As noted above, even though her indifference probabilities may be inexact, a breakeven analysis like that in Figure 10.5 can reveal that they lead to robust conclusions.

Realistic Choices and Consistency Checks

Figure 10.6 reveals a subtle difficulty of decision analysis. The mere act of specifying an indifference probability can require an unnatural decision, one that cannot arise in life. No woman ever chooses between aborting one sort of fetus and giving birth to a different sort of fetus. Nothing in a woman's experience prepares her for the choice that she is required to make in Figure 10.6. There is good reason to be dubious of subjective probabilities that are based on unnatural choices like these.

One way around this difficulty is to employ realistic choices and consistency checks in the assessment of indifference probabilities. The amniocentesis example illustrates this. Figure 10.7 presents Ms. C with a choice that is more realistic—to deliver or to abort the fetus that is in her womb.

Clearly, when $q = 1$, Ms. C prefers to abort, and when $q = 0$, she prefers to deliver. The probability q for which Ms. C is indifferent between the lotteries in Figure 10.7 can be estimated by bracketing. If a decision maker is indifferent between lotteries L and L', then $W(L) = W(L')$. The breakeven value of q satisfies

$$qw(4) + (1 - q)w(1) = qw(2) + (1 - q)w(3). \tag{10.3}$$

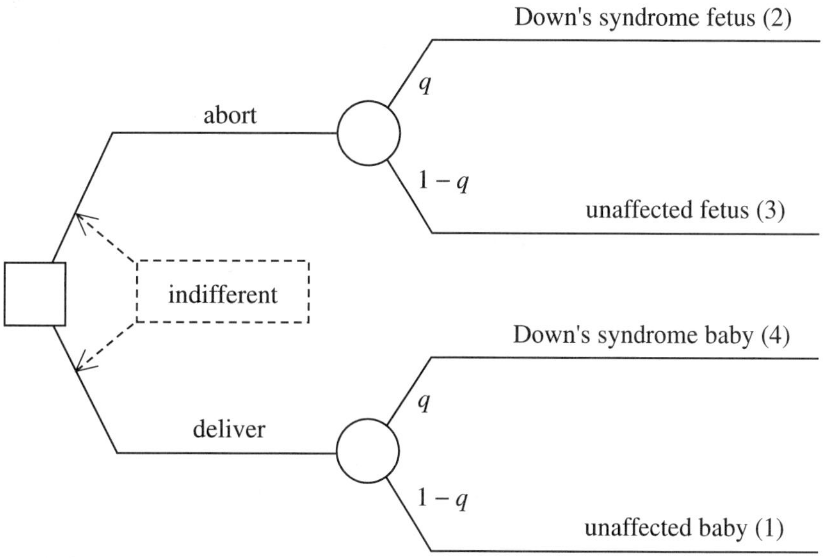

Figure 10.7 The decision to abort or deliver a fetus.

Substituting $w(4) = 1$ and $w(1) = 0$ into Equation (10.3) and then solving it for q gives

$$q = \frac{w(3)}{1 - w(2) + w(3)}. \tag{10.4}$$

With $w(2) = 0.05$ and $w(3) = 0.30$, Equation (10.4) gives $q = 0.30/1.25 = 0.24$. Thus, if Ms. C's indifference probability q differs substantially from 0.24, she should revise her estimate of $w(2)$ or $w(3)$.

10.8. THE MAIN THEOREM

We have presented the utility theory of John Von Neumann and Oskar Morgenstern for the (simplifying) case of decision trees that have finitely many end-points. Let us review this theory. We number the decision tree's end-points 1 through n. Axioms 1–4 guarantee that there is at least one end-point that the decision maker most prefers and at least one end-point that the decision maker least prefers. These axioms assign to each end-point k probabilities $w(k)$ and $b(k) = 1 - w(k)$ such that the decision maker is indifferent to substitution for end-point k of a lottery that gets a least favored (worst) end-point with probability $w(k)$ and a most favored (best) end-point with probability $b(k)$.

Each strategy has a probability distribution L over the end-points of the tree. The decision maker is indifferent between any probability distribution L over the tree's end-points and a lottery that gets a least-favored end-point with probability $W(L)$ and a most-favored end-point with probability $B(L)$, where

$$\begin{aligned} W(L) &= P(L = 1)\,w(1) + P(L = 2)\,w(2) + \cdots + P(L = n)\,w(n), \\ &= \sum_{k=1}^{n} P(L = k)\,w(k), \\ B(L) &= P(L = 1)\,b(1) + P(L = 2)\,b(2) + \cdots + P(L = n)\,b(n), \\ &= \sum_{k=1}^{n} P(L = k)\,b(k). \end{aligned}$$

The decision maker most prefers the strategy for which $W(L)$ is lowest and, equivalently, the strategy for which $B(L)$ is highest. Thus, we can assign each end-point k a "cost" $w(k)$ and minimize expected cost. Alternatively, we can assign each end-point k a "value" $b(k)$ and maximize expected value.

Utility and Disutility

The Von Neumann/Morgenstern theory is usually stated in the language of "utilities" or "disutilities" rather than indifference probabilities. To do so, interpret $b(k)$ as the **utility** of end-point k, and pick the lottery whose expected utility is largest. Alternatively, interpret $w(k)$ as the **disutility** of end-point k, and pick the lottery whose expected disutility is smallest.

Assuming that there are finitely many end-points led us to utilities and disutilities that lie between 0 and 1. There is nothing magic about 0 and 1 . With a as any number (positive or negative) and with s as any *positive* number, everything works if, for each end-point k, we replace $b(k)$ by $u(k) = a + s\,b(k)$ and $w(k)$ by $d(k) = a + s\,w(k)$. In this context, a lottery L has $a + s\,B(L)$ as its expected utility and has $a + s\,W(L)$ as its expected disutility. As before, one lottery is preferred to another if its expected utility is larger, and equivalently, if its expected disutility is smaller. When there can be infinitely many outcomes, the mathematics becomes more delicate, and we may need to employ a utility function $u(\cdot)$ that is unbounded.

Two Utility Functions

When the end-points are monetary amounts, particular utility functions have seen wide use. We now mention two of the most common ones. The simplest is the **linear** utility function, $u(x) = x$. When the utility function is linear, maximizing the expected utility is the same as maximizing the expected net revenue because $Eu(X) = E(X)$ for every random variable X.

The **exponential** utility function $u(x)$ has a single parameter, λ, and it is given by

$$u(x) = \begin{cases} -e^{-\lambda x} & \text{when } \lambda > 0 \\ e^{-\lambda x} & \text{when } \lambda < 0 \end{cases}.$$

When λ is positive, this function is concave, in which case the exponential utility function models risk aversion. When λ is negative, this function is convex, in which case the exponential utility function models risk seeking.

A key property of the exponential utility function is now developed. We shall see that, with an exponential utility function, the advisability of accepting an investment is independent of the decision maker's wealth. Let us consider the risk-averse case, $\lambda > 0$, of the exponential utility function. Let w be the wealth level of the decision maker, and let the random variable X be the net return (possibly negative) that the decision maker could earn from an investment opportunity. If this person accepts the investment opportunity, his wealth level changes from w to $w + X$. Let us compute the expected utility.

$$\begin{aligned} E(u(w + X)) &= E(-e^{-\lambda(w+X)}) = E(-e^{-\lambda w}e^{-\lambda X}), \\ &= -e^{-\lambda w}E(e^{-\lambda X}) = u(w)E(e^{-\lambda X}). \end{aligned}$$

For any level w of wealth, this risk-averse exponential utility function has $u(w) < 0$, and the equation,

$$E(u(w + X)) = u(w)E(e^{-\lambda X}),$$

shows that expected utility is improved (made less negative) by accepting an investment if and only if its net revenue X has $E(e^{-\lambda X})$ below 1. With this utility function, the decision maker accepts or rejects investment opportunities independent of his or her level w of wealth! A similar result holds for the risk-seeking case.

Subjective Assessments

To execute a decision analysis, the decision maker must make two different types of subjective assessment. The underlying data can include subjective probabilities, for example, $P(D)$ in the amniocentesis example. Each outcome has an indifference probability. The Von Neumann/Morgenstern postulates (Axioms 1–4) describe a rational process for assessing only the indifference probabilities. These axioms take the subjective probabilities as fixed.

A different axiom system, due to Leonard L. Savage, describes a rational process for *simultaneous* assessment of subjective probabilities and indifference probabilities. Savage's axiom system lies beyond the scope of this book. But we will soon discuss the "Ellsberg paradox," which points to inconsistencies that can occur in the simultaneous selection of subjective probabilities and indifference probabilities.

10.9. CRITIQUE OF UTILITY THEORY

Axioms 1 through 4 seem plausible, but situations have been found in which thoughtful people act in ways that violate them. This section illustrates some famous challenges to these axioms.

Incomparable Outcomes

Earning a few dollars is hard to compare with the loss of health or life. To dramatize this point, we consider:

Example 10.1 (Incomparable Outcomes). A lottery has three outcomes: you get \$50, you get \$10, and you get shot dead.

Getting killed is so onerous as to question the "continuity" axiom. Can you state a probability p that is strictly between 0 and 1 such that you are indifferent between receiving \$10 and a lottery that gets you shot dead with probability p and gets you \$50 with probability $1 - p$? Many of us would find any such indifference probability to be extremely problematic.

Some writers point out, however, that people assess these probabilities regularly, albeit implicitly, for instance, when they risk a highway fatality to enjoy an evening at a friend's home.

The Ellsberg Paradox

Decision analysis requires subjective assessment of indifference probabilities and subjective probabilities. Daniel Ellsberg noted that inconsistencies arise when people must make these assessments simultaneously; they do so by making pessimistic estimates of probabilities that they do not know.

Example 10.2 (Pessimistic Assessment of Subjective Probabilities). There are three outcomes, A, B, and C. One of them will occur. You know that $P(A) = 1/3$, and that is all you know about their probabilities.

(a) Suppose you win \$100 if you select the outcome that occurs. Which do you choose?

(b) Suppose you win \$100 if you select an outcome that does not occur. Which do you choose?

You know that $P(B) + P(C) = 2/3$, but that is all you know about $P(B)$ and $P(C)$. In part (a), many people prefer outcome A, which provides a known probability of $1/3$ of winning \$100. Implicitly, these people do not choose B because of a (pessimistic) estimate $P(B) < P(A)$. And in part (b), the same people again choose outcome A; now they avoid B because they estimate that $P(B) > P(A)$. That is inconsistent.

Framing

Two distinguished psychologists, Daniel Kahneman and Amos Tversky, found that the way in which a decision is "framed" can make a difference.

Example 10.3 (Framing). A new flu will arrive next winter.

(a) If you do nothing, 600 citizens will die. If you administer vaccine A, you will save 400 of these lives for certain. If you administer vaccine B, you will save none of these lives with probability of $1/3$, and you will save all 600 lives with probability of $2/3$. You can administer only one of these two vaccines. Which will it be?

(b) If you administer vaccine C, exactly 200 deaths will occur. If you administer vaccine D, 0 deaths will occur with probability of $2/3$, and 600 deaths will occur with probability $1/3$. You can administer only one of these two vaccines. Which will it be?

Kahneman and Tversky reported that, empirically, most medical professionals are risk averse as concerns lives saved but are risk seeking as concerns lives lost. For instance:

- In part (a), most medical professionals choose vaccine A over vaccine B; they prefer to save 400 lives for certain over a gamble that has the same expectation.
- In part (b), the same medical professionals choose vaccine D over vaccine C; rather than causing the death of 200 people for certain, they prefer a gamble with the same expectation.

But when parts (a) and (b) are represented as decision trees, they become identical. Vaccines A and C are indistinguishable, as are vaccines B and D. Did you spot that?

Kahneman and Tversky have found many examples of framing. They report that a person, when arriving at a theater, is more likely to buy a \$50 ticket if he discovers that he has lost \$50 in cash than if he had misplaced a ticket that he had previously bought and paid \$50 for. The framing paradox challenges the assumption that only the outcomes and their probabilities matter.

Regret and Elation

Earlier, we indicated that regret and elation can affect decision making. Here's an instance of that.

Example 10.4 (Regret and Elation). In a hospital, pathologists analyze tissues and organs that have been removed during surgery. Suppose that, after an elective abortion of Ms. C's fetus, the hospital's pathologist tests it and learns whether or not it had Down's syndrome. Should Ms. C be told? Does it matter?

It could matter greatly. If Ms. C learns that the fetus that she elected to abort was healthy, she might feel deep remorse (regret). If she learns that it had Down's syndrome, she might feel great relief (elation). If, before she had an abortion, she knew that her fetus would be tested and that she would be told the outcome, her decision might be entirely different.

Regret and elation challenge the assumption that the sole measure of the value of a lottery is its risk profile. David Bell and others have shown how to quantify regret and elation and to incorporate them in certain decision trees. (See Problem 7, on page 340.)

The Allais Paradox

At a conference of experts on decision analysis, M. Allais asked each of several participants to choose among three lotteries. Their choices violated the axioms of rationality. The term *Allais paradox* describes the type of model in which this inconsistency arises. See Problem 6 on page 340 for an instance of it. This paradox entails small differences in probabilities of outcomes that are extremely desirable or undesirable. For that reason, it tends to arise in decision trees whose conclusions would fail a robustness test.

A Mundane Consideration

The main limitation of utility theory may not lie in the brilliant paradoxes of Allais, Ellsberg, or Kahneman and Tversky, but in a mundane consideration. Often, the decision maker cannot specify indifference probabilities precisely enough to achieve robust conclusions.

10.10. REVIEW

Utility theory is a theory of rational decision making in the face of uncertainty. This chapter describes how a decision maker whose preferences satisfy the postulates of rational behavior would reach a decision. Excluded from this chapter is a large literature on decision making by groups.

Utility theory applies to operational and strategic decisions. It also applies when the important consequences of decisions are nonmonetary, as is the case in medical decisions. Earlier, in Chapter 8, we had suggested that in operational decisions, maximizing expected profit can be a good idea. We had also suggested that, when the monetary stakes are large, it may be prudent to examine the tradeoff between the expectation of profit (higher is better) and its variance (higher is riskier). Mean-variance analysis is an inaccurate substitute for utility theory. Decision analysis requires the decision maker to assess his or her attitude toward risk.

Decision analysis is inherently subjective. For each outcome that is ranked between the best and the worst, the decision maker must estimate an indifference probability. Being subjective, these estimates are imprecise. As a consequence, sensitivity analysis is usually needed to determine whether or not the strategy obtained from a decision analysis is robust. Each strategy has a risk profile, which is its probability distribution over the outcomes. These risk profiles facilitate breakeven analyses that determine whether a strategy is robust.

The amniocentesis example illustrates both the potential and the limitations of decision analysis. This example indicates how different decision makers view the same decision. We have used it to see that:

- Even the list of "outcomes" can depend on the decision maker.
- A decision maker's indifference probabilities can be elicited by bracketing.
- Using realistic choices and consistency checks helps to elicit indifference probabilities.
- Regret and elation can affect the decision maker's assessment of the end-points.

We have seen how a strategy can be robust, even in so subjective a situation as the amniocentesis decision.

This chapter also surveys the ways in which informed people violate the axioms of "rational" decision making. Even so, the principal impediment to the use of decision analysis may be practical, not theoretical. A strategy that works best for particular parameter values may not be sufficiently robust to recommend.

10.11. HOMEWORK AND DISCUSSION PROBLEMS

1. In Example 10.1, what bracketing questions would you ask yourself to determine whether you preferred to receive \$48,000 or a lottery that gets you \$96,000 with probability of $3/4$ and gets \$0 with probability of $1/4$?

2. Use the spreadsheet in Table 10.3 to find the range of $P(D)$ for which Strategy AM remains the best.

3. Consider a rational decision maker who is indifferent between outcomes A and B. Lottery L obtains outcome A with probability p and obtains outcome C with probability $1 - p$. Lottery L' obtains outcome B with probability p and obtains outcome C with probability $1 - p$. Prove or disprove: This person must be indifferent between lotteries L and L'.

4. Consider a rational decision maker who prefers outcome A to outcome B and prefers outcome B to outcome C. Prove or disprove: This person can be indifferent between outcome B and lotteries that give outcome C with probability p and outcome A with probability $1 - p$ for two distinct values of p.

5. Consider two lotteries, L and L', over the same outcomes. Prove or disprove:

(a) If a rational decision maker prefers L to L', then $W(L) < W(L')$.

(b) If a rational decision maker is indifferent between L and L', then $W(L) = W(L')$.

6. **(The Allais Paradox)** Consider these lotteries:

Lottery A: Win \$1,000,000 for certain.

Lottery B: Win \$1,000,000 with probability 0.89, win \$5,000,000 with probability 0.10, and win \$0 with probability 0.01.

Lottery C: Win \$5,000,000 with probability 0.90 and win \$0 with probability 0.10.

Lottery D: Win \$1,000,000 with probability 0.89 and win \$0 with probability 0.11.

Allais reported that many individuals (including some famous decision theorists) prefer lottery A to lottery B and prefer lottery C to lottery D. Provide a verbal argument as to why people might have these preferences. Prove or disprove: No rational decision maker can have these preferences.

7. **(Regret and the Allais Paradox)** Might regret play a role in the Allais paradox (Problem 6)? Can you quantify regret in a way that removes the paradox?

8. **(Framing)** The framing example must violate at least one of the four axioms of rationality. Which one(s)?

9. (**Domination**) In Chapter 9, a notion of domination of one random variable over another was described. The current chapter concerns lotteries over ranked outcomes, rather than random variables.

(a) Adapt the definition of domination to lotteries over ranked outcomes.

(b) For the Amniocentesis problem, find a Strategy Z such that Strategy NP dominates Strategy Z in the sense in part (a). (Assume that if a miscarriage would occur without amniocentesis, it will occur with amniocentesis.)

10. (**Stochastic Domination**) In Chapter 9, a notion of stochastic domination of one random variable over another was described. The present chapter concerns lotteries over ranked outcomes rather than random variables.

(a) Adapt the definition of stochastic domination to lotteries over ranked outcomes.

(b) For the Amniocentesis problem, find a Strategy P such that Strategy AM stochastically dominates Strategy P but does not dominate in the sense in Problem 9. *Hint:* You may find it helpful to "reverse" Figure 10.4.

11. (**The Spouse**) After Ms. C's indifference probabilities were elicited, her husband has had second thoughts. He is equally loath to abort a healthy and a Down's syndrome fetus, with indifference probabilities $w(2) = w(3) = 0.3$. How would you, the decision analyst, reconcile their differences?

12. (**Earth Slides**)[2] You are chief budget officer for the small town of High Trees, California. Extensive logging has exposed a hillside to the possibility of a mudslide. Reforestation is underway, but it will be a year before the new vegetation will be mature enough to remove the danger. Should a slide occur in the interim, human injury will be averted because mud moves slowly. The damage from such a slide would be limited to the road that passes beneath the hill. Constructing a retaining wall on the uphill side of the road could prevent the damage.

An expert tells you that there is only 1 chance in 100 that a slide will occur within the year. He estimates that 5% of the slides break through walls like the one proposed. He points out that he can better assess the likelihood of a slide if he conducts a geological test of the igneous rock layer below the hillside. The test he recommends has a sensitivity of 90% and a specificity of 85%. This test would cost $12,000 to undertake; the retaining wall would cost $200,000 to build; the road would cost $5 million to repair if it were damaged by a mudslide.

The mayor is willing to "play the averages" for expenses below $250,000, which are operational. But a $5 million repair bill is in a different category. Paying it off would require a bond, supported by a large and unpopular tax hike, to which she is averse.

(a) List the undominated strategies. For each dominated strategy, say how.

(b) For each undominated strategy, specify: the risk profile; the expectation of the total cost; the expectation of the cost other than repair; and the probability of a $5 million payout.

(c) Prepare a presentation for the mayor. Help her to reconcile the $5 million repair bill with operational expenses.

---

[2] Adapted from a problem created by Professor Ludo Van der Heyden for the John F. Kennedy School of Government, Harvard University, and for the Yale School of Management.

Chapter 11

Continuous Random Variables

11.1. PREVIEW

The discrete random variables that we studied in Chapter 7 are uncertain quantities that take isolated values, with gaps between them. By contrast, **continuous** random variables are uncertain quantities that can take any value in some given range, with no gaps between adjacent values. Continuous random variables are natural models for uncertain times and for uncertain distances, as these quantities vary continuously.

A discrete random variable can be described as a list of values, each of which has a probability of occurrence. We will soon see that a continuous random variable takes each value with a probability of zero. As a consequence, we will need to describe a continuous random variable in a different way.

This chapter presents the information about continuous random variables that will be used later in this work. If you have had prior exposure to probability, you may find our approach a bit novel. Older texts that cover continuous random variables tend to fall in one of two categories.

- *Without calculus:* Many texts avoid calculus entirely. They rely on intuitive descriptions of the expectation and variance, for instance.
- *With calculus:* The more advanced texts use calculus to define the expectation and variance and to do computations.

Spreadsheets open a third path, which is followed here. Rather than doing the calculus, we approximate the continuous random variable by a discrete one and do whatever calculation is needed on a spreadsheet. Once you get the knack of this approximation scheme, it is easy—and potent. It lets you do computations that are not possible with college-level calculus.

11.2. WHAT CAN YOU LEARN FROM THIS CHAPTER?

The first few sections of this chapter consist of a general discussion of continuous random variables. From these sections, you can learn how to:

- Specify a continuous random variable via its "density function" and its cumulative distribution function.
- Compute the expectation and variance of a continuous random variable.
- Transcribe to continuous random variables the insights you obtained in Chapter 8 for discrete random variables. These insights include the expectation as a measure of central tendency, the variance as a measure of spread, a function of a random variable, the sum of random variables, independent random variables, and conditional probability.

Other sections of this chapter survey five particularly useful probability distributions: uniform, triangular, exponential, normal, and lognormal. From those sections you can learn the uses to which particular distributions will be put in later chapters. These sections can be read selectivity.

One section is devoted to the Central Limit Theorem, which we introduced in Chapter 8. In this section, you can learn when this theorem's hypothesis is satisfied and how to use it. Another section is devoted to Brownian motion. From that section, you can learn of other ways in which the normal and lognormal distributions arise.

This chapter, like the others, is sprinkled with problems. Some problems will help you learn how to grapple with issues of general importance; others will show you how to exploit the properties of particular probability distributions.

11.3. OUTCOMES AND THEIR PROBABILITIES

To see why a continuous random variable cannot be described by a list of outcomes and their probabilities, we pose the question in:

Problem A (Equally Likely Outcomes?)

How can we describe the probability distribution of a random variable X that is guaranteed to lie between 2 and 5 and is equally likely to fall anywhere within this range? What is the expectation of X? What is the variance of X?

The random variable X in Problem A is equally likely to take any value between 2 and 5. Let us pick any number b between 2 and 5. To conclude that $P(X = b)$ equals zero, we observe that

- Infinitely many numbers lie between 2 and 5.
- Since our random variable X is equally likely to take any value between 2 and 5, all numbers between 2 and 5 are equally probable.
- If $P(X = b)$ was positive, it would be easy to construct an event (set of numbers) whose probability exceeded 1. No event can have probability larger than 1, so it must be that $P(X = b) = 0$.

This simple random variable X confronts us with what seems to be a paradox. It takes *some* value between 2 and 5 with probability of 1, but it takes *each* value between 2 and 5 with probability of 0.

11.4. THE DENSITY FUNCTION

Evidently, a continuous random variable cannot be described by assigning probabilities to outcomes. Instead, we describe it by assigning probabilities to intervals. This can be done in either of two ways: through the density function or through the cumulative distribution function. It's important to learn both ways. We present the density function first because it leads us directly to spreadsheet calculation.

Density Function of the Random Variable X

The continuous random variable X in Problem A is equally likely to take any value between 2 and 5. This random variable's **density function** $f(t)$ is plotted in heavy lines in Figure 11.1. Evidently, $f(t)$ equals zero for values of t that are below 2 and above 5, and $f(t)$ equals $1/3$ for $2 \le t \le 5$. In this example and throughout probability and statistics, the lower-case letter "f"is reserved for the density function.

For numbers a and b that satisfy $2 \le a \le b \le 5$, the area of the shaded region in Figure 11.1 is *defined* to be the probability $P(a \le X \le b)$. In other words,

$$P(a \le X \le b) = (1/3)(b - a) \quad \text{for } 2 \le a \le b \le 5, \tag{11.1}$$

because the shaded area is a rectangle with width $(b - a)$ and height $(1/3)$.

Let us check that the definition in Equation (11.1) has the properties that we desire of X. First, by substituting $a = 2$ and $b = 5$ in Equation (11.1), we see that $P(2 \le X \le 5)$ equals $(1/3)(5 - 2) = (1/3)(3) = 1$, as desired. Second, the expression $(1/3)(b - a)$ depends only the width $(b - a)$ of the interval between a and b. Thus, intervals of equal

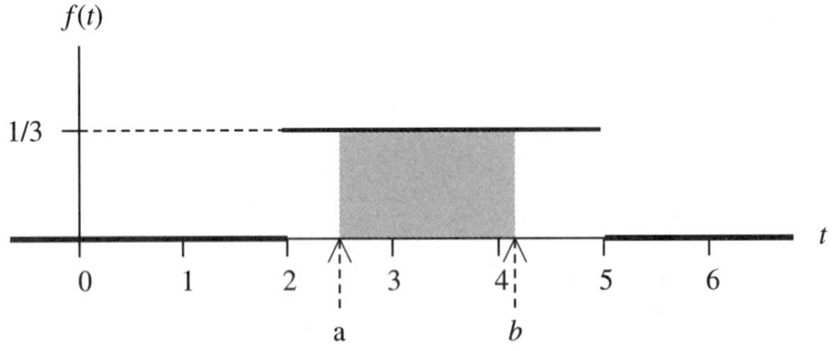

Figure 11.1 In thick lines, the density function $f(t)$ of a random variable X that takes a value between 2 and 5, all such values being equally likely.

width have equal probability; that is the sense in which X is "equally likely to fall anywhere" between 2 and 5.

Probability as Area

The shaded region in Figure 11.1 suggests how each density function assigns probabilities to intervals. Every continuous random variable Y has a density function $f(t)$, and the probability that Y takes a value between a and b equals the area of the region **under** the density function $f(t)$ and **between** the numbers a and b. This region lies:

- below the function $f(t)$,
- above the horizontal axis,
- to the right of the vertical line $t = a$,
- and to the left of the vertical line $t = b$.

The probability that a random variable takes a value between a and b is determined from its density function by the following rule:

> Let Y be any continuous random variable, and let $f(t)$ be its density function. For any real numbers a and b that satisfy $a \leq b$,
>
> $P(a \leq Y \leq b) =$ the area of the region under the density function $f(t)$ between a and b.

For the random variable X in Problem A, areas are easy to compute because the regions are rectangular.

A Different Example

Let us bring a curved density function into view. Figure 11.2 describes the density function of a continuous random variable, Y.

The probability that Y takes a value between 0 and 2 equals the area under its density function between these end-points, and Figure 11.2 indicates that $P(0 < Y \leq 2) = 0.593$. Figure 11.2 also indicates that $P(Y \geq 2) = 0.407$ and that

$$P(0 < Y < \infty) = 0.593 + 0.407 = 1.$$

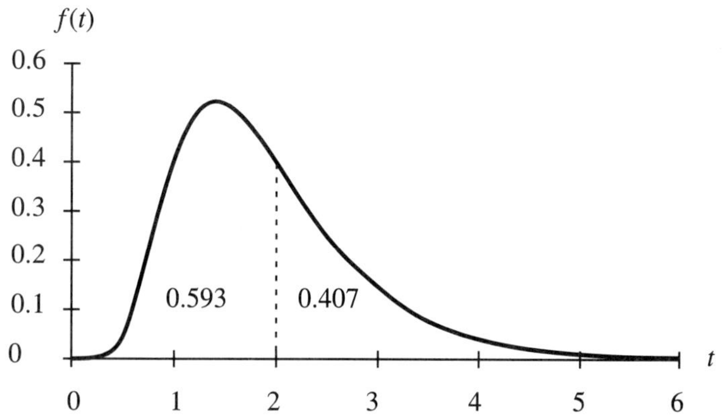

Figure 11.2 Density function $f(t)$ of a continuous random variable Y.

Common Properties

The density functions in Figures 11.1 and 11.2 illustrate properties that are shared by all density functions:

- Every density function is nonnegative; $f(t) \geq 0$ for each t.
- The area under the entire density function equals 1.
- The probability that a continuous random variable falls into an interval equals the area under its density function between that interval's end-points.

An interval whose width equals zero has 0 as its area. For this reason, it must be that every continuous random variable Y satisfies

$$P(Y = b) = 0 \quad \text{for each number } b. \tag{11.2}$$

Thus, a continuous random variable can assign a positive probability to an interval but not to a point. The probability that a continuous random variable Y takes any particular value b equals zero.

The Density Function as a Probability

With $f(t)$ as a density function, the quantity $f(x)$ is not a probability. But we can transfer to the density function the intuition we have developed about probabilities. To do so, we choose an interval that is narrow enough that the region under the density function between its end-points is very nearly rectangular.

Let us think of Δx as a number that is close to zero but positive, such as 0.01. Here is a way in which to interpret $f(x)\,\Delta x$ as a probability:

> Let Y be a continuous random variable whose density function is $f(t)$. The probability that Y takes a value between x and $x + \Delta x$ is approximately $f(x)\,\Delta x$.

Figure 11.3 illustrates this approximation. In this figure, $P(x \leq Y \leq x + \Delta x)$ equals the area under the density function $f(t)$ between x and $x + \Delta x$. This area consists of a shaded

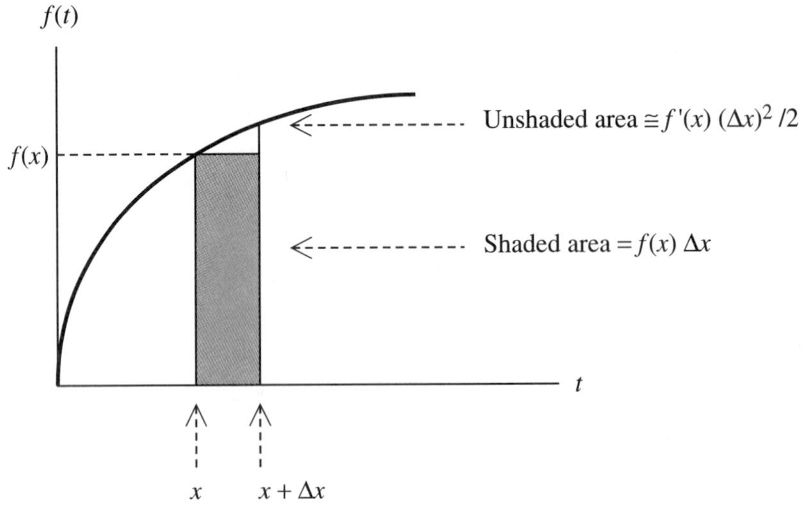

Figure 11.3 Approximation of $P(x \leq Y \leq x + \Delta x)$ by $f(x)\,\Delta x$.

rectangle and a smaller unshaded region, which is very nearly triangular. Visually, the area of the triangle is much smaller than the area of the rectangle.

The area of the shaded rectangle in Figure 11.3 equals its height times its base, namely, $f(x)\Delta x$. The unshaded region is nearly triangular, with base of Δx, with height of approximately $f'(x)\Delta x$ (where $f'(x)$ equals the slope of the density function at x), and with area of approximately $f'(x)(\Delta x)^2/2$, the triangle's height times its base times $(1/2)$.

The Accuracy of the Approximation

As Δx is reduced, the quality of the approximation in Figure 11.3 improves because the area of the rectangle is proportional to Δx, while the area of the triangle is proportional to $(\Delta x)^2$. This indicates that

$$P(x \leq Y \leq x + \Delta x) = f(x)\,\Delta x + \text{error term}, \tag{11.3}$$

where the "error term" is roughly proportional to $(\Delta x)^2$.

In brief, while the density $f(x)$ is not a probability, the product $f(x)\,\Delta x$ differs only slightly from the probability, $P(x \leq Y \leq x + \Delta x)$, that Y takes a value between x and $x + \Delta x$.

Improved Accuracy

The approximation in Figure 11.3 improves if we take x as the *midpoint* of an interval of width Δx rather than an end-point. Figure 11.4 indicates why. The height of the rectangle overestimates $f(t)$ in the shaded near-triangular region to the left, and it underestimates $f(t)$ in the unshaded near-triangular region to the right. Because their areas are very nearly equal, these regions very nearly cancel each other out.

Figure 11.4 indicates that

$$P(x - \Delta x/2 \leq Y \leq x + \Delta x/2) \cong f(x)\,\Delta x + \text{error term}, \tag{11.4}$$

where the error term in Equation (11.4) is roughly proportional to $(\Delta x)^3$.

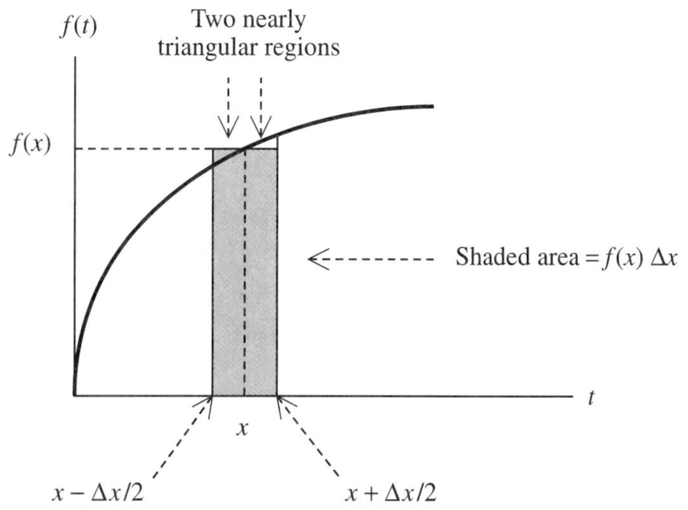

Figure 11.4 Approximation of $P(x - \Delta x/2 \leq Y \leq x + \Delta x/2)$ by $f(x)\,\Delta x$.

> Let Y be a continuous random variable whose density function is $f(t)$. The probability that Y takes a value within an interval of width Δx is approximately $f(x)\,\Delta x$, where x is the midpoint of the interval.

In brief, while the density $f(x)$ is not a probability, the product $f(x)\,\Delta x$ approximates the probability that the random variable falls within an interval of width Δx, with x an endpoint of the interval, or, more accurately, with x as the center of the interval.

Spreadsheet Computation of Areas

Figure 11.5 suggests how to approximate the probability that a random variable takes a value between the numbers a and b. To do so, divide the interval between a and b into narrow subintervals of equal width, evaluate the function at the midpoint of each subinterval, multiply it by the width of the subinterval, and take the sum.

In Figure 11.5, there are five subintervals, each having $(b - a)/5$ as its width. The approximation suggested by Figure 11.5 is

$$P(a \le Y \le b) \cong [f(x_1) + f(x_2) + f(x_3) + f(x_4) + f(x_5)](b - a)/5.$$

Visually, even with five intervals, the approximation in Figure 11.5 looks fairly close. The accuracy of this approximation improves as the *square* of the number of intervals. With 50 intervals, the approximation would be 100 times more accurate than it is with five intervals.

Figure 11.5 breaks the interval between a and b into five subintervals of equal width, which leads to the approximation of $P(a \le Y \le b)$ that is displayed above. Breaking the same interval into n subintervals of equal width gives

$$P(a \le Y \le b) \cong \sum_{k=1}^{n} f(x_k)\,\Delta x \quad \text{with } \Delta x = (b - a)/n, \tag{11.5}$$

where $x_1 = a + \Delta x/2$ and $x_{k+1} = x_k + \Delta x$ for each $k \ge 1$.

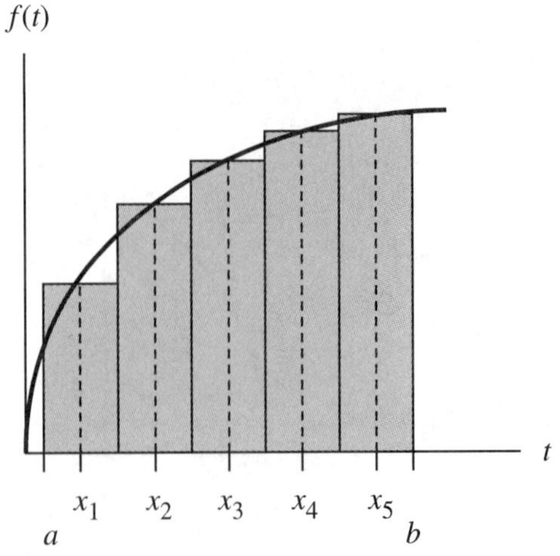

Figure 11.5 Estimating $P(a \le Y \le b)$ by adding the areas of five rectangles.

Even with a small number n, executing the computation in Equation (11.5) by hand would be gruesome. But on a spreadsheet, it is easy! You can compute the area of one rectangle in a way that lets you find the areas of the others with a drag. On a spreadsheet, using 50 subintervals is almost as easy as using 5.

Area as an Integral

Students of calculus will recognize the area of the region under the curve $f(t)$ between a and b as the **integral** of $f(t)$ from a to b. In the language of calculus, this fact is written as

$$P(a \leq Y \leq b) = \int_{x=a}^{b} f(x)dx \quad \text{for } a \leq b. \tag{11.6}$$

Equations (11.5) and (11.6) are intimately related. Equation (11.6) is a succinct way of writing what happens in Equation (11.5) as the number n of subintervals gets larger and larger. Comparing Equations (11.6) and (11.5) shows that the symbols in Equation (11.6) have these meanings:

- The numbers a and b are the end-points of the interval $a \leq x \leq b$ over which an area is being calculated.
- The function under which the area is being calculated is $f(x)$.
- The symbol "dx" recalls the fact that the limit is being taken as the grid width Δx on the variable x approaches zero.
- The "$\int$" reminds us that we are summing the areas of rectangles, each of width Δx and height $f(x)$.

A first course in integral calculus shows how to evaluate certain integrals exactly. In this book, you will *not* be required to know how to do that.

Numerical Integration

Using Equation (11.5) to approximate an area is known as **numerical integration**. Numerical integration has two virtues:

- It is easy to execute on a spreadsheet. To do so, compute the area of one rectangle in a way that lets you find the others with a drag.
- It lets us approximate areas and integrals that would be difficult (if not impossible) to evaluate exactly, using calculus.

In this text, we will find several uses for numerical integration.

Recap

Throughout probability and statistics, the symbol $f(t)$ is reserved for the density function. We have seen that:

- Each continuous random variable Y has a density function, $f(t)$.
- This density function $f(t)$ is nonnegative, and the area under the entire density function equals 1.
- The probability that Y takes a value in an interval equals the area under its density function between the end-points of that interval.

- For every number b, the probability that Y equals b is zero.
- For narrow intervals, the probability that Y falls into an interval is very nearly equal to $f(x)\ \Delta x$, where $f(x)$ is the value of the density function at the midpoint of the interval and where Δx is the width of the interval.
- On a spreadsheet, it's easy to compute the probability that Y falls into a broad interval by numerical integration, that is, by partitioning the broad region into narrow regions that are very nearly rectangular.

This completes our introduction to density functions.

11.5. THE CUMULATIVE DISTRIBUTION FUNCTION

A continuous random variable has been described in terms of the area under its density function. We now describe a continuous random variable in a different way—through its cumulative distribution function. In Chapter 8, the cumulative distribution function was introduced as a property of a discrete random variable. The definition is perfectly general, however. It applies to all random variables.

Specifically, every random variable Y has as its **cumulative distribution function** (CDF) the function $F(t)$ of t that is defined by

$$F(t) = P(Y \leq t) \quad \text{for each real number } t. \tag{11.7}$$

Throughout probability and statistics, the upper-case letter "F" is reserved for the cumulative distribution function, just as the lower-case letter "f" is reserved for the density function.

Examples

The random variable X in Problem A is equally likely to lie anywhere in the interval between 2 and 5. Figure 11.6 plots its CDF. This function $F(t)$ of t equals 0 for $t \leq 2$; it equals 1 for $t \geq 5$; and it increases linearly from 0 to 1 as t increases from 2 to 5. Evidently, $F(3) = 1/3$ and $F(4) = 2/3$. Why is its slope equal to $1/3$?

Similarly, Figure 11.7 plots the CDF of the random variable Y whose density function is depicted in Figure 11.2. From Figure 11.7, we see that $F(1) = P(Y \leq 1) = 0.109$, that $P(Y \leq 2) = 0.593$, and that $P(Y \leq 3) = 0.863$.

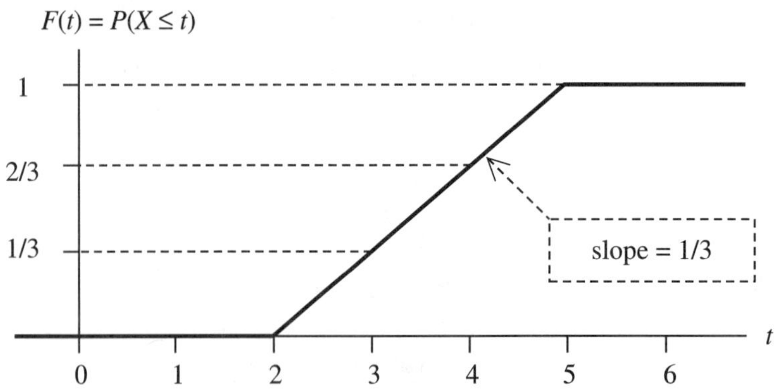

Figure 11.6 In a thick line, the CDF of the random variable X that takes a value between 2 and 5, all such values being equally likely.

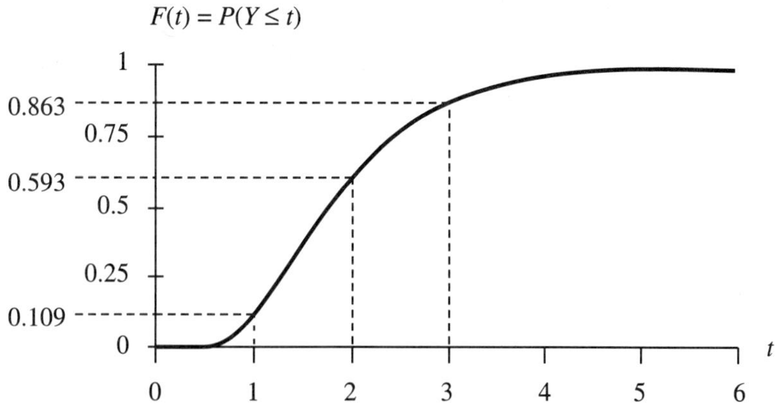

Figure 11.7 The CDF of the random variable Y whose density function is plotted in Figure 11.2.

Common Properties

Figures 11.6 and 11.7 exhibit properties that are shared by the CDF of every continuous random variable. Specifically:

- The CDF $F(t)$ is a nondecreasing function of t; $F(t)$ can increase as t increases, but it cannot decrease.
- As t gets more and more positive, $F(t)$ approaches (or equals) 1.
- As t gets more and more negative, $F(t)$ approaches (or equals) 0.
- $F(t)$ is a continuous function of t.

The CDF of a discrete random variable shares all of the above properties but the last. If a random variable is discrete, its CDF $F(t)$ jumps upward at the values that the random variable can take, and it is flat in between.

Intervals and Their Probabilities

As was the case in Chapter 8, the CDF lets us compute the probability that a random variable falls within an interval. Specifically, for any numbers $a < b$, the event $Y \le b$ is the union of the disjoint events $Y \le a$ and $a < Y \le b$. The probability of the union of two disjoint events equals the sum of their probabilities. Hence, for any numbers $a < b$,

$$P(Y \le a) + P(a < Y \le b) = P(Y \le b),$$

which rearranges itself as

$$P(a < Y \le b) = P(Y \le b) - P(Y \le a) = F(b) - F(a). \tag{11.8}$$

Equations (11.7) and (11.8) hold for every random variable Y, be it discrete or continuous.

If Y is continuous, we know from Equation (11.2) that $P(Y = a)$ equals 0 and that $P(Y = b) = 0$, which combines with Equation (11.8) to give

$$P(a < Y < b) = P(a \le Y \le b) = F(b) - F(a) \quad \text{for } a \le b. \tag{11.9}$$

Evidently, the CDF of Y lets us find the probability that Y falls into any given interval. It lets us compute this probability without having to estimate an area, that is, without doing any numerical integration. This fact can make the CDF easier to work with than the density.

Fractiles and the Median

The CDF also lets us compute "fractiles" of continuous probability distributions. Let Y be a continuous random variable whose CDF is the function $F(t)$. For any number p between 0 and 1, the **p fractile** of Y is a number t for which

$$p = P(Y \leq t) = F(t). \tag{11.10}$$

We will find several uses for fractiles.

A fractile is found by reading the CDF "backwards." To find the p fractile, start at the point p on the vertical axis of the CDF and find the value t such that $F(t) = p$. In Figure 11.7, for instance, the 0.863 fractile of Y is found by starting at the value 0.863 on the vertical axis, going "across" to the value x at which $F(x) = 0.863$, and dropping down to get $x = 3$. Fractiles are found by *inverting* the cumulative distribution function.

The **median** of a continuous random variable Y equals its 0.5 fractile. In other words, the median of Y is the number t such that $P(Y \leq t) = 0.5$. A continuous random variable Y is equally likely to lie on either side of its median. The random variable X whose CDF is plotted in Figure 11.6 has 3.5 as its median. The random variable Y whose CDF is plotted in Figure 11.7 has 1.8 as its median.

Fractiles are particularly useful when we describe population statistics. Let Y denote the annual income of a randomly selected American family. It's convenient to model Y as a continuous random variable (even though income is measured in discrete units, i.e., pennies). The median income is the number t such that half of the families make t or less and half make t or more. The 0.9 fractile is the number t such that $9/10$ of the families make t or less and $1/10$ of the families make t or more.

An Integral and a Derivative

For those readers who are conversant with calculus, we observe from Equation (11.9) that the density $f(a)$ is the derivative (slope) of the CDF $F(t)$ at a. Also, from Equations (11.7) and (11.8), the CDF $F(t)$ is the integral of the density $f(x)$ from $-\infty$ to t. We will not use these facts, however.

An Aside for Calculus Enthusiasts

Calculus demands that the density be undefined at points where the CDF lacks a derivative. To see why, we review Figure 11.6, which plots the cumulative distribution $F(t) = P(X \leq t)$ of the random variable X that is uniformly distributed on the interval between 2 and 5. This function $F(t)$ has a slope (derivative) for each value of t other than 2 and 5. We've defined the density of this random variable X as the function $f(t)$ that takes the value $1/3$ for all t between 2 and 5, inclusive.

Technically, that is erroneous! Our error makes areas easy to visualize. It gives no wrong answers because the value of a function at isolated points can have no effect on areas. We will repeat this error whenever it makes areas easy to visualize.

Any Other Random Variables?

Are there random variables that are neither discrete nor continuous? Yes. In Problem B, we adopt a convenient fiction—that an uncertain amount W of money is a random variable that can vary continuously. (Actually, W gets rounded off to the nearest penny, centime, or whatever.)

Problem B (Neither Discrete Nor Continuous)

On a quiz show, you have won \$1200 so far. Your total winnings W will be found as follows. The quiz-master spins a wheel that comes to rest at a point Z that is equally likely to be anywhere between 0 and 3. If $Z \leq 1$, you keep your earnings of \$1200. If $Z > 1$, you get (\$1200) Z instead. Is W a discrete random variable? Is W a continuous random variable? Does W have a density? If so, draw it. Does W have a CDF? If so, draw it.

The random variable W is neither discrete nor continuous. It cannot be described by a probability mass function or by a density function. It does have a CDF, $F(t) = P(W \leq t)$ for each t. It is not difficult to check that $F(t)$ jumps from 0 to 1/3 at $t = 1200$ and increases linearly from 1/3 to 1 as t increases from 1200 to 3600. In other words,

$$F(t) = \begin{cases} 0 & \text{for } t < 1200 \\ 1/3 + (t - 1200)/3600 & \text{for } 1200 \leq t \leq 3600. \\ 1 & \text{for } t > 3600 \end{cases}$$

This random variable W combines properties of discrete and continuous random variables. It assigns positive probability of 1/3 to the value $t = 1200$, and it distributes its remaining probability between 1200 and 3600 in a uniform manner.

11.6. THE EXPECTATION AND THE VARIANCE

We are not quite finished with Problem A. We've seen how to describe X through its density function and through its CDF. Either method allows us to compute the probability that X falls into a particular interval. We have not yet seen how to compute the expectation of X, or its variance. In fact, we have not yet defined the expectation and variance of a continuous random variable. Before doing so, we develop an intuitive feel for the expectation.

Repeated Trials

As in the discrete case, "repeated trials" motivates the expectation. The random variable X in Problem A is equally likely to fall anywhere between 2 and 5. Imagine that you observe 300 different random variables, each independent of each other and each having the same distribution as does X. Roughly 100 of these observations would take values between 2 and 3, roughly 100 would take values between 3 and 4, and roughly 100 would take values between 4 and 5. The average of the values that you will observe will be very close to 3.5, which suggests (correctly) that 3.5 is the expectation of X.

The Fulcrum

As was the case for a discrete random variable, the fulcrum (balance point) turns out to equal the expectation. Figure 11.8 presents the density function of X. In Figure 11.8, the area under its density function is shaded. To assign mass to this function, imagine that the shaded area has a constant density, say, of 1 pound per square inch. The fulcrum (balance point) is at the center of the shaded interval, so 3.5 is the expectation of X.

Review the density function in Figure 11.2 and try to guess the expectation of Y by "eyeballing" its fulcrum. Its fulcrum (expectation) equals 2.

Symmetric Densities

Before defining the expectation of a continuous random variable, we interject a handy observation. A function $f(t)$ is said to be **symmetric** about the number m if $f(m + x) = f(m - x)$

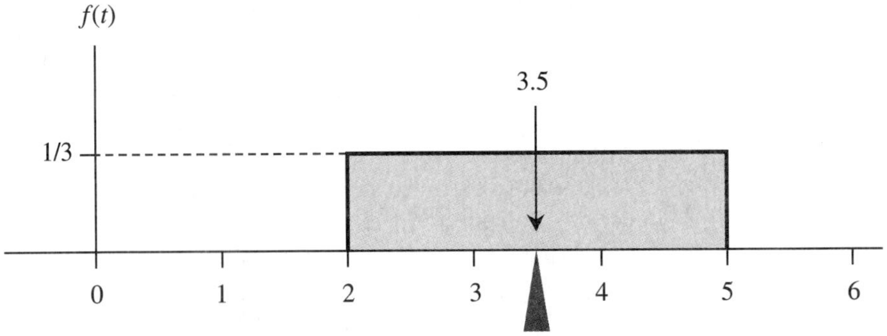

Figure 11.8 The density of X and its expectation (fulcrum).

for each positive number x. The uniform density function in Figure 11.1 is symmetric about 3.5. By contrast, the density function in Figure 11.2 is not symmetric.

If the density function of a random variable Z is symmetric about m, two things are easy to see. First, symmetry guarantees that $P(Z \leq m) = P(Z \geq m)$, and hence that m is the median of Z. Second, m is the fulcrum of the density function of Z, so m is the expectation of Z.

> If the density function of the random variable Z is symmetric, then the point about which it is symmetric is both the median of Z and the expectation of Z.

Defining the Expectation and the Variance

The question remains—how do we define the expectation and the variance of a continuous random variable? When Y is a continuous random variable, the definitions of $E(Y)$ and Var(Y) employ an approximation that is reminiscent of Figure 11.5. The **range** of a continuous random variable is the interval between the smallest and largest number for which its density is positive.

Procedure for Computing the Expectation and Variance of a Continuous Random Variable Y:

1. Partition the range of Y into nonoverlapping subintervals of width Δx.
2. Equate the discrete random variable W to the midpoint of the subinterval into which Y falls.
3. Compute $E(W)$ and Var(W).
4. $E(Y)$ and Var(Y) equal, respectively, the limiting value of $E(W)$ and Var(W) as the width Δx of the subintervals approaches 0.

The aforementioned approximation procedure *defines* the expectation and variance of a continuous random variable.

A Sample Calculation

To illustrate this procedure, we apply it to the uniform random variable X whose density is given in Figure 11.1. The range of X is the interval between 2 and 5. Let us partition this range into intervals of width $\Delta x = 0.1$. Table 11.1 shows the result of this partition. There are 30 such subintervals, each is equally likely to occur, and the probability that X falls into a particular subinterval equals $1/30$. The discrete random variable W is assigned the midpoint of the interval into which X falls.

Table 11.1 Discrete approximation W of the uniform random variable X, with subintervals of width $\Delta x = 0.1$.

Event	Value of W	Probability
$2.0 \le X < 2.1$	2.05	1/30
$2.1 \le X < 2.2$	2.15	1/30
$\vdots$	$\vdots$	$\vdots$
$4.9 \le X \le 5.0$	4.95	1/30

The random variables X and W are fairly close in value. Under no circumstance is the difference between W and X larger than 0.05. Computing $E(W)$ and $\text{Var}(W)$ is easy on a spreadsheet. One gets:

$$\begin{aligned} E(W) &= (2.05)(1/30) + (2.15)(1/30) + \cdots + (4.95)(1/30) \\ &= 3.5. \\ \text{Var}(W) &= (2.05 - 3.5)^2(1/30) + (2.15 - 3.5)^2(1/30) + \cdots + (4.95 - 3.5)^2(1/30) \\ &= 0.74916667. \end{aligned}$$

This computation indicates that $E(X)$ is approximately 3.5 and that $\text{Var}(X)$ is approximately 0.749. The former is no surprise; we had seen that 3.5 is the fulcrum of the density of X.

To get a closer approximation of W to X, use a smaller grid width, say, 300 intervals, each having width of 0.01. No matter how fine a grid we chose, we would get $E(W) = 3.5$. Hence, $E(X) = 3.5$. Incidentally, as Δx gets smaller and smaller, $\text{Var}(W)$ converges to 0.75, which equals $\text{Var}(X)$.

The quality of these approximations can be quite good. Even with a grid width $\Delta x = 0.1$, the error as a fraction of the variance equals $0.0001666/0.75 = 0.00022$, which is roughly 2 parts in 1000.

An Admission

It is time to admit to having done calculus. The above procedure finds $E(W)$ by **integration**. Conceptually, nothing is complicated. The expectation and variance of a continuous random variable Y equal the limits of the expectation and variance of the discrete approximation W to Y as the grid width Δx approaches zero.

To approximate the continuous random variable Y whose density is plotted in Figure 11.2 by a discrete random variable W, we proceed exactly as suggested in Table 11.1. Specifically, we place a grid of width Δx on the horizontal axis and represent each interval into which Y might fall by its midpoint. The expectation of W equals the sum of its values times their probabilities. Evidently,

$$E(W) \cong \sum_k x_k f(x_k)\Delta x. \tag{11.11}$$

As the grid width Δx approaches zero, three things happen in Equation (11.11): the difference between W and Y vanishes; the sum becomes an integral; and "$\cong$" becomes an equation. In the notation of calculus, this is written as

$$E(Y) = \int_{x=-\infty}^{+\infty} x f(x)dx, \tag{11.12}$$

where the symbol dx records the fact that the limit has been taken as the grid width Δx on the variable x approaches zero. Equation (11.12) defines the **expectation** of a continuous random variable Y.

The expectation of a continuous random variable Y is often called the **mean** of Y and is often denoted as μ, just as in the discrete case. We often write $\mu = E(Y)$. The variance of a continuous random variable is often written as σ^2, as is the case for discrete random variables.

Spreadsheet Computation

Equation (11.12) was approached gingerly because it entails integration. It is not necessary that you be familiar with integration. It suffices to interpret (well, actually, slightly misinterpret) dx as a "width" and the integral as a "sum." In this context, Equation (11.12) says we should interpret $f(x)\,dx$ as the probability that the random variable Y takes a value in an interval of width dx that is centered at x, multiply that probability by x, and sum, exactly as in the case of a discrete random variable.

The approximations in Equations (11.5) and (11.11) look strikingly similar, but there is one important difference. If the range of Y is infinite, the approximation in Equation (11.11) partitions the range into *infinitely* many intervals of a given width, Δx. When we execute this approximation on a spreadsheet, we need to drop the addends that are so tiny that their contribution to the sum is miniscule. Having dropped them, we can do the approximation easily, with a drag.

11.7. A FUNCTION OF A RANDOM VARIABLE

In Chapter 8, we described a function $g(X)$ of a discrete random variable X. The same intuition (and formulas) will now be used to describe a function of a continuous random variable. When Y is a continuous random variable, what is meant, for instance, by the function $g(Y) = (Y - 19)^3$ of Y? The answer is the same as in the discrete case; if Y takes a particular value, say 43.2, then the function $g(Y) = (Y - 19)^3$ takes the value $g(43.2) = (43.2 - 19)^3$.

The continuous random variable Y takes a value that is approximately x with probability $f(x)\,\Delta x$, and, when this occurs, the random variable $g(Y)$ takes a value that is approximately $g(x)$. The product $f(x_k)\,\Delta x$ equals the probability of the event in which Y falls in an interval of width Δx that is centered at x_k; when this event occurs, the function $g(Y)$ takes a value that is approximately $g(x_k)$. Thus, to approximate the expectation of $g(Y)$, we multiply $g(x_k)$ by $f(x_k)\,\Delta x$ and sum over x, getting

$$E[g(Y)] \cong \sum_k g(x_k) f(x_k)\,\Delta x. \qquad (11.13)$$

Again, the equation,

$$E[g(Y)] = \int_{x=-\infty}^{+\infty} g(x) f(x)\,dx, \qquad (11.14)$$

describes the limit in (11.13) as the width Δx approaches zero. Again, Equation (11.13) can be computed, numerically, on a spreadsheet.

Equation (11.13) is familiar from Chapter 8. There, to compute $E[g(Y)]$ for a discrete random variable Y, we multiplied $P(Y = x)$ by $g(x)$ and summed over x. In Equation (11.13), the quantity $f(x_k)\,\Delta x$ plays the role of $P(Y = x_k)$, so the addend $g(x)P(Y = x)$ is replaced by $g(x_k)f(x_k)\,\Delta x$.

The Function $(Y - \mu)^2$

With $\mu = E(Y)$, the **variance** of Y is defined to be the expectation of the function $g(Y) = (Y - \mu)^2$, namely, of the square of the difference between Y and its mean. By substituting $(Y - \mu)^2$ in Equation (11.14), we see that

$$\mathrm{Var}(Y) = E[(Y - \mu)^2] = \int_{x=-\infty}^{+\infty} (x - \mu)^2 f(x)dx. \tag{11.15}$$

By expanding $(x - \mu)^2 = x^2 - 2x\mu + \mu^2$ and substituting in Equation (11.15), we get the formula,

$$\mathrm{Var}(Y) = E[(Y)^2] - \mu^2, \tag{11.16}$$

which is familiar from the discrete-variable case.

The unit of measure of the variance equals the square of the unit of measure of the variable itself. If Y is a time, measured in seconds, then Var(Y) is measured in (seconds)^2. The **standard deviation** of a continuous random variable equals the square root of the variance.

$$\mathrm{StDev}(Y) = \sqrt{\mathrm{Var}(Y)} = \sigma. \tag{11.17}$$

The unit of measure of the standard deviation σ is the same as that of the variable itself.

The Function $aY + b$

Let a and b be fixed numbers, and consider the function $g(Y) = aY + b$ of the continuous random variable Y. Substituting this formula into Equation (11.14) gives

$$E(aY + b) = aE(Y) + b. \tag{11.18}$$

This too is familiar from Chapter 8. Adding b to each value of Y adds b to its expectation. Multiplying each value of Y by the factor a multiplies its expectation by a. As in Chapter 8,

$$\mathrm{Var}(aY + b) = a^2\,\mathrm{Var}(Y). \tag{11.19}$$

Once again, adding b to Y has no effect on its variance, but multiplying Y by a multiplies its variance by a^2. Since the standard deviation is the square root of the variance, Equation (11.19) gives

$$\mathrm{StDev}(aY + b) = |a|\,\mathrm{StDev}(Y) \tag{11.20}$$

The Expectation of a Nonnegative Random Variable

If a random variable Y is nonnegative, there is a second way in which to compute its expectation. The expectation of Y equals the area to the right of the vertical axis, above its CDF, and below the horizontal line that lies one unit above the horizontal axis. Areas are integrals, and

$$E(Y) = \int_{x=0}^{\infty} [1 - F(x)]dx \qquad \text{if } P(Y \geq 0) = 1. \tag{11.21}$$

If you know how to "integrate by parts," you can verify that the right-hand side of Equation (11.21) computes $E(Y)$ by taking $u = [1 - F(x)]$ and $dv = dx$.

11.8. CONDITIONAL PROBABILITY, INDEPENDENCE, AND SUMS

Nearly everything that we learned in the discrete case about conditional probability, independent random variables, and sums of random variables transcribes directly to the continuous case. This section makes the connections, but it avoids the mathematics that a rigorous presentation would require.

Conditional Probability

For continuous random variables Y and Z, the **conditional probability** $P(Y \le y | Z \le z)$ is defined in a familiar way; if $P(Z \le z)$ is positive, then

$$P(Y \le y | Z \le z) = \frac{P(Y \le y, Z \le z)}{P(Z \le z)}. \tag{11.22}$$

The analog of expression (11.22) holds with the event $\{Z \le z\}$ replaced by any event that occurs with positive probability, for instance, by the event $\{Z > z\}$ or by the event $\{Z \ge z\}$.

On the other hand, conditioning on an event whose probability equals zero leads to trouble; if we replace the event $\{Z \le z\}$ by $\{Z = z\}$ in Equation (11.22), its right-hand side becomes the ratio of two 0's, which is nonsense. The conditional probability $P(Y \le y | Z = z)$ does make intuitive sense, and mathematical sense can be made of it, but we shall not need to do so.

Independent Random Variables

The intuition behind dependent and independent random variables remains the same as in the discrete case. The continuous random variables Y and Z are said to be **independent** of each other if

$$P(Y \le y | Z \le z) = P(Y \le y) \quad \text{for all } y \text{ and } z \text{ having } P(Z \le z) > 0. \tag{11.23}$$

As in Chapter 8, two random variables are independent if information about either provides no information about the other. Again, there are several ways to test for independence. Again, the continuous random variables Y and Z are said to be **dependent** if they are not independent.

The Sum of Continuous Random Variables

The expectation of the sum of several random variables equals the sum of their expectations, whether or not they are independent. In other words, for any continuous random variables Y and Z,

$$E(Y + Z) = E(Y) + E(Z). \tag{11.24}$$

With $\mu_Y = E(Y)$ and with $\mu_Z = E(Z)$, the **covariance** of Y and Z is again given by

$$\text{Cov}(Y, Z) = E[(Y - \mu_Y)(Z - \mu_Z)]. \tag{11.25}$$

Again, variance of the sum of two random variables is given by

$$\text{Var}(Y + Z) = \text{Var}(Y) + \text{Var}(Z) + 2\text{Cov}(Y, Z). \tag{11.26}$$

If the continuous random variables Y and Z are independent, then $\text{Cov}(Y, Z) = 0$, as was the case in Chapter 8. Consequently,

$$\text{Var}(Y + Z) = \text{Var}(Y) + \text{Var}(Z) \quad \text{if } Y \text{ and } Z \text{ are independent.} \tag{11.27}$$

Equations (11.16) through (11.22) hold for discrete and continuous random variables. In fact, these equations hold for every random variable, including those that are neither discrete nor continuous. Briefly put, the understanding you obtained for discrete random variables applies to continuous random variables when you interpret $f(x)dx$ as the probability that a continuous random variable Y takes a value that is "approximately" x.

11.9. THE UNIFORM DISTRIBUTION

This and the next few sections describe several continuous probability distributions, along with the uses to which each is put. The first of these distributions is familiar from Figure 11.1. The random variable X is said to be **uniformly** distributed between a and b if its density function $f(t)$ is

$$f(t) = \begin{cases} 0 & \text{for } t < a \\ 1/(b - a) & \text{for } a \leq t \leq b. \\ 0 & \text{for } t > b \end{cases} \tag{11.28}$$

The fact that $f(t)$ is constant for t in the interval between a and b means that X is equally likely to fall anywhere in that interval.

Density Function of the Uniform

Figure 11.9 plots the density function $f(t)$ and the CDF $F(t)$ of this uniform random variable. Their shapes are familiar from Figures 11.1 and 11.6. The region under this density function is a rectangle whose width equals $(b - a)$, whose height equals $1/(b - a)$, and whose area equals 1, as must be. As t increases from a to b, this CDF increases linearly from 0 to 1, with a slope of $1/(b - a)$.

The density function in Figure 11.9 is symmetric about $m = (a + b)/2$, so we know that m equals both the expectation of X and its median. A bit of calculus (that we omit) computes its variance from Equation (11.15). We report:

$$E(X) = \frac{a + b}{2}, \quad \text{StDev}(X) = \frac{b - a}{\sqrt{12}}. \tag{11.29}$$

Uses of the Uniform Distribution

The uniform distribution has helped us understand the properties of continuous random variables. But is it of any practical value? Yes, it has two principal uses.

First, the uniform distribution is the natural model of an uncertain quantity that is known to fall within a particular interval and is equally likely to fall anywhere

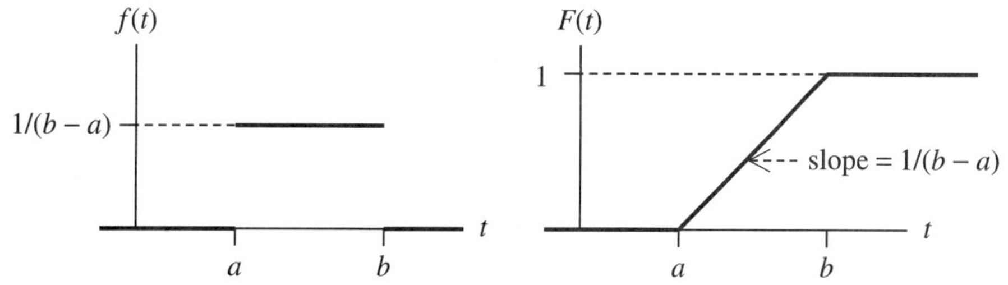

Figure 11.9 Density $f(t)$ and CDF $F(t)$ of a random variable that is uniformly distributed on the interval between a and b.

within it. If you spin an unbiased pointer, it is equally likely to come to rest pointing anywhere on the circle; its distribution is uniform on the interval between 0 and 360 degrees.

Second, the uniform is the basis for simulation. We mention how. Suppose you wish to simulate a random variable Y whose CDF $F(t)$ is plotted in Figure 11.7. First, use the Excel function RAND() to simulate a random variable X that is uniformly distributed on the interval between 0 and 1. Then find the value Y such that $F(Y) = X$. Effectively, in Figure 11.7, pick a uniform random variable X on the vertical axis, and read the CDF backwards.

11.10. THE TRIANGULAR DISTRIBUTION

The triangular distribution is a handy model for an uncertain quantity that is known to lie somewhere within a given interval and to have a most probable value within that interval. To illustrate, suppose that an engineer estimates an uncertain manufacturing cost, C. She might feel that C will be at least \$2.50 per unit and at most \$4.00 per unit, and moreover, that the most probable value of C is \$3.00 per unit. It's reasonable to model the distribution of C by a density function that has the triangular shape in Figure 11.10, with $a = 2.5$, $b = 4.0$, and $m = 3.0$.

The density function $f(t)$ is said to have the **triangular** distribution with parameters a, b, and m if $f(t)$ increases linearly from 0 to $2/(b - a)$ on the interval between a and m and if $f(t)$ decreases linearly from $2/(b - a)$ on the interval between m and b. Figure 11.10 graphs this distribution.

The area under the entire density function equals 1. For this reason, the peak value must equal $2/(b - a)$. Do you see why?

If $m = (a + b)/2$, this density is symmetric, in which case $E(T) = (a + b)/2$. To allow a comparison with the uniform random variable, we report the mean and standard deviation of T for the symmetric case.

$$E(T) = \frac{a + b}{2}, \quad \text{StDev}(T) = \frac{b - a}{\sqrt{24}} \quad \text{if } m = (a + b)/2. \tag{11.30}$$

As compared with the uniform density on the same interval, the symmetric triangular density concentrates the probability toward the middle. These two random variables have the same mean, but the triangular reduces the standard deviation by a factor of $\sqrt{2}$, as equations (11.29) and (11.30) demonstrate.

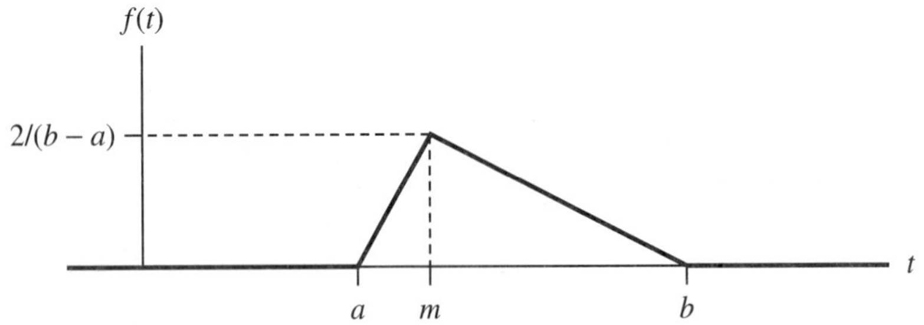

Figure 11.10 Density of a random variable T whose distribution is triangular, with parameters a, b, and m.

11.11. THE EXPONENTIAL DISTRIBUTION

Typically, the "exponential" distribution measures an interval of time whose duration is uncertain. The random variable T is said to have the **exponential** distribution with parameter λ (where λ is a fixed positive number) if its density function $f(t)$ is given by

$$f(t) = \begin{cases} 0 & \text{for } t < 0 \\ \lambda e^{-\lambda t} & \text{for } t \geq 0 \end{cases}. \tag{11.31}$$

Here, e is Euler's constant; $e \cong 2.718$.

Let the random variable T have the exponential distribution with parameter λ. Clearly, T cannot take a negative value. A bit of calculus (which we omit) verifies that

$$E(T) = 1/\lambda, \quad \text{StDev}(T) = 1/\lambda. \tag{11.32}$$

If, for instance, T is exponentially distributed with a mean of 10 minutes, then $E(T) = 1/\lambda = 10$ minutes, and $\lambda = 1/10$ per minute. Evidently, λ is a *rate*. The unit of measure of T and the unit of measure of λ are reciprocals of each other.

A Graph of the Density

Let us graph the density of this random variable. Note from Equation (11.31) that $f(0) = \lambda$ and that $f(t) = f(0)e^{-\lambda t}$, so the density takes its maximum value at 0 and drops exponentially, as t increases. Figure 11.11 plots this density function.

Each number in Figure 11.11 is the probability that T falls into the indicated interval. Since $E(T) = 1/\lambda$, Figure 11.11 shows that T takes a value below its mean with probability of 0.632. Evidently, if we observe an exponentially distributed random variable over and over, independently, about two-thirds of the observations will be somewhat below the mean, balanced by one-third that are considerably farther above the mean. And about 13.5% of the observations will be at least twice the mean.

The CDF and Tail Probabilities

The CDF $F(t) = P(T \leq t)$ of the exponential is found by calculating the area under its density function between 0 and t. The result is

$$F(t) = P(T \leq t) = \begin{cases} 0 & \text{for } t < 0 \\ 1 - e^{-\lambda t} & \text{for } t \geq 0. \end{cases} \tag{11.33}$$

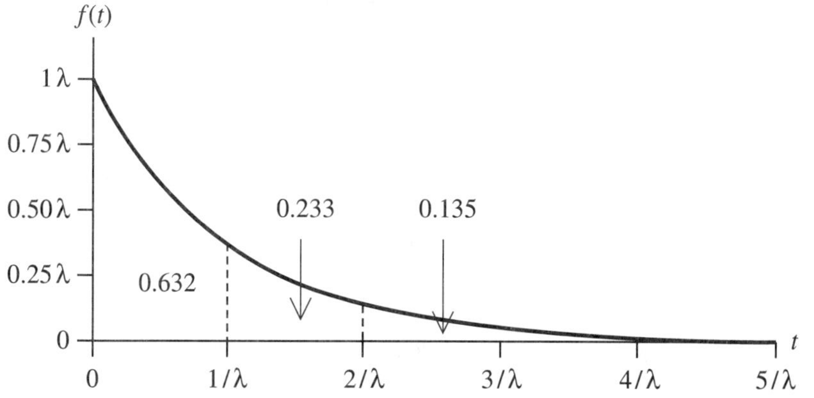

Figure 11.11 Density $f(t)$ of a random variable T whose distribution is exponential with parameter λ, so that $E(T) = 1/\lambda$.

Equation (11.33) gives a handy formula for the "tail" probability, $P(T > t)$.

$$P(T > t) = 1 - P(T \leq t) = e^{-\lambda t} \quad \text{for } t \geq 0 \tag{11.34}$$

No Memory

To bring into focus an important property of the exponential distribution, we pose the rhetorical question in:

Problem C (No Memory)

The interval T of time between the moments at which successive buses pass a particular corner is exponentially distributed with a mean of 10 minutes. You arrive at this corner the moment after a bus went by. You wait for 25 minutes, during which time the next bus does not come. Do the 25 minutes that you have waited in vain affect the time between now and the moment that the bus arrives?

No! We will soon see that, if the time between buses is exponential, the time that you have waited in vain has no influence on the remaining time until the bus comes. To verify that this is so, we compute the probability $P(T > x + t \mid T > x)$ that you will have to wait at least t more minutes given that the bus did not come during the first x minutes. The law of conditional probability gives

$$P(T > x + t \mid T > x) = \frac{P(T > x + t, T > x)}{P(T > x)} = \frac{P(T > x + t)}{P(T > x)}.$$

Equation (11.34) specifies the numerator and the denominator of this ratio.

$$P(T > x + t \mid T > x) = \frac{e^{-\lambda(x+t)}}{e^{-\lambda x}} = e^{-\lambda t} = P(T > t),$$

the last equation from a third application of Equation (11.34). Thus, the probability that you will need to wait at least t more minutes is independent of the number x of minutes that you have already waited. Probabilities sum to 1, so the above gives:

$$\begin{aligned} P(T \leq x + t \mid T > x) &= 1 - e^{-\lambda t} \\ &= P(T \leq t) \quad \text{for all } t \geq 0 \text{ and all } x \geq 0. \end{aligned} \tag{11.35}$$

Equation (11.35) states that the number x of minutes that you have been waiting has no effect on the probability that the bus will come within the next t minutes. This "memoryless" property of the exponential distribution is highlighted as follows.

Memoryless property: If a random time T has the exponential distribution, then the probability that T occurs in the next t units of time is independent of the time that has elapsed without its occurrence.

Incidentally, in Chapter 8, we saw that a discrete probability distribution—the geometric—has the same memoryless property.

Interpretation of λ as an Arrival Rate

We will express the memoryless property in a second way. Let us compute the probability that the bus is about to arrive, given that it has not yet come. That is, we compute the probability $P(T \leq x + \Delta t \mid T > x)$ that the bus will arrive within Δt units of time given that it has not arrived by time x. Substituting Δt for t in Equation (11.35) gives

$$P(T \leq x + \Delta t \mid T > x) = P(T \leq \Delta t).$$

Since $f(0) = \lambda$, we see that when Δt is small, the probability $P(T \leq \Delta t)$ is approximately the area of the rectangle with base Δt and height λ, so that

$$P(T \leq x + \Delta t \mid T > x) = \lambda \Delta t + \text{error term}, \tag{11.36}$$

where the "error term" is proportional to $(\Delta t)^2$, just as in Equation (11.3).

Expression (11.36) interprets λ as the *rate* at which T will occur, given that it has not previously occurred. The fact that λ is independent of the elapsed time x is another way of articulating the memoryless property of the exponential distribution. We emphasize:

Memoryless property. If a random time T has the exponential distribution with parameter λ, then the probability that T will occur within Δt units of time given that T has not yet occurred differs from $\lambda \Delta t$ by an error term that is proportional to $(\Delta t)^2$.

The Exponential as an Approximation

The exponential distribution has two important uses, both of which stem from its memoryless property. This property is so convenient that we often assume that an uncertain time has the exponential distribution even when we know that it does not. Typically, the time S that it takes to "serve" a customer is uncertain but has a standard deviation that is well below its mean: $\text{StDev}(S) < E(S)$. We often approximate S with an exponential random variable T whose expectation is that of S. Being exponential, T has $\text{StDev}(T) = E(T)$. Replacing S with T *overestimates* the level of uncertainty in the service time. But, as we shall see in later chapters, this approximation can render an otherwise difficult situation easy to analyze.

The Natural Model for Inter-arrival Times

Let us consider the stream of arriving customers at a service facility. The Superposition Theorem (in Chapter 8) shows that the Poisson distribution is the natural model for the number of customers who will arrive during a given period of time. We will soon see that the exponential distribution is the natural model for the length of time between the arrival of successive customers.

At a service facility, let the random variable $N(t)$ denote the number of customers who will arrive by time t. Thus, $N(s + t) - N(s)$ equals the number of customers who arrive between time s and $s + t$. The role of exponential inter-arrival times is evident in the:

Theorem (exponential inter-arrival times and Poisson arrivals). For an arrival process at a service facility, the following are equivalent:

1. The times between successive arrivals are independent and identically distributed, and each has the exponential distribution with parameter λ.
2. For every nonnegative number s and every positive number t, the number $[N(s + t) - N(s)]$ of customers who will arrive between time s and time $s + t$ has the Poisson distribution with mean of λt.

This theorem states that if inter-arrival times are exponential with parameter (rate) λ, then the number of arrivals during any interval t of time is Poisson with mean equal to the product λt, and conversely.

To dovetail this theorem with the Superposition Theorem, suppose that the arrival process at a service facility satisfies the hypothesis of the Superposition Theorem. If the aggregate arrival rate that is constant over time, say, 25 arrivals per hour, the Superposition Theorem guarantees that the number of customers who will arrive during an interval of t hours is Poisson with mean $25t$. Hence, the above theorem shows that the inter-arrival times are exponential with parameter 25, equivalently, with a mean of $1/25$ hour.

Periods of Intense Activity Punctuated by Periods of Inactivity

The fact that inter-arrival times are exponential warrants a closer look at the exponential distribution. Let us examine its fractiles. The p-fractile of the exponential distribution equals the value of t for which

$$p = P(T \leq t) = 1 - e^{-\lambda t}.$$

To solve this equation for t, we rearrange it as $1 - p = e^{-\lambda t}$ and then take logarithm, getting

$$t = -\ln(1 - p)/\lambda = -\ln(1 - p)\,E(T). \tag{11.37}$$

The p-fractile of T is the value of t that satisfies Equation (11.37). The spreadsheet in Table 11.2 reports the p-fractile for several values of p.

To interpret Table 11.2, suppose that inter-arrival times are exponential with mean of 10 minutes. Table 11.2 instructs us that roughly 20% of the inter-arrival times will be below 2.23 minutes, and roughly 20% will be above 16.09 minutes. Similarly, roughly 5% of the inter-arrival times will be below 0.51 minutes, and 5% will be above 29.96 minutes. We conclude:

> When inter-arrival times are exponential, there will be many short intervals between successive arrivals, punctuated by a few very long intervals.

This behavior is familiar to merchants. Their customers seem to appear in rapid succession, with long periods of inactivity.

The Exponential Distribution on a Spreadsheet

The exponential distribution is easy to use on a spreadsheet. Table 11.3 tells how. As this table reports, the density, CDF, and fractile are simple enough that you can enter them directly. You can also take advantage of the fact that its density and CDF are Excel functions.

Table 11.2 Fractiles of the exponential distribution, as a fraction of its mean.

	A	B	C	D	E	F	G	H
2	p =	**0.05**	**0.10**	**0.20**	**0.50**	**0.80**	**0.90**	**0.95**
3	p fractile =	0.051	0.105	0.223	0.693	1.609	2.303	2.996
4		↑						
5		= - ln(1 - B2)						

Table 11.3 Excel functions for the exponential random variable with parameter λ.

Quantity	Excel Function(s)
Density $f(t)$	=λ*EXP(−t*λ) or =EXPONDIST(t, λ, FALSE)
CDF $F(t)$	=1−EXP(−t*λ) or =EXPONDIST(t, λ, TRUE)
p-fractile	=−LN(1−p)*(1/λ)

Suppose, for instance, that you wish to compute $P(T \le 20)$ where T is exponential with parameter $\lambda = 0.1$. To do so, you could enter the function

=EXPONDIST(20, 0.1, TRUE)

into a cell of an Excel spreadsheet. In that cell, Excel will report the value 0.865, so $0.865 = P(T \le 20)$. Similarly, to find the 0.9 fractile of this distribution, you can enter the function

=−LN(1 − 0.9)*(1/0.1)

into a cell of a spreadsheet. In that cell, Excel will report the value 23.0, so $P(T \le 23.0) = 0.9$.

Recap

Let us review what we have learned about the exponential distribution. It has a single parameter, λ. A random variable T whose distribution is exponential with parameter λ has these properties:

- The value of T can be any nonnegative number.
- $E(T) = \text{StDev}(T) = 1/\lambda$.
- The unit of measure of T and of λ are the reciprocals of each other. For instance, if T measures time in minutes, then λ is measured in (1/minutes).
- The density of T is $f(t) = \lambda e^{-\lambda t}$, and the Excel function =EXPONDIST(t, λ, FALSE) returns $f(t)$.
- The CDF of T is $F(t) = 1 - e^{-\lambda t}$, and the Excel function =EXPONDIST(t, λ, TRUE) returns $F(t)$.
- T is memoryless; for any nonnegative numbers x and t, the conditional probability $P(T > x + t | T > x)$ is independent of x.
- $P(T \le x + \Delta x | T > x) \cong \lambda \Delta x$, with error proportional to $(\Delta x)^2$. This interprets λ as an arrival rate.
- This random variable T is the natural model for the time between successive arrivals at a facility that serves many customers, each of which uses the facility infrequently and independently of the others.
- The p-fractile of T equals $-\ln(1 - p)/\lambda$. When inter-arrival times are exponential, customers appear to arrive in clusters, punctuated by quiet periods.

11.12. THE NORMAL DISTRIBUTION

The random variable X is said to have the **normal** distribution with parameters μ and σ if its density function has the ungainly form

$$f(t) = \frac{1}{\sqrt{2\pi}\sigma} e^{-\frac{1}{2}\left(\frac{t-\mu}{\sigma}\right)^2} \quad \text{for } -\infty < t < \infty \tag{11.38}$$

where e is Euler's constant ($e \cong 2.718$) and π is the ratio of the circumference of the circle to its diameter ($\pi \cong 3.142$). The normal distribution is sometimes referred to as the **Gaussian** distribution; these terms are synonyms. Eventually, we shall see that the normal distribution gets its name from the fact that it is, in a sense, the *norm.*

The normal is the most important of the continuous probability distributions. Three sections of this chapter are devoted to it. The current section describes the normal distribution, the next section focuses on the Central Limit Theorem, and a later section discusses a class of "diffusion" models that give rise to the normal distribution.

The Bell-shaped Curve

Figure 11.12 graphs the normal density function; its "bell" shape may be familiar to you. This density function $f(t)$ stays positive for all t, but Figure 11.12 shows that values farther from μ are increasingly unlikely to occur.

As Figure 11.12 indicates, the normal density is symmetric about μ, which guarantees that μ equals the median and the mean of X. And the notation correctly connotes that the standard deviation of X equals σ. In short,

$$E(X) = \mu, \quad \text{StDev}(X) = \sigma. \tag{11.39}$$

The numbers in Figure 11.12 are the probabilities that a normal random variable falls within the indicated intervals. For instance, the normal random variable takes a value within one standard deviation of its mean approximately 68% of the time because $0.6826 = 0.3413 + 0.3413$. Similarly, the normal random variable takes a value within two standard deviations of its mean approximately 95% of the time because $0.9544 = 1 - 0.0228 - 0.0228$.

The normal distribution is familiar to takers of standardized tests, such as the Scholastic Aptitude Test (SAT). Typically, test results are scaled so that the scores of a target population of test-takers have the normal distribution with a mean of 500 and a standard deviation of 100, except that SAT scores are truncated at 200 and 800. A score of 700 is two standard deviations above the mean. Thus, Figure 11.12 indicates that roughly 2.28% of the target population receive scores of 700 or more. A score of 600 is one standard deviation above the mean. Figure 11.12 shows that scores of 600 or more are received by roughly 15.87% of this population ($0.1587 = 0.1359 + 0.0228$). And so forth.

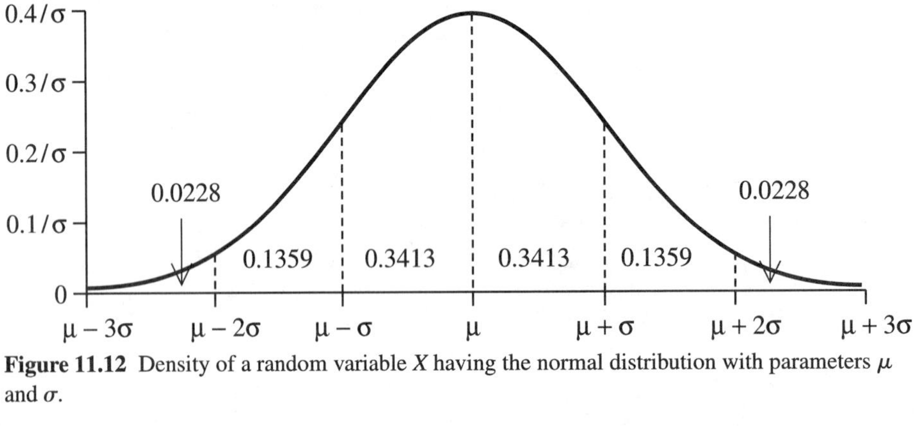

Figure 11.12 Density of a random variable X having the normal distribution with parameters μ and σ.

Table 11.4 Excel functions for the normal probability distribution having mean μ and standard deviation σ.

Quantity	Excel Function
Density $f(t)$	=NORMDIST(t, μ, σ, FALSE)
CDF $F(t)$	=NORMDIST(t, μ, σ, TRUE)
p-fractile	=NORMINV(p, μ, σ)

The Normal Distribution on a Spreadsheet

There is no simple functional form for the CDF of the normal random variable. The CDF has been approximated numerically, however, and is available as a spreadsheet function. Table 11.4 reports the Excel functions for the density, CDF, and fractile of the normal distribution.

To illustrate the use of spreadsheet calculations for the normal distribution, we pose the rhetorical questions in:

Problem D (A Gasoline Station)

Suppose that the daily demand X for gasoline at your filling station has the normal distribution with mean of 1500 gallons and standard deviation of 300 gallons. Find:

(a) The probability p that customers will demand more than 1650 gallons during a particular day.

(b) The number g of gallons such that the demand X exceeds g one day in 10, on average.

A handy way in which to attack problems like these is to sketch the normal density function and annotate it with what is known and what is unknown. The normal densities in Figure 11.13 depict the quantities that are known and sought in parts (a) and (b) of Problem D.

For part (a), we wish to compute $p = P(X > 1650)$. This probability is depicted in the left-hand sketch in Figure 11.13, as is $1 - p = P(X \leq 1650)$. The CDF of the normal distribution specifies $P(X \leq 1650)$, which equals the area to the left of the dashed line in the same sketch. Hence, for part (a), we can enter the function

=NORMDIST(1650, 1500, 300, TRUE)

into a cell of an Excel worksheet. In that cell, Excel reports 0.691, so $1 - p = P(X \leq 1650) = 0.691$. Hence, $p = 1 - 0.691 = 0.309$.

For part (b), we want the value of g such that $P(X > g) = 0.1$, which the right-hand sketch in Figure 11.13 depicts. The fractile of the normal distribution is an Excel function.

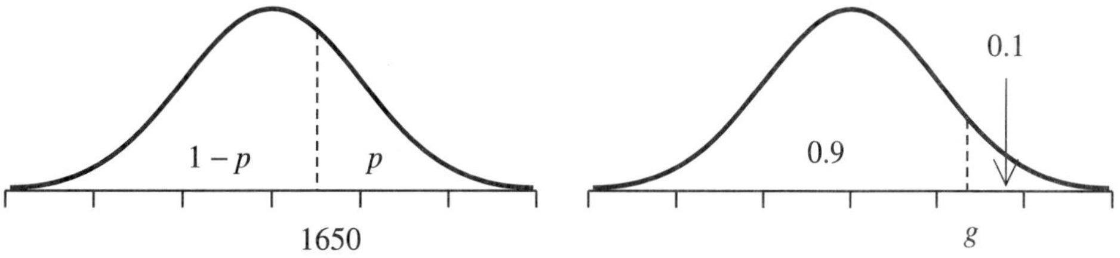

Figure 11.13 Parts (a) and (b) of Problem D posed graphically.

For whatever probability p we wish, the fractile reports the number g such that $p = P(X \leq g)$. This probability is the area to the left of the dotted line in the right-hand sketch. Evidently, the number g is the 0.9 fractile of X. To compute g, we enter the function

=NORMINV(0.9, 1500, 300)

into a cell of an Excel worksheet. In that cell, Excel reports 1884. Thus, $P(X \leq g) = 0.9$ for $g = 1884$ gallons.

The Sum of Normal Random Variables

What can be said about the sum of independent random variables, each of which has the normal distribution? The sum also has the normal distribution. We emphasize:

> The sum of independent random variables whose distributions are normal is a random variable having the normal distribution.

This property is handy but not surprising. If it did not hold, the Central Limit Theorem could not be true. Do you see why? Incidentally, the normal is not the only distribution that is preserved under summation. In Chapter 8, we saw that if independent random variables have the Poisson distribution, so does their sum.

Linear Function of a Normal Random Variable

Suppose that we take a linear function of a random variable whose distribution is normal. What sort of random variable do we get? Its distribution is normal. We highlight:

> Let the random variable X have the normal distribution, and let a and b be fixed numbers. Then the random variable $(aX + b)$ also has the normal distribution.

This property is handy, but it too is hardly surprising. The random variable $(X + b)$ adds b to whatever value X takes; $(X + b)$ has the same density as does X, except that it is centered at $E(X) + b$. Similarly, the random variable aX effects a change in the unit of measurement of X; its density remains "bell" shaped but with $E(aX) = aE(X)$ and with $\text{StDev}(aX) = |a|\text{StDev}(X)$.

For instance, if X has the normal distribution, then so does $-X$, with $E(-X) = -E(X)$ and with $\text{StDev}(-X) = \text{StDev}(X)$.

The Standard Normal Random Variable

Let the random variable X have the normal distribution with mean μ and standard deviation σ, and consider the random variable Z that is defined by

$$Z = \frac{X - \mu}{\sigma} = \frac{1}{\sigma}X - \frac{\mu}{\sigma}. \tag{11.40}$$

Evidently, Z is obtained from X by multiplying X by the constant $a = 1/\sigma$ and then adding the constant $b = -\mu/\sigma$. Hence, Z has the normal distribution. It is easy to check, using Equations (11.18) and (11.20), that

$$E(Z) = 0, \quad \text{StDev}(Z) = 1. \tag{11.41}$$

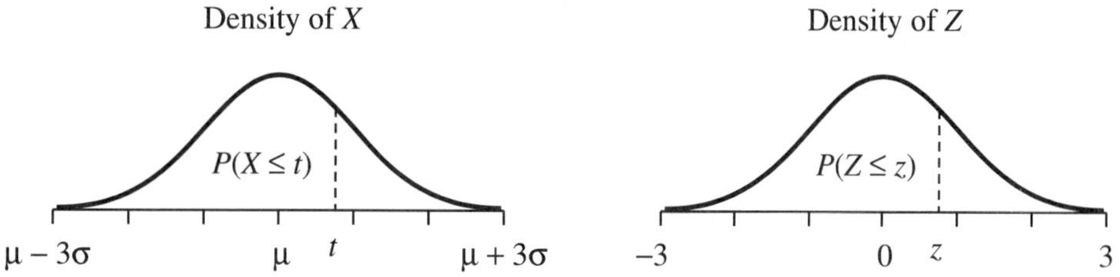

Figure 11.14 Equal areas have equal probability.

A random variable Z is said to have the **standard** normal distribution if its distribution is normal, if $E(Z) = 0$, and if $\text{StDev}(Z) = 1$. Any statement we might wish to make about a normal random variable X can be translated into an equivalent statement about a standard normal random variable Z, and conversely. Figure 11.14 suggests why. Pictorially, the areas to the left of the dashed lines in Figure 11.14 equal each other.

With Figure 11.14 in view, we translate $P(X \leq t)$ into an equivalent statement about Z. Let us begin with the inequality $X \leq t$. From both sides of this inequality, we subtract $\mu = E(X)$ and then divide by $\sigma = \text{StDev}(X)$. We obtain

$$P(X \leq t) = P\left(\frac{X - \mu}{\sigma} \leq \frac{t - \mu}{\sigma}\right) = P\left(Z \leq \frac{t - \mu}{\sigma}\right), \qquad (11.42)$$

the last from Equation (11.40). In brief,

$$P(X \leq t) = P(Z \leq z) \quad \text{with} \quad z = \frac{t - \mu}{\sigma}. \qquad (11.43)$$

Equation (11.43) translates a statement about a normal random variable X into an equivalent statement about a standard normal random variable Z.

To translate $P(Z \leq z)$ into an equivalent statement about X, we solve the equation $z = (t - \mu)/\sigma$ for t. This gives $t = \sigma z + \mu$, which lets us rewrite Equation (11.43) as

$$P(Z \leq z) = P(N \leq t) \quad \text{with } t = \sigma z + \mu. \qquad (11.44)$$

Equation (11.44) translates a statement about a standard normal random variable Z into an equivalent statement about a normal random variable X. Both statements comport with Figure 11.14 because the number t is $(t - \mu)/\sigma$ standard deviations above the mean of X.

The Standard Normal Table

We have seen how to transform statements about X and Z into each other. A **standard normal table** is a tabulation of the CDF of the standard normal random variable, Z. Table 11.5 is a standard normal table. In an earlier era, the standard normal table was vital to use of the normal distribution. That is no longer so. Spreadsheet functions can be far more convenient. The standard normal table remains important as a reference, however, and so you may wish to know how to use it.

Using the Standard Normal Table

To illustrate the use of the standard normal table, we solve Problem D a second time. In place of Excel functions, we will use the standard normal table. For part (a), we still want

Table 11.5 Standard normal table; $F(z) = P(Z \le z)$ for the standard normal random variable Z.

z	−0.00	−0.01	−0.02	−0.03	−0.04	−0.05	−0.06	−0.07	−0.08	−0.09	z	0.00	0.01	0.02	0.03	0.04	0.05	0.06	0.07	0.08	0.09
−3.0	.0013	.0013	.0013	.0012	.0012	.0011	.0011	.0011	.0010	.0010	0.0	.5000	.5040	.5080	.5120	.5160	.5199	.5239	.5279	.5319	.5359
−2.9	.0019	.0018	.0018	.0017	.0016	.0016	.0015	.0015	.0014	.0014	0.1	.5398	.5438	.5478	.5517	.5557	.5596	.5636	.5675	.5714	.5753
−2.8	.0026	.0025	.0024	.0023	.0023	.0022	.0021	.0021	.0020	.0019	0.2	.5793	.5832	.5871	.5910	.5948	.5987	.6026	.6064	.6103	.6141
−2.7	.0035	.0034	.0033	.0032	.0031	.0030	.0029	.0028	.0027	.0026	0.3	.6179	.6217	.6255	.6293	.6331	.6368	.6406	.6433	.6480	.6517
−2.6	.0047	.0045	.0044	.0043	.0041	.0040	.0039	.0038	.0037	.0036	0.4	.6554	.6591	.6628	.6664	.6700	.6736	.6772	.6808	.6844	.6879
−2.5	.0062	.0060	.0059	.0057	.0055	.0054	.0052	.0051	.0049	.0048	0.5	.6915	.6950	.6985	.7019	.7054	.7088	.7123	.7157	.7190	.7224
−2.4	.0082	.0080	.0078	.0075	.0073	.0071	.0069	.0068	.0066	.0064	0.6	.7257	.7291	.7324	.7357	.7389	.7422	.7454	.7486	.7517	.7549
−2.3	.0107	.0104	.0102	.0099	.0096	.0094	.0091	.0089	.0087	.0084	0.7	.7580	.7611	.7642	.7673	.7704	.7734	.7764	.7794	.7823	.7852
−2.2	.0139	.0136	.0132	.0129	.0125	.0122	.0119	.0116	.0113	.0110	0.8	.7881	.7910	.7939	.7967	.7995	.8023	.8051	.8078	.8106	.8133
−2.1	.0179	.0174	.0170	.0166	.0162	.0158	.0154	.0150	.0146	.0143	0.9	.8159	.8186	.8212	.8238	.8264	.8289	.8315	.8340	.8365	.8389
−2.0	.0228	.0222	.0217	.0212	.0207	.0202	.0197	.0192	.0188	.0183	1.0	.8413	.8438	.8461	.8485	.8508	.8531	.8554	.8577	.8599	.8621
−1.9	.0287	.0281	.0274	.0268	.0262	.0256	.0250	.0244	.0239	.0233	1.1	.8643	.8665	.8686	.8708	.8729	.8749	.8770	.8790	.8810	.8830
−1.8	.0359	.0351	.0344	.0336	.0329	.0322	.0314	.0307	.0301	.0294	1.2	.8849	.8869	.8888	.8907	.8925	.8944	.8962	.8980	.8997	.9015
−1.7	.0446	.0436	.0427	.0418	.0409	.0401	.0392	.0384	.0375	.0367	1.3	.9032	.9049	.9066	.9082	.9099	.9115	.9131	.9147	.9162	.9177
−1.6	.0548	.0537	.0526	.0516	.0505	.0495	.0485	.0475	.0465	.0455	1.4	.9192	.9207	.9222	.9236	.9251	.9265	.9279	.9292	.9306	.9319
−1.5	.0668	.0655	.0643	.0630	.0618	.0606	.0594	.0582	.0571	.0559	1.5	.9332	.9345	.9357	.9370	.9382	.9394	.9406	.9418	.9429	.9441
−1.4	.0808	.0793	.0778	.0764	.0749	.0735	.0721	.0708	.0694	.0681	1.6	.9452	.9463	.9474	.9484	.9495	.9505	.9515	.9525	.9535	.9545
−1.3	.0968	.0951	.0934	.0918	.0901	.0885	.0869	.0853	.0838	.0823	1.7	.9554	.9564	.9573	.9582	.9591	.9599	.9608	.9616	.9625	.9633
−1.2	.1151	.1131	.1112	.1093	.1075	.1056	.1038	.1020	.1003	.0985	1.8	.9641	.9649	.9656	.9664	.9671	.9678	.9686	.9693	.9699	.9706
−1.1	.1357	.1335	.1314	.1292	.1271	.1251	.1230	.1210	.1190	.1170	1.9	.9713	.9719	.9726	.9732	.9738	.9744	.9750	.9756	.9761	.9767
−1.0	.1587	.1562	.1539	.1515	.1492	.1469	.1446	.1423	.1401	.1379	2.0	.9772	.9778	.9783	.9788	.9793	.9798	.9803	.9808	.9812	.9817
−0.9	.1841	.1814	.1788	.1762	.1736	.1711	.1685	.1660	.1635	.1611	2.1	.9821	.9826	.9830	.9834	.9838	.9842	.9846	.9850	.9854	.9857
−0.8	.2119	.2090	.2061	.2033	.2005	.1977	.1949	.1922	.1894	.1867	2.2	.9861	.9864	.9868	.9871	.9875	.9878	.9881	.9884	.9887	.9890
−0.7	.2420	.2389	.2358	.2327	.2296	.2266	.2236	.2206	.2177	.2148	2.3	.9893	.9896	.9898	.9901	.9904	.9906	.9909	.9911	.9913	.9916
−0.6	.2743	.2709	.2676	.2643	.2611	.2578	.2546	.2514	.2483	.2451	2.4	.9918	.9920	.9922	.9925	.9927	.9929	.9931	.9932	.9934	.9936
−0.5	.3085	.3050	.3015	.2981	.2946	.2912	.2877	.2843	.2810	.2776	2.5	.9938	.9940	.9941	.9943	.9945	.9946	.9948	.9949	.9951	.9952
−0.4	.3446	.3409	.3372	.3336	.3300	.3264	.3228	.3192	.3156	.3121	2.6	.9953	.9955	.9956	.9957	.9959	.9960	.9961	.9962	.9963	.9964
−0.3	.3821	.3783	.3745	.3707	.3669	.3632	.3594	.3557	.3520	.3483	2.7	.9965	.9966	.9967	.9968	.9969	.9970	.9971	.9972	.9973	.9974
−0.2	.4207	.4168	.4129	.4090	.4052	.4013	.3974	.3936	.3897	.3859	2.8	.9974	.9975	.9976	.9977	.9977	.9978	.9979	.9979	.9980	.9981
−0.1	.4602	.4562	.4522	.4483	.4443	.4404	.4364	.4325	.4286	.4247	2.9	.9981	.9982	.9982	.9983	.9984	.9984	.9985	.9985	.9986	.9986
−0.0	.5000	.4960	.4920	.4880	.4840	.4801	.4761	.4721	.4681	.4641	3.0	.9987	.9987	.9987	.9988	.9988	.9989	.9989	.9989	.9990	.9990

$P(X > 1650) = 1 - P(X \leq 1650)$. To translate $P(X \leq 1650)$ to a statement about a standard normal random variable, we use $z = (t - \mu)/\sigma$ to get

$$P(N \leq 1650) = P\left(Z \leq \frac{1650 - 1500}{300}\right) = P(Z \leq 0.5) = 0.6915,$$

the last from the standard normal table. Hence, $P(X > 1650) = 1 - 0.6915 = 0.3085$.

For part (b), we still want g such that $P(X > g) = 0.1$, equivalently, $P(X \leq g) = 0.9$. By reading the standard normal table "backwards," we see that $P(Z \leq z) = 0.9$ for $z = 1.28$. To translate $P(Z \leq z)$ into an equivalent statement about X, we use $t = \mu + z\sigma = 1500 + (1.28)(300) = 1884$ gallons. Thus, $P(X \leq 1884) = 0.9$.

Using the Standard Normal Distribution

Some problems can be hard to solve without using the standard normal distribution. Here is one of them.

Problem E (The Gas Guzzler Tax)

You plan to manufacture one model of automobile for the U.S. market. Your engineers can control the mean mpg μ of your vehicles. Due to manufacturing variations, the mpg from an individual car is uncertain; it has a normal distribution with a mean μ and a standard deviation of 1.8 mpg. The mpg ratings of different cars are independent. For the U.S. market, you want your engineers to design for a low value of μ, which will make your cars peppier, hence more attractive to U.S. buyers. But you want to avoid the U.S. "gas guzzler" tax. To determine whether or not you will have to pay this tax, a government agency will pick six newly manufactured cars at random, measure the mpg of each, and impose this tax if the average is below 27 mpg. You are willing to run 1 chance in 100 of having to pay the gas guzzler tax. Subject to this constraint, you seek to minimize μ so as to make your car as peppy as possible. At what value do you set μ?

To reduce the probability of having to pay the "gas guzzler" tax to 0.01 (1 chance in 100), you need to set μ far enough above 27 mpg to provide a suitable margin of safety. To see how far, we investigate the average A of the mpg ratings of six randomly selected cars. Specifically,

$$A = (X_1 + \cdots + X_6)/6,$$

where X_1 through X_6 are independent random variables, each having the normal distribution with mean of μ and standard deviation of 1.8.

What do we know about $(X_1 + \cdots + X_6)$? The sum of independent normal random variables is normal, so $(X_1 + \cdots + X_6)$ is normal. We know that

$$E(X_1 + \cdots + X_6) = 6\mu, \quad \text{StDev}(X_1 + \cdots + X_6) = \sqrt{6}(1.8)$$

because each of the means and variances of independent random variables add and because the standard deviation equals the square root of the variance.

Finally, the random variable $A = (X_1 + \cdots + X_6)/6$ has the normal distribution because A multiplies a normal random variable $(X_1 + \cdots + X_6)$ by the number $1/6$. We recall from Equations (11.18) and (11.20) that multiplying a random variable by $1/6$ multiplies its expectation by $(1/6)$ and multiplies its standard deviation by $1/6$. Hence,

$$E(A) = 6\mu(1/6) = \mu, \quad \text{StDev}(A) = \sqrt{6}(1.8)(1/6) = (1.8)/\sqrt{6}.$$

Table 11.6 Using Solver to minimize μ such that $P(A \leq 27) \leq 0.01$.

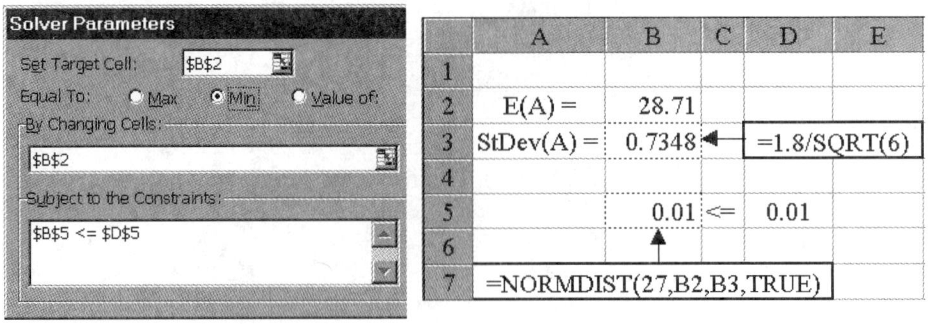

We need to set μ large enough that $P(A \leq 27) = 0.01$. We'll do this in two different ways, once with Solver, once without.

Table 11.6 indicates how to use Solver to find the smallest value of μ that keeps $P(A \leq 27)$ from exceeding 0.01. The function in cell B3 computes StDev(A) $= 1.8/\sqrt{6}$. The function in cell B5 uses the normal CDF to compute $P(A \leq 27)$ where $E(A)$ and StDev(A) are in cells B2 and B3, respectively. Solver selects the smallest value in cell B2 that keeps $P(A \leq 27) \leq 0.01$. Table 11.6 reports the optimal solution, $\mu = 28.71$. Evidently, the margin of safety is 1.71 miles per gallon, as $1.71 = 28.71 - 27$.

The preceding method is quick but may be a bit "slick." Let us solve Problem E again, this time without using Solver. A pair of sketches will help us figure out what to do. The left-hand sketch in Figure 11.15 shows that we wish to pick μ such that $P(A \leq 27)$ equals 0.01. We know that $\sigma = 1.8/\sqrt{6} = 0.735$. It isn't obvious from the left-hand sketch how to compute μ. But from the right-hand sketch, it is clear that z is the 0.01 fractile of Z, which we can easily compute.

The Excel function **=NORMINV(0.01, 0.1)** returns the value $z = -2.33$, which we could also find from the standard normal table (Table 11.5). Since $z = -2.33$, we substitute in the equation $t = \mu + z\sigma$ and get

$$27 = \mu + (-2.33)(0.735) = \mu - 1.71.$$

So we set $\mu = 27 + 1.71 = 28.71$. Your margin of safety is 1.71 mpg.

An Advantage of the Normal Distribution

Why is the normal distribution important? The normal distribution earns its fame primarily from the Central Limit Theorem, which was introduced in Chapter 8 and will be studied in the next section.

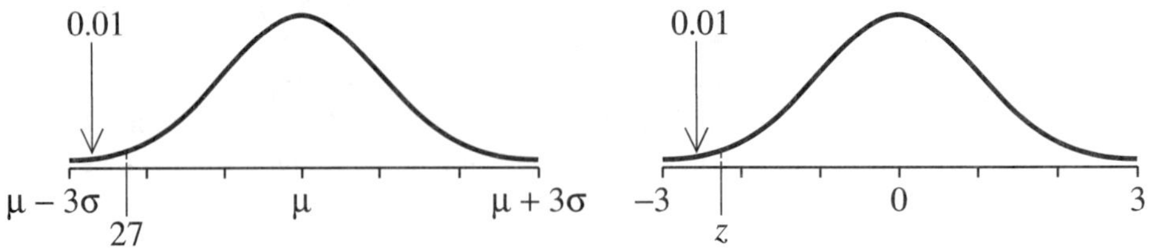

Figure 11.15 Setting μ so that $P(A \leq 27) = 0.01$.

Without invoking the Central Limit Theorem, we can point out one advantage of the normal distribution: It is flexible. By adjusting the parameters μ and σ, we can fit a normal random variable to any mean and any standard deviation. (That cannot be said, for instance, of either the exponential distribution or the Poisson distribution.) Thus, the normal fits any density that is roughly symmetric and is roughly bell-shaped.

The normal is so handy that it is often used to model quantities that cannot take negative values. Suppose a nonnegative random variable Y is known to have a symmetric distribution with a mean of 100 and a standard deviation of 30. As a model of Y, we might take a normal random variable X with $\mu = 100$ and $\sigma = 30$, ignoring the possibility that $X < 0$, as $P(X < 0) = 0.0004$, which is miniscule.

An important property of the normal distribution is that it has **skinny tails**, namely, a tiny probability of taking values that are several deviations away from its mean. For example, the probability that a normal random variable exceeds its mean by four or more standard deviations equals 0.00003, which is 3 chances in 100,000. Why the tail probabilities are small is evident in the normal density function, which behaves as e^{-t^2} for large t. By contrast, the exponential distribution has a fatter tail; its density drops off as e^{-t}.

Recap

Let us review what we have learned about the normal distribution. It has two parameters, μ and σ. A random variable X whose distribution is normal with parameters μ and σ has these properties:

- The value of X can be any number, positive, negative, or zero.
- $E(X) = \mu$ and $\text{StDev}(X) = \sigma$.
- The density $f(t)$ of X is symmetric about μ, and the Excel function =NORMDIST (t, μ, σ, FALSE) returns the density $f(t)$.
- The CDF $F(t)$ of X has no convenient formula, but the Excel function =NORMDIST(t, μ, σ, TRUE) returns the CDF $F(t)$.
- If b and c are fixed numbers, the random variable $(bX + c)$ also has the normal distribution.
- In particular, the random variable $Z = (X - \mu)/\sigma$ has the normal distribution with $E(Z) = 0$ and $\text{Var}(Z) = 1$. This random variable Z is called the **standard** normal random variable, and a table of its CDF appears on page p. 370.
- Facts about the normal random variable X and the standard normal random variable Z relate to each other through the transformation $z = (t - \mu)/\sigma$ because the number t is $(t - \mu)/\sigma$ standard deviations above the expectation of X.
- The sum, $X + Y$, has the normal distribution if the random variables X and Y are independent of each other and if both have the normal distribution.
- The normal distribution is a convenient model of a random variable whose density is roughly symmetric and bell-shaped, with skinny tails.

These facts explain why the normal is convenient but not why the normal plays a central role in probability and statistics.

11.13. THE CENTRAL LIMIT THEOREM

This section focuses on the uses of the Central Limit Theorem, which was first mentioned in Chapter 8. Let us begin by restating this theorem. Let $X_1, X_2, \ldots$ be a sequence of independent and identically distributed (i.i.d.) random variables, and consider the sum,

$$S_n = X_1 + X_2 + \cdots + X_n, \tag{11.45}$$

of the first n of these random variables. The X_i's have the same distribution, and expectations add. The X_i's are independent, so their variances also add. The standard deviation is the square root of the variance. Hence,

$$E(S_n) = nE(X_1), \quad \text{StDev}(S_n) = \sqrt{n}\,\text{StDev}(X_1). \tag{11.46}$$

Thus, the fact that the X_i's are i.i.d. determines the mean and standard deviation of their sum.

The Central Limit Theorem is a deeper result. It states that for sufficiently large n, the sum distribution of the sum S_n is approximately normal. The Central Limit Theorem establishes the normal distribution as the *norm*, the standard to which the sum converges. Let us summarize:

> **Central Limit Theorem.** For sufficiently large values of n, the sum S_n of n i.i.d. random variables is well-approximated by a random variable N whose distribution is normal with parameters $\mu = E(S_n)$ and $\sigma = \text{StDev}(S_n)$.

To understand the Central Limit Theorem and to use it intelligently, we must grapple with the following questions.

- How large must n be for S_n to be approximately normal?
- If S_n is a discrete random variable, how can its distribution be approximated by that of a continuous random variable N?
- Is it necessary that the X_i's have the same distribution?
- To which practical settings does the Central Limit Theorem apply?

This section probes each of these questions.

For n to be large enough that the distribution of S_n has become normal, two things must have occurred: (1) the distribution of S_n must have become symmetric, and (2) it must have become "bell-shaped." The speed with which this happens depends crucially on whether the distribution of the X_i's is symmetric or asymmetric. An asymmetric probability distribution is said to be **skewed**.

We will see that if the distribution of X_1 is symmetric, the distribution of S_n is guaranteed to be symmetric, and it can become bell-shaped for very low values of n. On the other hand, if the distribution of X_1 is highly skewed, the extent to which it is skewed attenuates only slowly with n, and it may be that n must be quite large for S_n to be well-approximated by the normal distribution.

Round Off

Let us first probe the question of approximating a discrete random variable by a continuous one. To do so, we consider the case in which the random variable X_1 is integer-valued. In this case, S_n is a sum of integers. It too is integer-valued. The normal is a continuous distribution. To approximate the discrete random variable S_n by the continuous random variable N, we round N off to the nearest value that S_n can take. To illustrate this procedure, consider:

Problem F (Round Off)

How well is the number S_7 of heads in seven tosses of a fair coin approximated by the normal variable N whose parameters μ and σ are the mean and standard deviation of S_7?

Here, S_7 is the sum of i.i.d. Bernoulli random variables X_1 through X_7 having parameter $p = 1/2$. We recall from Chapter 8 that the mean and standard deviation of this Bernoulli random variable X_1 are

$$E(X_1) = p = 1/2, \quad \text{StDev}(X_1) = \sqrt{p(1-p)} = 1/2.$$

Hence, Equation (11.46) shows that the mean and standard deviation of N are

$$\mu = E(S_7) = (7)E(X_1) = (7)(1/2) = 3.5,$$
$$\sigma = \text{StDev}(S_7) = (\sqrt{7})\,\text{StDev}(X_1) = (\sqrt{7})(1/2) = 1.3229.$$

The random variable S_7 takes the values 0, 1, . . . , 7. When we round N off to the nearest value that S_7 can take:

- All values of N that are between $-\infty$ and 0.5 are rounded off to 0.
- All values of N that are between 0.5 and 6.5 are rounded off to the nearest integer.
- All values of N that are between 6.5 and $+\infty$ are rounded off to 7.

Table 11.7 compares the probability distribution of S_7 with this round off of N. The bottom three lines of Table 11.7 record the functions that execute its calculations. Evidently, column B computes the CDF of N at the values 0.5, 1.5, . . . , 6.5. Column C uses the numbers in column B to compute the probability that N rounds off to each value that S_7 can take. Column D computes the probability distribution of S_7, which is binomial with parameters $n = 7$ and $p = 0.5$.

Table 11.7 Distribution of the number S_7 of heads in seven tosses of a fair coin and of a normal approximation N to S_7.

	A	B	C	D
2	mu =	3.5		
3	sigma =	1.3229		
4				
5	**k**	**P(N < k + 0.5)**	**P(N → k)**	**P(S7 = k)**
6	0	0.0117	0.0117	0.0078
7	1	0.0653	0.0536	0.0547
8	2	0.2248	0.1596	0.1641
9	3	0.5000	0.2752	0.2734
10	4	0.7752	0.2752	0.2734
11	5	0.9347	0.1596	0.1641
12	6	0.9883	0.0536	0.0547
13	7		0.0117	0.0078
14				
15	Cell B6 contains **=NORMDIST(A6+0.5,B$2,B$3,TRUE)**			
16	Cell C6 contains **=B6**		Cell C7 contains **=B7-B6**	
17	Cell D6 contains **=BINOMDIST(A6,7,0.5,FALSE)**			

Columns C and D of Table 11.7 are close to each other. The error is in the third decimal place. It would take many observations to distinguish S_7 from the normal approximation to it. Evidently, when the X_i's are Bernoulli with $p = 1/2$, taking $n = 7$ is enough for S_n to the sum to be approximately normal.

The Bernoulli as a Worst Case

If the range of X_1 is finite, a worst-case bound on n can be obtained by replacing X_1 by the Bernoulli-like random variable that takes only its extreme values and has the same mean, as in:

- Pick as s and L the smallest and the largest values that X_1 can take.
- Pick p so that $E(X_1) = ps + (1 - p)L$.
- Let n be the smallest integer such that the sum S_n of n Bernoulli random variables with parameter p is approximately normal.

Thus, by studying the case in which X_1 is Bernoulli with the parameter p determined as above, we find an upper bound on the value of n for which S_n is approximately normal.

Convergence When X_1 Is Symmetric

Let us consider the case in which the range of X_1 is finite and its distribution is symmetric. The Bernoulli with $p = 0.5$ is the least bell-shaped of these distributions. Even so, Table 11.7 shows that the normal is a pretty good approximation to the sum S_7 of seven symmetric Bernoulli random variables. For symmetric distributions that are more bell-shaped than the Bernoulli, convergence to the normal will occur even faster.

To illustrate, we consider the case in which X_1 is the number of pips on one fair die. Its distribution is symmetric and is closer to bell-shaped than the Bernoulli. Convergence to the normal occurs more quickly. Figure 11.16 compares the distribution of the total number S_6 of pips on six fair dice and with the normal approximation to it. Visually, they very nearly

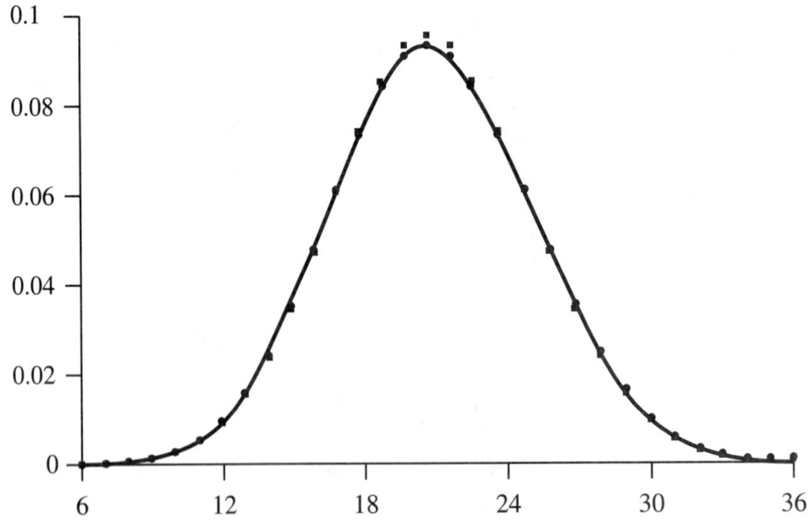

Figure 11.16 Probability distribution of total number S_6 of pips on six balanced dice (connected diamonds) and the normal approximation to it (unconnected squares).

coincide, the normal approximation being slightly more concentrated at the middle. Incidentally, Problem 27 on page 397, indicates how to use *one* "drag" to compute the probability distribution of the total number of pips on n dice.

Thus, when X_1 is symmetric and has a finite range, S_n is approximately normal for n equal to 7 and, perhaps, for n even lower. In general, when X_1 is symmetric, the quality of the fit improves *linearly* with n; doubling n decreases the error by half.

Convergence When X_1 Is Skewed

We now treat the case in which the range of X_1 is finite and its distribution is skewed. To focus on the worst case, we assume that X_1 is Bernoulli whose parameter p has been determined by the prior procedure. Using the Bernoulli has two advantages:

- Excel can be used to compute the probability distribution of S_n, which is binomial with parameters n and p.
- Solver can be used to find out how big n must be for the normal random variable N to provide a fit of given quality to S_n.

Let us begin with a picture. Figure 11.17 plots the distribution of the sum S_{80} of 80 Bernoulli random variables with parameter $p = 0.1$, and it plots the normal approximation N to this distribution.

The fit in Figure 11.17 is worse than that in Figure 11.16, even though many more random variables are being summed. Note in Figure 11.17 that the distribution of S_{80} is somewhat skewed; S_{80} takes the values 4 through 7 with higher probability than its normal approximation, and S_{80} takes the values 9 through 12 with lower probability than its normal approximation. Taking $n = 80$ has attenuated the skewness in X_1, but it remains noticeable.

The Rule of Seven

In Chapter 8, we asserted that the normal distribution is an adequate fit to the binomial random variable S_n when n and p are large enough that $np \geq 7$ and $n(1 - p) \geq 7$. The first of these inequalities requires the expected number of "successes" to be at least seven. The second, requires the expected number of "failures" be at least seven.

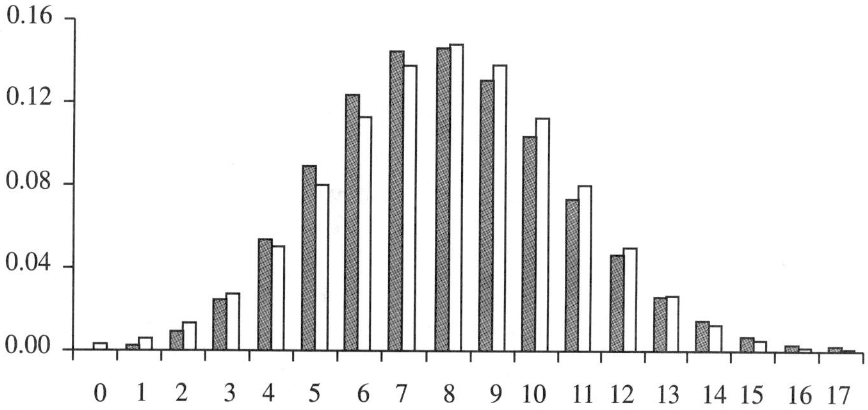

Figure 11.17 Distribution (shaded) of the sum S_{80} of 80 Bernoulli random variables with $p = 0.1$ and (unshaded) the normal approximation N to it.

Table 11.8 For given p, the smallest n for which the binomial with parameters n and p has gap $\Delta \le 0.02$.

p	0.1	0.5	0.2	0.01	0.005
n	80	19.4	53.72	1084	2192
np	8	9.7	10	10.84	10.96

This "rule of seven" works well for many purposes. Seven successes and seven failures are enough to cause the distribution of S_n to be roughly bell-shaped, but its skewness remains detectable. Figure 11.17 depicts the distribution of a binomial random variable S_{80} that has $np = 8$; when compared with the normal, the distribution of S_{80} has a slight bulge to the left.

A Measure of Fit

We seek to measure the difference between the distribution of a binomial random variable S_n and the normal random variable N that has the same mean and standard deviation, when N is rounded off to the nearest value that S_n can take. There are many ways in which to measure the difference between two probability distributions. We seek a measure that assesses the skewness of the distribution of S_n. As Figure 11.17 suggests, this distribution has a "bulge" in its CDF; more probability lies to one side of its mean than to the other.

As a measure of fit, we use the "gap" between two cumulative distributions, this being the largest difference between their CDFs. Specifically, the **gap** Δ between the CDF of S_n and the normal approximation N to it is defined by

$$\Delta = \max\{|P(S_n \le k) - P(N \le k + 0.5)| : 0 \le k \le n\} \tag{11.47}$$

where $E(N) = np$ and $\text{StDev}(N) = \sqrt{np(1-p)}$. Equation (11.47) defines Δ as the maximum value of the difference between the CDF of S_n and the CDF of the normal approximation to S_n.

For several values of p, Solver has been used to find the smallest integer n such that the binomial with parameters n and p reduces the gap Δ to 0.02 or less. Table 11.8 reports the results. Evidently, for p between 0.005 and 0.1, picking n so that the expected number np of successes is approximately 10 reduces the gap to 0.02.

For the same values of p as are in Table 11.8, Solver was used to find the smallest integer n that reduces the gap Δ to 0.01 or less. Table 11.9 presents the results. This table shows that for p between 0.005 and 0.1, we need to have the expected number np of successes approximately equal to 40 to reduce the gap below 0.01.

Let us summarize the insights obtained from Tables 11.8 and 11.9. For p in the interval between 0.005 and 0.1, taking $n = 10/p$ reduces the gap to approximately 0.02. This value of n is far larger than for the symmetric case. In order to improve the gap by a factor of 2, we need to increase n by a factor of 4. This reflects a general property. When X_1 is skewed, the quality of the fit improves as the *square root* of n.

Table 11.9 For given p, the smallest n for which the binomial with parameters n and p has gap $\Delta \le 0.01$.

p	0.1	0.5	0.2	0.01	0.005
n	346	794	2089	4407	8755
np	34.6	37.9	41.2	44.1	43.8

Convergence When the X_k's Are Not i.i.d.

Is it necessary for the Central Limit Theorem that the individual random variables have the same distribution? No! It is enough that the individual random variables be roughly independent and that each has a small effect on the total.

This is important in practice. Often, the quantity of interest is the sum of several random variables whose distributions are different, each making a small contribution to the total. The individual random variables are independent, or nearly so, but they have different probability distributions. The sum is normal if enough of them are added up.

Indeed, it can be helpful for the X_k's to have different distributions. The asymmetries can offset each other; if some X_k's are skewed to one side, and some to the other, S_n becomes symmetric more quickly.

Must the aggregate S_n be the *sum* of the individual random variables? Not exactly. In biology, a number of quantities have been found to have a normal distribution. These include:

- The length of the ear of a particular strain of corn at maturity,
- The length of a cuckoo egg,
- The height of a childbearing woman in England at a particular time in its history.

Each of these quantities is an aggregate of several factors but is not their sum. For instance, the height of a full-grown English woman is influenced by her gene composition, by the health of her mother during pregnancy, by her diet, and by environmental factors.

Settings in Which the Normal Distribution Occurs

In brief, then, we can anticipate that the normal distribution describes the aggregate of random fluctuations that may or may not be the sum of i.i.d. random variables. There are five practical settings in which this occurs:

- Measurement error.
- Engineering and manufacturing.
- Marketing and inventory control.
- Aggregate profit.
- Brownian motion.

Measurement Error

Typically, the error in a measurement is the sum of several independent random factors whose distributions are roughly symmetric, with no predominant factor. As suggested above, the distribution of the aggregate measurement error is often very nearly normal. In fact, an early and important use of the normal distribution was in the estimation of errors of measurement in astronomical observations.

Engineering and Manufacturing

The **fluctuation** in a dimension is the difference between an actual dimension and its intended value. This fluctuation is often the aggregate of several small random factors whose distributions are symmetric, so it often has the normal distribution. A prime concern of manufacturing is to control and reduce these fluctuations. Off-size components fit less well and quicken wearout. The field of **quality control** concerns the design of manufacturing processes whose random fluctuations are small and the design of measures that determine whether or not the manufacturing process is in control (operating as designed). The normal distribution plays a key role in quality control because fluctuations often have the normal distribution.

Marketing and Inventory Control

The total demand for a product is normal when it is the sum of many individual requests that are small and are independent of each other. For instance, the number of tourist-class seats requested on a particular flight is normal if it is uncontaminated by the unavailability of seats on other flights.

Aggregate Profit

The total net income that an entity will earn from many operational-scale activities is normally distributed when the individual activities are independent, or nearly so. For instance, the net revenue that an automobile insurer will earn over a year tends to be normal because the payoffs on individual automobile insurance policies are very nearly independent of each other. By contrast, the annual net revenue of a hurricane insurer tends not to be normal, unless the insurer is exceedingly diversified, geographically. Net income that is normally distributed is a consequence of successful diversification.

Brownian Motion

Roughly speaking, Brownian motion describes the aggregate displacement of a particle that has been subjected to many tiny "bumps." Brownian motion has found application to many areas, including the position of a molecule of a liquid or a gas, thermal noise, the behavior of certain queueing systems, and the price of a financial asset. A later section in this chapter is devoted to Brownian motion.

Recap

How quickly the sum of i.i.d. random variables becomes normal depends on whether the individual random variables are symmetric or skewed. If the individual random variables have a symmetric distribution, the sum can be approximately normal for very low values of n. Figure 11.16 shows that S_6 and the normal approximation to it are almost indistinguishable when X_1 is the number of pips on one fair die, for instance.

For random variables whose ranges are finite, the Bernoulli is the worst case, the most severely skewed. When X_1 is Bernoulli, Solver can be used to find the smallest n that attains a given quality of fit between S_n and the normal approximation N to it. For a broad range of the parameter p of the Bernoulli, the gap between the CDF of S_n and the approximation based on N falls below 0.02 when the number np of "successes" and the number $n(1 - p)$ of "failures" are at least 10. Skewness attenuates as the square root of n; to reduce this gap below 0.01, we need to have np and $n(1 - p)$ at least 40.

For the sum to be normal, the individual random variables need not have the same distribution. It can help if they have different distributions, the skewnesses in one offsetting another. Because this is so, the normal distribution is useful in marketing, engineering, manufacturing, quality control, and inventory control, and in the analysis of errors in measurement.

11.14. THE LOGNORMAL DISTRIBUTION

The random variable Y is said to have the **lognormal** distribution with parameters μ and σ if the natural logarithm of Y has the normal distribution with the same parameters. In other words, a lognormal random variable Y and a normal random variable X with the same parameters relate to each other through the equations

$$X = \ln(Y) \quad \text{and} \quad Y = e^X, \tag{11.48}$$

where, once again, e is Euler's constant ($e \cong 2.718$). A normal random variable X can take any value, positive, negative, or zero. By contrast, the value taken by a lognormal random variable Y can be any positive number; Y cannot equal zero, and Y cannot be negative.

The parameters μ and σ of a lognormal random variable Y are the mean and standard deviation of the natural logarithm of Y, *not* of Y itself. Later in this section, we will see how to fit μ and σ to specific values of $E(Y)$ and StDev(Y).

The Density, CDF, and Fractile

Throughout this section, Y is a random variable whose distribution is lognormal with parameters μ and σ. Let $F(t)$ denote the CDF of Y, so $F(t) = P(Y \leq t)$ for each t. The events $\{Y \leq t\}$ and $\{\ln(Y) \leq \ln(t)\}$ include the same values of Y because $\ln(t)$ is an increasing function of t. Hence,

$$F(t) = P(Y \leq t) = P[\ln(Y) \leq \ln(t)] = P[X \leq \ln(t)], \tag{11.49}$$

the last from Equation (11.48), where X has the normal distribution with parameters μ and σ. In brief, the CDF of Y, evaluated at t, equals the CDF of $X = \ln(Y)$, evaluated at $\ln(t)$.

The lognormal density $f(t)$ is given by

$$f(t) = \frac{1}{t\sqrt{2\pi}\,\sigma} e^{-\frac{1}{2}\left(\frac{\ln(t)-\mu}{\sigma}\right)^2} \quad \text{for } t > 0, \tag{11.50}$$

with $f(t) = 0$ for $t \leq 0$. (Equation (11.50) can be verified by differentiation in Equation (11.49), invoking the chain rule.)

The p-fractile of Y equals the number t for which

$$p = P(Y \leq t) = P[X \leq \ln(t)], \tag{11.51}$$

where, once again, X has the normal distribution with parameters μ and σ.

Spreadsheet Computation

Equations (11.49), (11.50), and (11.51) let us compute the CDF, density, and p-fractile of the lognormal distribution as Excel functions. Table 11.10 indicates how.

When using Table 11.10, we must recall that μ and σ are the mean and standard deviation of the logarithm of Y, *not* the mean and standard deviation of Y itself.

The Mean and Standard Deviation

Computing the mean and standard deviation of a lognormal random variable Y is an exercise in calculus that we omit. The results are:

$$E(Y) = e^{\mu+\sigma^2/2}, \quad \text{StDev}(Y) = E(Y)\sqrt{e^{\sigma^2} - 1}. \tag{11.52}$$

Table 11.10 Excel functions for the density, CDF, and fractile of the lognormal distribution having parameters μ and σ.

Quantity	Excel Function(s)
Density $f(t)$	=(1/t)*NORMDIST(LN(t), μ, σ, FALSE)
CDF $F(t)$	=NORMDIST(LN(t), μ, σ, TRUE) or =LOGNORMDIST(t, μ, σ)
p-fractile	=EXP(NORMINV(p, μ, σ))

Suppose we know that Y has the lognormal distribution, and suppose we know its mean and standard deviation; we know the numbers m and s, where

$$m = E(Y) \quad \text{and} \quad s = \text{StDev}(Y). \tag{11.53}$$

To find the parameters μ and σ of this distribution, we solve the right-hand equation in (11.52) for σ and then solve the left-hand equation in (11.52) for μ. The result is

$$\sigma = \sqrt{\ln(1 + (s/m)^2)}, \quad \mu = \ln(m) - \sigma^2/2. \tag{11.54}$$

In brief, if we are given the mean m and standard deviation s of random variable Y whose distribution is lognormal, we can compute its parameters σ and μ from Equation (11.54). We can then use the Excel functions in Table 11.10 to compute its density, CDF, and fractile.

Graph of the Lognormal Density

A lognormal density was plotted early in this chapter, but it was not identified as such. Figure 11.2 plots the density of a random variable Y whose distribution is lognormal with $E(Y) = 2$ and StDev$(Y) = 1$. Figure 11.2 shares a property with every lognormal density. Its median lies below its mean. To verify that this is so, we use Equations (11.49) and (11.52) in

$$P[Y < E(Y)] = P[X < \ln(E(Y))] = P(X < \mu + \sigma^2/2) > 0.5.$$

Figure 11.18 plots three lognormal density functions, all of which have the same mean. The figure demonstrates that, as the standard deviation increases, the density becomes increasingly skewed.

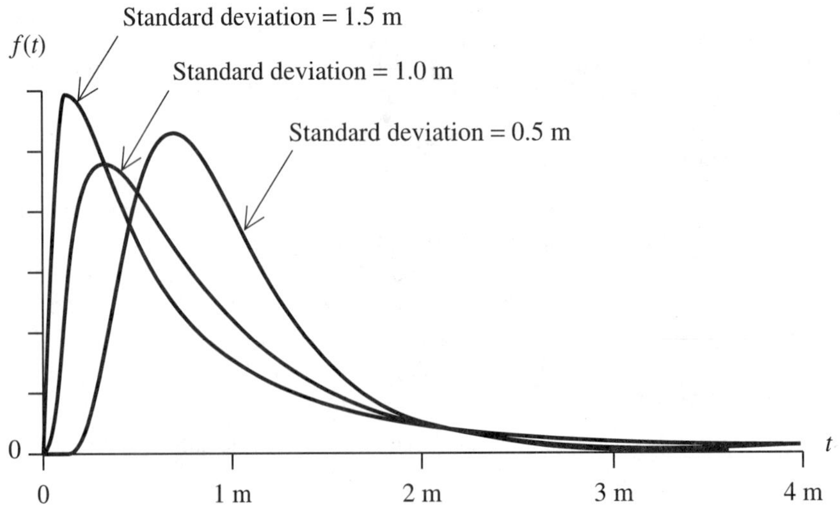

Figure 11.18 Three lognormal density functions $f(t)$, each having m as its expectation.

Table 11.11 The probability that the random variable Y takes a value at least four standard deviations above its mean, with $m = E(Y)$ and $s = \text{StDev}(Y)$.

Distribution of Y	$P(Y > m + 4s)$
Normal	0.00003
Exponential	0.007
Lognormal with $s = 0.5$ m	0.005
Lognormal with $s = 1.0$ m	0.009
Lognormal with $s = 1.5$ m	0.021

Tail Proabilities

Figure 11.18 suggests—correctly—that a lognormal random variable can have a fat tail. Table 11.11 compares the probability that various random variables take values that are at least four standard deviations above their means.

Of the tail probabilities reported in Table 11.11, the normal has the smallest by a wide margin. The exponential and lognormal have comparable tail probabilities when their standard deviations are similar. When the lognormal is very skewed, its tail probabilities dwarf the others.

One Use of the Lognormal

One use of the lognormal distribution is to model a positive random variable whose density has the general shape of Figure 11.18. Such a random variable:

- can take any value that is positive,
- takes values very close to 0 with miniscule probability,
- takes a value below its mean with probability in excess of 0.5,
- can have any mean and any standard deviation that are positive.

If a random variable's probability distribution has these characteristics, Equation (11.54) can be used to fit the parameters μ and σ of the lognormal distribution to its mean and standard deviation.

Importance of the Lognormal

The importance of the lognormal stems mainly from the role it plays for products of *positive* random variables. To bring this role into focus, we let $X_1, X_2, \ldots$ be a sequence of independent and identically distributed (i.i.d.) random variables, each of which is positive, and consider the product,

$$Y_n = (X_1)(X_2) \cdots (X_n), \qquad (11.55)$$

of the first n of these random variables.

Since the X_k's assume positive values, the product Y_n is a positive number, and we can take its logarithm. The logarithm of the product equals the sum of the logarithms, so

$$\ln(Y_n) = \ln(X_1) + \ln(X_2) + \cdots + \ln(X_n). \qquad (11.56)$$

In other words, the random variable $\ln(Y_n)$ is the sum of i.i.d. random variables, each having the same distribution as does $\ln(X_1)$. The Central Limit Theorem tells us that for large enough values of n, the random variable $\ln(Y_n)$ is approximately normal with mean $n\mu$ and standard deviation $\sqrt{n}\,\sigma$, where

$$\mu = E(\ln(X_1)) \quad \text{and} \quad \sigma = \text{StDev}(\ln(X_1)). \tag{11.57}$$

Since $\ln(Y_n)$ is approximately normal, Equation (11.48) shows that Y_n is approximately lognormal with the same parameters.

Central Limit Theorem for products. For sufficiently large values of n, the product Y_n of n positive i.i.d. random variables X_1 through X_n is approximately lognormal with parameters $n\mu$ and $\sqrt{n}\,\sigma$, where

$$\mu = E(\ln(X_1)) \quad \text{and} \quad \sigma = \text{StDev}(\ln(X_1)).$$

In the next section, we will see that the normal distribution is the appropriate model for the sum of many small random displacements and that the lognormal distribution is the appropriate model for the product of many small random displacements.

11.15. DIFFUSION

Early in the nineteenth century, the botanist Robert Brown observed that when a grain of pollen was suspended in water, it underwent repeated jerky motion. The particle seemed to be displaced by many bumps. These bumps were later attributed to collisions with molecules. The cumulative motion of the grain of pollen over a period of time is known to follow a normal distribution that is called Brownian motion in his honor. Brownian motion has found application in areas as diverse as particle physics, queueing, and finance.

Bumps

Brownian motion is based on a model of "diffusion" in which a particle gets bumped again and again. We restrict ourselves to one-dimensional diffusion; each bump displaces the position of the particle by a small amount, randomly, to the left or the right. Specifically, in a one-dimensional **diffusion** process, the position of a particle is altered by a sequence of *bumps* that have these properties:

- Many bumps occur in each unit of time.
- The intervals of time between successive bumps can be constant or random. If these intervals are random, they are i.i.d., and they have the exponential distribution.
- Each bump displaces the particle by a random amount, which can be to the right or the left. These displacements are i.i.d. random variables.
- The cumulative displacement of the particle is the sum of the individual displacements.

Let us interpret the present moment as time 0. Let us designate as $X(0)$ the position of the particle at this moment, which is time 0. Similarly, for any future time $t > 0$, we designate as $X(t)$ the position that the particle will occupy at time t. When t is positive, $X(t) - X(0)$ is the sum of many random displacements and hence is a random variable. The term **Brownian motion** describes the net displacement $X(s + t) - X(s)$ that will occur between time s and time $s + t$.

Since $X(s + t) - X(s)$ is the sum of many random displacements that are i.i.d., the Central Limit Theorem guarantees that $X(s + t) - X(s)$ has the normal distribution. And since the displacements are i.i.d. and occur at a uniform rate, the mean and variance of $[X(s + t) - X(s)]$ grow linearly in t. This motivates the:

Theorem (Diffusion). A one-dimensional diffusion has these properties:

1. For all $t \geq 0$ and all $s \geq 0$, the random variable $[X(s + t) - X(s)]$ is normal with mean μt and variance $\sigma^2 t$.
2. The random variables $[X(d) - X(c)]$ and $[X(b) - X(a)]$ are independent of each other whenever $d > c \geq b > a \geq 0$.

Part 1 of the diffusion theorem states that the aggregate displacement $[X(s + t) - X(s)]$ between times s and $s + t$ is normal, with mean and variance that grow linearly in t. Moreover, this displacement is independent of the position of the particle at time s. Part 2 indicates that the displacements in nonoverlapping intervals of time are independent of each other.

A remarkable feature of one-dimensional Brownian motion is that it is governed by two parameters, μ and σ. The diffusion theorem demonstrates that if you know μ and σ, you know all that can be known about this process. In Brownian motion, μ is naturally called the **drift rate**.

Stationary Independent Increments

In Brownian motion, the displacement $X(s + t) - X(s)$ is called the **increment** between times s and $s + t$, with $s \geq 0$ and $t \geq 0$. A random process $X(t)$ is said to have **stationary increments** if the distribution of the increment $X(s + t) - X(s)$ is independent of $X(s)$. A random process $X(t)$ is said to have **independent increments** if the increments in nonoverlapping intervals are mutually independent. The diffusion theorem shows that Brownian motion has stationary independent increments.

Other processes have stationary independent increments. Suppose, for instance, that the inter-arrival times at a service facility are exponential with parameter λ. The number $N(t)$ of arrivals by time t also has stationary independent increments.

Very Frequent Bumps

Brownian motion inherits its parameters μ and σ from the tiny displacements that it sums. To see how, we assume that exactly n bumps (displacements) occur in each unit of time. These n displacements are i.i.d., and we let the random variable Y_n have their common probability distribution.

The mean and variance of the total displacement during one unit of time must equal μ and σ^2. The mean and variance of this sum equals the sum of the means and variances, so $\mu = nE(Y_n)$ and $\sigma^2 = n\,\text{Var}(Y_n)$. Thus, the individual displacements, whose distribution is that of Y_n, must have

$$E(Y_n) = \mu/n \quad \text{and} \quad \text{StDev}\,(Y_n) = \sigma/\sqrt{n}. \tag{11.58}$$

Table 11.12 presents a two-outcome distribution of Y_n whose expectation and standard deviation satisfy Equation (11.58). (Problem 30 on page 398, describes another two-outcome distribution that satisfies Equation (11.58).)

Our presentation of the diffusion theorem was *imprecise*. We had described it as the result of many bumps per unit time. The diffusion theorem actually describes the limit, as n approaches infinity, of a displacement process like the one in Table 11.12.

Table 11.12 Distribution of each displacement Y_n in a diffusion whose bumps occur at the rate of n per unit time.

Probability	Value of Y_n
0.5	$\mu/n + \sigma\sqrt{1/n}$
0.5	$\mu/n - \sigma\sqrt{1/n}$

Equation (11.58) and Table 11.12 indicate that Y_n has a "drift" component that is proportional to $1/n$ and an "uncertain" component that is proportional to $\sqrt{1/n}$. The ratio $\text{StDev}(Y_n)/E(Y_n)$ equals, $(\sigma/\mu)\sqrt{n}$ which approaches *infinity* as n gets large. As n increases, the drift component approaches zero more rapidly than does the uncertain component, and the mathematics becomes delicate, to put it mildly. For our purposes, it suffices to think of the time t as large enough that the aggregate of many displacements between time 0 and time t has become normal.

The Normal versus the Binomial

The normal distribution emerges in the limit, as n approaches infinity. Both in theory and in practice, it can be simpler not to pass to the limit. Suppose that exactly one bump occurs at each unit of time. In Table 11.12, each displacement takes the value $\mu + \sigma$ with probability 0.5 and the value $\mu - \sigma$ with probability of 0.5. We call a displacement an *uptick* if it takes the value $\mu + \sigma$ and a *downtick* if it takes the value $\mu - \sigma$. In t units of time, the number $U(t)$ of upticks is binomial with parameters $n = t$ and $p = 0.5$, and the number of downticks equals $t - U(t)$. Thus, the position $X(t)$ of the particle at time t is given by

$$\begin{aligned} X(t) &= X(0) + (\mu + \sigma)U(t) + (\mu - \sigma)[t - U(t)] \\ &= X(0) + (\mu - \sigma)t + 2\sigma U(t). \end{aligned} \tag{11.59}$$

Equation (11.59) gives a binomial approximation to the normal. If Equation (11.59) is used, t should be large enough that $X(t)$ is approximately normal. Otherwise, the distribution is dependent on that of the individual displacements, which, as we have observed (by reference to Problem 30, page 398, is ambiguous.

A Financial Asset

Brownian motion is the basis for a family of models of market prices of financial assets. One can imagine that the market price $W(t)$ of an asset at time t reflects many small displacements,each of which changes the price by a random amount whose expectation is very small.

Many researchers and practitioners in finance question the assumption of additive displacements. They prefer a model in which the market price $W(t)$ of the asset at time t reflects many small displacements, each of which *multiplies* the price by a random factor that is close to 1.0. Specifically, if the bumps occur at each unit of time, then

$$W(t) = W(0)Y_1Y_2 \cdots Y_t, \tag{11.60}$$

where the Y_k's are i.i.d. and positive, with expectation that is approximately 1.0. To focus on the individual displacements, we rewrite Equation (11.60) as

$$W(t) = W(0)Y(t) \quad \text{with} \quad Y(t) = Y_1Y_2 \cdots Y_t. \tag{11.61}$$

In the preceding section, we saw that for large t, the product $Y(t) = Y_1Y_2 \cdots Y_t$ of t positive random variables is lognormal with parameters $t\mu$ and $t\sigma$, where

$$\mu = E[\ln(Y_1)] \quad \text{and} \quad \sigma = \text{StDev}[\ln(Y_1)] \tag{11.62}$$

Thus, the lognormal is the natural model for the aggregate displacement $Y(t)$ of a financial asset, the aggregate being the result of many bumps, which multiply the price by factors that are random, positive, independent, and identically distributed.

This multiplicative model has a discrete analog, just as the additive one did. Let us suppose that exactly one bump occurs at each unit of time. Each displacement multiplies the price of the asset by $e^{\mu+\sigma}$ with probability 0.5 and multiplies the price of the asset by $e^{\mu-\sigma}$ with probability of 0.5. Again, we call a displacement an *uptick* if it takes the value $e^{\mu+\sigma}$ and a *downtick* if it takes the value $e^{\mu-\sigma}$. In t units of time, exactly t displacements occur. Of these displacements, $U(t)$ are upticks and $[t - U(t)]$ are downticks, where $U(t)$ is binomial with parameters $n = t$ and $p = 0.5$. Thus, the market price $W(t)$ at time t is given by

$$\begin{aligned} W(t) &= W(0)e^{(\mu+\sigma)U(t)}e^{(\mu-\sigma)[t-U(t)]}, \\ &= W(0)e^{(\mu-\sigma)(t)}e^{(2\sigma)U(t)}. \end{aligned} \tag{11.63}$$

Equation (11.63) is not new; its logarithm describes the binomial model in Equation (11.59).

Volatility

Using a diffusion model to describe the fluctuation of an asset price can seriously underestimate its volatility. As noted in Table 11.11, the normal has a "skinny" tail. This difficulty is only slightly ameliorated by using a multiplicative model, as it amounts to a deformation of the time scale.

Historically, individual asset prices have undergone huge fluctuations, as have the aggregate stock market averages. During the 1980s, for instance, 40% of the change in the close-of-day S&P stock index occurred on *10* trading days. In a diffusion process, this level of turbulence would occur once in a million years, if then.

In brief, using a diffusion process to model equity prices works well during tranquil epochs and catastrophically during turbulent times, which do occur. Needless to say, perhaps, investing on the "wrong side" of a diffusion model during a turbulent time can be a certain path to financial ruin.

As Benoit Mandelbrot points out, the basic defect in Brownian motion is that $\text{StDev}(Y_n)/E(Y_n)$ proportional to n^p for $p = 0.5$. (See Equation (11.58).) He points out that this is *insufficiently* volatile, and he proposes a fractal with $p > 0.5$. He also proposes a fractal on the time axis, to measure the volatility in the daily trading volume. For a scientist, the goal would be to fit these fractals to market conditions.

11.16. RETAILING AND ASSET PRICING

This section presents a pair of applications or, more precisely, the same application in two different settings. These applications illustrate the use of a spreadsheet to approximate a computation that cannot be done with college-level calculus. Let us consider:

Problem G (A Retailer)

Imagine that you are a retailer. At the beginning of the day you possess 120 units of inventory of a particular item. You will receive no more until tomorrow. The demand D for this product that will occur today is uncertain. Its distribution is normal with expectation of 90 units and standard deviation of 40 units. You will satisfy as much of today's demand as you can. The demand that you cannot meet will vanish; it will not wait for you to restock. The excess, if any, of D over 120 equals the number of **lost sales**, namely, the number of sales opportunities that you were forced to forego due to insufficient inventory. Compute the mean and standard deviation of the number of lost sales.

Problem H (An Investor)

From a specific individual, you have purchased a **European call**, which has these characteristics:

- At a precise time (say 90 days in the future), you will have the option to buy from that person 100 shares of Company M's stock at a price of \$120 per share. (This is called the **striking** price.)
- The market price S of Company M's stock at the moment you can exercise your option is uncertain. Its distribution is normal, with a mean of \$90 per share and a standard deviation of \$40 per share.

If $S \leq \$120$, you will let the option expire because you could purchase the shares more cheaply at the market. On the other hand, if $S > \$120$, you will exercise your option, sell the shares at market, and earn $(S - \$120)100$, less transaction costs, which we are ignoring. Compute the mean and standard deviation of the income that you will receive from this European call. (Incidentally, an **American call** differs only in the fact that you can exercise it at any moment up to the expiration date.)

Problems G and H are virtually identical. Did you notice that? We will study the retailer's problem and transcribe the results to the investor. For specificity, we have assumed that the random variables D and S have the normal distribution. The general approach in this section applies to any probability distribution, be it normal, lognormal, Poisson, binomial, or whatever.

The Positive Part

First, let us present some notation. Let q denote the inventory position at the start of the period: $q = 120$. The number of lost sales equals the excess, if any, of D over q. To describe this excess, we introduce the function $(x)^+$ of the number x, where

$$(x)^+ = \begin{cases} x & \text{if } x \geq 0 \\ 0 & \text{if } x < 0 \end{cases}.$$

In the inventory context, the demand D that will occur for this item is random, you have q units, and you will receive no more until tomorrow. Note that

$$(D - q)^+ = \begin{cases} D - q & \text{if } D \geq q \\ 0 & \text{if } D < q \end{cases}.$$

In other words, the random variable $(D - q)^+$ equals the number of lost sales opportunities. Problem G asks us to compute the mean and standard deviation of the random variable $(D - q)^+$.

Two other random variables that are also of interest to you, the retailer, are:

$(q - D)^+ =$ the number of units that remain on hand at the end of the day.

$\min\{D, q\} =$ the number of units that you will sell today.

These three quantities are related to each other. To verify that

$$D = \min\{D, q\} + (D - q)^+, \tag{11.64}$$

check that Equation (11.64) holds in the case $D \le q$ and in the case $D > q$. To see that

$$(D - q)^+ = (q - D)^+ + D - q, \tag{11.65}$$

check that Equation (11.65) holds in the case $D \le q$ and in the case $D > q$. Equations (11.64) and (11.65) hold for any numbers D and q. We are employing these equations for the case in which D is a random variable and q is a specific number, that is, $q = 120$. After calculating the mean and variance of any one of these quantities, we could use Equations (11.64) and (11.65) to calculate the mean and variance of the others.

The Expectation of $(D - q)^+$

We wish to compute the expectation of the function $(D - q)^+$ of the continuous random variable D. We will do this in two ways. One method works for the normal distribution, and the other works for any distribution.

The diskette that accompanies this text contains a few Excel add-in functions. The Appendix tells how to copy them into your library and activate them. The function **=NL(q, μ, σ)** calculates $E[(D - q)^+]$ for the case in which D is normal with mean μ and standard deviation σ.

For Problem G, the random variable D has $\mu = 90$ and $\sigma = 120$. We wish to compute $E[(D - 120)^+]$. To do so, we select a cell, enter

=NL(120, 90, 40) (11.66)

into that cell, and observe that the number 5.246679 appears there. Thus,

$$E[(D - q)^+] = 5.25, \tag{11.67}$$

when rounded to three significant digits.

The Standard Normal Loss Function

For the reader who wishes to know how this computation works, we report the following: Let the random variable D have the normal distribution with parameters μ and σ. An exercise in calculus (that we omit) demonstrates that the expectation of $(D - q)^+$ can be found from the equations:

$$z = \frac{q - \mu}{\sigma}, \tag{11.68}$$

$$L(z) = \frac{e^{-0.5z^2}}{\sqrt{2\pi}} - z[1 - Pr(Z \le z)], \tag{11.69}$$

$$E[(D - q)^+] = \sigma L(z), \tag{11.70}$$

where Z has the standard normal distribution. The function $L(z)$ of z in Equation (11.69) is known as the **standard normal loss function**.

To illustrate Equations (11.68), (11.69), and (11.70), we apply them to the case in which D is normal with mean $\mu = 90$ and standard deviation $\sigma = 40$. With $q = 120$, Equation (11.68) computes $z = (120 - 90)/40 = 0.75$. To compute $L(0.75)$, we could place the number 0.75 in a cell and, say, cell B2, select some other cell, and type into that cell the (ghastly) function that Equation (11.69) represents, namely,

=EXP(-B2*B2/2)/SQRT(2*PI()) - B2*(1-NORMDIST(B2, 0, 1, TRUE))

and learn that $L(0.75) = 0.131167$. Then, we use Equation (11.70) to get $E[(D - 120)^+] = 40L(0.75) = 5.246679$. The Excel function **=NL(120, 90, 40)** automates this calculation.

A **standard normal loss table** is a table that reports the values of $L(z)$ for various values of z. In an earlier era, this table was important. Spreadsheet computation makes it obsolete.

Expected Sales and End-of-Period Inventory

Since the expectation of the sum equals the sum of the expectations, Equation (11.64) shows us that with $q = 120$ and our demand distribution D, the expectation of the end-of-day stock position is given by

$$\begin{aligned} E[(120 - D)^+] &= 120 - E(D) + E[(D - 120)^+], \\ &= 120 - 90 + 5.25 = 35.25. \end{aligned}$$

For the retailer in Problem G, the expectation of the end-of-period inventory is 35.25 units.

Similarly, Equation (11.65) shows us that the expectation of the number of sales is given by

$$\begin{aligned} E(\min\{D, 120\}) &= E(D) - E[(D - 120)^+], \\ &= 90 - 5.25 = 84.75. \end{aligned}$$

The preceding equations obtain the expectation of the number of sales and the safety stock from the expected number of lost sales.

The Investor

For the investor in Problem H, the situation is precisely the same, except that all computations have been made on a per-share basis. With k as the striking price, the net profit that you can earn from the option equals $100(S - k)^+$. Equation (11.67) shows that its expectation equals \$525, when rounded to the nearest dollar. In finance, the **value** of a call equals the expectation of the profit that a person who owns it can obtain. The value of this European call equals \$525. A risk-neutral decision maker is indifferent to buying this call at its value, which equals \$525.

The Standard Deviation

The number X of lost sales is the function $X = (D - 120)^+$ of the random variable D. Problem G asks for the standard deviation of X as well as its mean. And Problem H asks for the standard deviation of $100(S - k)^+$, which has the same distribution as $100X$. The normal loss function helps us to compute the mean but not the standard deviation. Not to worry. We will deploy a method that should now be familiar to you. It is to approximate X by a discrete random variable W, and compute $E(W)$ and $E(W^2)$, thereby obtaining (discrete) approximations of $E(X)$ and $E(X^2)$. The fancy phrase for this method is **numerical integration**.

A Discrete Approximation

On a spreadsheet, it is a simple matter to approximate the continuous random variable $X = (D - 120)^+$ by a discrete random variable W and compute the mean and variance of W. To do so, we:

- Ignore all values of D that are below 120 because these values assign X the value of zero.
- Ignore all values of D that are at least four standard deviations above its mean because of the miniscule probability D is this large.
- Divide the remainder of the range of D into intervals that are narrow enough that the density of D does not curve a great deal within each interval.
- Use the normal CDF to compute the probability that D falls into each interval.
- For each interval into which D can fall, set $W = m - 120$ where m is the midpoint of the interval.
- Use the SUMPRODUCT function to compute $E(W)$ and $E(W^2)$ and, from them, StDev(W).

A spreadsheet that accomplishes this is presented as Tables 11.13 and 11.14. This spreadsheet contains only four elements of data, which are in cells B6 through B9. The number 0.1 in cell B9 indicates that we are dividing the range of D into intervals whose width equals 0.1 standard deviations. The functions at the top of this spreadsheet compute the width of each interval (which equals 4), the value of D at its midpoint (which equals 122, 126, . . . ,), the value that the interval assigns to W (which equals 2, 6, 10, . . .), and the probability that D falls into the interval (which is found from the normal CDF).

Table 11.14 displays the bottom of the same spreadsheet. Rows 48–53 of this spreadsheet describe functions that it uses to compute the probability that X is positive, the expectation of W, the expectation of W^2, the variance of W, the standard deviation of W, and the expectation of W given that W is positive. Cell B44 reports $E(W) = 5.26$, which differs only slightly from the value $E(X) = 5.25$ that we had computed using the normal loss function. Cell B47 reports StDev(W) $= 13.33$, which we could not obtain from the normal loss function.

Table 11.13 Top of a spreadsheet that approximates $X = (D - 120)^+$ by the discrete random variable W.

	A	B	C	D
1	Cell B10 contains =**B7*B9**		Cell A13 contains =**B8+B10/2**	
2	Cell A14 contains =**A13+B$10**		Cell B13 contains =**A13-B$8**	
3	Cell C13 contains =**NORMDIST(A13-B$10/2, B$6, B$7, TRUE)**			
4	Cell D13 contains =**C14-C13**			
5				
6	mu =	90		
7	sigma =	40		
8	q =	120		
9	grid width =	0.1		
10	interval width =	4		
11				
12	**midpoint**	**discrete approx. W**	**CDF at lower end point**	**probability**
13	122	2	0.7734	0.0290
14	126	6	0.8023	0.0266
15	130	10	0.8289	0.0242

Table 11.14 Bottom of a spreadsheet that approximates $X = (D-120)^+$ by the discrete random variable W.

	A	B	C	D
41	234	114	0.9998	0.0001
42	238	118	0.9999	0.0001
43	P(X > 0) =	0.2266		
44	E(W) =	5.26	E(W | X > 0) =	23.19
45	E(W^2) =	205.24	E(W^2 | X > 0) =	905.65
46	Var(W) =	177.62	Var(W | X > 0) =	783.76
47	StDev(W) =	13.33	StDev(W | X > 0) =	58.81
48				
49	Cell B43 contains **=SUM(D13:D42)**			
50	Cell B44 contains **=SUMPRODUCT(B13:B42, D13:D42)**			
51	Cell B45 contains **=SUMPRODUCT(B13:B42, B13:B42, D13:D42)**			
52	Cell B46 contains **=B45 - B44*B44**			
53	Cell B47 contains **=SQRT(B46)**		Cell D44 contains **=B44/B$43**	

The European Call

Evidently, the European call in Problem H is a risky proposition. Table 11.14 indicates that the income that the investor earns from it has expectation of approximately \$526 and standard deviation of approximately \$1,333. Hence, it has a ratio of standard deviation to mean of approximately 2.6 = 1,333/526. A risk-averse decision maker would not pay \$526 for this call.

In business, calls and other options can reduce a company's exposure to uncertainty. Suppose, for instance, that an American company needs to pay 2 million Eurodollars 90 days in the future. If this company is concerned about losses due to adverse currency fluctuations, it might purchase a call for 2 million Eurodollars 90 days in the future. Purchasing this call eliminates the possibility of a big loss due to untoward currency fluctuation.

11.17. REVIEW

A continuous random variable assigns probability to intervals, not to points. The way in which this occurs is determined by the random variable's density function and by its CDF. Specifically, the probability that a continuous random variable takes a value between the numbers a and b equals:

- The area under its density function $f(t)$ between a and b.
- The quantity $F(b) - F(a)$, where $F(t)$ is this random variable's CDF.

Here, as in Chapter 8, the emphasis has been on the expectation as a measure of central tendency and on the variance as a measure of spread.

The expectation of a continuous random variable Y is found by making a discrete approximation W to Y, computing $E(W)$, and setting $E(Y)$ equal to the limiting value of $E(W)$ as the quality of the approximation gets better and better. The quality of these approximations improves if we use the midpoint of the interval rather than either end-point. Without passing to the limit, it is an easy matter to compute $E(W)$ on a spreadsheet. This computation and others like it let us approximate quantities that are beyond the methods of college-level calculus.

Five common models of continuous probability distributions have been discussed in this chapter. The uses of each are restated as follows.

- The uniform distribution is a natural model of an uncertain quantity that is known to fall within a certain range but about which nothing else is known.
- The triangular distribution describes an uncertain quantity that is known to fall within a certain range and has a most likely value within that interval.
- The exponential distribution describes an uncertain quantity that has the memoryless property. The exponential is the natural model for inter-arrival times at a service facility.
- The normal distribution is a convenient model of a random variable whose distribution is symmetric and roughly bell-shaped. The sum of sufficiently many i.i.d. random variables has the normal distribution. The sum can be approximately normal even if the individual random variables have different distributions. For these reasons, the normal distribution arises in measurement, manufacturing, marketing, inventory control, aggregate income, and finance.
- The lognormal distribution is a convenient model of a random variable whose value must be positive and whose distribution is skewed. The product of sufficiently many positive i.i.d random variables is lognormal. The lognormal distribution is a popular model of the market price of a financial asset, which is thought to evolve as the product of many random displacements, but the lognormal distribution may understate the volatility of this price.

In closing, we point to a difference between our presentations of discrete and continuous random variables. In our discussion of discrete random variables, each outcome is assigned a probability and the random variable is an attribute of the outcome. For example, let X be the number of heads in the first three tosses of a fair coin, and let Y be the number of heads that appear before the first tails. One outcome is (H, T, H). This outcome occurs with probability of $1/8 = (1/2)^3$. It assigns X the value 2, and it assigns Y the value 1. This example illustrates the way in which each outcome assigns values to several random variables, which may or may not be independent.

By contrast, in our discussion of a continuous random variable X, probability has been assigned *directly* to the intervals in which X can lie. We have provided no proper foundation on which to describe dependence or independence among continuous random variables. Why did we fail to lay this foundation? It would be abstruse, it would ensnare us deeply in mathematics, and it would add little to your ability to use continuous random variables. In place of the foundation, we have relied on your intuition, particularly on your ability to reason by analogy with the discrete case.

11.18. HOMEWORK AND DISCUSSION PROBLEMS

1. Draw the density function of the random variable V that is uniformly distributed on the interval between 3 and 9.

2. Let T have the symmetric triangular density, with parameters a, b, and $m = (a + b)/2$.

(a) For each number t between a and m, the probability $P(T \leq t)$ equals the area of a triangle. What is this triangle's height? What is its area?

(b) For each number t between m and b, the probability $P(T \geq t)$ equals the area of a triangle. Compute this probability.

(c) For each value of t, specify the CDF $F(t) = P(T \leq t)$ of this random variable.

3. Adapt Figure 11.5 to show, graphically, why approximating at midpoints improves on the accuracy of approximation by endpoints.

4. Figure 11.6 specifies the CDF of a random variable X. Plot the slope of this CDF as a function of t. Do you recognize this slope?

5. Suppose that 12 customers arrived during the past hour and that their arrival times were independent and uniformly distributed over the past hour. Compute the probability $P(t)$ that all 12 customers arrived at least t minutes ago. *Remark:* If inter-arrival times are exponential with a constant rate, then, given a number (such as 12) of arrivals during a specific period of time (such as one hour), each of these arrivals *is* uniformly distributed over that period.

6. For the random variable W in Table 11.1, use a spreadsheet to compute $E(W)$ and $E(W^2)$. On this spreadsheet, calculate $\text{Var}(W) = E(W^2) - [E(W)]^2$. Repeat the calculation with the interval between 2 and 5 broken into 60 smaller intervals, each of whose width is 0.05.

7. (**A Symmetric Density**) Let Z be a continuous random variable whose density function is symmetric about the number m. Pick any positive number Δx, and approximate Z by a discrete random variable W whose intervals have width Δx and are centered at m, at $m + \Delta x$, at $m - \Delta x$, at $m + 2\Delta x$, at $m - 2\Delta x$, and so forth.

(a) Show that $E(W) = m$.

(b) Conclude from part (a) that $E(Z) = m$.

8. The time you will wait in a bank line is exponential with a mean of 10 minutes. Compute:

(a) The probability that you will wait not more than 2 minutes.

(b) The probability that you will wait not more than 8 minutes.

(c) The probability that you will wait between 5 and 15 minutes.

(d) The probability that you will wait at least 12 minutes.

(e) The probability that you will wait not more than 8 more minutes if you are not served in the first 10 minutes.

9. The median of an exponential distribution equals 10 minutes. What is its mean?

10. Customers arrive at a service facility independently of each other, at the rate of 20 per hour. Compute:

(a) The probability that the next customer will appear within 2 minutes.

(b) The probability that the next customer will not appear within 4 minutes.

(c) The mean time between the arrival of successive customers.

(d) The median time between the arrival of successive customers.

11. (**Selection Bias**) The intervals of time between the moments at which successive buses pass a particular corner are exponential with a mean of 15 minutes. People queue up at this corner at the rate of 1 per minute, waiting for the next bus. You are one of these people. At a particular moment in time, you arrive at the corner and join the queue. The following is a fact (that we have not proved) about exponential inter-arrival times; the time between the moment the prior bus passed the corner and the moment that you appear at the corner (also) has the exponential distribution with a mean of 15 minutes. Let T denote the time that elapses between the prior bus and the one you get on.

(a) True or false; T has the exponential distribution with a mean of 15 minutes.

(b) True or false: $T = V + W$ where V and W are independent random variables whose distributions are exponential, each with a mean of 15 minutes.

(c) True or false: The expectation of the number of people who get on the average bus equals 15.

(d) True or false: The expectation of the number of people who get on your bus equals 30.

(e) Comment on the consistency between your answers to parts (c) and (d).

12. **(Selection Bias, continued)** Your supermarket has 12 check-out lines. To analyze congestion, you want to learn the probability distribution of the "check-out time" T, this being the length of time that it takes a server to serve a customer by tallying up the customer's order, bagging it, and getting paid. Each server's computer records the time at which each tally starts. To find a representative sample of check-out times, you visit each of the 12 lines at predetermined moments in time. If a line's server is engaged, you remain until the service is complete, and record the line's number and the time of completion. Later, from the line's computer, you determine the time at which service commenced, and you take as the check-out time the difference between the two.

(a) Argue for or against this proposition; the check-out times you record are representative of T.

(b) If, in part (a), you argued against the proposition, revise your sampling plan to obtain check-out times that are representative of T.

13. The annual income of graduates from college H in their first year of employment is normally distributed with mean of $45,000 and standard deviation of $10,000. A graduate is picked at random. Compute the probability that this graduate earns:

(a) Less than $35,000.

(b) Between $35,000 and $45,000.

(c) Between $45,000 and $60,000.

(d) Exactly $46,000.

(e) Greater than $65,000.

14. The annual income of graduates from college H in their first year of employment is normally distributed with mean of $45,000 and standard deviation of $10,000. Find:

(a) The 0.1 fractile of this distribution.

(b) The 0.95 fractile of this distribution.

15. **(Graph of the Normal Density and CDF)** Open a spreadsheet. In Column A, use a drag to insert the numbers $-6.0, -5.9, \ldots, +5.9, +6.0$.

(a) In columns B, C, and D, respectively, record the density of the normal probability distribution with mean of 0 and with standard deviation of 0.5, 1.0, and 2.0. Use an Excel chart to plot these densities.

(b) In columns E, F, and G, respectively, record the CDFs of the same three normal distributions. Use an Excel chart to plot them.

16. The random variable X has the normal distribution with $\text{StDev}(X) = 40$ and with $P(X \leq 100) = 0.15$. What is the expectation of X?

17. Suppose that 40% of all families own at least two cars. Out of a population of 10,000 families, what is the probability that at least 5000 families own two or more cars?

18. The random variables X and Y are independent, and their distributions are normal, with $E(X) = 1000$, $E(Y) = 900$, $\text{StDev}(X) = 300$, and $\text{StDev}(Y) = 400$.

(a) What can you say about the random variable $(X - Y)$?

(b) Compute $P(X < Y)$.

19. In the Gas Guzzler tax problem (Problem E within the chapter), suppose that the U.S. government offers to select 12 newly manufactured vehicles at random (rather than 6), test them, and charge you the gas guzzler tax if their average mpg is below 27.

(a) Intuitively, why do you prefer that they test 12 vehicles?

(b) Re-compute the mean mpg μ that reduces to 1 chance in 100 the probability of having to pay the gas guzzler tax, again with $\sigma = 1.8$ mpg. Comment on how the "margin of safety" has changed.

20. (**In Control**) An industrial process is said to be **in control** if it is operating as designed. If this process is thought to be out of control, production ceases while a team of engineers determines what is wrong and fixes it. While a process is in control, random fluctuations occur, but they are small, and they have known characteristics. While an automotive manufacturing process is in control, the width W of a certain gap is normally distributed with a mean of 2 millimeters and a standard deviation of 0.2 millimeters, and the gaps on different vehicles are independent. To indicate whether or not the process is in control, a sample of five vehicles is selected, and the gap on each of these vehicles is measured. If the average of their gaps differs from 2 millimeters by an amount whose absolute value exceeds K, the process is declared to be out of control, and the engineers are summoned. The smaller the value of K, the greater the danger that the engineers are summoned while the process is actually in control. The larger the value of K, the greater the danger that the engineers are not summoned while the process is out of control.

(a) Suppose that the probability distribution of W was roughly symmetric and roughly bell-shaped, but not exactly normal. Would it still be reasonable to use the normal distribution for this problem? Why or why not?

(b) Determine the number K such that the chance is only 1 in 100 that the engineers are summoned while the process is in control.

21. (**In Control, continued**) In the preceding problem, suppose (as seems natural) that gaps cannot be measured exactly. The error E in measurement of each gap is normally distributed with a mean of 0 millimeters and a standard deviation of 0.1 millimeters. Errors in the measurement of gaps are independent of each other and of the gap itself. Hence, while the process is in control, each sample measures $W + E$ rather than W. The sampling protocol is exactly as it was in the preceding problem, with the same goal, which is to determine the number K such that there is only 1 chance in 100 that the engineers are summoned while the process is in control.

(a) In words, say why the value of K should be larger or smaller than in the preceding problem.

(b) Compute K.

22. (**Planned Replacement**) In its classrooms, a college has 120 fluorescent light bulbs. Its classrooms have high ceilings, and these bulbs are easier and cheaper to replace all at once than one at a time. For this reason, the college has a policy of **planned replacement**: On September 1 of each year, the college replaces all 120 bulbs. In addition, each bulb that fails during the year is replaced immediately. On a per-bulb basis, unplanned replacement is five times as costly as planned replacement. Fluorescent bulbs do not age until they are put in service. Once a bulb is put into service, its lifetime has a normal distribution with a mean of 400 days and a standard deviation of 100 days.

(a) Consider a particular socket. What is the probability that the bulb in this socket will fail before planned replacement?

(b) Consider the same socket. What is the probability that both the bulb in this socket and its replacement will fail before planned replacement? *Hint:* The sum of two i.i.d. normal random variables is ____.

(c) Consider the same socket. What is the probability that the bulb in this socket, its replacement, and that bulb's replacement will all fail before planned replacement?

(d) Let X denote the number of unplanned replacements in a particular socket that will occur during the year. Specify $P(X \geq k)$ for $k = 1, 2$, and 3, and specify $P(X = k)$ for $k = 0, 1, 2$, and 3. Compute $E(X)$ and $\text{Var}(X)$.

(e) Let T denote the total number of unplanned replacements that the college will need to make in all 120 sockets. What can you say about T?

(f) How many spare bulbs should the college stock so as to have only 1 chance in 10 that they run out of spares between planned replacements?

(g) The college wishes to minimize its expected annual expenditure on bulb replacement. How frequently should it do planned replacement? (When solving this, ignore the inventory of bulbs they might need to keep on hand *between* planned replacements.)

23. (**Normal Approximation to the Binomial**) Use Excel (and, if you choose, Solver) to build the analog of Table 11.8 for the values $p = 0.2, 0.3, 0.4$, and 0.5. Repeat for Table 11.9.

24. (**Normal Approximation to the Poisson**) On a spreadsheet, compute two Poisson distributions, one with parameter $\lambda = 2$, the other with $\lambda = 12$. On the same spreadsheet, compute the normal approximation to each distribution. Plot them. How close do they look? Describe the discrepancy between them.

25. (**Normal Approximation to the Poisson**) On a spreadsheet, reserve cell B2 for the parameter λ of the Poisson distribution. Place a positive number in that cell.

(a) In a column of your spreadsheet, compute the CDF of the Poisson whose parameter is in cell B2. In another column, compute the CDF of the normal approximation to this distribution. In a third column, take the difference of the former CDF over the latter. In a fourth column, take the difference of the latter CDF over the former.

(b) Use Solver to find the smallest value of λ such that gap between the CDF of the Poisson and the CDF of the normal approximation to this Poisson is at most 0.02. *Hint:* It helps to initialize Solver with a feasible (large) value of λ.

(c) Repeat part (b) with a gap of 0.01. *Hints:* (1) Again, initialize Solver with a feasible value of λ. (2) If necessary, omit constraints that will obviously be satisfied.

(d) In part (c), did the value of λ increase by a factor of roughly 4. If so, why do you suppose that happened?

26. (**Normal Approximation to the Poisson**) The preceding problems demonstrate, graphically, that the normal provides a good approximation to the Poisson when its parameter λ is large. Why must this be so? *Hint:* What do you know about the sum of independent random variables whose distributions are Poisson?

27. (**A Convolution**) Let S_n denote the total number of pips on n fair dice.

(a) On an Excel spreadsheet, compute the probability distributions of S_n for the values $n = 2, 3, 4, 5$, and 6. Plot it. *Big hint:* You might:

- Enter the numbers 1, 2, . . . , in cells A7, A8, ...
- Enter the number 1/6 in cells B1 through B6.
- Enter the number 1/6 in cells C7 through C12.
- Enter the function **=SUMPRODUCT($B$1:$B$6, C2:C7)** in cell C8.
- Drag this function across and down.

(b) Adapt part (a) to compute the same probabilities for the case in which each die has $P(6) = 1/5$ and $P(k) = 4/25$ for $k = 1, 2, 3, 4$, and 5. *Hint:* Change only two cells. *Remark:* Parts (a) and (b) compute the **convolution** of probability distributions.

28. The annual income of graduates from college H in their first year of employment has the lognormal distribution with mean of \$45,000 and standard deviation of \$10,000. A graduate is picked at random. Compute the probability that this graduate earns:

(a) Less than \$35,000.

(b) Betweeen \$35,000 and \$45,000.

(c) Between \$45,000 and \$60,000.

(d) Exactly \$46,000.

(e) Greater than \$65,000.

(f) Compare part (e) with the result you would obtain if the distribution was normal with the same mean and standard deviation.

29. (**A European Call**) Suppose that the market price S of a share of Company M's stock 90 days in the future has the lognormal distribution with mean of \$90 and standard deviation of \$40. You have purchased the option to buy 100 shares of Company M's stock exactly 90 days from now at a striking price of \$120 per share. Compute the mean and standard deviation of income you can earn from this option. *Hint:* Adapt the spreadsheet in Tables 11.13 and 11.14, but be aware that the lognormal can have a fatter tail than the normal.

Remark: The diffusion model in Table 11.12 has equal "uptick" and "downtick" probabilities, and it fits the amounts of the displacements to match the desired μ and σ. Some researchers and practitioners prefer a model with fixed displacements but unequal uptick and downtick probabilities. The next problem develops such a model.

30. **(Diffusion—Displacements with Unequal Probabilities)** Let Y be a discrete random variable that takes the value $+K$ with probability p and the value $-K$ with probability $(1 - p)$. For fixed numbers $\mu > 0$ and σ, we seek values of p and K such that $E(Y) = \mu$ and $\text{Var}(Y) = \sigma^2$.

(a) Show that these requirements are satisfied by setting $s = \sigma/\mu$,

$$p = \left(\frac{1}{2}\right)\left(1 + \frac{1}{\sqrt{1 + s^2}}\right) \quad \text{and} \quad K = \mu\sqrt{1 + s^2}.$$

(b) Now suppose that Y equals the sum of n i.i.d. random variables Y_1 through Y_n, that Y_1 is the discrete random variable that takes the value $+K_n$ with probability p_n and the value $-K_n$ with probability $(1 - p_n)$, and that these values are selected so that $\mu = nE(Y_1)$ and $\sigma^2 = n\,\text{Var}(Y_1)$. Show that these requirements are satisfied by

$$p_n = \left(\frac{1}{2}\right)\left(1 + \frac{1}{\sqrt{1 + ns^2}}\right) \quad \text{and} \quad K_n = \frac{\mu}{n}\sqrt{1 + ns^2}.$$

(c) As n becomes large, does this model succeed in keeping the uptick and downtick probabilities unequal?

31. **(Normal Loss Function)** At this moment, your only asset is \$200 in your checking account. The revenue R that you will earn this month is normally distributed with a mean of \$2500 and a standard deviation of \$600. The expense S that you will incur this month is normally distributed with a mean of \$2300 and a standard deviation of \$800. You can defer expenses to the end of the month but not beyond. Revenue and expense are independent random variables.

(a) What can you say about your net worth at the end of the month?

(b) If your net worth is negative, you will need to take a loan to restore it to zero. What is the probability that you will need to take a loan?

(c) What is the expectation of the amount that you will have to borrow?

(d) Given that you do have to borrow, what is the expectation of the amount that you will need to borrow?

32. **(Variance of the End-of-Period Stock)** Let D be a random variable, and let q be a fixed number. Define the random variables X and Y by

$$X = (D - q)^+ \quad \text{and} \quad Y = (q - D)^+.$$

The retailer's problem (Problem G within the chapter) presented a general approach to computing the $E(X)$ and StDev(X). The spreadsheets in Tables 11.13 and 11.14 compute $E(X)$ and $E(X^2)$ and StDev(X) for these data; D is normal with mean 90 and standard deviation 40, and q equals 120. Here, you will be asked to relate the mean and variance of Y to the mean and variance of X.

(a) Prove or disprove: $Y - X = q - D$.

(b) Prove or disprove: $E(Y) = E(X) + q - E(D)$.

(c) Prove or disprove: $XY = 0$.

(d) Prove or disprove: $E(Y^2) = E(X^2) + q^2 - 2qE(D) + E(D^2)$.

(e) Create a formula that describes Var(Y) in terms of the mean and variance of X and of D.

(f) For the retailer's problem (Problem G within the chapter), compute the variance of the end-of-day stock position. *Hint:* Use part (e) and information in Table 11.14.

Part D

Stochastic Systems

Chapter 12

Inventory

12.1. PREVIEW

The goal of a production and distribution system is to deliver a high-quality product to customers at low cost and in a way that adapts quickly to changing market conditions. **Inventory** appears at each stage of these systems; inventory can take the form of a stock of raw materials, subassembles, work in process, and finished goods. A well-designed system of inventories lubricates these systems, enabling them to run smoothly without interruption and without unnecessary expense or waste.

Modern Production Systems

In recent years, interest in inventory has reawakened as engineers and managers have learned that well-designed production and distribution systems are essential to successful competition. Innovations in these systems have been reported in the press, and several terms have entered common usage. A **pull** system is a replenishment scheme in which each element of the production chain is triggered by demand (e.g., by the withdrawal of a product from a shelf). A **just-in-time** system aims to coordinate inventories and production so that parts and subassemblies appear very shortly before they are needed. **Flexible manufacture** describes a system in which production of small batches is economically viable. **Supply chain management** describes the process of coordination by the participants in a production and distribution system. Modern production systems strive to be lithe—to adapt with seeming ease to changes in market conditions and to do so without a surfeit of inventory.

A revolution in supply chain management has been made possible by the laser scanner and by modern multiaccess computer systems. Prior to the laser scanner, inventory needed to be counted, which was inaccurate and expensive. Information was scarce and difficult to share. Now, laser scanners routinely track the flow of material through a production and distribution system. Modern computer systems allow each participant in the supply chain to know what is selling, what is not, and what actions must be taken to react to the customer's needs.

The Roles of Inventory

Inventory plays four distinct roles in a production and distribution system. First, a **cycle stock** describes inventory that exists because an item is produced intermittently or is consumed intermittently. Farm goods are produced during the summer but are consumed all year round. Toys tend to be produced throughout the year but are sold mainly during the holidays. In these cases and others, a cycle stock grows and dwindles periodically.

Second, **pipeline** inventory exists for items that are produced a distance from their markets and must be transported. Examples include natural gas that is piped from well to market, clothes that are made in Asia for sale in the United States, and crops that are grown in California for sale nationwide. An inventory of goods (e.g., lettuce) must be maintained in the "pipeline" in order for goods to reach the market at a prescribed rate.

Third, **safety stock** is an inventory whose existence is due to an inherent uncertainty in the demand or the supply. Manufacturers and suppliers maintain safety stocks, without which production could be interrupted for lack of a subassembly or raw material. Retailers maintain safety stocks, without which consumers' needs could not be satisfied when they occurred.

A fourth role for inventory is **speculative**, as a hedge against price changes. The need for speculative inventory has increased as manufacturing and distribution have become more international. Forward contracts for currency swaps are one example of this type of inventory.

The inventory that a consumer perceives in a store is, typically, the sum of a safety stock and a cycle stock. The cycle stock reflects the fact that it is economical to restock periodically. The safety stock buffers the uncertainty in demand. Models of the uncertainty in demand were built in Chapters 8 and 11. We have seen that the Poisson distribution is the natural model for the number of consumers who will demand goods during a given period of time and that the normal distribution is a natural model for the aggregate demand of many customers.

12.2. WHAT CAN YOU LEARN FROM THIS CHAPTER?

Effective control of inventory is a challenge. A chapter like this can introduce you to the subject but cannot make you an expert on pull systems, just-in-time production, or supply chain management, for instance. The focus in this chapter is on safety stock and cycle stock. Our aim is to use simple problems to introduce themes that pervade inventory control. Collectively, the problems in this chapter introduce you to the range of issues that inventory addresses as well as to the methods used to resolve these issues.

Marginal analysis is used again and again in this chapter. It is an incisive tool in any setting that exhibits increasing marginal cost or, equivalently, decreasing marginal return. This chapter could be viewed as training in how to think at the margin.

12.3. THE NEWSVENDOR

Why should you study a model whose financial stakes are as puny as that of the newsvendor? This model appears frequently and in many disguises, often with significant financial consequences. The insights that it provides pervade inventory control. For these reasons, it may be the most important model of inventories.

Yesterday's newspaper has scant value, in which sense newspapers are a perishable commodity. Each day, the newsvendor must buy his or her newspapers before the demand materializes. The **newsvendor** model is as follows:

- The demand D for an item is uncertain. The probability distribution of D is known, however.
- Before the demand D is observed, the newsvendor must decide on the number q of units of this item to purchase.
- The newsvendor pays w dollars for each unit that she purchases. She receives r dollars for each unit that she sells and s dollars for each unit that she cannot sell. These data satisfy $s < w < r$.
- The newsvendor aims to maximize the expectation of the net revenue that she earns.

Timing is crucial here. If the newsvendor could observe the demand D before placing her order, she would purchase exactly D units and earn a profit of $(r - w)D$.

The Tradeoff

The newsvendor must select her order quantity q before the demand D materializes. Selecting a low value of q increases the risk of failing to satisfy the demand, with a loss of profit of $(r - w)$ on each missed sales opportunity. Selecting a high value of q increases the risk of failing to sell all that she orders, with a loss of $(w - s)$ on each unsold unit. The newsvendor faces a tradeoff. She wishes to select the order quantity that maximizes her expected net revenue. We designate the number q, the random variable $Y(q)$, and the quantity $R(q)$ by:

$$\begin{aligned} q &= \text{the number of units that the newsvendor buys,} \\ Y(q) &= \text{the net revenue that the newsvendor earns if she buys } q \text{ units} \\ &\quad\ \text{from the wholesaler,} \\ R(q) = EY(q) &= \text{the expectation of the newsvendor's net revenue if she buys} \\ &\quad\ q \text{ units from the wholesaler.} \end{aligned}$$

She seeks to order the quantity q that maximizes $R(q)$, thereby striking a balance between the risks of purchasing too few and too many.

Opportunity Costs

In the newsvendor problem, the symbols C_u and C_o have special meanings:

$$\begin{aligned} C_u &= \text{the loss of profit for ordering one unit fewer than (\textit{under}) the demand,} \\ C_o &= \text{the loss of profit for ordering one unit more than (\textit{over}) the demand.} \end{aligned}$$

If the newsvendor orders one fewer than the demand, she loses the profit on that unit, so $C_u = r - w$. If the newsvendor orders one more than the demand, she must scrap that extra unit, so $C_o = w - s$. In the jargon of inventory control, the quantity C_u is called the **underage cost**, and C_o is called the **overage cost**, the subscripts u and o being short for "under" and "over" the demand.

A Recipe

The numbers C_u and C_o are data in a recipe for selecting the order quantity q that maximizes expected net revenue, $R(q)$. The ratio $C_u/(C_u + C_o)$ is known as the **critical fractile**.

One can show (and we shall) that the profit-maximizing order quantity is the smallest value of q for which $P(D \le q)$ is at least as large as the critical fractile. In brief:

> When the demand D is integer-valued, the expectation of net revenue is maximized by ordering q^* units where q^* is the smallest integer q for which
>
> $$P(D \le q) \ge \frac{C_u}{C_u + C_o}.$$

The boxed-in recipe will be justified later in this section. To illustrate it and to gain insight into it, we turn our attention to a true newsboy problem.

Art's Newsstand

Problem A (Art's Newsstand)

At his newsstand, Art sells the daily newspaper. He cannot foretell the number D of papers that he can sell in a given day. Based on past experience, Art has estimated the probability distribution of D. Table 12.1 records this probability distribution, along with its cumulative distribution function. Before Art's customers appear, he must buy his newspapers from the wholesaler at 30 cents apiece. For each paper that he sells, he receives 50 cents, its retail value. For each paper that he does not sell, he receives 2 cents, its scrap value. Art wants to maximize the expectation of the net profit that he makes from newspaper sales. How many newspapers should he buy each day? How much does he expect to earn? How sensitive is Art's expected net profit to the number q of newspapers that he purchases?

Problem A is the newsvendor model with these data: $w = 30$, $r = 50$, $s = 2$, and demand D whose probability distribution is specified in Table 12.1. If Art purchases one unit fewer than the demand, he loses the profit on that unit, so $C_u = (50 - 30) = 20$. If Art purchases one unit more than the demand, he must scrap that unit, so $C_o = (30 - 2) = 28$. Thus, for Problem A, the critical fractile is given by

$$\frac{C_u}{C_u + C_o} = \frac{20}{20 + 28} = 0.417.$$

From the CDF of demand in Table 12.1, we see that the smallest number q for which $P(D \le q)$ is at least as large as 0.417 equals 25. Thus, q^* equals 25. Art should order 25 newspapers.

What's Been Accomplished

The solution to the newsvendor model seems to be rather simple. It consists of these steps:

- Find the opportunity cost C_u of ordering one unit too few.
- Find the opportunity cost C_o of ordering one unit too many.
- From these opportunity costs, compute the critical fractile, $C_u/(C_u + C_o)$ and pick as q^* the smallest value of q for which $P(D \le q)$ is at least as large as $C_u/(C_u + C_o)$.

Table 12.1 Probability distribution and CDF of the demand for Art's newspapers.

x	19	20	21	22	23	24	25	26	27	28	29	30
$P(D = x)$	0.00	0.01	0.03	0.06	0.10	0.18	0.22	0.17	0.12	0.06	0.04	0.01
$P(D \le x)$	0.00	0.01	0.04	0.10	0.20	0.38	0.60	0.77	0.89	0.95	0.99	1.00

But our analysis of the newsvendor model is not complete. We have not answered the following questions:

- What is Art's expected profit $R(q^*)$ from ordering q^* newspapers?
- How sensitive (or insensitive) is Art's expected profit $R(q)$ to the number of papers he purchases?
- Why does the critical-fractile method work?

The third of these questions is particularly important. We will encounter the newsvendor model in disguise—that is, in settings where it isn't obvious what C_u and C_o should be and where it isn't even clear what random variable plays the role of D. In order to make good use of the newsvendor model, we will need to understand it thoroughly.

A Frontal Assault

To mount a frontal assault, we account for the net revenue $Y(q)$ that the newsvendor earns if she orders q units. To justify the equation

$$Y(q) = C_u \min(q, D) - C_o \max(0, q - D), \tag{12.1}$$

note that:

- The number of newspapers that she sells equals the smaller of q and D.
- She earns $C_u = r - w$ dollars on each paper that she sells.
- The number of newspapers that she scraps equals the excess, if any of q over D.
- She loses $C_o = w - s$ dollars on each paper that she scraps.

Thus, Equation (12.1) indicates that the newsvendor's expected net revenue $R(q)$ if she orders q papers is given by:

$$\begin{aligned} R(q) &= E[Y(q)] \\ &= E\{C_u \min(q, D) - C_o \max(0, q - D)\}. \end{aligned} \tag{12.2}$$

where the expectation in Equation (12.2) is taken with respect to the random variable D.

Equation (12.2) is concise but unwieldy. The value q^* that maximizes $R(q)$ is not apparent from this equation, for instance. Should you wish to pursue a frontal attack on Equation (12.2), please refer to Problem 9 on page 435, where a spreadsheet calculation of $R(q)$ is sketched.

A Marginal Analysis

We will set Equation (12.2) aside because this is an instance in which it is more insightful—and simpler—to think at the margin. Our marginal analysis studies the effect of increasing the order quantity from q to $q + 1$. Each newspaper that the newsvendor buys is increasingly likely to go unsold. Hence, the marginal profit from the $(q + 1)$th paper decreases with q. At some point, the marginal profit switches from positive to negative, and the value q^* at which the switch occurs is the most profitable order quantity. Thus, to find q^*, we will compute:

$$\begin{aligned} \text{EMP}(q \to q + 1) &= E[Y(q + 1) - Y(q)] \\ &= \text{the expectation of the marginal profit that the newsvendor} \\ &\quad \text{receives if she increases her order from the wholesaler} \\ &\quad \text{from } q \text{ to } q + 1 \text{ newspapers.} \end{aligned}$$

Evidently, the expected marginal profit, EMP($q \to q + 1$), relates to the expected net revenue $R(q)$ through

$$\text{EMP}(q \to q + 1) = R(q + 1) - R(q). \tag{12.3}$$

Equation (12.3) holds for a familiar reason: the expectation of the difference equals the difference of the expectations.

Equation (12.3) reflects the strategy that we are using. Rather than computing the expected net revenue, $R(q)$, directly from Equation (12.2), we will compute the expected marginal profit, EMP($q \to q + 1$) for each q. We can then find $R(q)$ for each q from Equation (12.3) and the "boundary condition,"

$$R(0) = 0,$$

which holds because the newsvendor earns no revenue and incurs no losses if she sets $q = 0$.

Art's Marginal Profit Tree

The key to computing the expected marginal profit is the probability tree in Figure 12.1. This tree depicts the **marginal profit** (or incremental profit), $Y(q + 1) - Y(q)$, for increasing the order from q to $q + 1$ papers. This tree accounts for the fate of the marginal newspaper, that is, for the ($q + 1$)st newspaper. For specificity, we study this tree for the data in Problem A, namely, for Art's newsstand. In Figure 12.1:

- The marginal paper, (the $q + 1$)st, must be purchased at a price of 30 cents, so a revenue of -30 appears on the stem of this tree. Here, as always, a cost is converted to a revenue by multiplying by -1.
- What happens to the marginal paper depends on whether or not the demand D exceeds q. This accounts for the chance node in Figure 12.1.
- If $D \geq q + 1$, the marginal paper, the ($q + 1$)st, is sold, and 50 cents are received, so a revenue of 50 appears on the branch in which the marginal paper is sold.
- If $D \leq q$, the marginal paper is scrapped, and 2 cents are received, so a revenue of 2 appears on the branch in which the marginal paper is scrapped.

The upper branch of Figure 12.1 has a net total revenue of $50 - 30 = 20 = C_u$. The lower branch has a net total profit of $2 - 30 = -28 = -C_o$.

Figure 12.1 describes the marginal profit that Art receives if he increases the order from q papers to $q + 1$ papers. Let us recall that EMP($q \to q + 1$) is the expectation of Art's marginal profit. From Figure 12.1, we see that

$$\text{EMP}(q \to q + 1) = P(D \leq q)(-28) + P(D \geq q + 1)(20).$$

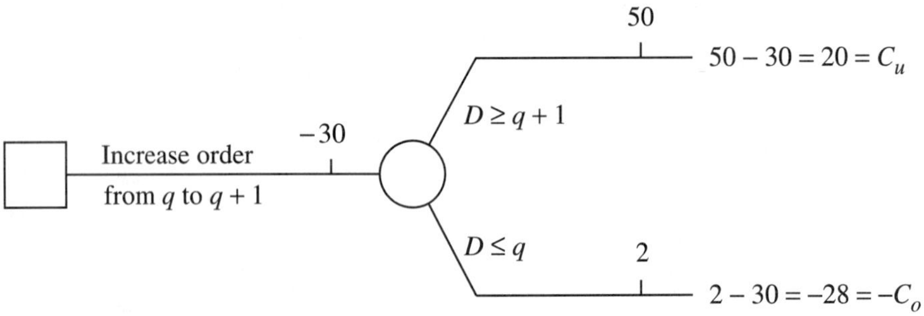

Figure 12.1 The marginal profit, $Y(q + 1) - Y(q)$, for increasing Art's order from q newspapers to $q + 1$.

Since $P(D \geq q + 1) = 1 - P(D \leq q)$, the preceding equation gives

$$\begin{aligned} \text{EMP}(q \rightarrow q + 1) &= P(D \leq q)(-28) + [1 - P(D \leq q)](20) \\ &= 20 - P(D \leq q)(20 + 28). \end{aligned} \qquad (12.4)$$

Equation (12.4) measures the benefit of increasing the order quantity from q to $q + 1$. To study Equation (12.4), we note that:

- As q increases, the probability $P(D \leq q)$ that the demand does not exceed q also increases.
- Hence, as q increases, Equation (12.4) shows that the expected marginal return of the $(q + 1)$st paper decreases.
- At some number q, the marginal value of the $(q + 1)$st paper switches from positive to negative, and the value q^* at which this occurs equals the optimal order quantity.

In brief:

> To maximize the expectation of net revenue, equate the order quantity q^* to the smallest value of q for which $\text{EMP}(q \rightarrow q + 1)$ is not positive.

The Recipe, Revisited

We will soon compute and $\text{EMP}(q \rightarrow q + 1)$ and $R(q)$ on a spreadsheet. Before doing so, we pause to justify the use of the critical fractile. Toward that end, we substitute C_u for 20 and C_o for 28 in Equation (12.4). The result is

$$\text{EMP}(q \rightarrow q + 1) = C_u - P(D \leq q)(C_u + C_o). \qquad (12.5)$$

As just noted, the optimal order quantity q^* is the smallest value of q for which $\text{EMP}(q \rightarrow q + 1)$ is not positive. By manipulating Equation (12.5), we see that

$$\text{EMP}(q \rightarrow q + 1) \leq 0 \quad \text{if and only if} \quad P(D \leq q) \geq \frac{C_u}{C_u + C_o}. \qquad (12.6)$$

The ratio $C_u/(C_u + C_o)$ has been dubbed the **critical fractile**. We have just seen that the profit-maximizing order quantity q^* equals the smallest value of q for which $P(D \leq q)$ is at least as large as the critical fractile, $C_u/(C_u + C_o)$, which justifies the recipe that was given earlier.

A Spreadsheet

The spreadsheet in Table 12.2 uses Equation (12.4) to compute $\text{EMP}(q \rightarrow q + 1)$ for each value of q between 20 and 30. This spreadsheet also uses Equation (12.3) to compute $R(q)$ for the same values of q. For the latter, it uses the "boundary condition" $R(20) = 400$ cents rather than $R(0) = 0$. To verify that $R(20) = 400$, we note in Table 12.1 that $P(D \geq 20) = 1$. If Art sets $q = 20$, he sells all 20 newspapers and earns a profit of 20 cents on each (because $20 = 50 - 30$), for a total profit of $(20)(20) = 400$ cents. Table 12.2 identifies the functions whose values are in particular cells. Specifically:

- The function in cell C6 sets $P(D \leq 21) = P(D \leq 20) + P(D = 21)$.
- The function in cell B7 sets $\text{EMP}(20 \rightarrow 21) = 20 - (20 + 28)P(D \leq 20)$, which implements Equation (12.4).
- The function in cell C8 equates $R(21) = \text{EMP}(20 \rightarrow 21) + R(20)$, which implements Equation (12.3).

Needless to say, perhaps, each of these three functions has been dragged across its row.

Table 12.2 Spreadsheet computation of $P(D \le q)$, of $\text{EMP}(q \to q + 1) = R(q + 1) - R(q)$ and of $R(q)$.

	A	B	C	D	E	F	G	H	I	J	K	L
1	Cu =	20										
2	Co =	28		= B6 + C5								
3												
4	q	20	21	22	23	24	25	26	27	28	29	30
5	P(D = q) =	0.01	0.03	0.06	0.1	0.18	0.22	0.17	0.12	0.06	0.04	0.01
6	P(D ≤ q) =	0.01	0.04	0.10	0.20	0.38	0.60	0.77	0.89	0.95	0.99	1
7	R(q+1) - R(q) =	19.5	18.1	15.2	10.4	1.76	-8.8	-17	-23	-26	-28	-28
8	R(q) =	400	420	438	453	463	465	456	439	416	391	363
9												
10	= $B1 - B6*($B1 + $B2)			= B7 + B8								

Scan row 7 of Table 12.2. Observe that the expected marginal profit of the $(q + 1)$st newspaper decreases with q, as had been predicted. For quantities $q \le 24$, row 7 shows that the expected marginal profit of the $(q + 1)$st paper is positive. For quantities $q \ge 25$, row 7 shows that the expected marginal profit of the $(q + 1)$st paper is negative. The numbers in row 7 demonstrate that 25 is the most profitable order quantity; $q^* = 25$.

The Excel chart in Figure 12.2 uses row 8 to plot $R(q)$ versus q. The function in Figure 12.2 is concave; it exhibits decreasing marginal return, as had been predicted. Its maximum occurs at $q^* = 25$, as has just been calculated.

Figure 12.2 and Table 12.2 indicate that Art's expected profit varies only slightly for order quantities q that are between 23 and 26. On an expected-profit basis, Art is nearly indifferent to these order quantities. What about the standard deviation of Art's profit? How does it vary with his order quantity q? That question is explored in Problem 9, on page 435.

Recap

We have found the optimal order quantity q^* in two different ways:

- *The recipe:* Determine the underage cost C_u and the overage cost C_o. Then compute the critical fractile, $C_u/(C_u + C_o)$. Equate q^* to the smallest value of q for which the cumulative distribution function $P(D \le q)$ equals or exceeds the critical fractile.

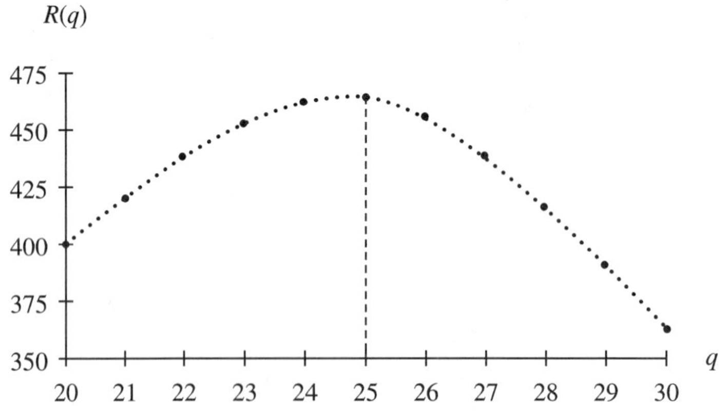

Figure 12.2 Art's expected net revenue $R(q)$ (in pennies) versus number q of newspapers that he buys from the wholesaler.

- *The marginal analysis:* Draw the marginal profit tree. From it, specify EMP($q \rightarrow q + 1$). On a spreadsheet, compute EMP($q \rightarrow q + 1$) for each q and $R(q)$ for each q. Equate q^* to the smallest q for which EMP($q \rightarrow q + 1$) is not positive.

The recipe is easy. It does not require a spreadsheet. To use it, all you need is an understanding of the opportunity costs, C_u and C_o. But you are urged to draw the marginal profit tree and to execute a marginal analysis. It provides more information. In subtle applications, it can help you to get the correct values for C_u, C_o, and q^*.

12.4. THE NEWSVENDOR WITH CONTINUOUS DEMAND

Let us now adapt the newsvendor model to the case in which the demand D is a continuous random variable. As before, Equation (12.1) specifies the random variable $Y(q)$, which describes the net revenue that the newsvendor obtains if she orders q units. As before, $R(q) = E[Y(q)]$, and $R(q)$ denotes the expectation of her net revenue if she orders q units.

Since q varies continuously, a marginal analysis concerns the slope (derivative) $R'(q)$ of $R(q)$. The equation for this slope turns out to be very familiar. Later in this section, we will verify that

$$R'(q) = C_u - (C_u + C_o)P(D \leq q), \tag{12.7}$$

where, as before, the **underage** cost C_u equals the reduction in net profit for ordering one unit fewer than (under) the demand, and the **overage** cost C_o equals the reduction in net profit for ordering one unit more than (over) the demand.

Let us see what Equation (12.7) implies. The probability $P(D \leq q)$ increases with q. Thus, Equation (12.7) shows that the slope of $R(q)$ decreases as q increases. In other words, the profit function $R(q)$ exhibits decreasing marginal return. And the optimal order quantity q^* is the value of q for which $R'(q)$ equals zero. With $R'(q) = 0$, Equation (12.7) assumes a familiar form; it equates $P(D \leq q)$ to the critical fractile.

> When the demand D is a continuous random variable, expected net revenue is maximized by ordering q^* units, where
>
> $$P(D \leq q^*) = \frac{C_u}{C_u + C_o}.$$

In other words, when the demand is a continuous random variable, the optimal order quantity q^* *equals* the critical fractile of the demand. There is no need to "round up," as had been necessary when the demand was discrete.

Adirondack Life

To investigate the case of continuous demand distribution, we pose the rhetorical question in:

Problem B (Art's Calendars)

At Art's chi-chi newsstand, he also sells each year's Adirondack Life calendar. Art must order these calendars in August, well before his customers start to request them. These calendars cost Art \$4.50 each, and he sells them for \$12.00 each. Any that are left over after New Year's Day are "remaindered" at a price of \$2.00 apiece. From past experience, Art estimates the demand D for next year's calendars to be normally distributed with a mean of 150 units and a standard deviation of 40 units. (Technically, this random variable D can take a negative value, but the probability of that event is small enough to ignore.) How many calendars should Art buy, and how sensitive is his profit to this quantity?

The underage cost C_u equals the loss of profit for having ordered one calendar too few: $C_u = \$12.00 - \$4.50 = \$7.50$. The overage cost C_o equals the loss of profit for ordering one calendar too many: $C_o = \$4.50 - \$2.00 = \$2.50$. Thus, Art should order q^* calendars, where

$$P(D \leq q^*) = \frac{C_u}{C_u + C_o} = \frac{7.50}{7.50 + 2.50} = \frac{7.50}{10.00} = 0.75.$$

For Problem B, the critical fractile equals 0.75. The demand D has the normal distribution with parameters $\mu = 150$ and $\sigma = 40$. To find the critical fractile of the demand, we can open an Excel worksheet, select a cell, enter into that cell the function

=NORMINV(0.75, 150, 40),

and observe that Excel reports the value 176.98 in that cell. Rounding off to three significant digits, Art should order $q^* = 177$ calendars.

A Tree

For Problem B, it is easy to determine the overage cost and the underage cost without drawing a tree. That is not always true. We now provide a recipe for determining the overage and underage cost from a marginal profit tree. This recipe takes 1 as a surrogate for a number that is small and positive. It *ignores* the possibility that the demand D could fall between q and $q + 1$. It gives rise to the tree in Figure 12.3.

The idea is to use the tree in Figure 12.3 to specify the number $-w$ that goes on the trunk of the tree, the number r that goes on the upper branch, and the number s that goes on the lower branch. Figure 12.3 indicates that:

- The number w equals the cost of increasing the order quantity from q to $q + 1$.
- The number r equals the marginal revenue that is earned from the $(q + 1)$st unit in the event that D exceeds $d + 1$.
- The number s equals the marginal revenue that is earned from the $(q + 1)$st unit in the event that D does not exceed q.
- The numbers C_u and C_o are given by $C_u = r - w$ and $C_o = w - s$.

Having found C_u and C_o from the tree in Figure 12.3, we select the optimal order quantity q^* so that $P(D \leq q^*) = C_u/(C_u + C_o)$. Later in this section, calculus can be used to justify this approximation.

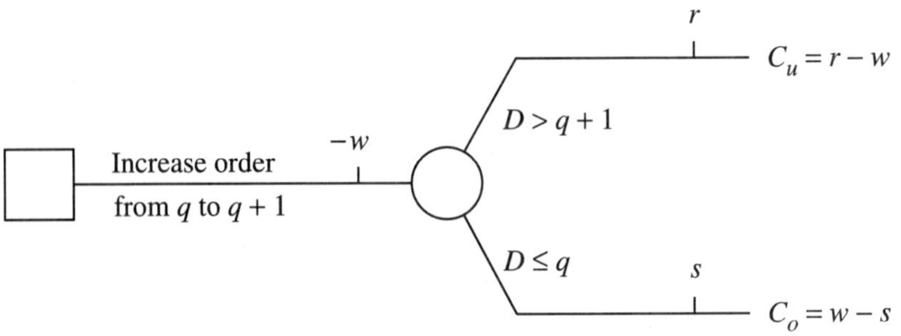

Figure 12.3 Approximate marginal profit $Y(q + 1) - Y(q)$ with continuous demand.

What Remains to Be Done?

For Problem B, we have seen that Art's profit-maximizing order quantity q^* equals 177 calendars, but we have not yet shown how to compute Art's expected net revenue $R(q)$ if he orders q calendars, and we have not yet seen how sensitive $R(q)$ is to the order quantity q. We'll do this twice, once with a method that works because the demand D has the normal distribution, and once with a method that works for any demand distribution.

Using the Normal Loss Function

The demand D has the normal distribution, and this fact will allow us to express $R(q)$ in terms of the "normal loss function." Let us begin by reviewing ideas that were introduced in Section 11.16 of Chapter 11. The number of "lost sales opportunities" equals the excess, if any, of the demand over the supply. With D as the demand and q as the supply, the number of lost sales opportunities is given by

$$(D - q)^+ = \max(D - q, 0).$$

Note that if the demand D exceeds q, then $(D - q)^+ = D - q$, which is the excess of D over q. And if D does not exceed q, then $(D - q)^+ = 0$.

Equation (12.1) expresses the newsvendor's net profit $Y(q)$ in terms of the number of units that she sells and the number of units that she scraps. Both of these quantities can be written in terms of $(D - q)^+$, as follows:

$$\min\{D, q\} = D - (D - q)^+,$$
$$\max\{0, q - D\} = q - D + (D - q)^+.$$

Substituting the above into Equation (12.1) rewrites that equation as

$$Y(q) = (C_u + C_o)D - C_o q - (C_u + C_o)(D - q)^+. \tag{12.8}$$

Finally, since $R(q)$ is the expectation of $Y(q)$, Equation (12.8) gives

$$\begin{aligned} R(q) &= E[Y(q)] \\ &= (C_u + C_o)\,E(D) - C_o\,q - (C_u + C_o)\,E[(D - q)^+] \end{aligned} \tag{12.9}$$

In the case of Art's calendars, the demand D has the normal distribution, and the expectation of $(D - q)^+$ is the **normal loss function**. The normal loss function is an Excel add-in on the CD that accompanies this text. After you install that function in your library and activate it, the Excel function **=NL(q, μ, σ)** computes $E[(D - q)^+]$, where $\mu = E(D)$ and $\sigma = \text{StDev}(D)$.

A Spreadsheet

Table 12.3 contains the first few lines of a spreadsheet that uses Equation (12.9) and the normal loss function to evaluate $R(q)$ for Art's calendar sales. The function in cell B7 computes $R(100)$. Dragging this function down the sheet produces $R(q)$ for higher values of q.

Figure 12.4 was constructed from the same spreadsheet. This figure plots $R(q)$ versus q. The function $R(q)$ is concave; it exhibits decreasing marginal return, which was ordained by Equation (12.7). Figure 12.4 demonstrates that $R(q)$ is fairly flat for order quantities q that are between 160 and 190. Any order quantity in this range will do a pretty good job of maximizing Art's expected profit from Adirondack Life calendar sales.

Table 12.3 Using the normal loss function to compute $R(q)$ for q = 100, 105, 110, ...

	A	B	C	D	E	F	G	H	I
2	μ =	150		σ =	40				
3	C_u =	7.50		C_o =	2.50				
4									
5			= (B$3 + E$3)*B$2 - E$3*A7 - (B$3 + E$3)*NL(A7, B$2, E$2)						
6	q	R(q)							
7	100	729.77							
8	105	761.38							
9	110	791.67							

An Aside

We have focused on the case in which the demand D has the normal distribution, but Equation (12.9) specifies $R(q)$ for any demand distribution. Calculus buffs are invited to differentiate Equation (12.9) with respect to q. What results is the expression in Equation (12.7) for $R'(q)$.

What Remains?

Our discussion of Newsboy with continuous demand is incomplete. We've seen how to compute $R(q)$ for the case in which D has the normal distribution. Can we compute $R(q)$, at least approximately, for the case in which D has a continuous distribution other than the normal? Yes. We can approximate $R(q)$ in either of these ways:

- Approximate D by a discrete random variable W as was done in Table 12.1 of Chapter 11. For the discrete random variable W, use marginal analysis to compute $R(q)$.
- Approximate $R(q)$ directly from Equation (12.7), which specifies its slope (derivative) $R'(q)$.

The second method is easier to execute. It is sketched in Figure 12.5.

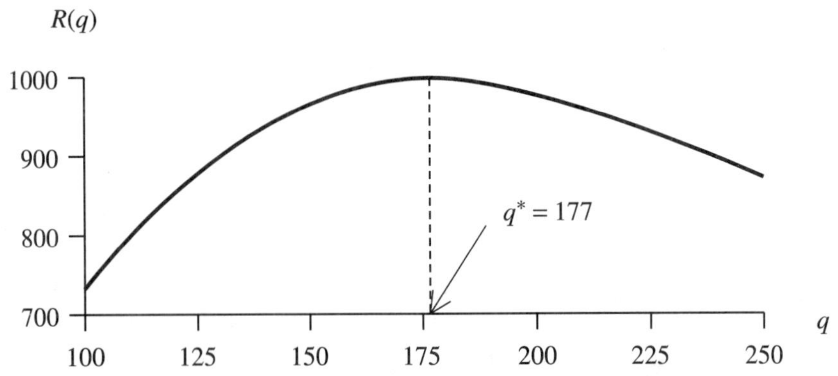

Figure 12.4 Art's expected net revenue $R(q)$ (in dollars) from Adirondack Life calendar sales versus the number q that he buys from the wholesaler.

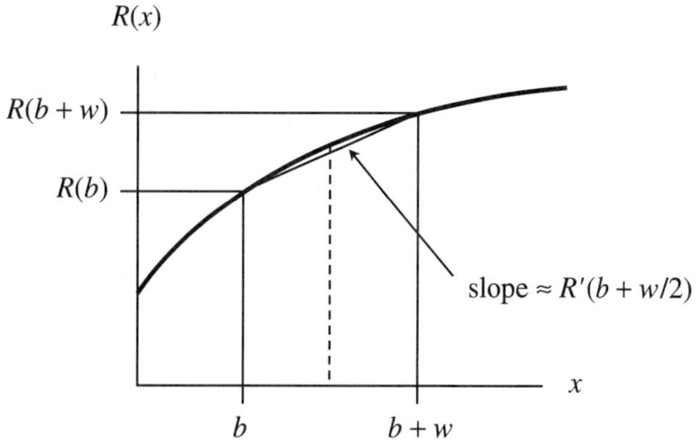

Figure 12.5 The increment between $R(b)$ and $R(b + w)$.

Numerical Approximation

Equation (12.7) specifies the slope (derivative) $R'(q)$ of the function $R(q)$. The key to recovering $R(q)$ from its slope is evident in Figure 12.5.

Figure 12.5 indicates how $R(x)$ changes as x increases from b to $b + w$. The slope of the line segment in Figure 12.5 is very close to the slope of the function $R(x)$ at the midpoint of the interval between b and $b + w$. In other words,

$$\text{slope of line segment} = \frac{R(b + w) - R(b)}{w} \approx R'(b + w/2),$$

which reorganizes itself as

$$R(b + w) - R(b) \approx wR'(b + w/2). \tag{12.10}$$

Equation (12.10) uses $R'(x)$ at the midpoint of an interval to estimate the change in $R(x)$ over the course of the interval. To cement your understanding of Figure 12.5 and Equation (12.10), ask yourself, "Why was the midpoint chosen; why not the left end-point, for instance?"

Equation (12.10) estimates the change in the function $R(x)$ over an interval. To recover $R(q)$ Equation (12.10), we need to know its value for any single value of q. If Art orders no calendars, he earns nothing, so

$$R(0) = 0. \tag{12.11}$$

Equations (12.7), (12.10), and (12.11) will enable us to approximate the function $R(q)$ of q on a spreadsheet. The method by which they accomplish this is called **numerical integration**.

A Spreadsheet

Table 12.4 displays the first few lines of a spreadsheet that approximates $R(q)$ for q in the interval between 0 and 200. We have aimed for an excellent approximation. Cell B3 reports a grid width of $1/20$th of a standard deviation. Cell B4 reports an "interval" width of 2 because $2 = (1/20)\sigma = (1/20)(40)$. With $c - b = 2$, Equation (12.10) becomes

$$R(b + 2) \cong R(b) + (2)R'(b + 1). \tag{12.12}$$

Table 12.4 Approximation of $R(0)$, $R(2)$, . . . , $R(200)$ by numerical integration on a spreadsheet.

	A	B	C	D	E	F	G
1	μ =	150		Cu =	7.5		
2	σ =	40		Co =	2.5		
3	grid width	1/20		=B2*B3			
4	interval width	2					
5			=E$1-(E$1+E$2)*NORMDIST(A7,B$1,B$2, TRUE)				
6	**center**	**R′(center)**	**left**	**R(left)**			
7	1	7.4990	0	0.0000			
8	3	7.4988	2	14.9980		=D7+B$4*B7	
9	5	7.4986	4	29.9957			

In Table 12.4, the width of each interval equals 2. Column C records the left end-point of each interval. Column A records the midpoint of each interval. In addition:

- Column B computes $R'(q)$ at each midpoint q. For instance, the function in cell B7 uses Equation (12.7) to compute

$$R'(1) = C_u - (C_u + C_o)P(D \leq 1).$$

- Column D computes $R(q)$ at each left end-point. For instance, the function in cell D8 uses Equation (12.12) to compute

$$R(2) = R(0) + 2R'(1).$$

How accurate is the approximation in Table 12.4? The exact values of the function $R(q)$, as computed from Equation (12.9), are virtually identical to the function plotted in Figure 12.4. For no value of q does the difference between the exact and approximate value of $R(q)$ exceed 0.03 dollars.

Recap

We've provided three separate approaches to the newsvendor problem with continuous demand.

- *The recipe:* Determine the underage cost C_u and the overage cost C_o. Then compute the critical fractile, $C_u/(C_u + C_o)$. Equate q^* to the value of q for which the cumulative distribution function $P(D \leq q)$ equals the critical fractile.
- *The frontal assault for the case in which D has the normal distribution:* Use the normal loss function to compute $R(q)$ directly from Equation (12.9).
- *Numerical approximation:* Equation (12.7) specifies the slope $R'(q)$. Use Equation (12.10) to approximate $R(q)$ by numerical integration.

This completes our introduction to the newsvendor with a continuous distribution of demand.

12.5. A NEWSVENDOR IN YIELD MANAGEMENT

Art's newsstand has motivated two newsvendor computations, one with a discrete random variable, the other with a continuous random variable. In both cases, the underage cost C_u and the overage cost C_o have been easy to calculate. Both cases have entailed small sums of money. Let us now adapt our understanding of the newsvendor problem to a setting that is more subtle and entails substantial sums.

Airlines sell seats at a variety of prices and terms. Hotels rent rooms at a variety of prices and terms. Auto rental companies lease the same car at different prices under different circumstances. The process of setting these prices and terms is known as **yield management**. Yield management reacts to market uncertainties in an attempt to maximize profit. If done well, yield management contributes significantly to corporate profit.

In a sense, each seat on a flight is a perishable commodity. If a seat is not filled when the flight departs, the opportunity to earn money from that seat is irrevocably lost. This connotes that the newsvendor model may be relevant. To see how, we study **no shows**, namely, individuals who reserve seats but fail to appear to claim the seats that they reserved. Airlines compensate for no shows by overbooking, that is, by accepting more reservations than the number of available seats. To study an instance of this, we pose the rhetorical question in:

Problem C (Business Class No Shows)

You have been asked to determine your airline's policy for accepting Business class reservations. You begin with a particular flight. You have ascertained that:

- The Business class section of this flight contains 32 seats.
- The number D of individuals who will seek to reserve these Business class seats is uncertain. Its distribution is Poisson with mean of 44.
- Each individual who is granted a reservation for a Business class seat appears to claim it with probability of 0.8. Different individuals decide whether to show up independently of each other.
- Each occupied seat in the Business class section earns the company a contribution (price of a Business class ticket less variable cost) of $600.
- If more than 32 Business class ticketholders appear to claim the seats they reserved, the airline compensates each customer that it cannot seat at a cost to the airline of $300 apiece.
- Business class customers pay no penalty for reserving a seat and then failing to show up.

You need to determine the "quota" q of reservations to accept, where the first q requests for Business class seat reservations are accepted and any requests excess of q are denied. Picking a quota q that is only slightly above 32 forgoes the opportunity to profit from filling Business class seats. Setting q far above 32 increases the likelihood that Business class customers cannot be seated and will need to be compensated. A balance must be struck.

A Marginal Analysis

To strike this balance, we do a marginal analysis. What happens, we ask ourselves, if the quota q is increased by 1? This increases the probability that the plane is overbooked, so it increases the probability that the marginal customer cannot be seated and must be compensated. Consequently, the marginal profit for increasing the quota by 1 decreases with q. Marginal analysis can identify the profit-maximizing quota, q^*. Following a familiar pattern, we will compute:

$$\text{EMP}(q \rightarrow q+1) = \text{the expectation of the marginal profit for increasing the quota from } q \text{ to } q+1.$$

Because it will prove useful, we pause to introduce the random variable N_k whose definition is:

N_k = the number of customers who appear to claim the seats that they reserved, from a group of k customers who made reservations.

This random variable N_k is familiar from Chapter 8. Exactly k people have reserved seats, each person shows up with probability of 0.8, independent of the others, and N_k is the total number who show up. Thus, N_k has the binomial distribution with parameters $n = k$ and $p = 0.8$.

A Marginal Profit Tree

The tree in Figure 12.6 describes the marginal profit for increasing the quota from q to $q + 1$. This tree ascertains the effect of increasing the quota on the marginal customer, the $(q + 1)$st. To animate the discussion, we give the $(q + 1)$st customer to request a seat the name Ludo. If q or fewer customers request seats, none of them is named Ludo. If more than q customers request seats, Ludo is the $(q + 1)$st, and he is granted a reservation when the quota is increased from q to $q + 1$.

The marginal profit tree in Figure 12.6 accounts for what happens to Ludo. To interpret this tree, we consider these cases:

- Suppose $D \le q$. In this case, all reservation requests were granted when the quota was q. Increasing the quota has no effect on any reservation requests, and hence no effect on income. This explains a "0" in Figure 12.6.
- Suppose $D \ge q + 1$ and suppose that Ludo (who got a reservation when the quota was increased) is a no show. Ludo pays no penalty for being a no show, so no change occurs in revenue, which explains another "0."
- Suppose $D \ge q + 1$ and suppose Ludo appears to claim a seat. Being the marginal customer, Ludo will get a seat if there is room. Exactly q other customers have made reservations. Of them, N_q appear to claim their seats.
  - **(1)** If $N_q \le 31$, Ludo gets a seat, and the airline earns \$600, which explains the one branch in Figure 12.6.
  - **(2)** If $N_q \ge 32$, there is no room for Ludo, and the airline loses \$300, which explains the remaining branch in Figure 12.6.

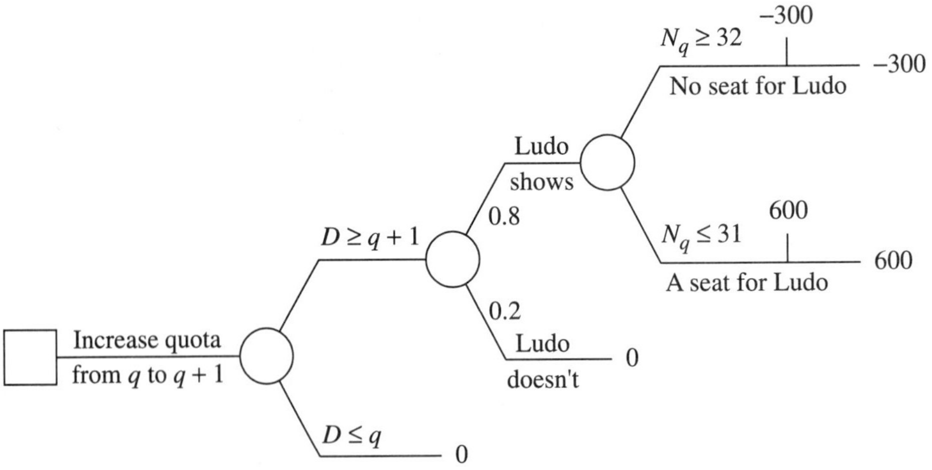

Figure 12.6 Marginal profit for increasing the quota from q to $q + 1$.

To compute the expected marginal profit from Figure 12.6, we multiply the probability of reaching each end-point by the profit earned if that end-point is reached and take the sum, obtaining

$$\text{EMP}(q \to q+1) = 0 + 0 + P(D \geq q+1)(0.8)P(N_q \leq 31)(600)$$
$$+ P(D \geq q+1)(0.8)P(N_q \geq 32)(-300).$$

To simplify the above, we substitute $1 - P(N_q \leq 31)$ for $P(N_q \geq 32)$ and factor out the term $P(D \geq q+1)(0.8)$, getting

$$\text{EMP}(q \to q+1) = P(D \geq q+1)(0.8)[-300 + 900P(N_q \leq 31)]. \quad (12.13)$$

From Equation (12.13), we see that:

- The term $P(D \geq q+1)(0.8)$ is positive, independent of the value of q.
- Hence, the quantities $\text{EMP}(q \to q+1)$ and $[-300 + 900P(N_q \leq 31)]$ have the same sign.
- In other words, the marginal profit of the $(q+1)$st seat is positive if and only if the expression $[-300 + 900P(N_q \leq 31)]$ is positive.
- Since $P(N_q \leq 31)$ equals the probability that q reservations result in 31 or fewer customers, this probability decreases as q increases.
- Hence, Equation (12.13) exhibits decreasing marginal return, and expected profit is maximized by selecting the smallest value of q for which the quantity $[-300 + 900P(N_q \leq 31)]$ is not positive.

In brief, the profit-maximizing quota q^* equals the smallest number q such that the expression $[-300 + 900P(N_q \leq 31)]$ is not positive.

A Spreadsheet

Table 12.5 shows how to find the profit-maximizing quota q^* on a spreadsheet, using the Excel function for the CDF of the binomial. Table 12.5 also indicates that $q^* = 41$.

A surprising insight in Equation (12.13) is that q^* does not depend on the demand D. The fact that D has the Poisson distribution with a mean of 44 has no effect on q^*. Had you guessed that?

The expected net revenue $R(q)$ does depend on D. The expected net revenue $R(q)$ from a quota of size q can be found by marginal analysis, using $R(0) = 0$ and

$$R(q+1) = R(q) + \text{EMP}(q \to q+1),$$
$$= R(q) + P(D \geq q+1)(0.8)[-300 + 900P(N_q \leq 31)]. \quad (12.14)$$

for $q = 1, 2, \ldots$. On a spreadsheet, computing $R(q)$ by the above recursion is straightforward.

Table 12.5 Spreadsheet computation of $[-300 + 900P(N_q \leq 31)]$ for q equal to 40 through 44.

	I	J	K	L	M	N
2	q =	40	41	42	43	44
3	−300 + 900 P(N$_q$ ≤ 31) =	66.19	-33.64	-115.50	-178.01	-222.81
4		↑				
5	= - 300 + 900 * BINOMDIST(31, J2, 0.8, TRUE)					

Recap

This yield management problem illustrates the value of thinking at the margin. The marginal profit tree in Figure 12.6 is the key to its analysis. From this tree, we saw that D is not relevant, and we saw how to compute the profit-maximizing quota, q^*.

This marginal analysis was a bit subtle. To execute it, we had to think carefully about what happens to the marginal customer when the quota is increased by 1. If you are not yet convinced of the value of marginal analysis, mount a frontal assault. Try to compute $R(q)$ directly on a spreadsheet rather than from Equation (12.14).

12.6. THE BASE STOCK MODEL

The newsvendor is a **one-period** model in the sense that inventory "perishes" at the end of each period; it cannot be used to satisfy the next period's demand. In a **multiple-period** model, the inventory that is not used during a period remains available for use in subsequent periods. This section concerns a simple multiple-period inventory control model.

In a multiperiod model, much depends on those customers whose demands cannot be satisfied immediately. In the jargon of inventory theory, **backlogging** occurs when a customer whose demand cannot be satisfied immediately waits for the merchandise to appear. Similarly, a **lost sale** occurs when a customer whose demand cannot be satisfied disappears, perhaps to buy the item elsewhere. Backlogs and lost sales add to expense but in different ways.

This section focuses on the case of lost sales. For it, the **base stock** model is as follows:

- Time is broken into periods of equal length.
- The demand D for an item during a period is uncertain. The demands in different periods are independent of each other, and they have the same probability distribution. This probability distribution is known.
- Any inventory that remains at the end of a period is available for use in future periods.
- Inventory can be replenished at the beginning of each period, and only then.
- There is an economic motive to avoid excess inventory.

In the base stock model, the key decision is to determine the level q to which inventory is replenished at the beginning of each period. This level q is known as the **base stock level** and the **order-up-to quantity**. These terms are synonyms.

In the base stock model, the excess, $q - E(D)$, of the start-of-period inventory over the expected demand is called the **safety stock**. This safety stock protects against the uncertainty in the demand. It controls the likelihood of lost sales.

Achieving a Service Level

The **service level** is the fraction (or percentage) of the periods in which the customers' demands can be fully satisfied. In a base stock model, a service level of 95% is achieved if $P(D \le q)$ is 0.95. To bring the base stock model into focus, we pose the rhetorical question in:

Problem D (Base Stock with a Service Level)

Your job is to manage inventory for a large number of products in each of several stores. Each store is restocked each evening, after the close of the business day. Your directive is to provide a service level for each product in each store of 99%. Subject to this requirement, your company wants you to avoid excess inventory. Here is what you have learned about the demands:

- The demand for each product in each store on each day is uncertain.
- The demands for a particular product in a particular store on different days are independent of each other and have the same distribution. This distribution is Poisson, and you know its mean.
- The demands for different products are mutually independent.

What is your inventory control policy?

Let D denote the demand for a particular product in a particular store during a particular day. The distribution of D is Poisson. Its mean $E(D)$ is particular to the product and to the store. Your goal is to set the smallest base stock level q such that $P(D \leq q)$ is at least 0.99. We shall see that an effective (close-to-optimal) base stock policy is to compute

$$q = E(D) + 2.35\sqrt{E(D)} \tag{12.15}$$

and then round q off to the nearest integer.

An Economy of Scale

Before justifying Equation (12.15), we interpret it. The safety stock, $q - E(D)$, equals the excess of the start-of-period inventory level q over the expected demand $E(D)$ during the period. Equation (12.15) shows that the safety stock $q - E(D)$ equals $2.35\sqrt{E(D)}$. This safety stock is proportional to the *square root* of the mean demand. Thus, if the mean demand doubles, the size of the safety stock increases by the factor of $\sqrt{2} \cong 1.414$, which is well below 2. This manifests an economy of scale, one that does not depend on the details in this model. We emphasize:

> In nearly every inventory control model, the safety stock grows as the square root of the mean demand.

A Spreadsheet

To justify Equation (12.15), we must determine how the start-of-day stock position q should vary with the mean daily demand, $E(D)$, or, conversely, how the expected demand $E(D)$ must change with the start-of-day stock position q in order to provide the desired service level. The latter computation is easy to execute on a spreadsheet, using Solver. The spreadsheet in Table 12.6 accomplishes this. In the table, row 4 contains data, and row 5 contains changing cells for Solver. We note that:

- Cell B4 specifies 5 as a base stock level.
- Cell B5 is reserved for a value of $E(D)$ that Solver determines.
- The function in cell B6 computes $P(D \leq 5)$ where D is Poisson with mean in cell B5.
- Solver has been used to select the largest value in cell B5 that satisfies the constraint B6 $\geq$ B2, thereby keeping $P(D \leq 5) \geq 0.99$.
- Columns C through G execute similar computations for larger base stock levels.

Table 12.6 also reports the solutions that Solver provided.

To see how the safety stock varies with the mean demand, we compare rows 7 and 4. Cells B4 and B7 show that when $E(D) = 1.79$, a safety stock of 3.21 units is needed, for a

Table 12.6 For base stock levels between 5 and 120, the value of $E(D)$ for which the base stock provides a service level of 99%.

	A	B	C	D	E	F	G
2	service level =	0.99					
3							
4	**base stock level, q =**	**5**	**10**	**20**	**40**	**80**	**120**
5	E(D) =	1.79	4.77	11.82	27.59	61.54	96.89
6	P(D ≤ q) =	0.99	0.99	0.99	0.99	0.99	0.99
7	safety stock, q − E(D) =	3.21	5.23	8.18	12.41	18.46	23.11
8	$(q - E(D))/\sqrt{E(D)}$ =	2.41	2.39	2.38	2.36	2.35	2.35
9							
10	=POISSON(B4, B5, TRUE)		=(B4-B5)/SQRT(B5)				

ratio of 1.8 = 3.21/1.79 of safety stock to mean demand. By contrast, cells G4 and G7 show that when $E(D) = 96.89$, a safety stock of 23.11 units is needed, for a ratio of 0.24 = 23.11/96.89. As the demand increases, less safety stock is needed, as a fraction of the mean demand. An economy of scale is evident.

In fact, row 8 shows that the ratio of the safety stock to the *square root* of the mean demand varies only slightly, from 2.41 to 2.35. When the mean demand is 1.8 units per day, a safety stock equal to 2.41 times the square root of the mean demand provides a service level of 99%. When the mean demand is 97 units per day, a safety stock equal to 2.35 times the square root of the mean demand provides the same service level. In Equation (12.15), we have used the coefficient of 2.35. Virtually no difference would occur if we used the coefficient of 2.41, or something that is in between.

Profit Maximization

The basic stock model has just been used to achieve a prescribed level of service with as little inventory as is possible. In that setting, the safety stock increased as the square root of the mean demand.

Profit was not measured directly. The base stock model will now be used in a profit-maximization setting, and the same economy of scale will recur. Let us consider:

Problem E (Base Stock with Profit Maximization)

Again, your job is to manage inventory for a variety of products in each of several stores. Each store is restocked once each week. As in Problem D, the demand for each product in each store during each week is Poisson. The demands for a particular product in different weeks are independent of each other and are identically distributed, and the demands for different products are independent of each other. Again, you know the mean weekly demand for each product in each store. If a customer demands a product that you cannot supply, you lose the sale because he or she goes elsewhere. The cost of keeping each item in inventory for all or any part of one week equals 3% of the contribution your company receives for selling it. What inventory control policy maximizes expected profit?

To analyze Problem E, we consider the demand D for a particular product in a particular store during a particular week. This demand is uncertain. Its distribution is Poisson. You know its mean. Let us denote by K the contribution your company earns for each unit of this product that is sold. We must determine the number q of units of this product that should be on hand at this store at the beginning of each week.

As in the newsvendor model, we compute EMP($q \rightarrow q + 1$), the expected marginal profit for increasing the base stock from q to $q + 1$ for one week. Keeping the extra unit in inventory for one week costs $0.03K$. The extra unit is sold if $D \geq q + 1$, in which case K is earned. Thus,

$$\begin{aligned} \text{EMP}(q \rightarrow q + 1) &= -0.03K + P(D \geq q + 1)K, \\ &= -0.03K + [1 - P(D \leq q)]K, \\ &= [0.97 - P(D \leq q)]K. \end{aligned} \tag{12.16}$$

Equation (12.16) shows that EMP($q \rightarrow q + 1$) is positive if and only if $P(D \leq q)$ is below 0.97. Thus, the profit-maximizing base stock level q^* is the smallest value of q for which $P(D \leq q)$ is at least 0.97.

In brief, the profit-maximizing inventory control policy is to minimize the inventory needed to obtain a service level of 97%. Problem E is just like Problem D, except that the service level is 97% rather than 99%. It is solved in the same way. Table 12.7 reports the solution that was obtained from Solver.

Table 12.7 evinces an economy of scale. The safety stock grows as the square root of the mean demand. In this case, an effective base stock policy is to compute

$$q = E(D) + 1.87\sqrt{E(D)} \tag{12.17}$$

and then round q up to the nearest integer.

Continuous Demand

Let us now consider a base stock model in which the demand D has the normal distribution rather than Poisson. We recall from Chapter 11 that the distribution of a normal random variable D is specified by two parameters—its mean $\mu = E(D)$ and its standard deviation

Table 12.7 For base stock levels between 5 and 120, the values of $E(D)$ for which the base stock policy maximizes expected profit.

	A	B	C	D	E	F	G
2	service level =	0.97					
3							
4	**base stock level, q =**	**5**	**10**	**20**	**40**	**80**	**120**
5	E(D) =	2.30	5.66	13.26	29.83	64.94	101.17
6	P(D ≤ q) =	0.97	0.97	0.97	0.97	0.97	0.97
7	safety stock, q − E(D) =	2.70	4.34	6.74	10.17	15.06	18.83
8	(q − E(D))/√E(D) =	1.78	1.83	1.85	1.86	1.87	1.87
9							
10			=POISSON(B4,B5,TRUE)				

$\sigma = \text{StDev}(D)$. In Problem D, we sought a base stock level q such that $P(D \leq q) = 0.99$. In other words, we sought the 0.99 fractile of the demand.

In Chapter 11, we learned how to compute fractiles of the normal distribution. Let the random variable Z have the normal distribution with $E(D) = 0$ and $\text{StDev}(D) = 1$. To compute the number z such that $P(Z \leq z) = 0.99$, we can open a spreadsheet, select a cell, enter the function

=NORMINV(0.99, 0, 1)

and learn $z = 2.33$. (We could get the same information by reading the Standard Normal Table backwards.) Thus, by a familiar transformation,

$$P(D \leq q) = 0.99 \quad \text{for} \quad q = \mu + 2.33\sigma, \quad \text{when } D \text{ is normal.} \tag{12.18}$$

Similarly, in the profit-maximization case, we sought a base stock level that satisfies 97% of the demand. This is the 0.97 fractile. To compute it, we observe that the function **=NORMINV(0.97, 0, 1)** reports $z = 1.88$, and obtain

$$P(D \leq q) = 0.97 \quad \text{for} \quad q = \mu + 1.88\sigma, \quad \text{when } D \text{ is normal.} \tag{12.19}$$

The numbers 2.33 and 1.88 are strikingly close to the constants in Equations (12.15) and (12.17), respectively. Is that a coincidence?

A Normal Approximation

In Chapter 8, we learned that the sum of independent Poisson random variables is Poisson. In Chapter 11, we learned that the sum of sufficiently many i.i.d. random variables is approximately normal. Thus, the sum of n i.i.d. Poisson random variables with mean of 1 is Poisson with mean of n and is approximately normal for n large enough.

Let us recall from Chapter 8 that a Poisson random variable D has variance equal to its mean; it has $\sigma = \text{StDev}(D) = \sqrt{E(D)}$. Thus, Equations (12.18) and (12.19) show that for large values of $E(D)$, the safety stocks in Tables 12.6 and 12.7 should approximate $2.33\sqrt{E(D)}$ and $1.88\sqrt{E(D)}$ respectively. And they do! The safety stock reacts to the tail probability, but the normal approximation is working well, even there.

12.7. THE ECONOMIC LOT SIZE MODEL

We have studied two models in which a safety stock protects against the uncertainty in the demand. Our attention now turns to a model that requires a cycle stock but no safety stock. F. W. Harris of the Westinghouse Corporation developed this model in 1915. The **economic lot size** model is as follows:

- The demand for a product occurs at a constant rate of A units per year.
- The objective is to minimize the annual cost of satisfying all of the demand without allowing any backlog.
- Any quantity q can be ordered at any time. A fixed replenishment interval of L years elapses between the moment at which the order is placed and the moment at which delivery occurs. (This number L is often well below 1, e.g., $L = 14/365$ years.)
- There is a cost of K dollars for placing each order; this cost is independent of the size of the order.
- Keeping units in inventory accrues cost at the rate of H dollars per unit per year.

The value of the product has been omitted from this model. Why? Demand must be satisfied, so the value of satisfying it is not relevant. The economic lot size model is often called the **economic order quantity model**; these terms are synonyms.

In the economic lot size model, the demand is certain, and there is no need for a safety stock. Each order incurs a fixed charge, so a cycle stock will be needed. To illustrate the economic lot size model and to explore it, we turn our attention to the example in:

Problem F (Refrigerator Production)

A production line makes refrigerators and nothing else. The demand is sufficient to run this line without interruption, at the rate of 120,000 refrigerators per year. Each refrigerator requires a compressor. Compressors are manufactured on a different production line, one that serves several other purposes and is capable of making compressors at a rate far in excess of 120,000 units per year. Switching that line to the manufacture of compressors requires it to be shut down for several hours, and the opportunity cost (lost of profit) of each such shutdown equals \$1600. Each order for compressors must be placed two weeks before delivery. The inventory-carrying cost is \$6 per compressor per year. When should compressors be ordered, and in what quantity?

Problem F is the economic lot size model with these data: $A = 120{,}000$ units/year, $H = 6$ dollars per unit per year, $K = 1600$ dollars per order, and $L = 14/365$ years.

Reorder Point and Order Quantity

There is no uncertainty in the demand for compressors and none in the supply, so there is no need for a safety stock. Each order should be timed to appear at the moment that the inventory runs out. This is reflected in Figure 12.7, which plots the inventory level $I(t)$ versus time, t. The plot has a "saw-toothed" shape. Inventory $I(t)$ decreases at the rate of $-A$ until it reaches 0, at which point the receipt of an order causes the inventory to jump upward, whereupon it commences another descent.

This inventory control policy is described in terms of a **reorder point** r and an **order quantity** q. The reorder point is the inventory level at which an order is placed. The reorder point is chosen so that inventory runs out at the moment of replenishment. Figure 12.7 shows that the "rise" (of r) equals the "slope" (of A) times the "run" (of L); $r = AL$. The order quantity q equals the height of the saw-toothed function in Figure 12.7.

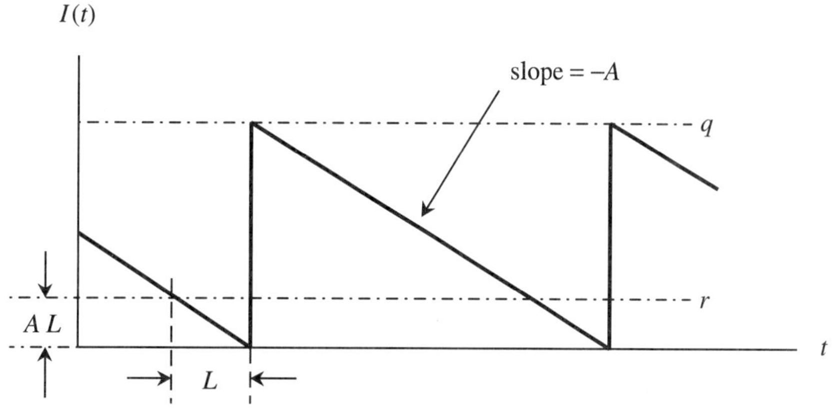

Figure 12.7 Plot of inventory $I(t)$ versus t, with reorder quantity q and reorder point chosen so that the order is received at the moment of stockout.

In the economic lot size model, the crucial question is how many to order. Let us designate q and $C(q)$ by

q = the size of each order,

$C(q)$ = the average cost per year of a policy that orders in lots of q units, timing each order for receipt at the moment inventory is exhausted.

A small value of q means frequent orders, at a cost of K apiece. A large value of q means a large volume of inventory, at a cost of H apiece. A tradeoff exists.

Ordering Cost and Holding Cost

To measure this tradeoff, we compute the annual ordering cost, the annual holding cost, and their total, $C(q)$. It is evident from Figure 12.7 that:

- The average inventory level over a long period of time equals $q/2$. Since each unit of inventory costs H dollars per year, the annual holding inventory is the product,

$$\text{holding cost} = \frac{Hq}{2}, \quad \text{measured in dollars per year.}$$

- The average number of orders placed per year equals the ratio A/q of the annual demand to the number q of units in each order. Since each order incurs a fixed charge of K dollars, the annual ordering cost is the product,

$$\text{ordering cost} = \frac{AK}{q}, \quad \text{measured in dollars per year.}$$

For instance, if $A = 120$ units per year and $q = 10$ units per order, then $A/q = 120/10 = 12$ orders per year.

The sum, $C(q)$, of the average annual inventory carrying cost and ordering cost is given by

$$C(q) = \frac{Hq}{2} + \frac{AK}{q}. \tag{12.20}$$

The order quantity q represents a **cycle stock**; this inventory owes its existence to the fact that it pays to order intermittently. As mentioned earlier, no safety stock is needed because there is no uncertainty in the demand.

The Economic Order Quantity

We aim to select the order quantity q^* whose annual cost is smallest. For the data in Problem F, Figure 12.8 plots the annual cost $C(q)$ versus q. This figure also plots the annual inventory holding cost $(Hq)/2$ and the annual ordering cost $(AK)/q$ versus q. Figure 12.8 measures the tradeoff between small q (low holding cost) and large q (low ordering cost).

It is evident from Figure 12.8 that $C(q)$ is convex, and that hence this is a case of increasing marginal cost. The slope $C'(Q)$ increases as q increases, so the optimal order quantity q^* satisfies $C'(Q^*) = 0$. By differentiating Equation (12.20), we get

$$C'(q) = \frac{H}{2} - \frac{AK}{(q)^2}.$$

Setting $C'(q^*) = 0$ and then solving for q^* gives

$$q^* = \sqrt{\frac{2AK}{H}}. \tag{12.21}$$

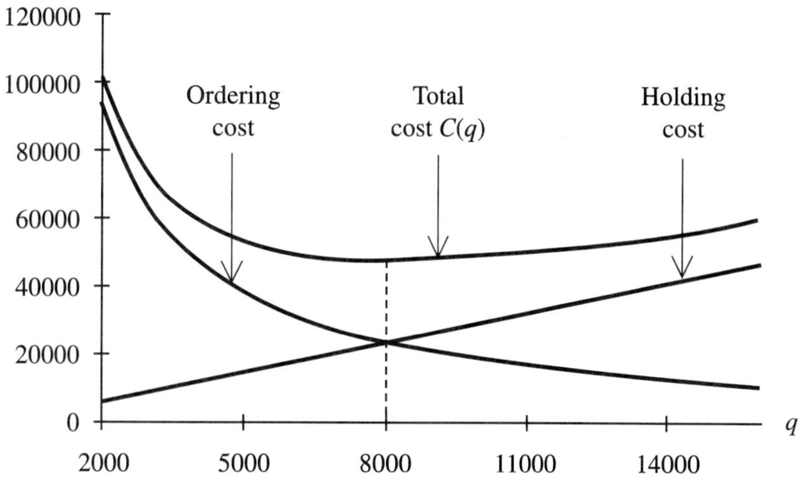

Figure 12.8 Annual cost $C(q)$ and its components for Problem F.

The number q^* given by Equation (12.21) is called the **economic order quantity** and is often abbreviated as the **EOQ**. The EOQ balances the tradeoff between inventory-carrying cost and ordering cost.

Let us compute the EOQ for Problem F. Its data are $A = 120{,}000$, $K = 1600$, and $H = 6$. Substituting these values into Equation (12.21) gives

$$q^* = \sqrt{\frac{2 \times 120{,}000 \times 1600}{6}} = \sqrt{64{,}000{,}000} = 8000,$$

as is indicated in Figure 12.8. For this example, the number A/q^* of orders per year is given by $120{,}000/8000 = 15$, and the time q^*/A between orders is given by $8000/120{,}000 = 1/15$ years.

Figure 12.8 also shows that q^* causes the annual carrying cost $Hq^*/2$ to equal the annual ordering cost KA/q^*. This is not peculiar to the data in Problem F. For every economic lot size model, the EOQ happens to equate the annual carrying cost to the annual ordering cost.

Substituting the value for q^* into Equation (12.20) and doing a little algebra (that we omit) results in

$$C(q^*) = \sqrt{2AKH}. \tag{12.22}$$

An Economy of Scale

How do the economic order quantity q^* and the annual cost $C(q^*)$ vary with the annual demand A? Equations (12.21) and (12.22) show that q^* and $C(q^*)$ are proportional to the *square root* of A, not to A itself. Doubling the demand A causes q^* and $C(q^*)$ to increase by the factor of $\sqrt{2} \cong 1.414$ not by the factor of 2. We emphasize:

> In the economic lot size model, the economic order quantity q^* and the annual cost $C(q^*)$ increase as the square root of the demand.

In the base stock model, we saw that the safety stock increases as the square root of the demand. Here, in the economic lot size model, we see that the cycle inventory increases as

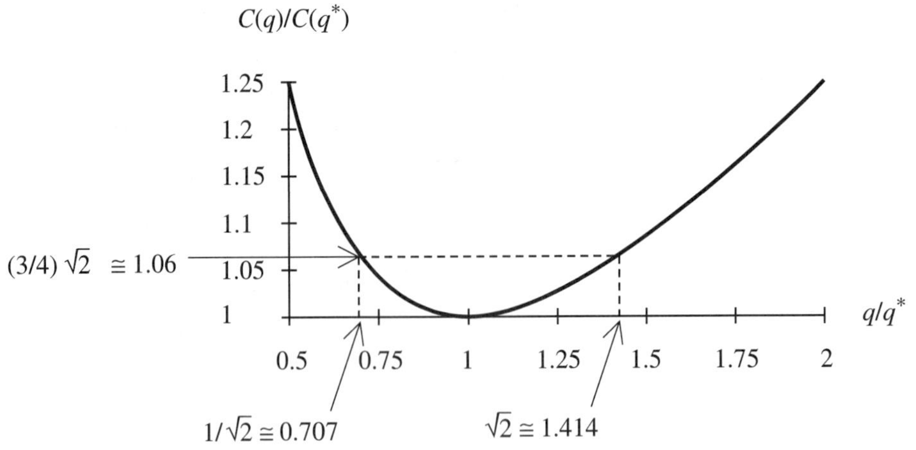

Figure 12.9 The ratio $C(q)/C(q^*)$ of costs versus the ratio q/q^* of quantities, with q^* as the economic order quantity.

the square root of the demand. This "square root" economy of scale is a recurring theme in inventory control.

A Flat Bottom

Figure 12.8 suggests that for values of q that are close to the economic order quantity, the cost $C(q)$ does not exceed $C(q^*)$ substantially. To measure the extent to which $C(q)$ varies with q, we investigate the ratio $C(q)/C(q^*)$ of costs. Algebraic manipulation of Equations (12.20) and (12.22) (that we omit) produces

$$\frac{C(q)}{C(q^*)} = \frac{1}{2}\left(\frac{q}{q^*} + \frac{q^*}{q}\right). \tag{12.23}$$

Equation (12.23) is startlingly simple. The ratio $C(q)/C(q^*)$ of costs depends solely on the ratio q/q^* of order quantities, and not on the data K, A, and H. The ratio $C(q)/C(q^*)$ achieves its minimum when $q = q^*$, of course. Figure 12.9 plots the ratio $C(q)/C(q^*)$ versus q.

Checking the computations in Figure 12.9 requires a bit of algebra, which is sketched in Problem 21 on page 439. Since $(3/4)\sqrt{2} \cong 1.06$, Figure 12.9 reports that $C(q)$ increases by only 6% for q as low as $q^*/\sqrt{2} \cong 0.707q^*$ and as high as $q^*\sqrt{2} \cong 1.414q^*$.

> **Flat bottom:** The cost $C(q)$ exceeds $C(q^*)$ by not more than 6% as q varies by a factor of 2, between $0.707q^*$ to $1.414q^*$.

The fact that $C(q)$ has a "flat bottom" is extremely useful. An order quantity q that is close to q^* is good enough. For instance, little may be lost by rounding q^*/A off to the nearest week, or by coordinating ordering intervals for different items.

12.8. THE ECONOMIC LOT SIZE MODEL WITH UNCERTAIN DEMAND

The economic lot size model is now generalized to include uncertain demand. The resulting model combines features of the economic lot size model and the base stock model. It will require both a safety stock and a cycle stock. In this section, we study the version of this model in which unsatisfied demand is backlogged. Problem 27 on page 441, suggests how to adapt the analysis to the case of lost sales.

The data in the **economic lot size model with uncertain demand** include the numbers A, b, L, K, and H, where:

- Demand for a product is random but has stationary independent increments,[1] with an expected demand of A units per year.
- Unsatisfied demand is backlogged, at a cost of b dollars for each unit placed in backlog.
- Any quantity q can be ordered at any time. A fixed replenishment interval of L years elapses between the moment at which the order is placed and the moment at which delivery occurs. The demand D that occurs during this replenishment time is uncertain. The probability distribution of D is known, however.
- There is a cost of K dollars for placing each order; this cost is independent of the size of the order.
- Keeping units in inventory accrues cost at the rate of H dollars per unit per year.

Again, the value of the product has been omitted from this model because demand must be satisfied.

The main difference between this model and the economic lot size model is that the demand D that occurs during the replenishment interval is random. Since demand occurs at a constant average rate throughout the year, the expectation $E(D)$ of the demand during the replenishment interval still equals LA.

Blood

To illustrate this model, we pose the rhetorical question in:

Problem G (Blood Resupply)

You have been asked to advise the local hospital as to its policy for inventorying blood. You have decided to focus, initially, on the most common blood type, which is O positive. These are the relevant considerations:

- The hospital draws its blood supply from the regional bank, which fills routine orders for each blood type in five calendar days at a cost of \$50 per order, independent of the size of the order.
- The blood bank also loans small quantities of blood at a cost of \$2 per unit borrowed. Borrowed blood is delivered with negligible delay and is debited from the next order.
- For the hospital's inventory, carrying cost is \$3 per unit of blood per year.
- The demand for type O positive blood has stationary independent increments, with an expected demand of 36,500 units per year (100 units per day).
- The demand D for type O positive blood during the five-day resupply time is normal, with a mean of 500 units and a standard deviation of 110 units.

For rare blood types, there is a possibility that blood will become "outdated," that is, too old to use. This has never occurred with type O positive blood, however, and you have decided to ignore the possibility of outdating, at least initially.

---

[1] The term *stationary independent increments* was discussed in Chapter 11. It implies that the demands in nonoverlapping periods of the same length are i.i.d. random variables, among other things.

Problem G has these data: a demand rate $A = 36{,}500$ units per year, an ordering cost $K = 50$ dollars per order, a holding cost $H = 3$ dollars per unit per year, a replenishment time L of 5 days or $5/365$ years, a backorder cost $b = 2$ dollars per unit, and demand D during replenishment time that has the normal distribution with parameters $\mu = 500$ and $\sigma = 110$. Borrowed blood is backordered because the amount that the hospital borrows is deducted from the next order.

The Reorder Point and the Order Quantity

To an extent, the saw-toothed curve in Figure 12.7 continues to describe the fluctuation in inventory level. What remains the same, and what has changed? As in Figure 12.7, the inventory position $I(t)$ jumps abruptly on receipt of an order and dwindles between replenishment points, but now $I(t)$ decreases at random, with A as the expectation of the rate of decrease. There is still a reorder point, r, but the demand D during the replenishment time is random, and it may be prudent to set $r > E(D)$ to protect against backorders.

The economic lot size model had a single decision, the order quantity. With uncertain demand, there are two decisions to make: the order quantity q and the reorder point r. We designate:

$$q = \text{the number of units that are ordered.}$$

$$r = \text{the inventory on hand at the moment an order is placed.}$$

$$C(q, r) = \text{the expected annual cost of an policy that orders } q \text{ units each time the inventory level is reduced to } r \text{ units.}$$

To measure $C(q, r)$, we must account for three types of annual cost—the inventory carrying cost, the ordering cost, and cost of backorders.

Annual Ordering Cost and Holding Cost

Let us first specify the annual ordering cost. The expected demand per year equals A units. Each order is for q units. So

$$\frac{A}{q} = \text{expected number of orders per year.}$$

Each order incurs a fixed cost of K dollars, so

$$\frac{A}{q}K = \text{expected annual ordering cost.}$$

Let us next specify the annual backorder cost. On average, A/q orders are received each year. At the moment an order is received, the number of units backordered equals $(D - r)^+$, this being the excess of the demand D during the replenishment interval over the inventory level r at the moment the decision to replenish is made. Thus, $E[(D - r)^+]$ equals the expectation of the number of units that are backordered at the moment an order is received. Each backorder costs b, and so

$$\frac{A}{q}bE[(D - r)^+] = \text{expected annual backorder cost.}$$

Since D has the normal distribution with parameters $\mu = 500$ and $\sigma = 110$, we know from Chapter 11 that $E[(D - r)^+]$ equals the normal loss function, which can be computed by the Excel add-in **=NL(r, 500, 110)**.

Annual Inventory Carrying Cost

It remains to specify the annual inventory carrying cost. The physical inventory is the number of units on hand, which equals zero when there is a backlog. To measure the size of the backlog, when one exists, we introduce the concept of the inventory position, defined as follows. The **inventory position** x is a number that can be positive, negative, or zero: when x is nonnegative, there are x units on hand; when x is negative, a backlog of $-x$ units exists. Of what use is the inventory position? Receipt of an order of q units increases the inventory position by q units. In particular,

$$r - D = \text{the inventory position at the moment before replenishment,}$$
$$r - D + q = \text{the inventory position at the moment after replenishment.}$$

Suppose, for instance, that $r - D = -12$. A backlog of 12 units exists. Receipt of an order of $q = 60$ units increases the inventory position from -12 to $-12 + 60 = 48$. The backlog has been exhausted, and 48 units remain on hand. Similarly,

$$r - E(D) = \text{the expected inventory position at the moment before replenishment,}$$
$$r - E(D) + q = \text{the expected inventory position at the moment after replenishment.}$$

Figure 12.10 plots the expected inventory position versus the time t since the last replenishment. At time 0, the moment of replenishment, the expected inventory position jumps from $r - E(D)$ to $r - E(D) + q$. The expected inventory position then drops at the rate of A. When the inventory position decreases to r, an order is placed, and the expected inventory position equals $r - E(D)$ at the moment just before that order is received. At that point, the cycle is renewed.

It is evident from Figure 12.10 that $r - E(D)$ is a **safety stock** and that q is a **cycle stock**. Of these q units, on average, half are present. Thus,

$$\left[r - E(D) + \frac{q}{2}\right] = \text{average inventory position.}$$

In the following, we follow standard practice, which introduces an error:

$$\left[r - E(D) + \frac{q}{2}\right]H = \text{expected annual inventory carrying cost.}$$

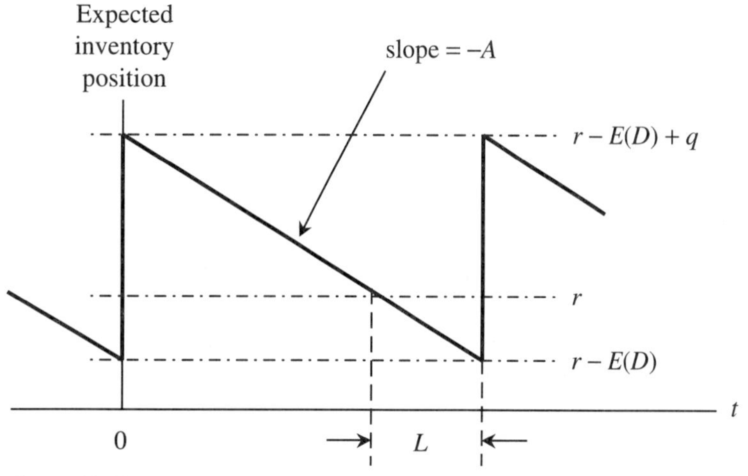

Figure 12.10 Expected inventory position versus time t since replenishment.

Did you spot the error? We have just applied the holding cost to the inventory position, *not* to the physical inventory.

We have underestimated the inventory carrying cost because the physical inventory exceeds the inventory position when a backorder exists. Backlogs are infrequent and of short duration, so the error is small. This error can be fixed. In fact, in most instances, the error is minuscule and so is not worth fixing.

We have now accounted for the annual ordering, backorder, and holding costs. The total $C(q, r)$ of these annual costs is given by

$$C(q, r) = \left[r - E(D) + \frac{q}{2}\right]H + \frac{A}{q}K + \frac{A}{q}bE[(D - r)^+]. \tag{12.24}$$

For Problem G, the demand D during the replenishment interval has the normal distribution with mean of 500 and standard deviation of 110. For this example, the normal loss function computes $E[(D - r)^+]$. It remains to find values q^* and r^* that minimize $C(q, r)$. The function $C(q, r)$ is nonlinear, and the problem of minimizing it may seem daunting. We will approximate its solution, and then we will compute it exactly.

An Approximation

Let us begin with an approximation to the values q^* and r^* that minimize $C(q, r)$. This approximation sets

$$q^* = \sqrt{\frac{2AK}{H}} \quad \text{and} \quad P(D \leq r^*) = 1 - \frac{q^*H}{bA}. \tag{12.25}$$

For the data in Problem G, the approximation in Equation (12.25) gives

$$q^* = \sqrt{\frac{(2)(36{,}500)(50)}{3}} = 1103.03,$$

$$1 - \frac{q^*H}{bA} = 1 - \frac{(1103.03)(3)}{(2)(36{,}500)} = 0.955,$$

r^* **=NORMINV(0.955, 500, 110)** = 686.1.

How does this approximation work? It consists of two steps. Step 1 imagines that there is no uncertainty in D, in which case the model reduces to a deterministic economic lot size model, whose optimal order quantity q^* is given by the square root formula in Equation (12.25). Step 2 accepts this value of q^* as a datum (fixed number) and uses a marginal analysis to compute the least costly reorder point, r^*.

To execute this marginal analysis, we denote as EMC$(r \to r + 1)$ the expectation of the change in the annual cost if the reorder point is increased from r to $r + 1$, with the reorder quantity q^* given by Equation (12.25). The tree in Figure 12.11 measures this change. A cost of H goes on its stem because increasing r by 1 increases the annual carrying cost by H. If, during a reorder interval, the demand D is at least as large as $r + 1$, increasing r by 1 has reduced the number of backorders by 1, and b dollars have been saved. On average, the number of reorders per year equals A/q^*, for a total savings of bA/q^* in the event $D \geq r + 1$, which explains the cost $-bA/q^*$ on the top branch of Figure 12.11.

From Figure 12.11, we see that EMC$(r \to r + 1)$, the expected marginal cost of increasing the reorder point from r to $r + 1$, is given by

$$\text{EMC}(r \to r + 1) = H - P(D > r + 1)\frac{bA}{q^*}.$$

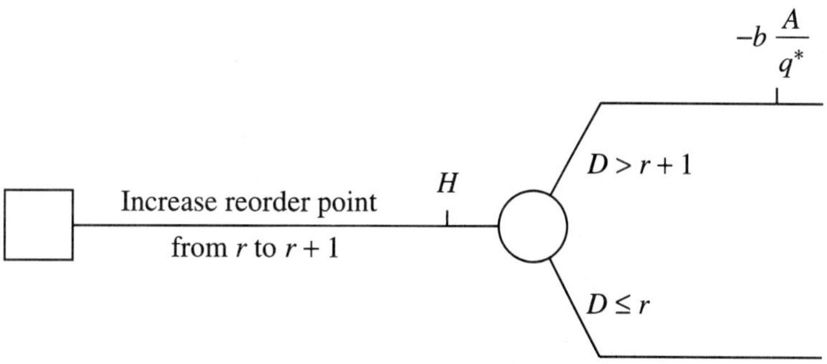

Figure 12.11 The marginal cost of increasing r by 1, with fixed value of q^*.

Substituting $1 - P(D \le r)$ for $P(D > r + 1)$ in the preceding equation and setting $\text{EMC}(r^* \to r^* + 1) = 0$ produce

$$P(D \le r^*) = 1 - \frac{q^*H}{bA},$$

which is the expression for r^* in Equation (12.25).

Successive Approximation

Having estimated q^* and then r^* from Equation (12.25), we could improve our estimate of q^* by re-solving the deterministic lot size model with the value of r^* that was just obtained as a datum (fixed number). The effect of this is to replace K in the square root formula by $[K + bE[(D - r)^+]]$. With this new value of q^*, we could then improve the estimate of r^* by re-computing the critical fractile in Equation (12.25). Iterating this process is known as **successive approximation**. For the problem at hand, it works well, but there's a better idea.

A Nonlinear Program

A **nonlinear program** differs from a linear program in that one or more of the expressions is not linear. Solver finds solutions to many nonlinear programs but not to all of them. There can be no panacea for nonlinear programs, but there is good reason to believe that Solver will work well for the problem at hand. The cost function $C(q, r)$ exhibits increasing marginal cost in q and r. Solver uses this sort of convexity to "home in" on optimal values of q and r.

The spreadsheet in Table 12.8 computes $C(q, r)$. For whatever values of q and r are placed in cells D3 and D4, this spreadsheet computes $C(q, r)$ and places that value in cell D11; rows 7 through 11 explain how. Solver has been used to minimize the value in cell D11, with D3 and D4 as changing cells, subject to the constraint $D3 \ge 0$. Table 12.8 reports the solution that Solver has found.

Table 12.8 reports the optimal solution q^* and r^* and its annual cost $C(q^*, r^*)$. Specifically,

$$q^* = 1150, \quad r^* = 684, \quad C(q^*, r^*) = 4000.$$

Thus, expected annual cost is minimized by ordering 1150 units of O positive blood each time the inventory position goes below 684 units. This optimal solution is not far from the two-stage approximation that we obtained earlier.

Table 12.8 Spreadsheet for Problem G, Solver having minimized the value in cell D11, with cells D3 and D4 as changing cells and with the constraint D3 ≥ 0.

	A	B	C	D
1	A =	36,500	μ =	500
2	K =	50	σ =	110
3	H =	3	q =	1149.52
4	b =	2	r =	683.94
5				
6	**quantity**	**formula**	**Excel function**	**value**
7	# backordered	$E[(D-r)^+]$	= NL(D4, D1, D2)	2.15
8	holding cost	$H(r - \mu + q/2)$	= B3 * (D4 - D1 + D3/2)	2276.11
9	ordering cost	$(A\,K)/q$	= B2 * B1 / D3	1587.62
10	backorder cost	$b\,E[(D-r)^+]\,A/q$	= B4 * D7 * B1 / D3	136.66
11	total cost	$C(q, r)$	= D8 + D9 + D10	4000.39

By varying the model in Table 12.8, one can see how the optimal policy varies with the demand rate and, therefore, with the blood type. To "scale" the demand by the factor *f*, multiply the entries in cells B1 and D1 by *f*, and multiply the entry in cell D2 by the square root of *f*.

Earlier, we observed that Equation (12.24) underestimates the inventory carrying cost. Problem 25 on page 440 uses information in Table 12.8 to sketch an argument that the amount by which cost has been underestimated is miniscule.

12.9. REVIEW

Inventory control problems exist in a great many varieties, several of which this chapter has introduced. For instance:

- The newsvendor model is a *single-period* model because decisions made in each period have no impact on the next period.
- The base stock model is a *multiperiod* model because inventory that is not consumed during a period remains available for later periods.
- The base stock model has *periodic review* because its inventory position is reviewed periodically.
- By contrast, the economic lot size model has *continuous review* because the inventory position is monitored continuously.
- The version of the base stock model that we studied has *lost sales* because customers whose needs cannot be met go elsewhere.
- In the EOQ model with uncertain demand, we studied the case of *backlogging*, in which customers whose demands cannot be met wait for the order to be received.
- For the base stock model, we used a *service criterion*, for example, no stock-out in 99% of the periods.
- And we have used an *economic criterion*, that is, maximize expected profit.

Which combination of features best fits a particular situation depends on its details.

Throughout this chapter, we have focused on the control of a single item of inventory, treating this item as distinct from the production and distribution system in which it has been embedded. The models we have introduced reveal themes that hold broadly and enable analysis of more complex systems. These themes include:

- *Marginal analysis:* In a sense, this entire chapter is an exercise in thinking at the margin. Marginal analysis works well in any setting that exhibits decreasing marginal return, equivalently, increasing marginal cost.
- *Tradeoffs:* Inventory manages the tradeoff between conflicting goals. We have used a cycle stock to balance the fixed charge for placing an order with the inventory carrying cost. We have used a safety stock to balance the cost of stockout with the inventory carrying cost.
- *Square root economy of scale:* Typically, the cycle stock and the safety stock increase in rough proportion to the square root of the demand.
- *Flat bottom:* Typically, cost increases only slightly as the cycle stock q varies by a factor of 2 from its optimal value.
- *Normal approximation:* Another theme, only lightly touched, is that the normal distribution can provide a good fit and can be especially tractable.

From a computational perspective, we have employed three tools that are particularly well suited to spreadsheets—marginal analysis, numerical integration, and nonlinear programming.

12.10. HOMEWORK AND DISCUSSION PROBLEMS

1. **(The Nighthawks)**[2] City residents deposit their refuse on the curb at the start of each working day. The amount of refuse that appears each day is an uncertain quantity whose probability distribution is as follows:

Refuse (in tons)	15	20	25	30	35	40	45	50	55	60
Probability	0.025	0.075	0.15	0.20	0.175	0.15	0.10	0.075	0.025	0.025

The amounts of refuse produced on different days are independent of each other. Refuse is collected by two-person crews, with each crew collecting 5 tons of refuse per 8-hour working day. Each crew earns $256 per day. Whatever refuse remains at the end of the work day *must* be collected by a special team, called the Nighthawks, who pick up the surplus at a cost of $520 per ton. The city aims to minimize the expected cost of picking up the trash.

(a) What number q^* of regular-time crews should the city employ?

(b) Designate as $C(q)$ the expected cost of using q crews. Compute $C(q)$ and plot it versus q.

2. **(Good Will)** In Problem A within the chapter, Art's primary income comes from the supplementary items that customers purchase when they drop by to buy his newspapers. If Art cannot sell them a newspaper, they may switch their patronage to a competitor. Art estimates that failing to satisfy a customer's need for a newspaper costs him $2.00, on average, in future earnings. How many newspapers should Art order? Why? (This $2.00 is called a **good will** cost.)

3. **(Planning for Electrical Generation Capacity)** Energy has been deregulated, and your company has contracted to supply the local grid with electrical power at a rate of between 8 and 10 trillions (10^{12}) of BTUs. The demand for this energy varies with the season and the time of day, as indicated below. These data show, for instance, that demand occurs at the peak rate of 10 trillion BTUs/year for 1% of the year.

[2] Adapted from a problem written by Arthur J. Swersey and the author.

Demand Rate	8.0	8.2	8.4	8.6	8.8	9.0	9.2	9.4	9.6	9.8	10.0
Fraction of year	0.01	0.02	0.06	0.10	0.22	0.20	0.15	0.10	0.08	0.05	0.01

You must decide what plant to build. Base-load plant is more capital intensive but burns fuel more efficiently. The annualized capital cost of base-load plant is \$3.657 per million BTUs of output. The annualized capital cost of peak-load turbines is \$3.070 per million BTUs of output. The fuel cost is \$12.33 per million BTUs of output for base-load plant and is \$16.71 per million BTUs of output for peak-load plant. Capital costs are incurred for all purchased capacity, whether or not it is used. Fuel costs are incurred for the fuel burned, of course. Designate as q the capacity of the base-load plant that the company builds, measured in millions of BTUs. The remaining capacity consists of peak-load plant.

(a) What value q^* minimizes annual cost?

(b) For each value of q compute the cost $C(q)$ of a design that builds a capacity q units of base-load plant and the remainder of peak-load plant. Plot $C(q)$ versus q.

(c) You could supply the peak-load demand for which you have contracted by purchasing electricity from other suppliers at market prices. Find the largest market price p for which it would be profitable to build no peak-load capacity.

(d) Identify the risk, if any, of supplying any of the demand for which you have contracted by buying it at the market price rather than by building production capacity.

4. (Noxious Emissions)[3] A company operates a plant that smelts lead and produces sulfur dioxide (SO_2) as a noxious byproduct. Because of variations in ore composition and product mix, the quantity of SO_2 produced each month is uncertain. Its distribution is normal, with a mean and standard deviation of 6 and 2 tons per month, respectively. The Environmental Protection Agency (EPA) wants this company to reduce its emissions by purchasing scrubbers with a capacity of 7.5 tons per month. The company can purchase any scrubbing capacity it wishes at an amortized cost of \$20,000 per month of SO_2 removal capacity; this cost is incurred whether or not the scrubbers are used. Operating scrubbers cost an additional \$8000 per ton of SO_2 removed, up to the scrubbing capacity. Any SO_2 produced in excess of scrubber capacity is discharged into the environment.

(a) Suppose the tax is \$75,000 per ton of SO_2 released into the environment. What amount of scrubbing capacity minimizes the company's expected cost?

(b) What is the smallest tax that will induce the company to install scrubber capacity sufficient to remove 7.5 tons/month of SO_2?

(c) Suppose the tax in part (b) is imposed. What is the probability that SO_2 emissions occur in a given month? What is the expectation of the number of tons of SO_2 emitted during a month? For a month in which emissions do occur, what is the expectation of the number of tons of emission?

5. (Fleet Size) The operations manager of a national auto rental company must determine the size of the fleet to purchase each year, and she must allocate that fleet among the company's rental locations. Each car that she purchases incurs a fixed charge of \$4000 per year. The demand for cars varies with the location, as does the contribution.

(a) Cars are rented for one day. The daily demand at a particular location is Poisson with a mean of 60. Demands on different days are independent, with the same distribution. The contribution at that location is \$30 per car rented. How many cars should be located there?

(b) What is the company's expected contribution from car rental at that location?

(c) Suppose that, due to one-way rentals, the supply of cars at this location fluctuates at random by an amount X whose expectation equals zero, with $|X| < 6$. By how much can the profit be degraded?

(d) Describe a procedure for determining the fleet size that maximizes contribution.

[3] Adapted from a problem written by Kurt Anstreicher.

6. (For Problem B, Art's Calendar Sales) On reflection, Art thinks that the demand D for his Adirondack Life calendars has the lognormal distribution with the same mean and standard deviation, that is, with $E(D) = 150$ and $\text{StDev}(D) = 40$.

(a) On a spreadsheet, compute the parameters μ and σ of this distribution. *Hint:* review Equation (12.54) in Chapter 11.

(b) The overage and underage cost in Problem B are unaffected, so the critical fractile is unchanged. Art seeks the order quantity q^* for which $P(D \leq q^*)$ equals 0.75. On your spreadsheet, compute q^*. *Hint:* Solver does a nice job of computing a fractile.

(c) Is q^* lower or higher than for the case of normal demand? Why?

(d) On your spreadsheet, use Equation (12.12) to approximate the function $R(q)$. Draw a graph of your approximation.

7. (Prepaid Medical Expense) U.S. tax law allows an employee to prepay for uninsured medical expense in a way that can save money by reducing federal and state income tax. To see how this works, we consider a particular taxpayer, Joe. His marginal income tax rate is 40%. Joe's uninsured medical expense X for the forthcoming year is uncertain. Based on past experience, Joe estimates that X is normally distributed with a mean of \$2000 and a standard deviation of \$400. Before the beginning of each calendar year, Joe (like any employee) can declare an amount q of uninsured *prepaid* medical expense. Joe's salary for the year is *reduced* by q dollars. Joe's company reimburses Joe for the first q dollars of his medical expense for the forthcoming year. Joe pays any excess, $(X - q)^+$, out of after-tax income. Prepaying q dollars of medical expenses saves Joe $0.4q$ in after-tax income because he does not pay taxes on the amount reduced from his salary. But there is a downside. The law requires Joe's employer to keep any excess $(q - X)^+$ of Joe's prepaid medical expenses over his actual medical expense. Joe's goal is to prepay the amount q^* that maximizes the expectation of his after-tax income.

(a) What amount q^* should Joe choose? Justify your answer.

(b) For the value q^* selected in part (*a*), compute:

(i) The probability that Joe will need to pay for any uninsured medical expense with after-tax dollars.

(ii) The probability that Joe's company will pay fewer than q^* dollars toward Joe's uninsured medical expense.

(iii) The expectation of the amount of uninsured medical expense that Joe will pay with after-tax dollars.

(iv) The expectation of the amount by which q^* exceeds the amount of Joe's uninsured medical expense that his company will pay.

(v) The expectation of the increase in after-tax income that Joe garners from prepaying q^* dollars of medical expense.

8. (Prepaid Medical Expense, continued) The tax rules are as stated in the prior problem. Again, Joe estimates that his uninsured medical expense X for the forthcoming year has the normal distribution with a mean of \$2000 and a standard deviation of \$400. Now, X excludes the expense of a \$300 pair of prescription glasses that Joe will order in December of the coming year for delivery in the following January. After Joe learns the value that X takes, he determines how much (if any) of this \$300 expense to pay in December. He pays the remainder of this cost in January of the following year. Joe knows what he will pay in January *before* he determines the amount of uninsured medical benefits to deduct for the following year. Thus, the sole benefit of paying in December is to avoid prepaying more than X during the current year. Redo parts (a) and (b) of the prior problem to account for these eyeglasses. What changes? by how much? Why? In your answer, account for $E(G)$ where G is the amount Joe pays this year toward the cost of these eyeglasses.

9. (Computing $R(q)$ Directly) For Art's newsstand (Problem A within the chapter), the random variable $Y(q) = 20 \min\{D, q\} - 28 \max\{0, q - D\}$ denotes Art's net profit when he orders q newspapers. This problem shows how to compute the mean and variance of $Y(q)$ directly rather than by marginal analysis.

(a) Open a spreadsheet. In cells B1 through L1, enter the integers 20, 21, . . . , 30. For each number k that you placed in row 1, enter the probability $P(D = k)$ below it in row 2.

(b) Place the order quantities $q = 20, 21, \ldots, 30$ in cells A3 through A13. In cells B3 through L13, specify the value taken by Y_q for the value of q and each value of D. *Hint:* One function and one drag should suffice.

(c) In columns M, N, and O, compute $E(Y_q)$ and $E[(Y_q)^2]$ and StDev(Y_q) for the values of q in column A. *Hint:* One SUMPRODUCT function will have a single colon, and the other will have two colons.

(d) Plot the mean and standard deviation of Y_q versus q. Of the order quantities 23 through 27, which entails this least risk?

10. **(Numerical Integration)** In Figure 12.5, the change in the function's value over an interval is approximated by the product of the width of the interval and the slope of the function at the midpoint of the interval. What is the advantage of using the midpoint rather than one of the end-points?

11. **(Stocking CDs)** A music store is restocked weekly. The weekly demand for a best-selling CD is normal, with mean of 200 and standard deviation of 50. These CDs cost \$8.00 apiece, and the store earns a contribution of \$4.50 per CD sold. The store estimates its inventory carrying cost as \$0.20 per CD per week. Excess demand is lost; customers go to a competitor's store rather than wait for resupply.

(a) What order policy maximizes expected profit?

(b) Suppose that exactly 10% of the start-of-week inventory of this CD disappears from the store without being sold; this is euphemistically known as a 10% *shrinkage* rate. Now, what order quantity maximizes expected profit?

12. **(College Admissions)** The director of Admissions of your college has asked for your advice. She needs to determine the number of offers of admission to make, as well as the number of wait list offers to make. The target size for the freshman class is 1230 persons (freshmen). By agreement among a large number of colleges, including yours, the following procedure is employed by each college in the group:

- On a particular date in the spring, each college informs each applicant to it that one of three actions has been taken on that person's application—Admit, Wait List, or Decline.
- Within several weeks, all applicants who were admitted must accept or reject their offers. In addition, all applicants who were offered positions on Wait Lists must indicate whether or not they wish to accept those positions.
- Subsequently, applicants who elect to remain on each Wait List are ranked. The ranked Wait List candidates are offered admission, one by one: the process stops when the Wait List is exhausted or when the college has attained the class it wishes, whichever comes first.

It is understood that an applicant can accept admission to a college and, simultaneously, place himself or herself on one or more Wait Lists, hoping to gain admission later on to a college that he or she prefers.

The percentage of the candidates who accept offers of admission *and* join the entering class for your college has been stable at 55% for the past several years. There is no reason to anticipate that the current year will differ from earlier years. About 60% of the applicants who are offered positions on the Wait List decline those positions. Of the 40% who accept positions on the Wait List, 80% will accept offers of admission and join the entering class if offered the opportunity to do so.

Your college has no shortage of qualified applicants from which to select a balanced class of 1230 persons. But dorm space is tight. The college dean is willing to accept 1 chance in 20 that the admissions process will result in an entering class of more than 1230 persons.

The provost reports that the contribution per student (tuition plus board less variable cost) is \$10,000. The provost also reports that the cost of offering each person a position on the Wait List is \$100. This includes the variable cost of dealing with these people and their relatives as well as the loss of good will due to friends' and relatives' threats to withhold contributions. The director of Admissions has asked you to advise her as to the number t of Admit offers to make and the number w of Wait List offers to make.

(a) Without (as yet) specifying the best number t of Admit offers to make, what can you say about the number $Y(t)$ of students who will join the class if t applicants receive Admit offers? Is $Y(t)$ random?

If so, what is its mean? its variance? its distribution? Is there a convenient approximation to its distribution?

(b) How large should t be? Interpret $[Y(t) - 1230]^+$. Compute or approximate its expectation.

(c) Without (as yet) determining the best number w of Wait List offers to make, what can you say about the number $X(w)$ of students who will join the class from a pool of w applicants who are offered positions on the Wait List, assuming that *all* of those who accept positions on the Wait List are subsequently offered admission?

(d) How large should w be? *Hint:* to do a marginal analysis, ask yourself, "Under what circumstance has the director of admissions any interest in the $w + 1$st person to be offered a position on the Wait List?"

(e) For the values of t and w that you have selected in parts (b) and (d), interpret $Y(t) + X(w)$. Say what you can about its distribution. Interpret $[1230 - Y(t) - X(w)]^+$. Compute or approximate its expectation.

(f) Describe the probability distribution of the size C of the entering class.

13. (Business Class No Shows) For Problem C within the chapter, compute the expected profit $R(q)$ for the values $q = 0, 1, \ldots, 50$. *Hint:* Combine the calculation in Table 12.5 with Equation (12.13), and note that $P(N_q \leq 31) = 1$ for $q = 0, 1, \ldots, 31$. Plot $R(q)$ versus q.

14. (Business Class No Shows) In Problem C within the chapter, Business class customers could cancel their reservations without notice and without fee. Adapt the prior analysis of Problem C to encompass the case in which each Business class customer who makes a reservation pays a nonrefundable fee of \$50.

15. (Yield Management) An airplane has 200 seats, each of which is identical to the others. The airline sells these seats to two populations of travelers, at different prices. Economy class travelers buy their seats at least 14 days in advance of the flight on a nonrefundable basis. Business class travelers make their travel plans within six days of the flight. The airline earns a contribution of \$125 for each Economy class seat it sells. It earns a contribution of \$400 for each Business class seat that it sells. The demand D_E for economy class seats on this flight is normally distributed with mean and standard deviation of 180 and 50, respectively. The demand D_B for Business class seats on this flight is normally distributed with mean and standard deviation of 50 and 20, respectively. These demands are mutually independent.

(a) Is it reasonable that the random variables D_E and D_B have the normal distribution? If so, why? Is it reasonable that these random variables be independent of each other? If so, why?

(b) The airline wishes to maximize expected contribution from this flight. Of the 200 seats, how many should be set aside for Business class customers? Justify your answer.

(c) On what data does your answer to part (b) depend?

16. (Training Sessions) A company needs to train service representatives at the rate of 40 per month. Each training class lasts one month and costs \$15,000, excluding the salaries of the trainees. Each trained service representative earns \$1500 per month, including benefits. Each trainee earns \$750 during the month of training.

(a) How frequently should the company conduct training classes, and how many should it train in each class?

(b) Suppose that, due to a shortage of teachers, the company cannot run two training classes concurrently. Does this make a difference? If so, what should the company do?

17. (Cash Withdrawals) Joe spends cash at a uniform rate of \$12,000 per year. He withdraws this cash from his checking account, which earns interest at the rate of 4% per year. Each trip to the bank to withdraw cash takes 30 minutes, and Joe values his time at \$30 per hour.

(a) How frequently should he go to the bank? How much should he withdraw each time? Is Joe maintaining a cycle stock, and, if so, how large is it?

(b) Suppose Joe values his time at \$60 per hour, not \$30 per hour. Redo part (a).

(c) Suppose Joe withdraws cash twice a month. What value does he place on his time?

18. (**EOQ with Backorders Allowed**) Augment the deterministic economic lot size model by allowing backorders; the company incurs cost at the rate of B dollars per year per unit backordered. An inventory control policy is now governed by two parameters—an order quantity Q and a shortage level S at which the order is placed. The inventory position has the saw-toothed shape given below; note that the length of each cycle is Q/A years.

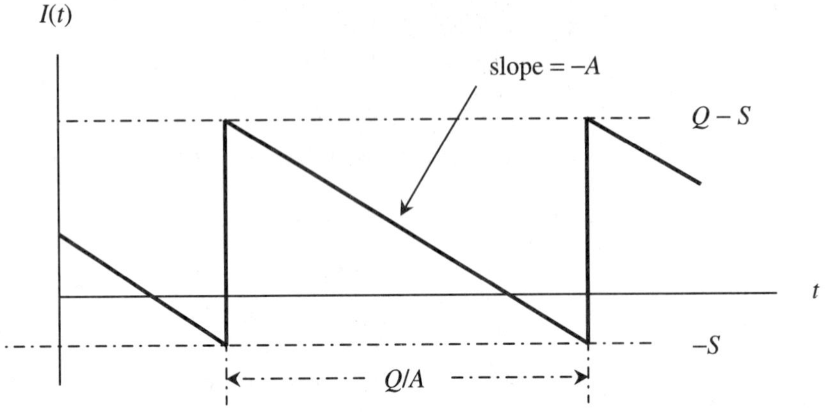

(a) Demonstrate that the annual cost $C(Q, S)$ of this policy is as given below. *Hint:* compute the cost per cycle, and multiply by the number of cycles per year.

$$C(Q, S) = \frac{A}{Q}\left[K + \frac{H(Q - S)^2}{2A} + \frac{BS^2}{2A}\right].$$

(b) Show that for any order quantity Q, the cost-minimizing shortage amount S_Q satisfies

$$S_Q = Q\frac{H}{H + B},$$

and, as a consequence,

$$C(Q, S_Q) = \frac{A}{Q}K + \hat{H}\frac{Q}{2} \quad \text{with} \quad \hat{H} = \frac{HB}{H + B}.$$

(c) Specify the order quantity Q^* and shortage level S^* that minimize annual cost.

19. (**EOQ with Price Breaks**) The unit ordering cost has been excluded from the economic lot size model. Now, let us consider a schedule of **price breaks** whose data are $0 < d_1 < d_2 < d_3$ and $0 < q_1 < q_2 < q_3$, with

Size of Order	Unit Cost
$0 < Q < q_1$	C
$q_1 \leq Q < q_2$	$C - d_1$
$q_2 \leq Q < q_3$	$C - d_2$
$q_3 \leq Q$	$C - d_3$

For instance, the price is reduced by d_1 dollars per unit for order quantities that are at least q_1 units.

(a) Show that for Q in the range $q_1 \leq Q < q_2$, annual cost is reduced by d_1A and that for Q in the range $q_2 \leq Q < q_3$, annual cost is reduced by d_2A, and so forth.

(b) True or false: With Q^* as the EOQ with no price breaks, these price breaks can cause the optimal order quantity to increase but not to decrease.

(c) Pick k such that $q_{k-1} \leq Q^* < q_k$. True or false: the optimal order quantity is either Q^* or q_p for some $p \geq k$.

(d) Redo Problem F with these price breaks: 0.10 dollars per unit for order quantities of at least 10,000 compressors and 0.20 dollars per unit for order quantities of at least 20,000 compressors.

20. (EOQ with Coordinated Orders) A company manufactures a large number of different models of printers on the same production line. Each switch in production between one model and another incurs a fixed charge because the line must normally be shut down for two hours. However, printers B, C, and D are similar enough that it only takes 0.5 hour to switch the line from production of one of these three to another. Shutting this line down costs \$1200 per hour in lost contribution. The demand for printers B, C, and D occurs at the constant rates of 3400, 4600, and 50 per year, respectively. The holding cost for each of these printers is \$130 per year. Should the orders for any or all of these printers be coordinated? If so, how and why?

21. (Flat Bottom for the EOQ Model) For the economic lot size problem:

(a) Verify that $C(q^*) = 2(q^*H/2) = 2(KA/q^*)$.

(b) Use part (a) to verify Equation (12.23).

(c) In Equation (12.23), substitute $q = q^*\sqrt{2}$ and $q = q^*/\sqrt{2}$ in Equation (12.23), and observe that $C(q^*)/C(q) \cong 1.06$ in both cases.

22. (Multiple Warehouses) A company satisfies demand for a particular product from a warehouse that is co-located with its production facility. This product fits the hypothesis of the economic lot size model with backlogging and uncertain demand. Its data are: $A = 36{,}500$ units/year, $H = 6$ dollars per unit per year, $K = 1600$ dollars per order, and $L = 14/365$ years, $E(D) = 1400$, $\text{StDev}(D) = 500$, and $b = 25$ dollars per unit.

(a) Find the parameters q^* and r^* of the inventory control policy that minimizes expected annual cost, and compute that cost. *Hint:* Use a nonlinear program.

(b) To reduce shipping costs, the company plans to replace its central warehouse with four regional warehouses. Each regional warehouse would handle one-fourth of the business. Each regional warehouse would act independently of the others. As concerns this product, each warehouse would experience these data: $A = 36{,}500/4$, $H = 6$, $K = 1600$, $L = 14/365$ years, $E(D) = 1400/4$, $\text{StDev}(D) = 500/2$, and $b = 25$. Find the parameters q^* and r^* for each warehouse, as well as the annual cost of satisfying its demand.

(c) Compare the solutions in parts (a) and (b). Roughly speaking, what happens? Under what circumstances are regional warehouses beneficial?

23. (Fixed Time between Orders) In Problem G within the chapter, the blood bank accepted orders for a fixed quantity q but with an uncertain time between orders. The blood bank has changed its policy. To help it plan its blood drives, the blood bank now insists on a fixed time between orders, but it allows each hospital to order an uncertain quantity Q. The data are as in Problem G. One parameter of an ordering policy is the number t of days between orders. The other parameter is the expectation s of the inventory position at the moment after the order is received. The delivery lag is five days; so the inventory position is observed five days before the order will be received, and enough blood is ordered to restore to s the expected inventory position on receipt of order.

(a) Prove or disprove: the demand D_t for type O positive blood that occurs over an interval of t days is normal with parameters $E(D_t) = 100t$ and $\text{StDev}(D_t) = 110\sqrt{t/5}$.

(b) Let the random variable S denote the inventory position at the moment after the order is received. Describe S. *Hint:* Is S normal? If so, what are its mean and its standard deviation?

(c) Describe the inventory position F at the moment before the order is received. *Hint:* Does $F = S - D_t$?

(d) The safety stock equals the expectation of F. Express the safety stock in terms of s and t. Express the cycle stock in terms of these variables.

(e) The expected number of backorders per period equals $E(-F)^+$. Specify this expectation in terms of the normal loss function.

(f) Use parts (d)–(e) to specify the expected annual cost $c(s, t)$ of a policy whose parameters are s and t.

(g) Use Solver to find the values s^* and t^* that minimize $c(s, t)$.

(h) Is this policy more expensive than the one in Problem G? If so, what accounts for the added expense?

24. (**Lateral Resupply**) In military logistics, "lateral resupply" refers to the practice of seeking goods from other bases rather than from the depot. In *M*A*S*H*, lateral resupply was one of Radar's many talents. Your city has two hospitals that do not engage in lateral resupply. Both hospitals obtain their blood supplies from the regional blood bank. Hospital A is as described in Problem G within the text. Hospital B handles twice as many patients as does Hospital A. Otherwise, the characteristics of these hospitals are identical.

(a) What are the mean and standard deviation of the demand for type O positive blood in a five-day period in Hospital B? Justify your answer.

(b) For Hospital B, compute the values of the reorder quantity q and of the reorder point r that minimize expected annual cost. Compute its expected annual cost. Use $E(D - q)^+$ to compute P_B = the fraction of the time that Hospital B is out of blood. Compute the comparable number P_A for Hospital A.

(c) Compare the reorder policies for the two hospitals.

At a recent conference, Hospitals A and B have agreed to change their blood inventory systems. They will continue to purchase individually from the blood bank, but they have agreed to the following terms for lateral resupply:

- Should only one hospital be out of type O positive blood, the other will loan it all it needs,
- This loan will be repaid when the hospital that is out of blood receives its next shipment from the blood bank.

The cost of ferrying blood from one hospital to another is \$0.30 per unit; this includes the cost of repaying the loan. With this lateral resupply policy, hospitals will borrow from the blood bank only when both are out.

(d) Designate as b_A and b_B the effective backorder costs for the two hospitals. Is there a sense in which

$$b_A = 2P_B + 0.3(1 - P_B) \quad \text{and} \quad b_B = 2P_A + 0.3(1 - P_A)?$$

(e) Find the policies for the two hospitals that minimize their total cost, using lateral resupply. Use Solver, with coordinated backorder cost.

(f) What fraction of each hospital's blood is borrowed from the other? What fraction is borrowed from the blood bank? How, if at all, does lateral resupply improve the hospital's blood inventory systems?

(g) To what extent, if any, does lateral resupply aggravate or ameliorate the blood bank's inventory control problem?

25. (**Underestimate of the Holding Cost?**) In Section 12.8, we underestimated the holding cost by applying it to the "inventory position," which is below the physical inventory when there is a backorder. Let's estimate the error in the holding cost for the data in Table 12.8, which has $q^* = 1150$, an average of 2.15 units backordered at each moment at which q^* is received.

(a) Estimate the fraction of the ordering cycle for which any of these 2.15 units are backordered.

(b) Estimate the average size of the number backordered over the course of the year.

(c) Estimate the amount by which we underestimated the holding cost. Was it miniscule?

26. (**Alternate Verification of the Critical Fractile in Equation (12.25)**) In Equation (12.24), for any fixed value of $\hat{Q}$, we seek the value $\hat{r}$ that minimizes the expression

$$rH + bE[(D - r)^+]\frac{A}{\hat{Q}}.$$

This can be found by marginal analysis or by calculus. To sketch the latter approach, let $f(x)$ be the density function of D, and write the preceding expression as

$$rH + \frac{bA}{\hat{Q}}\int_r^{\infty}(x - r)f(x)dx.$$

Differentiate with respect to r. Does the derivative increase as r increases? Is the expression minimized by setting the derivative to zero? Show that

$$H - \frac{bA}{\hat{Q}} P(D > \hat{r}) = 0,$$

which verifies the critical fractile.

27. (**Economic Lot Size with Uncertain Demand and Lost Sales**) This problem asks you to adapt the analysis in Section 12.8 to the case of lost sales rather than backlogging. The model's data are exactly as in Section 12.8 except that b is omitted, and each lost sale incurs a penalty (loss of profit) of p. The policy remains the same: when the inventory position is reduced to r units, order q. Let α denote the expectation of the inventory position at the moment before the order is filled, and let β denote the expectation of the number of lost sales per ordering cycle.

(a) How, if at all, is α related to $E[(r - D)^+]$ or to $E[(D - r)^+]$?

(b) How, if at all, is β related to $E[(r - D)^+]$ or to $E[(D - r)^+]$?

(c) Show that α and β relate to each other through $\alpha = \beta + r - \mu$.

(d) In this model, what plays the role of safety stock? And what plays the role of cycle stock?

(e) In this model, which is a better measure of the number of orders per year, A/q or $A/(q + \alpha)$? Explain your answer.

(f) Let $C(q, r)$ denote the expected annual loss of profit for this policy, where "loss of profit" includes inventory carrying cost, ordering cost, and lost sales. Write an expression for $C(q, r)$.

(g) On a spreadsheet, find values of q and r that minimize $C(q, r)$ for the data in Table 12.8, with p = \$3 per unsold unit.

Chapter 13

Markov Chains

13.1. PREVIEW

The term *state* has appeared in several previous chapters. In a sequential decision process, a **state** is a summary of the prior history of the process that suffices to determine its law of motion and to evaluate the costs and revenues of alternative actions, if any. In Chapter 4, identifying the states helped us to formulate the tire fabrication example as a linear program. In Chapter 5, the nodes played the role of states in shortest-path problems. In Chapter 9, we saw that, in a decision tree, the choice nodes are states.

In this chapter, decision making is suppressed. Here, the focus is on a **Markov chain**—namely, on a process that evolves from state to state at random, the law of motion being determined solely by the current state. Each Markov chain is defined in terms of a set S of states and an array P of transition probabilities. This chapter treats the case in which S consists of finitely many states. The transition probabilities form the square array P in which

$$P(i, j) = P(\text{the next transition will occur to state } j \mid \text{state } i \text{ is observed now}).$$

This notation reflects the premise that the law of motion (transition probabilities) depends *solely* on the current state, and not on any prior states. Being probabilities, the $P(i, j)$'s are nonnegative. The transition probabilities are conditional probabilities.

Throughout this chapter, transitions are assumed to occur from each state to some state, so that the sum of the transition probabilities out of state i equals 1:

$$\sum_{j \in S} P(i, j) = 1, \quad \text{for each state } i.$$

In a Markov chain, the set S of all states is called the **state space**.

Markov chains exist in several varieties, two of which are treated here. In a **discrete-time** Markov chain, each transition takes a fixed amount of time, which is allowed to vary with the state i that is observed now and the state j to which the transition will occur. In a so-called **continuous-time** Markov chain, the times between transitions have the exponential distribution, with a mean that can depend on the state i that is currently occupied and the state j to which transition will occur.

Markov chains are an enormous subject, with many applications and a vast literature. This introductory chapter treats the following three questions:

- *The probability of reaching a particular state.* Given that state i is now observed, what is the probability that at least one transition to state j will occur?
- *The first passage times.* Given that state i is now observed and given that at least one transition to state j will occur, how much time will elapse until the first transition to state j does occur?
- *The long-term behavior.* Over a long period of time, how frequently will transitions to state j occur?

Each of these questions is introduced in the context of a specific problem. The discrete-time case is discussed first, and the basic results for the discrete-time case are then adapted to the continuous-time case. This chapter includes applications of Markov chains to two operational issues.

13.2. WHAT CAN YOU LEARN FROM THIS CHAPTER?

This chapter deals directly with the three problems we have described. It also serves several other purposes. From it, you can learn about:

- *Embedding.* The computation of the probability of a complicated event can be eased by use of a Markov chain to "embed" the problem of interest in a family of related problems, one per state, and build a system of equations that links their solutions.
- *Steady-state probabilities.* Often, one is interested in the long-term behavior of a probabilistic system. In the case of a Markov chain, the long-term behavior can be described in terms of a "steady-state" probability distribution over its states.
- *Transition diagrams and flux.* A "transition diagram" is a handy way to describe a Markov chain. This diagram also helps analyze a Markov chain because, in steady state, the "flux" (rate of flow) into any group of states must equal the flux out of that group of states.
- *The Erlang loss model.* This model is a continuous-time Markov chain that plays a central role in telecommunications. Studying it can clarify the role of transition diagrams, flux, and steady-state probabilities.
- *The needle exchange model.* This is a much-lauded application of operations research, one that illustrates the use of continuous-time Markov chains.

- *Reversibility.* If a condition called reversibility is satisfied, the steady-state probabilities are particularly easy to calculate.

If you wish to master all of the material in this chapter, read its sections in their normal sequence. This chapter is also written to enable you to skip around. If, for instance, your main goal is to prepare yourself for the chapter on queues, you might plunge directly into Section 13.10 (Telecommunications Traffic), using it to learn about transition diagrams and steady-state probabilities, referring back to earlier sections as needed, and then skim Section 13.12 (Reversible Markov Chains).

13.3. THE PROBABILITY OF REACHING A PARTICULAR STATE

The so-called Gambler's Ruin problem is used to introduce the problem of computing the probability of reaching a particular state in a Markov chain. This problem has several states, two of which are particularly important. The gambler hopes to reach the "good" state and to avoid the "ruinous" state. Let us consider:

Problem A (Gambler's Ruin)

A gambler possesses one dollar. His goal is to increase his asset position to six dollars. He can bet one dollar at a time on a game in which he wins one dollar with probability $p = 0.6$ and loses one dollar with probability $1 - p = 0.4$. The outcomes of different bets are independent of each other. If the gambler runs out of cash, he can bet no more and is ruined. He continues to bet until he reaches his goal of six dollars or is ruined, whichever event comes first. What is the probability that the gambler reaches his goal of six dollars?

It's convenient to model this problem as a Markov chain that has one state for each asset position that the gambler might reach. Thus, its state space S consists of the integers 0 through 6. If the gambler reaches state 0, he remains there; if he reaches state 6, he remains there. Whenever he is at an intermediary state i, he bets and observes a transition to state $i + 1$ with probability of 0.6 and a transition to state $i - 1$ with probability of 0.4.

A Graph of Transition Probabilities

Figure 13.1 depicts each state of this game as a node and each nonzero transition probability as an arc, with an arrow pointing to the state to which transition occurs.

As is indicated in Figure 13.1, $P(0, 0) = 1$ because the gambler must stop betting if his asset position is reduced to 0. Also, $P(6, 6) = 1$ because he elects to stop betting if he reaches his goal of 6 dollars. And, for each state i between 1 and 5, he bets and observes a

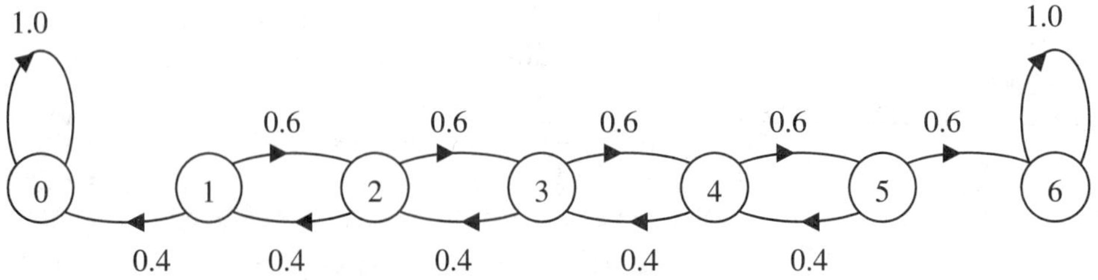

Figure 13.1 A graph of the transition probabilities for the Gambler's Ruin problem.

transition according to the probabilities

$$P(i, i+1) = 0.6 \quad \text{and} \quad P(i, i-1) = 0.4.$$

There are infinitely many ways in which the gambler can reach state 6 (his goal). Fortunately, we needn't enumerate them because the notion of a state will simplify the calculation of his probability of reaching state 6.

To gain familiarity with arrays (matrices), we record the array P of transition probabilities for the gambler's ruin problem. This matrix has seven rows and seven columns. Each row corresponds to a wealth position that is between 0 and 6 dollars, as does each column. The sum of the transition probabilities across each row of P equals 1. The five "below diagonal" entries record the fact that $P(i, i-1) = 0.4$ for i equal to 1 through 5, and the five "above-diagonal" entries record the fact that $P(i, i+1) = 1$ for the same values of i.

$$P = \begin{bmatrix} 1 & 0 & 0 & 0 & 0 & 0 & 0 \\ 0.4 & 0 & 0.6 & 0 & 0 & 0 & 0 \\ 0 & 0.4 & 0 & 0.6 & 0 & 0 & 0 \\ 0 & 0 & 0.4 & 0 & 0.6 & 0 & 0 \\ 0 & 0 & 0 & 0.4 & 0 & 0.6 & 0 \\ 0 & 0 & 0 & 0 & 0.4 & 0 & 0.6 \\ 0 & 0 & 0 & 0 & 0 & 0 & 1 \end{bmatrix}$$

This array P records the same information as does the transition diagram in Figure 13.1. Arrays like this will be handy when we get to the mathematics.

Embedding

The probability that the gambler will reach state 6 depends solely on the current state (asset position), and not on how that state was reached. Let us define $w(i)$ as follows:

$$w(i) = P(\text{state 6 will be reached} \mid \text{state } i \text{ is occupied now}).$$

Problem A asks us to compute $w(1)$, the probability that the gambler reaches his goal of six dollars given an initial asset position of one dollar. This problem has now been **embedded** in a family of problems, one per state. Embedding seems to be a step backward, for it replaces one problem with several.

But we will soon find that the problems are linked in a way that facilitates their solution. It's easy to see that

$$w(0) = 0 \quad \text{and} \quad w(6) = 1 \tag{13.1}$$

because the gambler has no probability of reaching his goal if his asset position is 0 and because he has reached his goal if his asset position is 6.

If the gambler's asset position is some number i between 1 and 5, he bets and reaches his goal with probability $w(i)$. We shall justify the equation,

$$w(i) = 0.6w(i+1) + 0.4w(i-1) \quad \text{for } 1 \leq i \leq 5. \tag{13.2}$$

Given state i, there are two ways in which to reach state 6. The gambler could win the next bet and then reach state 6; that event occurs with probability $0.6w(i+1)$. The gambler could lose the next bet and still reach state 6; that event occurs with probability $0.4w(i-1)$. These two events are disjoint, and they account for all of the ways in which the gambler can reach his goal, which justifies Equation (13.2).

Table 13.1 Spreadsheet computation of $w(i)$ for each state i.

	A	B	C	D	E	F	G	H	I
2	p =	0.6		Solver selects values in cells C11: G11 that satisfy I6:I10 = 0					
3	1 - p =	0.4							
4									
5	state i =	0	1	2	3	4	5	6	
6	1	-0.4	1	-0.6					0.00
7	2		-0.4	1	-0.6				0.00
8	3			-0.4	1	-0.6			0.00
9	4				-0.4	1	-0.6		0.00
10	5					-0.4	1	-0.6	0.00
11	w(i) =	0	0.365	0.609	0.771	0.88	0.952	1	
12									
13				=SUMPRODUCT(B10:H10,B$11:H$11)					

Embedding may have seemed a step backward. It places the problem of interest in a family of related problems, one per state. But expression (13.2) links the solutions to the problems in this family, making them easy to solve.

A Spreadsheet

Equations (13.1) and (13.2) are easy to solve on a spreadsheet. To do so, we rewrite the latter with its variables on its left-hand side, as in:

$$-0.4w(i-1) + w(i) - 0.6w(i+1) = 0 \quad \text{for } 1 \le i \le 5.$$

We then build the spreadsheet in Table 13.1, and use Solver as is indicated there. In this table, cell B11 contains the datum 0, cell H11 contains the datum 1, and cells C11 through G11 are changing cells that contain values of $w(1)$ through $w(5)$ that Solver computes. Solver has selected values of the changing cells that equate the numbers in cells I6 through I10 to zero. In this application of Solver, its "target cell" has been made blank because Solver is being used to solve a system of equations, not to optimize something.

Evidently, $w(1)$ equals 0.365, when rounded off to three decimal places. The probability that the gambler will reach state 6 before reaching state 0 equals 0.365.

A similar calculation would show that the probability that the gambler will be ruined (reach state 0 before reaching state 6) equals $0.635 = 1 - 0.365$. Incidentally, the probability of the event in which the gambler plays forever equals zero.

A Recursion*

In this subsection, the Gambler's Ruin problem is used to introduce an "uptick" trick that can prove useful, as it does in Problem 29 on page 479.

*This subsection is not essential to a basic understanding of Markov chains. It can be skipped with no loss of continuity.

The Gambler's Ruin problem is that of computing the probability of a complicated event. In this case and in many others, the probability of a complicated event can be computed by a recursion. To see how, consider the event in which state $i + 1$ is reached given that state i is occupied now. Specifically, we define an *uptick* as the event in which state $i + 1$ is reached, given that state i is occupied now. This uptick cannot occur if state 0 is reached first. The probability that this uptick occurs is defined as $u(i)$.

$$u(i) = P(\text{state } i + 1 \text{ is reached} \mid \text{state } i \text{ is occupied now}).$$

Initially, state 1 is occupied. In order for state 6 to be reached, there must be an uptick to state 2, followed by an uptick to state 3, and so forth. In other words,

$$w(1) = u(1)\, u(2)\, u(3)\, u(4)\, u(5).$$

The preceding equation parses the event in which the gambler reaches his goal into a sequence of five upticks. But can we compute the uptick probabilities? With $u(0) = 0$, we shall argue that

$$u(i) = p + (1 - p)\, u(i - 1)\, u(i) \quad \text{for } i \geq 1. \tag{13.3}$$

To see why Equation (13.3) holds, suppose state i is observed, and consider the next bet. An uptick can occur in two disjoint ways. The next bet can win, in which case the uptick has occurred; the probability of that event equals p. Alternatively, the next bet can lose, in which case state $i - 1$ is observed, so an uptick must occur to state i and an uptick must then occur to state $i + 1$; the probability of that event equals $(1 - p)\, u(i - 1)\, u(i)$. This justifies Equation (13.3).

To express $u(i)$ in terms of $u(i - 1)$, we collect terms containing $u(i)$ on the left-hand side of Equation (13.3), as follows.

$$u(i)[1 - (1 - p)\, u(i - 1)] = p.$$

Solving the above for $u(i)$ gives

$$u(i) = \frac{p}{1 - (1 - p)\, u(i - 1)}, \quad \text{for } i \geq 1, \tag{13.4}$$

with $u(0) = 0$. On a spreadsheet, the recursion in Equation (13.4) is easy to compute in ascending i.

Once you understand the embedding trick, discovering the equation for $w(i)$ is fairly routine. By contrast, the equation for $u(i)$ is subtle, not something a beginner would be likely to dream up, unless her probabilistic intuition was phenomenal.

That said, the $u(i)$'s are more trenchant. From them, the probability of reaching an asset position of n dollars from an initial wealth level i that is below n equals the product $u(i)\, u(i + 1) \cdots u(n - 1)$ of the uptick probabilities from states i through $n - 1$.

Incidentally, Problem 14 on page 477 provides a different solution method for the Gambler's Ruin problem.

13.4. FIRST PASSAGE TIME

This section illustrates the computation of the time that elapses until a particular state of a Markov chain is observed. To introduce this topic, consider:

Problem B (Punxsutawney Pete)

Punxsutawney Pete, the groundhog, is in room A of his modest three-room burrow, which is depicted in Figure 13.2. It is February 2, and Pete is eager to leave the burrow. Being dim

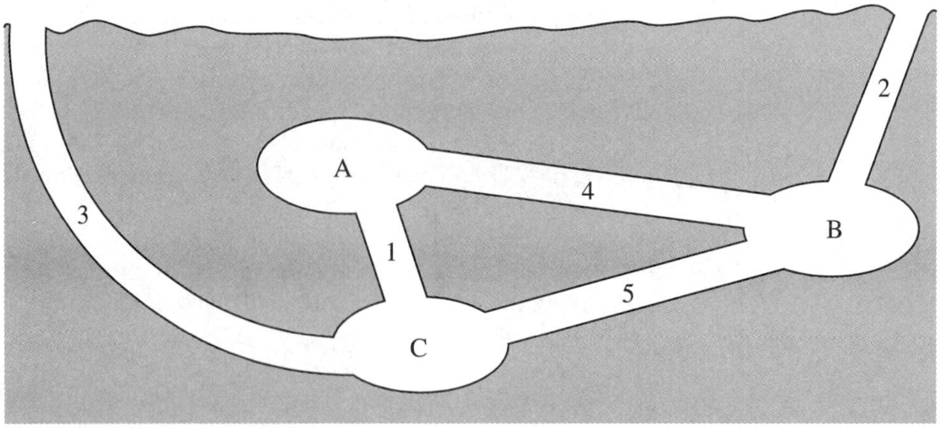

Figure 13.2 A three-room burrow, with travel times in minutes.

of wit, Pete has lost his sense of direction. He is equally likely to leave each room by any of its connecting tunnels, independent of how he entered that room. Shown in Figure 13.2 is the travel time in each tunnel, measured in minutes. Let T denote the time that elapses from now until Pete emerges from his burrow. Compute the expectation of T.

A direct computation of $E(T)$ would require the enumeration of each route that leads from room A to the outside. There are infinitely many such routes, so enumerating them would be daunting. A simple way to compute $E(T)$ is to recognize that we are dealing with a discrete-time Markov chain.

An Embedding

Since Pete is "memoryless," the room to which he moves depends solely on the room that he now occupies. For Pete, there are four states—room A, room B, room C, and above ground—with transition probabilities from state to state. Pete will eventually reach the "above ground" state, and our concern lies with the length of time that it will take him to get there.

Rather than dealing directly with the random variable T, we treat three random variables, one per state. The time that remains before Pete emerges depends solely on the room he now occupies. Let the random variable $T(A)$ denote the time that remains until Pete emerges from his burrow, given that he has reached room A. Define $T(B)$ and $T(C)$ similarly.

Figure 13.3 shows how these three random variables are linked to each other. This figure presents a probability tree for each random variable. Each colon in Figure 13.3 means "has the same distribution as" rather than "equals." If, for instance, Pete is in room A and chooses the tunnel to room B, the time it will take him to emerge from his burrow has the

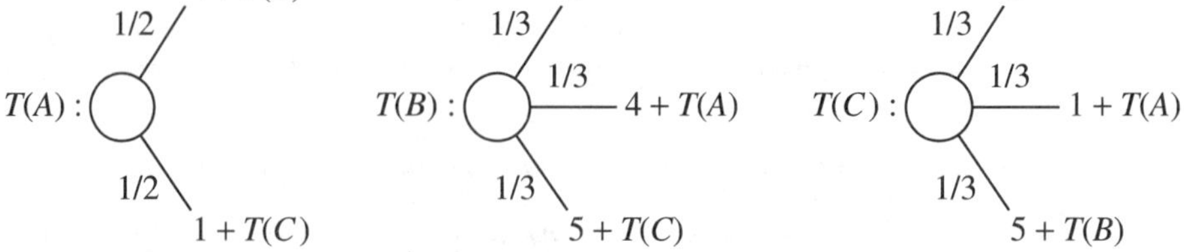

Figure 13.3 Probability trees for time to emergence.

distribution of $[4 + T(B)]$, where the random variable $T(B)$ describes the time from occupancy of room B to emergence.

Mean Times to Emergence

Figure 13.3 relates the random variables $T(A)$, $T(B)$, and $T(C)$ to each other. Let us define $\mu(A)$, $\mu(B)$, and $\mu(C)$ as their expectations. Specifically,

$$\mu(A) = E[T(A)], \quad \mu(B) = E[T(B)], \quad \mu(C) = E[T(C)].$$

Recall that the expectation of the sum equals the sum of the expectations. Thus, it is evident from the leftmost tree in Figure 13.3 that

$$\begin{aligned}\mu(A) &= (1/2)[4 + E[T(B)]] + (1/2)[1 + E[T(C)]],\\ &= (1/2)[4 + \mu(B)] + (1/2)[1 + \mu(C)].\end{aligned}$$

Similarly, the middle and the rightmost trees in Figure 13.3 indicate that

$$\mu(B) = (1/3)(2) + (1/3)[4 + \mu(A)] + (1/3)[5 + \mu(C)]$$
$$\mu(C) = (1/3)(3) + (1/3)[1 + \mu(A)] + (1/3)[5 + \mu(B)]$$

A Spreadsheet

Above are three linear equations in the three variables, $\mu(A)$, $\mu(B)$, and $\mu(C)$. Let us reorganize these equations so that the variables are on their left-hand sides, the constants on the right, as in:

$$\begin{aligned}\mu(A) - (1/2)\mu(B) - (1/2)\mu(C) &= 5/2,\\ -(1/3)\mu(A) + \mu(B) - (1/3)\mu(C) &= 11/3,\\ -(1/3)\mu(A) - (1/3)\mu(B) + \mu(C) &= 9/3.\end{aligned}$$

These equations are easy to solve on a spreadsheet. Table 13.2 indicates how. In it, column E contains the usual "sumproduct" functions, and cells B7 through D7 are changing cells that contain values of the variables $\mu(A)$, $\mu(B)$, and $\mu(C)$ that Solver has calculated. Solver's "target cell" has been left blank because Solver is finding a solution to a system of equations rather than optimizing anything.

Table 13.2 Spreadsheet computation of Pete's expected time to emergence, given his current state.

	A	B	C	D	E	F	G	H
3		μ(A)	μ(B)	μ(C)				
4	state A	1	-0.5	-0.5	2.5	=	2.5	
5	state B	-0.3333	1	-0.3333	3.6667	=	3.6667	
6	state C	-0.3333	-0.3333	1	3	=	3	
7	values	15	12.75	12.25	↑			
8								
9		=SUMPRODUCT(B6:D6,B$7:D$7)						
10								
11	Solver selects values in cells B7:D7 that satisfy E4:E6 = G4:G6							
12								

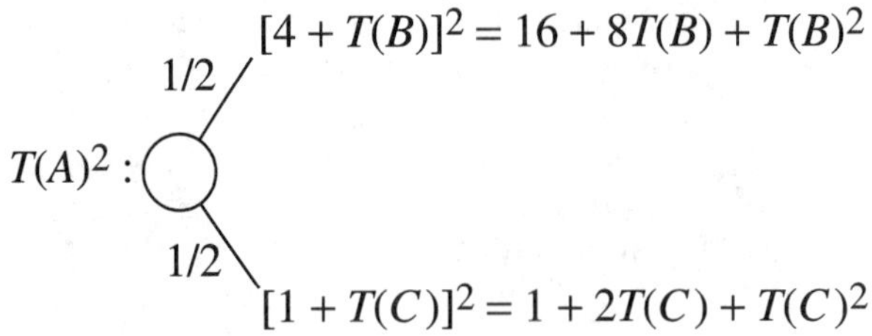

Figure 13.4 Probability tree for the random variable $T(A)^2$.

Pete begins in room A. Table 13.2 shows that the mean length of time until Pete emerges is 15 minutes.

The Variance of T

The same method can be used to compute the variance of the time T that it takes Pete to emerge from his burrow. Let us see how. Figure 13.4 presents the analog of the tree in Figure 13.3 for the random variable $T(A)^2$.

Recall that the expectation of the sum equals the sum of the expectations. Hence, the tree in Figure 13.4 shows that

$$E[T(A)^2] = 17/2 + 4\mu(B) + \mu(C) + (1/2)E[T(B)^2] + (1/2)E[T(C)^2].$$

Substituting the value $\mu(B) = 12.75$ and $\mu(C) = 12.25$ gives

$$E[T(A)^2] = 71.75 + (1/2)E[T(B)^2] + (1/2)E[T(C)^2].$$

Figure 13.4 presents the analog of Figure 13.3 for $[T(A)]^2$. Similar trees specify $T(B)^2$ and $T(C)^2$. Taking expectations, as above, produces three equations in three unknowns, which are easy to solve on a spreadsheet. Having solved them, you can get the variances from the standard trick,

$$\text{Var}[T(A)] = E[T(A)^2] - [E(T(A)]^2.$$

It turns out that the standard deviation of T is approximately 12.5 minutes.

13.5. MULTIPLE TRANSITIONS

Our attention now turns to the question of how frequently the various states will be observed over a long period of time. The next several sections will address that question. We begin by studying the likelihood that the various states are observed after several transitions. This topic is introduced in the context of:

Problem C (Auto Rental)

An auto rental firm has four locations, which are numbered 1 through 4. Each of its rental cars can be returned to the location at which it was picked up or, at an extra charge, can be returned at a different location. Table 13.3 displays the probability that a car that is rented in a particular location will be returned to each of the four locations. A car is presently in location 1. After several rentals of this car, with what probability will it be returned to each of the four locations?

The transition probabilities appear in cells C5 through F8 of Table 13.3. Because they are probabilities, these entries are nonnegative numbers. Being conditional probabilities, the

Table 13.3 Transition probabilities from location to location.

	A	B	C	D	E	F
3			ending location			
4			1	2	3	4
5	starting location	1	0.7	0.3	0	0
6		2	0.2	0.6	0.2	0
7		3	0	0.3	0.5	0.2
8		4	0	0	0.6	0.4

sum across each row equals one. For each pair (i, j) of states, we designate $P(i, j)$ as follows:

$$P(i, j) = \text{the probability that a car rented in location } i \text{ will be returned to location } j.$$

For instance, row 7 of this table contains the transition probabilities from location 3 to the various locations, and

$$\begin{aligned} P(3, 1) + P(3, 2) + P(3, 3) + P(3, 4) &= 0 + 0.3 + 0.5 + 0.2 \\ &= 1, \end{aligned}$$

as must be.

Problem C poses the question of where this car might be returned after several rentals. This car begins at location 1. To account for the possibilities, we let the symbol k denote the number of rentals that have occurred, and we let the symbol j denote the location to which the car is returned. For each state j and each nonnegative integer k, the quantity $q(k, j)$ is defined by

$$q(k, j) = P(\text{the car will be returned to location } j \text{ after exactly } k \text{ rentals}).$$

Initially, after 0 rentals have occurred, the car is in location 1, so that

$$q(0, 1) = 1, \quad q(0, 2) = 0, \quad q(0, 3) = 0, \quad q(0, 4) = 0.$$

A Recursion

The $q(k, j)$'s can be computed recursively, in increasing k. To suggest how, we account for the probability $q(6, 3)$ that the car is returned to location 3 after six rentals. This car must have been returned to some location after five rentals, and it must have been returned to location 3 after the sixth rental. The joint probability that the car is returned to location i after five rentals and is returned to location 3 after the sixth rental equals $q(5, i)\, P(i, 3)$. By accounting for each possible location i of the car after five rentals, we get

$$q(6, 3) = \sum_{i=1}^{4} q(5, i)\, P(i, 3).$$

Once $q(5, i)$ has been found for each i, the preceding equation can be used to compute $q(6, 3)$.

The equation for $q(6, 3)$ illustrates the general result, which is that

$$q(k + 1, j) = \sum_{i=1}^{4} q(k, i)\, P(i, j). \tag{13.5}$$

Equation (13.5) expresses the probability $q(k + 1, j)$ as the sum of the probabilities of four mutually exclusive events, each of which is that the car is returned to a particular location i at the end of the kth rental and proceeds from there to location j.

A Solution Method

In this subsection, we sketch a method that uses the SUMPRODUCT function to solve Equation (13.5). The arguments in Excel's SUMPRODUCT function can be two rows of equal length or two columns of equal length. But its arguments *cannot* be a row of a given length with a column of the same length. If, for instance, you enter into a cell the function

=SUMPRODUCT(B3:D3,B5:B7)

the error message "#VALUE!" will appear in that cell.

The equation in (13.5) for $q(k+1, j)$ multiplies the numbers $q(k, 1)$ through $q(k, 4)$ times the numbers $P(1, j)$ through $P(4, j)$, which form the jth *column* of numbers in the array in Table 13.3. Thus, in order to execute equation (13.5) conveniently on the spreadsheet in Table 13.3, we can place the numbers $q(k, 1)$ through $q(k, 4)$ in a *column* and use the SUMPRODUCT function to put the numbers $q(k+1, 1)$ through $q(k+1, 4)$ in the next column to the right. Then we could drag this column rightward to compute the $(k+2)$nd column, and so forth.

This method works, but it turns out to be a bit awkward. For technical reasons, the functions for $q(k+1, 1)$ through $q(k+1, 4)$ must be entered individually; the copy-and-paste trick will not work.

Matrix Arithmetic

An accounting system known as matrix arithmetic provides an easier way to solve Equation (13.5). A rectangular block of numbers is called a **matrix**. Matrix arithmetic is straightforward and is easy to learn; there is nothing mysterious about it. If you are not comfortable with matrix arithmetic, please read the sections in the Appendix that describe array formulas and matrix arithmetic. And if you are conversant with matrix arithmetic, read those sections *anyhow.* Why? Excel's scheme for dealing with matrices contains booby traps that you need to be aware of!

The rules for taking the product AB of two matrices are recapitulated as follows:

- The product AB of the matrices A and B cannot be taken unless the number of columns in A equals the number of rows in B.
- The product AB is a matrix that has the number of rows in the matrix A and the number of columns in the matrix B.
- The entry in the ith row and jth column of AB is found by multiplying each element in the ith row of A by the corresponding element in the jth column of B and taking the sum.

Interpreting Equation (13.5) as a Matrix Product

We shall identify Equation (13.5) with the product of two matrices. The array of transition probabilities form the matrix P that has four rows and four columns, where

$$P = \begin{bmatrix} P(1,1) & P(1,2) & P(1,3) & P(1,4) \\ P(2,1) & P(2,2) & P(2,3) & P(2,4) \\ P(3,1) & P(3,2) & P(3,3) & P(3,4) \\ P(4,1) & P(4,2) & P(4,3) & P(4,4) \end{bmatrix} = \begin{bmatrix} 0.7 & 0.3 & 0 & 0 \\ 0.2 & 0.6 & 0.2 & 0 \\ 0 & 0.3 & 0.5 & 0.2 \\ 0 & 0 & 0.6 & 0.4 \end{bmatrix}$$

Similarly, the numbers $q(k, 1)$ through $q(k, 4)$ can be formed into the matrix that has one row and four columns, as follows:

$$q(k) = [q(k, 1) \quad q(k, 2) \quad q(k, 3) \quad q(k, 4)].$$

Equation (13.5) states that the jth entry in $q(k + 1)$ is found by multiplying each element in $q(k)$ by the corresponding element in the jth column of P and taking the sum. In brief, Equation (13.5) is the matrix product

$$q(k + 1) = q(k)P. \tag{13.6}$$

Table 13.4 uses the matrix multiplication capability of Excel to compute $q(2)$ through $q(12)$. To compute $q(2)$ as indicated in Table 13.4, execute these steps:

- Select cells I6:L6.
- Type **=MMULT(I5:L5,C$5:F$8)** but do *not* hit the Enter key. Instead, type **Ctrl+Shift+Enter**

Then select cells I6:L6 and drag downward to compute $q(3)$ through $q(12)$. Again, you are urged to read the discussion of matrices in the Appendix before you attempt any matrix calculations with Excel.

The probabilities in Table 13.4 seem to be settling down; $q(10, 1)$ differs only slightly from $q(12, 1)$, for instance. These probabilities *are* settling down. Moreover, the probability distribution to which they are settling down is independent of the initial location of the car.

Intuitively, this behavior is what we might anticipate: After many transitions have occurred, the probability that state i will be observed settles down to a **steady-state** value $\pi(i)$ that does not depend on the state that was occupied initially. In this sense, the Markov chain loses memory of its starting state.

Table 13.4 Computation of the probability distribution over the states after k transitions.

	A	B	C	D	E	F	G	H	I	J	K	L
3			ending location									
4			1	2	3	4		k	q(1.k)	q(2,k)	q(3,k)	q(4,k)
5	starting location	1	0.7	0.3	0	0		0	1.00	0.00	0.00	0.00
6		2	0.2	0.6	0.2	0		1	0.70	0.30	0.00	0.00
7		3	0	0.3	0.5	0.2		2	0.55	0.39	0.06	0.00
8		4	0	0	0.6	0.4		3	0.46	0.42	0.11	0.01
9								4	0.41	0.42	0.14	0.03
10								5	0.37	0.42	0.17	0.04
11		=MMULT(I5:L5,C$5:F$8)						6	0.34	0.41	0.19	0.05
12								7	0.32	0.41	0.21	0.06
13								8	0.31	0.41	0.22	0.07
14								9	0.30	0.40	0.23	0.07
15								10	0.29	0.40	0.24	0.07
16								11	0.28	0.40	0.24	0.08
17								12	0.28	0.40	0.25	0.08

13.6. COMPUTING THE STEADY-STATE PROBABILITIES

How can we compute these steady-state probabilities? Being probabilities, they sum to one, so

$$\pi(1) + \pi(2) + \pi(3) + \pi(4) = 1. \tag{13.7}$$

Let us build the 1 by 4 matrix π in which

$$\pi = [\pi(1), \pi(2), \pi(3), \pi(4)].$$

Equation (13.6) is that $q(k+1) = q(k)P$. Since $q(k)$ and $q(k+1)$ are settling down to π, it must be that

$$\pi = \pi P. \tag{13.8}$$

Equations (13.7) and (13.8) specify the steady-state probabilities for the auto rental problem.

A Spreadsheet

The spreadsheet in Table 13.5 solves equations (13.7) and (13.8). To understand how it does so, we rewrite the latter as

$$\pi - \pi P = 0,$$

and we note that in Table 13.5:

- Cells C3 through F6 contain the matrix P.
- Cells C7 through F7 are changing cells that contain the values of $\pi(1)$ through $\pi(4)$ that Solver computes.
- The array formula in cells C8: F8 computes $\pi - \pi P$.
- The function in cell G7 computes the sum of the steady-state probabilities.
- Solver has selected values of cells C7 through F7 that satisfy Equation (13.8) by equating the values in cells C8 through F8 to 0 and satisfy Equation (13.7) by equating the value in cell G7 to 1.

The calculation in Table 13.5 shows that the steady-state probabilities are

$$\pi(1) = 0.261, \quad \pi(2) = 0.391, \quad \pi(3) = 0.261, \quad \pi(4) = 0.087,$$

Table 13.5 Steady-state probabilities for the auto rental example.

	A	B	C	D	E	F	G	H	I	J
1			ending location							
2			1	2	3	4				
3	starting location	1	0.7	0.3	0	0				
4		2	0.2	0.6	0.2	0				
5		3	0	0.3	0.5	0.2				
6		4	0	0	0.6	0.4				
7		π =	0.261	0.391	0.261	0.087	1			
8	π − π P =		3E-15	-0	4E-14	-0				
9								=SUM(C7:F7)		
10		=C7:F7-MMULT(C7:F7,C3:F6)								

Solver selects values in cells C7:F7 that satisfy C8:F8=0 and G7=1

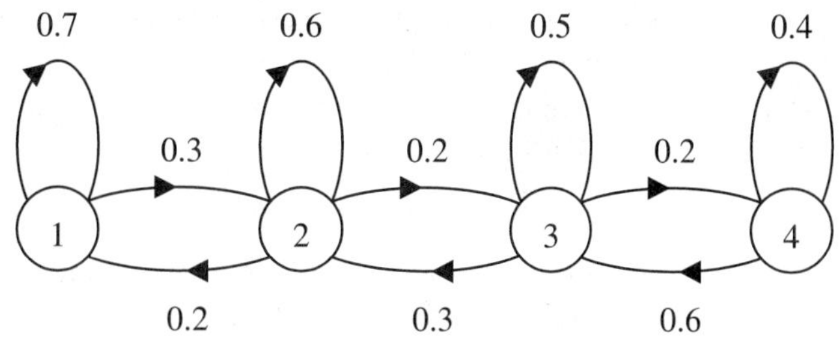

Figure 13.5 Transition probabilities for the auto rental example.

when rounded off to three decimal places. These probabilities state that, over a long period of time, the car will be returned to state 1 after roughly 26% of the rentals, to state 2 after roughly 39% of the rentals, and so forth.

Transition Diagram

The steady-state probabilities can be computed by a different method, one that is worth learning because it can be much simpler. To bring this method into view, we display the transition probabilities for this example in Figure 13.5.

In Figure 13.5, each state is represented as a circle with a number inside. Each nonzero transition probability is represented as an arrow that points from state to state. For instance, two arrows point "away" from state 1, and they represent the probabilities

$$P(1, 1) = 0.7, \quad P(1, 2) = 0.3,$$

which sum to 1, as they must because they are the nonzero probabilities of transition from state 1 to the various states.

Flux

Let A denote any subset of the states. The **flux** into the set A of states equals the probability of transition from a state j that is not in A to a state i that is in A. Similarly, the **flux** out of the set A equals the probability of transition from a state i that is in A to a state j that is not in A. Intuitively, it seems clear that, in steady state, the flux into A must equal the flux out of A. This intuition is correct; Equation (13.8) can be used to verify it.

Let us illustrate. The dashed line in Figure 13.6 partitions the states into the sets $\{1\}$ and $\{2, 3, 4\}$. In Figure 13.6, flux is represented as a transition across the dashed line. In steady state, the fluxes in the two directions across this line must balance each other. In other words,

$$\pi(1)(0.3) = \pi(2)(0.2)$$

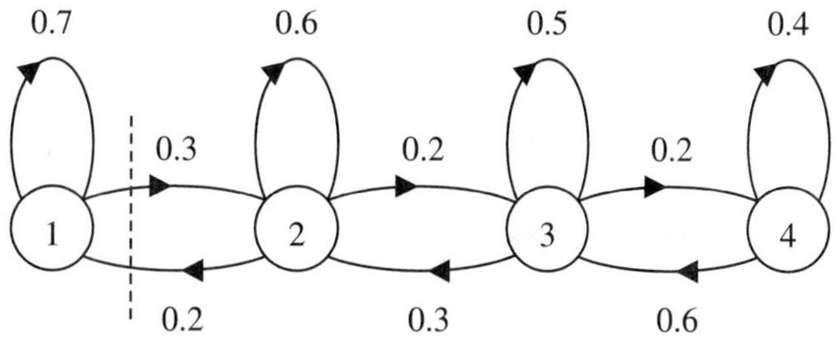

Figure 13.6 Transition probability diagram, with a dashed line separating state 1 from the others.

because the probability $\pi(1)(0.3)$ of crossing this line from left to right must equal the probability $\pi(2)(0.2)$ of crossing it from right to left.

Similarly, a dashed line that partitions the state space into sets $\{1, 2\}$ and $\{3, 4\}$ demonstrates that

$$\pi(2)(0.2) = \pi(3)(0.3)$$

because the probability $\pi(2)(0.2)$ of crossing that line from left to right equals the probability $\pi(3)(0.3)$ of crossing it from right to left. And the line that separates state 4 from the others is responsible for the equation

$$\pi(3)(0.2) = \pi(4)(0.6).$$

The steady-state probabilities are the unique solution to the preceding three equations and

$$\pi(1) + \pi(2) + \pi(3) + \pi(4) = 1.$$

A Simpler Equation Set

These four equations are easy to solve on a spreadsheet. Indeed, they are simple enough to solve by hand. We use the first three equations to express $\pi(2)$, $\pi(3)$, and $\pi(4)$ in terms of $\pi(1)$, getting

$$\pi(2) = \pi(1)(3/2), \quad \pi(3) = \pi(1), \quad \pi(4) = \pi(1)(1/3),$$

and substitute into the fourth equation to obtain

$$\pi(1)[1 + 3/2 + 1 + 1/3] = 1, \quad \text{so} \quad \pi(1) = 6/23,$$

from which it's easy to see that the complete solution is

$$\pi(1) = 6/23, \quad \pi(2) = 9/23, \quad \pi(3) = 6/23, \quad \pi(4) = 2/23.$$

This is the same solution that was obtained in Table 13.5, but Solver has not been needed. Instead, we used the idea of flux to express each probability in terms of $\pi(1)$, and then we solved for the value of $\pi(1)$ for which the probabilities sum to 1.

Generic Properties?

The auto rental example has illustrated the following features:

- The probability distribution over the states settles down to steady-state values as the number of transitions becomes large.
- These steady-state values are independent of the starting state.
- These steady-state values are the unique solution to Equations (13.7) and (13.8).

Are these features particular to the auto rental example, or do they hold for every Markov chain?

These features are shared by many Markov chains, though not all of them. To describe what can go wrong, we now introduce some nomenclature.

13.7. CLASSIFYING MARKOV CHAINS

The Markov chain in Figure 13.7 will be used to illustrate several definitions. This Markov chain has four states, which have been labeled 1, 2, 3, and 4. Cells C4 through F7 contain its transition probabilities, each of which is the conditional probability $P(i, j)$ of a transition

	A	B	C	D	E	F
2			to state			
3			1	2	3	4
4	from state	1	1	0	0	0
5		2	0.3	0.6	0.1	0
6		3	0	0	0.5	0.5
7		4	0	0	0.6	0.4

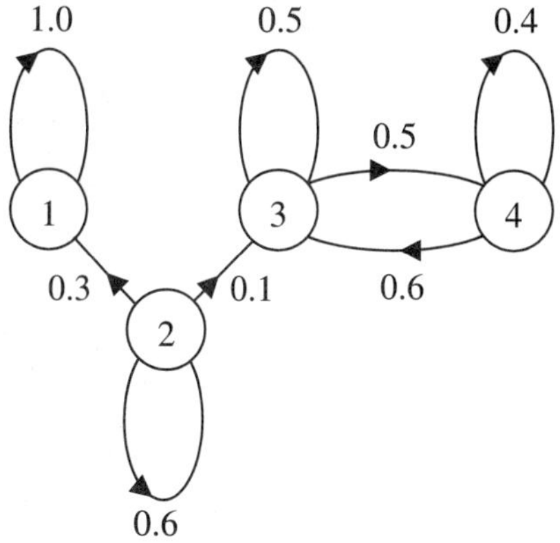

Figure 13.7 Transition probabilities and diagram for a four-state Markov chain.

to state j given that state i is now observed. By summing these probabilities across a row, we get 1, as we must.

In a Markov chain, a nonempty set C of states is now said to be **closed** if $P(i, j) = 0$ for each i in C and each j that is not in C. Escape from a closed set is impossible; if we begin at a state i in a closed set C, no sequence of transitions will lead us to any state j that is not in C. The Markov chain in Figure 13.7 has three closed sets of states, which are

$$\{1, 2, 3, 4\}, \quad \{1\}, \quad \text{and} \quad \{3, 4\}.$$

Invariably, the set S of all states in a Markov chain is closed.

Again, escape from a closed set of states is impossible. For example, if, in Figure 13.7, you begin at state 3, no sequence of transitions will lead you to any state that is not in the set $\{3, 4\}$. Similarly, if you begin at state 1, you will remain at state 1 after any sequence of transitions. Thus, for the Markov chain in Figure 13.7, the probability distribution over the states cannot settle down to values that are independent of the starting state.

An Ergodic Set of States

In a Markov chain, a set C of states is said to be **ergodic** if C is closed and if no subset of C is closed. An ergodic set is a *smallest* set from which escape is impossible. The Markov chain in Figure 13.7 has two ergodic sets. One ergodic set consists of state 1, and the other consists of states 3 and 4.

A state is said to be **recurrent** if it is a member of an ergodic set, and it is said to be **transient** if it is not recurrent. In Figure 13.7, states 1, 3, and 4 are recurrent, and state 2 is transient. The terms *recurrent* and *transient* earn their names from the following properties:

- If we begin at a state i that is recurrent, the probability of visiting state i again equals 1.
- Alternatively, if we begin at a state i that is transient, the probability of visiting state i again is less than 1.

In a Markov chain, state i is said to be **absorbing** if the set $\{i\}$ is itself an ergodic set. In Figure 13.7, state 1 is absorbing. In the Gambler's Ruin problem (Figure 13.1), states 0 and 6 are absorbing, and states 1 through 5 are transient. In the Punxsutawney Pete example (Figure 13.2), states A, B, and C are transient, and the above-ground state is absorbing.

	A	B	C	D	E	F
3			to state			
4			1	2	3	4
5	from state	1	0	1/3	2/3	0
6		2	0	0	0	1
7		3	0	0	0	1
8		4	1	0	0	0

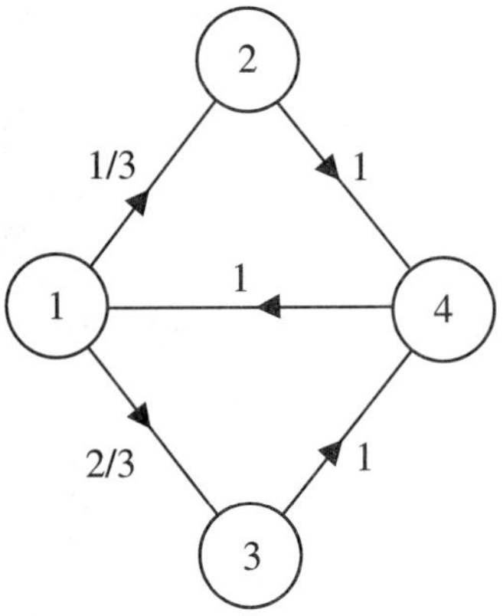

Figure 13.8 Transition probabilities and transition diagram for an ergodic set.

The Period of an Ergodic Set

We seek conditions that can keep a Markov chain from having steady-state probabilities. Evidently, if a Markov chain has more than one ergodic set, the probability distribution over its states cannot settle down to values that are independent of the starting state.

The probability distribution over the states in a single ergodic set *can* fail to settle down to steady-state values. To see how, we present the transition matrix and transition diagram in Figure 13.8.

In Figure 13.8, the set $\{1, 2, 3, 4\}$ consisting of all four states is ergodic. Suppose you begin at state 2 of this ergodic set. After how many transitions might you revisit state 2? You can *only* revisit state 2 after a multiple of three transitions, that is, after 3, 6, 9, 12, . . . , transitions.

An ergodic set has p as its **period** if p is the *largest* integer such that you can only revisit each state in it after a multiple of p transitions. The ergodic set in Figure 13.8 has 3 as its period. To see this, we note that its states partition themselves into these three groups,

$$\{1\}, \quad \{2, 3\}, \quad (4\},$$

with transition from each group to the next, and with transition from the third group back to the first. If an ergodic set had 2 as its period, then its set of states would partition itself into two groups, transitions occurring from each group to the other. By contrast, the ergodic sets in Figure 13.7 have 1 as their period.

An ergodic set is said to be **periodic** if its period equals 2 or more, and it is said to be **aperiodic** if its period is equal to 1. Figures 13.7 and 13.8 illustrate the types of finite-state Markov chain for which the probability distribution over the states fails to settle down to steady-state values that are independent of the starting state. (This chapter does not deal with infinite-state Markov chains, for which other pathologies can occur.)

13.8. LONG-TERM BEHAVIOR

This section concerns the long-term behavior of a discrete Markov chain that has a finite set S of states and a matrix (array) P of transition probabilities, with $P(i, j)$ denoting the probability that transition will occur to state j, given that state i is observed now. Being conditional probabilities, they sum to 1, so that

$$1 = \sum_{j \in S} P(i, j) \quad \text{for each state } i \text{ in } S.$$

Our prior examples indicate that this Markov chain cannot have a unique set of steady-state probabilities if it has two or more ergodic sets.

If the Markov chain has only one ergodic set, it has a unique solution to the analog of Equations (13.7) and (13.8). The properties of this solution are presented in:

Theorem 13.1. Suppose that a discrete-time Markov chain has only one ergodic set. Then:

- **(a)** (**stationary distribution**) There exists exactly one solution π to the equation system

$$\sum_{j \in S} \pi(j) = 1 \quad \text{and} \quad \pi(i) = \sum_{j \in S} \pi(j)P(j, i) \quad \text{for each } i \in S. \qquad (13.9)$$

- **(b)** (**time average**) Over a long period of time, the fraction of the transitions that occur to state i will be approximately equal to $\pi(i)$.
- **(c)** (**time average**) Over a long period of time, the fraction of the transitions that occur from state i to state j will be approximately equal to $\pi(i)P(i, j)$.
- **(d)** (**expected recurrence times**) If state i is recurrent, then $\pi(i)$ is positive and $1/\pi(i)$ equals the expectation of the number of transitions that occur between successive occurrences of state i.
- **(e)** (**steady state**) If, in addition, this ergodic set is aperiodic, then the probability that state i is observed after k transitions approaches $\pi(i)$ as k becomes large.

To interpret Theorem 13.1, we return to the auto rental example. It has one ergodic set, and that set is aperiodic. For it:

- Part (a) demonstrates that there exists a unique solution to system (13.9). We've calculated this solution and found, for instance, that $\pi(1) = 6/23$.
- Part (b) states that, over a long period of time, the fraction of the returns that will bring the car to location i is approximately $\pi(i)$.
- Part (c) states that, over a long period of time, the fraction of the rentals that will be picked up in location i and returned to location j is approximately $\pi(i)P(i, j)$.
- Part (d) states that the expectation of the number of rentals between successive visits to location i equals $1/\pi(i)$.
- Part (e) states that the probability that the kth rental will be returned to location i settles down to $\pi(i)$ as k becomes large.

Thus, the behavior that we observed in the auto rental example holds whenever the Markov chain has a single ergodic set whose period equals 1.

Parts (b) and (c) of Theorem 13.1 describe an "ergodic" property that holds for many systems. This **ergodic** property is a conditional statement: *if* the behavior of a time-varying system settles down to steady-state values as the elapsed time becomes large, then the values to which it settles down also describe the time-average behavior of the system.

This ergodic property provides a second interpretation of the steady-state probabilities. To illustrate it, we recall from the auto rental example (Problem C) that the probability that a car is returned to location 1 settles down to the value 6/23. The ergodic property states that, over a long period of time, approximately 6/23 of the returns will occur to location 1.

13.9. A CONTINUOUS-TIME MARKOV CHAIN

The prior sections dealt with discrete-time Markov chains, for which each transition took a fixed length of time. Our attention now turns to a continuous-time Markov chain in which the times between transitions are uncertain and will turn out to have the exponential distribution.

Again, the Markov chain is defined in terms of a finite set S of states, and its law of motion depends only on the current state, not on prior states. A "continuous"-time Markov chain gets its name from the fact that we think of state i as being occupied continuously for a (random) length of time, followed by an instantaneous jump from state i to some state j.

We shall see that the data of a continuous-time Markov chain are a set of transition rates rather than a set of transition probabilities. These rates will:

- imply that the time until transition has the exponential distribution and determine that distribution's expectation,
- determine the probability of transition from state to state,
- lead to an analog of Theorem 13.1, one that describes the steady-state behavior of a continuous-time Markov chain.

Data

The data of a continuous-time Markov chain consist of a finite set S of states and an array Q of nonnegative numbers, where $Q(i, j)$ is interpreted as the *rate* of transition from state i to state j. Specifically,

$$P\left(\begin{matrix}\text{transition will occur to} \\ \text{state } j \text{ by time } t + \Delta t\end{matrix} \,\middle|\, \begin{matrix}\text{state } i \text{ is observed} \\ \text{at time } t\end{matrix}\right) = Q(i, j)\Delta t + \text{error term}, \qquad (13.10)$$

where the "error term" is proportional to $(\Delta t)^2$, just as in Equation (11.3) of Chapter 11.

Let us designate as $\nu(i)$ (the Greek letter ν is pronounced "nu") the sum of the transition rates from state i. Specifically,

$$\nu(i) = \sum_{j \in S} Q(i, j), \quad \text{for each state } i \text{ in } S. \qquad (13.11)$$

By summing Equation (13.10) over j, we see that

$$P\left(\begin{matrix}\text{transition will occur} \\ \text{by time } t + \Delta t\end{matrix} \,\middle|\, \begin{matrix}\text{state } i \text{ is observed} \\ \text{at time } t\end{matrix}\right) = \nu(i)\Delta t + \text{error term}, \qquad (13.12)$$

where the "error term" is proportional to $(\Delta t)^2$.

Each transition rate $Q(i, j)$ must be nonnegative. The sum of the transition rates out of each state must be positive, but these rates need not sum to 1. In other words, each state i must have $\nu(i) > 0$, but $\nu(i)$ can be any positive number. In this model, we allow transitions from state i to itself, so $Q(i, i)$ can be positive.

Exponential Times until Transition

Equation (13.12) is familiar. The probability that transition will occur within the next Δt units of time is approximately $\nu(i)\Delta t$, and this probability is independent of the length of time that state i has been occupied. This is the memoryless property of the exponential distribution, as it appears in Equation (11.36) of Chapter 11.

In other words, Equation (13.12) guarantees that, given that state i is occupied, the time that will elapse before the next transition has the exponential distribution with rate $\nu(i)$, equivalently, with mean of $1/\nu(i)$. Similarly, given that state i is occupied now, the probability $P(i, j)$ that the next transition will occur to state j is given by

$$P(i, j) = \frac{Q(i, j)}{\nu(i)}, \quad \text{for each pair } (i, j) \text{ of states.}$$

In this way, the transition rates imply that the time to transition given state i has the exponential distribution with mean $1/\nu(i)$, and they determine the probability $P(i, j)$ that this transition will occur to state j.

Classification

Continuous-time Markov chains are classified in much the same way as are discrete-time Markov chains. Specifically, in a continuous-time Markov chain:

- A nonempty set C of states is said to be **closed** if $Q(i, j) = 0$ for each i in C and each j that is not in C. (The set S of all states is closed, for instance.)
- A set C of states is said to be an **ergodic set** if C is closed and if no subset of C is closed.
- State i is said to be **recurrent** if state i is part of an ergodic set.
- State i is said to be **transient** if state i is not a member of an ergodic set.

Again, escape from an ergodic set is impossible. The set S of all states is closed, so a continuous-time Markov chain must have at least one ergodic set.

In a continuous-time Markov chain, the times between transitions are exponential. This makes it impossible that the states in an ergodic set be visited periodically. In this sense, a continuous-time Markov chain avoids an annoying pathology (periodicity) of a discrete-time Markov chain.

One Ergodic Set

Let us focus on a continuous-time Markov chain that has only one ergodic set. Theorem 13.2 shows that the probability distribution over its states settles down to steady-state values that are independent of the starting state. Hence, the probability that state i will be observed settles down to the steady-state value $\rho(i)$ as the elapsed time t becomes large. (The Greek letter ρ, pronounced "rho," is used for the steady-state probabilities of a continuous-time Markov chain.)

Being a probability distribution, the $\rho(i)$'s must sum to 1; they satisfy

$$\sum_{i \in S} \rho(i) = 1. \tag{13.13}$$

Theorem 13.2 also implies that, when a continuous-time Markov chain is in steady state, the **flux** (rate of flow) into any group of states equals the flux (rate of flow) out of that group of states. In particular, in steady state, the flux (rate of flow) into state i must equal the flux (rate of flow) out of state i. In other words,

$$\sum_{j \in S} \rho(j)\, Q(j, i) = \sum_{j \in S} \rho(i)\, Q(i, j), \quad \text{for each } i,$$

because the left-hand side of the above equation equals the rate of transition, into state i and its right-hand side equals the rate of transition out of state i. Equation (13.11) defines

$\nu(i)$ as the sum over j of $Q(i, j)$, which lets us rewrite the above as

$$\sum_{j \in S} \rho(j)\, Q(j, i) = \rho(i)\, \nu(i), \quad \text{for each } i. \tag{13.14}$$

Theorem 13.2, is the analog of Theorem 13.1 for continuous-time Markov chains.

Theorem 13.2. Suppose that a continuous-time Markov chain has only one ergodic set. Then:

(a) **(stationary distribution)** There exists exactly one solution ρ to Equations (13.13) and (13.14).

(b) **(steady state)** The probability that state i will be occupied at time t settles down as t becomes large to $\rho(i)$, independent of the state that was occupied initially.

(c) **(time average)** Over a long interval t of time, the fraction of the time that state i will be observed is approximately equal to $\rho(i)$.

(d) **(time average)** Over a long interval t of time, the rate at which transition from state i to state j will be observed is approximately equal to $\rho(i)\, Q(i, j)$.

(e) **(mean recurrence times)** If state i is recurrent, then $\rho(i)$ is positive, and the expectation of the time that elapses between successive transitions to state i equals $1/[\rho(i)\, \nu(i)]$.

In a continuous-time Markov chain, the steady-state probabilities have a somewhat different interpretation than in the discrete-time Markov chain. We emphasize:

- In a discrete-time Markov chain, the state is observed intermittently, at the moment just after transition, and $\pi(i)$ equals the steady-state probability that state i is observed.
- In a continuous-time Markov chain, the state is observed continuously, except at moments of transition, and $\rho(i)$ equals the steady-state probability that state i is observed.

In a continuous-time Markov chain, states are occupied for intervals of time; by contrast, in a discrete-time Markov chain, states are occupied for instants. This distinction accounts for the differences between Theorems 13.2 and 13.1.

13.10. TELECOMMUNICATIONS TRAFFIC

By 1917, A. K. Erlang, a Danish mathematician, had built a model of congestion in the traffic between telephone offices. Both his model and his insights into that model proved to be prescient, and they remain fundamental to the analysis of modern telecommunication systems. Erlang's work also helped stimulate the development of the field of queueing, to which Chapter 14 of this text is devoted. To introduce his model, we consider:

Problem D (The Erlang Loss System)

The circuit that connects a pair of central offices has a capacity of n simultaneous conversations. In this system:

- Requests for service (call attempts) between these central offices form a Poisson process with a steady rate of λ call attempts per minute.

- Each request for service is accepted if fewer than n conversations are in progress when it appears. Each request for service is denied access to the system if it appears while n conversations are in progress.
- If a request for service is accepted, the call duration (interval of time that elapses between the instant a call is accepted and the moment when it is completed) has the exponential distribution with a rate of μ per minute, and hence a mean of $1/\mu$ minutes. Call duration times of different calls are independent of each other.

Under steady-state conditions, what fraction of the calls must be denied access to the system for the following values of n, λ, and μ?

$$n = 8, \quad \lambda = 0.5, \quad \mu = 1/6. \tag{13.15}$$

In the telecommunications literature, the model in Problem D is referred to as an **Erlang loss system**, the "loss" referring to the requests for service that are lost because they cannot be satisfied. Equation (13.15) describes a circuit with a capacity of 8 simultaneous calls, which arrive at a rate of 0.5 calls per minute and have a mean call duration time of 6 minutes.

Modeling

Let us first comment on Erlang's model. His assumption of Poisson arrivals is sound. As many as 10,000 customers are served by a single central office. If each customer calls another central office intermittently and occasionally, the Superposition Theorem guarantees that the aggregate call process will be Poisson. Problem D studies this system for an interval of time in which the aggregate call rate is steady.

Erlang's assumption of exponential call duration times is a convenience that simplifies the analysis. There is no reason to believe that call duration times are exponential, that is, that the conditional probability that the call will terminate in the next second is independent of the length of time the parties have been conversing. Later in this section, we'll remark on the extent (if any) to which the assumption of exponential call duration times limits the applicability of Erlang's results.

A Recursion

The operating characteristics of an Erlang loss system can be specified in terms of its capacity of n simultaneous conversations and its ratio

$$R = \lambda/\mu,$$

of the arrival rate λ to its service rate μ. In the telecommunications literature, the constant $R = \lambda/\mu$ measures the traffic load in **Erlangs**. (This honors Erlang, even though the unit of measure is dimensionless.)

The steady-state probability that n conversations are in progress in an Erlang loss system with parameters n and R is denoted $B(n, R)$. In the telecommunications literature, the quantity $B(n, R)$ is now known as the **Erlang probability**. Call attempts that arrive while n conversations are in progress must be denied access to the system, so $B(n, R)$ equals the fraction of the calls that the system cannot serve.

For a fixed value of R, there turns out to be a simple recursion for the reciprocals of the Erlang probabilities. This recursion is

$$\frac{1}{B(n, R)} = \frac{1}{B(n-1, R)}\frac{n}{R} + 1, \tag{13.16}$$

with $B(0, r) = 1$. Later in this section, we will see why Equation (13.16) holds.

Table 13.6 A recursion for the Erlang probability $B(k, R)$.

	A	B	C	D	E	F
2	λ =	0.5		n	1/B(n, r)	B(n, r)
3	μ =	1/6		0	1	1.0000
4	R =	3		1	1.333333	0.7500
5				2	1.888889	0.5294
6				3	2.888889	0.3462
7		=1+E3*D4/B$4		4	4.851852	0.2061
8				5	9.08642	0.1101
9				6	19.17284	0.0522
10				7	45.73663	0.0219
11				8	122.9643	0.0081

The spreadsheet in Table 13.6 uses the data in Equation (13.15) and the recursion in Equation (13.16) to compute $B(k, R)$ for the values $k = 1$ through 8.

From Table 13.6, we see that, in steady state, a circuit with a capacity of eight conversations is fully utilized with probability 0.0081. Requests for service that arrive while this system is fully utilized must be denied access to it. Thus, slightly less than 1% of the requests for service are denied access to the system.

Incidentally, before the spreadsheet era, many telecommunication engineers had tables of Erlang probabilities atop their desks. Spreadsheets have made those tables obsolete.

States

Equation (13.16) is an end-point of an analysis that we now begin. Let us recall a theorem on page 363 of Chapter 11, which shows that a Poisson arrival process with arrival rate of λ per unit time implies exponentially distributed inter-arrival times with a rate of λ per minute and hence a mean inter-arrival time of $1/\lambda$ minutes.

Thus, the inter-arrival time has the exponential distribution with λ as its rate. The call duration time has the exponential distribution with μ as its rate. The exponential distribution is memoryless. Thus, all we need to know in order to predict the law of motion is the number of calls in progress. In brief, the number k of calls in progress is a *state.* With a capacity of 8 simultaneous calls, the state k is an integer between 0 and 8, inclusive. Since transition times are exponential, we are dealing with a continuous-time Markov chain. Its transition diagram is presented as Figure 13.9.

To justify the transition diagram in Figure 13.9, consider a small interval Δt of time. The probability that a call attempt will occur during this interval is approximately $\lambda\Delta t$. If the

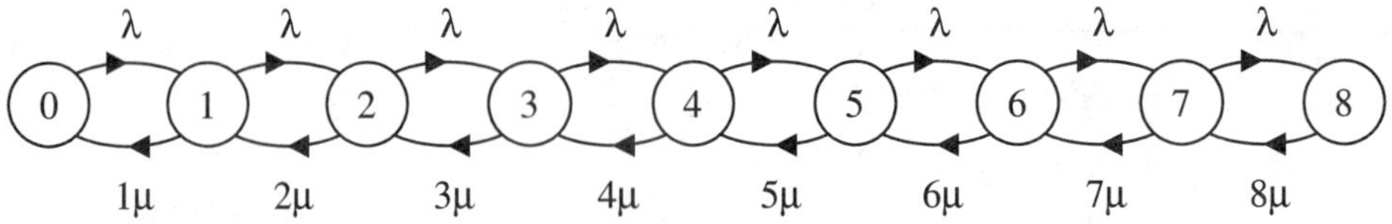

Figure 13.9 Transition rate diagram for the Erlang loss model with $n = 8$, hence with a capacity of 8 simultaneous calls.

number k of calls in progress is below 8, this call attempt will be accepted, and it will increase k by 1, which explains the right-pointing arrows in Figure 13.9. To justify the left-pointing arrows, suppose that k calls are in progress, where $k > 1$. In a small interval Δt of time, each of these k calls is completed with a probability of approximately $\mu\Delta t$, so the probability that one of them is completed is roughly $k\mu\Delta t$, which explains the left-pointing arrow from state k. (Here, as always, we ignore terms proportional to higher powers of Δt, as they become immaterial for small values of Δt.)

Flux and Steady-State Probabilities

It is evident from Figure 13.9 that the set of all states is ergodic. Theorem 13.2 applies. The steady-state probabilities sum to 1, so they satisfy

$$\rho(0) + \rho(1) + \cdots + \rho(8) = 1. \qquad (13.17)$$

Imagine, in Figure 13.9, a dashed line that separates states 1 through $k - 1$ from states k through 8. In steady state, the net flux across this line must equal zero; thus,

$$\lambda\rho(k - 1) = k\mu\rho(k) \quad \text{for } k = 1, 2, \ldots, 8, \qquad (13.18)$$

because the rate of flow (flux) from state $k - 1$ to state k equals $\lambda\rho(k - 1)$ and because the rate of flow (flux) from state k to state $k - 1$ equals $k\mu\rho(k)$.

Together, Equations (13.17) and (13.18) specify the probability distribution over the states. These equations are easy to solve on a spreadsheet. Solving them would demonstrate that $\rho(8) = 0.0081$, as that information had been gleaned from Table 13.6.

A Famous Formula

Equations (13.17) and (13.18) lead to a famous formula for the steady-state probability $\rho(k)$ that k calls are in progress. Let us recall that $R = (\lambda/\mu)$. We rewrite Equation (13.18) as

$$\rho(k) = \frac{R}{k}\rho(k - 1), \quad \text{for } k = 1, 2, \ldots, 8. \qquad (13.19)$$

Substituting once in Equation (13.19) gives

$$\rho(k) = \frac{R}{k}\rho(k - 1) = \frac{R^2}{(k)(k - 1)}\rho(k - 2).$$

Repeated substitution produces

$$\rho(k) = \frac{R^k}{k!}\rho(0), \quad \text{for } k = 0, 1, \ldots, 8, \qquad (13.20)$$

where $k!$ (pronounced k factorial) is defined by $0! = 1$ and, for $k \geq 1$, by $k! = (k)(k - 1)\cdots(2)(1)$.

Equation (13.20) expresses the probability that each state is occupied in terms of the probability that state 0 is occupied. Equation (13.17) records the fact that the sum of these probabilities equals one. Solving Equations (13.20) and (13.17) gives

$$\rho(k) = \frac{\dfrac{R^k}{k!}}{1 + \dfrac{R}{1!} + \cdots + \dfrac{R^7}{7!} + \dfrac{R^8}{8!}}, \quad \text{for } k = 0, 1, \ldots, 8. \qquad (13.21)$$

Equation (13.21) is the formula we had sought. This formula is famous because Erlang showed how little information was needed to determine the probability distribution over the states—only capacity (8 calls in this case) and the ratio R of the arrival rate to the service rate matter.

On a spreadsheet, Equation (13.21) is easy to evaluate because its numerator is the probability that a random variable having the Poisson distribution takes the value k and its denominator is the probability that the same random variable is not greater than 8. In other words, $p(k)$ is given by

=POISSON(k, R, 0)/POISSON(8, R, 1)

In the above equation, the capacity n of the circuit is eight simultaneous calls. This ratio of Excel functions is handy when n is small but not when n is in the hundreds. To see why, try computing 200!.

The Erlang Probabilities

At the beginning of this section, the Erlang probability $B(n, R)$ was defined to be the steady-state probability that all n conversations are in progress in a model with a capacity of n simultaneous conversations and with a traffic load of R Erlangs. To specify $B(n, R)$ from Equation (13.21), we replace k and 8 by n and get

$$B(n, R) = \frac{\dfrac{R^n}{n!}}{1 + \dfrac{R}{1!} + \cdots + \dfrac{R^{n-1}}{(n-1)!} + \dfrac{R^n}{n!}}. \tag{13.22}$$

Computing $B(n, R)$ from Equation (13.22) is possible when $n = 8$. Often, however, the number n of simultaneous telephone conversations that a circuit can carry is in the dozens or hundreds. Computing $B(n, R)$ directly from Equation (13.22) would be daunting when $n = 100$, for instance. To manipulate Equation (13.22) into a form that is tractable, numerically, we take the reciprocal and organize the numerator and denominator as

$$\frac{1}{B(n, r)} = \frac{1 + \dfrac{R}{1!} + \cdots + \dfrac{R^{n-1}}{(n-1)!}}{\dfrac{R^{n-1}}{(n-1)!}\,\dfrac{R}{n}} + 1 = \frac{1}{B(n-1, R)}\,\frac{n}{R} + 1,$$

which verifies Equation (13.16).

Insensitivity

Equation (13.21) describes the steady-state probability distribution over the states. To obtain Equation (13.21), we used the fact that the call duration time has the exponential distribution. But Equation (13.21) specifies the steady-state probability distribution for *any* call duration time whose mean equals $1/\mu$. The call duration time could be constant, exponential, lognormal, or whatever.

In the literature, this is called an **insensitivity** result. In the Erlang loss model, the steady-state distribution of the number of in-progress calls depends on the mean call duration time but is independent of (insensitive to) its distribution. If this insensitivity result astounds you, you are in good company. It seems, however, that Erlang had been aware of it. Problem 28 on page 479 provides a hint as to why it is true.

13.11. AIDS AND NEEDLE EXCHANGE

Within the United States, the epidemic of human immunodeficiency virus (HIV) was established by 1978. Beginning in 1982, the end states of the HIV infection began to be called acquired immune deficiency syndrome (AIDS). Tests for the HIV antibody were marketed in the United States in 1985 and were used thereafter to diagnose HIV infection.

Barring treatment, the median time between HIV infection and the onset of AIDS is about 10 years. AIDS can develop in as little as two years. A small fraction of HIV positive individuals have survived as long as 18 years without developing AIDS. The median survival time of an untreated AIDS patient is roughly 9 months.

Unlike influenza, the HIV virus cannot be transmitted by kiss, cough, or touch. It can be transmitted from an infected to an uninfected person in these ways:

- Anal or vaginal sex, without the use of a protective condom.
- Injection with an infected needle and syringe, without adequate cleansing prior to use.
- During pregnancy and delivery, from an infected pregnant woman to her unborn fetus.
- From an infected mother to her baby via breast milk.
- By injection or transfusion of infected human blood or blood products. (This route of transmission has been all but eliminated in industrialized nations by screening potential donors.)
- Uncommonly, by oral-genital contact.
- Rarely, by accidental exposure of an uninfected person to an infected person's blood via a needlestick or equivalent accident.
- Very rarely, by the inadvertent transplant of infected tissues or organs.

AIDS and Medical Science

As of this writing, no cure for an established HIV infection has been identified.

Some medical success has been obtained in precluding effective transmission. Specifically, for a HIV-positive woman who becomes pregnant, administration of zidovudine (ADZ) during pregnancy and delivery and to the baby for six weeks after birth reduces the likelihood of transmission from 25% to 8% and, with best care, even lower. In addition, swift administration of AZT (and probably other retroviral regimens) can inhibit transmission after accidental exposure in health care and, possibly, after sexual or needle-sharing events.

Administration of a well-designed regimen of anti-retroviral drugs to HIV positive individuals slows the progression to AIDS and alleviates each stage of the illness. However, these drugs can have significant toxic side effects that limit their utility for some patients. Patients must also be able and willing to adhere to the prescribed regimen; intermittent adherence may severely compromise efficacy.

AIDS in New Haven, Connecticut

By 1986, it was becoming clear that the primary cause of HIV transmission in New Haven was the sharing of contaminated needles. At that time, supplying intravenous drug users (IDUs) with new and uncontaminated needles was illegal in Connecticut. Evidence that needle exchange in Holland and Australia had retarded the spread of HIV among IDUs was encouraging but anecdotal, relying as it did on IDUs' self-reported surveys of behavior.

Responding to an initiative by Yale Professor Alvin Novick, in 1987 New Haven Mayor Ben DeLieto established the New Haven Mayor's Task Force on AIDS. An intense multi-year lobbying and educational campaign by task force members Novick, Sher Horasko, and Elaine O'Keefe bore fruit in June of 1990. On the final day of the legislative session, with vital support from then-mayor John Daniels, the Connecticut General Assembly voted to authorize a pilot needle exchange program in New Haven, with the participation of the Health Department of the city of New Haven and Yale faculty members and staff. This program's goals were to serve clients by reducing HIV transmission and to do so in ways that generated experimental data that could be used to design and evaluate other needle exchange programs.

The New Haven Needle Exchange Experiment

In this experiment, a van donated by the Yale Medical School visited four New Haven neighborhoods frequented by IDUs. Each adult IDU could voluntarily become a client. To guarantee anonymity, each client selected a pseudonym. Each client could trade needles on a one-for-one basis with "program" needles. Each program needle bore a unique label. Records were kept on the date of issue of each needle, the pseudonym of the client to whom it was issued, the date of return of each program needle, and the pseudonym of the client who traded it in. A sample of the needles that were traded in was tested for the HIV antibody. Robert Heimer of the Yale School of Public Health coordinated these studies of the contamination of needles.

A major insight obtained from the New Haven needle exchange experiment was that measures of the value of needle exchange could be obtained directly from the needles themselves. With very conservative—and therefore defensible—modeling assumptions, it was concluded that needle exchange reduces the rate of exposure of uninfected participants to infected needles by a minimum of 33% . This result was obtained from two separate lines of analysis, one reported by Elaine O'Keefe and Edward H. Kaplan[1] and the other by Kaplan.[2] This experiment and its analyses have altered the perception of needle exchange at the local, national, and international levels. Needles and syringes can now be purchased legally in Connecticut, for instance.

A Simplified Experiment

Problem E, which sketches the New Haven needle exchange experiment, provides less data than was used in the aforementioned studies. Our analysis of this data is simple enough to use the basic methodology of this chapter. But our analysis is less conservative than the aforementioned; it leads to conclusions that are more contentious.

Problem E (Needle Exchange)

In a needle exchange experiment, a van visits the regions of a city that are frequented by intravenous drug users (IDUs). Each IDU can voluntarily become a client. Each client is guaranteed anonymity and can trade needles on a one-for-one basis with "program" needles. Each program needle is new, uncontaminated, and uniquely identified. Each needle that was

---

[1] E. H. Kaplan and E. O'Keefe, "Let the Needles Do the Talking! Evaluating the New Haven Needle Exchange," *Interfaces* 23 (1993), pp. 7–26.

[2] E. H. Kaplan, "Probability Models of Needle Exchange," *Operations Research* 34 (1995), pp. 558–569.

traded for a new one is tested for the HIV antibody. In the early months of this experiment:

- Roughly 60% of the nonprogram needles that were traded in were HIV positive.
- Roughly 40% of the program needles that were traded in were HIV positive.
- Of the needles that clients traded in, 68% had been issued to the client who returned them, and virtually all of the remaining 32% were "street" needles, that is, needles that had not been issued to other clients.
- Needles that were traded in by the client to whom they had been issued had spent an uncertain length of time with the client, with a mean of roughly seven days.

On the basis of this information, what can be said about the effect of needle exchange on the rate h at which uninfected clients inject themselves with infected needles?

Problem E differs from most of the problems in this text. It describes an experiment, but it omits the modeling assumptions that will be used to analyze the data that the experiment produces.

A Model

Each client possesses a population of needles. In our model, clients can swap needles with other IDUs, and they can return needles to the program in exchange for new (and uninfected) needles. In our analysis of Problem E, each client is assumed to:

(a) swap needles with other IDUs at rate s,

(b) trade in needles at rate r,

(c) select each needle in his or her possession with equal probability when making an injection.

Here, the word "rate" connotes a continuous-time Markov chain. For instance, the probability that a needle is swapped during Δt units of time equals $s\Delta t$ plus an error term that is small with respect to Δt. Program needles return in seven days, on average, so $r = 1/7$. The data in Problem E do not include the parameter s, but we will find a way around this difficulty. To the assumptions listed above, we make one other, which is that:

(d) Each needle that is turned in by the client to whom it was issued has never been swapped.

Later, assumption (d) will be criticized.

States

A client's needle is now said to be in *state 1* if it was issued by the program and has not been swapped with another IDU. A client's needle is said to be in *state 2* if it is the result of at least one swap of a state 1 needle. Each needle that a client possesses is either in state 1 or state 2.

Figure 13.10 presents the transition rate diagram between the two states. Transition from state 1 to state 2 occurs at rate s, in accord with assumption (a), above. Transitions from each state to state 1 occur at rate r, in accord with assumption (b), because each needle that is traded in is replaced by a new one. (This diagram remains correct if it also contains an arc pointing from state 2 to itself with rate s, but we cannot observe these swaps, and their occurrence does not affect the steady-state probability distribution.)

Let $\rho(1)$ and $\rho(2)$ denote the steady-state probability of observing states 1 and 2, respectively. These probabilities sum to 1, and the (flux) net rate of flow across the dashed gray

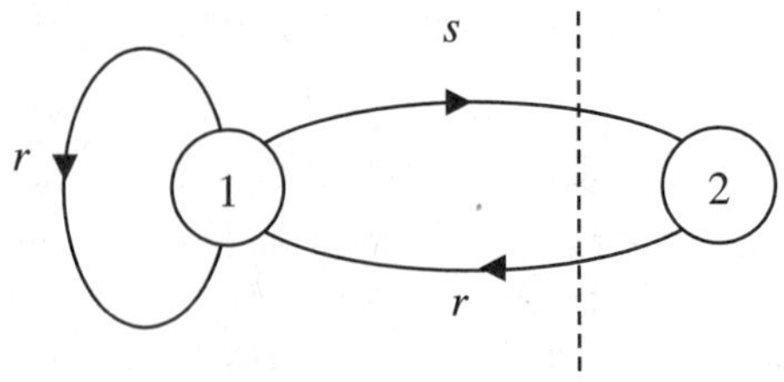

Figure 13.10 Transition rate diagram for needle swapping and trade-in.

line equals zero. Hence, the steady-state probabilities satisfy

$$\rho(1) + \rho(2) = 1 \quad \text{and} \quad \rho(1)s = \rho(2)\,r. \tag{13.23}$$

Expression (13.23) is two equations in two unknowns. One of these equations gives $\rho(1) = \rho(2)\,r/s$, and substituting $\rho(2)\,r/s$ for $\rho(1)$ in the other gives

$$\rho(2) = \frac{1}{1 + r/s} \quad \text{and} \quad \rho(1) = \frac{r/s}{1 + r/s} \tag{13.24}$$

The needle exchange data include $r = 1/7$, but not s, which might seem to impede our use of Equation (13.24).

Return Rates

The data in Problem E include the fraction f of needles that were returned but had not been issued to the client to whom they were returned: $f = 0.32$. Let us see what f determines. In steady state, state 2 needles are returning at the rate $\rho(2)\,r$. Thus, over a long period of time, part (d) of Theorem 13.2 states that the rate of return of state 2 needles is approximately equal to $\rho(2)\,r$. Part (d) also states that, over a long period of time, the rate of return of state 1 needles is approximately $\rho(1)\,r$. Since f is defined to be the fraction of the returning needles that are in state 2, it must be that f equals $\rho(2)\,r$ divided by the sum of $\rho(1)\,r$ and $\rho(2)\,r$. Algebraically,

$$0.32 = f = \frac{\rho(2)\,r}{\rho(1)\,r + \rho(2)\,r} = \frac{\rho(2)}{\rho(1) + \rho(2)} = \rho(2). \tag{13.25}$$

Evidently, f equals 0.32. From this fact, we could compute s, but we need not do so, as neither s nor r will affect our analysis.

Assumption (c) is that each needle in the client's possession is equally likely to be selected for an injection, so the probability h that the client chooses a needle that has been swapped equals $\rho(2)$. If an uninfected client injects with a state 1 needle, no infection can occur. Thus, the probability that an uninfected client injects with a street needle decreases from 1 to $\rho(2) = h$. Since $\rho(2) = 0.32$, this is a decrease of 68% in the rate of exposure to needles that can be contaminated. Theorem 13.2 has allowed us to conclude:

> The reduction in the rate by which uninfected clients inject themselves with infected needles equals the fraction of program needles that are returned by the client to whom they were issued.

Curiously, for our model, only one datum matters. This datum, 0.68, is the fraction of the program needles that are returned by the client to whom they were issued. Our model uses a continuous-time Markov chain to infer a 68% reduction in the rate at which uninfected clients inject themselves with infected needles.

Critique of the Model

Our simplified model overlooks two factors. First, IDUs share needles as well as swapping them. For instance, a pair of IDUs may "shoot up" with a needle and syringe that one of them provides. This challenges assumption (d); not all of the state 1 needles have been used solely by the person to whom they were issued. Second, as favors, some IDUs brought their friends' street needles to the program van for trade-in without ever using them. This challenges assumption (c); some of the state 2 needles had never been candidates for injection. To an extent, these considerations offset each other.

In the more sophisticated models, mentioned earlier, only *pessimistic* assumptions are made, therefore obtaining a lower bound on the benefit of needle exchange. Those models make use of information about whether returned needles were contaminated or uncontaminated to assay the effect of needle exchange.

13.12. REVERSIBLE MARKOV CHAINS

This section introduces a topic that is known as reversibility. If a Markov chain is reversible, we will see that its steady-state probabilities are particularly easy to compute. And, in Chapter 14, a remarkable consequence of reversibility will be described. Reversibility will be introduced in the context of a continuous-time Markov chain and will then be adapted to discrete-time Markov chains.

A Transition Rate Diagram

Figure 13.11 presents the transition rate diagram of a continuous-time Markov chain that has six states. In the figure, the symbols a through f stand for known positive numbers, each of which is a transition rate.

A Directed Network

A transition rate diagram can be interpreted as a directed network, using the language introduced in Chapter 5. The states of the Markov chain are the *nodes* of this network. Each transition rate $Q(i, j)$ that is positive is identified with a *directed arc* that points from node i to node j.

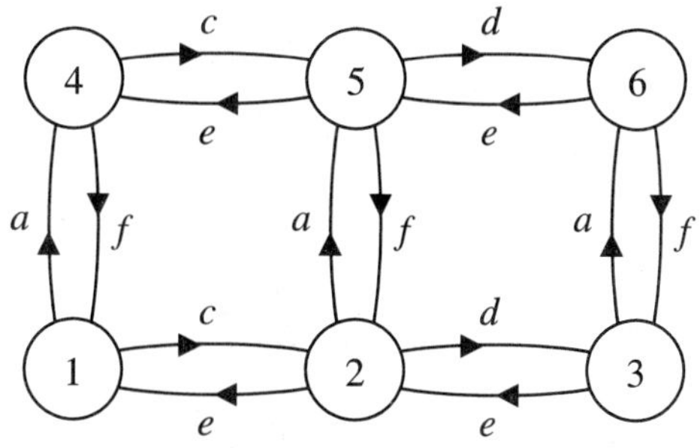

Figure 13.11 Transition rate diagram for a continuous-time Markov chain.

As in Chapter 5, a *path* is a sequence of nodes whose arcs can be "traversed" in sequence, each in the direction of its arrow. For instance, the directed network in Figure 13.11 contains path (1, 4, 5, 6); this path consists of arc (1, 4), arc (4, 5), and arc (5, 6), which are traversed in sequence.

As in Chapter 5, a *cycle* is a path that ends at the node where it started. This directed network contains many cycles, one of which is (1, 4, 5, 6, 3, 2, 1).

The network in Figure 13.11 has two special properties that are not shared by all networks, notably:

- First, if (i, j) is an arc, then so is (j, i). As a consequence, each path has a **reversed** path, which visits the same nodes in the reverse sequence.
- Second, the product of the transition rates of the arcs in each cycle equals the product of the transition rates of the arcs in the reversed cycle.

It is evident from Figure 13.11 that the first property is satisfied. To illustrate the second property, we consider the cycle (1, 2, 5, 4, 1). The product of the transition rates of its arcs equals $c\,a\,e\,f$. The reversed cycle is (1, 4, 5, 2, 1), and the product of the transition rates of its arcs equals $a\,c\,f\,e$. These products are the same, and that fact is true of every cycle.

A Puzzle

Let us set Figure 13.11 aside for a moment and introduce a puzzle. Suppose that you know the transition rates for a continuous-time Markov chain. Suppose that this chain has only one ergodic set and that all states are recurrent. You are given a **sample path**, which is a plot like that in Figure 13.12. Its y-axis records the state that is occupied at a particular time, and its x-axis records the length of time each state is occupied. In Figure 13.12, state 1 is occupied for a length of time that is proportional to the width of the interval A. State 2 is occupied for two intervals of time whose lengths are proportional to intervals B and E.

What you are *not* told is whether, in Figure 13.12, time increases from left to right or from right to left. If time increases from left to right, a transition has occurred from state 1 to state 2. If time increases from right to left, a transition has occurred from state 2 to state 1.

A continuous-time Markov chain is said to be **reversible** if, knowing its transition rates, you cannot tell from any sample path, however long, whether time increases to the left or to the right.

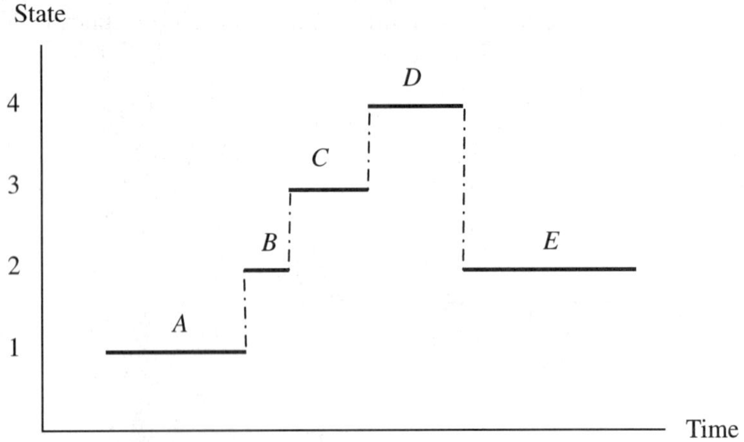

Figure 13.12 A sample path of a continuous-time Markov chain.

The puzzle is this: Are there any reversible Markov chains? If so, what do they look like? And why would we care? Each of these questions will be answered.

Necessary Conditions

Let us first identify conditions that a reversible Markov chain must satisfy. Suppose that the Markov chain whose sample path is depicted in Figure 13.12 has $Q(2, 1)$ positive and $Q(1, 2)$ equal to zero. In this chain, we can observe transitions from state 2 to state 1 but not from state 1 to state 2. In Figure 13.12, time must increase from right to left, as a transition from state 1 to state 2 cannot occur. This chain cannot be reversible. The property that this illustrates holds in general as is highlighted in the following.

> **A necessary condition for reversibility.** If $Q(i, j)$ is positive, then $Q(j, i)$ must also be positive.

We can improve upon this condition.

As usual, the $\rho(i)$'s denote the steady-state probability distribution over the states. Consider a Markov chain for which $\rho(i)\, Q(i, j) = 2.2$ and $\rho(j)\, Q(j, i) = 1.6$. Given an infinitely long sample path, part (d) of Theorem 13.2 assures you that the time-average rate of transition from state i to state j will equal 2.2 and that the time-average rate of transition from state j to state i will equal 1.6. Thus, from such a sample path, you could tell which way time increases. This chain cannot be reversible.

Thus, in order for a continuous-time Markov chain to be reversible, it must be the case that $\rho(i)\, Q(i, j)$ and $\rho(j)\, Q(j, i)$ equal each other.

> **A necessary condition for reversibility.** The steady-state probabilities satisfy $\rho(i)\, Q(i, j) = \rho(j)\, Q(j, i)$ for every pair i and j of states.

Reversed Paths and Cycles

The preceding condition has a lovely implication for paths. To explore it, we consider the path (i, j, k). In a reversible Markov chain, the steady-state probabilities must satisfy the equations

$$\rho(i)\, Q(i, j) = \rho(j)\, Q(j, i) \quad \text{and} \quad \rho(j)\, Q(j, k) = \rho(k)\, Q(k, j).$$

Solving both equations for $\rho(j)$, eliminating $\rho(j)$, and then clearing denominators produces

$$\rho(i)\, Q(i, j)\, Q(j, k) = \rho(k)\, Q(k, j)\, Q(j, i),$$

which relates the path (i, j, k) to the reversed path (k, j, i). The analog of this equation holds for paths having any number of nodes. And if such a path is a cycle, $\rho(i)$ cancels out. What remains on the left-hand side of the equation is the product of the transition rates in the cycle. And what remains on the right-hand side is the product of the transition rates in the reversed cycle. These products must equal each other, as is highlighted in the following statement.

> **A necessary condition for reversibility.** The product of the transition rates in any cycle must equal the product of the transition rates in the reversed cycle.

A Necessary and Sufficient Condition

This condition that we have just identified is not only necessary but can be shown to be sufficient as well. So we have:

Theorem 13.3 (Reversibility in a Continuous-Time Markov Chain). In a continuous-time Markov chain, an ergodic set is reversible if and only if the product of the transition rates in each of its cycles equals the product of the transition rates in the reversed cycle.

Evidently, the continuous-time Markov chain in Figure 13.11 is reversible. (By the way, the continuous-time Markov chain in Figure 13.9 is also reversible.)

Who Cares?

Reversibility can ease the computation of steady-state probabilities. We have already seen that, if a continuous-time Markov chain is reversible, then the steady-state probabilities must satisfy

$$\rho(i)\, Q(i,j) = \rho(j)\, Q(j,i) \quad \text{for every pair } \{i,j\} \text{ of states.} \tag{13.26}$$

Equation (13.26) can simplify the computation of the steady-state probabilities. In Figure 13.11, for instance, reversibility assures us that

$$\rho(2) = \frac{c}{e}\rho(1), \quad \rho(4) = \frac{a}{f}\rho(1), \quad \rho(5) = \frac{c}{e}\rho(4) = \frac{ca}{ef}\rho(1),$$

and so forth.

Reversibility in a Discrete-Time Markov Chain

In a discrete-time Markov chain, an ergodic set is said to be **reversible** if, knowing the transition probabilities, you cannot tell from any discrete-time sample path whether time is increasing from left to right or from right to left.

The prior discussion applies to discrete-time Markov chains when we replace the term *rate* by *probability*. The analog of Theorem 13.3 is:

Theorem 13.4 (Reversibility in a Discrete-Time Markov Chain). In a discrete-time Markov chain, an ergodic set is reversible if and only if the product of the transition probabilities in each of its cycles equals the product of the transition probabilities in the reversed cycle.

If an ergodic set is reversible, its steady-state distribution is easy to compute from the analog of Equation (13.26), which is

$$\pi(i)\, P(i,j) = \pi(j)\, P(j,i) \quad \text{for each pair } \{i,j\} \text{ of states.} \tag{13.27}$$

By the way, the discrete-time Markov chain in Figure 13.6 is reversible.

13.13. REVIEW

The early sections of this chapter posed and solved the following two problems:

- Given a starting state s, what is the probability of reaching state t?
- Given a starting state s, how long will it take to reach state t?

To solve each of these problems, we first embedded it in a class of problems, one per starting state. We developed a system of equations, one per starting state.

Later sections of this chapter studied the long-term behavior of a Markov chain. Those sections focused on its steady-state probability distribution. Conditions were provided under which such a distribution exists, and methods were found for computing it. An easy method is to observe that, in steady state, the flux (rate of flow) into each group of states equals the flux out of that group of states.

The ergodic property gives a second interpretation of the steady-state probabilities. In a discrete-time Markov chain, the steady-state probability $\pi(i)$ that state i is occupied is the fraction of the transitions that occur to state i, over a long period of time. In a continuous-time Markov chain, the steady-state probability $\rho(i)$ that state i is occupied is the long-run fraction of the time that state i will be occupied.

A set of states is said to be ergodic if it is a smallest set from which escape is impossible. An ergodic set of states can be reversible. If it is reversible, then its steady-state probabilities are especially easy to compute because they satisfy Equation (13.26) in the case of a continuous-time Markov chain and Equation (13.27) in the case of a discrete-time Markov chain.

The Erlang loss model in Section 13.10 is the pioneering use of Markov chains in telecommunications. The needle exchange model in Section 13.11 applies a Markov chain to an important issue in the public sector.

In Chapter 14, the ideas that have been introduced here will be applied to Markov chains that have infinitely many states. This will work because the Markov chains we will encounter in Chapter 14 satisfy a **positive recurrence** condition, which is that for each state s, the probability of revisiting state s equals 1, and the expectation of the time until state s is revisited is finite.

13.14. HOMEWORK AND DISCUSSION PROBLEMS

1. Pete, the groundhog, finds himself in room 2 of the nine-room maze shown below. A snack is in room 9, and a predator skulks in room 4. From each room, Pete is equally likely to move to any adjacent room to which there is an open passage.

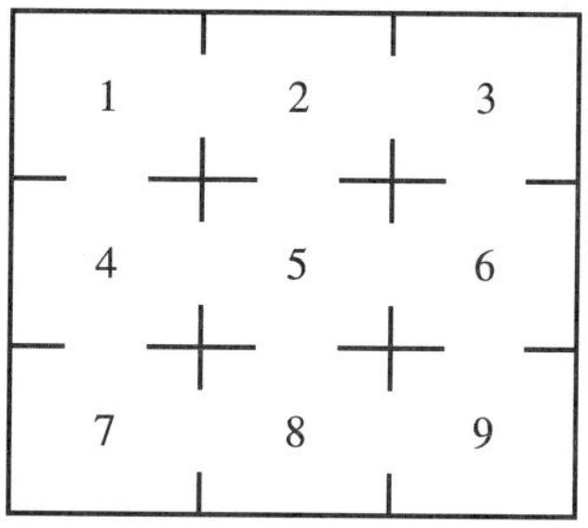

(a) Compute the probability that Pete reaches room 9 (the snack) before he reaches room 4 (the predator).

(b) Compute the expectation of the number of moves from room to room that occur before termination in part (a).

2. Eliminate the predator from the nine-room maze in the preceding problem. Suppose Pete moves from room to room as above and moves repeatedly. This problem concerns Pete's steady-state probability distribution (the $\pi(i)$'s) over the nine rooms.

(a) Symmetry guarantees that $\pi(i)$ equals $\pi(j)$ for certain values of pairs i and j. Partition the set $\{1, 2, \ldots, 9\}$ into three groups, so that each state within a particular group has the same value of $\pi(i)$. Label your groups A, B, and C.

(b) Draw a transition diagram showing the probability of transition from group to group.

(c) Use the diagram in part (b) to find the steady-state probability distribution over the groups.

(d) Find the steady-state probability distribution over the rooms.

3. **(The Absent-minded Professor)** Vince, the absent-minded professor, keeps a stock of five umbrellas, some at home and the rest at the office. If it's raining when he leaves either location for the other, he takes an umbrella with him, if there is one. If it's not raining when he makes a trip, he leaves the umbrellas where they are. It rains with probability p each time he makes a trip in either direction, and these events are independent of each other.

(a) For a Markov chain that represents this situation, identify the states and draw the transition diagram.

(b) Compute the steady-state probability distribution over the states.

(c) With what probability does Vince get wet on the way to work?

(d) With what probability does Vince get wet at least once during the day, either on his way to work or on his way home?

4. **(Random Swapping)** Three white balls and three black balls are distributed between urns A and B in such a way that each urn contains exactly three balls. At each step, one ball is drawn at random and simultaneously from each urn; these two balls are then placed in the other urn.

(a) For a Markov chain that represents this situation, identify the states and draw the transition diagram.

(b) Compute the steady-state probability distribution over the states.

5. Compute the steady-state probability distribution for the three-state Markov chain whose transition matrix appears below.

$$P = \begin{bmatrix} 0.3 & 0 & 0.7 \\ 0.1 & 0.7 & 0.2 \\ 0.6 & 0 & 0.4 \end{bmatrix}$$

6. An automobile insurance company charges higher premiums to customers who have had accidents in the prior two years. Each year, you have an accident with probability of 0.2. Accidents in different years are independent events. Next year's premium is determined according to these rules:

- If you had accidents in each of the prior two years, you pay an extra \$800.
- If you had an accident this year but not last year, you pay an extra \$400.
- If you had an accident last year but not this year, you pay an extra \$200.

Over a long period of time, how much extra do you pay per year?

7. Consider the six-state Markov chain whose nonzero transition probabilities are marked below with an asterisk. Identify its transient states, its ergodic set or sets, and the period of each ergodic set.

	1	2	3	4	5	6
1	*		*			
2			*			
3	*			*		
4						*
5		*		*		
6		*		*		

8. Consider the Six-state Markov chain whose nonzero transition probabilities are marked below with an asterisk. Does this Markov chain have any transient states? How many ergodic sets does it have? What is the period of each ergodic set? Partition each ergodic set whose period equals p into p groups of states, with transition from each group to the next and from the pth group back to the first.

	1	2	3	4	5	6
1			*		*	
2			*			
3				*		
4					*	*
5		*				
6	*	*				

9. In the Gambler's Ruin problem (Problem A in the text), let the symbol $r(i)$ denote the probability that the gambler will be ruined (by reaching state 0 before state 6) given that he is in state i. Write a system of equations for $r(0)$ through $r(6)$. Use Solver to compute their solution. For each state i, compute the probability of the event in which the gambler plays forever, without winning or losing. *Remark:* Some events do have zero as their probabilities.

10. In the Gambler's Ruin problem (Problem A in the text), let the random variable N denote the number of times he bets before reaching either state 0 or state 6. Compute the expectation of N. *Hint:* Embedding and an analog of Figure 13.3 may be helpful.)

11. (**Problem 10, continued**) Compute the standard deviation of N. *Hint:* Review Figure 13.4.

12. (**Timid Play**) In the Gambler's Ruin problem (Problem A in the text):

(a) Use a spreadsheet to solve the recursion in Equation (13.4).

(b) Suppose the gambler could bet one half dollar at a time rather than one dollar. If he did so, would the probability of reaching his goal increase? If so, by how much? *Hint:* Adapt part (a) to answer part (b).

13. (**Bold Play**) In the Gambler's Ruin problem (Problem A in the text), the odds were favorable in the sense that the probability p of winning was greater than 0.5. Suppose $p = 0.4$, and suppose that the gambler can bet any number of dollars at a time, up to his current asset position. On a spreadsheet, compute the probability that he wins if he bets **boldly**, that is, when he observes state i, he bets the smaller of i and $6 - i$ dollars. When $p = 0.4$, does bold play improve the probability that he reaches his goal?

14. (**Gambler's Ruin**) A gambler begins with i dollars. At each play, the gambler wins \$1 with probability p and loses \$1 with probability $q = 1 - p$. Successive plays are independent, and the gambler keeps playing until his capital increases to N dollars or decreases to 0 dollars, whichever comes first. Let $w(i)$ denote the probability that the gambler reaches N dollars before he reaches 0 dollars.

(a) Write the analog of Equations (13.1) and (13.2) for $w(i)$.

(b) Show that these equations are satisfied by

$$w(i) = \begin{cases} \dfrac{1 - (q/p)^i}{1 - (q/p)^N} & \text{if } p \neq q \\ \dfrac{i}{N} & \text{if } p = q \end{cases} \qquad \text{for each } i.$$

15. For the Punxsutawney Pete problem (Problem B in the chapter), compute the standard deviation of the time T that it takes Pete to emerge from his burrow.

16. For the Punxsutawney Pete problem (Problem B in the chapter), suppose that Pete has limited recall; Pete can remember the tunnel he took into the room he currently occupies, and he chooses each of the other tunnels with equal probability. How does this change the state space? Re-compute his mean time to emergence. Does it increase? decrease?

17. (**Craps**) The game of "craps" is played with two fair dice. The "shooter" rolls the dice repeatedly, until he wins or loses. The shooter wins if the total number of pips on the first roll is either 7 or 11. The shooter loses if the total number of pips on the first roll is 2, 3, or 12. If the number of pips on the first roll is some number other than these, that number becomes the shooter's "point." He rolls repeatedly until a 7 appears or the point reappears, whichever comes first. He wins if the point reappears before a 7. He loses if a 7 appears before the point.

(a) What is the probability that the shooter wins on the first roll?

(b) Suppose that the shooter rolls a 5 on his first roll. Given this point, let the random variable X equal 1 if he makes this point, and let X equal 0 if he does not. Use a probability tree like those in Figure 13.3 to describe the probability distribution of X. Compute $P(X = 1)$.

(c) For the numbers 3 through 6 and 8 through 10, compute the joint probability that his point is this number and that he makes his point.

(d) Compute the probability that the shooter wins at craps.

18. For the auto rental example (Problem C in the text):

(a) Use the SUMPRODUCT function to compute $q(2, j)$ for each j. Do this in a way that will let you find $q(k, j)$ for each $k > 2$ without extra typing.

(b) From part (a), compute $q(k, j)$ for $k = 3, 4, \ldots, 12$.

19. For the auto rental problem (Problem C in the text), suppose that the car begins at location 4. Compute $q(k, j)$ for each j and for $k = 0$ through 12. Is there a familiar pattern?

20. For the Markov chain in Figure 13.7, compute:

(a) The steady-state probability distribution over the states given that you begin in state 3. *Hint:* Is $0.5\,\pi(3) = 0.6\,\pi(4)$.

(b) The steady-state probability distribution over the states given that you begin in state 1.

(c) The probability of reaching state 1 given that you begin in state 2.

(d) The steady-state probability distribution over the states given that you begin in state 2.

21. Consider a discrete-time Markov chain that has only one ergodic set. For each state j that is recurrent, define:

$m(i, j) =$ the expectation of the number of transitions that will occur prior to the next transition to state j given that state i is observed now.

(a) Justify the equation

$$m(i, j) = 1 + \sum_{k=1}^{n} P(i, k)\, m(k, j) - P(i, j)\, m(j, j).$$

Hint: An analog of the probability tree in Figure 13.3 may be helpful.

(b) With $\pi(i)$ as the steady-state probability of observing state i, multiply the equation for $m(i, j)$ by $\pi(i)$. Then sum over i and cancel certain terms to demonstrate that $m(j, j) = 1/\pi(j)$.

(c) Have you proved part of Theorem 13.1? If so, which part?

22. Consider a continuous-time Markov chain that has only one ergodic set. For each state i, set

$$\nu(i) = \sum_{k} Q(i, k) \quad \text{and} \quad P(i, k) = Q(i, k)/\nu(i).$$

Note that the expectation of the time to transition given state i is exponential with rate $\nu(i)$. Moreover, note that $P(i, k)$ is the probability that this transition occurs to state k. For each state j that is recurrent, define:

$m(i, j) =$ the expectation of the time that elapses until the next transition to state j given that state i is observed now.

(a) Justify the equations

$$m(i, j) = 1/\nu(i) + \sum_{k} P(i, k)\, m(k, j) - P(i, j)\, m(j, j),$$

$$\nu(i)\, m(i, j) = 1 + \sum_{k} Q(i, k)\, m(k, j) - Q(i, j)\, m(j, j).$$

(b) With $\rho(i)$ as the steady-state probability of observing state i, multiply the preceding equation by $\rho(i)$. Then sum over i and cancel certain terms to demonstrate that $m(j, j) = 1/[\rho(j)\, \nu(j)]$.

(c) Have you proved part of Theorem 13.2? If so, which part?

23. **(Flux)** Consider a discrete-time Markov chain that has only one ergodic set. Let A be any subset of the states. Use Equation (13.9) to verify that

$$\sum_{i \in A,\, j \notin A} \pi(i)\, P(i, j) = \sum_{j \notin A,\, i \in A} \pi(j)\, P(j, i).$$

Hint: Sum an expression in Equation (13.9) over i in A and use the fact that $1 = \sum_{j \in A} P(i, j) + \sum_{j \notin A} P(i, j)$ to cancel certain terms in that sum.

24. **(Reversibility)** Consider a discrete-time Markov chain. Let S denote its set of states. Suppose that S is an ergodic set, and suppose that this chain's transition probabilities satisfy these special conditions:.

- If $P(i, j)$ is positive, then so is $P(j, i)$.
- If $P(i, j)$ is positive, then $P(i, j) = 1/b(i)$.

(a) Show that this chain is reversible. *Hint:* Does $b(i)$ equal the number states j for which $P(i, j)$ is positive?

(b) Use the fact that the chain is reversible to show that its stationary distribution is

$$\pi(i) = \frac{b(i)}{\sum_{k \in S} b(k)} \quad \text{for each } i \text{ in } S.$$

25. **(Punxsutawney Pete, again)** Reconsider Problem 2 of this chapter. Show that its Markov chain is reversible. Use reversibility to compute its stable probability distribution. *Hint:* Review the prior problem.

26. **(Knight's Move)** A 64-state Markov chain is described as follows: A knight hops from square to square on a chessboard, repeatedly. At each hop, the knight chooses each legal move with equal probability.

(a) Is this Markov chain reversible? Why? *Hint:* Review the prior hint(s).

(b) Compute the stable probability that the knight is in the lower right-hand square of the chessboard. *Hint:* Review the prior hint(s).

27. A service organization has a call rate of 200/hour and a mean call duration of 4 minutes. The telephone company has been asked to provide them with a number of lines such that not more than 1% of the calls cannot be answered immediately. How many lines should be provided?

Remark: In the Erlang loss model, we noted that the steady-state probability distribution of the number of customers in the system is given by Equation (13.21), independent of the distribution of the call duration time, with $1/\mu$ as its mean. The next problem verifies (13.21) for the case of a call duration time that is the mixture of two exponential probability distributions. (The result remains true for the sum of exponential probability distributions and, more generally for **phase** distributions, which are mixtures of sums of exponential distributions. Reversibility plays a role in the development.)

28. **(Insensitivity)** Let us alter the Erlang loss model (Problem D in the text) by changing the probability distribution of the call duration, as follows: Each call is Type A with probability $p(A)$ and Type B with probability $p(B) = 1 - p(A)$. Type A calls have an exponential duration with rate $\mu(a)$. Type B calls have an exponential duration with rate $\mu(b)$. The types of different calls are mutually independent.

(a) Compute the mean call duration, and specify the rate μ as its reciprocal.

(b) The state of the system is now a pair (A, B) of integers, A being the number of Type A calls that are in progress and B being the number of Type B calls in progress. For $n = 8$, build a transition diagram akin to that in Figure 13.9.

(c) Is this chain reversible?

(d) Show that the steady-state probability distribution of this Markov chain equates $\rho(A, B)$ to the product of $\rho(A + B)$, as given by Equation (13.21), and the probability of A successes in $A + B$ Bernoulli trials, each having probability $p(A)$ of success.

(e) Show that $\rho(A + B)$ equals the steady-state probability that $A + B$ calls are in progress.

29. **(Upticks and Infinite-State Markov Chains)** You can flip a coin repeatedly. At each flip, this coin has probability p of coming up heads. The outcomes of different flips are independent of each other. You win one dollar each time the coin comes up heads, and you lose one dollar each time it comes up tails. The net amount X that you have won or lost can be described by an infinite-state Markov chain.

(a) What are the states of this Markov chain? What are its transition probabilities?

(b) Suppose $p = 0.6$. You plan to stop as soon as X is positive. Let s denote the probability that you stop. Does s satisfy the equation $s = 0.6 + 0.4s^2$? Compute s.

(c) Redo part (b) for the case $p = 0.5$.

(d) Redo part (a) for the case $p = 0.4$.

(e) Denote as T the number of flips that occur until you stop. Is $E(T) = 1 + (1 - p)\, 2E(T)$? Why? For which p?

Remark: The next several problems presume a familiarity with matrices. They use the idea of a "transient" matrix to develop the so-called fundamental matrix of Kemeny and Snell, and they show how to use it in the analysis of a Markov chain.

30. For the transition matrix and initial condition in Table 13.4, let $Q(i, k)$ denote the expectation of the number of times state i is observed up to and including the kth transition.

(a) Demonstrate that

$$Q(i, k) = \sum_{m=0}^{k} q(i, m).$$

(b) Augment the spreadsheet in Table 13.4 to compute $Q(i, k)$ for values of k between 0 and 20.

(c) Define the matrices P, π, P^*, and I, letting

- P be the 4×4 transition matrix in Table 13.4.
- π be the 1×4 matrix of steady-state probabilities for the matrix P (so that $\pi(1) = 6/23$ and so forth).
- P^* be the 4×4 matrix each of whose rows equals π.
- I be the 4×4 identity matrix (so that I has 1's on the diagonal and 0's elsewhere).

In columns C through F of the same spreadsheet, use Excel to compute the matrix Z that is given by

$$Z = (I - P + P^*)^{-1}.$$

(d) To the right of the columns in which you computed $Q(1, k)$ through $Q(4, k)$, compute the sum of the top row of Z and $k\pi$. What is happening as k increases?

Remarks: The matrix Z that is defined in part (c) of the preceding problem was dubbed the **fundamental matrix** by Kemeny and Snell. Part (d) observes, empirically, that the expectation of the number of times that state j will be observed up to and including the kth transition given that state i was observed initially approaches $Z(i, j) + k\pi(j)$ as k becomes large. The next two problems verify this mathematically.

Definition: An $n \times n$ matrix Q is said to be **transient** if each entry in the matrix Q^k approaches zero as k becomes large.

31. Let the $n \times n$ matrix Q be transient.

(a) Verify that

$$(I + Q + \cdots + Q^k)(I - Q) = I - Q^{k+1}.$$

(b) Show that the matrix $(I - Q)$ is invertible. *Hints:* The determinant of the product equals the product of the determinants. A matrix whose determinant is not zero is invertible. Is the determinant of $(I - Q^{k+1})$ positive for large values of k?

(c) Show that:

$$(I + Q + \cdots + Q^k) = (I - Q)^{-1}(I - Q^{k+1}),$$

$$\sum_{k=0}^{\infty} Q^k = (I - Q)^{-1}.$$

32. Let the $n \times n$ matrix P be the transition matrix of a discrete-time Markov chain that has only one ergodic set, and let the $1 \times n$ matrix π be the unique solution to Equation (13.9). Let **1** denote the $n \times 1$ matrix each of whose entries equals 1, and define the $n \times n$ matrices P^* and Q by

$$P^* = \mathbf{1}\pi \quad \text{and} \quad Q = P - P^*.$$

Observe that each row of P^* equals π.

(a) Demonstrate that $P^* = P^*P = PP^* = P^*P^*$.

(b) Demonstrate that

$$Q^k = P^k - P^* \quad \text{for } k = 1, 2, \ldots.$$

Hint: This is clear when $k = 1$, so prove it by induction on k.

(c) Is this matrix Q transient?

(d) Let T^k denote the $n \times n$ array (matrix) whose ijth entry equals the expectation of the number of times state j will be observed up to and including the kth transition, given that state i is observed initially. Is it true that:

$$T^k = (I + P + \cdots + P^k) = (I + Q + \cdots + Q^k) + kP^*?$$

(e) The **fundamental matrix** Z of Kemeny and Snell is defined to be

$$Z = (I - Q)^{-1} = [I - P + P^*]^{-1}.$$

Show that as $T^k - (Z + kP^*) \to 0$ as $k \to \infty$.

Remark: The next problem unravels a "paradox" that is attributed to the Spanish physicist, Juan Parrando. It blends two money-losing betting schemes into a profitable one.

33. **(Parrando's Paradox)** A gambler can place either of two types of bet. Each bet either wins \$1 or loses \$1, and the outcomes of different bets are mutually independent. Take ε as a small positive number, say, $\varepsilon = 0.01$. Each bet of type A wins with probability of $0.5 - \varepsilon$. Each bet of type B wins with probability of $0.7 - \varepsilon$ if the gambler's net winnings to date is an even number and wins with probability of $0.1 - \varepsilon$ if the gambler's net winnings to date is an odd number.

(a) Suppose that the gambler places type A bets repeatedly. Draw a transition diagram of the probability that betting shifts his wealth level between even and odd numbers. What is the steady-state probability distribution of his wealth level over these states? What is his expected net profit per bet?

(b) Redo part (a) for the case in which the gambler places type B bets repeatedly.

(c) Redo part (a) for the case in which the gambler places type A bets when his wealth level is an odd number and places type B bets when his wealth level is an even number.

Chapter 14

Queueing

This chapter is dedicated to Carl M. Harris, a leader in queueing, who died in the spring of 2000. Carl read drafts of nearly all of the chapters of this text and provided superb feedback. He helped me unstintingly and enthusiastically, as he had done for so many others in so many ways.

14.1. PREVIEW

A **queue** is a line of customers who are waiting for service. These "customers" can be people, vehicles, aircraft, packages, or work-in-process inventory. These customers can also be electronic. On the Internet, for instance, messages are sent from node to node, each node holding a queue of messages that are waiting for transmission to other nodes.

The most familiar examples of queues are customers arriving at a fast-food counter, a bank, or a toll booth. Queues play important roles throughout the manufacturing and service sectors of modern economies. At an airport, for instance, the ease with which people, freight, and aircraft move through the facility is strongly influenced by the queues that they encounter. The customers in the queue needn't be physically lined up. Examples are the queue of patients waiting for donations of vital organs, the queue of couples waiting to adopt children, and the queue of people waiting for permission to immigrate.

Performance Measures

A customer is said to be **in the queue** from the moment at which the customer appears until the moment at which the customer enters service. Thus, a customer is in the queue while he or she waits for service to begin. Similarly, a customer is said to be **in the system** from the moment at which he or she appears until the moment at which his or her service is completed.

Queueing is the mathematical analysis of queues. The fruit of this mathematics includes the measures of performance that follow:

- The expectation of L_q of the number of customers in the queue.
- The expectation L of the number of customers in the system.
- The expectation W_q of the time that a customer spends in the queue.
- The expectation W of the time that a customer spends in the system.
- The fraction of the time that each server is idle.

An individual customer is concerned with the expectation W_q of the time until service begins and with the expectation W of the time until service is complete. Because these times affect customer satisfaction, they are also important to management. In addition, managers and facility designers are concerned with the expected number L_q of customers in the queue, as this determines the room that must be made available to accommodate them. Managers may also be concerned with the fraction of the time that the servers are idle.

Describing Queues and Their Users

The customers that a queue serves can behave in several ways. A customer may **balk**, which is to refuse to join a queue if the line is too long. A customer may **renege**, which is to leave a queue if he or she has been waiting too long. In a queue with multiple servers, a customer may **jockey**, that is, shift from one line to another. A queue may have **unlimited capacity** (room for any number of waiting customers) or **fixed capacity** (room for only a given number of customers to wait).

There are many varieties of queues. A familiar type of queue is the **S-line**, which is a line of customers who wait for their turn to be served by one of a group of servers. An S-line forms in front of a group of bank tellers and at a check-in area at an airport. At Disneyland, S-lines are omnipresent. An S-line of telephone callers waits for the "next available service representative" to become available. An S-line establishes equity. In an S-line, each customer experiences roughly the same waiting time. When compared with individual queues with jockeying, an S-line reduces the variance of the time in the queue.

The rule used to determine which customer is served next is known as the **queue discipline**. An S-line implements the **first-come-first-served** rule. This rule is equitable, but it is not always the most effective. In the **long-waits-for-short** rule, the customer whose service time is shortest is served next. Customers may have **priority classes**, as when "first-class" customers "step to the head of the line." In a supermarket, express checkout lines (10 items or less, say) are designed to allow priority access to customers with short service times. Thus, an express line implements a surrogate for the long-waits-for-short rule.

Systems of Queues

Queues can occur individually or in systems of interacting queues. In a system of queues, the terms *bottleneck* and *gridlock* have special meaning. A **bottleneck** occurs where service capacity is least adequate. In a manufacturing system, great effort is made to identify and relieve bottlenecks, for they limit the productivity of the system. **Gridlock** occurs when the queue of users of one facility blocks access to another. At a supermarket, gridlock occurs when the queue of people waiting to check out blocks access to the shopping aisles. Gridlock can quickly propagate in ways that cripple a system. Relieving bottlenecks and avoiding gridlock are important facets of traffic engineering. Short-sighted planning or inept facility design can easily result in bottlenecks and gridlock of the sort experienced at the New York metropolitan airports.

14.2. WHAT CAN YOU LEARN FROM THIS CHAPTER?

Queueing is an enormous subject, one that an introductory chapter can only probe. From this chapter, you can learn why queues form. You will see how easy it is to calculate the fraction of the time that each server is busy. You will also discover a simple relationship between the performance measures L_q, L, W_q, and W.

This chapter presents three different models of queues. For each of these models, a formula is provided for L_q. From these formulas, it will be easy to determine L, W_q, and W. Listed below are insights that these formulas provide into the expectation W_q of the time that a customer spends waiting for service to begin.

- *Idle time:* If the servers are rarely idle, the waiting time W_q can be huge.
- *Pooling:* Pooling queues by consolidating their arrival processes and their servers can reduce W_q markedly.
- *Variability:* With fixed arrival and service rates, reducing the uncertainty in the inter-arrival time and in the service time decreases W_q.

This chapter closes with a model of a network of interacting queues. You will see that analysis of this network of queues will require no new tools, merely that you reassemble information learned previously.

This chapter focuses on queues with Poisson arrivals. It was noted in Chapter 8 that the Poisson process is a natural model of an arrival stream, and it was noted in Chapter 11 that the exponential distribution is a natural model of the time between successive arrivals. The reason why this is so is recorded below.

Suppose that:

- Many customers share the same service facility.
- Each customer acts independently of the others.
- No customer accounts for a significant fraction of the arrivals.
- Their aggregate arrival rate is λ per unit time.

Then the Superposition Theorem guarantees that:

- The number $N(t)$ of customers who arrive during t units of time has the Poisson distribution with mean of λt.
- The inter-arrival times are exponentially distributed with rate λ.

Our analysis of queues draws on material in Chapter 13, particularly on continuous-time Markov chains, transition rate diagrams, flux, and reversibility.

14.3. WHY QUEUES FORM

A "rush hour" queue forms at a bridge, tunnel, or airport in periods when customers arrive more rapidly than they can be served. Suppose, for instance, that during the evening rush hour vehicles arrive at a bridge's toll booth at the rate of 25 per minute and can only be served at the rate of 20 per minute. While this condition persists, 5 more vehicles arrive

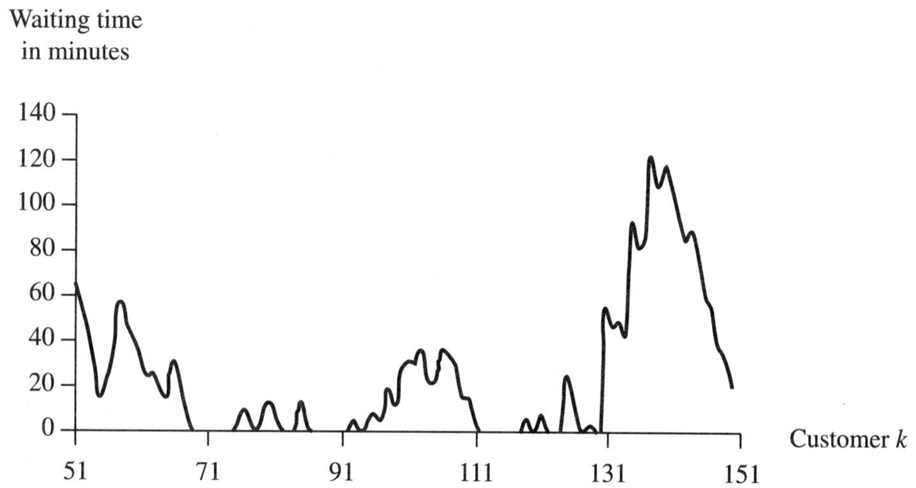

Figure 14.1 Simulated waiting times (in minutes) of 100 customers who arrive every 15 minutes and have exponential service times with mean of 13 minutes.

each minute than can be served. They form a queue that keeps growing until the arrival rate falls below the service rate.

For a queue to form, must the mean rate at which customers arrive exceed the mean rate at which they can be served? No! *Uncertainty* in the service times or in the inter-arrival times can cause a queue to form even if the mean arrival rate is below the mean service rate. To see why, let us suppose that:

- Customers arrive at a service facility every 15 minutes, so there is no uncertainty in their inter-arrival times. Specifically, a customer appears at 9:00 o'clock, the next at 9:15, the one after that at 9:30, and so forth.
- Let there be one server, and let the expected service time be 13 minutes.

First, suppose that the service time is exactly 13 minutes. In this case, no queue forms. The server works the first 13 minutes of every 15-minute period and is idle for the remaining 2 minutes.

Now let the service time be uncertain, with the same mean. Specifically, suppose that the service time has the exponential distribution with a mean of 13 minutes, hence with parameter $\mu = 1/13$. If a customer's service time exceeds 15 minutes, the next customer must wait. If a customer is forced to wait, the likelihood increases that the following customer must also wait, and so forth. Excel has been used to simulate the waiting times of 150 customers in such a queue. Figure 14.1 reports the waiting times of customers 51 through 150. Some of these customers did not wait at all. Others waited as long as 120 minutes. A description of how Figure 14.1 was constructed is postponed to Chapter 15. Problem 34 on page 519, shows how to analyze this queue.

For the simulation in Figure 14.1, there is no uncertainty in the inter-arrival times of the customers. If the inter-arrival times were uncertain, the waiting times would tend to lengthen.

14.4. PRELIMINARY OBSERVATIONS

This introductory chapter studies only a few of the many types of queues. We will focus on the behavior of queues whose mean arrival rates are below their mean service rates. Thus, we will not deal directly with the evanescent behavior of "rush hour" queues.

A "Plain Vanilla" Queue

In this chapter, our primary concern lies with a single queue whose characteristics are as follows:

- There is only one class of customer, not several.
- The mean rate at which customers arrive is constant, not time varying. The inter-arrival times of the customers are mutually independent.
- If there are several servers, they are identical. The service times are independent of each other and of the inter-arrival times.
- If a customer appears while all servers are busy, that customer joins the queue and stays there until service can begin.
- No server is allowed to be idle while any customers are in the queue.

This model does not allow balking, reneging, or jockeying. It does not apply to the evanescent behavior of rush-hour queues. Although this model is relatively simple, it will reveal insights that hold broadly.

The Symbols λ, c, μ, and ρ

Throughout this chapter, the letters λ, c, and μ are reserved for the following meanings:

- Customers arrive at the rate of λ customers per unit time, so the expectation of the inter-arrival time equals $1/\lambda$ units of time.
- There are c servers, working in parallel.
- Each server's service rate is μ customers per unit time, so the expectation of the service time equals $1/\mu$ units of time.

The rates λ and μ must have the *same* unit of measure. If, for instance, λ is measured in customers per minute, then μ must be measured in customers per minute.

Throughout, the symbol ρ is reserved for the ratio of the arrival rate λ to the aggregate service rate $c\mu$ while all servers are working; that is,

$$\rho = \frac{\lambda}{c\mu}. \tag{14.1}$$

Since λ and μ are required to have the same unit of measure, the ratio ρ is dimensionless. We will soon observe that ρ equals the fraction of the time that each server is busy. That result lets ρ be interpreted as the server's **utilization rate**.

An Illustration

To illustrate this notation, suppose that customers arrive at the rate of 15 per hour and that there are 3 servers, each with a mean service time of 10 minutes.

Let us express the arrival and service rates as customers per hour. The arrival rate λ equals 15 customers per hour. To measure the service rate μ in customers per hour, we note that the mean service time is 10 minutes, and

$$10 \text{ minutes} \times (1/60) \text{ hours/minute} = (1/6) \text{ hours},$$

so the mean service time $1/\mu$ equals $(1/6)$ hours, and the service rate μ equals 6 customers per hour. The aggregate service rate for three servers equals $3 \times 6 = 18$ customers per hour.

Substituting into Equation (14.1) gives

$$\rho = \frac{\lambda}{c\mu} = \frac{15/\text{hour}}{(3) \times (6/\text{hour})} = \frac{15}{18} = \frac{5}{6}.$$

Thus, each server is busy $5/6$th of the time and is idle $1/6$th of the time, on average.

Throughout, we study the behavior of queues for which the utilization rate ρ is below 1. If ρ exceeded 1, the queue would grow without limit.

The Nomenclature *X*/*Y*/*c*

A queue is described by three symbols that are separated by slashes. The first symbol specifies the distribution of the inter-arrival times, the second symbol specifies the distribution of the service time, and the third is the number c of servers. Here, the letter D is an abbreviation for deterministic, the letter M for memoryless (exponential), and the letter G for general.

For instance, the *M*/*G*/1 queue has exponential inter-arrival times, general service times that are mutually independent, and one server. The *M*/*M*/*c* queue has exponential inter-arrival times, exponential service times, and c servers. Figure 14.1 simulates the *D*/*M*/1 queue; it has fixed inter-arrival times, exponential service times, and one server.

Throughout this chapter, we study models in which inter-arrival times and service times are independent. For that reason, we have dropped the *I* from *GI*. When we speak of an *M*/*G*/1 queue, we assume that the service times are mutually independent; in a more general taxonomy, this would be called an *M*/*GI*/1 queue.

The *G*/*G*/*c* Queue

The *G*/*G*/*c* queue can have any distribution of inter-arrival times, any distribution of service time, mutually independent arrival and service times, and any number c of servers. Can anything be said of this queue? Yes. In fact, we have already observed that ρ is the fraction of the time that each server is busy. And, as there are c servers, the time-average number of busy servers equals $c\rho$.

> In the *G*/*G*/*c* queue, each server is working the fraction ρ of the time and is idle the fraction $1 - \rho$ of the time. Over a long period of time, the average number of busy servers equals $c\rho = \lambda/\mu$.

Intuitively, it is easy to grasp the idea that ρ is the fraction of the time that each server is busy. But what is the underlying mathematics? This interpretation of ρ reflects an "ergodic" property that was introduced in Chapter 13 and will be used repeatedly in this chapter. An **ergodic** property is the statement that *if* a property of a time-varying system settled down, as the elapsed time becomes large, then the value x to which it settles down equals the time-average value of that property.

For instance, if the probability that a server is busy settles down to some number x as the elapsed time becomes large, then x equals the long-run fraction of the time that the server is busy. That the latter equals ρ follows, basically, from the fact that the expectation of the sum equals the sum of the expectations. It holds even if the inter-arrival times and service times are dependent.

A Pair of Examples

Two examples appear here. As the chapter unfolds, these examples will be used to illustrate various properties of queues.

Problem A (An *M/M/*1 Queue)

Arrivals at a clinic are Poisson with a mean of 4 per hour. This clinic is served by one medical doctor. The time that this doctor takes to serve each patient has the exponential distribution with a mean of 13 minutes. Under steady-state conditions, what can be said of the number of patients in the queue? of the time that a patient spends in the queue?

In Problem A, it is convenient to measure the arrival rate λ in customers per hour; $\lambda = 4$/hour. The mean service time equals 13 minutes or $13/60$ hours, so the service rate $\mu = 60/13$ per hour, and the utilization rate ρ is given by

$$\rho = \frac{\lambda}{\mu} = \frac{4 \text{ per hour}}{(60/13) \text{ per hour}} = \frac{4 \times 13}{60} = \frac{13}{15} = 0.8666$$

Thus, for Problem A, the doctor is busy approximately 87% of the time and is idle approximately 13% of the time.

Problem B (An *M/M/*3 Queue)

Three clinics like the one in Problem A have merged. Arrivals at the consolidated clinic are Poisson with a mean of 12 per hour. All three doctors serve the consolidated clinic. The time that each doctor takes to serve each patient (still) has the exponential distribution with a mean of 13 minutes. Under steady-state conditions, what can be said of the number of patients in the queue? of the time that a patient spends in the queue?

In Problem B, each doctor's service rate is unchanged, but the arrival rate triples, and there are three doctors. For the consolidated system, the utilization rate ρ is given by

$$\rho = \frac{\lambda}{3\mu} = \frac{12 \text{ per hour}}{(3)(60/13) \text{ per hour}} = \frac{12 \times 13}{3 \times 60} = \frac{13}{15} = 0.8666.$$

It is hardly surprising that each doctor is still busy approximately 87% of the time and is idle approximately 13% of the time.

Pause to mull over the differences between the unconsolidated and the consolidated systems. Try to guess the answers to these questions:

- Why might the consolidated system provide quicker service?
- How does the length of the consolidated queue compare with the lengths of the individual queues?
- How does the time that customers spend in the consolidated queue compare with the time they spend in the individual queues?

As the chapter unfolds, these questions will be answered. The answers may surprise you.

Four Random Variables

Throughout this chapter, the symbols N_q, N, T_q, and T are reserved for random variables whose meanings are highlighted below:

$$\begin{aligned} N_q &= \text{the number of customers in the queue,} \\ N &= \text{the number of customers in the system,} \\ T_q &= \text{the time that a customer spends in the queue,} \\ T &= \text{the time a customer spends in the system.} \end{aligned}$$

The value of N can be any nonnegative integer. If no customers are in the system, then $N = 0$. The value of N_q can also be any nonnegative integer; $N_q = k$ if k customers are waiting for service to commence. Being times, the values of T_q and T can be any nonnegative numbers.

Each of these four random variables has as its expectation a performance measure that was introduced earlier. Specifically:

$$L_q = E(N_q) \quad \text{and} \quad L = E(N),$$
$$W_q = E(T_q) \quad \text{and} \quad W = E(T).$$

We will soon see that these four performance measures are closely linked so that if we can determine any one of them, we can easily find the rest.

Four Interpretations of N_q and N

We have just been a bit sloppy. To see why, consider a queue that has settled down to steady-state conditions. For this queue, what, exactly, is N_q? And what is N? The random variable N_q can have any of the following interpretations:

1. *Steady state:* Perhaps the random variable N_q describes the steady-state probability distribution of the number of customers in the queue.
2. *Time average:* Perhaps, over a long period of time, the probability distribution of N_q specifies the fraction of the time during which each number of customers has been in the queue.
3. *What the arriving customer sees:* Perhaps the random variable N_q is the number of customers in the queue at the moment before an arrival occurs.
4. *What the departing customer sees:* Perhaps the random variable N_q is the number of customers in the queue at the moment after a customer begins service.

Potentially, at least, all four of these random variables could have different probability distributions. But:

- The first two random variables have the same distribution.
- The final two random variables have the same distribution.
- If arrivals are Poisson, all four have the same distribution.

Let us explain why.

Theorems 13.1 and 13.2 of Chapter 13 illustrate an "ergodic" property that holds for a great many systems. This ergodic property is that, *if* the behavior of a system settles down to steady-state values, then the values to which it settles down also describe the time-average behavior of the system. This property holds for N_q.

> **Ergodic property:** The steady-state probability distribution of N_q is the time-average distribution of N_q.

For instance, the steady-state probability that three customers are in the queue equals the fraction of the time that three customers are in the queue.

To relate the third and fourth interpretations of N_q, we call an arrival that occurs while there are k customers in the system an "uptick" from k to $k + 1$ customers. Similarly, we call a service completion that occurs while there are $k + 1$ customers in the system a "downtick" from $k + 1$ to k customers. Between every pair of upticks from k to $k + 1$ customers, there must occur *exactly one* downtick from $k + 1$ customers to k customers.

Thus, in steady state, the uptick probabilities must balance the downtick probabilities. In brief:

> **Upticks balance downticks:** The steady-state distribution of the number N_q of customers in the queue at the moment before an arrival equals the steady-state distribution of the number N_q of customers in the queue at the moment after a departure.

PASTA is an acronym for an amazing result, which is that "Poisson arrivals see time averages."

> **PASTA:** In a *M/G/c* queue, the steady-state distribution of the number N_q of customers in the queue at the moment before an arrival equals the time-average distribution of the number in the queue.

Thus, if arrivals are Poisson, all four interpretations of N_q have the same probability distributions. If arrivals are not Poisson, the first two interpretations of N_q have the same probability distribution, and the final two interpretations of N_q have the same probability distribution, but these distributions can be different.

These results hold not just for N_q, but for N as well. In fact, they hold for any property of a queue that settles down to steady-state values.

Who Cares?

An arriving customer cares about the number N_q of customers in the queue at the moment of arrival, for this determines the customer's waiting time. A facility designer is interested in the steady-state and time-average distributions of N_q because these distributions determine the amount of room that must be set aside to accommodate the queue. The remaining interpretation of N_q (as the number in the queue just after a customer departs) may seem esoteric, but uses will be found for it.

Relating the Performance Measures

The performance measures L_q, L, W_q, and W are now related to each other. Let us recall that W is the expectation of the time in the system and that W_q is the expectation of the time in the queue. The equation,

$$W = W_q + 1/\mu, \tag{14.2}$$

holds because the expected time in the system equals the sum of the expected time in the queue and the expected time in service, the latter being $1/\mu$.

Let us also recall that L equals the expected number of customers in the system and that L_q equals the expected number in the queue. The equation,

$$L = L_q + c\rho = L_q + \frac{\lambda}{\mu}, \tag{14.3}$$

holds because $c\rho$ is the time-average number of busy servers.

In 1961, John D. C. Little published a relationship that, over time, has become known as **Little's law**; it is that

$$L_q = \lambda W_q \quad \text{and} \quad L = \lambda W. \tag{14.4}$$

Thus, once L_q has been determined, the other three performance measures can be found as follows:

- Set $W_q = L_q/\lambda$,
- Set $W = W_q + 1/\mu$,
- Set $L = \lambda W$.

After any of these four performance measures has been calculated, determination of the others is routine.

14.5. LITTLE'S LAW

But what does the equation $L = \lambda W$ actually mean? If arrivals are Poisson, there is no ambiguity in this equation because PASTA shows that arrivals see time averages. But if the arrival process is not Poisson, there are two possible values of W and two of L.

For which interpretations of L and W does Little's law hold? For the answer to this question, you may skip directly to the unstarred subsection entitled "The Bottom Line." The starred subsections that precede it suggest why Little's law holds. Skipping them entails no loss of continuity.

A Picture*

Proofs of Little's law focus on the "sample path" of the number of customers that are actually present in the system as its random variables are realized (take values). The **sample path** of the number of customers in the system is the function that reports, for each value of t, the number $N(t)$ of customers in the system at time t due to realizations of all random variables. A sample path is deterministic. We will account for the area under the sample path in two ways, one of which relates to W, the other to L.

An **idle/busy cycle** is an interval of time that begins at a moment when the server become idle and ends at the moment when the server next becomes idle. Figure 14.2 plots $N(t)$ versus t for a single idle/busy cycle. In this figure, time t is measured from the moment that this cycle begins. This particular cycle consists of four arrivals and four departures, the

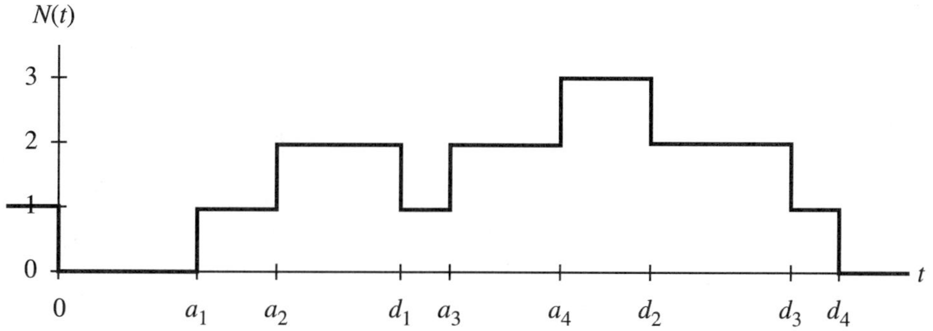

Figure 14.2 An idle/busy cycle that consists of four arrivals and four departures.

arrivals occurring at times a_1 through a_4, the departures at times d_1 through d_4. The fourth departure reduces $N(t)$ to zero, which completes this cycle.

The Area under $N(t)$*

We shall relate the area under $N(t)$ to the quantities $\overline{L}$ and $\overline{W}$, defined as follows:

$\overline{L}$ = the time-average number of customers in the system during k consecutive idle/busy cycles.

$\overline{W}$ = the average of the times that the customers who arrived during these k idle/busy cycles spent in the system.

Figure 14.2 plots $N(t)$ for a single idle/busy cycle, the case $k = 1$. The length of this cycle equals d_4, and the area under the function $N(t)$ in Figure 14.2 equals the product of d_4 and $\overline{L}$, so

$$(d_4)(\overline{L}) = (a_2 - a_1)1 + (d_1 - a_2)2 + (a_3 - d_1)1 + (a_4 - a_3)2 + (d_2 - a_4)3 + (d_3 - d_2)2 + (d_4 - d_3)1.$$

The preceding equation reorganizes itself as

$$(d_4)(\overline{L}) = d_1 + d_2 + d_3 + d_4 - (a_1 + a_2 + a_3 + a_4). \tag{14.5}$$

But the sum $4\overline{W}$ of the times in system of the four customers who arrive during this idle/busy cycle is given by

$$4\overline{W} = (d_1 - a_1) + (d_2 - a_2) + (d_3 - a_3) + (d_4 - a_4). \tag{14.6}$$

The right-hand sides of Equations (14.5) and (14.6) are identical. Evidently,

$$4\overline{W} = (d_4)(\overline{L}) \tag{14.7}$$

for a single idle/busy cycle that is comprised of four arrivals and departures.

Equation (14.7) was developed under the tacit hypothesis of a first-come first-served rule. But it holds for any service discipline. And if the cycle consisted of p arrivals and departures, Equation (14.7) would hold with the number 4 replaced by p. Furthermore, the comparable computation for any number k of consecutive idle/busy cycles gives

$$n\overline{W} = (d_n)(\overline{L}),$$

where n is the departure that ends the kth idle/busy cycle and where d_n is the time that elapses until this departure occurs. Let's rearrange the preceding equation as

$$\overline{W} = (\overline{L})(d_n/n). \tag{14.8}$$

Verifying Equation (14.8) has been easy. This equation suggests the form of Little's law; $\overline{W}$ is the average of the waiting times that the customers experience, while $\overline{L}$ is the time-average number of customers in the system.

The Nitty Gritty*

Equation (14.8) also hints at a proof of Little's law. An actual proof develops an analog of Equation (14.8) for an arbitrary interval of time of length T and examines what happens as T becomes large. We'll avoid the nitty gritty. But we do note what happens to d_n/n as k becomes large. Recall that a_n is the time of the nth arrival. As k becomes large, so does n, and it's clear that

$$\frac{d_n - a_n}{n} \to 0, \quad \text{so} \quad \frac{d_n}{n} \to \frac{a_n}{n} \to \frac{1}{\lambda},$$

the last from the law of large numbers. Thus, as k becomes large, Equation (14.8) motivates the form of Little's law that is highlighted below.

The Bottom Line

In Little's law, as presented below, the ergodic property has been used to provide two interpretations of W and two of L.

> **Little's law.** The equation $L = \lambda W$ holds when W and L have these interpretations:
>
> - In steady state, W is the expectation of the time that an *arriving* customer spends in the system. Equivalently, W is the average of the times that the arriving customers spend in the system.
> - In steady state, L is the expectation of the number of customers in the system. Equivalently, L is the time-average number of customers in the system.

Thus, a general statement of Little's law requires different interpretations of W and L. Although W is the expectation of the time that an arriving customer will spend in the system, L may *not* be the expectation of the number of people that an arriving customer will observe at the moment of arrival.

The same line of analysis shows that $L_q = \lambda W_q$, with W_q as the expectation of the time that an arriving customer spends in the queue and L_q as the expectation of the number of customers in the queue, in steady state.

Generality

The proof of Little's law that we've outlined is very general. At the heart of this proof lies an exercise in *accounting*, one that interprets the integral of $N(t)$ over a long period of time in two different ways, producing Equation (14.8).

Little's law has been cast as a result in queueing, but it applies to a great many systems. The equation $L = \lambda W$ states that, in steady state,

$$\begin{bmatrix}\text{the average number of}\\ \text{customers in a system}\end{bmatrix} = \begin{bmatrix}\text{the rate at which}\\ \text{new customers arrive}\end{bmatrix} \times \begin{bmatrix}\text{the average time an arriving}\\ \text{customer spends in the system}\end{bmatrix}.$$

This equation is familiar to epidemiologists. In steady state, the average number of people who have a particular disease equals the product of the rate at which that disease is acquired by the population and the average time that each person who acquires this disease has it.

For another illustration of this result, suppose that physical therapists obtain licenses to practice in Connecticut at the rate of 160 per year and that those who do practice in Connecticut for an average of 7.3 years during their careers. In steady state, how many licensed physical therapists are practicing in Connecticut?

A Bit of History

The equations $L = \lambda W$ and $L_q = \lambda W_q$ are not due to Little. For an *M*/*M*/*c* queue, these equations appear in a 1958 book by Philip Morse.[1] Morse was a pioneer in operations research,

---

[1] Philip M. Morse, *Queues, Inventories and Maintenance* (New York: Wiley, 1958).

and he was an international figure in physics. Little was his Ph.D. student, the first in operations research at MIT. Little's 1961 paper[2] aimed to show that these equations hold generally, and not just for $M/M/c$ queues. Although Little's proof was flawed, the ideas he introduced launched a fleet of researchers on quests for general conditions and valid proofs. In 1974, Shaler Stidham, Jr., published a sample path argument that is correct, simple, and general.[3]

The Development of PASTA*

This starred subsection focuses on PASTA. With B as a predetermined set of integers, PASTA examines those times t for which $N(t)$ is in B. This subsection poses and answers three questions.

QUESTION 1: Does the long-run fraction of the time that $N(t)$ is in B equal the long-run fraction of the time that an arriving customer finds $N(t)$ to be in B?

ANSWER: Yes, if the arrival process is Poisson.

QUESTION 2: What makes PASTA work?

ANSWER: The key is a lack-of-anticipation assumption, namely, that for any time x, the set of times up to x for which $N(t)$ has been in B provides no information about (is independent of) the stream of arrivals that will come after time x. A Poisson arrival stream has this property, even if its arrival rate $\lambda(x)$ varies with x, and so do some others.

QUESTION 3: When was PASTA developed and by whom?

ANSWER: Considering how fundamental PASTA is, the result is surprisingly recent. Ronald Wolff published the key paper in 1982,[4] fully two decades after Little's law. In his paper, Wolff acknowledges Ward Whitt for the lack-of-anticipation assumption. Wolff's proof is brief and elegant, relying as it does on the deft use of a martingale. Its antecedents include a 1970 paper by Ralph Strauch.[5]

14.6. THE $M/M/1$ QUEUE

This and the next two sections study specific queues. Each of these two sections has an unstarred portion and a starred portion. The unstarred portions present formulas for L_q and use these formulas to reveal themes that pervade queueing. The starred portions present the mathematics that leads to the formulas for L_q. The starred portions can be skipped or skimmed.

As its title indicates, this section concerns the $M/M/1$ queue, which has Poisson arrivals and exponential service times. As noted earlier, the Poisson process is a natural model for arrivals at a service facility. There is no particular reason why a service time should have the exponential distribution; this assumption is a mathematical convenience, one that, as we shall see, often gives a conservative (high) estimate of L_q.

---

[2] John D. C. Little, "A Proof for the Queueing Formula: $L = \lambda W$," *Operations Research* 9 (1961), pp. 383–387.

[3] Shaler Stidham, Jr., "A Last Word on $L = \lambda W$," *Operations Research* 22 (1974), pp. 417–421.

[4] Ronald W. Wolff, "Poisson Arrivals See Time Averages," *Operations Research* 30 (1982), pp. 223–231.

[5] Ralph E. Strauch, "When a Queue Looks the Same to an Arriving Customer as to an Observer," *Management Science* 17 (1970), pp. 140–141.

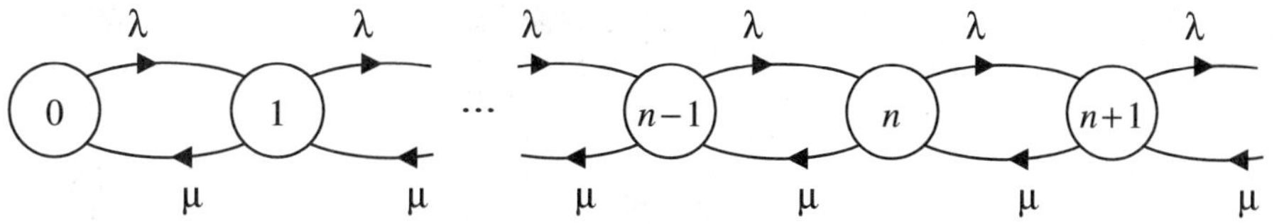

Figure 14.3 Transition rate diagram for the *M/M/*1 queue.

A Transition Rate Diagram

Figure 14.3 is the transition rate diagram for the *M/M/*1 queue. Arrivals occur at the rate λ, and each arrival increases the number of customers in the system by 1, which explains the right-pointing arrows in Figure 14.3. As long as the number of customers in the system is positive, service completions occur at the rate μ, and each service completion decreases the number of customers in the system by 1, which explains the left-pointing arrows in Figure 14.3.

A Formula for L_q

Reported below are formulas for the mean and variance of the number N_q of customers in the *M/M/*1 queue.

$$L_q = E(N_q) = \frac{\rho^2}{1-\rho}, \quad \text{Var}(N_q) = \frac{\rho^2}{(1-\rho)^2}(1 + \rho - \rho^2) \tag{14.9}$$

These formulas will be verified in the starred portion of this section.

Equation (14.9) implies that $\text{StDev}(N_q) > E(N_q)/\rho$. Thus, the standard deviation of N_q exceeds the mean by a factor in excess of $1/\rho$. In this case and in general, the queue length N_q is quite variable. Evidence of this variability can be found in Figure 14.1.

The Folly of Full Utilization

The formula for L_q has the term $(1 - \rho)$ in its denominator. As ρ gets closer and closer to 1, the mean queue length L_q grows without bound, as is indicated in Figure 14.4.

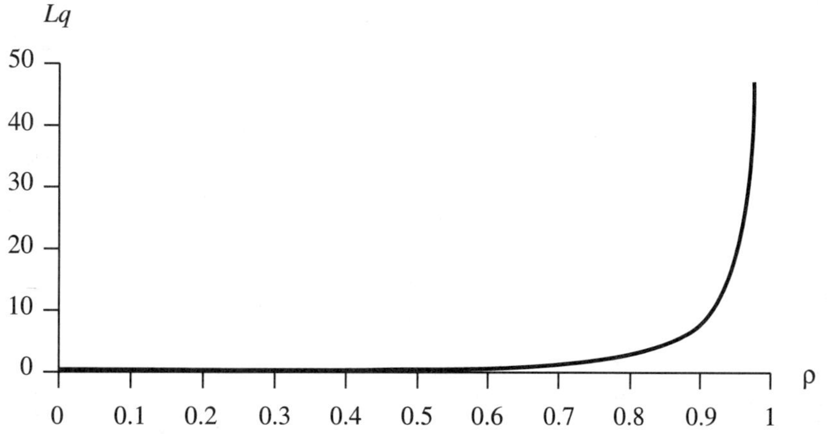

Figure 14.4 Plot of L_q versus ρ for the *M/M/*1 queue.

Let us recall from Little's law that the expectation W_q of the time in the queue is given by $W_q = L_q/\lambda$, so W_q exhibits the same behavior as does L_q as the utilization ρ approaches 1. We emphasize:

> As the utilization ρ approaches 1, the mean queue length L_q and the mean time in the queue W_q approach infinity.

This property is shared by more complicated queues. In each queue that we shall consider, the formula for L_q has the term $(1 - \rho)$ in its denominator. In each case, as ρ approaches 1, the mean queue length approaches infinity; L_q grows as the hyperbola, $1/(1 - \rho)$.

The curve in Figure 14.4 begins to rise steeply as the utilization rate approaches 90%. This fact is important for systems designers and managers. They must understand how vital it is for the servers to have some idle time. Full utilization or near-full utilization of the servers results in enormous queues and in huge waiting times for customers.

In addition, if servers have subsidiary tasks (such as paperwork) that can be postponed, it is important that these tasks be postponed to idle periods. Failure to do so can result in enormous waiting times for the customers.

Problem A, Revisited

With these formulas in view, we return to Problem A, which has λ = 4/hour, $\mu = 60/13$ per hour, and $\rho = 13/15$. For Problem A, we now know that

$$L_q = \frac{(13/15)^2}{(1 - 13/15)} = 5.6333,$$

$$W_q = \frac{L_q}{\lambda} = \frac{5.6333}{4/\text{hour}} = 1.41 \text{ hours},$$

$$W = W_q + 1/\mu = 1.41 + 13/16 = 1.625 \text{ hours},$$

$$L = \lambda W = (4 \text{ customers/hour})(1.625 \text{ hours}) = 6.5 \text{ customers}.$$

These computations can be executed on a spreadsheet, of course. Table 14.1 shows how. This spreadsheet has the usual advantage of enabling a sensitivity analysis.

Table 14.1 Spreadsheet computation of performance measures for an *M*/*M*/1 queue.

	A	B	C	D
2	λ =	4		
3	μ =	4.615385	←	=60/13
4	ρ =	0.866667	←	=B2/B3
5	Lq =	5.633333	←	=B4*B4/(1-B4)
6	Wq =	1.408333	←	=B5/B2
7	W =	1.625	←	=B6+1/B3
8	L =	6.5	←	=B7*B2

Table 14.1 shows that, on average, a customer spends 1.41 hours (that's 84.5 minutes) in the waiting room before being seen by the doctor. The mean waiting time is high because the utilization rate ρ is given by $\rho = 13/15 = 0.8666$, which is close to 1.

The mean time in queue is long and is quite sensitive to the service time. If the mean service time was 14 minutes rather than 13, the utilization ρ would increase to $14/15 = 0.9333$, and W_q would jump to 3.2666 hours.

Steady-State Probabilities*[6]

In what follows, we verify Equation (14.9) and specify the probability distribution of the random variables N_q, N, T_q, and T. The distribution of T turns out to be exponential, by the way.

The transition rate diagram in Figure 14.3 lets us compute the steady-state distribution of the number N of customers in the system. It also allows us to calculate the probabilities $\pi(0), \pi(1), \pi(2), \ldots,$ where

$$\pi(n) = P(N = n) \quad \text{for } n = 0, 1, \ldots.$$

The symbol $\pi(n)$ is being used here rather than $\rho(n)$ because we are reserving ρ for the utilization rate.

Flux*

We recall from Chapter 13 that, in steady state, the flux (rate of flow) out of any group of states equals the flux into that group of states. Figure 14.5 reproduces Figure 14.3, but shades states 0 through $n - 1$.

The dashed line in Figure 14.5 separates the shaded states from the others. One arc points from each group to the other. In steady state, the fluxes on these arcs must be equal. In other words,

$$\lambda\pi(n - 1) = \mu\pi(n) \quad \text{for each } n \geq 1, \tag{14.10}$$

because the rate of flow $\lambda\pi(n - 1)$ on the arc pointing from node $n - 1$ to node n must equal the rate of flow $\mu\pi(n)$ on the arc pointing from node n to node $n - 1$.

This queue has $c = 1$. Its utilization rate ρ equals λ/μ, so Equation (14.10) can be rearranged as

$$\pi(n) = \frac{\lambda}{\mu}\pi(n - 1) = \rho\,\pi(n - 1) \quad \text{for each } n \geq 1.$$

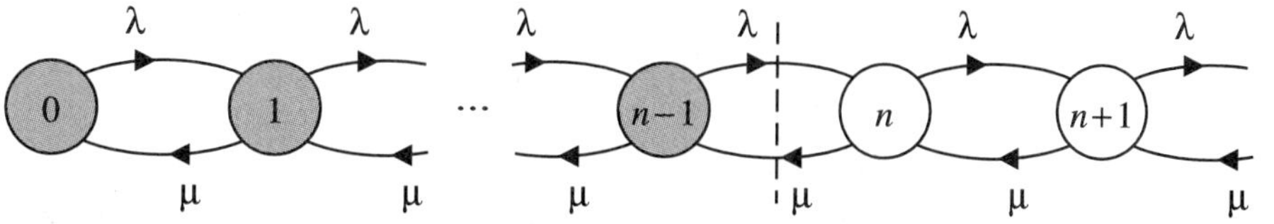

Figure 14.5 The *M/M/*1 queue, showing the flux between shaded and unshaded states.

[6]The remainder of this section is starred and can be skipped with no loss of continuity.

Repeated substitution in this equation results in

$$\pi(n) = \rho^n \pi(0) \quad \text{for each } n \geq 1. \tag{14.11}$$

Equation (14.11) makes it easy to compute the steady-state probabilities and to prove:

Theorem 14.1 (Number in System). For an *M*/*M*/1 queue whose utilization factor $\rho = \lambda/\mu$ is below 1, the steady-state probability distribution of the number N of customers in the system is given by

$$\pi(n) = P(N = n) = (1 - \rho)\rho^n \quad \text{for } n = 0, 1, 2, \ldots. \tag{14.12}$$

Proof. The probabilities sum to 1.

$$1 = \pi(0) + \pi(1) + \pi(2) + \cdots.$$

Substituting Equation (14.11) in the above gives

$$1 = \pi(0)(1 + \rho + \rho^2 + \cdots).$$

Since the absolute value of ρ is below 1, the geometric series $(1 + \rho + \rho^2 + \cdots)$ converges and has $1/(1 - \rho)$ as its sum, so $1 = \pi(0)/(1 - \rho)$, and $\pi(0) = 1 - \rho$. Substituting $(1 - \rho)$ for $\pi(0)$ in Equation (14.11) verifies Equation (14.12), completing the proof. ◆

If you are unfamiliar with the geometric series, you might refer to Problem 30 on page 518, which sums it and related series. Incidentally, we did not need to sum this series. Why? Recall our earlier observation that ρ equals the fraction of the time that the server is busy. Since $\pi(0)$ equals the fraction of the time that the server is idle, we must have $\pi(0) = 1 - \rho$.

The Mean and Variance of N*

Computing the mean and variance of N directly from Equation (14.12) is an exercise in summing series. We could take that approach, but instead, we take a shortcut. The random variable N in Equation (14.12) reminds us of the geometric distribution in Chapter 8. The smallest value that N can take equals 0. The smallest value that a geometric random variable can take equals 1. Except for that fact, their distributions are similar. In fact, $N + 1$ has the geometric distribution; $N + 1$ equals the number of independent Bernoulli trials that are needed to obtain the first success, each trial having probability $p = 1 - \rho$ of success.

Formulas for the mean and variance of the geometric distribution were provided in Chapter 8. With $p = 1 - \rho$, these formulas give

$$E(N) = \frac{\rho}{1 - \rho} \quad \text{and} \quad \text{Var}(N) = \frac{\rho}{(1 - \rho)^2}. \tag{14.13}$$

The Mean and Variance of N_q*

With a bit of effort, we could relate N_q to the geometric distribution and use the formulas in Chapter 8 to compute its mean and variance. Instead, we sum series, as that technique is well worth learning.

Let us recall that the *M*/*M*/1 queue has $c = 1$. For every positive integer k, there are k customers in queue if $N = c + k$. In other words,

$$P(N_q = k) = P(N = k + c) \quad \text{for } k = 1, 2, 3, \ldots, \tag{14.14}$$

which combines with Equation (14.11) to give

$$P(N_q = k) = \pi(c)\rho^k \quad \text{for } k = 1, 2, 3, \ldots. \tag{14.15}$$

From Equations (14.14) and (14.15), we see that

$$E(N_q) = \pi(c)[1\rho + 2\rho^2 + 3\rho^3 + 4\rho^4 + \cdots] \tag{14.16}$$

$$E[(N_q)^2] = \pi(c)[1^2\rho + 2^2\rho^2 + 3^2\rho^3 + 4^2\rho^4 + \cdots]. \tag{14.17}$$

The series in Equations (14.16) and (14.17) are easy to sum by term-by-term differentiation of the geometric series. To learn how, please review Problem 30 on page 518. The results are

$$E(N_q) = \pi(c)\frac{\rho}{(1-\rho)^2}, \tag{14.18}$$

$$E[(N_q)^2] = \pi(c)\frac{(1+\rho)\rho}{(1-\rho)^3}. \tag{14.19}$$

Since $c = 1$, we have $\pi(c) = \pi(1) = \rho(1-\rho)$, so Equation (14.18) gives

$$L_q = E(N_q) = \frac{\rho^2}{(1-\rho)}. \tag{14.20}$$

And, since $\text{Var}(N_q) = E[(N_q)^2] - (L_q)^2$, Equations (14.19) and (14.20) give

$$\text{Var}(N_q) = \frac{\rho^2}{(1-\rho)^2}(1 + \rho - \rho^2). \tag{14.21}$$

Equation (14.9) has now been verified.

The Distribution of T_q and T^*

Our attention now turns to the distribution of the time T_q that a customer spends in the queue and the distribution of the time T that a customer spends in the system. Theorem 14.2 relates these random variables to a random variable X that has the exponential distribution with parameter $(\mu - \lambda)$, so that

$$P(X \le t) = 1 - e^{-(\mu-\lambda)t} \quad \text{for } t \ge 0. \tag{14.22}$$

Theorem 14.2 (An Exponential Distribution). Consider an *M/M/*1 queue whose utilization factor $\rho = \lambda/\mu$ is below 1 and whose queue discipline is first-come first-served. Let X have the exponential distribution with parameter $(\mu - \lambda)$. Under steady-state conditions:

(a) The time T spent in the system and X have the same probability distribution.

(b) The time T_q spent in the queue relates to X as indicated in Figure 14.6.

A direct proof of Theorem 14.2 is an exercise in calculus, which we omit.

Theorem 14.2 is remarkable. To see why, we interpret part (a). Imagine that you arrive at an *M/M/*1 queue with first-come first-served (FCFS) queue discipline. When you arrive, N people are ahead of you. The time T that will elapse until your service is complete is the sum of $N + 1$ i.i.d. random variables, each having the exponential distribution with parameter μ. The sum of a fixed number of i.i.d. exponential random variables is *not* exponential. But N is random, not fixed, and its distribution is such that the sum of $N + 1$ of these random variables does have the exponential distribution.

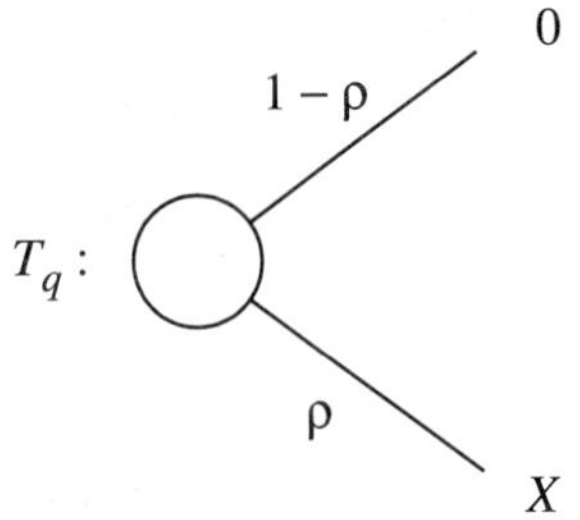

Figure 14.6 Relating T_q to an exponential random variable X with parameter $(\mu - \lambda)$.

14.7. THE *M/M/c* QUEUE

The *M/M/c* queue is similar to the *M/M/*1 queue, the lone difference being that now there are c servers, rather than 1. For the *M/M/c* queue, arrivals are Poisson and service times are exponential. Figure 14.7 presents its transition diagram. Arrivals occur at the rate λ, and each arrival increases the number N of customers in the system by 1, which explains the right-pointing arrows. While N is at least c, each of c servers is at work, and each of these c servers is completing service at the rate μ. Thus, while $N \geq c$, the aggregated service completion rate equals $c\mu$. While $N < c$, exactly N servers are at work, and the aggregated service completion rate equals $N\mu$. Each service completion decreases N by 1.

Figure 14.7 differs from Figure 14.3 in that the aggregate service rate dwindles for N below c. This complicates the formula for L_q but not the method by which it is obtained.

A Formula for L_q

The formula for L_q entails the number c of servers and the utilization rate $\rho = \lambda/(c\mu)$. The formula itself is the ungainly expression,

$$L_q = \frac{\dfrac{1}{c!}\left(\dfrac{\lambda}{\mu}\right)^c}{\dfrac{1}{c!}\left(\dfrac{\lambda}{\mu}\right)^c + (1-\rho)\sum_{k=0}^{c-1}\dfrac{1}{k!}\left(\dfrac{\lambda}{\mu}\right)^k}\left(\frac{\rho}{1-\rho}\right). \tag{14.23}$$

Although complicated, this formula for L_q shares a feature with the single-server case, namely, the factor $(1 - \rho)$ in its denominator. Thus, as the utilization rate approaches 1, the mean queue length L_q grows without bound, exactly as it does in the single-server case.

You do *not* need to wrestle with Equation (14.23). The value of L_q that it computes is the Excel add-in function **=LQP(ρ, c)** on the diskette that accompanies this text. Table 14.2 uses this function to report the values of L_q for utilization rates between 0.7 and 0.95 and for numbers of servers between 1 and 5.

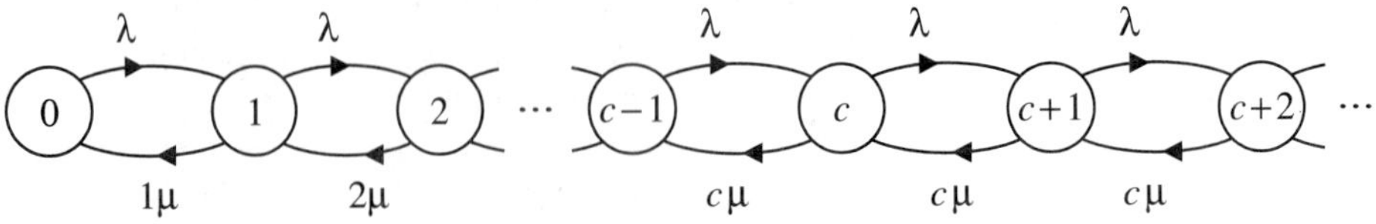

Figure 14.7 Transition rate diagram for the *M/M/c* queue.

Table 14.2 Mean queue length L_q for the *M/M/c* queue versus its utilization ρ and number c of servers.

	A	B	C	D	E	F
1		number c of servers				
2	ρ	1	2	3	4	5
3	0.7	1.6333	1.3451	1.1488	1.0002	0.8816
4	0.8	3.2	2.8444	2.5888	2.3857	2.2165
5	0.9	8.1	7.6737	7.3535	7.0898	6.8624
6	0.95	18.05	17.5872	17.2332	16.9370	16.6782
7						
8						
9		=LQP($A6,B$2)				

By looking down the columns of Table 14.2, we see that as the utilization ρ approaches 1, the mean queue length grows rapidly, as anticipated. This is due to the factor $(1 - \rho)$ in the denominator of the formula for L_q.

By looking across the rows of Table 14.2, we see evidence of a result that holds in general but is not obvious from Equation (14.23).

> For any fixed value of ρ, the expected number of customers in the queue *decreases* as the number c of servers increases.

Problem B, Revisited

To interpret these results, we review Problem B. In this problem, three identical clinics have been merged. The consolidated clinic has these properties:

- The pooled arrival rate λ equals 12 customers per hour.
- The service rate μ remains equal to $60/13$ customers per hour.
- The number c of servers equals 3.
- The utilization rate ρ remains equal to $13/15$.

The Excel function **=LQP(13/15, 3)** reports the value

$$L_q = 4.933 \text{ customers},$$

and the remaining three performance measures can be found from the equations

$$W_q = \frac{L_q}{\lambda} = \frac{4.933}{12/\text{hour}} = 0.411 \text{ hours},$$

$$W = W_q + 1/\mu = 0.411 + 0.217 = 0.628 \text{ hours},$$

$$L = W\lambda = 0.628 \times 12 = 7.53 \text{ customers}.$$

The pooled queue in Problem B has a mean queue length $L_q = 4.933$. For the unpooled queues, each of three queues has a mean queue length $L_q = 5.633$. Thus, the consolidated queue is shorter than *any one* of the individual queues. A surprise?

Not only is the pooled queue shorter than any one of the individual queues, but it moves three times as fast because the aggregate service rate is three times as large. Thus, in the aggregated system, customers spend less than one-third the time in queue than in the individual systems. Surprised?

Pooling

Let us now discuss pooling in general terms. To **pool** queues is to consolidate their arrival processes and grant all of the customers equal access to all of the servers. Consider what happens when we pool k *identical M/M/c* queues. This consolidation:

- increases the arrival rate λ by the factor of k,
- increases the number of servers by the factor of k,
- has no effect on the utilization rate ρ,
- reduces the mean queue length L_q below that for any one of the unpooled queues,
- and from the equation $W_q = L_q/\lambda$, reduces the mean time in the queue by a factor of at least k.

We emphasize:

> Pooling k identical *M/M/c* queues produces a consolidated queue whose expected queue length L_q is lower than any of the unconsolidated queues and whose mean waiting time W_q is less than $1/k$ times the mean waiting time for any of the individual queues.

If queues are not identical, the effect of pooling is more complex. If two queues have widely different values of W_q, pooling *all* of their servers can result in a queue whose value of W_q lies between the values for the individual queues. But it is always possible to share servers in a way that reduces the waiting times for the customers in each of the individual queues. Again, all customers can be made better off by pooling.

The Math*[7]

The remainder of this section presents the mathematics of the *M/M/c* queue. The formulas developed here are more intricate than those for the *M/M/*1 queue, but the methods are exactly the same.

Let the random variable N have the steady-state probability distribution for the Markov chain in Figure 14.7. As in the *M/M/*1 queue, we will compute the steady-state probabilities $\pi(0), \pi(1), \pi(2), \ldots,$ where

$$\pi(n) = P(N = n) \quad \text{for } n = 0, 1, \ldots .$$

In Figure 14.7, imagine a gray line that separates states 1 through $n - 1$ from the others. The net flux across this line equals zero. Thus

$$\lambda\,\pi(n-1) = \begin{cases} n\mu\,\pi(n) & \text{if } n \le c \\ c\mu\,\pi(n) & \text{if } n \ge c \end{cases}. \tag{14.24}$$

[7] The remainder of this section is starred and can be omitted with no loss of continuity.

This queue has utilization rate $\rho = \lambda/(c\mu)$. Let us set $r = c\rho = \lambda/\mu$ and reorganize Equation (14.24) as

$$\pi(n) = \begin{cases} \dfrac{r}{n}\pi(n-1) & \text{if } n \le c \\ \rho\pi(n-1) & \text{if } n \ge c \end{cases}. \tag{14.25}$$

Repeated substitution in the above produces

$$\pi(n) = \begin{cases} \dfrac{r^n}{n!}\pi(0) & \text{if } n \le c \\ \rho^{n-c}\pi(c) & \text{if } n \ge c \end{cases}. \tag{14.26}$$

The bottom line of Equation (14.26) shows that

$$\begin{aligned} P(N \ge c) &= \pi(c) + \pi(c+1) + \pi(c+2) + \cdots \\ &= \pi(c)(1 + \rho + \rho^2 + \cdots) = \frac{\pi(c)}{1-\rho}, \end{aligned} \tag{14.27}$$

the last by summing a geometric series. The fact that the $\pi(i)$'s sum to 1, coupled with Equations (14.26) and (14.27), produces

$$\pi(0) = \frac{1}{1 + \cdots + \dfrac{r^{c-1}}{(c-1)!} + \dfrac{r^c}{c!(1-\rho)}} \quad \text{with } r = c\rho = \frac{\lambda}{\mu}. \tag{14.28}$$

The top line of Equation (14.26) shows that

$$\pi(c) = \pi(0)\frac{r^c}{c!} \quad \text{with } r = c\rho = \frac{\lambda}{\mu}. \tag{14.29}$$

For every positive integer k, there are k customers in queue if $N = c + k$. In other words,

$$P(N_q = k) = P(N = k + c) \quad \text{for } k = 1, 2, 3, \ldots, \tag{14.30}$$

which combines with the bottom line in Equation (14.26) to give

$$P(N_q = k) = \pi(c)\rho^k \quad \text{for } k = 1, 2, 3, \ldots. \tag{14.31}$$

Equation (14.31) is identical to Equation (14.15). Thus, when expressed in terms of $\pi(c)$, the probability distribution of the queue length is the same for the single-server and multiple-server cases. Hence, the analysis in the prior section computes the mean and variance of N_q. In particular, Equation (14.18) gives

$$L_q = E(N_q) = \pi(c)\frac{\rho}{(1-\rho)^2}. \tag{14.32}$$

Substituting Equations (14.29) and (14.28) into the above verifies the formula for L_q that was presented in Equation (14.23).

The distribution of the time T_q in the queue relates to the exponential random variable X given in Equation (14.22) in exactly the same way as in Figure 14.6, provided the symbol ρ in Figure 14.6 is replaced by $P(N \ge c) = \pi(c)/(1-\rho)$.

14.8. THE *M/G/1* QUEUE

The *M/G/1* queue has one server, exponential inter-arrival times, and a general service time distribution. This queue is not a continuous-time Markov chain because the service time need not have the exponential distribution. We'll see how to view it as a discrete-time Markov chain.

The State of an *M/G/1* System

The term *state* has appeared in earlier chapters. For a system that evolves over time, the **state** of the system is enough information about its history to describe the law of motion that governs its evolution. In an *M/G/1* system, the state normally consists of the following pieces of information:

- The number of customers in the system.
- If any customers are in the system, the time that has elapsed since the customer who is being served entered service.

The arrival process is Poisson (memoryless), so the state need not include the time that has elapsed since the last arrival occurred. The service distribution is not assumed to be exponential (memoryless). Thus, in order to determine the distribution of the time that remains until service is complete, the state must include the time that has elapsed since service began.

The State at Service Completion Moments

For the *M/G/1* queue, the state of the system at each service completion moment simplifies; it consists of a single piece of information, which is the number of customers that remain in the system.

Suppose, for instance, that two customers remain in an *M/G/1* system at the moment after a service was completed. Of these two customers, one has just entered service and has a service time whose distribution is known; the other customer remains in the queue, and the time until the next customer arrives has the exponential distribution with known parameter.

A Discrete-time Markov Chain

Thus, an *M/G/1* queue becomes a discrete Markov chain if we agree to observe it *only* at service completion moments. Consequently, for an *M/G/1* queue, the natural interpretation of N_q is as the number of customers in the system at the moment after a service completion. This is the fourth of the interpretations of N_q given in Section 14.4. Since upticks balance downticks, N_q can also be interpreted as the number of customers in the system at the moment before an arrival occurs. And since arrivals are Poisson, N_q is also the time average number of customers in the system. Thus, for the *M/G/1* queue, all four of the prior interpretations of N_q have the same probability distribution.

A Formula for L_q

In the case of the *M/G/1* queue, the formula for L_q contains one new parameter, namely,

$$\sigma^2 = \text{the variance of the service time.}$$

The formula for L_q is

$$L_q = E(N_q) = \frac{\rho^2 + \lambda^2\sigma^2}{2(1-\rho)}, \quad \text{with } \rho = \frac{\lambda}{\mu}. \tag{14.33}$$

The expression for L_q in Equation (14.33) is called the **Pollaczek-Khinchine** formula in honor of the mathematicians who discovered it.

The Pollaczek-Khinchine formula has the familiar factor of $(1 - \rho)$ in its denominator. As the utilization rate approaches ρ, the mean number L_q of customers in the queue and the mean time W_q that they spend in the queue increase without limit.

A Rule of Thumb

The Pollaczek-Khinchine formula brings into view a useful approximation. As was indicated in Chapter 11, the exponential distribution has a "long tail"; its standard deviation equals its mean.

Let us contrast the cases of an exponential service time and a constant service time. First, suppose that the service time is exponential. In this case, we have $\sigma = 1/\mu$, so $\lambda^2\sigma^2 = \lambda^2/\mu^2 = \rho^2$, and the Pollaczek-Khinchine formula in Equation (14.33) boils down to

$$L_q = E(N_q) = \frac{\rho^2 + \rho^2}{2(1 - \rho)} = \frac{\rho^2}{(1 - \rho)}, \quad \text{for the } M/M/1 \text{ queue,}$$

which is no surprise, for this is the formula for L_q for an *M*/*M*/1 queue.

Now suppose that the service time is constant. In this case, we have $\sigma = 0$, and the Pollaczek-Khinchine formula reduces to

$$L_q = E(N_q) = \frac{\rho^2 + 0}{2(1 - \rho)} = \frac{\rho^2}{2(1 - \rho)}, \quad \text{for the } M/D/1 \text{ queue.}$$

Evidently, the mean queue length with constant service times is *half* of the mean queue length with exponential service times.

When the number c of servers exceeds 1, the *M*/*G*/*c* queue is very hard to analyze. And the *G*/*G*/*c* queue is all but impossible to analyze. Highlighted below is a rule of thumb that the Pollaczek-Khinchine formula suggests.

> As a rule of thumb:
>
> - In an *M*/*M*/*c* queue, half of the queue length L_q is due to the uncertainty in the service time and half to the uncertainty in the inter-arrival time.
> - A *G*/*G*/*c* queue has a lower value of L_q than the *M*/*M*/*c* queue with the same arrival and service rates if the standard deviation of its inter-arrival and service time distributions are below their means.

In cases where the standard deviation is below the mean, using this rule of thumb avoids a difficult calculation or a lengthy simulation. As an approximation, this rule of thumb works pretty well but not universally. Typically, in manufacturing, the service time S has a probability distribution whose standard deviation σ is positive but is below the mean service time. On the other hand, in telecommunication systems, the standard deviation of S can be well in excess of its mean.

An Equation That *N* Satisfies*[8]

The remainder of this section presents the mathematics behind the Pollaczek-Khinchine formula. This math is *definitely* of the "extra for experts" variety. It entails two tricks, along with some bookkeeping.

---

[8] The remainder of this section can be skipped with no loss of continuity.

The first trick is to find a way to work with the fourth interpretation of N, namely, as the number of customers in the system at the moment after a service completion. Let us designate the random variables N_k, S_k, and A_k by:

N_k = the number of customers in the system at the moment after the kth service completion,

S_k = the service time of the k + 1st customer to enter service,

A_k = the number of customers who arrive between the kth and the k + 1st service completion.

We first show that random variable N_{k+1} relates to N_k, S_k, and A_k as follows:

$$N_{k+1} = \begin{cases} N_k - 1 + A_k & \text{if } N_k > 0 \\ N_k + A_k & \text{if } N_k = 0 \end{cases}. \tag{14.34}$$

Justifying Equation (14.34) is a matter of counting. For its top line, we account for the case in which at least one customer remains in the system after the kth service completion. In this case, the number N_{k+1} of customers in the system after the k + 1st service completion equals the number N_k there after the kth completion minus 1 (the k + 1st customer) plus the number A_k of arrivals during the service time S_k of the k + 1st customer. For the bottom line of Equation (14.34), we account for the case in which no customer remains in the system after the kth service completion. In this case, the number N_{k+1} in the system after k + 1st service completion equals the sum of N_k and the number A_k who arrive during the service time S_k of the k + 1st customer.

To consolidate the lines of Equation (14.34), we employ the **indicator** random variable $\mathbf{1}_E$ that equals 1 if the event E occurs and equals zero if the event E does not occur. The indicator random variable lets us write Equation (14.34) compactly as

$$N_{k+1} = N_k + A_k - \mathbf{1}_{\{N_k>0\}}, \tag{14.35}$$

as this indicator random variable equals 1 if N_k is positive.

In steady state, the variables N_k and N_{k+1} have the same probability distributions and, in particular, have L as their expectation. In Equation (14.35), we take expectations, recall that the expectation of the sum equals the sum of the expectations, and drop subscripts, getting

$$L = L + E(A) - E(\mathbf{1}_{\{N>0\}}).$$

Since $E(\mathbf{1}_{\{N>0\}}) = P(N > 0) = \rho$, the above shows that

$$E(\mathbf{1}_{\{N>0\}}) = \rho = E(A), \tag{14.36}$$

which is doubly disappointing. Nothing has been learned about L, and Equation (14.36) could have been obtained more easily.

But let us square Equation (14.35) before taking expectations. When taking the square, we drop subscripts in the case of products of independent random variables, and we simplify two of the terms that entail $\mathbf{1}_{\{N>0\}}$. The result is

$$N^2 = N^2 + A^2 + \mathbf{1}_{\{N>0\}} + 2NA - 2N - 2A\mathbf{1}_{\{N>0\}}.$$

Now, we take expectations in the above. The expectation of the product of independent random variables equals the product of their expectations. Two terms cancel each other, and Equation (14.36) shows that what remains is

$$0 = E(A^2) + \rho + 2L\rho - 2L - 2\rho\rho. \tag{14.37}$$

Progress! We can compute L from Equation (14.37) if we can evaluate $E(A^2)$.

The Conditional Variance

Evaluating $E(A^2)$ is the second trick. A conditional variance calculation (see Problem 33 page 519) shows that for any pair A and S of random variables,

$$E(A^2) = \text{Var}_S E(A|S) + E_S E(A^2|S). \tag{14.38}$$

We will use Equation (14.38) with the random variable S as the service time.

Specifically, with S as the service time, we will evaluate both of the addends on the right-hand side of Equation (14.38). The random variable $A|S$ equals the number of arrivals that occur during the service time S. Arrivals are Poisson with rate λ per unit time, so formulas given in Chapter 8 for the mean and variance of the Poisson distribution give

$$E(A|S) = \lambda S \quad \text{and} \quad \text{Var}(A|S) = \lambda S. \tag{14.39}$$

The service time S has $\text{Var}(S) = \sigma^2$, so the left-hand expression in (14.39) gives

$$\text{Var}_S E(A|S) = \lambda^2 \sigma^2. \tag{14.40}$$

To evaluate the other addend in (14.38), we note, from the standard formula for the expectation of the square and Equation (14.39),

$$\begin{aligned} E(A^2|S) &= \text{Var}(A|S) + [E(A|S)]^2 \\ &= \lambda S + (\lambda S)^2. \end{aligned} \tag{14.41}$$

Since $E(S) = 1/\mu$ and $E(S^2) = \text{Var}(S) + [E(S)]^2 = \sigma^2 + (1/\mu)^2$, taking expectations in expression (14.41) produces

$$\begin{aligned} E_S E(A^2|S) &= \lambda/\mu + \lambda^2(\sigma^2 + (1/\mu)^2) \\ &= \rho + \lambda^2\sigma^2 + \rho^2. \end{aligned} \tag{14.42}$$

By substituting expressions (14.40) and (14.42) into Equation (14.38), we get

$$E(A^2) = 2\lambda^2\sigma^2 + \rho + \rho^2. \tag{14.43}$$

Finally, substituting the expression for $E(A^2)$ in Equation (14.43) into Equation (14.37) and shifting terms that entail L to its left-hand side produce

$$L = \frac{\lambda^2\sigma^2 + \rho^2}{2(1 - \rho)} + \rho. \tag{14.44}$$

Since ρ is the expectation of the number of customers in service, the equation $L = L_q + \rho$ combines with Equation (14.44) to verify Equation (14.33).

14.9. REVERSIBILITY OF THE *M/M/c* QUEUE

The next section describes a network of queues. In preparation, we recall from Chapter 13 that a continuous-time Markov chain is reversible if these two conditions are satisfied:

- All states must be members of the same ergodic set.
- The product of the transition rates around each cycle must equal the product of the transition rates around the reversed cycle, the latter being the cycle that visits the states in reverse order.

Consider an *M/M/c* queue whose utilization rate $\rho = \lambda/(c\mu)$ is below 1. Its chain has one ergodic set, which consists of all the states. What do its cycles look like? Each transition either increases the state by 1 or decreases the state by 1, so its cycles have a simple form. Consider, for instance, the cycle (3, 4, 5, 6, 5, 4, 3). The reverse of this cycle is the cycle itself.

In this queue, the reverse of *each* cycle is the cycle itself. Hence, the $M/M/c$ queue is reversible provided that its utilization rate is below 1.

So what? Let us recall from Chapter 13 that, in a reversible Markov chain, we cannot tell from any sample path the direction in which time increases. That fact is the second in this line of reasoning:

- The input to this queue is a Poisson process with rate λ.
- Since the process is reversible, the input to the time-reversed process must also be a Poisson process with rate λ.
- The input to the time-reversed process is the output of the process itself.
- Hence, the output of this queue is a Poisson process with rate λ.

> If the utilization rate of an $M/M/c$ queue is below 1, its output process is Poisson with the same rate as its input process.

This result is astonishing. To see why, consider an $M/M/1$ queue whose arrival rate λ equals 1 per hour and whose service rate μ equals 10 per hour. Suppose that 15 customers have emerged from this system during the past hour. This is a rare event, but it does occur. Given that it has just occurred, it seems to be all but certain that a queue has formed and that at least one customer remains in the system, in which case the mean time to emergence of the next customer is $1/10$ hour, not 1 hour. *Wrong.* The output process is Poisson, so the time until the next output is exponential with a mean of 1 hour, independent of what has happened before.

14.10. A NETWORK OF QUEUES

Prior sections have studied individual queues. In many settings, customers experience a network of related queues rather than a single queue. For instance, an item that is undergoing manufacture may visit several workstations, and it may need to wait for service at each. Similarly, a message that is being transmitted through a telecommunication system visits several nodes of its network en route to its ultimate destination, and it may need to wait at each node for transmission toward another.

The results in prior sections adapt almost effortlessly to networks of queues. To see how, we turn our attention to:

Problem C (An Emergency Room)[9]

Figure 14.8 depicts the flow of patients through the Surgery unit of an emergency room. As the figure suggests, arrivals at the emergency room are Poisson with a rate of 12 per hour. Of these arrivals, one-sixth proceed directly to the Surgery unit. The remainder proceed to the Diagnosis unit, which is staffed by four physicians. The time required for diagnosis has the exponential distribution with a mean of 15 minutes. Diagnosis results in referral of one patient in three to the Surgery unit, the rest to other units. The Surgery unit is staffed by three surgeons. The total time that a surgeon spends with each patient has the exponential distribution with a mean of 30 minutes. Surgeons may order blood work or X rays, which lengthens the patient's stay in the surgery unit. Excluding extra time spent waiting for blood work

[9]This example stems from a project done by Yale College seniors under the direction of Arthur J. Swersey, Jr.

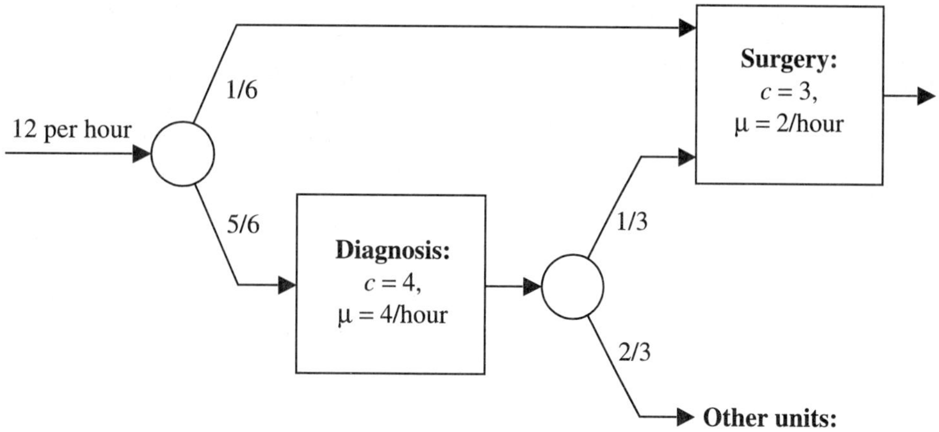

Figure 14.8 Flow of surgery patients through an emergency room.

and X rays, how long do surgery patients spend in this emergency room? What would happen if a fourth surgeon was added to the Surgery unit?

Analysis of Problem C will entail no new ideas. To solve it, we will cobble together material from this chapter and Chapter 8. This material includes the formula for L_q for the *M*/*M*/*c* queue and the earlier formulas for L and W.

Random Splitting of a Poisson Process

The customer arrival process at the emergency room is Poisson with rate $\lambda = 12$ per hour. Each arrival goes directly to the Surgery unit with probability $1/6$ and goes first to the Diagnosis unit with probability $5/6$. This is a random splitting, of the sort discussed in Chapter 8. In that chapter, we observed that a random split of a Poisson process produces Poisson processes that are independent *of each other*, though they are not independent of the process from which they had been split. Thus, in Problem C, the stream of patients who go directly to surgery is Poisson with a rate of 2 per hour, the stream of patients who go first to diagnosis is Poisson with a rate of 10 per hour, and these streams are independent of each other.

Reversibility of an *M*/*M*/*c* Queue

Let us examine the Diagnosis unit. Its input process is Poisson with a rate of 10 per hour. This unit has four servers, each with a service rate of 4 per hour. Its utilization rate ρ equals $10/(4 \times 4) = 5/8$, which is well below 1.

The Diagnosis unit is an *M*/*M*/4 queue whose utilization rate equals $5/8$. Like any *M*/*M*/*c* queue whose utilization rate is below 1, this one is reversible, so the *output* of the Diagnosis unit is a Poisson process with a rate of 10 per hour. This output is subject to a random split, with $1/3$ going to surgery.

Superposition of Poisson Processes

The input to the Surgery unit is the superposition of two Poisson processes that are independent of each other. One of these Poisson processes has a rate of 2 per hour, and the other has a rate of $10/3$ per hour. In Chapter 8, we observed that the superposition of independent Poisson processes is a Poisson process, and hence, that the input to the Surgery unit is a

Poisson process with rate $\lambda = 2 + 10/3 = 16/3$ per hour. With three surgeons, this unit has a utilization rate $\rho = (16/3)/(3 \times 2) = 16/18 = 8/9$.

In brief, then, the Diagnostic unit is an *M*/*M*/4 queue having utilization rate $\rho = 5/8$, and the Surgery unit is an *M*/*M*/3 queue with utilization rate $\rho = 8/9$. These facts enable computation of L_q and W_q for each queue, as well as the expectation of the time that each customer spends in the network of queues. This computation is left for you as a homework problem.

A Jackson Network

Let us now begin to build a model that Problem C illustrates. This model is a network of K queues. We number these queues 1 through K. The characteristics of this network are listed below. For $k = 1, 2, \ldots, K$:

- The external input to queue k is a Poisson process with rate $r(k)$ that is fixed and nonnegative.
- Queue k has $c(k)$ identical servers who work in parallel; their service times have the exponential distribution with rate $\mu(k)$.
- Each customer who completes service at queue k next visits queue j with probability $P(k, j)$ and departs from the system with probability $d(k)$, where

$$d(k) = 1 - \sum_{j=1}^{K} P(k, j) \quad \text{for } k = 1, \ldots, K.$$

- All random variables are mutually independent.

The model that has just been described is called a **Jackson network** in honor of the first of two path-breaking papers published by J. R. Jackson, this one in 1957. The nodes in this network correspond to the queues, and the directed arcs in this network correspond to the transition probabilities that are positive. Thus, the network contains nodes $1, 2, \ldots, K$, and it contains arc (k, j) if and only if $P(k, j)$ is positive.

In a network of queues, the time that a customer spends in the (entire) system of queues is said to be that customer's **sojourn** time. We will see what can be said of the sojourn times in a Jackson network.

Utilization Rates

The departure probabilities—the $d(k)$'s—do not appear explicitly in anything that follows, but some of these departure probabilities must be positive. As we shall see, they must be sufficiently positive that the aggregate arrival rate at each queue is below its service rate.

For $k = 1, 2, \ldots, K$, we designate as $\lambda(k)$ the aggregate (total) arrival rate at queue k, as given by

$$\lambda(k) = r(k) + \sum_{j=1}^{K} \lambda(j)\, P(j, k) \quad \text{for } k = 1, 2, \ldots, K. \tag{14.45}$$

In Equation (14.45), the addends to the left and the right of the "+" sign are, respectively, the rates at which external and internal arrivals occur at node k.

We will study this system of queues under the assumption that the utilization rate $\rho(k)$ of each queue is below 1, that is, that

$$\rho(k) = \frac{\lambda(k)}{c(k)\, \mu(k)} < 1 \quad \text{for } k = 1, 2, \ldots, K. \tag{14.46}$$

A famous result of Jackson requires nothing more than that each utilization rate $\rho(k)$ be below 1. His result is presented later in this section. Our initial goal is to study a special class of Jackson networks, namely, those for which the input to each queue is a Poisson process.

An Acyclic Jackson Network

As in Chapter 5, a directed network is said to be **acyclic** if it has no cycle, equivalently, if the transition probabilities are such that no queue can ever be revisited. For Problem C, the Jackson network has two nodes, and it is acyclic. Customers can proceed from the triage node to the surgery node, but not from the surgery node to the triage node.

An acyclic network may take the form of a series of queues, with transition from each queue to the next and with departure after the Kth queue. Alternatively, customers, can "hop" forward from queue k to any higher-numbered queue, and different customers can enter at different queues. An acyclic Jackson network is sometimes called a **feed-forward** network in order to distinguish it from the "feedback" case in which nodes can be revisited.

We have seen that the output of an *M/M/c* queue is a Poisson process, that random splitting of a Poisson process produces independent Poisson processes, and that the superposition of independent Poisson processes is Poisson. The amalgam of these properties proves:

Theorem 14.3 (An Acyclic Network of Queues). Suppose that an acyclic network of *M/M/c* queues has utilization rates that satisfy expression (14.46). Then, for each k, the input to queue k is a Poisson process whose arrival rate $\lambda(k)$ is given by Equation (14.45).

Theorem 14.3 lets us use the Excel function **=LQP(ρ(k), c(k))** to compute the expectation L_q of the number of customers in queue k. Once a queue's expected length L_q has been calculated, its quantities L, W_q, and W can be found in the usual way. Since arrivals are Poisson, PASTA applies.

Sojourn Times in an Acyclic Jackson Network

We recall that a customer's sojourn time is that person's total time in the system of queues. Let us consider the sojourn times in an acyclic Jackson network. Theorem 14.3 describes conditions under which the input to each queue is a Poisson process. In this system, the inputs to the different queues are *not* independent of each other. The expectation of the sum equals the sum of the expectations, whether or not the random variables are independent. Thus, we can compute a customer's expected sojourn time by multiplying the probability that the customer visits each queue by the expected time in system for that queue and taking the sum.

On the other hand, since the times spent in different queues are dependent, the variances cannot be added. Thus, for an acyclic network of queues, we can offer no convenient way to compute the variance of the sojourn time. In this system, it is difficult to measure the uncertainty of the sojourn time.

Jackson's Result

Now suppose that the Jackson network is not acyclic, that one or more of its nodes can be revisited. In this case, the inputs to the queues are not Poisson processes, and we seem to be stymied. But Jackson demonstrated:

Theorem 14.4 (A Jackson Network of Queues). Suppose that a Jackson network of queues has utilization rates that are below 1. Then:

(a) For each k, the steady-state probability distribution of the number of customers in queue k is identical to that of an $M/M/c(k)$ queue whose arrival rate is $\lambda(k)$.

(b) The steady-state probability distributions of the numbers of customers in different queues are mutually independent.

This is another of the startling surprises of applied probability. The proof of Theorem 14.4 is not insightful. It is of the "plug and chug" variety: plug the alleged solution into the equation system that the true solution must satisfy, chug along, and see that it does.

Theorem 14.4, Jackson's result, shows that the *steady-state* distribution of the number N_q of customers in a queue is what we would obtain for this queue if its arrivals were Poisson. It also shows that the steady-state distributions of the numbers of customers in the various queues are mutually independent.

Dovetailing

Part (a) shows that the Excel function **=LQP(ρ(k), c(k))** computes the expectation L_q of the steady-state number of customers in the kth queue. Arrivals are not Poisson, so PASTA does not apply, and we cannot interpret L_q as the expectation of the number of customers in the queue at the moment when an arrival occurs.

But Jackson's result dovetails perfectly with Little's law, which does *not* require Poisson arrivals. Little's law states that, with L_q as the expected number of customers in the kth queue under steady-state conditions, the equation $W_q = L_q/\lambda(k)$ specifies W_q as the expectation of the length of time that a customer who arrives at queue k must wait. Serendipity?

Again, the expectation of the customer's sojourn time is easily computed. To do so, multiply the expectation of the number of visits that the customer makes to each queue by the expectation of the time in system for that queue, and sum the result. This works because the expectation of the sum of random variables equals the sum of their expectations, even when they are dependent. Again, variances of dependent random variables do not add, and we can offer no way to compute the variance of the sojourn time.

14.11. REVIEW

The unstarred portions of this chapter build on rather simple ideas. First, as concerns performance measures, once L_q has been found, it's easy to compute $W_q = L_q/\lambda$ and $W = W_q + 1/\mu$ and $L = \lambda W$. Formulas for L_q are given for the $M/M/1$, $M/M/c$, and $M/G/1$ queues. In the case of the $M/M/c$ queue, the formula is unwieldy, but the Excel function **=LQP(ρ, c)** specifies it.

Four different interpretations have been provided of the random variable N_q, and each is important. Luckily, in the case of Poisson arrivals, all four interpretations have the same probability distribution, in which case $L_q = E(N_q)$ equals:

- The expectation of the number of customers present in the queue at the moment before a customer arrives
- The expectation of the number of customers present in the queue at the moment a customer enters service
- The expectation of the number of customers in the queue under steady-state conditions
- The time-average number of customers in the queue

The first interpretation of L_q is of interest to the arriving customer; all are of interest to the manager and the facility designer.

The presentation has emphasized the importance of keeping the servers' utilization rate $\rho = \lambda/(c\mu)$ well below 1. For each queue, the formula for L_q contains the term $(1 - \rho)$ in the denominator, and L_q begins to grow rapidly as the utilization rate ρ approaches 0.9.

The presentation has also highlighted an advantage of pooling queues. Pooling k identical queues produces a lower value of L_q than any of the unpooled queues. Moreover, as the formula $W_q = L_q/\lambda$ demonstrates, the pooled queue not only shortens the mean line length, L_q, but also causes the line to move k times as fast.

The Pollaczek-Khinchine formula indicates that in an *M*/*M*/1 queue, half of the expected queue length L_q is due to the uncertainty in the service time, and half is due to the uncertainty in the inter-arrival time. This gives a rule of thumb for more complicated queues whose inter-arrival and service times have less uncertainty than the exponential.

If a network of *M*/*M*/*c* queues is acyclic, its analysis proves to be easy. But Jackson's result and Little's law show how to compute the expectation of the time in system even if the network is cyclic. Simply compute the aggregate arrival rate at each queue. Then use the standard formula to compute L_q for each queue, use Little's law to compute W_q, and add expectations.

The starred segments of this chapter present the mathematics that supports its main results. The *M*/*M*/1 and *M*/*M*/*c* queues are continuous-time Markov chains that are easy to analyze using the methods developed in Chapter 13. The *M*/*G*/1 queue is a discrete-time Markov chain when it is viewed only at service completion moments. Its analysis is more involved, resting as it does on the "conditional variance" formula.

It is also possible to analyze the *G*/*M*/1 queue, which is a discrete-time Markov chain when it is viewed only at customer arrival moments. For a sketch of that analysis, see Problem 34 on page 519.

14.12. HOMEWORK AND DISCUSSION PROBLEMS

1. Over the course of a 24-hour period, make a log of each queue that you experience.

2. **(Air Travel)** Imagine (or experience) air travel. List each queue that you experience from the moment you leave home until your aircraft takes off.

3. **(Long Waits for Short)** Customers X, Y, and Z enter a single-server queue in alphabetical order. Their service times are, respectively, 12 minutes, 8 minutes, and 6 minutes. Which service sequence minimizes the sum of the times that they wait? Which service sequence minimizes the average of the times that they wait?

4. True or false: For a queue in steady state, the random variable N_q equals the number of customers in the queue at the moment after a customer has arrived. Support your answer.

5. **(The Number in System)** List four different interpretations of the number N of customers in the system. Under what circumstances do the various interpretations have the same probability distribution?

6. In an *M*/*M*/*c* queue, which of the following interpretations applies to the random variable T? Briefly indicate why.

(a) T is the time an arriving customer will spend in system.

(b) With a FCFS queue discipline, T is the remaining time that the most recent arrival will spend in the system, given that at least one customer is in the system.

(c) With a FCFS queue discipline, T is the time that a departing customer spent in the system.

7. **(à la Little)** The university's Economics Department enrolls Ph.D. students at the rate of 25 per year. The expectation of the time that a student remains in the Ph.D. program is 4.3 years. In steady state, how many students are in the program, on average?

8. **(Work)** In queueing, the amount Z of **work** in a system equals the sum of the service times of the customers who are in the queue plus the sum of the remaining service times of the customers who are in service. Which of the following is true? Why?

(a) For an $M/M/1$ queue in steady state and with a FCFS queue discipline, Z and T have the same probability distribution.

(b) For an $M/M/c$ queue in steady state and with a FCFS queue discipline, Z and T have the same probability distribution.

9. Suppose you are presented with a formula for L, the expected number of customers in the system. Express L_q, W_q, and W in terms of L and state the conditions, if any, under which these equations are valid.

10. **(Expected Busy Period)** In an $M/G/1$ queue, an **idle period** I is an interval of time that begins when the server becomes idle and ends when the server next becomes busy, that is, when the next customer appears. A **busy period** B of this queue is an interval of time that begins at the moment when the server becomes busy and ends when the server next becomes idle. Evidently, each busy period is followed by an idle period, and each idle period is followed by a busy period.

(a) Justify the equation

$$\rho = \frac{E(B)}{E(B) + E(I)}.$$

(b) Show that $E(B) = 1/(\mu - \lambda)$.

11. To earn spending money, Jane "moonlights" in the food court at the student union during the evenings. She is the only person who works at the food counter while she is there. She has noted that customers arrive at a rate of 10 per hour and that it takes her four minutes to serve a customer, on average. She adopts the Poisson as a model of the arrival process and the exponential as a model of her service time distribution.

(a) What fraction of the time is she serving customers?

(b) What is the average queue length?

(c) What is the average time that a customer is in the system?

(d) What fraction of the time is at least one person waiting for service to commence?

(e) What is the probability that a customer waits at least four minutes in the queue before being waited on?

(f) Jane's other job is grading calculus assignments. It takes her about five minutes to grade each assignment. She spends her idle time at the food court grading. How many papers per hour can she grade while working at the food court?

12. The local utility company has its own service center for routine maintenance of its vehicles. The service center works around the clock. The arrival process of vehicles at this service center is Poisson with a rate of three per day. The service center can work on only one vehicle at a time. Its service time is exponential, with a mean of 7 hours. The opportunity cost (loss of profit) of each vehicle that is undergoing service is $65 per hour.

(a) What is the expectation of the opportunity cost of the vehicles that are in the service center?

(b) At an increase in variable cost of $75 per hour, you can decrease the mean service time from 7 hours to 6, keeping its distribution exponential. Would this change be profitable? If so, how profitable would it be?

13. **(Engine Overhaul)** An airline owns a fleet of two-engine aircraft. Each engine requires periodic overhaul, which occurs at the airline's repair depot. The repair depot can overhaul only one engine at a time. The time needed to overhaul one engine is exponential with a rate of three per day. Current policy is to overhaul only one engine on each visit of an aircraft to the depot. Arrivals of aircraft whose engines require overhaul occur according to a Poisson process with a rate of two per day.

(a) Compute and interpret L_q, L, W_q, and W.

(b) The company is contemplating a change in policy in which both engines are overhauled each time a plane comes to the depot. This would halve the rate at which aircraft arrive at the depot but would double the mean and variance of the repair time. Re-compute and re-interpret L_q, L, W_q, and W for this new policy.

(c) In what ways, if any, are the policies in parts (a) and (b) preferable to each other?

14. (**Street Lamps**) A city has 24,000 street lamps, each of which lasts an average of 400 days before burning out. This city has hired a contractor to replace burned-out street lamps. The contract stipulates that street lamps are to be replaced within five days of burning out, on average. To assure compliance with this contract, the town auditor counts burned-out street lamps at random intervals. She has just done so and has found that 1200 street lamps are burned out.

(a) On a typical day, how many lamps should be burned out? Why?

(b) If the contractor is complying with his contract, what is the probability that at least 1200 lamps will be burned out under steady-state conditions?

15. An efficiency expert has observed that a bank's customers make deposits or withdrawals on a single visit, but not both. Customers arrive at the rate of 30 per hour, and each customer is equally likely to make a withdrawal or a deposit. This expert has observed that the bank's tellers can serve customers at a rate of 20 per hour if they handle either deposits or withdrawals, but at a rate of 18 per hour if they handle both. Service times are exponential.

(a) Based on the expert's assessment, the bank has two tellers, one of whom handles withdrawals and the other deposits. How long do its customers wait for service to begin, on average? How long do its customers spend in the bank, on average? What is the probability that a customer waits five minutes or longer?

(b) Would service improve or degrade if the bank had two tellers, each of whom served both classes of customer? Support your answer.

16. (**Pipeline Repair**) A gas pipeline company maintains 8 skilled teams that it dispatches to emergency repairs on its pipelines. Demands for pipeline repair form a Poisson process with a rate of 0.8 per day. Repair time is exponential with a mean of two weeks (14 days).

(a) On average, how long does a repair wait for a team to be dispatched to it?

(b) Each repair team costs the company \$350 per day, excluding the cost of the repairs themselves. Assume that the company has sized its emergency repair force so as to minimize its expected cost. What can you say about its cost per pipeline that requires repair but is not receiving it?

17. (**Emergency Vehicles**) A city locates emergency vehicles at each firehouse. A particular firehouse experiences Poisson demand for emergency vehicles with a rate of 30 per day. The time that it takes each emergency vehicle to handle a request for service is exponential with a mean of 40 minutes. The city's policy is to locate enough emergency vehicles at each firehouse to have only a 1% chance that a vehicle is not present when a call arrives. How many vehicles are needed at this firehouse? *Hint:* To calculate $P(N > c)$ on a spreadsheet, note that for an *M*/*M*/*c* queue, $P(N > c) = \pi(c)\rho/(1 - \rho)$ and, from Equation (14.32), that $\pi(c) = L_q(1 - \rho)^2/\rho$.

18. (**Balking**) The local service station has one gasoline pump. The arrival process of customers who want gasoline is Poisson with a rate of 15 per hour. The service time is exponential with a rate of 13 per hour. Arriving customers may "balk" by going to a different station. If k customers are in the queue when an arrival occurs, the arriving customer balks with probability $k/3$ for $k \leq 3$ and with probability 1 with $k > 3$.

(a) Draw a transition rate diagram for this system.

(b) On a spreadsheet, compute the steady-state probability distribution of the number of customers in the system.

(c) At what rate do customers balk? What fraction of the arriving customers balk?

(d) What is the expectation of the time that a customer who uses the system waits for service to commence? Does your answer to part (d) employ Little's law? If so, how?

19. (**Requires Calculation**) At the local grocery store, the customer arrival process at the checkout station is Poisson with a rate of 20 per hour. This checkout station is staffed by a single server whose service time is exponential with a mean of three minutes. But while three or more customers are in the queue, the manager appears and "bags" groceries, which increases the service rate by 50%.

(a) Draw a transition rate diagram for this system.

(b) Compute the steady-state probability distribution of the number of customers in this system. *Hint:* Adapt Equation (14.26) with $c = 3$.

(c) Compute the expected number of customers in the queue. *Hint:* To avoid a nasty computation, you can adapt Equation (14.16) with $c = 3$ to account for some of the customers in the queue, so you need only account for the others.

(d) Compute the expected time that an arriving customer spends in the queue.

(e) Suppose that, to avoid insulting his clientele, the manager arrives as soon as at least three customers are in the queue but does not leave until the system next empties out. Revise the transition diagram accordingly.

20. The arrival process of tankers at a major crude-oil unloading port is Poisson with a rate of eight per day. The port has six berths, and but it has the capability to unload only four tankers simultaneously. It takes an average of 10 hours to unload each tanker, and the unloading time is approximately exponential. When all six berths are full, incoming tankers are routed to another facility whose operating characteristics are similar to this one's. Routing a tanker to the other facility lengthens its time at sea by 15 hours, on average.

(a) Draw a transition rate diagram for the number of tankers that are being berthed at this facility, and compute the steady-state probability $\rho(n)$ that n tankers are berthed here.

(b) At what rate are tankers diverted to the other facility? On average, how long must a tanker that will be unloaded here have to wait for unloading to commence?

(c) The company is thinking about adding the capacity to unload a fifth tanker at this port. Redo parts (a) and (b) to determine the circumstances under which this would be worthwhile.

21. **(Problem C)** For Problem C, compute the expectation of the time that a surgery patient spends in the system when the Surgery unit has three doctors. What happens if a fourth surgeon is added to the Surgery unit?

22. **(Rework)** In a quality-oriented production facility, each manufacturing operation ends with a test. The item may pass the test, fail it, or require rework. The external arrival process of items at operation Q is Poisson with a rate of 5 per hour. Each visit of each item to operation Q has these characteristics—a processing time that is exponential with a mean of 8 minutes (including the test), a probability 0.7 of passing the test, a probability 0.05 of failing it, and a probability 0.25 of requiring rework. If an item requires rework at operation Q, it revisits operation Q. If an item fails inspection, it is discarded.

(a) Compute the aggregate arrival rate at operation Q.

(b) Compute the utilization rate at operation Q.

(c) Is the aggregate arrival process at operation Q Poisson? Why?

(d) In steady state, what is the expected number of items that are waiting in the queue at operation Q?

(e) For each item that reaches operation Q, what is the expectation of the time that it waits for processing to commence?

(f) What can you say about the expected sojourn time of a unit at operation Q?

23. **(A Network of Queues)** A regional TV repair center fixes those sets it can and sends the rest to the manufacturers. The arrival process of sets requiring repair is Poisson with a rate of 6 per hour. A diagnostician examines each set that appears. She spends an average of 9 minutes with each set, and she determines where to send it. She sends 15% of the sets to their manufacturers, 50% to the general repair section, and the remaining 35% to the expert repair section. The general repair section has three repair people who work in parallel; it takes each of them an average of 50 minutes per set. The expert repair section has four repair people who work in parallel; it takes each of them an average of 90 minutes per set. The experts fix everything that comes their way. The general repair people fix 90% of the sets sent to them and forward the rest to the expert group. Each set (including those sent to the manufacturers) goes to shipping and billing, which is staffed by two clerks who work in parallel, each taking an average of 15 minutes per set. All times are exponential, approximately.

(a) Draw a transition rate diagram for this network. Compute the utilization rate at each node.

(b) Is the arrival process at each node Poisson? Briefly, say why.

(c) For each node, compute the expected time that a set reaching that node will spend there.

(d) Compute the expectation of the time that a TV set spends at this repair center.

24. (**What the Departing Customer Sees**) The fourth interpretation of N_q was as the number of customers who remain in the queue at the moment after a customer enters service. Within Chapter 14, has any use been found for this interpretation? If so, where?

25. (**Simulating a *D/M/1* Queue**) Figure 14.1 plots the result of simulating a *D/M/1* queue whose inter-arrival times are exactly 15 minutes and whose service time is exponentially distributed with parameter $\mu = 1/13$ per minute, hence with a mean service time of 13 minutes. To indicate how this figure is constructed, we designate

$$W_k = \text{the time that the } k\text{th customer waits for service to begin,}$$
$$S_k = \text{the service time of the } k\text{th customer.}$$

(a) Does W_k satisfy the recursion, $W_{k+1} = \max\{0, W_k + S_k - 15\}$? If so, why?

(b) Verify that $P(S_k \le t) = -\ln(Z) \times 13$ where Z is uniformly distributed on the interval $0 \le Z \le 1$. Use this fact to simulate W_1 through W_{150} on a spreadsheet. *Remarks:*

- The function **=RAND()** simulates a random variable whose probability distribution is uniformly distributed on the interval between 0 and 1.
- But every time you do anything to a spreadsheet, this function simulates anew, which can be frustrating.
- After simulating 150 uniform random variables, you can "freeze" their values by selecting them all, copying them, and using the "Paste Special" option to paste (only) their values.

(c) Make a chart of W_{51} through W_{150}.

26. (**The Economics of Queues**) Suppose that each customer in the queue incurs cost (e.g., an opportunity cost of time) of K dollars per unit time spent in the queue. Show that the expectation of the cost of entering an *M/M/1* queue equals $K\rho/(\mu - \lambda)$. *Hint:* Does W_q equal $\rho/(\mu - \lambda)$?

27. (**The Economics of Queues, continued**) Consider an *M/M/1* queue in steady state. This problem asks you to compute b, the expectation of the time that elapses from the moment a customer enters the system until the queue first becomes empty.

(a) Consider a moment at which the number n of customers in the queue is positive. Let t denote the expectation of the time that elapses from this moment until the first time at which $n - 1$ customers are in the queue. Show that

$$t = \frac{1}{\lambda + \mu} + \frac{\mu}{\lambda + \mu} \times 0 + \frac{\lambda}{\lambda + \mu} \times (2t) = \frac{1}{\mu - \lambda}.$$

(b) Consider a customer who enters the system in steady state. If there are n customers in the system at the moment just before that customer enters, the expectation of the time until the queue next becomes empty equals nt. Why is that?

(c) Show that

$$b = \sum_{n=1}^{\infty} P(N = n)nt = t(1 - \rho)\rho(1 + 2\rho + 3\rho^2 + \cdots)$$
$$= \frac{t\rho}{1 - \rho} = \frac{W_q}{1 - \rho}.$$

(d) Is it reasonable that b exceeds W_q? If so, why?

28. (**The Economics of Queues, concluded**) Suppose, as above, that each customer in the queue incurs cost (e.g., an opportunity cost of time) of K dollars per unit time spent in the queue. In steady state, how much cost does each user impose on the other customers? *Hint:* The customer's expected service time equals $1/\mu$ and, during the remaining busy period, the expectation of the number of customers who arrive equals λb.

Remark: The preceding economic analysis has been proposed, sporadically, as a proper basis for setting landing fees at airports. The idea is to set the fee equal to the marginal cost of using the facility.

An airport is an extremely expensive facility, and the demand for takeoff and landing slots fluctuates with the time of the day. A fee structure that smoothes these fluctuations could use the facility more efficiently. But in this environment, equating the landing fee to the marginal cost can be catastrophic. Why? The demand is cyclic, not steady, and the fee at the beginning of the busy period can be astronomical.

29. (**Summing a Random Number of Random Variables**) Let μ and λ be positive numbers with $\lambda < \mu$ and set $\rho = \lambda/\mu$. Let N be the integer-valued random variable whose probability distribution is

$$P(N = n) = (1 - \rho)\rho^n \quad \text{for } n = 0, 1, \ldots.$$

Let $X_1, X_2, \ldots$ be a sequence of i.i.d. random variables, each having the exponential distribution with parameter μ. With $S_0 = 0$, set

$$S_N = X_1 + X_2 + \cdots + X_N.$$

Notice that S_N equals the sum of a random number of random variables. What is the probability distribution of S_N? *Hint:* Do not even dream of computing this. Instead, review Theorem 14.2.

30. (**Summing the Geometric Series**) Let x be any number whose absolute value is less than 1.

(a) Show that $(1 + x + x^2 + \cdots + x^n)(1 - x) = (1 - x^{n+1})$.

(b) Show that the infinite sum $(1 + x + x^2 + \cdots)$ equals $1/(1 - x)$.

(c) Differentiate in part (b) to show that

$$(1x + 2x^2 + 3x^3 + \cdots) = \frac{x}{(1 - x)^2}.$$

(d) Differentiate in part (c) to show that

$$(1^2x + 2^2x^2 + 3^2x^3 + \cdots) = \frac{(1 + x)x}{(1 - x)^3}.$$

31. (**Variability of an *M/M/c* Queue**) Equations (14.14)–(14.19) are stated for the *M/M/*1 queue, but these equations are correct for the *M/M/c* queue. From these equations, it is possible to use the function **=LQP(ρ, c)** for L_q to recover $\pi(c)$ and $\text{Var}(N_q)$ for the *M/M/c* queue.

(a) Is it true that an *M/M/c* queue has $E[(N_q)^2] = L_q(1 + \rho)/(1 - \rho)$. If so, why?

(b) On a spreadsheet, create a table whose columns record the values $c = 1, 2, 3, 4, 5$, whose rows record the values $\rho = 0.7, 0.8, 0.9$, and 0.95, and whose entries record the value of the standard deviation of N_q for the utilization rate ρ in its row and the number c of servers in its column.

(c) In your table, how does the standard deviation of N_q vary with c and with ρ? How does the standard deviation of N_q compare with the expectation of N_q?

Remark: Let X be a continuous random variable whose density function is $f(x)$, let Y be any random variable, and suppose that X and Y are independent of each other. The probability distribution of their sum is given by the formula

$$P(X + Y \le t) = \int_{x=-\infty}^{+\infty} P(Y \le t - x) f(x)\, dx,$$

and this formula is said to take the **convolution** of the distributions of X and Y. To see why this formula is true, interpret $f(x)\,dx$ as the probability that X is "approximately" x. The random variables are independent, so $P(Y \le t - x)\, f(x)dx$ equals the joint probability that X is approximately x and $Y \le t - x$, so that $X + Y \le t$. The convolution sums (integrates) over all x to compute $P(X + Y \le t)$.

32. (**The Distribution of *T* for an *M/M/*1 Queue**) Assume that T_q has the distribution given in Figure 14.6. Evidently, $T = T_q + S$, where the random variable S has the exponential distribution with parameter μ. Demonstrate that T has the exponential distribution with parameter $(1 - \rho)\lambda$. *Hint:* Use (only) the convolution formula (see above) and the fact that

$$\int_0^t ae^{-ax}\,dx = 1 - e^{-at}.$$

33. **(Conditioning the Square and the Variance)** Let X and Y be discrete random variables that need not be independent. For each outcome x that X can take, set

$$\mu(x) = E(Y|X = x) \quad \text{and} \quad \sigma^2(x) = \mathrm{Var}(Y|X = x).$$

(a) Show that

$$\begin{aligned} E(Y^2) &= \sum_x \sum_y [y - \mu(x) + \mu(x)]^2 P(Y = y|X = x)P(X = x), \\ &= \sum_x \sum_y [(y - \mu(x))^2 + \mu(x)^2]\,P(Y = y|X = x)P(X = x), \\ &= E_X\mathrm{Var}(Y|X) + E_X[E(Y^2|X)]. \end{aligned}$$

(b) From part (a), show that

$$\mathrm{Var}(Y) = E_X\,\mathrm{Var}(Y|X) + \mathrm{Var}_X\,E(Y|X).$$

Remark: The formula in part (b) of the preceding problem is known as the **conditional variance formula**. This formula also holds for continuous random variables, with a similar proof.

34. **(The *D/M/*1 queue)** Figure 14.1 simulates the queue in which a customer arrives every 15 minutes, there is one server, and service times are exponential with a mean of 13 minutes. This problem outlines an analysis of this queue. Let $X_1, X_2, \ldots$ be a sequence of i.i.d. random variables, each of which has the exponential distribution with mean $m = 13$. Set $S_n = X_1 + X_2 + \cdots + X_n$. This random variable S_n has the **Gamma** distribution with parameters n and m, and the Excel function **=GAMMADIST(t, n, m, 1)** reports $P(S_N \le t)$.

(a) Let the random variable N denote the number of service completions that can take place within 15 minutes. True or false:

$$\begin{aligned} P(N \ge n) &= P(S_n \le 15) \\ P(N = k) &= P(S_n \le 15) - P(S_{n+1} \le 15). \end{aligned}$$

(b) Open a spreadsheet and compute, for the values $k = 0, 1, \ldots, 18$,

$$q(k) = P(N = k).$$

(c) The number S of customers in the system the moment before a customer appears is the state of a discrete-time Markov chain. Why? This state S can take the values 0, 1, 2, ... Why? The transition matrix P for this Markov chain is given below, where each "*" is the number that causes the row sum to equal 1. Why?

$$P = \begin{bmatrix} * & q(0) & 0 & 0 & 0 & \cdots \\ * & q(1) & q(0) & 0 & 0 & \cdots \\ * & q(2) & q(1) & q(0) & 0 & \cdots \\ * & q(3) & q(2) & q(1) & q(0) & \cdots \\ * & \vdots & \vdots & \vdots & \vdots & \end{bmatrix}$$

(d) Let $\pi(0), \pi(1), \pi(2), \ldots$ be the steady-state probabilities for this Markov chain, so $\pi(k) = P(S = k)$ for $k = 0, 1, \ldots$. These steady-state probabilities sum to one. They must satisfy the equation that is displayed below. Why?

$$\pi(k) = \pi(k - 1)q(0) + \pi(k)q(1) + \pi(k + 1)q(2) + \cdots.$$

Remark: The preceding equation has so much symmetry that we can find a number z such that $0 < z < 1$ and

$$\pi(k) = (1 - z)z^k \quad \text{for } k = 0, 1, 2, \ldots.$$

(e) If a solution of the above form does exist, show that z must satisfy

$$z = q(0) + q(1)z + q(2)z^2 + q(3)z^2 + \cdots.$$

(f) Aiming for such a z, we introduce the function $f(z)$ that is defined by

$$f(z) = q(0) + q(1)z + q(2)z^2 + \cdots.$$

Which of the following statements are true?

- The function $f(z)$ of z is increasing on the interval $0 \leq z \leq 1$.
- The function $f(z)$ of z is convex on the interval $0 \leq z \leq 1$.
- $f(0)$ is positive.
- $f(1) = 1$.
- $f'(1) = E(N) > 1$.

(g) Use your answer to part (f) to plot the function $f(z)$ of z. Is there a value of z that satisfies $z < 1$ and $z = f(z)$. Why? Is this value unique? Why?

(h) On the spreadsheet in part (b), use Solver to minimize z subject to the constraint $f(z) \geq z$. (You should get a number $z^* = f(z^*)$.)

(i) Compute the expectation of S.

(j) Compute the expectation of the time that an arriving customer must wait for service to begin.

Chapter 15

Simulation

15.1. PREVIEW

To **simulate** something is to emulate its behavior, either in a laboratory or on a computer. There are many types of simulation; for example:

- A wind tunnel simulates the flow of air around an aircraft wing.
- An underwater environment simulates the weightlessness of space and is used to train astronauts to perform mechanical tasks in a zero-gravity environment.
- A cockpit simulator exposes pilots to dangerous situations and evaluates their responses.
- A man–machine simulation trains radar operators to distinguish flocks of birds from aircraft.
- A war game simulates the clash of enemy forces, aiming to evaluate equipment, tactics, logistics, and communication systems.
- A computer model simulates the dynamics of nuclear fission.

A simulation can entail a physical device, such as a wind tunnel. It can include human beings, either as an essential part of the system being modeled or for training purposes. It can even model competition.

Discrete Event Digital Simulation

This chapter does not survey the full range of simulations; rather, our attention is limited to discrete event digital simulation. **Digital simulation** means that the simulation occurs entirely within a digital computer, and **discrete event** means that the system under study can be viewed intermittently rather than continuously.

A discrete event digital simulation can model systems that evolve in continuous time, provided their laws of evolution are governed by discrete events. A queue illustrates this point. It evolves in continuous time, but we can describe it in terms of a sequence of customer arrivals, service initiations, and service completions, for each of these events occurs at a discrete point in time.

The Output of a Simulation

A simulation imitates the behavior of a system. It does not tell how the system should behave. Nor does it tell what settings of the system's parameters cause it to perform best.

In nearly every digital simulation, uncertainty plays a central role. When a simulation samples uncertain quantities, its output emulates the behavior of a system in a *particular instance.* As a consequence:

- If you wish to see how the system behaves on average (in expectation), you may need to run the simulation many times.
- If you wish to see how the system reacts to different settings of its parameters, you may need to simulate repeatedly and to experiment with a variety of parameter settings.

As a general rule, simulation is less powerful than analysis. A mathematical analysis measures performance rather than estimating it. An optimization model evaluates a range of parameters and selects the values whose performance is best.

Why Simulate?

Regardless of the advantages of analysis, discrete event digital simulation is a vital part of the science of decision making. It is, in fact, the most frequently used tool of operations research. There are three good reasons to simulate.

- The mathematical description of the system may be too complex to solve. Alternatively, mathematics may provide only part of the desired information.
- Experimenting with the actual system may be too expensive, too risky, or simply impossible.
- Digital simulation may be quick and easy.

The last of these reasons is a modern development, for until recent times, simulation was difficult and expensive. Building a simulation model once took person-years, and each run of the simulation model could require hours of time on an expensive mainframe computer.

Computer memories have grown by orders of magnitude, and clock-speeds are faster by comparable factors. User-friendly simulation software has greatly eased the problem of constructing a simulation. Today, simulation is often the tool that is properly tried *first.* A simulation can provide insight that would be hard to garner from an analytic model. It can do this quickly.

15.2. WHAT CAN YOU LEARN FROM THIS CHAPTER?

At the heart of digital simulation lies a method for sampling from a probability distribution. From this chapter, you can learn the principle that underlies this sampling.

Although a few simulations can be run with Excel itself, nearly every simulation is easier to build and run using an Excel add-in. A simulation add-in called RiskSim, written by

Michael Middleton, is included on the CD that accompanies this text. From this chapter, you can learn how to make good use of RiskSim.[1]

This chapter formulates two problems for simulation. The first of these problems is a queue, and from its simulation, you can obtain a surprising insight. It is that the expectation W of the time in the system can be a poor indicator of the average of the times in the system of a cohort of 100 customers who arrive one after another.

The second problem is a yield management problem that is a bit too complex to solve analytically. From it, you can learn how to use simulation to home in on the parameters of a policy whose expected contribution is largest.

Some simulations are easy to execute on spreadsheets. Others require commercial simulation software that is not spreadsheet-friendly. A section of this chapter helps you to discern which are which.

15.3. MONTE CARLO

In the present context, **Monte Carlo** refers not to a gambling casino on the French Riviera but to the sampling of a probability distribution. Nearly every discrete event digital simulation models uncertain quantities, and thus, nearly every simulation samples from probability distributions. This section shows how to sample uncertain quantities inside a computer.

Sampling a Uniform Random Variable

As we shall soon see, the key to simulating any probability distribution is to sample from the uniform distribution. A great many computer programs, including Excel, contain a feature that generates numbers that *appear* to be mutually independent and uniformly distributed on the interval between 0 and 1. Table 15.1 shows how to simulate these random variables on an Excel spreadsheet. In the table, the Excel function =RAND() was entered in cell B2 and was then dragged across to cell G2. Note that different numbers appear in cells B2 through G2, and that each of these numbers is between 0 and 1.

The Excel function **=RAND()** is actually a **pseudorandom number** generator. If you use it to generate a list of 1000 random variables, they will pass every known test for independence and for the uniform distribution, but they will not be truly random. Most pseudorandom number generators actually **cycle**; for some integer n, these random number generators produce the same number every 2^n iterations. The random number generator in Excel cycles with a period of not less than $2^{16} = 65{,}536$. The random number generator in Visual Basic cycles with a period of $2^{24} = 16{,}777{,}216$.

Table 15.1 Realizations of six independent random variables, each uniformly distributed between 0 and 1, each generated by the =RAND() function.

	A	B	C	D	E	F	G
1							
2		0.025944	0.667875	0.414939	0.323582	0.105693	0.008803
3		↑					
4		=RAND()					

[1]Other Excel add-ins that can help you to execute simulations on spreadsheets include @RISK and RISKOptimizer, which are products of Palisade Corporation, and Crystal Ball, which is a product of Decisioneering, Inc.

Simulating a Discrete Random Variable

Suppose that you want to simulate a discrete random variable X whose CDF (cumulative distribution function) you know. How can you do that? More precisely, how can you use the **=RAND()** function to do that? Figure 15.1 makes the answer to this question easy to grasp. It displays the probability distribution of a discrete random variable X, the cumulative distribution of this random variable, and, in heavy lines, a graph of this CDF.

Let Z be a random variable that is uniformly distributed on the interval between 0 and 1. The arrows in Figure 15.1 show how to use Z to simulate the discrete random variable X; these arrows transform the simulated value $Z = 0.563$ into $X = 4$. To execute the procedure suggested by these arrows, we:

- Set Z **=RAND()**.
- Interpret Z as a value on the ordinate (vertical axis).
- At the ordinate Z, move horizontally until the CDF is encountered.
- Then move vertically to the value X on the abscissa (horizontal axis).

Why does this work? Basically, since Z is uniformly distributed on the interval between 0 and 1,

$$P(Z \le t) = t \quad \text{for } 0 \le t \le 1. \tag{15.1}$$

From Figure 15.1 and Equation (15.1), we see that

$$\begin{aligned}
P(X = 1) &= P(0.00 < Z \le 0.15) = 0.15 - 0.00 = 0.15,\\
P(X = 3) &= P(0.15 < Z \le 0.45) = 0.45 - 0.15 = 0.30,\\
P(X = 4) &= P(0.45 < Z \le 0.65) = 0.65 - 0.45 = 0.20,\\
P(X = 7) &= P(0.65 < Z \le 1.00) = 1.00 - 0.65 = 0.35,
\end{aligned}$$

exactly as desired.

Mathematically, the dashed line in Figure 15.1 uses the uniform random variable Z to simulate X by *inverting* the CDF of X, that is, by reading it backwards.

From a computational viewpoint, simulating this random variable X on a spreadsheet would be awkward, but we'll soon describe an Excel add-in that does away with the drudgery.

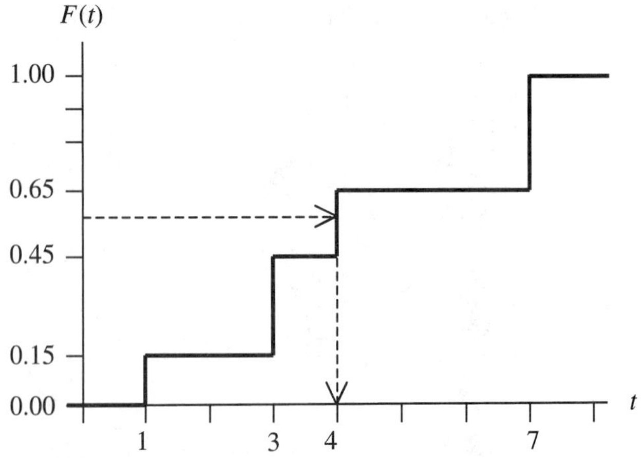

Figure 15.1 Simulating a discrete random variable X.

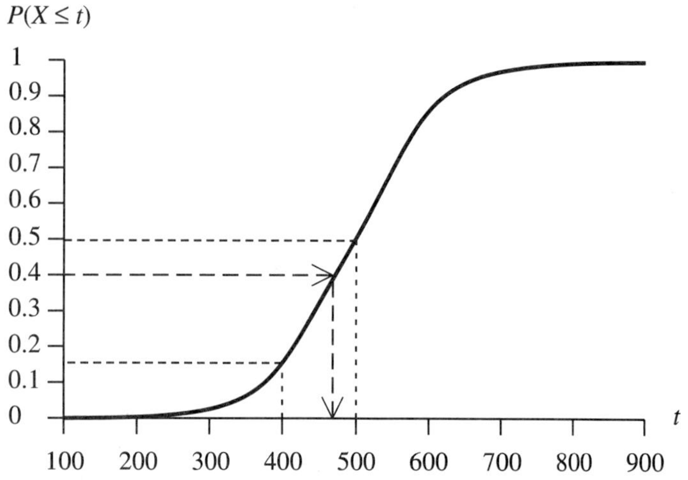

Figure 15.2 Simulation of a random variable X whose distribution is normal with mean of 500 and standard deviation of 100.

Simulating a Continuous Random Variable

The same principle can be used to simulate a continuous random variable. To illustrate, let X have the normal distribution with mean of 500 and standard deviation of 100. Figure 15.2 plots the CDF of X. To simulate X, we can begin with a random variable Z that is uniformly distributed on the interval between 0 and 1 and follow the procedure that is indicated by the arrows in Figure 15.2, which transform the value $Z = 0.398$ into the value $X = 474$. Again, we:

- Set Z =**RAND()**.
- Interpret Z as a value on the ordinate (vertical axis).
- At the ordinate Z, move horizontally until the CDF is encountered.
- Then move vertically to the value X on the abscissa (horizontal axis).

Figure 15.2 also indicates why this works; it illustrates the computation of $P(a < X \leq b)$ for the values $a = 400$ and $b = 500$. For these or any other numbers $a < b$, the simulation in Figure 15.2 couples with Equation (15.1) to give

$$\begin{aligned} P(a < X \leq b) &= P[F(a) < Z \leq F(b)], \\ &= P[Z \leq F(b)] - P(Z \leq F(a)], \\ &= F(b) - F(a), \end{aligned}$$

exactly as desired. As Figure 15.2 indicates, X is being simulated by inverting its CDF.

Happily, the inverse of the CDF of a normal random variable is an Excel function. For this reason, Figure 15.2 shows us how to simulate a normal random variable on a spreadsheet. Let the random variable X have the normal distribution with mean μ and standard deviation σ. The Excel function **=NORMINV(RAND(), μ, σ)** simulates X.

15.4. SIMULATING RANDOM VARIABLES WITH RISKSIM

But how can we simulate other garden-variety random variables, such as the Poisson? The inverse of its CDF is not an Excel function; for it, the procedure outlined in Figure 15.1 is not that easy to execute on a spreadsheet.

Table 15.2 RiskSim functions for seven random variables.

Distribution of the Random Variable X	Excel Function
Normal with mean μ and standard deviation σ	**=randnormal(μ, σ)**
Uniform on the interval between a and b	**=randuniform(a, b)**
Triangular between a and b with m being the most likely value	**=randtriangular(a, m, b)**
Exponential with parameter (rate) λ	**=randexponential(λ)**
Poisson with parameter (mean) λ	**=randpoisson(λ)**
Binomial with parameters n and p	**=randbinomial(n, p)**
Equally likely to take any integer value between a and b, with $a < b$	**=randinteger(a, b)**

An Excel add-in called RiskSim lets you simulate a variety of random variables. Before you can use RiskSim, however, it must be installed and activated. The Appendix tells how to accomplish these steps. Installing RiskSim and activating it causes several "user-defined" functions to appear on your Excel "Insert" menu. These functions include the seven that are listed in Table 15.2.

For instance, to simulate a random variable whose distribution is Poisson with parameter (expectation) of 6.2, click on a cell and then type the function

=randpoisson(6.2)

into that cell. To simulate six independent random variables each having this distribution, select the same cell and then drag it across five other cells.

Simulating a Specified Discrete Random Variable

A different RiskSim function lets you simulate a discrete random variable whose probability distribution you specify, and the spreadsheet in Table 15.3 indicates how to do that.

In Table 15.3, the function **=randdiscrete(A$3:B$6)** has been entered in cell F3 and then dragged down to cell F8. This creates six independent random variables, each having the probability distribution specified by cells A3 through B6. As Table 15.3 connotes, the randdiscrete function has two *adjacent columns* as its arguments, where:

- The left column specifies the values that the random variable can take.
- Each entry in the right column specifies the probability that the random variable takes the value to its left.

Table 15.3 The randdiscrete function.

	A	B	C	D	E	F
2	t	P(X = t)				
3	-1.4	0.23				-1.4
4	2	0.14				3
5	3	0.36				6
6	6	0.27				2
7						6
8		=randdiscrete(A$3:B$6) →				3

Table 15.4 The randcumulative function.

	A	B	C	D	E	F
2	t	P(X ≤ t)				
3	-4	0				
4	-2	0.08				
5	0	0.24				
6	3	0.56				1.071
7	6	0.8				2.497
8	9	0.94				4.950
9	11	1				-3.967
10						-0.194
11		=randcumulative(A$3:B$9) →				2.837

The probabilities in the right column must sum to 1. If they do not, the randdiscrete function will give an error message.

Simulating a Specified Continuous Random Variable

Another RiskSim function simulates a continuous random variable whose CDF you specify, approximately. Table 15.4 suggests how.

In Table 15.4, the function **=randcumulative(A$3:B$9)** has been entered in cell F6 and dragged down to cell F11. This creates six independent random variables, each having a CDF specified by the values in cells A3 through B9 and by "straight-line interpolation." For instance, on the interval between 9 and 11, the CDF increases linearly from 0.94 to 1.00, with slope (density) equal to $0.06/2$.

As the table suggests, the randcumulative function has two adjacent columns as its arguments, where

- The left column contains values of the random variable, in ascending order.
- Each entry in the right column equals the probability that the random variable does not exceed the value to its left.
- The top probability must equal zero, and the bottom probability must equal 1.

A Speed-up

Each RiskSim function checks for valid parameters. If you find that your simulation runs slowly (and many do), you can speed it up a bit by suppressing the checks. To do so, replace the letters "rand" with "fast." For instance, the function **=fastexponential(10)** simulates a random variable whose distribution is exponential with rate of 10, but it does not check that 10 is a positive number.

An Unpleasant Feature

Whenever you change any entry in any cell of a spreadsheet, Excel recalculates each function whose value can change. Normally, that is what you want it to do, but whenever you change any entry in any cell of your spreadsheet, a new number will appear in each cell in which the function **=RAND()** has been entered.

That can be unpleasant because new values of the random variables can spawn changes throughout the spreadsheet, as well as on any charts that summarize it. Re-computation can be time-consuming, inconvenient, and aggravating. A splitting headache can come on fast.

Ideally, Excel should include a "toggle" that allows you to turn random number regeneration on and off. At this writing, it doesn't. As an expedient, you can "freeze" values of your random variables by using the Paste Special feature, notably

- Select the cells in which the function **=RAND()** appears explicitly or implicitly, for example, as an argument in a RiskSim function.
- Copy them.
- Then use the Paste Special option on the Edit menu to paste only their values, and not the functions themselves.

This eliminates the **=RAND()** functions but preserves the values that these functions have most recently computed.

Freezing values is perilous, however. It's likely that, at a later time, you will want new values of the random variables. To get them, you must re-insert the appropriate function in each cell where you used Paste Special to freeze values.

15.5. A QUEUE

Figure 14.1 of Chapter 14 graphs the simulation of a queue. In that chapter, we did not indicate how to use Excel to produce that chart. We do so now, and in preparation, we first restate the problem that had been simulated.

Problem A (A Clinic)

Patients arrive at a clinic every 15 minutes, with no uncertainty in their inter-arrival times. Specifically, a patient appears at 9:00 o'clock, the next at 9:15, the one after that at 9:30, and so forth. Patients wait their turn to be treated by a single medical doctor (MD). The time that the MD spends treating each patient is exponentially distributed, with a mean of 13 minutes. The times that the MD spends with different patients are mutually independent. Use a simulation to describe the lengths of time that patients spend at the clinic, that is, from the moment of arrival to the moment at which treatment is complete.

Before reading further, you are invited to pause to design your own simulation of this clinic and to compare it with what appears below.

Design of a Simulation

Each patient experiences a time in the queue (possibly zero), followed by a time being treated. Let us number the patients 1, 2, 3, . . . , and let us designate:

$Q(n)$ = the time that the nth patient waits for treatment to begin, measured in minutes.

$S(n)$ = the time that the nth patient spends being treated, measured in minutes.

$T(n)$ = the time that the nth patient spends in the clinic, measured in minutes.

How can we simulate these quantities? The service times $S(1)$, $S(2)$, . . . are independent and identically distributed random variables, each having the exponential distribution with rate $1/13$. The function **=randexponential(1/13)** simulates them. The time that a patient spends at the clinic equals the sum of the time in the queue and the time being treated. In other words,

$$T(n) = Q(n) + S(n) \quad \text{for } n = 1, 2, \ldots. \tag{15.2}$$

It remains to describe the time $Q(n)$ that the nth patient spends in the queue. We will verify that

$$Q(n) = \max\begin{cases} 0 \\ T(n-1)-15 \end{cases} \quad \text{for } n = 2, 3, \ldots. \tag{15.3}$$

To show why Equation (15.3) holds, we note that:

- The nth patient appears exactly 15 minutes after the $(n-1)$st patient does, and the $(n-1)$st patient spends a total of $T(n-1)$ minutes in the clinic.
- If $T(n-1)$ does not exceed 15 minutes, the nth patient is served immediately upon arrival.
- If $T(n-1)$ does exceed 15 minutes, the nth patient spends $T(n-1) - 15$ minutes in the queue.

Equations (15.2) and (15.3) illustrate a point that holds in general. A simulation program—like any other computer program—benefits from careful thought. It can be simple and transparent, or it can be complex and opaque.

An Excel Simulation

Table 15.5 simulates the times $T(1)$ through $T(5)$ that the first five patients spend in the clinic. In this table, the functions in column B simulate five independent random variables, each having the exponential distribution with rate $1/13$, and equivalently, with mean of 13 minutes. We assume that the MD is idle at 9:00, when the first patient appears, for which reason cell C4 contains the number 0. The remaining entries in column C reflect Equation (15.3), and the entries in column D reflect Equation (15.2).

Table 15.5 Simulation of patients 1 through 5 in Problem A.

	A	B	C	D	E
1					
2	μ =	0.076923 ←			=1/13
3	n	S(n)	Q(n)	T(n)	
4	1	25.796	0.000	25.796	
5	2	15.767	10.796	26.563	
6	3	4.780	11.563	16.342	
7	4	9.700	1.342	11.043	
8	5	0.538	0.000	0.538	
9		↑	↑	↑	
10	=randexponential(B$2)			=$B8+$C8	
11					
12			=MAX(0, $D7-15)		

For the simulated treatment times in Table 15.5, patients 2, 3, and 4 were required to wait. Patients 1 and 5 saw the doctor immediately upon arrival.

Simulated Average Time in the System

As in Chapter 14, the symbol W denotes the expectation of the time that a customer spends in the system. For this $D/M/1$ queue, it is possible to compute W analytically; Problem 34 in Chapter 14 indicates how. For the data in Problem A, this computation gives $W = 51.13$ minutes.

Table 15.5 reports the simulated times $T(1)$ through $T(5)$ that the first five customers spend in the system. From the simulated times $T(1)$ through $T(n)$ of the first n customers, we can compute the average $\overline{W}(n)$ of the times they spend in the system.

$$\overline{W}(n) = [T(1) + T(2) + \cdots T(n)]/n$$

This quantity $\overline{W}(n)$ is a **statistic**, namely, a realization of a random variable. For large enough values of n, the value taken by $\overline{W}(n)$ is very likely to be close to 51.13 minutes, the expectation of W.

How large? During the course of an eight-hour shift, exactly 32 patients appear (because $32 = 4 \times 8$). Is the average $\overline{W}(32)$ of their times in the system likely to be close to W?

An Experiment

To gain insight into this issue, we used Excel to simulate $\overline{W}(n)$ for various values of n and observed how quickly it settles down. Figure 15.3 reports the results of this experiment. The figure indicates that the average time in the system of the first n customers settles down *very slowly* as n increases. It's clear from the figure that the first significant digit of W is a 5, but the second significant digit could be a 1 or a 2, and that's with $n = 40{,}000$.

Why is it that $\overline{W}(n)$ approaches W so slowly? If the times $T(n)$ and $T(n + 1)$ that successive patients spend in the clinic were independent of each other, then $\overline{W}(100)$ would be very close to W.

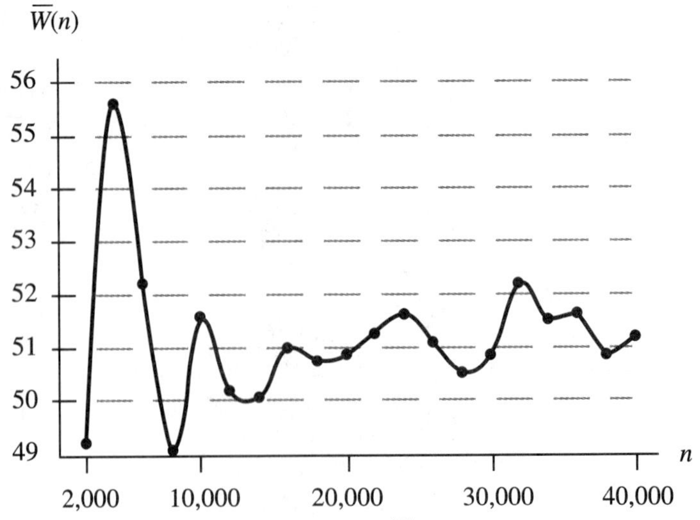

Figure 15.3 Average time in the system $\overline{W}(n)$ versus n.

$\overline{W}(n)$ is converging slowly because $T(n)$ and $T(n+1)$ are very strongly correlated. One long service time can delay many patients.

A "Typical" Day at the Clinic

Figure 15.3 suggests that the operation of the clinic during a single day may be quite volatile. This leads us to pose the question in:

Problem A (A Clinic, continued)

Suppose that the clinic accepts patients during one eight-hour shift each day. It schedules their arrivals at 15-minute intervals, the first arrival occurring at 9:00 A.M., the second at 9:15 A.M., and the last at 4:45 P.M. The MD and her staff begin work at 9:00 A.M. and remain until all patients have been served. What can be said of the following random variables?

- The average of the times in the system of one day's patients.
- The average of the times in the queue of one day's patients.
- The time at which the day's work is done.

These questions would be hard to answer by mathematical analysis, but they can be answered by simulation. We'll show how to answer the first of them; the other two are left for you as homework problems. The first question seeks information about the random variable

$$T_{\text{avg}} = \text{the average of the times in the system of the 32 customers who will arrive during a single day.}$$

We emphasize that T_{avg} is a random variable; it has a probability distribution. This random variable summarizes the experiences of the cohort of patients who are served during a single day.

Estimating a Probability Distribution

By contrast, $\overline{W}(32)$ is the average of the times in the system of the 32 patients who actually arrived during a particular day; $\overline{W}(32)$ is a realization of the random variable T_{avg}. By simulating the operation of the clinic for one day, we can obtain a value for $\overline{W}(32)$. Thus, by simulating the clinic repeatedly and recording the value of $\overline{W}(32)$ obtained in each simulation, we can estimate the probability distribution of T_{avg}.

The spreadsheet in Table 15.5 simulates the times in the system of the first five patients. To simulate 32 patients, select row 8 and drag it down to row 35. (Row 35 reports the service time, queueing time, and time-in-system of the 32nd patient.) Then, in cell D36, enter the function

=SUM(D4: D35)/A35

whose value is the average $\overline{W}(32)$ of the times in the system of these 32 patients.

Simulating with RiskSim's "One Output" window

To estimate the probability distribution of T_{avg}, we wish to simulate the clinic repeatedly and record the value of $\overline{W}(32)$ for each simulation. A RiskSim window is ideal for this purpose. On the Excel Tools menu, scroll down to the Risk Simulation entry, shift right to the One Output window, and click on it. The window in Table 15.6 will appear but with blanks in its Output Formula cell and Number of Trials boxes.

Table 15.6 RiskSim's One Output window for Problem A.

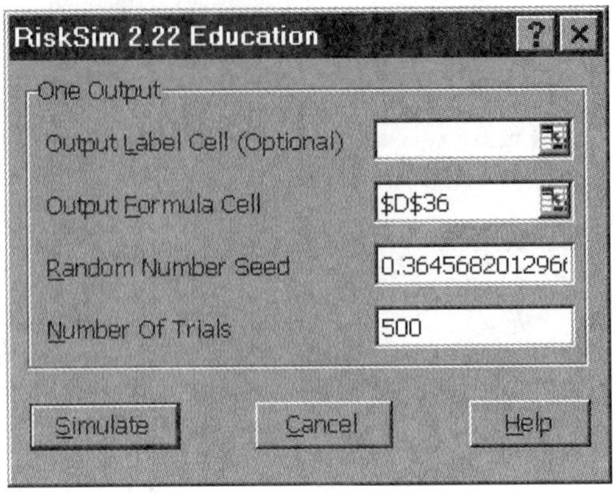

Cell D36 contains the statistic $\overline{W}(32)$ whose distribution we wish to study, so D36 has been entered in the Output Formula cell. To estimate the probability distribution of T_{avg}, we simulate for 500 replications by entering 500 in the Number of Trials cell. Without messing with the Random Number Seed cell, we then click on the Simulate button.

Random Number Seed

In the language of simulation, the **seed** is the starting point of a sequence of pseudorandom numbers. If a pseudorandom number generator is re-initialized with the same seed, it will produce the same sequence of random numbers. As Table 15.6 suggests, RiskSim lets you control the seed of its random number generator. That feature will soon prove useful.

Simulation Output

The window in Table 15.6 instructs RiskSim to accumulate 500 values of the quantity in the Output Formula cell. After it does so, RiskSim computes the empirical CDF of these quantities and estimates the mean and standard deviation of the random variable that is being simulated. In also computes the empirical fractiles and a wealth of other information, some of which is reported in Figure 15.4, including the following:

- The best available "point" estimate of $E(T_{avg})$ is 31.27 minutes.
- The probability that $E(T_{avg})$ lies within 0.88 of 31.27 is roughly 0.68, from a normal approximation.
- The best available "point" estimate of StDev(T_{avg}) is 19.60 minutes.
- The lowest observed value of T_{avg} was 7.28 minutes; the highest was 183.74 minutes.
- The density of T_{avg} is spread out (fat) rather then being concentrated about its mean and is heavily skewed to the left.

As just noted, T_{avg} has a fat density, so there is considerable day-to-day variation in the service times experienced by the customers. And every now and then, the clinic experiences a horrendously busy day.

Figure 15.4 reports a mean time in the system of 31.27 minutes, which is well below the steady-state value of 51.13 minutes. This occurs because the system starts each day with no patients in the queue and because it does not reach steady state quickly.

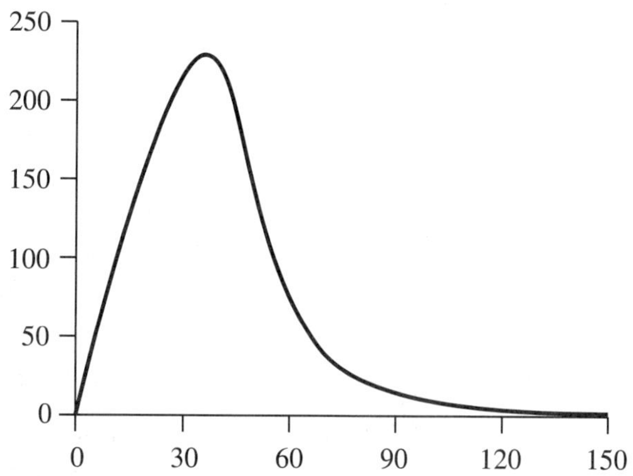

Q	R
Mean	31.27
St. Dev.	19.60
Mean St. Error	0.88
Minimum	7.28
First Quartile	18.05
Median	24.80
Third Quartile	38.10
Maximum	183.74
Skewness	2.22

Figure 15.4 Estimating the distribution of T_{avg}, the average of the times in the system of one day's patients, measured in minutes.

To answer the other two questions posed by Problem A, you could use the One Output window with different cells as the Output Formula cell. To use the same random variables for each simulation, begin them with exactly the same Random Number Seed in the One Output window.

15.6. INSIGHTS INTO QUEUES

Figure 15.3 exhibits a property that holds for nearly every queue and has important implications, not just for queues but for models that include queues. In Problem A and in general, $\overline{W}(n)$ converges slowly to W.

Queues lie at the center of a wide range of models. In the simulation of a production system, queues of items wait for machines to become available. In a simulation of an airport, planes queue up as they wait for runways, gates, and service crews, and so forth. These simulations inherit the slow-convergence feature of queues.

Slow Convergence

To investigate the slowness of convergence, we turn our attention from $\overline{W}(n)$ to a related performance measure, which is:

$$\overline{Q}(t) = \text{the time-average number of customers in queue during the first } t \text{ units of time.}$$

As t becomes large, $\overline{Q}(t)$ approaches the mean queue length, L_q. Figure 15.3 suggests that $\overline{Q}(t)$ converges slowly to L_q. We will see how slowly this occurs.

The **absolute error** in $\overline{Q}(t)$ equals $|\overline{Q}(t) - L_q|$, the absolute value of the difference between $\overline{Q}(t)$ and its long-run value, L_q. The **relative error** in $\overline{Q}(t)$ equals $|\overline{Q}(t) - L_q|/L_q$, namely, the absolute error as a function of the quantity L_q that is being estimated.

Let us assume that customers arrive at the rate of one per unit time, on average, so that t is a surrogate for the number of customers who have arrived. Consider an $M/M/1$ queue whose utilization rate ρ equals 0.8. For this queue, we learned in Chapter 14 that $L_q = \rho^2/(1 - \rho) = (0.64)/(0.2) = 3.2$. Ask yourself, "How large must t be to have only one chance in ten that the relative error in $\overline{Q}(t)$ exceeds 5% of its long-term value?" Before reading further, pause to guess a value of t.

With $\beta = 0.1$ and with $\varepsilon = 0.05$, the question just posed is to find the value of t for which

$$P\left\{\frac{|\overline{Q}(t) - L_q|}{L_q} > \varepsilon\right\} = \beta.$$

In a lovely paper, Ward Whitt[2] approximates t for a wide variety of queues. When equation (19) of his paper is applied to the question posed above, t is found as follows. First, with $\beta = 0.1$, compute the $\beta/2$ fractile of the standard normal distribution, getting

$$z = \textbf{NORMINV}(\boldsymbol{\beta}/2, 0, 1) = \textbf{NORMINV}(0.05, 0, 1)\ -1.96.$$

Then, with $z = -1.96$, $\rho = 0.8$ and $\varepsilon = 0.1$, approximate t by the formula

$$t = \left[\frac{4z}{\rho(1-\rho)\varepsilon}\right]^2 = \left[\frac{(4)(-1.96)}{(0.8)(0.2)(0.1)}\right]^2 = \left[\frac{-7.84}{0.016}\right]^2 \approx 240{,}000.$$

It takes a simulation of approximately 240 *thousand arrivals* to achieve a relative error of 5% with probability of 0.9. More than you had guessed?

Whitt's formula shows how t varies with allowable errors and the queue's utilization. For the cognoscenti, we mention that, in more general queues, the "4" in the numerator of his formula is replaced by $(1 + C_A^2)(1 + C_S^2)$, where C_A equals the ratio of the standard deviation of the inter-arrival time to its mean, and C_S is the ratio of the standard deviation of the service time to its mean.

In a nutshell, large values of t occur because extraordinarily long lines form, infrequently. For users of simulation, this is somber news. Models are often used to evaluate alternative designs and to determine which design performs best. An improvement of only a few percent can be an enormous benefit. A cost reduction of 3% can double net profit, for instance. If such a model has queues at its core and if the difference between alternative designs is only a few percent, very long simulation runs can be necessary to detect the difference with any confidence.

A Cohort of Customers

What can be said about the *average* of the queue lengths experienced by a cohort of, say, 100 customers who arrive in sequence? If the queue lengths they experienced were independent, the standard deviation of the average would equal StDev(N_q)/10 because $10 = \sqrt{100}$. Figure 15.3 suggests, correctly, that it is barely lower than StDev(N_q). We emphasize:

> **The experience of a cohort:** The average of the queue lengths experienced by a cohort of, say, 100 customers who arrive in sequence has a mean of $L_q = E(N_q)$ and, to a first approximation, has a standard deviation of StDev(N_q) rather than StDev(N_q)/10.

This too is because long lines are infrequent but persistent, providing added incentive to compute the standard deviation of N_q, as well as its mean.

---

[2]Ward Whitt, "Planning Queueing Simulations," *Management Science*, 35 (1989): 1341–1366.

15.7. A YIELD MANAGEMENT PROBLEM

In the current deregulated era, airlines are free to set the fares that they charge for seats. The process by which they do so is known as **yield management**. In Chapter 12, we studied yield management problems that were simple enough to solve analytically. Here we investigate a yield management problem that requires simulation.

Airlines accomplish yield management by having many different fare classes on each flight and by adjusting the "seat availability" of these fare classes in response to demand. Suppose, for instance, that the reservations for a particular flight are being made more slowly than the airline had anticipated. Aiming to compete for customers, the airline may cause previously unavailable low-fare-class seats to become available. As a consequence, a person who buys the "cheapest available seat" the day after you do may pay *less* for it.

A Two-Stage Model

An attempt to simulate the evolution of seat availability over time would lead us to an elaborate model and to a complicated simulation. To simplify the discussion, we assume in Problem B, that the uncertain demand reveals itself in two stages rather than continuously. This model has only two fare classes, and the demand for the lower-priced seats materializes first.

Problem B (Yield Management)

A particular flight has 120 seats, all of which are identical. It sells these seats at two prices—full fare and discount. Full fare seats are aimed at business travelers, who need flexible schedules, whereas discount fare seats are aimed at tourists, who can fix their travel plans well in advance of their travel dates. Customers who pay the full fare price make their reservations within seven days of takeoff, and they are not penalized if they change their reservations. Customers who pay the discount price must make their reservations more than seven days before takeoff, and they are penalized if they change their reservations.

Table 15.7 presents data on each fare class. In the table, "normal" means that the actual demand is normal with the indicated parameters rounded off to the nearest nonnegative integer. The airline wishes to maximize the expectation of the net profit it earns from this flight. Here, "net profit" equals the contribution earned from passengers who are *seated* on the flight, plus any rebooking fees paid by people who reserved seats on this flight but do not claim them, less the overbooking cost that the airline incurs for those passengers it cannot seat. What policy maximizes expected net profit?

Table 15.7 indicates that discount fare customers incur a re-booking fee of $75 if they reserve a seat and fail to claim it. By contrast, full fare customers incur no fees if they book seats but fail to show up. The cost to the airline of "bumping" a customer to another flight

Table 15.7 Data by fare class.

Attribute	Full Fare	Discount Fare
Demand D is "normal" with $E(D) = \mu$ and StDev$(D) = \sigma$	$\mu = 50$ $\sigma = 20$	$\mu = 100$ $\sigma = 30$
Contribution per passenger	$400	$150
P(no show)	0.15	0.05
Fee for re-booking	$0	$75
Cost of overbooking	$125	$125

is \$125. If the airline cannot seat everyone, it elects to "bump" discount fare customers first because their contributions are lower.

Decision Variables

The airline's yield management problem has two decision variables, defined as follows.

q = the number of reservations to accept for discount fare seats.

t = the total number of reservations to accept.

The airline's goal is to select values of q and t that maximize the expectation of the net profit it earns from this flight. When selecting q and t, the airline faces the usual tradeoffs. Increasing q risks loss of revenue from full fare purchases; increasing t risks loss of income due to overbooking fees; and decreasing q or t risks loss of revenue from empty seats. The spreadsheet in Table 15.8 simulates a single flight, with specific values of the parameters q and t.

Columns B and C contain data and functions, nearly all of which are described in columns E through G. Evidently, in this simulation:

- The flight's quota q of discount fare reservations equals 110.
- The flight's quota t of discount and full fare reservations equals 170.
- For this flight, 121 customers request discount fare seats, and 110 of these customers receive discount fare reservations because $110 = \min\{121, 110\}$.
- For this flight, 82 customers request full fare seats, and 60 of them receive full fare reservations because $60 = \min\{82, 170 - 110\}$.

Table 15.8 Simulation of one flight.

	A	B	C	D	E	F	G
2	number of seats =	120		overbooking penalty =			125
3		FF	DF				
4	mu =	50	100				
5	sigma =	20	30				
6	contribution/pass. =	400	150				
7	P(no show) =	0.15	0.05				
8	penalty/no show =	0	75				
9	demand (normal) =	82.18	121.39	←	= randnormal(C4, C5)		
10	demand (rounded) =	82	121	←	= MAX(0, INT(C9 + 0.5))		
11	q =		110		= MIN(B10, B12 - C13)		
12	t =	170					
13	# reservations =	60	110	←	= MIN(C10, C11)		
14	# no shows =	13	8	←	= randbinomial(C13, C7)		
15	# shows =	47	102	←	= C13 - C14		
16	# overbooked =		29	←	= MAX(0, B15 + C15 - B2)		
17	# passengers =	47	73	←	= MAX(0, C15 - C16)		
18	contribution =	18800	10950	←	= C17 * C6		
19	no show fees =		600	←	= C14 * C8		
20	overbooking cost =		3625	←	= G2 * C16		
21	net profit =		26725	←	= B18 + C18 + C19 - C20		

- There are 13 full fare no shows, so 47 = 60 − 13 customers appear to request full fare seats.
- There are 8 discount fare no shows, so 102 = 110 − 8 customers appear to claim discount fare seats.
- The plane is overbooked by 29 seats because 29 = 47 + 102 − 120, so 73 = 102 − 29 discount fare customers are seated.
- Net profit equals $26,725 because

$$26{,}725 = 400 \times 47 + 150 \times 73 + 75 \times 8 - 125 \times 29.$$

A Debugging Tip

A simulation can be notoriously difficult to debug because a simulation must account correctly for each of a host of contingencies. A well-designed spreadsheet helps. It lays bare the details of the calculation, which can help you check that each contingency is correctly accounted for.

The data in Table 15.8 have allowed us to verify that the computation is correct for certain of the contingencies. To check the others, we could—and should—insert different values of the random variables in cells B9 and D9, and verify that the computations are correct. It is important to perform this type of check.

> To debug your simulation, use the spreadsheet to check that it makes correct calculations under *each* contingency.

By the way, not all of the functions in column B are found by dragging from column C. One of them has been left for you to discover.

The Distribution of Net Profit

The flight simulated in Table 15.8 has a net profit of $26,725. The net profits earned from different simulations of this flight are independent random variables. To estimate the distribution of net profit, we can run this simulation repeatedly. To simulate 1000 flights, scroll down the Excel Tools menu to the Risk Simulation entry, shift right, click on the One Output entry, fill in the window as indicated in Table 15.9, and then click on the Simulate button.

Table 15.9 Arranging to simulate 1000 flights, each using the data in Table 15.8.

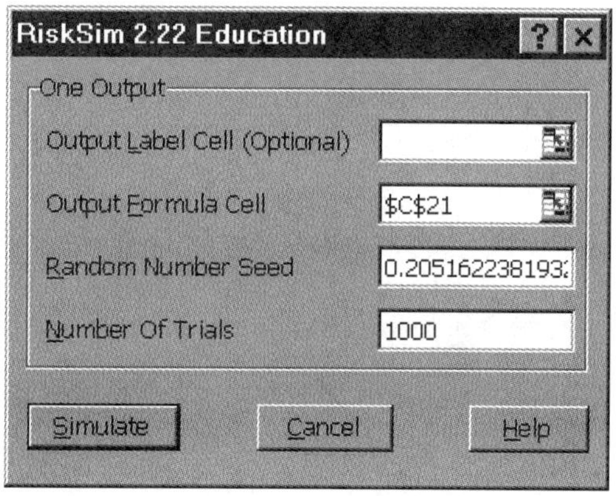

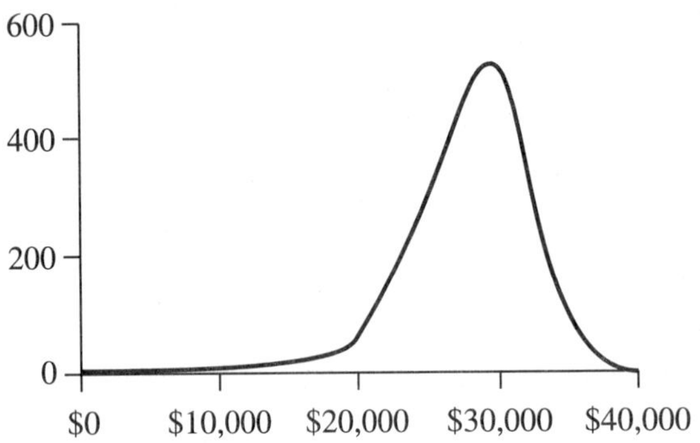

Q	R
Mean	$25,531
St. Dev.	$3,945
Mean St. Error	124,77
Minimum	$4,650
First Quartile	$24,000
Median	$26,013
Third Quartile	$27,700
Maximum	$37,700
Skewness	−1,0409

Figure 15.5 Estimated profit per flight, with $q = 110$ and $t = 170$.

Simulation Output

Figure 15.5 summarizes information obtained from this simulation. As Table 15.9 indicates, a good estimate of the mean net profit is $25,531, and a good estimate of its standard deviation is $3945. The most likely value of net profit is about $30,000, and its distribution is skewed to the right.

If we ran this simulation again with a different random number seed, we would get a different estimate of mean net profit. Figure 15.5 indicates that mean net profit lies within $125 of $25,531 with probability of 0.68, from a normal approximation.

A Variability Reduction Trick

The airline aims to select the parameters q and t that maximize the expectation of its net profit. To do that, we can run the simulation repeatedly, each time with different values of q and t. If we do so, there are two sources of variation: that due to random fluctuations in the demand for airline seats and that due to the values of q and t that have been chosen.

To suppress the variation due to random fluctuations, we use the *same* sequence of random variables for each simulation. A RiskSim window lets us do this. To open it, scroll down the Excel Tools menu to the Risk Simulation entry, shift right, and click on the Two Non-Random Inputs box. The window in Table 15.10 will appear, with all but two of its cells blank.

As Table 15.10 indicates, we elect to estimate net profit for 16 cases, namely, for each pair (q, t) with $q = 60$, 80, 100, and 120 and with $t = 160$, 180, 200, and 220. As soon as we click on the Simulate button in this dialog box, RiskSim simulates 1000 flights for each pair of parameters. It uses the same sequence of random variables for each simulation because the Same Seed box has been checked. For each of these 16 runs, RiskSim computes the empirical CDF, plots a two-dimensional histogram, and provides various summary statistics, including those in Table 15.11.

Evidently, setting a quota q between 80 and 100 is a good idea, and there is no economic merit in limiting t. The reason for the latter is that when the plane is overbooked, full fare customers displace discount fare customers, with a net profit of $125 apiece (because $125 = 400 - 150 - 125$), unless all of the discount fare customers have been bumped.

Let us reemphasize the importance of using the same random number seed for each simulation. That way, Table 15.11 estimates the change that is due to the values of the

Table 15.10 The "Two Non-Random Inputs" dialog box in RiskSim.

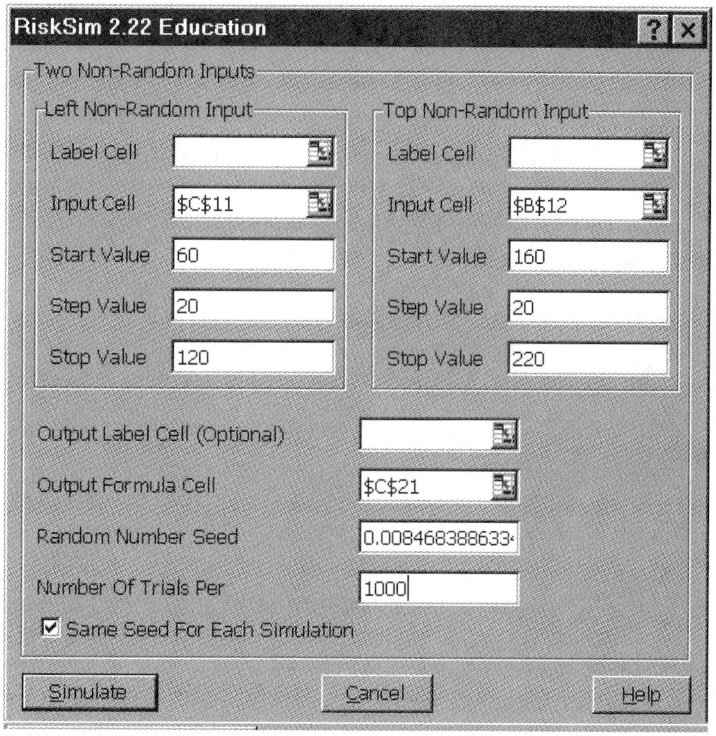

Table 15.11 Estimate of expected profit per flight for each of 16 pairs of quotas.

F	G	H	I	J	K
		Quota t on total reservations			
		160	180	200	220
Quota q on discount fare reservations	60	$25,438	$25,444	$25,444	$25,444
	80	$26,636	$26,697	$26,702	$26,702
	100	$26,148	$26,390	$26,436	$26,439
	120	$25,117	$25,592	$25,736	$25,766

parameters q and t, suppressing the variability (and it is large) that is due to the inherent uncertainty in the demand for seats.

15.8. STATES AND REGENERATION POINTS

Throughout this text, the "state" of the system is a summary of what has transpired to date that contains enough information to specify the law that governs the evolution of the system. A simulation emulates a particular realization of a system. Hence, the **state** of a simulation includes the realizations (values) of enough of the random variables that the system's law of evolution can be specified.

Roughly speaking, a **regeneration point** of a simulation is a moment in time at which the state simplifies. Problem A illustrates both definitions. Normally, the state of its simulation keeps track of two pieces of information:

- The time that remains until the next customer appears.
- The total of the remaining service times of the customers who are present.

Problem A is a $D/M/1$ queue, so the moments when customers appear are regeneration points; at these times, the state need record only one piece of information, which is the total of the remaining service times of the customers who are present.

Problem B also illustrates these definitions. For it, the regeneration point is the moment in time at which discount fare customers can no longer be accepted. At that moment, its state consists of a single piece of information, which is the number of discount fare class reservations that have been accepted.

Both simulations have been organized so that their states have recorded only one piece of information. That makes these simulations easy to execute on a spreadsheet.

A Measure of Complexity

A key measure of the complexity of a simulation is the number of pieces of information that its states must record. Problems A and B were particularly simple; their states consisted of a single piece of information.

If the state records two or three items, use of IF statements can make the simulation reasonably easy to simulate on a spreadsheet. In many simulations, the state can record dozens of pieces of information, or hundreds. For them, simulation on a spreadsheet is impossible; it's time for a commercial simulation software package.[3]

Simulating Telecommunications Traffic

Let us briefly mention how to organize a simulation that is much too complex to execute on a spreadsheet. Imagine that we simulate the traffic experienced by a telephone network that links a group of central offices (nodes). Requests for service (call attempts) occur between each node pair. Each call attempt must be rejected or accepted. If it is accepted, it must be assigned to a route between its node pair. In a simulation of this system, a state includes the following information:

- For each pair of central offices, the time at which the next call attempt will occur.
- For each in-progress call, the path in the telecommunications network to which it has been assigned and the time at which it will be completed.

Evidently, the state of this simulation records a great many pieces of information.

In a simulation, the term **event epoch** describes a moment in time at which the state changes, often by the realization of one or more random variables. In the telecommunications simulation, each event epoch is either a moment at which a new request for service occurs or a moment at which an in-process call is completed. Each event epoch changes the state.

The times at which event epochs will occur are kept on an **event list**, and **simulated time** is advanced to the earliest time on the event list. The event that occurs at this time changes the state, which can trigger the occurrence of other events, whose times must be appended to the event list. In Problems A and B, we did without the event "lists" because they consisted of a single item.

---

[3]The journal *ORMS Today* periodically surveys simulation software packages. Among the more popular commercial packages are ARENA, GPSS, ProModel, SIMSCRIPT, SIMFACTORY, SIMAN, and SLAM.

15.9. THE PERILS OF SIMULATION

A general discussion of simulation would be incomplete without mention of the trio of concerns described here.

It's important to remember that *insight* comes from deft modeling, not from detail. KISS (Keep It Simple, Stupid) applies with equal force to simulations as to other models. Beginners may be tempted to construct a "realistic" simulation model that includes a wealth of detail and pray that it will tell them something important. Those prayers go unanswered, however. Simulation is no substitute for thought. The history of operations research is strewn with examples of large-scale simulations that missed the main point.

A simulation can be devilishly difficult to *debug* because a simulation entails a host of contingencies, each of which must be coded properly. If a simulation is not debugged, its output will be confused and misleading.

A simulation can be difficult to *audit.* Its computer code may be so intricate that it is nearly impossible to judge whether it is executing the desired computation. This makes it possible for nefarious people to "tilt" a simulation model toward the conclusions that they espouse.

Those simulations that can be run on a spreadsheet are relatively transparent and relatively easy to debug and audit. The spreadsheet itself helps you to trace the computation and to verify that it is correct. More complex simulation languages may lack these features.

15.10. REVIEW

At the heart of digital simulation is a pseudorandom number generator, which produces a sequence $Z_1, Z_2, \ldots$ of numbers that seem to be the values taken by independent and identically distributed random variables, each having the uniform distribution on the interval between 0 and 1. These uniform random variables are translated into random variables of interest by inverting their CDFs. The diskette that accompanies this text includes an Excel add-in called RiskSim, which accomplishes these inversions and facilitates simulation in other ways. One of its dialog boxes lets you repeat a simulation and obtain evidence about the distribution of a random variable. Another dialog box lets you vary one or two parameters and determine the extent to which the performance of the system depends on their values.

In our simulation of the $D/M/1$ queue, the state records the amount of work left for the server at the moment when the nth customer arrives. This led to the simple recursion in Equations (15.2) and (15.3). The times that successive customers spend in the system are so dependent that the mean time in the system may be a poor proxy for the times experienced by one day's customers.

In our simulation of a yield management problem, the spreadsheet for a single flight was a bit involved. In this case and in others like it, you are advised to check the computations by hand. In this simulation, the profits earned from different flights are mutually independent, so relatively few flights need be simulated to get an accurate estimate of the expected profit per flight.

When simulation is used to determine the best values of parameters, try to use the same random variables with different parameter values. That way, your simulation focuses more on the variation due to the value of the parameter, and less on the inherent uncertainty in the system.

The simulations that can be executed on spreadsheets have states that record one, two, or perhaps three pieces of information. For more complex simulations, a commercial software package will be needed.

When compared with analysis, what are the advantages of simulation? Simulation can be much easier. The output of a simulation can provide an empirical probability distribution of a quantity of interest, not merely its expectation. Simulation can evaluate models that are too difficult to analyze, and, occasionally, simulation provides insights that are difficult to discover mathematically.

What are the disadvantages of simulation? A simulation models a particular design; it does not select a design that performs best. A simulation provides no sensitivity analysis, at least not without repeated simulation runs. A simulation can be very difficult to debug. It can fail to reveal insights that are readily obtained from analysis.

The most effective uses of simulation may be in tandem with analyses. Use the analysis to obtain insight into the system and to discover efficient operating rules. Use simulation to check the approximations and simplifications made by the analysis and to improve on the rules found by analysis.

15.11. HOMEWORK AND DISCUSSION PROBLEMS

1. The random variables $T(1)$ and $T(2)$ are independent, and each has the exponential distribution, with expectations of 10 and 5 minutes, respectively.

(a) On a spreadsheet, simulate $T = T(1) + T(2)$.

(b) On the same spreadsheet, replicate part (a) 200 times, and use the information you have learned to estimate the median of T and its 0.9 fractile.

2. Customers are of two types, Type A and Type B. Customers arrive at random, Type A being twice as likely to arrive as Type B. The service time of each Type A customer is the sum of two independent exponential random variables whose expectations are 10 and 5 minutes, respectively. The service time of each Type B customer is exponential with a mean of 12 minutes. The service times of different customers are mutually independent.

(a) On a spreadsheet, simulate the service time T of a customer.

(b) On the same spreadsheet, replicate part (a) 200 times, and use the information you have gleaned to estimate the median of T and its 0.9 fractile.

3. **(Problem A, The Clinic)** From the simulation output of Problem A, is it possible to estimate the average number of patients a customer finds in the queue when she or he arrives? If so, how?

4. **(Problem A, The Clinic, continued)** Figure 15.4 reports on simulations of the average of the times in the system of one day's patients. Construct and run the analog of this simulation for the time in the system of the day's final patient, who arrives at 4:45 P.M. Compare your results with Figure 15.4.

5. **(Problem B, Yield Management)** In Problem B, the airline did not "credit" itself with the contribution made by discount fare customers who reserved seats but did not appear to claim them. These customers did not get their money back. Should the airline have credited the flight with the revenue or contribution it earned from these customers? Why?

6. **(An *M/M/1* Queue)** In Problem A, suppose that the inter-arrival time of patients is exponential with a mean of 15 minutes rather than being fixed. Designate as $A(n)$ the time that elapses between the arrival of the $n - 1$st and the nth patient.

(a) On a spreadsheet, how would you simulate $A(2), A(3), A(4) \ldots$?

(b) How does Equation (15.3) change?

(c) Alter the spreadsheet in Table 15.5 to simulate the times in the queue and in the system experienced by the first five patients of this clinic.

(d) Complete this sentence. In this *M/M/*1 queue, the average time in the system $\overline{W}(n)$ settles down ________ rapidly than for the *D/M/*1 queue in Problem A because ________.

7. **(An *M/M/2* Queue)** Consider a clinic whose arrivals are Poisson with a rate of 8 per hour, whose service times are exponentially distributed with a mean of 13 minutes, and with two servers. Patients wait their turn for the first available server; the service discipline is FCFS (first come, first served).

(a) Designate as $A(n)$ the time that elapses between the arrival of the $n-1$st and the nth patient. On a spreadsheet, how would you simulate $A(2), A(3), A(4) \ldots$?

(b) To calculate the time $Q(n)$ that the nth patient spends in the queue, we must account for the server to whom this patient will be assigned. Use the quantities $F(1, n)$ and $F(2, n)$ that appear below to calculate $Q(n)$.

$F(i, n)$ = the time that will elapse from the moment that the nth customer appears and server i completes service on those among customers 1 through $n-1$ that have been assigned to it, measured in minutes.

(c) Create a spreadsheet that computes $A(n)$, $Q(n)$, $S(n)$, $T(n)$, $F(1, n)$, and $F(2, n)$ for the first 10 customers, assuming that both servers are idle initially.

8. **(A Newsvendor)** Let us reconsider Problem A in Chapter 12. We recall that Art pays 30 cents for each newspaper he buys from the wholesaler, that he receives 50 cents for each newspaper he sells retail, and that he receives 2 cents for each newspaper that he scraps. Art must select his order quantity q before the demand D materializes, and D has the following probability distribution.

x	19	20	21	22	23	24	25	26	27	28	29	30
$P(D = x)$	0.00	0.01	0.03	0.06	0.10	0.18	0.22	0.17	0.12	0.06	0.04	0.01
$P(D \leq x)$	0.00	0.01	0.04	0.1	0.20	0.38	0.60	0.77	0.89	0.95	0.99	1.00

(a) Let the random variable $X(q)$ denote Art's profit if he orders q newspapers. Show that

$$X(q) = 50 \min\{D, q\} + 2 \max\{0, q - D\} - 30\, q.$$

(b) On a spreadsheet, simulate $X(q)$ for 100 different days, and compute the average net profit for these 100 different days.

(c) Use an Excel table to repeat part (b) for each value of q between 19 and 30, employing the same (random) variables for each value of q.

(d) Make a chart that plots average net profit versus q.

9. For the spreadsheet in Table 15.8, what function goes in cell B17?

10. Without using RiskSim, find a way in which to simulate X in Figure 15.1 on a spreadsheet. *Hint:* The most straightforward approach has IF statements in a series of cells.

11. **(A Recursion for the *G/G/*1 Queue)** For Problem A, a recursion was used to simulate a *D/M/*1 queue with a first-come first-served queue discipline. The same idea simulates a *G/G/*1 queue with first-come first-served queue discipline. To suggest how, let us designate:

$A(n)$ = the time that elapses between the arrival of the $n-1$st customer and the nth.

$S(n)$ = the service time of the nth customer.

$Q(n)$ = the time that the nth customer waits for service to begin.

$T(n)$ = the time that the nth customer spends in the system.

Is it true that

$$Q(n) = \max\{0, T(n-1) - A(n)\} \quad \text{and} \quad T(n) = Q(n) + S(n)?$$

12. **(A Recursion for the *G/G/*2 Queue)** The recursion for the *G/G/*1 queue requires only one state variable, which is the amount of work that remains in the system when an arrival occurs. For a *G/G/*2 queue with first-come first-served queue discipline, two state variables will be needed.

(a) Adapt the recursion for the *G/G/*1 queue to compute the times in the system of the customers in a *G/G/*2 queue with a first-come first-served queue discipline. *Hint:* Keep track of the remaining work before each server at the moment a customer arrives.

(b) On a spreadsheet, simulate 15 arrivals in a *M/M/*2 queue with arrival rate of 10 customers per hour and with service rates of 6 per hour. *Hint:* The IF statement should be handy.

Part E

Game Theory

Chapter 16

Game Theory

16.1. PREVIEW

In all previous chapters, we have modeled the actions of a single decision maker who sought to maximize his or her well-being. **Game theory** is the study of decision-making models of two or more decision makers whose actions affect each other's well-being. The models of game theory encompass competitive behavior, cooperative behavior, and mixtures of the two. Games that have three or more types of players can model the formation of coalitions, that is, of groups of participants who band together to improve their own welfare at the expense of the rest.

Game theory plays important roles in politics, business, and military affairs and in political science, economics, and operations research. The terminology from game theory has entered common usage. Two terms that quickly come to mind are win-win situation and zero-sum situation. A **win-win** situation is one in which both players can benefit by cooperating, and a **zero-sum** situation is purely competitive.

With a single decision maker, the idea of an optimal solution is unambiguous. In contrast, in a game theoretic model, there are several valid notions of an optimal solution, each appropriate for particular settings. This chapter illustrates three notions of optimality:

- A set of strategies, one per player, is said to form an **equilibrium** if no single participant can benefit by changing his or her strategy.[1] Thus, in an equilibrium, each player's strategy is a *best response* to the actions of the others.

[1]Economists distinguish between different types of equilibrium; what we are calling an equilibrium is often referred to as a **Nash equilibrium**, after John Nash.

- A set of strategies, one per player, is said to be **stable** if no group of participants can all benefit from changing their strategies simultaneously.[2]
- A player's strategy is said to be **dominant** for that player if it performs the best for that player, independent of the actions of the others.

The idea of an equilibrium is central to economics: Each agent (player) acts in his or her own self-interest, and the agents (players) interact with each other through a market in ways that inhibit coalition-forming.

The idea of a stable solution is apt for models in which agents have motives to form coalitions and in which they have the opportunity to do so. A stable solution must be an equilibrium, of course, but an equilibrium need not be stable.

A dominant strategy is a best response to all actions that the other players might take.[3] A few games do have dominant strategies, but most do not. A dominant strategy may seem to be ideal, but the Prisoner's Dilemma (see Problem 13 on page 568) indicates that another solution can be better still, if it can be enforced.

16.2. WHAT CAN YOU LEARN FROM THIS CHAPTER?

This introductory chapter cannot present a systematic exploration of game theory, for that would require volumes. Instead, this chapter presents a few models that have been selected to illustrate the aforementioned solution concepts and to reveal themes that pervade game theory. The most fundamental of these themes is **empathy**; look at the situation from the viewpoints of the other participants as well as your own.

The first model that we present is that of a lawsuit. For this model, we'll use decision trees to evaluate the actions of the adversaries and to find a range of out-of-court settlements which both sides prefer to a trial. Each settlement in this range forms an equilibrium; neither side can benefit from rejecting it.

The second model is a "marriage game" whose players form two groups, with each member of either group (e.g., women) having a preference over the members of the other group (e.g., men). For this model, we'll see that the concept of an equilibrium is vapid because there is little that a man or woman can do singly. We'll find a stable solution for this game, namely, a pairing such that no group of individuals can all benefit from defecting.

The third model is a sealed-bid auction. For this model, each bidder will be seen to have a dominant strategy, namely, a strategy that is best for that bidder, independent of what the others do. This notion of dominance recovers the idea of optimality of a one-decision-maker situation.

The fourth model is a game in which two players are in direct opposition. For this model, we'll use a linear program to compute equilibrium strategies for both players. In this game, there can be multiple equilibria, but they all have the same economic benefit to each player.

---

[2]Our use of the term *stable* is a bit nonstandard. In cooperative game theory, a set of *outcomes* is said to be stable if no group of participants can all benefit from changing their strategies simultaneously. We've used *stable* to describe strategies, not outcomes. In economics, a set of strategies, one per player, has been called a **strong equilibrium** if no group of participants can all benefit from changing their strategies simultaneously. Since the term *strong equilibrium* never caught on, we'll stick with *stable.*

[3]Unfortunately, "dominance" carries different meanings in decision analysis and game theory. In decision analysis, one strategy dominates another if it performs at least as well under every chance outcome and if it performs better under at least one chance outcome. In game theory, a strategy is dominant for a player if it maximizes that player's *expected* utility for all strategies that the other players might take.

The fifth and final model is a game between two players who need not be in direct opposition and who may benefit by cooperating. This game also has an equilibrium, but that equilibrium cannot be found from the solution to a linear program.

Collectively, these five models suggest the ways in which game theory contributes to the science of decision making.

16.3. A LAWSUIT

Our first game theoretic model, a lawsuit, is an apt example of the merit of empathy, that is, of the value of examining the situation from the perspectives of the other participants.

Problem A (The AB/C Patent Infringement Suit)[4]

Ray, the president of Company C, has just hired you as an analyst. After welcoming you, he informs you that your first assignment is to review a patent infringement suit that has been brought against the company. Ray will brief you himself on the legal aspects of the case. He has arranged for you to meet later in the day with his brother Tom, who is responsible for production and marketing. Ray expects you to report to him tomorrow on your assessment of the lawsuit.

Over coffee and donuts, Ray explains that Companies A and B jointly brought a suit against Company C for infringing on their patent for the manufacture of compound V. Company A is a large pharmaceutical manufacturer whose scientists discovered and patented a process for the manufacture of compound V. Company A did not wish to manufacture this compound. It had granted an exclusive license for the manufacture of compound V to Company B, under these terms:

- Company B pays to Company A a royalty of 4% of its sales of compound V.
- Company B can sublicense the manufacture of compound V to other companies; under these sublicenses, the first 4% of the royalty would go to Company A, and the excess over 4% to Company B.

To remedy the infringement, Companies A and B seek a royalty of 10% of past and future sales of compound V. Ray mentions that Company B, being an exclusive licensee, could have sued by itself. Company C's scientists had discovered the same production method somewhat later but had not thought it to be patentable. No one in Company C had known that Company B was paying a royalty, so the lawsuit was a complete surprise.

Because patent law is a specialized field, Ray engaged the Cambridge patent attorney, Ann Howe. She advised them that a challenge to this patent was likely to succeed in court. She also suggested a token offer of a 2% royalty to avoid the expense and headache of litigation. When this offer was presented to Companies A and B, it was promptly declined. Ray then asked Ann to prepare for trial. She has spent \$50,000 so far, and she estimates that the trial will require another \$180,000 in legal expenses.

After lunch, you meet with Tom to discuss the marketing and production aspects of this suit. Tom informs you that Company B has about 50% of the market for compound V and that the rest of the market is divided between five producers, Company C with 25% being the largest. Compound V accounts for over half of Company B's sales but for under one-tenth of Company C's sales. Small quantities of compound V are used in many different products, and the aggregate market for compound V is stable and not price-sensitive. Working together,

---

[4]Problem A derives from the Harvard Business School Case, *C. K. Coughlin, Inc.*, prepared by Donald L. Wallace under the direction of John S. Hammond.

you and Tom estimate the present value of a 10% royalty on sales of compound V at $1.2 million. Future sales account for two-thirds of this amount, and past sales for one-third.

Tom reports that the other small producers of compound V use a different manufacturing process, so only your company is subject to this patent infringement suit. Production of compound V is capital intensive. For each producer, the variable cost of manufacture of compound V is below one-fourth of the sales price. Each producer of compound V had tooled up for an optimistic share of the market. As a result, the aggregate production capacity is substantially in excess of demand.

With this information in hand, you turn your attention to tomorrow's briefing. The options facing Ray are clear: They are to pursue the lawsuit and to settle for a royalty payment that is at or below 10%.

A Decision Tree

Your first thought is that a decision tree might help to bring this issue into focus. Most of the information needed for this decision tree is available; the information that is not available consists of:

P = the probability that Company C would win a lawsuit.

S = the royalty rate (in percent) for which Company C should offer to settle.

You construct the decision tree shown in Figure 16.1, which evaluates Ray's options. Needless to say, the $50,000 spent so far on attorney's fees is a sunk cost and is not relevant. Fighting the lawsuit incurs legal fees of 180 thousand and a royalty payment of 1200 thousand if the suit is lost. Settling at S% costs 120 S thousands, which comes to 1200 thousands when S = 10 (percent).

A Breakeven Analysis

Having built this tree, it occurs to you to do a breakeven analysis, namely, to determine the values of P and S for which Company C is indifferent between pursuing the lawsuit and settling it. These values satisfy

$$120\,S = 180 + (1 - P)1200.$$

Solving this equation for S gives

$$S = 11.5 - 10\,P. \qquad (16.1)$$

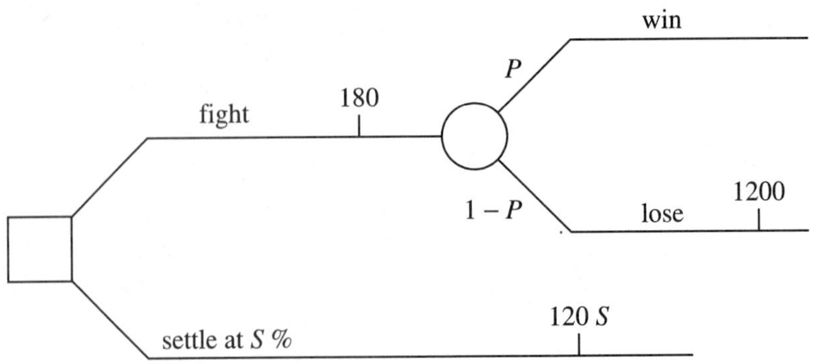

Figure 16.1 A decision tree for Company C's options in the patent infringement suit, with costs in thousands of dollars.

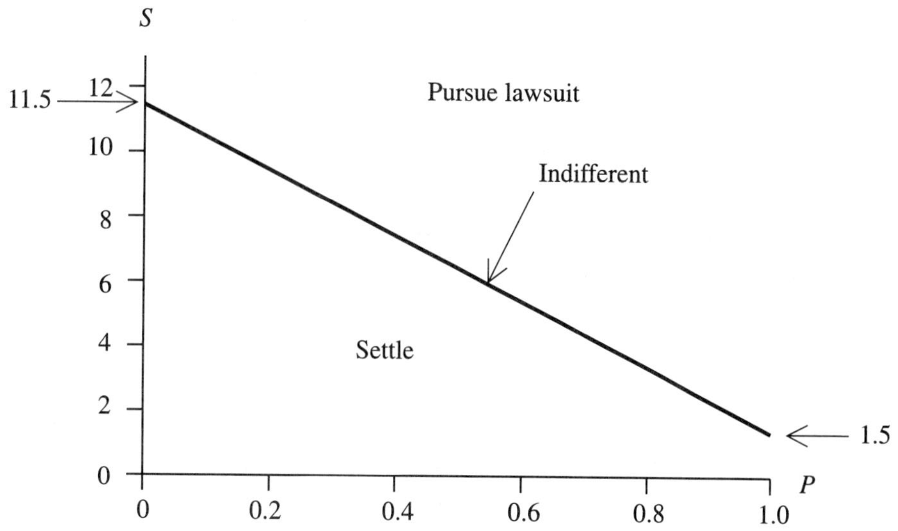

Figure 16.2 Breakeven between fighting and settling from Company C's perspective.

Equation (16.1) is linear, and Figure 16.2 plots its solution. In this figure, the probability P of winning the lawsuit is measured on the horizontal axis, and the settlement percent S is measured on the vertical axis. If $P = 0$, Equation (16.1) gives $S = 11.5$. If $P = 1$, Equation (16.1) gives $S = 1.5$. This indifference line divides the plane into two regions. For pairs P and S in the lower region, Company C prefers to settle. For pairs in the upper region, it prefers to fight.

Your presentation to Ray includes Figures 16.1 and 16.2. He is delighted. Your data seem sound to him, and your analysis has clarified a great deal. Ray schedules an appointment that afternoon at the law office of Ann Howe, the patent attorney. He orders you to come along and to bring your charts.

The Patent Attorney

At Ann Howe's office, Ray has you present your charts to her. He then presses Ann for her assessment of the probability of winning the lawsuit. This question discomfits her; after much hesitation and contemplation, she estimates P as 0.6.

While Ray was pressing Ann, it occurred to you to look at the lawsuit from the other side's viewpoint. You ask Ann for her guess as to the value that the patent attorney for Companies A and B would place on P. After some thought, she suggests that their attorney might estimate P at a lower value but not lower than 0.4. You also inquire as to the cost of the litigation to the other side. She reports that both sides do similar research and incur similar costs. At this point, Ray thanks Ann for her help, and he asks her to pause in her legal research for a day or two while he decides how to proceed.

Empathy

On the way back to the office, you show Ray the other side's decision tree, which is similar to Figure 16.1. They win when you lose. You estimate that their cost of going to court is \$180,000, just like yours. They break even for values of P and S that satisfy

$$120\,S = -\,180 + (1 - P)1200.$$

With a probability $P = 0.6$, Equation (16.1) shows that you prefer to settle for values

$$S \leq 11.5 - 10 \times 0.6 = 5.5. \tag{16.2}$$

Similarly, with probability $P = 0.4$, their breakeven equation shows that they prefer to settle for values

$$S \geq 8.5 - 10 \times 0.4 = 4.5. \tag{16.3}$$

Ray is ecstatic. He recalls from his economics course that settlement for a royalty rate between 4.5% and 5.5% is an equilibrium—that neither side can do better by refusing to settle, given each side's assessment of the situation.

Multiple Equilibria

In all previous chapters, an optimization problem could have more than one optimal solution, but they all had the same optimal value (payoff). That is not typical of game theory. Often, in a game theoretic model, there can be multiple solutions, each having different payoffs for the respective players.

This patent infringement suit has multiple equilibria because *any* royalty rate between 4.5% and 5.5% is preferable to both sides to going to court. In this instance, the equilibria have taken the form of an interval. A quantitative analysis has established a basis for negotiation, not an exact solution.

Postscript

At a late supper with your significant other, you cannot resist the urge to relate your success. After she listens intently to your entire tale, she congratulates you for your ability to gather data and to construct a quantitative model. Then she advises you that a royalty rate in the range you advocate is likely to be rejected. What had you overlooked?

16.4. A TWO-SIDED MARKET

The next game theoretic model has an interesting history. In 1962, David Gale and Lloyd Shapley published a seminal paper describing how a certain family of two-sided markets can be organized. Over two decades elapsed before Alvin Roth discerned in 1984 that the essence of their method had been used with great success, since 1951, to pair medical interns with hospitals.[5]

Gale and Shapley's approach applies to a variety of two-sided markets, including these:

College admissions: Suppose each of several colleges has a quota (maximum class size) and a strict preference ranking of the applicants that it is willing to accept for this year's entering class. Each applicant has a strict preference ranking of the colleges he or she is willing to attend.

Graduates and firms: Each of several graduates has a strict preference ranking of the firms that she or he is willing to join. Each firm has a quota (maximum number) of new hires and a strict preference ranking of the graduates that it is willing to hire.

[5]This observation and the literature to which it relates are lucidly recorded in A. Roth and M. A. O. Sotomayer, *Two-Sided Matching: A Study in Game-Theoretic Modeling and Analysis* (New York: Cambridge University Press, 1990).

Sorority rush: Each of several sororities has a quota (maximum pledge class size) and has a strict preference ranking of the rushees. Each rushee has a strict preference ranking of the sororities she is willing to join.

In each case, the problem is to match members of one group (e.g., rushees) with members of the other group (e.g., sororities). Gale and Shapley proposed an attractive procedure for constructing such matches. In the game theory literature, the models to which their methods apply are known loosely as **marriage games**.

An Illustration

To introduce this class of matching problem, we turn to the whimsical setting in:

Problem B (The Dance)

For Saturday night's ballroom dance, each of four women might date (pair up with) any of five men. The four women are labeled A through D, and the five men are labeled a through e. Each woman has a strict preference ranking of the men, and each man has a strict preference ranking of the women. Table 16.1 lists their rankings. It indicates, for instance, that woman A's first choice is man b, that her second choice is man c, and so forth. Similarly, man e's first choice is woman A, his second choice is woman B, his third choice is woman D, and he'd rather stay home than be partnered with woman C. You have been asked to suggest a way in which they might reasonably pair up for the dance.

For Problem B, you seek to pair (or match) some or all of the women with men. In this context, the idea of an equilibrium isn't very helpful. The only action that an individual can take by herself or himself is to refuse to go to the dance with the person to whom she or he has been matched. In fact, the only matchings that fail to be equilibria are those that pair man e with woman C. He will refuse to go to the dance with her because he'd rather stay home.

Stability

For Problem B, equilibrium is a weak measure of desirability. You need something more trenchant. You recall the notion of stability. A matching is stable if no group of participants can *all* improve by defecting from it. In particular, a matching is stable if no pair consisting of a woman W and a man m prefer each other to the assignments that the matching has given them.

Consider, for instance, a matching that includes the pairs (B, b) and (C, a). Woman B is paired with man b, but Table 16.1 indicates that she prefers man a. Man a has been paired with woman C, but the table shows that he prefers woman B. This matching is unstable because the pair (B, a) prefer each other to the assignments it gives them. The pair (B, a) can benefit by breaking the dates that have been assigned to them and going to the dance with each other.

Table 16.1 Preferences of each woman and man.

Woman	Preference	Man	Preference
A	b, c, a, d, e	a	A, B, C, D
B	c, a, b, d, e	b	D, B, C, A
C	e, d, a, b, c	c	D, C, A, B
D	a, d, e, b, c	d	A, D, C, B
		e	A, B, D

DAP/M

Does there exist a stable matching, and, if so, can you find one? Gale and Shapley's paper answers both of these questions in the affirmative. That paper contains an elegant algorithm called *DAP/M* that constructs a stable matching, thereby guaranteeing that one exists. *DAP/M* is an acronym for *d*eferred *a*cceptance *p*rocedure, *m*en proposing. This algorithm goes as follows:

1. Each man proposes to the woman he ranks highest.
2. Each woman who has multiple offers rejects all but the man whom she ranks most highly. (No woman has yet accepted any offer.)
3. Each man who has been rejected then proposes to the woman he ranks highest among those who have not yet rejected him.
4. Step 2 is repeated, and then Step 3 is repeated. This cycle iterates until each rejected man has proposed to all the women whom he wishes to take to the dance. At that point, each woman accepts the proposal, if any, that she has received but has not yet rejected.

Table 16.2 illustrates the progress of this algorithm. In Round 1, each man proposes to the woman he ranks first. Thus, woman A receives proposals from men a, d, and e, and woman D receives proposals from men b and c. Woman A rejects proposals from men d and e because she ranks man a higher. Woman D rejects the proposal from man c because she ranks man b higher. An overstrike (×) records each rejection.

In Round 2, men c, d, and e, who were rejected in Round 1, propose to the women they rank second. At this point, woman D has two proposals. She rejects man b because she ranks man d higher.

In Round 3, man b proposes to the woman he ranks second. At this point, woman B has two proposals. She rejects man e because she ranks man b higher.

In Round 4, man e proposes to the woman he ranks third. Now, woman D has two proposals. She rejects man e because she ranks man d higher.

At the close of Round 4, the procedure stops. Only man e has been rejected in Round 4, and he has been rejected by each woman whom he wishes to take to the dance. Each woman now accepts a proposal from the man whom she has not yet rejected. The matching that results is

$$(A, a) \quad (B, b) \quad (C, c) \quad (D, d),$$

with man e unmatched.

Proof of Stability

To demonstrate that *DAP/M* constructs a stable matching, we examine a pair (W, m) consisting of a woman W and a man m who are *not* matched to each other by *DAP/M*. This matching is stable if we can guarantee that woman W and man m do not prefer each other

Table 16.2 Four rounds of proposals for *DAP/M*.

Man	Round 1	Round 2	Round 3	Round 4
a	A	A	A	A
b	D	~~D~~	B	B
c	~~D~~	C	C	C
d	~~A~~	D	D	D
e	~~A~~	B	~~B~~	~~D~~

to their current assignments. There are two cases to consider:

- Suppose that woman W has received a proposal from man m. She has rejected man m. When she rejected man m, she had an assignment that she preferred to him. And she ended either with that assignment or with one she finds even more preferable to man m. Thus, woman W prefers her current assignment to man m.
- Alternatively, suppose that woman W has received no proposal from man m. He proposed to women in decreasing order of preference, stopping before he got to her. The proposal with which he stopped was accepted. Hence, man m prefers his current assignment to woman W.

In brief, it cannot be that an unmatched pair (W, m) prefer each other to the assignments they are provided by *DAP/M*. This demonstrates that *DAP/M* constructs a stable matching.

DAP/W

The algorithm *DAP/W* exchanges the roles of the men and women. In *DAP/W*, the women propose and the men reject. In the first step of *DAP/W*, each woman proposes to the man she ranks highest. Next, each man who has multiple proposals rejects all women but the one whom he ranks highest. After that, each woman who has been rejected proposes to the highest-ranked of the men who have not yet rejected her, and so forth. It's easy to see that *DAP/W* ends with a stable matching, just as *DAP/M* does.

Uniqueness?

Is there a unique stable matching? No. There can be several stable matchings, as can be seen when you execute *DAP/W* for the data in Table 16.1. You will get a different matching, which will be stable.

Optimality?

In a sense that will soon be made precise, *DAP/M* is best for the men and worst for the women. Table 16.2 shows that *DAP/M* assigns men a, b, c, and d the options that they rank 1, 2, 2, 2, and 4, respectively. The same matching assigns women A, B, C, and D the options that they rank 3, 3, 5, and 2, respectively. It can be shown that no other stable matching assigns any man a woman whom he ranks higher, and no other stable matching assigns any woman a man whom she ranks lower. Similarly, *DAP/W* is best for the women and worst for the men. In summary:

> There can be several stable matchings. Among all stable matchings:
>
> - The one created by *DAP/M* grants each man the outcome he most prefers and grants each women the outcome she least prefers.
> - The one created by *DAP/W* grants each women the outcome she most prefers and grants each man the outcome he least prefers.

Again, as was the case in the patent infringement suit, game theory does not prescribe a unique solution. There can be several stable matchings. For example, *DAP/M* constructs one that is optimal for all of the men, and *DAP/W* constructs one that is optimal for all of the women. Game theory has restricted the range of negotiation but has not prescribed a unique solution.

Misrepresentation?

Can a player benefit by misrepresenting his or her preferences? When *DAP/M* is used, no man can benefit. When *DAP/M* is used, a woman can benefit from listing the men in a sequence that fails to correspond to her true preferences, but she must know a lot of information in order to do so. Similarly, when *DAP/W* is employed, no woman can benefit by misrepresenting her preferences, but a man can benefit.

Medical Interns

Let us turn our attention to the problem of matching new MDs with hospital internships. We suppose that each MD has a strict preference as to the hospitals in which she or he is willing to intern. Similarly, each hospital has a strict preference as to the interns it is willing to have as interns. Each hospital also has a quota—a maximum number of interns that it needs.

Prior to 1951, MDs and hospitals negotiated directly with results that were chaotic and pairings that were unstable. In 1951, an algorithm that we would recognize as *DAP/H* was implemented. This algorithm proceeds as does *DAP/M*, except that each hospital keeps proposing until its quota is filled or it runs out of acceptable candidates, whichever comes first.

DAP/H reorganized the market for medical interns. By 1955, about 95% of the new MDs who graduated from American medical schools elected to have their internships determined by *DAP/H*. In later decades, this algorithm became somewhat less successful. It failed to allow pairs of MDs to negotiate internships that co-located them in one geographic region or another. This algorithm has been revised, and the current one is close to *DAP/I* (Interns proposing), with provision for pairs to co-locate.[6]

16.5. A SEALED-BID AUCTION

Auctions are used to sell commodities as diverse as art, flowers, oil rights, telecommunication bandwidth, and U.S. government bonds.

At the major auction houses, art is sold by an *ascending-bid* or *English* auction method in which each of several bidders has a numbered paddle or some other agreed-upon (possibly secret) method for making bids. The auctioneer states an opening bid. Each bidder can increase the bid at any time, usually by an amount determined by the auctioneer, and the item is sold to the person who places the highest bid.

In Amsterdam's wholesale flower marked, flowers are sold by a *descending-price* or *Dutch* auction method in which each potential purchaser has a button. The opening price for a lot of flowers is displayed visually, and the price decreases at a constant rate until someone buys the lot by pushing his or her button.

In 1961, William Vickery created a sensation with his analysis of an auction in which

- Each bidder submits a sealed bid for an item.
- The item goes to the bidder whose bid is highest.
- The winning bidder pays the *second-highest* bid for the item.

This type of auction is often called a **Vickery** auction, in his honor. For his work on sealed-bid auctions and for other ingenious research, Vickery was announced as winner

[6] A. E. Roth and Elliot Peranson, "The Redesign of the Matching Market for American Physicians: Some Engineering Aspects of Economic Design," *American Economic Review* 5, 89 (1999): 748–780.

of the Nobel Prize in Economics in 1996. Sadly, he died only a few days after the announcement. To introduce Vickery's work, we consider:

Problem C (A Vickery Auction)

The U.S. government is using a Vickery auction to lease an offshore oil parcel. You are to submit a bid on behalf of your company. Your company's geologists estimate that the value of this lease is $22.3 million. Others will bid for this lease, but you know nothing about the bids they will make. How much shall you bid?

A remarkable feature of the Vickery auction is that you can make a sensible decision based solely on the information in Problem C. Not only is this decision sensible, but it is the best possible decision, independent of whatever anyone else bids.

A Dominant Strategy

In the argot of game theory, a strategy that you select is said to be *dominant* for you if this strategy is best for you no matter what anyone else does. Are there any realistic games that have dominant strategies? Yes, there are a few. A Vickery auction is one of them. In Problem C, you have a dominant strategy, which is to bid the value of the lease to your company, namely, $22.3 million.

Proof of Dominance

We will now show that bidding $22.3 million is a dominant strategy. To do so, we introduce two items of nomenclature:

B = the amount that you bid on behalf of your company, in millions.

x = the amount of the second-highest bid, in millions.

Let us compare bids of B and 22.3. We consider two cases:

- Suppose B exceeds 22.3. What happens depends on the value of x.
  If x exceeds B, your company loses the lease and earns nothing with either bid.
  If x lies between 22.3 and B, a bid of B causes your company to get the lease and to lose $x - 22.3$ million, which would not have occurred if you had bid 22.3 million.
  And if x is below 22.3, your company wins the lease and earns $22.3 - x$ with either bid.
- Suppose B is below 22.3 million. Again, what happens depends on the value of x.
  If x exceeds 22.3, your company loses the lease and earns nothing with either bid.
  If x lies between B and 22.3, a bid of B causes your company to fail to get the lease and fail to earn $22.3 - x$, which would have occurred had you bid 22.3 million.
  If x is below B, your company wins the lease and earns $22.3 - x$ with either bid.

The first case shows that it is foolish to bid more than 22.3. The second case shows that, absent collusion between the bidders, it is foolish to bid less than 22.3.

In this example and in general, each participant in a Vickery auction does best by bidding the value V that he or she places on the item. In a Vickery auction, the winning bidder is the one whose value V is largest, and the winning bidder earns a profit of $(V - x)$, where x is the second-highest bidder's value.

Third-highest Bid?

Let's change the rules. Suppose, in a sealed-bid auction, the bidder whose bid is largest wins and pays the amount bid by the third-highest bidder rather than the second-highest bidder. What happens? The preceding argument still applies. Each bidder still has a dominant strategy, which is to bid the value V that she or he places on the item.

Empathy suggests that we look at the auction from the seller's viewpoint too. In a Vickery auction, the seller receives the second highest valuation placed on the item. In the third-highest-bid variant, the seller receives less. No seller would participate in such an auction if a Vickery auction was available.

16.6. A TWO-PERSON ZERO-SUM MATRIX GAME

Our attention now turns to a type of game that has a glorious history, to which a later section is devoted. We now focus on a game between two players who are in direct competition. What one player wins, the other loses. This is also a game of simultaneous play; both players select their strategies at the same time.

The data in this game are a rectangular array (or matrix) A of numbers, for example, the 3×3 array A:

$$A = \begin{bmatrix} 5 & 2 & 6 \\ 2 & 3 & 1 \\ 1 & 4 & 7 \end{bmatrix} \tag{16.4}$$

The rules of this **matrix game** are as follows:

- You and I know all of the entries in the array A.
- You choose a row. Simultaneously, I choose a column.
- I pay you the entry in A that lies at the intersection of the column I chose and the row you chose.
- You aim to maximize the expectation of the amount that you receive.
- I am to minimize the expectation of the amount that I lose.

If, for instance, you select row 1 and I choose column 3, then I pay you six units.

This is a **zero-sum** game because the sum of what you win and what I win equals zero. It is called a "matrix" game because the payoffs are the entries in a matrix.

An Illustration

We will see how to use linear programming to find equilibrium strategies for matrix games. This will be accomplished in the context of:

Problem D (A Matrix Game)

For the array in Equation (16.4), do you and I have equilibrium strategies? If so, how can we find them, and what are they?

Without (yet) reporting how to find them, we report the optimal strategies for Problem D. It turns out to be wise for each of us to pick a randomized strategy. We'll see that your best strategy $\hat{p}$ is given by the numbers

$$\hat{p}(1) = 1/2, \quad \hat{p}(2) = 0, \quad \hat{p}(3) = 1/2, \tag{16.5}$$

where $\hat{p}(i)$ denotes the probability that you play row i. Thus, you pick rows 1 and 3 with equal probability, and you avoid row 2. Similarly, my best strategy $\hat{q}$ is given by the numbers

$$\hat{q}(1) = 1/3, \quad \hat{q}(2) = 2/3, \quad \hat{q}(3) = 0, \tag{16.6}$$

where $\hat{q}(j)$ denotes the probability that *I* play column j. Thus, I play column 1 with probability of 1/3, I play column 2 with probability of 2/3, and I avoid column 3.

An Equilibrium for Problem D

Let us demonstrate that the strategies $\hat{p}$ and $\hat{q}$ given in Equations (16.5) and (16.6) form an equilibrium. Let us first suppose that you use strategy $\hat{p}$. When you do so, how much do I lose? That depends on the column I choose. By reading down the columns of the matrix A in Equation (16.4), we see that the expectation of my loss is

$$\begin{aligned} 5(1/2) + 2(0) + 1(1/2) &= 3 && \text{if I choose column 1,} \\ 2(1/2) + 3(0) + 4(1/2) &= 3 && \text{if I choose column 2,} \\ 6(1/2) + 1(0) + 7(1/2) &= 6(1/2) && \text{if I choose column 3.} \end{aligned}$$

Evidently, if you choose strategy $\hat{p}$, then:

- I lose 3 units if I select column 1,
- I lose 3 units if I select column 2,
- I lose 3 units if I randomize in any way over columns 1 and 2.
- I lose more than 3 units if I pick column 3 with positive probability.

Strategy $\hat{q}$, as specified by Equation (16.6), calls for me to play column 1 with probability 1/3 and column 2 with probability 2/3. This costs me 3 units, and I can do no better. In brief, if you play strategy $\hat{p}$, I have no economic motive to deviate from strategy $\hat{q}$.

Now, let's suppose that I play strategy $\hat{q}$. How much you win depends on the row that you choose. By reading across the rows of the matrix A in Equation (16.4), we see that the expectation of your gain equals

$$\begin{aligned} 5(1/3) + 2(2/3) + 6(0) &= 3 && \text{if you choose row 1,} \\ 2(1/3) + 3(2/3) + 1(0) &= 2\,2/3 && \text{if you choose row 2,} \\ 1(1/3) + 4(2/3) + 7(0) &= 3 && \text{if you choose row 3.} \end{aligned}$$

Thus, if I choose strategy $\hat{q}$, then:

- You gain 3 units if you select row 1 or row 3.
- You gain 3 units if you randomize over rows 1 and 3.
- You gain less than three units if you pick row 2 with positive probability.

Your strategy $\hat{p}$, as specified by Equation (16.5), randomizes over rows 1 and 3. In brief, if I play strategy $\hat{q}$, you have no economic motive to deviate from strategy $\hat{p}$. This demonstrates that the strategies $\hat{p}$ and $\hat{q}$ form an equilibrium.

A Matrix Having an Equilibrium in Nonrandomized Strategies

We've seen that the strategies $\hat{p}$ and $\hat{q}$ given by Equations (16.5) and (16.6) form an equilibrium. We have not yet shown how to compute this equilibrium.

Before tackling the payoff matrix A given by Equation (16.4), we will use two simpler examples to sharpen our intuition. Let's start with the 2×3 array A of numbers:

$$A = \begin{bmatrix} 5 & 6 & -1 \\ 7 & 8 & 3 \end{bmatrix} \tag{16.7}$$

For the array A in Equation (16.7), the optimal strategies for both players are easy to spot. You choose row 2 because it pays you more than row 1 no matter what column I select. I choose column 3 because it pays you less than columns 1 or 2 independent of the row that you select. This pair is an *equilibrium*; if you pick row 2, I have no economic motive to deviate from column 3. Similarly, if I pick column 3, you have no economic motive to deviate from row 2.

For the payoff matrix in equation (16.7), your best strategy is **nonrandomized** because you pick a particular row with probability of 1. My best strategy is also nonrandomized; I pick a particular column with probability of 1.

A Matrix Having No Equilibrium in Nonrandomized Strategies

Does every matrix A have an equilibrium in nonrandomized strategies? Definitely not. Consider this example.

$$A = \begin{bmatrix} 0 & 2 \\ 3 & 1 \end{bmatrix} \tag{16.8}$$

For this payoff matrix, the row you prefer depends on the column I choose, and conversely. For instance, if you choose row 1, I prefer column 1. If you choose row 2, I prefer column 2. This game does not have an equilibrium in nonrandomized strategies.

We now construct an equilibrium in "randomized" strategies. A **randomized** strategy for you, the row player, selects row i with probability $p(i)$, where the $p(i)$'s are nonnegative numbers that sum to 1. Similarly, a randomized strategy for me, the column player, selects column j with probability $q(j)$, where the $q(j)$'s are nonnegative numbers that sum to 1. Technically, the randomized strategies include the nonrandomized strategies; for instance, a randomized strategy for you allows you to pick row 2 with probability of 1.

A Trick

To find an equilibrium for the payoff matrix in Equation (16.8), use this trick: You choose a randomized strategy $\hat{p}$ for which your expected gain does *not* depend on the column that I choose. Similarly, I choose a randomized strategy $\hat{q}$ so that my expected loss is independent of the row that you choose.

Your randomized strategy selects row 1 with probability $\hat{p}(1)$, and it selects row 2 with probability $\hat{p}(2) = 1 - \hat{p}(1)$. By reading down the columns of the matrix A in Equation (16.8), we see that your expected gain does not depend on the column that I choose when $\hat{p}(1)$ is selected so that

$$0\,\hat{p}(1) + 3[1 - \hat{p}(1)] = 2\,\hat{p}(1) + 1[1 - \hat{p}(1)].$$

Solving this equation for $\hat{p}(1)$ gives

$$\hat{p}(1) = 1/2 \quad \text{and} \quad \hat{p}(2) = 1/2.$$

This strategy $\hat{p}$ calls for you to choose row 1 with probability $1/2$ and to choose row 2 with probability $1/2$.

My randomized strategy $\hat{q}$ selects column 1 with probability $\hat{q}(1)$, and it selects column 2 with probability $\hat{q}(2) = 1 - \hat{q}(1)$. By reading across the rows of the array A in Equation (16.8), we see that my expected loss does not depend on the row that you choose when $\hat{q}(1)$ satisfies

$$0\,\hat{q}(1) + 2[1 - \hat{q}(1)] = 3\,\hat{q}(1) + 1[1 - \hat{q}(1)].$$

Solving this equation for $\hat{q}(1)$ gives

$$\hat{q}(1) = 1/4 \quad \text{and} \quad \hat{q}(2) = 3/4.$$

Let us check that strategies $\hat{p}$ and $\hat{q}$ do form an equilibrium. Suppose I choose strategy $\hat{q}$. What do you win? By reading across the rows of the matrix A in Equation (16.8), we see that the expectation of your winnings is given by

$$\begin{aligned} 0(1/4) + 2(3/4) &= 1.5 \quad \text{if you choose row 1,} \\ 3(1/4) + 1(3/4) &= 1.5 \quad \text{if you choose row 2.} \end{aligned}$$

Your expected winnings equal 1.5 if you choose row 1, if you choose row 2, and if you randomize over these rows by choosing any strategy p, including strategy $\hat{p}$.

Now suppose that you choose strategy $\hat{p}$. By reading down the columns of the matrix A in Equation (16.8), we see that the expectation of my loss is given by

$$\begin{aligned} 0(1/2) + 3(1/2) &= 1.5 \quad \text{if I choose column 1,} \\ 2(1/2) + 1(1/2) &= 1.5 \quad \text{if I choose column 2.} \end{aligned}$$

Thus, the amount that I lose equals 1.5, independent of my strategy. If you choose strategy $\hat{p}$, my expected loss equals 1.5 for all strategies I might play, including strategy $\hat{q}$. In brief, the pair $(\hat{p}, \hat{q})$ is an equilibrium.

Dual Problems

Problem D asks us to construct an equilibrium for the array A in Equation (16.4). The examples in Equations (16.7) and (16.8) suggest how to proceed. In both examples:

- You seek a randomized strategy $\hat{p}$ that maximizes the least amount that you will earn if I pick any one of the columns at my disposal.
- I seek a randomized strategy $\hat{q}$ that minimizes the largest amount that I will lose if you pick any one of the rows at your disposal.

Each of these optimization problems can be formulated as a linear program. Not until Chapter 19 shall we demonstrate that—no matter what the matrix A—both linear programs have the same optimal value; moreover, the shadow prices for either linear program form an optimal solution to the other. But we can use the matrix in Equation (16.4) to illustrate these results, and we will do so.

A Linear Program

We will use a linear program to construct an equilibrium for the payoff matrix A given in Equation (16.4). For convenient reference, that array is reproduced as follows.

$$A = \begin{bmatrix} 5 & 2 & 6 \\ 2 & 3 & 1 \\ 1 & 4 & 7 \end{bmatrix}$$

Let's examine your optimization problem. The matrix A has three rows, so each randomized strategy at your disposal is a set $p(1)$, $p(2)$, and $p(3)$ of nonnegative numbers whose sum equals 1. These are three of the four decision variables in:

Program 16.1: Maximize $\{v\}$, subject to the constraints

$$
\begin{array}{llll}
\textit{column 1:} & v \le 5\,p(1) + & 2\,p(2) + & 1\,p(3), \\
\textit{column 2:} & v \le 2\,p(1) + & 3\,p(2) + & 4\,p(3), \\
\textit{column 3:} & v \le 6\,p(1) + & 1\,p(2) + & 7\,p(3), \\
 & p(1) + & p(2) + & p(3) = 1, \\
 & p(1) \ge 0, & p(2) \ge 0, & p(3) \ge 0.
\end{array}
$$

To study Program 16.1, we note that:

- Each randomized strategy p for you, the row player, consists of numbers $p(1)$, $p(2)$, and $p(3)$ that satisfy the final two constraints in Program 16.1.
- The right-hand side of the constraint labeled column j equals your expected payoff if you choose randomized strategy p and I pick column j.
- For each strategy p, the maximization operator in Program 16.1 equates v to the smallest expected payoff you get if you choose strategy p and I choose one of the columns.
- Moreover, the optimal solution to Program 16.1 selects the randomized strategy p that maximizes the minimum payoff that you can obtain if you choose a randomized strategy and I play a column.

In brief, the optimal solution to Program 16.1 selects a strategy $\hat{p}$ for you, the row player, that maximizes the least expected amount that you can earn if I pick one of the columns at my disposal.

Table 16.3 formulates this linear program for solution by Solver. The table also reports the optimal value that Solver has found, along with the shadow prices of the constraints.

Table 16.3 Optimal solution and shadow prices for Program 16.1.

	A	B	C	D	E	F	G	H	I
2	name	v	p(1)	p(2)	p(3)			RHS	shadow price
3	value	3	1/2	0	1/2				
4	col. 1	1	-5	-2	-1	0.0000	<=	0	1/3
5	col. 2	1	-2	-3	-4	0.0000	<=	0	2/3
6	col. 3	1	-6	-1	-7	-3.5	<=	0	0
7			1	1	1	1	=	1	3
8						↑			
9			=SUMPRODUCT(B$3:E$3,B7:E7)						
10									
11									
12			Solver maximizes the value in cell B3 subject to the constraints						
13			C3:E3 >= 0 , F4:F6 <= H4:H6, F7 = H7						

Finding an Equilibrium

Table 16.3 shows that the optimal solution to Program 16.1 is to select strategy $\hat{p}$ having

$$\hat{p}(1) = 1/2, \quad \hat{p}(2) = 0, \quad \hat{p}(3) = 1/2,$$

The table also shows that 3 is the optimal value of Program 16.1. The shadow prices for the first three constraints in Program 16.1 prescribe the strategy $\hat{q}$, where

$$\hat{q}(1) = 1/3, \quad \hat{q}(2) = 2/3, \quad \hat{q}(3) = 0.$$

Earlier, we had demonstrated that this pair $(\hat{p}, \hat{q})$ of strategies is an equilibrium.

It is *not* an accident that the shadow prices for the columns sum to 1. Nor is it an accident that the optimal solution and its shadow prices form an equilibrium. In Chapter 19, we will use "duality" to prove that these results hold for any matrix A, not just for the one illustrated in Program 16.1.

The Value of a Matrix Game

For a matrix game, can there be more than one equilibrium? Yes. That's no surprise because a linear program can have more than one optimal solution. Must every equilibrium $(\hat{p}, \hat{q})$ have the same expected payoff for you, the row player? Yes!

In brief, although a matrix game can have more than one pair $(\hat{p}, \hat{q})$ of equilibrium strategies, each pair of equilibrium strategies has the same expected payoff to the row player and hence has the same expected cost to the column player. That payoff is called the **value** of the game. The value of the game whose payoff matrix A is given by Equation (16.4) equals 3, for instance.

Thus, a matrix game shares a property with the one-decision-maker problems we've studied in earlier chapters; there can be multiple optimal solutions, but there is only one optimal value.

16.7. A BI-MATRIX GAME

The matrix game in the prior section models pure competition. What you win, I lose. By contrast, a "bi-matrix" game can have elements of competition and cooperation. The data in this bi-matrix game are a pair of matrices, each having the same number of rows and columns, such as the 2×3 arrays A of and B that appear below.

$$A = \begin{bmatrix} 5 & 8 & -1 \\ 2 & 0 & 3 \end{bmatrix} \qquad B = \begin{bmatrix} 7 & 12 & -3 \\ 1 & -5 & 6 \end{bmatrix} \tag{16.9}$$

The rules of the **bi-matrix game** are as follows:

- You and I know all of the entries in the arrays A and B.
- You choose a row. Simultaneously, I choose a column.
- You lose the amount in the matrix A that lies at the intersection of the row that you chose and the column that I chose.
- I lose the amount in the matrix B that lies at the intersection of the row that you chose and the column that I chose.
- You aim to minimize the expectation of your loss.
- I am to minimize the expectation of my loss.

This bi-matrix game boils down to the matrix game if $A = -B$, that is, if you win what I lose. We are prepared to pose the question in:

Problem E (A Bi-matrix Game)

For the matrices A and B in Equation (16.9), does there exist an equilibrium? If so, how can we find one?

For the matrices in Equation (16.9), an equilibrium is rather easy to spot. It is for you to play row 1 and for me to pay column 3. Note that:

- If you choose row 1, I have no motive to deviate from column 3 because I earn 3 units if I select column 3 and lose money if I select either of the other columns.
- And if I choose column 3, you have no motive to deviate from row 1 because you earn 1 unit if you select row 1 and lose 3 units if you select row 2.

In general, however, equilibrium strategies are randomized and are not that easily spotted. For every bi-matrix game:

- There does exist an equilibrium in randomized strategies.
- This equilibrium cannot be found by solving a known linear program.
- It can be found by an adaptation of Dantzig's simplex method that is known as the "complementary pivot method."

The complementary pivot method is not included within Solver. Unfortunately, it lies a bit beyond the scope of this text. A bi-matrix game must have at least one equilibrium, but it can have more than one, and different equilibria can have different expected costs to the two players.

For 2×2 matrices A and B, it's easy to find an equilibrium for a bi-matrix game. First, check whether the game has an equilibrium in nonrandomized strategies. If it does, fine; if not, arrange for each player to randomize so that his expected cost is independent of the action of the other player. In other words, use the same trick as for the 2×2 matrix A in Equation (16.8).

16.8. A BIT OF HISTORY

Matrix games and their generalizations have a fabulous history, beginning in 1910 with Brouwer's fixed-point theorem. The mathematical statement of this theorem is a bit abstruse, but the interpretation is not. Here goes. . .

Brouwer's Fixed-Point Theorem

Let X be a subset of $\Re^n$ that is nonempty, closed, bounded, and convex, and let f be a continuous function that maps X into itself. **Brouwer's fixed-point theorem** states that X contains at least one element x such that $x = f(x)$.

Here is the interpretation. Imagine that you have a cup of coffee on your table, and let X denote the surface of the coffee, the part that you can see looking down into the cup. This set X is a subset of $\Re^2$. Unless your coffee cup is oddly shaped, the perimeter of X is a circle. This set X is closed, bounded, and convex. Now imagine that you wiggle your coffee cup enough to cause waves but not breakers. Wait until the surface of your coffee settles down. Wiggling has mapped the surface onto itself. Finally, imagine the function f that maps the location of each point x on the surface before you wiggled the cup into a point $f(x)$ on the surface after it came to rest. This function f is continuous because there were no "breakers."

Brouwer's fixed-point theorem states that the net result of wiggling the cup leaves at least one point x where it began and so $x = f(x)$ for at least one point on the surface of the coffee.

Brouwer's fixed-point theorem lies within the field of nonlinear analysis; the function f must be continuous, but it need not be linear. Until the era of linear programming, all known proofs of Brouwer's fixed-point theorem were *existential*, which is to say, *nonconstructive.* They demonstrated the existence of a fixed point but offered no clue as to how to find it.

Fixed Points in Game Theory

In 1928, John Von Neumann used Brouwer's fixed-point theorem to show that each zero-sum matrix game has a "minimax" solution, effectively to demonstrate that each such game has an equilibrium. Resting as it did on Brouwer's fixed-point theorem, Von Neumann's result was nonconstructive; it provided no way in which to compute the fixed point.

In 1947, George Dantzig devised the simplex method, which quickly computes solutions to linear programs. Later that year, in a celebrated conversation at a railroad station in Princeton, Dantzig told Von Neumann about his work on linear programming. Von Neumann immediately realized that Program 16.1 constructs a solution to his zero-sum matrix game. This meant that solutions to these games not only existed but could be found. It also meant that, as an analytical tool, Brouwer's fixed-point theorem was obsolete as concerns zero-sum matrix games.

In 1950–1951, John Nash published some elegant papers that used Brouwer's fixed-point theorem to demonstrate the existence of several types of equilibria, including that for a bi-matrix game. For this work, Nash shared the Nobel Prize in Economics in 1994.

In 1964–1965, Carl Lemke published a pair of papers, the first coauthored with J. T. Howson, that introduced **complementary pivot methods**. These methods reorganize the pivot scheme in Dantzig's simplex method in ways that construct an equilibrium for a bi-matrix game, compute solutions to "quadratic" programs, and solve other problems.

Almost immediately, Herbert Scarf saw how to adapt the ideas of Lemke and Howson to construct an approximate solution to Brouwer's fixed-point problem and thereby compute economic equilibria, approximately.

This brought the analysis full circle. It began with a nonconstructive theorem in nonlinear analysis, namely, a demonstration that a certain function f maps at least one point into itself. It ended with the use of linear methods to approximate the fixed point.

16.9. REVIEW

This chapter has glimpsed game theory and has illustrated three solution concepts. A set of strategies, one per player, forms an *equilibrium* if no single player can benefit from changing his or her strategy unilaterally. A set of strategies, one per player, is *stable* if no group of players can all benefit from changing their strategies simultaneously. A strategy is *dominant* for a particular player if it is the best response to all strategies of the other players.

We've presented five game theoretic models. In the first, a lawsuit, an equilibrium takes the form of any out-of-court settlement that both sides prefer to a court fight. In the second, a marriage game, a stable solution is one in which no pair of participants can improve. Both models had multiple solutions, with different benefits to the participants. In our third example, a Vickery auction, each player has a dominant strategy.

The fourth example is a zero-sum two-person matrix game that has an equilibrium, which takes the form of a randomized strategy for each player. This game can have more than one equilibrium, but all of them have the same expected payoff to the profit-maximizing player and, consequently, the same expected payout from the cost-minimizing player. Solutions to this game can be found from linear programming.

Our final example is a non-zero-sum two-person matrix game. This game has at least one equilibrium, and each of its equilibria takes the form of a randomized strategy for each player. Different equilibria can have different expected costs to the two players. A solution to this game can be found by a variant of linear programming that is known as the complementary pivot method. For payoff matrices that have only a few rows and columns, solutions can be found by tricks introduced within the chapter.

Although this chapter provides only a glimpse of game theory, it does touch on landmarks in the field—the marriage game of Gale and Shapley, the sealed-bid auction of Vickery, the minimax theorem of Von Neumann, its generalizations by Nash, the solution of bi-matrix games by Lemke and Howson, and the computation of general equilibria by Scarf.

The notion of an equilibrium will be revisited in Chapter 19, where linear programming will be used to demonstrate the existence of a general equilibrium for a simplified model of an economy.

One limitation of this chapter warrants special mention. The games treated here are static; players pick strategies simultaneously. Many games are *dynamic*; they unfold over time as uncertainty is revealed and as the actions of each player affect the reactions of the others. The literature on dynamic games has not been probed.

16.10. HOMEWORK AND DISCUSSION PROBLEMS

1. For Problem A (**Patent Infringement**):

(a) Draw the decision tree that evaluates the options faced, collectively, by Company C's opponents.

(b) Draw their analog of Figure 16.2, find their indifference line, and justify expression (16.3) in the chapter.

(c) The indifference lines for the two sides are parallel. Would these lines remain parallel if the settlement from the present value of a 10% royalty differed somewhat from $1,200,000?

2. For Problem A (**Patent Infringement**), suppose that Companies A and B lose the suit. This relieves Company B of the 4% royalty it had been paying Company A, the patent being no longer valid. Should Company B pass this savings along to its customers in order to increase its market share? Why?

3. For Problem A (**Patent Infringement**), the analyst's significant other hinted that he had overlooked something important. What did she have in mind? What effect would it have on the settlement? *Hint:* Review the preceding problem.

4. For Problem A (**Patent Infringement**), Ann Howe is one of three partners in a widely known Cambridge law firm. Who are the other two? And what is Tom and Ray's surname? And what is the name of their chauffeur?

5. For Problem B (**The Dance**), execute *DAP/W*. Is each pairing different from the one obtained by *DAP/M*? Is each woman better off? Is each man worse off?

6. (**Misrepresentation in the Marriage Game**) For the marriage game with *DAP/M*, Claire's first proposal is from Brian, who is her second choice, her first choice being Adam. Diana's first proposal is from Adam, who is her second choice, her first choice being Brian. As for the men, Claire and Diana are Adam's and Brian's second choice, respectively. Does either woman have a motive to misrepresent her preference? Is it possible for the women to obtain a stable matching that they prefer? If so, how?

7. (**Further Misrepresentation in the Marriage Game**) For the marriage game with *DAP/W* (for which the women do the proposing), can you devise a way (albeit devious) in which the men get the stable matching that's best for them?

8. By the middle of the 1950s, nearly all of the new American MDs used *DAP/H* to obtain their internships; by the middle of the 1990s, some elected not to. In the interim, American society had evolved in a way that made *DAP/H* less attractive. How? Which postulate about preferences was better satisfied in the earlier period?

9. (**A Dominant Strategy**) The TV game, *Final Jeopardy*, has these two stages:

- In Stage 1, each of three contestants competes to accumulate as much "funny" money as possible.
- Stage 2 consists of a single question and proceeds as follows.
  First, the category of this question is announced.
  Next, each contestant elects to bet any amount of his or her funny money by writing it on a slate that is visible to no other contestant.
  After that, the question is posed. Each contestant writes his or her answer on the slate, which remains visible to no other contestants.
  The slates are then revealed. If a contestant gets the correct answer, her funny money is augmented by the amount she bet. If a contestant gets an incorrect answer, her funny money is decreased by the amount she bet.

The contestant who ends Stage 2 with the largest amount of funny money wins that amount of real money. The other two contestants win nothing.

The issue for Stage 2 is how much to bet. Your goal is to bet the amount that maximizes the expectation of your winnings. To determine your best betting strategy, you designate:

x = your asset position, in funny money, at the end of Stage 1.
q = the probability that you can correctly answer a question in the announced category.
b = the amount you bet in Stage 2, so $0 \le b \le x$.
$f(w)$ = the probability that you will win given that you accumulate w units of funny money at the end of Stage 2.

(a) This notation suppresses the fact that the probability $f(w)$ depends on the other players. Is it true that in every play of the game, no matter what the other players do, $f(w)$ cannot decrease as w increases? If so, why?

(b) Let $m(x, b)$ denote the expectation of the number of dollars you win if you possess x units of funny money at the end of Stage 1 and bet b units of funny money in Stage 2. Suppose that $q > 0.5$. Show that

$$\begin{aligned} m(x, b) &= f(x+b)\,q\,(x+b) + f(x-b)\,(1-q)\,(x-b), \\ &\le f(2x)\,q\,(x+b) + f(2x)\,(1-q)\,(x-b), \\ &= f(2x)\,[x + b\,(2q-1)], \\ &\le f(2x)\,[x + x\,(2q-1)] \\ &= f(2x)\,q\,(2x) = m(x, x). \end{aligned}$$

(c) Prove or disprove: Each contestant for whom q is at least 0.5 has a dominant strategy, which is bet all, independent of the amounts of funny money that the other contestants possess at the end of Stage 1 and independent of the other contestant's strategies in Stage 2.

10. Suppose that you have placed an item for sale in a Vickery auction, one that allows you (the seller) to place a sealed bid for the item. Let V denote the value that you place on this item. What is your bid? Does your bid represent a dominant strategy? Will you earn a profit? If so, how much?

11. For the matrix A in Equation (16.4), formulate the analog of Program 16.1 for the column player. This analog chooses the randomized strategy q for the column player that minimizes v, where v is the largest expected payout if the row player chooses one of the rows. Use Solver to find the optimal solution to this linear program. Record the shadow prices. Is this optimal solution familar? Are the shadow prices familiar?

12. True or false: For a matrix game, each player's equilibrium strategy is dominant. *Hint:* Review the optimal solution to Program 16.1.

The next problem concerns the **Prisoner's Dilemma**, which is as follows: You and I are in jail. The prosecuting attorney meets, individually, with each of us and delivers the following news:

- He has enough to put each of us in jail for 1 year.
- If one of us squeals on the other, he will be able to put the other in jail for 7 years, and he will let the squealer go free.
- If we both squeal, the judge will let him put us both in jail for 5 years.

Each of us can "squeal" or "clam." The Prisoner's Dilemma is summarized in the following table. Its rows correspond to your actions, its columns to mine. For each pair of actions, your cost (years in jail) appears to the left, mine to the right. For instance, if you squeal and I clam, you do no time, and I do 7 years.

Jail time for the Prisoner's Dilemma

	Squeal	Clam
Squeal	5, 5	0, 7
Clam	7, 0	1, 1

13. For the Prisoner's Dilemma:

(a) What is the equilibrium strategy for each of us?

(b) Does each of us have a dominant strategy, and if so, what is it?

(c) If we could enforce a cooperative solution, would we prefer it to the solution in part (a)? to the solution, if any, in part (b)? If so, why?

14. Find an equilibrium and value for the matrix game whose payoff matrix A is

$$A = \begin{bmatrix} 1 & 2 & 3 & 4 \\ 6 & 3 & 0 & -3 \\ 2 & 8 & -5 & -1 \end{bmatrix}.$$

15. (**Bi-matrix Game**) Find an equilibrium to the bi-matrix game whose cost matrices are as follows. Show that the strategies you picked do form an equilibrium. *Hint:* Pick randomized strategies so that what each player loses is independent of the other player's strategy.

$$A = \begin{bmatrix} 1 & 5 \\ 3 & 0 \end{bmatrix}, \quad B = \begin{bmatrix} 4 & 1 \\ 3 & 2 \end{bmatrix}$$

16. (**Steroids**) Many amateur and professional sports associations test athletes at random for steroids and other performance-enhancing drugs. This problem builds a simplified model of such a test. Consider a sprint between two athletes, Abel and Cain, who are equally matched. Each is equally likely to win the sprint if neither uses performance-enhancing drugs or if both use them. One of them is certain to win the race if only he uses performance-enhancing drugs. The entries in the following table display the probability that Abel and Cain win the race under each circumstance.

	Cain abstains	Cain uses drugs
Abel abstains	1/2, 1/2	0, 1
Abel uses drugs	1, 0	1/2, 1/2

(a) Without drug testing, are there dominant strategies? If so, what are they?

(b) Now suppose that, with probability p, *both* athletes will be tested for performance-enhancing drugs just before the race. If either athlete is found to be using performance-enhancing drugs, he will be disqualified from the race and will be penalized an amount equivalent to losing K races. For specificity take $K = 15$. How does drug testing affect the arrays of payoffs that appear above? Are there values of p that induce the athletes to abstain? If so, how large must p be?

Part F

Solving Linear Programs

Chapter 17

Solving Linear Systems

17.1. PREVIEW

This chapter probes ideas introduced by the great mathematicians, Carl Friedrich Gauss (1777–1855) and Camille Jordan (1838–1921), in their studies of sets of linear equations. They showed how to transform a set of linear equations into a form whose solution is self-evident. One of their methods is now known as Gaussian elimination, and the other as Gauss-Jordan elimination. You likely first encountered one of these methods in high school when you used it to solve two equations in two unknowns or three equations in three unknowns.

17.2. WHAT CAN YOU LEARN FROM THIS CHAPTER?

The focal point of this chapter is Gauss-Jordan elimination. Gauss-Jordan elimination finds a solution to a system of linear equations. You already know how to do that. Simply use Solver. Thus, from a pragmatic viewpoint, this chapter may seem pointless. Yet Gauss-Jordan elimination deserves to be mastered because it is one of the most important elements of linear algebra and it plays a central role in linear programming.

Gauss-Jordan elimination consists of a sequence of "pivots." The same pivots lie at the heart of Dantzig's simplex method, which employs a deft sequence of pivots to find the optimal solution to a linear program. Once you understand pivots, it will be easy to learn how the simplex method works. Knowing how the simplex method works will in turn help you to understand the optimal solutions that it reports.

17.3. GAUSSIAN OPERATIONS

Gauss-Jordan elimination wrestles a system of linear equations into a form for which a solution is obvious. This is accomplished by repeated and systematic use of two operations that now bear Gauss's name. These **Gaussian operations** are:

- To replace an equation by a nonzero constant c times itself.
- To replace an equation by the sum of itself and a constant d times another equation.

To replace an equation by a constant c times itself, we multiply each addend in that equation by the constant c. Suppose, for example, that we replace the equation $2\,x - 3\,y = 6$ by the constant -4 times itself. This yields the equation, $-8\,x + 12\,y = -24$. Every solution to the former equation is a solution to the latter, and conversely. In fact, the former equation can be re-created by replacing the latter by the constant $-1/4$ times itself.

These Gaussian operations are **reversible** because their effects can be undone (reversed). To undo the effect of the first Gaussian operation, replace the equation that it produced by the constant $(1/c)$ times itself. To undo the effect of the second Gaussian operation, replace the equation that it produced by the sum of itself and the constant $-d$ times the other equation. Because Gaussian operations are reversible, they preserve the set of solutions to an equation system. We emphasize:

> Each Gaussian operation preserves the set of solutions to the equation system; it creates no solutions, and it destroys none.

To introduce Gauss-Jordan elimination, we employ system (1), which consists of four linear equations. We have numbered them Equations (17.1.1) through (17.1.4). These equations have four variables or "unknowns": x_1, x_2, x_3, and x_4. The number p that appears on the right-hand side of Equation (17.1.3) is a datum, not a decision variable.

$$2\,x_1 + 4\,x_2 - 1\,x_3 + 8\,x_4 = 4 \qquad (17.1.1)$$
$$1\,x_1 + 2\,x_2 + 1\,x_3 + 1\,x_4 = 1 \qquad (17.1.2)$$
$$2\,x_3 - 4\,x_4 = p \qquad (17.1.3)$$
$$-1\,x_1 + 1\,x_2 - 1\,x_3 + 1\,x_4 = 0 \qquad (17.1.4)$$

We will attempt to solve system (1) for a variety of values of p. Before we do so, however, please pause to speculate as to how many solutions there are to system (1). Has it no solutions? or one? or many? Or does the number of solutions depend on p? We will find out.

17.4. A PIVOT

At the heart of Gauss-Jordan elimination lies the "pivot," which is designed to give a variable a coefficient of $+1$ in a particular equation and a coefficient of 0 in each of the other equations. This pivot "eliminates" the variable from all but one of the equations.

To **pivot** on a nonzero coefficient c of a variable x in equation (j), execute these Gaussian operations:

- First, replace Equation (j) by the constant $(1/c)$ times itself.
- Then, for each k other than j, replace Equation (k) by itself minus Equation (j) times the coefficient of x in Equation (k).

The language is a bit awkward, but an example will make everything clear. We will use system (1) to illustrate pivots. For clarity, we will *not* execute these pivots on a spreadsheet—at least not yet. In a later section, we will see how easy it is to pivot on a spreadsheet.

Let us pivot on the coefficient of x_1 in Equation (17.1.1). This coefficient equals 2. This pivot executes the following *sequence* of Gaussian operations:

- Replace Equation (17.1.1) with the constant $(1/2)$ times itself.
- Replace Equation (17.1.2) with itself minus 1 times Equation (17.1.1).
- Replace Equation (17.1.3) with itself minus 0 times Equation (17.1.1).
- Replace Equation (17.1.4) with itself minus -1 times Equation (17.1.1).

The first of these Gaussian operations changes the coefficient of x_1 in Equation (17.1.1) from 2 to 1. The second of these operations changes the coefficient of x_1 in Equation (17.1.2) from 1 to 0. The third operation keeps the coefficient of x_1 in Equation (17.1.3) equal to 0. The fourth changes the coefficient of x_1 in Equation (17.1.4) from -1 to 0.

This pivot transforms system (1) into system (2). This pivot consists of Gaussian operations, so it preserves the set of solutions to system (1). In other words, each set of values of the variables x_1, x_2, x_3, and x_4 that satisfies system (1) also satisfies system (2), and conversely.

$$\begin{aligned} 1\,x_1 + 2\,x_2 - 0.5\,x_3 + 4\,x_4 &= 2 &\quad (17.2.1)\\ 1.5\,x_3 - 3\,x_4 &= -1 &\quad (17.2.2)\\ 2\,x_3 - 4\,x_4 &= p &\quad (17.2.3)\\ 3\,x_2 - 1.5\,x_3 + 5\,x_4 &= 2 &\quad (17.2.4)\end{aligned}$$

This pivot has *eliminated* the variable x_1 from Equations (17.2.2), (17.2.3), and (17.2.4) because its coefficients in these equations equal zero.

17.5. A BASIC VARIABLE

A variable is said to be **basic** for an equation if its coefficient in that equation equals 1 and if its coefficients in the other equations equal zero. The pivot that we have executed made x_1 basic for Equation (17.2.1), exactly as planned. We emphasize:

> A pivot on a nonzero coefficient of a variable in an equation makes that variable basic for that equation.

Having made x_1 basic for Equation (17.2.1), we will pivot on a nonzero coefficient in Equation (17.2.2). The variables x_3 and x_4 have nonzero coefficients in this equation. Let's pivot on the coefficient of x_3 in Equation (17.2.2). This pivot consists of the following sequence of Gaussian operations:

- Replace Equation (17.2.2) by itself divided by 1.5.
- Replace Equation (17.2.1) by itself minus -0.5 times Equation (17.2.2).

- Replace Equation (17.2.3) by itself minus 2 times Equation (17.2.2).
- Replace Equation (17.2.4) by itself minus -1.5 times Equation (17.2.2).

These Gaussian operations transform system (2) into system (3). They create no solutions and destroy none.

$$\begin{aligned}
1\,x_1 + 2\,x_2 \qquad\quad + 3\,x_4 &= 5/3 &\quad (17.3.1)\\
+ 1\,x_3 - 2\,x_4 &= -2/3 &\quad (17.3.2)\\
0\,x_4 &= p + 4/3 &\quad (17.3.3)\\
3\,x_2 \qquad\quad + 2\,x_4 &= 1 &\quad (17.3.4)
\end{aligned}$$

This pivot made x_3 basic for Equation (17.3.2). It *kept* x_1 basic. This is no accident. Why? The coefficient of x_1 in Equation (17.2.2) had been set equal to zero, so replacing another equation by itself less some constant times Equation (17.2.2) cannot change its coefficient of x_1. The property that this illustrates holds in general. We emphasize:

> Pivoting on a nonzero coefficient of a variable x in an equation has these effects:
>
> - The variable x becomes basic for the equation whose coefficient has been pivoted upon.
> - Any variables that had been basic for other equations remain basic for those equations.

The idea that motivates Gauss-Jordan elimination is to keep pivoting until we create a basic variable for each equation.

17.6. TRITE AND REDUNDANT EQUATIONS

There is, a slight complication, however, and it is now within view. Equation (17.3.3) is

$$0\,x_1 + 0\,x_2 + 0\,x_3 + 0\,x_4 = p + 4/3.$$

Let us recall that p is a datum (number), not a variable. Equation (17.3.3) has a solution if $p = -4/3$, and it has no solution $p \neq -4/3$.

This motivates a pair of definitions. The equation

$$0\,x_1 + 0\,x_2 + \cdots + 0\,x_n = d$$

is said to be **trite** if $d = 0$. The same equation is said to be **inconsistent** if $d \neq 0$. A trite equation poses no restriction on the values taken by the variables. An inconsistent equation has no solution.

Gauss-Jordan elimination creates no solutions and destroys none. Thus, if Gauss-Jordan elimination produces an inconsistent equation, the original equation system can have no solution. In particular, system (1) has no solution if $p \neq -4/3$. For the remainder of this section, we assume that $p = -4/3$. In this case, Equations (17.3.1) and (17.3.2) have basic variables, and Equation (17.3.3) is trite.

Gauss-Jordan elimination continues to pivot, aiming for a basic variable for each nontrite equation. Equation (17.3.4) lacks a basic variable. The variables x_2 and x_4 have nonzero coefficients in Equation (17.3.4). We could make either of them basic for it. Let's make x_2 basic for Equation (17.3.4). To do so, we execute this sequence of Gaussian operations:

- Replace Equation (17.3.4) by itself divided by 3.
- Replace Equation (17.3.1) by itself minus 2 times Equation (17.3.4).

- Replace Equation (17.3.2) by itself minus 0 times Equation (17.3.4).
- Replace Equation (17.3.3) by itself minus 0 times Equation (17.3.4).

This pivot transforms system (3) into system (4).

$$\begin{aligned} 1\,x_1 & & & + (5/3)\,x_4 &= 1 & \quad (17.4.1)\\ & & + 1\,x_3 & - 2\,x_4 &= -2/3 & \quad (17.4.2)\\ & & & 0\,x_4 &= 0 & \quad (17.4.2)\\ & + 1\,x_2 & & + (2/3)\,x_4 &= 1/3 & \quad (17.4.4) \end{aligned}$$

In system (4), each nontrite equation has been given a basic variable. A solution to system (4) is evident. Equate each basic variable to the right-hand-side value of the equation for which it is basic, and equate any other variables to zero. That is, set:

$$x_1 = 1, \quad x_3 = -2/3, \quad x_2 = 1/3, \quad x_4 = 0.$$

These values of the variables satisfy system (4) and hence must satisfy system (1).

More can be said. Shifting the nonbasic variable x_4 to the right-hand side of system (4) expresses *every* solution to system (4) as a function of x_4. Specifically, for each value of x_4, setting

$$\begin{aligned} x_1 &= 1 - (5/3)\,x_4,\\ x_3 &= -2/3 + 2\,x_4,\\ x_2 &= 1/3 - (2/3)\,x_4, \end{aligned}$$

satisfies system (4) and, consequently, satisfies system (1).

17.7. A BASIC EQUATION SYSTEM AND ITS BASIC SOLUTION

With system (4) in view, we introduce two key definitions. A system of linear equations is said to be **basic** if each nontrite equation has a basic variable. System (4) is basic because Equation (17.4.3) is trite and because the remaining equations have basic variables.

A basic equation system's **basic solution** equates each nonbasic variable to zero and equates each basic variable to the right-hand-side value of the equation for which it is basic. The basic solution to system (4) is:

$$x_1 = 1, \quad x_3 = -2/3, \quad x_2 = 1/3, \quad x_4 = 0.$$

Let us summarize. Gauss-Jordan elimination pivots in search of a basic system or an inconsistent equation, as follows.

Gauss-Jordan elimination:

1. Stop if each nontrite equation has a basic variable. Otherwise, proceed to Step 2.
2. Select a nontrite equation that lacks a basic variable. Stop if this equation is inconsistent. Otherwise, select any variable whose coefficient in this equation is nonzero, and pivot on it. Return to Step 1.

Each pivot that is executed in Step 2 creates a basic variable for an equation that lacked one. If Gauss-Jordan elimination stops in Step 1, its basic solution satisfies the original equation system. Alternatively, if Gauss-Jordan elimination stops in Step 2, it has constructed an inconsistent equation, in which case the original equation system can have no solution.

17.8. THE WORK OF GAUSS-JORDAN ELIMINATION

Some computers are faster than others. A feature of a calculation that is common across computers is its **work**, which is defined to be the number of computer operations needed to execute the calculation. These computer operations include memory accesses, additions, multiplications, divisions, comparisons, and so forth. Less work connotes less time to execute a computation on a computer.

We aim to estimate the work needed by Gauss-Jordan elimination to make a system basic. As a surrogate for work, we count multiplications and divisions. This is reasonable because the total work of Gauss-Jordan elimination is roughly proportional to the number of multiplications and divisions.

Let us consider a system of m linear equations in n variables (unknowns). How many multiplications and divisions are needed to execute one pivot? Each equation has $n + 1$ data elements, including its right-hand-side value. The first Gaussian operation in a pivot divides an equation by the coefficient of one of its variables. This requires n divisions (not $n + 1$) because we needn't divide a number by itself. Each of the remaining Gaussian operations in a pivot replaces an equation by itself less a particular constant d times another equation. This requires n multiplications (not $n + 1$) because we needn't compute $d - d = 0$. We've seen that each Gaussian operation in a pivot requires n multiplications or divisions. Each pivot entails m Gaussian operations, one per equation. Also, Gauss-Jordan elimination requires as many as m pivots. From this, we conclude:

> Suppose that Gauss-Jordan elimination is used to find a solution to a system of m linear equations in n unknowns. Then:
>
> - Each pivot requires mn multiplications and divisions.
> - Gauss-Jordan elimination requires m^2n multiplications and divisions.

This work-bound of Gauss-Jordan elimination is m^2n. Multiplying m and n by 2 multiplies m^2n by $2^3 = 8$. Evidently, the work grows as the *cube* of the problem size. In later sections, we'll reduce the work of Gauss-Jordan elimination by nearly half, and we'll see how to reduce it further for systems that are "sparse."

17.9. PIVOTING ON A SPREADSHEET

In this section, we see how easy it is to pivot on a spreadsheet. First, we pivot by specifying a pair of Excel functions and using Excel's "drag" and "copy" features. Then we describe an Excel Add-In that pivots.

Detached-Coefficient Tableau

A spreadsheet is now used to solve system (1) for the case in which $p = -4/3$. Our first step is to introduce the **detached-coefficient tableau** in Table 17.1. This tableau is similar to the standardized spreadsheet, except that its column of "=" signs has been abbreviated to a (vertical) dashed line. Each variable has become a column heading, and the data in each equation have been placed in a row.

Table 17.1 Detached-coefficient tableau for system (1).

	A	B	C	D	E	F
1	equation	x1	x2	x3	x4	RHS
2	(1.1)	2	4	-1	8	4
3	(1.2)	1	2	1	1	1
4	(1.3)	0	0	2	-4	-1.33333
5	(1.4)	-1	1	-1	1	0

The First Pivot

As before, the first pivot will occur on the coefficient of x_1 in Equation (17.1.1). The functions in Table 17.2 indicate how to accomplish this on a spreadsheet.

Rows 8 through 11 of Table 17.2 contain the data in system (2). To create these rows, you can use the "drag" and "copy" features, as follows:

- To equate row 8 to row 2 divided by the number 2, drag the function **=B2/$B2** in cell B8 across row 8.
- To equate row 9 to row 3 minus (+1) times row 3, drag the function **=B3−$B3*B$3** in cell B9 across row 9.
- To equate row 10 to row 4 minus (0) times row 3, copy the function in cell B9, paste it into cell B10, and then drag it across row 10.
- To equate row 11 to row 5 minus (−1) times row 3, copy the function in cell B9, paste it into cell B11, and then drag it across row 11.

Why so complicated? We could have put the function **=B$2/2** in cell B8. Dragging this function across row 8 would give the correct answer. Similarly, we could have put the

Table 17.2 A spreadsheet that makes x_1 basic for Equation (17.2.1).

	A	B	C	D	E	F
1	equation	x1	x2	x3	x4	RHS
2	(1.1)	2	4	-1	8	4
3	(1.2)	1	2	1	1	1
4	(1.3)	0	0	2	-4	-1.33333
5	(1.4)	-1	1	-1	1	0
6						
7	equation	x1	x2	x3	x4	RHS
8	(2.1)	1	2	-0.5	4	2
9	(2.2)	0	0	1.5	-3	-1
10	(2.3)	0	0	2	-4	-1.33333
11	(2.4)	0	3	-1.5	5	2
12						
13	= B2/$B2		= B3 - $B3*B$8			

function **=B3 − (+1)*B3** in cell B11. Dragging this function across row 11 would give the correct answer. These simpler functions do the job. You may prefer them, and you are welcome to use them, but bear in mind that

- You will need four of these simpler functions rather than two.
- These simpler functions will not pivot correctly if you subsequently change the data in column B.

The Second Pivot

We are now ready for the second pivot. As before, the second pivot occurs on the coefficient of x_3 in Equation (17.2.2). Table 17.3 indicates how to execute this pivot on a spreadsheet.

In Table 17.3, rows 16–19 depict system (3). To create row 17, drag the function in cell D17 across row 17. To create row 19, drag the function in cell D19 across it. To create rows 16 and 18, copy the function in cell D17, paste it into cells D16 and D18, and then drag it across their rows.

The Third Pivot

Equation (17.3.3) is trite, and so Gauss-Jordan elimination skips this equation. As before, its third pivot occurs on the coefficient of x_2 in Equation (17.3.4). This pivot results in Table 17.4.

Three pivots have now been executed on a spreadsheet. Once you get the knack of fixed and relative addressing, pivoting on a spreadsheet is easy. Pivoting by hand is time-consuming, error-prone, repetitious, boring, and obsolete.

An Add-In That Expedites Pivoting

Pivoting on a spreadsheet is even easier than we have made it seem. An Add-In function on the CD that accompanies this text accomplishes the pivot with a single Array operation. To use it, you must remember that Array operations do *not* end by hitting the **Enter** key by

Table 17.3 The variable x_3 has been made basic for Equation (17.3.2).

	A	B	C	D	E	F
7	equation	x1	x2	x3	x4	RHS
8	(2.1)	1	2	-0.5	4	2
9	(2.2)	0	0	1.5	-3	-1
10	(2.3)	0	0	2	-4	-1.33333
11	(2.4)	0	3	-1.5	5	2
12						
13			= D9/$D9		= D8 - $D8*D$17	
14						
15	equation	x1	x2	x3	x4	RHS
16	(3.1)	1	2	0	3	1.666667
17	(3.2)	0	0	1	-2	-0.66667
18	(3.3)	0	0	0	0	0
19	(3.4)	0	3	0	2	1

Table 17.4 A basic equation system.

	A	B	C	D	E	F
15	equation	x1	x2	x3	x4	RHS
16	(3.1)	1	2	0	3	1.666667
17	(3.2)	0	0	1	-2	-0.66667
18	(3.3)	0	0	0	0	0
19	(3.4)	0	3	0	2	1
20						
21	equation	x1	x2	x3	x4	RHS
22	(4.1)	1	0	0	1.666667	1
23	(4.2)	0	0	1	-2	-0.66667
24	(4.3)	0	0	0	0	0
25	(4.4)	0	1	0	0.666667	0.333333

itself. To perform an Array operation, you:

- Select the block of cells in which you wish the result of the Array operation to appear.
- Type in the Array function.
- Depress the Ctrl and Shift keys and, while both are depressed, hit the Enter key.

In particular, to execute the pivot displayed in Table 17.5, you:

- Select the block B8:F11 in which you wish the result of this pivot to appear.
- Type **=PIVOT(B2: B2:F5)** to pivot on the coefficient in cell B2 of the array in cells B2:F5.
- Type **Ctrl+Shift+Enter** to tell Excel that this is an Array operation.

Table 17.5 Using the Array function =PIVOT(cell, range) to pivot.

	A	B	C	D	E	F
1	equation	x1	x2	x3	x4	RHS
2	(1.1)	2	4	-1	8	4
3	(1.2)	1	2	1	1	1
4	(1.3)	0	0	2	-4	-1.33333
5	(1.4)	-1	1	-1	1	0
6						
7	equation	x1	x2	x3	x4	RHS
8	(2.1)	1	2	-0.5	4	2
9	(2.2)	0	0	1.5	-3	-1
10	(2.3)	0	0	2	-4	-1.33333
11	(2.4)	0	3	-1.5	5	2
12		↑				
13		= PIVOT(B2, B2:F5)				

In Excel, Array functions are a bit quirky. When you use them, you may encounter difficulties. If you do have problems, check the Appendix, which discusses their peculiarities.

17.10. EXCHANGE OPERATIONS

A typical presentation of Gauss-Jordan elimination includes four Gaussian operations, of which we have presented only two. We have omitted these two **exchange operations**:

- Exchange the positions of a pair of equations.
- Exchange the positions of a pair of variables.

Like the others, these exchange operations can be undone. To recover the original equation system after doing an exchange operation, simply repeat it.

The exchange operations do not help us to construct a basis. They do serve a "cosmetic" purpose, one that lets us state results in simple language. For instance, the exchange operations let us place the basic variables on the diagonal and the trite equations at the bottom. To illustrate, reconsider Table 17.4. Exchanging rows 24 and 25 shifts the trite equation to the bottom. Then, exchanging columns C and D puts the basic variables on the diagonal.

In linear algebra, the two Gaussian operations that were introduced earlier and the first of the above two exchange operations are known as **elementary row operations**. Most texts on linear algebra *begin* with a discussion of elementary row operations and their properties. They do so because these Gaussian operations are the workhorse of linear algebra. For a glimpse of the fundamental role that Gauss-Jordan elimination plays in linear algebra, see Problem 13.

17.11. LOWER PIVOTS AND GAUSSIAN ELIMINATION*

The (full) pivots that we have described are crucial to an understanding of the simplex method and they are important to linear algebra. But they are inefficient. The method that's known as Gaussian elimination substitutes "lower pivots" for pivots. It solves an equation system with roughly half the effort.

To describe lower pivots, we identify the set S of equations on which lower pivots have not yet occurred. Initially, S consists of all the equations in the system that is being solved. Each lower pivot selects an equation in S, removes it, and executes certain Gaussian operations on the equations that remain in S. Specifically, each **lower pivot** consists of these steps:

- Select an Equation (j) in S and a variable x whose coefficient in Equation (j) is nonzero.
- Remove Equation (j) from S.
- For each Equation (k) that remains in S, replace Equation (k) by itself less the multiple of Equation (j) that equates the coefficient of x in Equation (k) to zero.

This verbal description of lower pivots is cumbersome, but an example will make everything clear.

A Familiar Example

To illustrate lower pivots, we return to system (1). We will re-solve this system with each "full" pivot replaced by the comparable lower pivot. For convenient reference, system (1)

*This section concerns efficient computation. It can be skipped.

is reproduced here as system (5).

$$2\,x_1 + 4\,x_2 - 1\,x_3 + 8\,x_4 = 4 \qquad (17.5.1)$$
$$1\,x_1 + 2\,x_2 + 1\,x_3 + 1\,x_4 = 1 \qquad (17.5.2)$$
$$2\,x_3 - 4\,x_4 = p \qquad (17.5.3)$$
$$-1\,x_1 + 1\,x_2 - 1\,x_3 + 1\,x_4 = 0 \qquad (17.5.4)$$

Initially, before any lower pivots have occurred, the set S consists of Equations (17.5.1) through (17.5.4).

The First Lower Pivot

In this illustration, we will select the same pivot elements that we did when implementing Gauss-Jordan elimination with full pivots. As before, the first lower pivot will occur on the coefficient of x_1 in Equation (17.5.1). This lower pivot eliminates (drives to zero) the coefficient of x_1 in Equations (17.5.2), (17.5.3), and (17.5.4). To execute this lower pivot, we remove Equation (17.5.1) from S and then:

- Replace Equation (17.5.2) by itself minus (1/2) times Equation (17.5.1).
- Replace Equation (17.5.3) by itself minus (0/2) times Equation (17.5.1).
- Replace Equation (17.5.4) by itself minus (−1/2) times Equation (17.5.1).

The three equations that remain in S become:

$$1.5\,x_3 - 3\,x_4 = -1 \qquad (17.6.2)$$
$$2\,x_3 - 4\,x_4 = p \qquad (17.6.3)$$
$$3\,x_2 - 1.5\,x_3 + 5\,x_4 = 2 \qquad (17.6.4)$$

The variable x_1 does not appear in Equations (17.6.2), (17.6.3), and (17.6.4). These three equations are *identical* to Equations (17.2.2), (17.2.3), and (17.2.4), as must be.

Equation (17.5.1) is set aside. Once Equations (17.6.2) through (17.6.4) have been solved for values of the variables x_2, x_3, and x_4, we will return to Equation (17.5.1) and solve it for the value of x_1 that is prescribed by these values of x_2, x_3, and x_4.

The Second Lower Pivot

As was the case in our initial presentation of Gauss-Jordan elimination, the second pivot element will be the coefficient of x_3 in Equation (17.6.2). A lower pivot on this coefficient will drive to zero the coefficient of x_3 in Equations (17.6.3) and (17.6.4). To execute this lower pivot, we remove Equation (17.6.2) from S and then:

- Replace Equation (17.6.3) itself minus (2/1.5) times Equation (17.6.2).
- Replace Equation (17.6.4) by itself minus (−1.5/1.5) times Equation (17.6.2).

This lower pivot replaces (17.6.3) and (17.6.4) by Equations (17.7.3) and (17.7.4).

$$0\,x_4 = p + 4/3 \qquad (17.7.3)$$
$$3\,x_2 \qquad + 2\,x_4 = 1 \qquad (17.7.4)$$

The variable x_3 has been eliminated from Equations (17.7.3) and (17.7.4). These two equations are identical to Equations (17.3.3) and (17.3.4), exactly as in the case for the first lower pivot.

Equation (17.6.2), on which this pivot occurred, is set aside. After solving Equations (17.7.3) and (17.7.4) for values of the variables x_2 and x_4, we will return to Equation (17.6.2) and solve it for the variable x_3 on which the lower pivot has occurred.

The next lower pivot is slated to occur on Equation (17.7.3). Again, there are two cases to consider. If p is unequal to $-4/3$, Equation (17.7.3) is inconsistent, so no solution can exist to the original equation system. Alternatively, if $p = -4/3$, Equation (17.7.3) is trite, and it has nothing to pivot upon.

We proceed on the assumption that $p = -4/3$. In this case, Equation (17.7.3) is trite, so we remove it from S, which reduces S to Equation (17.8.4), below.

$$3\,x_2 \qquad\qquad + 2\,x_4 = 1 \tag{17.8.4}$$

The Final Lower Pivot

Only Equation (17.8.4) remains in S. The next step calls for a lower pivot on Equation (17.8.4). The variables x_2 and x_4 have nonzero coefficients in Equation (17.8.4), so a lower pivot could occur on either of them. As before, we pivot on the coefficient of x_2 in this equation. But no equations remain in S after Equation (17.8.4) is removed. Hence, this lower pivot entails no arithmetic. As concerns lower pivots, we are finished.

Back-Substitution

It remains to construct a solution to system (5). To do so, we equate to zero each variable on which no lower pivot has occurred, and then we solve the equations on which lower pivots have occurred in "reverse" order. In our example, no lower pivot has occurred on the variable x_4. With $x_4 = 0$, the three equations on which lower pivots have occurred are:

$$\begin{aligned} 2\,x_1 + 4\,x_2 - 1\,x_3 &= 4 \\ 1.5\,x_3 &= -1 \\ 3\,x_2 &= 1 \end{aligned}$$

The first lower pivot eliminated x_1 from the bottom two equations, and the second lower pivot eliminated x_3 from the bottom equation. Thus, these equations can be solved for the variables on which their lower pivots have occurred by working from the bottom up. This process is aptly called **back-substitution**. For our example, back-substitution first solves the bottom equation for x_2, then solves the middle equation for x_3, and finally solves the top equation for x_1. This computation gives $x_2 = 1/3$ and $x_3 = -2/3$ and $x_1 = 1$, exactly as before.

Solving an equation system by lower pivots and back-substitution is known as **Gaussian elimination** and by the fancier label, **L-U decomposition**. By either name, it requires roughly half as many multiplications and divisions as does Gauss-Jordan elimination. This suggests that lower pivots are twice as good. Actually, lower pivots are vastly superior. This is so because lower pivots will let us take advantage of "sparsity" and help us to control "round-off" error.

17.12. SPARSENESS AND ROUND-OFF ERROR*

Typically, a large system of linear equations is **sparse**, which means that all but a tiny fraction of its coefficients are zeros. Sparseness accelerates Gauss-Jordan elimination because we can avoid multiplying by zero. Unfortunately, as pivoting proceeds, a sparse equation system tends to "fill in" as nonzero entries replace zeros. An adroit sequence of pivots can reduce the rate at which fill-in occurs. This section tells how.

*This section concerns efficient computation. It can be skipped.

Retarding Fill-in

A simple method for retarding fill-in requires us to count the number of nonzero entries in each "column" and "row." Specifically:

- For each variable x_k, we count the number c_k of different equations in which x_k has a nonzero coefficient.
- For each Equation (j), we count the number r_j of different variables whose coefficients in Equation (j) are nonzero.

Take a moment to convince yourself that a pivot on the coefficient of the variable x_k in Equation (j) will fill in (render nonzero) at most $(r_j - 1)(c_k - 1)$ zeros. This motivates the rule that's displayed below.

Myopic Pivot Element Picker:

1. Select a nonzero pivot element that minimizes $(r_j - 1)(c_k - 1)$ as j ranges over those equations on which pivots have not yet occurred and as k ranges over those indices such that x_k has not yet been made basic. Pivot on this element. Then go to Step 2.
2. Update r_j for each equation j on which no pivot has yet occurred. Update c_k for each variable x_k that has not yet been made basic. Return to Step 1.

This rule is **myopic** (near-sighted) in the sense that it aims to minimize the amount of fill-in at the moment, without looking ahead.

Fill-in with Lower Pivots

The myopic pivot element finder has been presented for normal (full) pivots. This lessens fill-in. The myopic pivot element finder further lessens fill-in when it is used in conjunction with lower pivots. To do so, we redefine c_k to be the number of equations on which lower pivots have not yet occurred in which x_k has nonzero coefficients. With this change in the definition of c_k, a lower pivot on the coefficient of the variable x_k in Equation (j) will fill in (render nonzero) at most $(r_j - 1)(c_k - 1)$ zeros.

Numerical Approximation

No pivot rule will work well unless it is qualified to avoid pivoting on coefficients that are close to zero. To indicate why, we need to discuss the bane of numerical computation: "round-off" error.

To facilitate arithmetic, spreadsheet programs store numbers in "floating-point" format. This causes certain fractions, such as $1/3$, to be approximated. The binary equivalent of $1/3$ is 0.010101 . . . . Here, the ellipsis (three dots) means that sequence 01 repeats without limit, that is, forever. If you place the function **=1/3** in a cell, the computer records a fixed number (perhaps 52) of these 0's and 1's, and it drops the rest. Dropping them introduces a miniscule amount of **round-off** error.

Round-off error is apparent in the output that Solver reports. Solver may report that the slack variable in a constraint equals −2.06E-13. Here, the symbol "E" stands for exponent, and −2.06E-13 is to be read as -2.06×10^{-13}. This number is tiny. It is almost surely round-off error. Round-off errors start tiny, but each arithmetic operation can introduce new round-off error and can magnify earlier round-off errors.

To see how round-off errors can magnify, we consider an extreme example. Imagine that we pivot on a coefficient that equals 10^{-6}, and imagine that all other nonzero coefficients are between 1 and 10 in absolute value. The first Gaussian operation multiplies the nonzero elements in the pivot row by $10^6 = 1{,}000{,}000$. As a result:

- Any round-off error in the pivot row gets multiplied by a factor of 1,000,000.
- The pivot element becomes 1. The other nonzero coefficients in the pivot row have magnitudes between 1,000,000 and 10,000,000.

This Gaussian operation has had two bad effects. First, it greatly amplifies any round-off errors in the pivot row. Second, the large coefficients in the pivot row *dwarf* the data in the other rows. When another equation is replaced by itself less a constant times the pivot row, the large coefficients in the pivot row will dominate that equation. Whatever information it held will be severely attenuated. Needless to say perhaps, computer codes that execute Gauss-Jordan elimination avoid pivoting on numbers whose magnitudes are close to zero.

In a commercial computer code that executes Gaussian elimination, the hard work lies in the control of round-off error. Inevitably, after enough arithmetic operations have occurred, the round-off error will have magnified to the point that it swamps the data. This limits the size of the problem that can be solved. Lower pivots and careful selection of pivot elements reduce the number of arithmetic operations, slow the growth of round-off error, and increase the size of the system that can be solved.

17.13. REVIEW

Gauss-Jordan elimination makes repeated and systematic use of two Gaussian operations. These operations are organized into pivots, with each pivot creating a basic variable for an equation that lacked one. In addition, each pivot keeps the variables that had been basic for the other equations basic for those equations. Gauss-Jordan elimination keeps pivoting until:

- Either it constructs an inconsistent equation,
- Or it creates a basic system, specifically a basic variable for each nontrite equation.

If Gauss-Jordan elimination constructs an inconsistent equation, the original equation system can have no solution. If Gauss-Jordan elimination constructs a basic system, its basic solution satisfies the original equation system. This basic solution equates each nonbasic variable to zero, and it equates each basic variable to the right-hand-side value of the equation for which it is basic.

In Chapter 18, we will see that the simplex method begins with Gauss-Jordan elimination and that it keeps on pivoting; it *continues* to pivot in search of an optimal solution to the linear program. Before studying that chapter, make sure that you are familiar with these terms: pivot, basic variable, basic system, basic solution, trite equation, and inconsistent equation. Also make sure that you can pivot on a spreadsheet.

A pair of starred sections has shown that, from a computational perspective, lower pivots are preferable to full pivots. Lower pivots require less work, they retard fill-in, and they create less opportunity for round-off error to accumulate.

17.14. HOMEWORK AND DISCUSSION PROBLEMS

1. To solve the following system of linear equations, implement Gauss-Jordan elimination on a spreadsheet. Turn your spreadsheet in, and indicate the functions that you have used in your computation.

$$\begin{aligned} 1\,A - 1\,B + 2\,C &= 10 \\ -2\,A + 4\,B - 2\,C &= 0 \\ 0.5\,A - 1\,B - 1\,C &= 6 \end{aligned}$$

2. Consider the following system of three equations in three unknowns.

$$\begin{aligned} 2\,A + 3\,B - 1\,C &= 12 \\ -2\,A + 2\,B - 9\,C &= 3 \\ 4\,A + 5\,B \qquad &= 21 \end{aligned}$$

(a) Use Gauss-Jordan elimination to find a solution to this equation system.

(b) Plot those solutions to this equation system in which each variable is nonnegative. Complete this sentence: The solutions that have been plotted form a ________.

(c) What would have happened if one of the right-hand-side values had been different from what it is? Why?

3. Use a spreadsheet to find *all* solutions to this system of linear equations:

$$\begin{aligned} 2\,x_1 + 4\,x_2 - 1\,x_3 + 8\,x_4 + 10\,x_5 &= 4 \\ 1\,x_1 + 2\,x_2 + 1\,x_3 + 1\,x_4 + 2\,x_5 &= 1 \\ 2\,x_3 - 4\,x_4 - 4\,x_5 &= -4/3 \\ -1\,x_1 + 1\,x_2 - 1\,x_3 + 1\,x_4 - 1\,x_5 &= 0 \end{aligned}$$

4. Redo the spreadsheet computation in Tables 17.1–17.4 using lower pivots in place of (full) pivots. Turn in your spreadsheet, and on it, indicate the functions that you used.

5. Tables 17.1–17.4 showed how to execute Gauss-Jordan elimination on a spreadsheet for the special case in which the datum p equals $-4/3$. Redo this spreadsheet for the general case in which the datum p can be any number. *Hint:* To Table 17.1, append column G whose heading (in row 1) is p and whose coefficients in rows 2, 3, 4, and 5 are 0, 0, 1, and 0, respectively.

6. **(Identical Columns)** A detached-coefficient tableau is said to have **identical columns** if two of its columns are identical, except for their column headings. For specificity, suppose that the column headings are labeled x_1 through x_{23} and that columns 5 and 12 are identical. Then the coefficient of x_5 in each equation that this tableau depicts equals the coefficient of x_{12} in that equation. Complete the following sentences.

(a) After one Gaussian operation, columns 5 and 12 ________.

(b) After any number of Gaussian operations, columns 5 and 12 ________.

(c) If a pivot makes x_5 basic for some equation, then x_{12} ________.

(d) Gauss-Jordan elimination, as we have specified it, is not quite precise because ________.

7. **(Opposite Columns)** Redo parts (a)–(c) of the preceding problem for the case of opposite columns.

Definition: A system of m linear equations in n variables (unknowns) is said to be **homogeneous** if the right-hand-side value of each equation equals zero. A solution to a homogeneous system is said to be **nontrivial** if at least one variable is not assigned the value zero.

8. **(Homogeneous Systems)** True or false?

(a) When Gauss-Jordan elimination is applied to a homogeneous system, it can produce an inconsistent equation.

(b) Every homogeneous system has at least one nontrivial solution.

9. **(Homogeneous Systems)** A homogeneous system has the "trivial" solution in which each decision variable is set equal to zero.

(a) Show how to use Gauss-Jordan elimination to determine whether a homogeneous system has a non-trivial solution and to construct one if it does.

(b) Apply your method to the homogeneous version of system (1).

(c) Prove or disprove: Every homogeneous system of four equations in five variables had a non-trivial solution.

10. (**Work for Lower Pivots and Back-Substitution**) Imagine that we solve a system of m equations in n variables by lower pivoting and back-substitution. Suppose that no trite or inconsistent equations are encountered.

(a) Show that the number of multiplications and divisions required by back-substitution equals $(1 + 2 + \cdots + m) = (m)(m + 1)/2$.

(b) For each $j < m$, show that the jth lower pivot requires $(m + 1 - j)(n)$ multiplications and divisions.

(c) How many multiplications and divisions are needed to execute Gauss-Jordan elimination with lower pivots and back-substitution? *Hint:* Summing part (b) gives $(2 + 3 + \cdots + m)(n) = (n)(m)(m + 1)/2 - n$.

11. (**Sparseness**) In the detached-coefficient tableau that follows, each nonzero number is represented by an asterisk (*). Specify a sequence of lower pivots that implements the myopic rule, with c_k equal to the number of nonzero coefficients of x_k in rows on which pivots have not yet occurred. How many Gaussian operations does this implementation require? How many multiplications and divisions does it require, assuming that you omit multiplication by zero?

Equation	x_1	x_2	x_3	x_4	x_5	RHS
(1)	*	*	*	*	*	*
(2)	*	*				*
(3)		*	*			*
(4)			*	*	*	*
(5)				*	*	*

12. (**Small Pivot Elements**) You are to solve the following system twice, each time by Gauss-Jordan elimination. Throughout each computation, you are to approximate each coefficient by three significant digits; this would round the number 0.01236 to 0.0124, for instance. Begin the first execution with a pivot on the coefficient of A in the topmost equation. Begin the second execution with a pivot on the coefficient of B in the topmost equation. Compare your solutions. What happens? Why?

$$0.001\,A + 1\,B = 10$$
$$1\,A - 1\,B = 0$$

Definitions: The next problem relates material in this chapter to linear algebra. It presumes a familiarity with vector spaces, particularly with the multiplication of a vector α by a scalar x and the addition of vectors α and β. The set $\{\alpha_k : k \in S\}$ of vectors in a vector space is said to be **linearly independent** if the only solution to the (homogeneous) equation system

$$\sum_{k:k\in S} \alpha_k x_k = 0$$

is $x_k = 0$ for each $k \in S$. This set $\{\alpha_k : k \in S\}$ of vectors is said to **span** the vector β if there exists a set $\{x_k : k \in S\}$ of scalars such that

$$\beta = \sum_{k:k\in S} \alpha_k x_k.$$

Finally, a set $\{\alpha_k : k \in S\}$ of vectors is called a **basis** for a vector space if this set $\{\alpha_k : k \in S\}$ is linearly independent and if it spans each vector in the vector space.

13. (**A Basis**) This problem concerns the four vectors that are listed below. Solve parts (a), (b), and (c) *without* doing any numerical computation. (Instead, adapt Tables 17.2–17.4.)

(a) Show that the leftmost three of these vectors are linearly independent.

(b) Show that the leftmost three of these vectors span the other one.

(c) Show that the leftmost three of these vectors are a basis for the vector space that consists of all linear combinations of these four vectors.

$$\begin{bmatrix} 2 \\ 1 \\ 0 \\ -1 \end{bmatrix}, \begin{bmatrix} 4 \\ 2 \\ 0 \\ 1 \end{bmatrix}, \begin{bmatrix} -1 \\ 1 \\ 2 \\ -1 \end{bmatrix}, \begin{bmatrix} 8 \\ 1 \\ -4 \\ 1 \end{bmatrix}.$$

Chapter 18

The Simplex Method

18.1. PREVIEW

This chapter presents the simplex method, which is the principal tool for computing solutions to linear programs. The simplex method provides a *complete* analysis of a linear program. When it is applied to a linear program:

- The simplex method determines whether or not the linear program has a feasible solution.
- If the linear program does have a feasible solution, the simplex method determines whether or not its objective is bounded.
- If the linear program is feasible and bounded, the simplex method finds an optimal solution to it.

We will see how the simplex method performs these tasks.

George B. Dantzig developed the simplex method in 1947. Over the ensuing decades, many other techniques for solving linear programs have been proposed. Nearly all of them have proved to be slower than the simplex method in solving practical problems. Only in the recent past has a family of "interior-point" methods been developed that compete in speed with the simplex method over broad ranges of practical linear programs. These newer techniques are far more complicated than the simplex method, however, and they may provide less insight into the problem that is being solved.

18.2. WHAT CAN YOU LEARN FROM THIS CHAPTER?

Computer codes that execute the simplex method are widely available, and they run on nearly every computer, large or small. Solver is one such code, and it comes with popular spreadsheet packages, including Excel. You can solve linear programs without knowing how the simplex method works. Why, then, should you learn it? We give three reasons.

First, at the heart of Dantzig's simplex method lies a beautiful idea that is known as the "feasible pivot." The feasible pivot is simple, insightful, and potent. It has a lovely geometric interpretation. We'll see that the simplex *method* describes a family of algorithms, each of which crochets feasible pivots into a scheme that solves linear programs. We'll also see how the feasible pivot helps us to understand the optimal solution, its shadow prices, its reduced costs, and the ranges over which they apply.

The feasible pivot plays a fundamental role in constrained optimization, much as Gaussian operations play a fundamental role in linear algebra. Feasible pivots adapt to solve optimization problems that are distinctly nonlinear. They also adapt to provide deep insight into analysis. Learning about feasible pivots is one good reason to study this chapter.

In addition, the simplex method provides economic insight into linear programs and into the problems they model. In Chapter 3, shadow prices were described as a property of the optimal solution. Here, we'll see that shadow prices are intrinsic to the simplex method; they are present at each step of the simplex method, and they guide its progress.

Finally, the simplex method sets the stage for the discussion of duality in Chapter 19. There, we will see that the simplex method solves a pair of linear programs, the one under attack and its "dual," and we will learn how duality broadens the class of problems that can be posed as linear programs.

This chapter dovetails with Chapter 17. To understand this chapter, you will need a firm grasp of these terms: trite equation, inconsistent equation, basic variable, nonbasic variable, basic system, basis, basic solution, and tableau. In addition, you will need to know how to execute a pivot on a spreadsheet. The first 16 sections of this chapter contain its basic insights. The remaining sections are more advanced and so can be skimmed or skipped.

18.3. AN EXAMPLE

Our introduction to the simplex method commences with an example that is familiar from earlier chapters. It is:

Program 18.1: Maximize $\{2A + 3B\}$ subject to the constraints

$$\begin{aligned} A \quad\quad &\le 6,\\ A + \quad B &\le 7,\\ 2B &\le 9,\\ -A + 3B &\le 9,\\ A \quad\quad &\ge 0,\\ B &\ge 0. \end{aligned}$$

Since Program 18.1 has only two decision variables, we will be able to use it to provide a simple visual interpretation of the simplex method.

A Format That Facilitates Pivoting

The simplex method consists of deftly selected pivots. Our first step is to place Program 18.1 in a format that facilitates pivoting. This format is dubbed "Form 1." A linear program is said to be written in **Form 1** if:

- Its objective is to maximize the quantity z or to minimize z.
- Each decision variable other than z is constrained to be nonnegative.
- All other constraints are equations.

Let us cast Program 18.1 in Form 1. To establish z as the quantity that we wish to maximize, we append to the program the "counting" constraint,

$$2A + 3B = z.$$

This constraint equates z to the value of the objective function. Program 18.1 has four "$\le$" constraints. To convert these constraints into equations, we insert a nonnegative slack variable on the left-hand side of each. This rewrites the program as:

Program 18.2: Maximize $\{z\}$, subject to the constraints

$$\begin{aligned} 2A + 3B \quad\quad\quad\quad - z &= 0, && (18.1.0)\\ 1A \quad\quad + s_1 \quad\quad\quad &= 6, && (18.1.1)\\ 1A + 1B \quad + s_2 \quad\quad &= 7, && (18.1.2)\\ 2B \quad\quad + s_3 \quad &= 9, && (18.1.3)\\ -1A + 3B \quad\quad\quad + s_4 &= 9, && (18.1.4)\\ A \ge 0, \quad B \ge 0, \quad s_i \ge 0 \quad &\text{for } i = 1, 2, 3, 4. \end{aligned}$$

Program 18.2 is written in Form 1. Program 18.2 has seven decision variables and five equality constraints. Each decision variable other than z is constrained to be nonnegative. The variable z has been shifted to the left-hand side of Equation (18.1.0) because we want all of the decision variables to be on the left-hand sides of the constraints.

Form 1 aims to treat the objective as closely as possible to the other decision variables. In Form 1, the variable z plays a special role because it measures the objective. We elect to think of $-z$ as a decision variable. In Program 18.2, the variable $-z$ is basic for Equation (18.1.0) because $-z$ has a coefficient of $+1$ in Equation (18.1.0) and coefficients of 0 in all other equations.

A Canonical Form

A **canonical form** for linear programs is any format into which each linear program can be cast. Form 1 is a canonical form, which we can verify through the following observations:

- Form 1 encompasses maximization problems and minimization problems.
- An equation can be included that equates z to the value of the objective.
- Each inequality constraint can be converted into an equation by insertion of a nonnegative (slack or surplus) variable.
- Each variable that is unconstrained in sign can be replaced by the difference of two nonnegative variables.

Since Form 1 is canonic, describing the simplex method for Form 1 shows how to solve *every* linear program. How easy is it to place a linear program in Form 1? Slack and surplus variables can be inserted automatically as the linear program is inserted into the computer. The same is true of the equation that defines z as the objective value. On the other hand, replacing a variable that is unconstrained in sign by the difference of two nonnegative variables is a cumbersome process. And it corrupts sensitivity analysis.

Form 1 is not ideal. Unconstrained variables do arise in applications. Later in this chapter, we will see how to handle unconstrained variables directly, without forcing the linear program into Form 1.

18.4. A FEASIBLE BASIS

Collectively, Equations (18.1.0) through (18.1.4) form system (1). System (1) is basic because each of its equations has a basic variable. The variable $-z$ is basic for Equation (18.1.0), and the slack variables are basic for the other equations. In system (1), the variables A and B are nonbasic. The basic solution to a basic system equates each nonbasic variable to zero, and it equates each basic variable to the right-hand-side value of the equation for which it is basic. The basic solution to system (1) is:

$$A = 0, \quad B = 0, \quad -z = 0, \quad s_1 = 6, \quad s_2 = 7, \quad s_3 = 9, \quad s_4 = 9.$$

A basis system for Form 1 is said to be **feasible** if its basic solution satisfies the linear program's nonnegativity constraints. Evidently, system (1) is feasible.

Let us recall from Chapter 17 that, in a basic system, the set of basic variables is called its basis. System (1) is basic. The basis for system (1) is the set $\{-z, s_1, s_2, s_3, s_4\}$ of variables. A basis for Form 1 is said to be **feasible** if its basic solution is feasible. Evidently, the basis $\{-z, s_1, s_2, s_3, s_4\}$ is feasible.

For Program 18.1, a feasible basis sprang immediately into view. All that we needed to do was to insert the slack variables and the equation that defined z as the objective. That is *not typical*. Casting a linear program in Form 1 does not automatically produce a basis, let alone a feasible basis. Normally, a feasible basis must be wrung out of the linear program by a procedure that is known as "Phase 1" of the simplex method. By using Program 18.1 to introduce the simplex method, we are beginning with "Phase 2" of the simplex method. Why did we omit Phase 1? It will turn out to be a minor adaptation of Phase 2.

Phase 2 of the simplex method begins with a feasible basis and with $-z$ basic for one of its equations. Phase 2 executes a series of pivots. None of these pivots occurs on any coefficient in the equation for which $-z$ is basic. Each of these pivots:

- Keeps $-z$ basic.
- Changes the basis.
- Keeps the basic solution feasible.

- Improves the basic solution's objective value or, barring an improvement, keeps it from worsening.

Phase 2 keeps on pivoting until it can no longer improve the basic solution's objective value. How this occurs will soon be explained.

18.5. ENTERING AND LEAVING VARIABLES

Let us recall from Chapter 17 that Gauss-Jordan elimination pivots in search of a basic system. When Gauss-Jordan elimination encounters a basic system, it stops because a solution to the equation system has been found. By contrast, the simplex method keeps on pivoting after it encounters a basic system. The simplex method pivots from one basic system to another, searching for an optimal solution to the linear program.

In any pivot on a basic system, the variable whose coefficient is pivoted upon is called the **entering variable**, and the variable that was basic for the equation whose coefficient is being pivoted upon is called the **leaving variable**. To illustrate these definitions, imagine (and this will turn out to be a *poor* choice) that we pivot on the coefficient of A in Equation (18.1.4). Executing this pivot has the following effects:

- The variable A becomes basic for Equation (18.1.4).
- The variable s_4 that had been basic for Equation (18.1.4) becomes nonbasic.
- The variables that had been basic for the other equations remain basic for those equations.
- The basis changes from $\{-z, s_1, s_2, s_3, s_4\}$ to $\{-z, s_1, s_2, s_3, A\}$.

Evidently, this pivot changes the basis. The "entering variable" A joins the basis, and the "leaving variable" s_4 departs.

What is wrong with this pivot? The coefficient of A in Equation (18.1.4) equals -1. The new basic solution would set $A = 9/(-1) = -9$. This basic solution equates A to a negative value, so it fails to keep the basic solution feasible. This pivot was a bad idea, but it reveals a good rule:

To keep the basic solution feasible, avoid pivoting on coefficients that are negative.

In system (1), we aim to pivot in a way that keeps the basic solution feasible, that keeps $-z$ basic for Equation (18.1.0), and that improves its objective value. This pivot must occur on a positive coefficient of A or on a positive coefficient of B, but not on their coefficients in Equation (18.1.0).

18.6. SELECTING A PIVOT ELEMENT

To see which pivots do the trick, we cast system (1) in a format that is so revealing that it has been dubbed a "dictionary."[1] To put system (1) in this format, we execute two steps.

- Shift the nonbasic variables A and B to the right-hand sides of the constraints.
- Multiply Equation (18.1.0) by -1, so that z (and not $-z$) appears on its left-hand side.

[1] The term *dictionary* is widely attributed to Vašek Chvátal, who popularized it in his lovely book, *Linear Programming*, published in 1983 (New York: W. H. Freeman and Co). In that book, he attributes the term to J. E. Strum's, *Introduction to Linear Programming*, published in 1972 (San Francisco: Holden-Day).

This rewrites system (1) as:

$$
\begin{aligned}
z &= 0 + 2\,A + 3\,B, &\quad (18.2.0)\\
s_1 &= 6 - 1\,A - 0\,B, &\quad (18.2.1)\\
s_2 &= 7 - 1\,A - 1\,B, &\quad (18.2.2)\\
s_3 &= 9 - 0\,A - 2\,B, &\quad (18.2.3)\\
s_4 &= 9 + 1\,A - 3\,B. &\quad (18.2.4)
\end{aligned}
$$

In system (2), the variable z (rather than $-z$) is basic for the topmost equation, and the slack variables are basic for the remaining equations. The basic solution to system (2) equates each nonbasic variable to zero and, consequently, equates each basic variable to the number on the right-hand-side of the equation for which it is basic.

A Perturbation

Imagine that we **perturb** the basic solution to system (2) by setting *exactly one* nonbasic variable positive and by adjusting the values of the basic variables so as to preserve a solution to system (2). We have a choice. We could perturb system (2) by setting B positive and keeping $A = 0$. Alternatively, we could perturb it by setting A positive and keeping $B = 0$.

We elect to perturb the basic solution to system (2) by setting only A positive. To see what this perturbation accomplishes, we set $B = 0$ in system (2) and obtain:

$$
\begin{aligned}
z &= 0 + 2\,A, & & \quad (18.3.0)\\
s_1 &= 6 - 1\,A & &\text{so } s_1 \text{ decreases to zero when } A = 6/1 = 6. \quad (18.3.1)\\
s_2 &= 7 - 1\,A, & &\text{so } s_2 \text{ decreases to zero when } A = 7/1 = 7. \quad (18.3.2)\\
s_3 &= 9 - 0\,A, & &\text{so } s_3 \text{ stays positive for all values of } A. \quad (18.3.3)\\
s_4 &= 9 + 1\,A, & &\text{so } s_4 \text{ increases as } A \text{ becomes positive.} \quad (18.3.4)
\end{aligned}
$$

Equation (18.3.0) shows us that this perturbation increases z by $2\,A$. This is good; we aim to maximize z. Equations (18.3.1) and (18.3.2) show us that setting A positive decreases the values of the basic variables s_1 and s_2. When A equals 6, the variable s_1 has decreased to zero, and s_2 remains positive. When A exceeds 6, the variable s_1 becomes negative. Evidently, the largest value of A that keeps this perturbation feasible is 6.

We aim to pivot in a way that keeps the basic solution feasible. With A selected as the entering variable, it's the *ratios* in system (3) that tell us what to pivot upon. The equation for which s_1 is basic has the *smallest* ratio. In system (1), a pivot on the coefficient of A in the equation for which s_1 is basic will cause s_1 to leave the basis. Without executing this pivot, what can we learn about the basic solution to the equation system that results from it? Since A enters the basis, this basic solution must satisfy system (3). Since s_1 leaves the basis, this basic solution will have $s_1 = 0$. In particular:

- Equation (18.3.1) and $s_1 = 0$ give $A = 6/1 = 6$.
- Equation (18.3.0) shows that z increases by $2\,A = 2 \times 6 = 12$.
- Equation (18.3.2) gives $s_2 = 7 - 1\,A = 1 = 7 - 1 \times 6 = 1$.
- Equation (18.3.3) gives $s_3 = 9 - 0\,A = 9$.
- Equation (18.3.4) gives $s_4 = 9 + 1\,A = 15$.

Evidently, this pivot keeps the basic solution feasible and improves its objective value by 12. By contrast, a pivot on the coefficient of A in Equation (18.1.2) would not keep the basic solution feasible. Do you see why?

Table 18.1 Detached-coefficient tableau for system (1), with entering variable, ratio column, and pivot element.

	A	B	C	D	E	F	G	H	I	J
1	equation	A	B	s1	s2	s3	s4	-z	RHS	ratio
2	(1.0)	2	3					1	0	none
3	(1.1)	1		1					6	6
4	(1.2)	1	1		1				7	7
5	(1.3)		2			1			9	***
6	(1.4)	-1	3				1		9	***
7										
8								= IF(B6>0, I6/B6, "***")		

A Spreadsheet

In this chapter, pivots will be executed on spreadsheets, exactly as was done in Chapter 17. Table 18.1 presents the first few rows of such a spreadsheet. Columns A through I form a detached-coefficient tableau for system (1). As in Chapter 17, the dashed line represents a column of "=" signs.

In Table 18.1, cell B1 is shaded because *A* has been selected as the entering variable. The ratios in column *J* are familiar from system (2). Specifically:

- Cell J2 contains the word "none" because we never pivot on a coefficient in the equation for which $-z$ is basic.
- Cell J3 contains the ratio $6/1 = 6$ of the right-hand-side value of Equation (18.1.1) to its coefficient of *A*.
- Cell J4 contains the ratio $7/1 = 7$ of the right-hand-side value of Equation (18.1.2) to its coefficient of *A*.
- Cell J5 contains no ratio because the coefficient of *A* in Equation (18.1.3) equals zero.
- Cell J6 contains no ratio because the coefficient of *A* in Equation (18.1.4) is negative.

To compute the ratios in column J, we have dragged the IF function in cell J6 across cells J5, J4, and J3.

Compare Table 18.1 with system (2). In Table 18.1, ratios are computed for *positive* coefficients of the entering variable. (The positive coefficients became negative when the nonbasic variables were switched to the right-hand side of system (2).) Each ratio gives the value of the entering variable for which the equation's basic variable decreases to zero. Thus, pivoting on a coefficient whose ratio is smallest preserves feasibility.

In Table 18.1 and throughout our discussion of the simplex method, the selection of the pivot element is recorded as follows:

- The entering variable is shaded.
- The smallest ratio is shaded.
- The pivot element is outlined in heavy lines.

In this way, Table 18.1 records the fact that the first pivot is slated to occur on the coefficient of *A* in Equation (18.1.1).

Table 18.2 The first pivot.

	A	B	C	D	E	F	G	H	I	J
1	equation	A	B	s1	s2	s3	s4	-z	RHS	ratio
2	(1.0)	2	3					1	0	none
3	(1.1)	1		1					6	6
4	(1.2)	1	1		1				7	7
5	(1.3)		2			1			9	***
6	(1.4)	-1	3				1		9	***
7										
8		= PIVOT(B3, B2:I6)					=IF(B6>0,I6/B6,"****")			
9										
10	equation	A	B	s1	s2	s3	s4	-z	RHS	ratio
11	(4.0)	0	3	-2	-0	-0	-0	1	-12	none
12	(4.1)	1	0	1	0	0	0	0	6	
13	(4.2)	0	1	-1	1	-0	-0	-0	1	
14	(4.3)	-0	2	-0	-0	1	-0	-0	9	
15	(4.4)	0	3	1	0	0	1	0	15	

Executing the Pivot

We have seen how rows 1–6 of the spreadsheet in Table 18.1 identify the pivot element. Table 18.2 reproduces these rows and displays the result of the pivot. We recall from Chapter 17 that executing pivots is eased by an Excel Add-In that accompanies this text. Before you can use this Add-In, you must install it (see the Appendix which tells how). To execute this pivot, select cells B11 through I15, then type **=PIVOT(B3, B2:I6)** followed by **Ctrl+Shift+Enter** rather than hitting the Enter key by itself.

Rows 10–15 of Table 18.2 describe the equation system that results from this pivot. This equation system is:

$$
\begin{array}{lllllll}
 & 3B - 2s_1 & & & -z = -12, & & (18.4.0) \\
A & \quad + 1s_1 & & & = 6, & & (18.4.1) \\
 & 1B - 1s_1 + s_2 & & & = 1, & & (18.4.2) \\
 & 2B + 0s_1 & + s_3 & & = 9, & & (18.4.3) \\
 & 3B + 1s_1 & & + s_4 & = 15. & & (18.4.4)
\end{array}
$$

System (4) indicates that the first pivot did everything we predicted it would. Specifically:

- The variable A became basic, and s_1 became nonbasic.
- The variables that had been basic for the other equations remained basic for them.
- The new basic solution is feasible.
- The new basic solution equates A to 6, which equals the ratio for the pivot row. This basic solution improves z by $2A = 2 \times 6 = 12$.

In system (4), we will wish to pick B as the entering variable for a second pivot because perturbing the basic solution to system (4) by setting B positive increases the basic solution's objective value.

18.7. FEASIBLE PIVOTS

Before executing the second pivot, we pause to abstract from the first pivot the properties that keep the basic solution feasible and that improve its objective value. These properties will be described in language that applies to every linear program, not just to Program 18.2.

Reduced Cost

As noted previously, no pivot ever occurs on a coefficient in the equation for which $-z$ is basic. Consequently, $-z$ stays basic for this equation. The coefficients in this equation have been given names. With $-z$ excluded, the coefficient of each variable in the equation for which $-z$ is basic is now called the **reduced cost** of that variable.

System (4) illustrates this definition. Each basic variable (other than $-z$) has zero as its reduced cost. In system (4), the variables B and s_1 are nonbasic, the reduced cost of B equals 3, and the reduced cost of s_1 equals -2.

Preserving Feasibility

This section focuses on pivots that keep the basic solution feasible, whether or not they improve its objective value. In this section, we allow *any nonbasic variable* to be the entering variable. Having selected an entering variable, we must pick a coefficient on which to pivot. The recipe that follows keeps the basic solution feasible:

- Begin with a basic feasible system for Form 1.
- Select any nonbasic variable as the entering variable.
- Compute no ratio for the equation in which $-z$ is basic. (This keeps $-z$ basic.)
- Compute no ratio for any equation whose coefficient of the entering variable is negative or is zero.
- For each equation whose coefficient of the entering variable is positive, compute the **ratio** of this equation's right-hand-side value to its coefficient of the entering variable.
- A **feasible** pivot is one that occurs on the coefficient of the entering variable in an equation whose ratio is smallest.

This precise recipe was executed in Table 18.2. All that we have done is to describe it in general terms and to define the terms *ratio* and *feasible pivot*.

To illustrate this recipe further, we select B as the entering variable in system (4). The variable B has positive coefficients in Equations (18.4.2), (18.4.3), and (18.4.4). The ratios for these equations are $1/1, 9/2$, and $15/3$, respectively. The smallest of these ratios equals 1, so the feasible pivot occurs on the coefficient of B in Equation (18.4.2). Here are the key properties of a feasible pivot.

> Each feasible pivot has these three effects:
>
> - It keeps the basic solution feasible.
> - It makes the entering variable x basic for a row whose ratio R is smallest.
> - It changes the basic solution's objective value by the product of R and the reduced cost of x.

To illustrate these properties, we consider a feasible pivot in system (4), with s_1 as the entering variable. The reduced cost of s_1 equals -2. The variable s_1 has ratios for

Equations (18.4.1) and (18.4.4); these ratios are 6/1 and 15/1, respectively. The smallest ratio equals 6, and this ratio occurs in Equation (18.4.1). A pivot on the coefficient of s_1 in Equation (18.4.1) keeps the basic solution feasible, and it changes the objective value by $6 \times (-2) = -12$, which equals the smallest ratio times the reduced cost of s_1.

Reconciliation

The term *reduced cost* first appeared in Chapter 3 where it was described as a property of the optimal solution that the simplex method found. In this chapter, reduced costs are defined for *every* basic system that the simplex method encounters.

The current definition encompasses the one given in Chapter 3, as we can see when we consider any basic system for a linear program in Form 1. For each variable x other than $-z$, we observe that:

- If x is basic, the reduced cost of x equals 0.
- If x is nonbasic, the reduced cost of x equals the change that occurs in the objective value if the basic solution is perturbed by setting $x = 1$ and by adjusting the values of the basic variables so as to satisfy the equations of Form 1.

18.8. SIMPLEX PIVOTS

Our aim is to pivot in a way that preserves feasibility and improves the objective value. Each feasible pivot changes the objective value by the product of two numbers—the reduced cost of the entering variable and the ratio R for the coefficient on which the pivot occurs. The denominator of R is positive, and the numerator is either positive or zero. Thus, R can be positive or zero.

A later section explores the case in which R can equal zero. Here, we assume that R is positive. Thus, the change in the objective value equals the product of the reduced cost of the entering variable and the positive number, R. We want to improve the objective value. Evidently, the reduced cost of the entering variable should be positive in the case of a maximization problem and negative in the case of a minimization problem. In brief, Dantzig's pivot rule is as follows:

> In a basic feasible system for Form 1, a **simplex pivot** is a feasible pivot in which the reduced cost of the entering variable is:
>
> - positive in the case of a maximization problem,
> - negative in the case of a minimization problem.

Phase 2 of the simplex method executes repeated simplex pivots, in search of an optimal solution to the linear program. For our example, Phase 2 began with system (1) where the variables A and B had positive reduced costs. Hence, in system (1), either A or B could have been the entering variable for the first simplex pivot. In system (4), only B can be the entering variable in a simplex pivot because B is the only nonbasic variable whose reduced cost is positive.

18.9. ILLUSTRATION OF PHASE 2

Let us now return to our example and resume Phase 2 of the simplex method. We continue to execute Phase 2 on a spreadsheet. The diskette that accompanies this text includes this spreadsheet; you may wish to explore it as you read this section.

Table 18.3 The second simplex pivot.

	A	B	C	D	E	F	G	H	I	J
10	equation	A	B	s1	s2	s3	s4	-z	RHS	ratio
11	(4.0)	0	3	-2	0	0	0	1	-12	none
12	(4.1)	1	0	1	0	0	0	0	6	***
13	(4.2)	0	1	-1	1	0	0	0	1	1
14	(4.3)	0	2	0	0	1	0	0	9	4.5
15	(4.4)	0	3	1	0	0	1	0	15	5
16										
17			= PIVOT(C13, B11:I15)					= IF(C15>0, I15/C15, "****")		
18										
19	equation	A	B	s1	s2	s3	s4	-z	RHS	ratio
20	(5.0)	0	0	1	-3	0	0	1	-15	none
21	(5.1)	1	0	1	0	0	0	0	6	
22	(5.2)	0	1	-1	1	0	0	0	1	
23	(5.3)	0	0	2	-2	1	0	0	7	
24	(5.4)	0	0	4	-3	0	1	0	12	

The Second Simplex Pivot

Table 18.3 describes the second simplex pivot. Rows 10–15 of this table identify the element on which the second simplex pivot occurs. Here, B must be the entering variable because only its reduced cost is positive. The functions in column J compute ratios for rows 13, 14, and 15 because their coefficients of B are positive. Row 13 has the smallest ratio. The pivot element is the coefficient of B in row 13. This coefficient is outlined in heavy lines.

In Table 18.3, rows 19–24 present the system that results from the second simplex pivot. The table includes the Array function that executes this pivot. The second simplex pivot keeps the basic solution feasible, causing B to enter the basis and s_2 to leave. The new basic solution equates B to 1, the ratio for its pivot row, and it improves the objective value by $3\,B = 3 \times 1$, from 12 to 15.

The Third Simplex Pivot

Table 18.4 describes the third simplex pivot. Row 20 of the table shows that s_1 must be the entering variable for this pivot because s_1 is the only nonbasic variable whose reduced cost is positive. The ratios in Table 18.4 show that this pivot will occur on the coefficient of s_1 in the equation for which s_4 is basic. This table also displays the system that results from this pivot.

Rows 29–33 of Table 18.4 describe the system that results from the third simplex pivot. The basic solution to this system is feasible. It equates z to 18. This basic solution sets

$$s_2 = s_4 = 0, \quad -z = -18, \quad A = 3, \quad B = 4, \quad s_3 = 1, \quad s_1 = 3.$$

Row 29 shows that both of the nonbasic variables have negative reduced costs; no reduced cost is positive. The simplex method cannot select an entering variable. It must stop.

Table 18.4 The third simplex pivot.

	A	B	C	D	E	F	G	H	I	J
19	equation	A	B	s1	s2	s3	s4	-z	RHS	ratio
20	(5.0)	0	0	1	-3	0	0	1	-15	none
21	(5.1)	1	0	1	0	0	0	0	6	6
22	(5.2)	0	1	-1	1	0	0	0	1	***
23	(5.3)	0	0	2	-2	1	0	0	7	3.5
24	(5.4)	0	0	4	-3	0	1	0	12	3
25										
26			= PIVOT(D24, B20:I24)				= IF(D24>0, I24/D24, "****")			
27										
28	equation	A	B	s1	s2	s3	s4	-z	RHS	ratio
29	(6.0)	0	0	0	-2.25	0	-0.25	1	-18	none
30	(6.1)	1	0	0	0.75	0	-0.25	0	3	
31	(6.2)	0	1	0	0.25	0	0.25	0	4	
32	(6.3)	0	0	0	-0.5	1	-0.5	0	1	
33	(6.4)	0	0	1	-0.75	0	0.25	0	3	

An Optimal Solution

When the simplex method encounters the system in rows 28–33 of Table 18.4, it stops pivoting because each nonbasic variable has a nonpositive number as its reduced cost. To see what this means, we multiply the equation that row 29 represents by -1 and transfer its nonbasic variables to its right-hand side, thereby obtaining

$$z = 18 - 2.25\, s_2 - 0.25\, s_4. \qquad (18.6.0)$$

As noted above, setting $s_2 = 0$ and $s_4 = 0$ produces a basic solution that is feasible and has $z = 18$. *Every* feasible solution equates s_2 and s_4 to nonnegative values. Consequently, Equation (18.6.0) demonstrates that every feasible solution has objective value $z \leq 18$. In brief, the basic solution to system (6) is optimal.

This observation holds in general. In a maximization problem, the **optimality test** is that the reduced cost of each nonbasic variable is nonpositive. Similarly, in a minimization problem, the **optimality test** is that the reduced cost of each nonbasic variable is nonnegative. When the optimality test is satisfied, the simplex method must stop because it cannot select an entering variable. The basic solution with which it stops is an optimal solution to the linear program. This is true for the reason that Equation (18.6.0) illustrates; perturbing the basic solution by setting one or more of the nonbasic variables positive can only worsen the objective. We emphasize:

> **Test for optimality.** The basic solution to a basic feasible system for Form 1 is optimal if the reduced costs of the nonbasic variables are:
>
> - nonpositive in the case of a maximization problem,
> - nonnegative in the case of a minimization problem.

The reduced costs that the Sensitivity Report presents are those that cause the simplex method to stop—no positive reduced costs for a maximization problem and no negative reduced costs for a minimization problem.

18.10. DETECTING AN UNBOUNDED LINEAR PROGRAM

What happens if Phase 2 is applied to an unbounded linear program? Phase 2 cannot find an optimal solution because none exists. To explore this issue, we reverse the sign of two coefficients of A in Program 18.1, transforming it into:

Program 18.3: Maximize $\{2\,A + 3\,B|$ subject to the constraints

$$\begin{aligned} -A \quad\quad &\le 6, \\ -A + 1\,B &\le 7, \\ 2\,B &\le 9, \\ -A + 3\,B &\le 9, \\ A \quad\quad &\ge 0, \\ B &\ge 0. \end{aligned}$$

To cast Program 18.3 in Form 1, we insert a slack variable on the left-hand side of each "≤" constraint, and we add an equation that defines z as the objective. This results in a linear program whose detached-coefficient tableau is presented as Table 18.5.

The reduced costs of A and B are positive. We could pick either of them as the entering variable for a simplex pivot. If we pick B, this pivot will occur on the coefficient of B in Equation (18.7.4). On the other hand, if we pick A as the entering variable, there will be *nothing to pivot upon* because the coefficients of A in Equations (18.7.1) through (18.7.4) are nonpositive. To see what this means, we perturb the basic solution to system (7) by setting A positive and keeping B equal to zero. In system (7), we set $B = 0$ and shift A to the right-hand side. We also multiply Equation (18.7.0) by -1. This produces:

$$z = 0 + 2\,A \tag{18.8.0}$$

$$s_1 = 6 + 1\,A \tag{18.8.1}$$

$$s_2 = 7 + 1\,A \tag{18.8.2}$$

$$s_3 = 9 + 0\,A \tag{18.8.3}$$

$$s_4 = 9 + 1\,A \tag{18.8.4}$$

In system (8), setting A positive increases z and decreases none of the basic variables. As A increases without limit, the perturbed solution remains feasible (nonnegative), and z (its objective value) increases without limit. Evidently, Program 18.3 is unbounded.

We have just learned how the simplex method detects an unbounded linear program. It does so by picking an entering variable that has no pivot row. Perturbing the basic solution

Table 18.5 Detached-coefficient tableau for Program 18.3.

	A	B	C	D	E	F	G	H	I	J
1	equation	A	B	s1	s2	s3	s4	-z	RHS	ratio
2	(7.0)	2	3					1	0	none
3	(7.1)	-1		1					6	
4	(7.2)	-1	1		1				7	
5	(7.3)		2			1			9	
6	(7.4)	-1	3				1		9	

by setting the entering variable positive preserves feasibility and improves the objective, without limit. We emphasize:

> **Test for unboundedness.** A linear program in Form 1 is unbounded if an entering variable for a simplex pivot has nonpositive coefficients in each equation other than the one for which $-z$ is basic.

In brief, Phase 2 of the simplex method begins with a basic feasible solution and executes simplex pivots until it encounters a basic feasible system that satisfies either the optimality test (in which case the basic solution is optimal) or the unboundedness test (in which case the basic solution can be improved without limit).

18.11. THE DICTIONARY

Detached-coefficient tableaus make it easy to execute pivots on spreadsheets, but for figuring out what to pivot upon, nothing beats the "dictionary." An equation system for Form 1 is said to be written as a **dictionary** if:

- Only the basic variables appear on the left-hand sides of the equations.
- Each basic variable has a coefficient of $+1$ on the left-hand side of the equation for which it is basic.
- The variable z (and not $-z$) is basic.

System (2) is a dictionary. This dictionary has helped us in two ways. First, it showed us what happens when we *perturb* the basic solution by setting one nonbasic variable positive. Second, it showed us how *ratios* identify pivot elements that keep the basic solution feasible. That's how the "dictionary" earns its name.

System (4) is not a dictionary because $-z$ is basic for its topmost equation and because the nonbasic variables are on its left-hand sides. To convert system (4) into a dictionary, multiply Equation (18.4.0) by -1 and switch the nonbasic variables B and s_1 to the right-hand sides of the equations. This rewrites system (4) as:

$$
\begin{aligned}
z &= 12 + 3\,B - 2\,s_1, && (18.9.0)\\
A &= 6 - 0\,B - 1\,s_1, && (18.9.1)\\
s_2 &= 1 - 1\,B + 1\,s_1, && (18.9.2)\\
s_3 &= 9 - 2\,B + 0\,s_1, && (18.9.3)\\
s_4 &= 15 - 3\,B - 1\,s_1. && (18.9.4)
\end{aligned}
$$

System (9) is a dictionary. From it, we can see at a glance what happens if the basic solution is perturbed by equating one nonbasic variable to a positive value. Setting B positive increases z and decreases the values of the basic variables s_2, s_3, and s_4, for instance. And it's clear from system (9) how the ratios determine the pivot element.

We could "pivot" from one dictionary to another, which would require us to shift variables left and right as they become basic and nonbasic. That is a cumbersome approach. We use dictionaries to uncover pivot rules, but we execute pivots on detached-coefficient tableaus.

18.12. GEOMETRIC INTERPRETATION

Program 18.1 is now used to provide a geometric interpretation of the simplex method. Figure 18.1 reproduces a diagram that we used in Chapter 3 to visualize its feasible solutions.

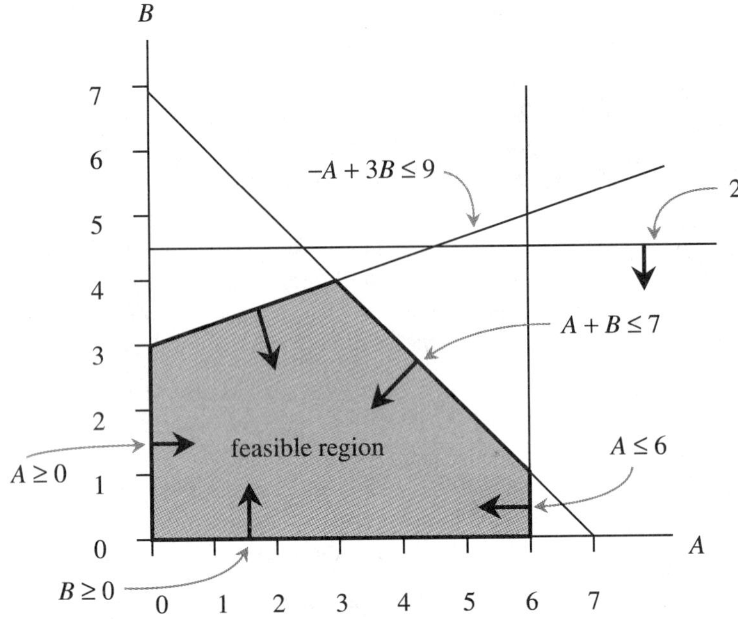

Figure 18.1 The feasible region for Program 18.1.

Program 18.1 has six inequality constraints. In Figure 18.1, each of these constraints accounts for a line and an attached arrow. For instance, the pairs (*A*, *B*) that satisfy the equation $A + B = 7$ form a line, and an arrow points from this line into the region that satisfies the constraint $A + B \leq 7$. The points (*A*, *B*) that satisfy all six constraints are shaded. These points form the feasible region for Program 18.1.

This feasible region has five extreme points, each of which lies at the intersection of two lines. For example, the extreme point (0, 3) is the intersection of the lines $A = 0$ and $-A + 3B = 9$. Similarly, the extreme point (3, 4) is the intersection of the lines $-A + 3B = 9$ and $A + B = 7$, and so forth. We will see how the simplex method pivots from extreme point to extreme point, improving the objective value with each pivot.

Feasible Region for Form 1

In order to apply the simplex method to Program 18.1, we placed it in Form 1. Slack variables converted its "≤" constraints to equations (see Figure 18.2). In Figure 18.2, each line is labeled with the variable in Program 18.2 that equals zero on it. For instance, the line on which the inequality $A + B \leq 7$ holds as an equation is relabeled $s_2 = 0$ because s_2 is the slack variable for the constraint $A + B + s_2 = 7$.

Table 18.6 traces the progress of the simplex method. We will see that the *nonbasic* variables in each feasible basis identify an extreme point in Figure 18.2. We will also see that each entering variable identifies an edge in which one of the extreme point's nonbasic variables becomes positive.

Bases and Extreme Points

When we applied the simplex method to Program 18.2, we encountered four different bases. For each of these bases, the nonbasic variables appear in the first column of Table 18.6.

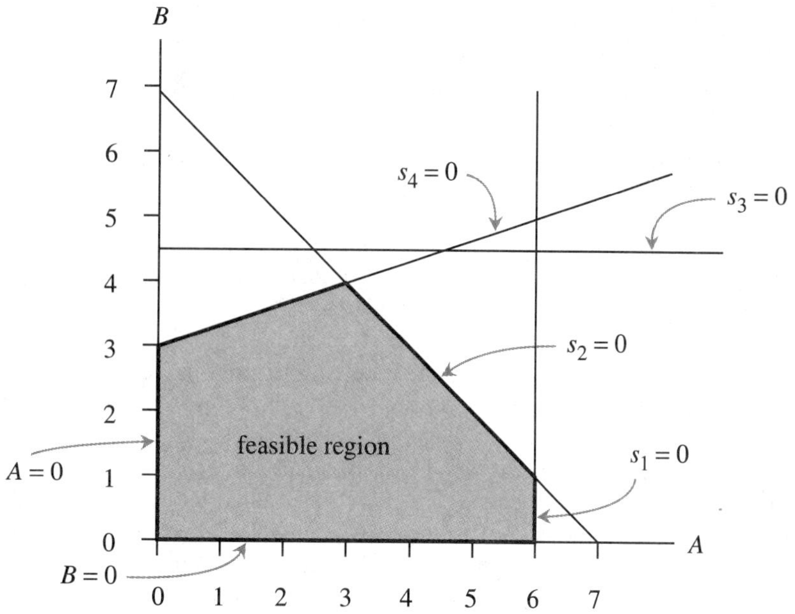

Figure 18.2 The feasible region for Program 18.2.

Initially, the variables A and B were nonbasic. The initial basic solution set $A = 0$ and $B = 0$. Figure 18.2 shows that the intersection of the lines $A = 0$ and $B = 0$ is the extreme point (0, 0). In the second basis, the variables s_1 and B were nonbasic. Figure 18.2 shows that the intersection of the lines $s_1 = 0$ and $B = 0$ is the extreme point (6, 0), and so forth. In this way, the nonbasic variables identify the extreme points.

Perturbations and Edges

Each of the first three rows of Table 18.6 describes a perturbation of a basic solution in which an entering variable becomes positive. From Figure 18.2 and the first row of Table 18.6, we see that:

- Initially, A and B are nonbasic. In Figure 18.2, the initial basic solution lies at the intersection of the lines $A = 0$ and $B = 0$.
- For the first pivot, A is the entering variable. The first perturbation moves away from the extreme point (0, 0) along the line $B = 0$; it moves in the direction of increasing A, which is to the east.
- Figure 18.2 shows that this perturbation encounters the lines $s_1 = 0$ and $s_2 = 0$ when B equals 6 and 7, respectively. Each ratio is the value of an entering variable at which

Table 18.6 Descriptions of three simplex pivots.

Nonbasic variables	Extreme point	Entering variable	Line of perturbation	Ratios	Leaving variable
A, B	(0, 0)	A	$B = 0$	6, 7	s_1
s_1, B	(6, 0)	B	$s_1 = 0$	1, 4.5, 5	s_2
s_1, s_2	(6, 1)	s_1	$s_2 = 0$	3, 3.5, 6	s_4
s_4, s_2	(3, 4)	none			

a new line is encountered. The variable s_1 leaves the basis because its line is encountered first.

- Thus, the first pivot occurs to extreme point (6, 0), for which the variables s_1 and B are nonbasic.
- As the entering variable B increases from 0 to 6, this perturbation traverses the *edge* (line segment) between extreme points (0, 0) and (6, 0).

In a similar way, the second line of Table 18.6 interprets the second perturbation. It begins at the extreme point (6, 0), which is the intersection of the lines at which the nonbasic variables s_1 and B equal zero. The entering variable is B, so the perturbation moves away from the extreme point (6, 0) along the line $s_1 = 0$ in the direction of increasing B, which is to the north. Figure 18.2 shows that this perturbation encounters the lines $s_2 = 0$, $s_3 = 0$, and $s_4 = 0$ when B equals 1, 4.5, and 5, respectively. The variable s_2 leaves the basis because its line is encountered first. As B increases from 0 to 1, this perturbation traverses the edge between extreme points (6, 0) and (6, 1).

Figure 18.2 is two-dimensional. Each variable in Program 18.2 equals zero on a line, and each extreme point is the intersection of two nonparallel lines. Each entering variable has a perturbation; this perturbation removes the entering variable's line and moves away from the extreme point along the other line. Each ratio is a value of the entering variable at which the perturbation encounters a new line. The smallest ratio determines the pivot element. Each perturbation traverses an edge of the feasible region.

A Three-Dimensional Example

Our attention now turns to the Recreational Vehicle example, which has three decision variables. Its feasible region can be visualized in three-dimensional space. With slack variables included, its linear program becomes:

Program 18.4: Maximize $\{840\,S + 1120\,F + 1200\,L\}$, subject to the constraints

$$
\begin{aligned}
3\,S + 2\,F + 1\,L + s_1 &= 120,\\
1\,S + 2\,F + 3\,L + s_2 &= 80,\\
2\,S + s_3 &= 96,\\
3\,F + s_4 &= 102,\\
2\,L + s_5 &= 40,\\
S \geq 0,\quad F \geq 0,\quad L \geq 0,\quad s_1 \geq 0,\quad s_2 &\geq 0, \ldots, \quad s_5 \geq 0.
\end{aligned}
$$

Cartesian coordinates identify each triplet (S, F, L) that satisfies the constraints of Program 18.4 with a point in three-dimensional space. The set of all such points forms the feasible region in Figure 18.3.

In Figure 18.3, each extreme point has been labeled with the values that it assigns to the variables S, F, and L. One of these extreme points, labeled (35, 0, 15), identifies the feasible solution to Program 18.4 in which $S = 35$, $F = 0$, and $L = 15$.

Each variable in Program 18.4 equals zero on a plane in Figure 18.3. For instance, s_1 equals zero on the plane that includes the triangle with vertical stripes. Each extreme point is the intersection of three such planes. For instance, the extreme point (35, 0, 15) is the intersection of the planes $L = 0$ and $s_1 = 0$ and $s_2 = 0$. This extreme point corresponds to the basis that excludes L, s_1, and s_2.

Each perturbation of an extreme point removes the entering variable's plane and shifts away from the extreme point along the *line* that is the intersection of the other two planes.

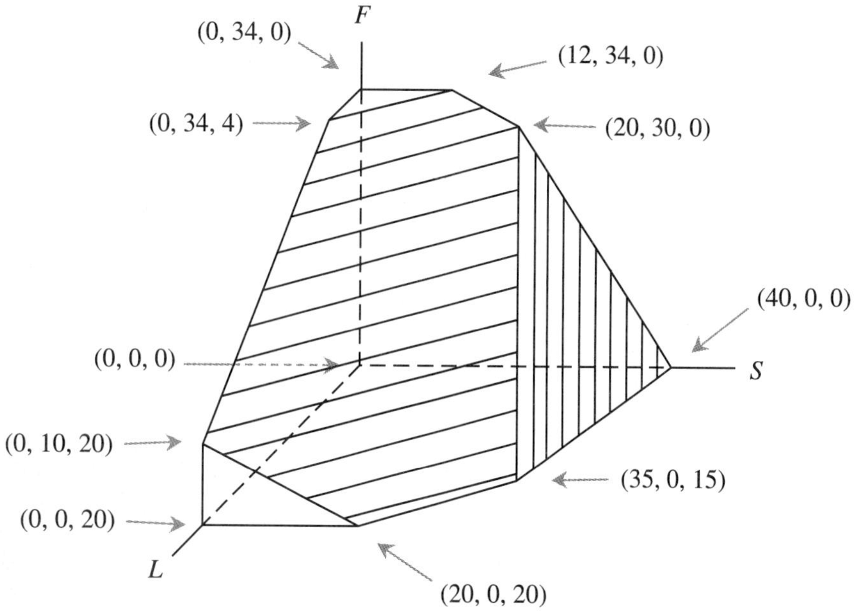

Figure 18.3 Feasible region for Program 18.4.

To illustrate, suppose we perturb the extreme point (35, 0, 15) by selecting F as the entering variable. As F increases, the solution shifts upward along the line that is the intersection of the striped planes, $s_1 = 0$ and $s_2 = 0$. When F increases to 30, this perturbation encounters the plane $L = 0$. As F increases from 0 to 30, the perturbation traverses the edge that connects between extreme points (35, 0, 15) and (20, 0, 30).

Higher Dimensions

What about-higher dimensions? The beauty of geometry is that the intuition we develop from two-dimensional and three-dimensional examples tends to hold up in higher dimensions, which we cannot visualize. In an m-dimensional example, each variable equals zero on an $(m - 1)$-dimensional "hyperplane." Each extreme point is the intersection of m hyperplanes. Each line is the intersection of $m - 1$ of these hyperplanes. Each perturbation moves away from an extreme point along the line on which one of its m hyperplanes is removed and $m - 1$ remain.

18.13. ECONOMIC PERSPECTIVE

The language of economics is now used to interpret the simplex method. For this purpose, we focus on a linear program whose object is profit maximization. We assume that this linear program has been cast in Form 1, so that it has equality constraints and nonnegative variables.

This linear program is now viewed in terms of the agents in an economy. These agents allocate scarce resources among competing activities. The **resources** are the right-hand-side values, and the **activities** are the variables. Since the linear program has been cast in Form 1, these activities *include* any slack and surplus variables that had been used to convert inequality constraints into equations. An activity is said to **participate** in the economy when it is basic, that is, when it is part of the basis. The participating activities establish a set of "shadow prices" according to rules that are highlighted as follows.

> A **shadow price** is assigned to each resource.
>
> - The unit of measure of each shadow price equals the unit of measure of the objective divided by the unit of measure of the resource.
> - The values of the shadow prices are determined from this rule: Each agent who engages in a participating activity is indifferent to perturbing the level of that activity and buying or selling the resources that this change requires or frees up at their shadow prices.

These "shadow prices" sound familiar, and they are. In Chapter 3, shadow prices were associated with an optimal solution to a linear program. In this section, shadow prices are associated with *each* basis. As was the case in Chapter 3, the shadow prices are breakeven prices. Agents who engage in participating activities are indifferent to varying the levels of their activities and buying or selling the requisite resources at their shadow prices. This establishes a **market** for the resources.

A *nonparticipating* activity will be profitable if its contribution exceeds the opportunity cost of the bundle of resources that it requires. If nonparticipating activity is profitable, an agent can earn a profit by engaging in it. This agent will buy the resources that are needed to engage in this activity. Buying these resources reduces the levels of those basic activities that compete with it for scarce resources. To maximize profit, the agent will engage in this activity at the largest possible level. This reduces to zero the level of some competing activity, thereby pivoting the economy to a new basis, with a new set of prices.

The computation of these shadow prices will be illustrated twice, first for a general linear program that has been cast in Form 1, then for a specific example. This computation will deepen our understanding of the simplex method.

General Description of Form 1

Interjected here is a general description of a profit-maximizing linear program that has been cast in Form 1. This linear program appears below as Program 18.5. In it, the integer m denotes the number of equations, excluding the equation that defines z as the objective value. Similarly, the integer n denotes the number of decision variables, excluding z. These n decision variables are labeled x_1 through x_n. The data in Program 18.5 are the c_j's, the b_i's, and the A_{ij}'s. Evidently, c_j is the contribution of the jth activity, b_i is the level of the ith resource, and A_{ij} is the number of units of the ith resource that are needed to engage in one unit of the jth activity.

Program 18.5: Maximize $\{z\}$, subject to the constraints

$$\begin{array}{lrcl}
 & c_1 x_1 + c_2 x_2 + \cdots + c_n x_n - z & = & 0, \\
y_1: & A_{11} x_1 + A_{12} x_2 + \cdots + A_{1n} x_n & = & b_1, \\
y_2: & A_{21} x_1 + A_{22} x_2 + \cdots + A_{2n} x_n & = & b_2, \\
\vdots & \vdots \qquad \vdots \qquad \vdots & & \vdots \\
y_m: & A_{m1} x_1 + A_{m2} x_2 + \cdots + A_{mn} x_n & = & b_m, \\
 & x_1 \geq 0, \quad x_2 \geq 0, \ldots, \quad x_n \geq 0. & &
\end{array}$$

Opportunity Costs and Shadow Prices for Form 1

The quantities y_1 through y_m that appear to the left of the equations in Program 18.5 will be the shadow prices. The unit of measure of each shadow price equals the unit of measure of the objective divided by the unit of measure of its resource.

Each basis (set of participating activities) assigns values to the shadow prices. Participants are *indifferent* to buying or selling resources at their shadow prices and adjusting the levels of the activities accordingly. This establishes a market for the resources, and it lets us compute the opportunity cost of engaging in one unit of each activity. Engaging in one unit of activity j consumes A_{1j} units of the first resource, A_{2j} units of the second resource, A_{3j} units of the third resource, and so forth. With shadow prices y_1 through y_m on the resources, the opportunity cost (reduction in contribution) of the bundle of resources that is needed to set $x_j = 1$ is given by:

$$\begin{bmatrix} \text{opportunity cost of the resources needed to} \\ \text{engage in one unit of activity } j \end{bmatrix} = y_1 A_{1j} + y_2 A_{2j} + \cdots + y_m A_{mj}$$

The contribution of each participating activity (basic variable) equals the opportunity cost of the resources that are needed to engage in one unit of it. In other words,

$$c_j = y_1 A_{1j} + y_2 A_{2j} + \cdots + y_m A_{mj}, \quad \text{for each participating activity } x_j. \qquad (18.10)$$

System (10) is a set of equations, one per basic variable. The shadow prices satisfy system (10). All m shadow prices can be found by solving system (10).

Reduced Cost

Consider any feasible basis for Program 18.5. Having found its shadow prices y_1 through y_m from system (10), we now define the numbers $\bar{c}_1$ through $\bar{c}_n$ by

$$\bar{c}_j = c_j - [y_1 A_{1j} + y_2 A_{2j} + \cdots + y_m A_{mj}], \quad \text{for } j = 1, 2, \ldots, n. \qquad (18.11)$$

Compare systems (10) and (11). Note that each basic variable x_j has $\bar{c}_j = 0$. In other words,

$$0 = \bar{c}_j = c_j - [y_1 A_{1j} + y_2 A_{2j} + \cdots + y_m A_{mj}] \quad \text{if } x_j \text{ is basic.} \qquad (18.12)$$

Moreover, for each nonbasic variable x_j, Equation (18.11) shows that $\bar{c}_j$ equals the change in the objective value if the basic solution is perturbed by setting $x_j = 1$.

Doesn't this sound familiar? The reduced cost of each basic variable x_j equals zero, as does $\bar{c}_j$. The reduced cost of each nonbasic variable x_j equals the change that occurs in the objective value when x_j is set equal to 1 and the values of the basic variables are adjusted so as to satisfy the equations of the linear program, as does $\bar{c}_j$. In other words, the $\bar{c}_j$'s *are* the reduced costs.

Equations (18.11) and (18.12) determine the shadow prices and the reduced costs. In earlier sections, we had found the reduced costs from a basic system. Equations (18.11) and (18.12) provide a different way in which to compute the reduced costs.

A Technical Difficulty

There is a technical difficulty in this computation of the shadow prices. Our economic argument is not quite airtight. From a mathematical viewpoint, it needs to be qualified slightly. *If* a basic system includes no trite equations, then:

- The basis includes $-z$ and *exactly* m other variables, one per equation.
- System (10) is a set of m equations in m unknowns.

- System (10) has exactly one solution, and this solution consists of the shadow prices.

But what happens if the basic system includes one or more trite equations? In this case:

- System (10) is a set of fewer than m equations with m unknowns.
- System (10) has multiple solutions.
- Even so, system (11) can be shown to have exactly one solution, and this solution correctly specifies the reduced costs.

Thus, the economic argument correctly computes the reduced costs, even when the shadow prices are ill-defined.

Shadow Prices for the Recreational Vehicle Problem

We have seen how systems (10) through (12) compute the shadow prices and the reduced costs for each basis. Let us now apply these results to the Recreational Vehicle problem. Program 18.4 is reproduced here, where the shadow prices y_1 through y_5 have been assigned to its constraints.

Program 18.4: Maximize $\{z\}$, subject to the constraints

$$
\begin{array}{lllllllllll}
 & 840\,S + & 1120\,F + & 1200\,L & & & & & & -z = & 0, \\
y_1: & 3\,S + & 2\,F + & 1\,L & +\,s_1 & & & & & = & 120, \\
y_2: & 1\,S + & 2\,F + & 3\,L & & +\,s_2 & & & & = & 80, \\
y_3: & 2\,S & & & & & +\,s_3 & & & = & 96, \\
y_4: & & 3\,F & & & & & +\,s_4 & & = & 102, \\
y_5: & & & 2\,L & & & & & +\,s_5 & = & 40,
\end{array}
$$

$$S \geq 0, \quad F \geq 0, \quad L \geq 0, \quad s_1 \geq 0, \quad s_2 \geq 0, \ldots, s_5 \geq 0.$$

Our goal is to compute the shadow prices and the reduced costs for a basis for Program 18.4. Figure 18.3 identifies each feasible basis with an extreme point; one of these extreme points is (35, 0, 15). Pictorially or algebraically, we can verify that this extreme point equates exactly three variables to zero. It sets $F = 0$, $s_1 = 0$, and $s_2 = 0$. Its basis excludes F, s_1, and s_2. Its basis consists of $-z$ and these five decision variables: S, L, s_3, s_4, and s_5. According to Equation (18.10), the contribution (objective coefficient) of each basic variable equals the opportunity cost of the resources that are needed to make one unit of it. In other words, Equation (18.10) demonstrates that

$$
\begin{array}{rlll}
840 = & 3\,y_1 + 1\,y_1 + 2\,y_3 & & \text{because } S \text{ is basic,} \\
1200 = & 1\,y_1 + 3\,y_2 & +\,2\,y_5 & \text{because } L \text{ is basic,} \\
0 = & 1\,y_3 & & \text{because } s_3 \text{ is basic,} \\
0 = & 1\,y_4 & & \text{because } s_4 \text{ is basic,} \\
0 = & & 1\,y_5 & \text{because } s_5 \text{ is basic,}
\end{array}
$$

Displayed above are five equations in five unknowns. Solving these equations is easy. The last three of them give $0 = y_3 = y_4 = y_5$; the other two give $y_1 = 165$ and $y_2 = 345$.

Letting Solver Do the Work

There is no need to compute the shadow prices and the reduced costs by hand: Solver can do it. Table 18.7 shows how to use a spreadsheet and Solver in order to compute the shadow

Table 18.7 Shadow prices that equate to zero the reduced cost of the basic variables S, L, s_3, s_4, and s_5.

	A	B	C	D	E	F	G	H	I	J	K
1	prices		S	F	L	s1	s2	s3	s4	s5	RHS
2	y1 =	165	3	2	1	1					120
3	y2 =	345	1	2	3		1				80
4	y3 =	0	2					1			96
5	y4 =	0		3					1		102
6	y5 =	0			2					1	40
7	contribution		840	1120	1200	0	0	0	0	0	
8	reduced cost		0	100	0	-165	-345	0	0	0	
9											
10											
11			= C7 - SUMPRODUCT($B2:$B6, C2:C6)								

With blank Target Cell, pick values of the changing cells B2:B6 subject to C8 = 0 E8 = 0 H8:J8 = 0

prices and reduced costs for the basis that consists of $-z$ and the variables S, L, s_3, s_4, and s_5. Dragging the function in cell C8 across row 8 equates each entry in row 8 to the reduced cost of its variable. Solver has been used to find values of the shadow prices that equate the reduced cost of each basic variable to zero.

Table 18.7 also indicates that the entering variable for a simplex pivot will be F because only its reduced cost is positive. Let us review what Table 18.7 accomplishes. This calculation

- computes the shadow prices for the basis that excludes F, s_1, and s_2;
- computes the reduced costs for this basis;
- identifies the entering variable for a simplex pivot.

In particular, we've identified an entering variable without constructing a basic system. That will turn out be advantageous.

A Defect

The simplex method has been presented as a scheme for pivoting from one tableau to the next. This is a great way to learn how the simplex method works, but as a method for solving large linear programs, it stinks.

To see why, imagine that we attempted to solve a large linear program on a digital computer. Suppose this linear program has 500 equations and 8000 variables. Its tableaus would have about $500 \times 8000 = 4{,}000{,}000$ entries. Pivoting from tableau to tableau would require lots of work. Controlling the round-off error on the entire tableau would be a horrendous task. Unless round-off error was controlled, it would accumulate to the point at which it swamped the data. For large linear programs, the accumulation of round-off error is all but certain to cause the tableau-based simplex method to fail.

A Remedy

The so-called **revised simplex method** keeps track of only $3\,m$ numbers per iteration rather than the entire tableau, m being the number of equations. This speeds up computation and greatly eases the control of round-off error.

Of these $3\,m$ numbers, m are the shadow prices for the current basis, which we can calculate from Equation (18.10). From these shadow prices, we can calculate the reduced cost of each nonbasic variable, using Equation (18.11). Having selected an entering variable, we need to compute its coefficients in each equation of the current tableau. (That's m more numbers.) We also need to compute the values of the basic variables in the current tableau. (Those are the final m numbers.) From these 2 m numbers, we can identify the leaving variable; it's the one with the smallest ratio.

This can be done efficiently. How? Because an answer to that question would embroil us in matrix notation that is a bit technical, we omit it. For a lucid presentation of the revised simplex method, you may refer to any of several standard texts on linear programming.[2]

18.14. DEGENERACY

In Program 18.2, each simplex pivot improved the basic solution's objective value. That is not always the case, however. A simplex pivot can "stall" by changing the basis without changing the basic solution. To see how this can occur, we alter Program 18.2 by changing the right-hand-side value of constraint (1.4) from 9 to 0. Table 18.8 presents the initial tableau for this variant of Program 18.2.

In Table 18.8, the reduced costs of A and B are positive. We could pick either A or B as the entering variable for a simplex pivot. We have selected B as the entering variable. In Table 18.8, ratios are computed for Equations (18.1.2), (18.1.3), and (18.1.4). Equation (18.1.4) has the smallest ratio. This ratio equals 0 because the right-hand-side value of Equation (18.1.4) equals zero. With B as the entering variable, a feasible pivot occurs on the coefficient of B in Equation (18.1.4). This pivot will make B basic, and it will make s_4 nonbasic, but it will change no right-hand-side values. This pivot will not change the basic solution, nor will it improve the basic solution's objective value.

A pivot is said to be **degenerate** if it occurs on the coefficient of a variable in an equation whose right-hand-side value is zero. If a pivot is degenerate, the ratio for its pivot row equals zero. A degenerate pivot "perturbs" the basic solution by equating a nonbasic variable to zero. This perturbation has no effect on the values of any of the basic variables. This pivot changes the basis, but it does not change the basic solution. We summarize:

> Each degenerate pivot changes the basis but does not change the basic solution; hence, it does not change the basic solution's objective value.

In Table 18.8, a pivot on the coefficient of B in Equation (18.1.4) is degenerate. This pivot changes the basis, but it changes neither the basic solution nor its objective value.

Table 18.8 Illustration of a degenerate pivot.

	A	B	C	D	E	F	G	H	I	J
1	equation	A	B	s1	s2	s3	s4	-z	RHS	ratio
2	(1.0)	2	3					1	0	none
3	(1.1)	1		1					6	***
4	(1.2)	1	1		1				7	7
5	(1.3)		2			1			9	4.5
6	(1.4)	-1	3				1		0	0

[2] See, for instance, *Linear Programming* by Vašek Chvátal (New York: W. H. Freeman and Co., 1983) or *Operations Research: Algorithms and Applications* by Wayne L. Winston (Boston: Duxbury, 1978), or *Linear Programming: Foundations and Extensions* by Robert J. Vanderbei (Boston: Kluwer, 1997).

A pivot is said to be **nondegenerate** if it occurs on the coefficient of a variable in an equation whose right-hand-side value is nonzero. A simplex pivot that is nondegenerate perturbs the basic solution by equating the entering variable to a positive value. This changes the basic solution, and it improves the objective value. We repeat:

> Each nondegenerate simplex pivot changes the basic solution and improves its objective value.

The simplex pivots in Tables 18.2, 18.3, and 18.4 were nondegenerate because, in each case, the right-hand-side value of the pivot row was positive. Each of these pivots improved the objective value, as it must.

A basic system for Form 1 is said to be **degenerate** if the right-hand-side value of some equation equals zero, other than the equation for which $-z$ is basic. Table 18.8 depicts a degenerate basic solution because the right-hand-side value of Equation (18.1.4) equals zero. If the basic solution is degenerate, must a simplex pivot be degenerate? Not necessarily. In Table 18.8, pick A as the entering variable.

18.15. IS PHASE 2 FINITE?

A degenerate pivot changes the basis but not the basic solution. This raises an unpleasant possibility. A **cycle** is a sequence of degenerate simplex pivots that ends with the basis from which it started. If the simplex method executes a cycle once, it can do so again and again, without limit. Not good!

Cycles will soon be investigated. Before doing so, we make two preliminary observations. Specifically:

- There are finitely many bases. Why is that? There are only finitely many decision variables, each basis is a subset of these decision variables, and a finite set has finitely many subsets.
- Each basis has exactly one basic solution. Why is that? Once the nonbasic variables have been equated to zero, there is no choice as to the values of the basic variables.

A Sack of Balls

To visualize the progress of the simplex method, imagine a sack that contains one ball for each feasible basis. Each ball has the objective value of its basic solution painted on it. There are finitely many bases, so there are finitely many balls in this sack. You begin with a ball in your hand. (Phase 1 puts it there.) Each simplex pivot replaces the ball in your hand with another. If this pivot is degenerate, the number on the new ball is the same as the number on the one it replaces. If this pivot is nondegenerate, the number on the new ball improves on the number on the old ball. Notice that each nondegenerate pivot places a *new* ball in your hand, one whose number improves on all numbers that you have previously seen. There are only finitely many balls in the sack. Thus, only finitely many nondegenerate pivots can occur prior to termination.

Consequently, the simplex method will terminate unless it executes a cycle of degenerate pivots. This cycle puts a ball in your hand that was there previously. If the simplex method uses a consistent rule for replacing one ball with another, such a cycle will repeat again and again, without limit.

Ambiguity

Is a cycle of degenerate pivots a theoretical possibility? Or can it actually occur? It can indeed occur. Can a cycle be kept from occurring? Yes. The simplex method contains some ambiguity, and whether or not it can cycle depends on how that ambiguity is resolved.

Let's identify this ambiguity. In a maximization problem, the entering variable can be *any* nonbasic variable whose reduced cost is positive. In Table 18.2, for instance, we could have chosen either A or B as the entering variable. Similarly, if there are ties for the smallest ratio, the leaving variable could be the basic variable for *any* of the equations whose ratios tie for the smallest.

An Unambiguous Simplex Pivot Rule

To describe a simplex pivot rule unambiguously, we must resolve this ambiguity. Let us consider a linear program that is written in Form 1. We denote as n the number of its decision variables, excluding z, and we number these decision variables x_1 through x_n. For specificity, we suppose that the objective is maximization. Stated below as Rule 1 is a very popular version of the simplex method.

Rule 1 (for maximization):

- The entering variable is a nonbasic variable whose reduced cost is most positive. If two or more variables tie for the most positive reduced cost, the entering variable is the lowest-numbered of them.
- The leaving variable is the lowest-numbered of the basic variables whose rows have the smallest ratio.

To illustrate Rule 1, we apply it to the maximization problem in Table 18.9. In that table, x_1 and x_3 have positive reduced costs, but the reduced cost of x_1 is larger, so Rule 1 selects x_1 as the entering variable. With x_1 as the entering variable, Equations (18.13.1) and (18.13.2) will tie for the smallest ratio, which equals zero. Of these two equations, (18.13.1) has the lower-numbered basic variable, which is x_5. Hence, Rule 1 calls for a pivot on the coefficient of x_1 in Equation (18.13.1).

Rule 1 can cycle. In fact, it *does* cycle when it is applied to the linear program in the table. After six degenerate pivots, the tableau in Table 18.9 reappears.

The data in Table 18.9 may seem innocuous, but this example is carefully contrived. Constructing examples in which Rule 1 cycles wasn't easy. Alan Hoffman found the first such example in 1952.

Anticycling Rules

An **anticycling rule** resolves the ambiguity in the simplex method in a way that precludes cycling and hence terminates finitely. Are there any anticycling rules? Two such rules were

Table 18.9 A maximization problem for which Rule 1 cycles.

	A	B	C	D	E	F	G	H	I	J	K
1	equation	x1	x2	x3	x4	x5	x6	x7	-z	RHS	ratio
2	(13.0)	0.75	-10	0.2	-6	0	0	0	1	0	none
3	(13.1)	0.25	-4	-0.4	9	1	0	0		0	
4	(13.2)	0.5	-6	-0.2	3	0	1	0		0	
5	(13.3)			1		0	0	1		1	

developed early in the history of linear programming. In 1952, A. Charnes published an anticycling rule that is based on perturbing the right-hand-side values of the constraints in a miniscule nonlinear way that makes all pivots nondegenerate and hence precludes cycling. Problem 26 on page 634 describes Charnes's scheme. In 1955, G. Dantzig, A. Orden, and P. Wolfe published an anticycling rule that is based on treating the constraints lexicographically. A controversy developed as to whether or not their work predated Charnes's. When these two anticycling rules are reduced to a computer code, they turn out to be equivalent. The computer code is identical, and it is complicated. Its execution slows down the simplex method.

In 1977, R. G. Bland created a sensation with an anticycling rule that is easy to state and implement. The research community was so surprised to learn that this rule precludes cycles that it immediately became known as **Bland's pivot rule**. His rule breaks ties as follows:

Bland's Pivot Rule:

- The entering variable is the lowest-numbered variable whose reduced cost is positive in a maximization problem, negative in a minimization problem.
- The leaving variable is the lowest-numbered of the basic variables whose rows have the smallest ratio.

Proving that Bland's rule does not cycle is a bit cumbersome. Even so, it is remarkable that it had been overlooked for a quarter century. Bland's rule need not be used at each pivot of the simplex method. It can be brought into play when a cycle is suspected, for example, after 25 pivots, each of which had been degenerate.

18.16. RECAP OF PHASE 2

Our attention will soon turn to Phase 1, but first we recapitulate what we have learned about Phase 2. Phase 2 has been described for linear programs whose decision variables are nonnegative. We've cast such a program in a format that's been dubbed Form 1, namely, as the problem of maximizing or minimizing z, subject to linear equality constraints, with nonnegative decision variables. Phase 2 is initialized with a basic feasible system, that is, with a basic system whose basic solution satisfies the nonnegativity constraints on the variables.

For a maximization problem that is cast in Form 1, Phase 2 is presented as a four-step procedure, with an italicized comment that interprets each step.

Phase 2 of the Simplex Method for Form 1, Maximization:

1. Begin with a basic feasible system. Go to Step 2. *Comment*: *Phase 1 will construct a basic feasible system if one exists.*
2. Stop if all reduced costs are nonpositive. Otherwise, go to Step 3. *Comment*: *If Phase 2 stops here, the basic solution to the system with which it stops is optimal.*
3. Select any nonbasic variable x whose reduced cost is positive. Stop if the coefficient of x in each equation other than the one for which $-z$ is basic is nonpositive. Otherwise, go to Step 4. *Comment*: *If Phase 2 stops here, the linear program is unbounded.*
4. Excluding the equation for which $-z$ is basic, for each equation whose coefficient of x is positive, compute the ratio of that equation's right-hand-side value to its coefficient of x. Pivot on the coefficient of x in an equation whose ratio is smallest. Go to Step 2. *Comment*: *This pivot keeps the basic solution feasible, improves the objective value if the smallest ratio is positive, and causes no change in the basic solution if the smallest ratio is zero.*

To adapt Phase 2 to a minimization problem, we need only switch the signs in Steps 2 and 3, as follows. In Step 2, stop if the reduced costs are nonnegative. In Step 3, pick x as any nonbasic variable whose reduced cost is negative.

As written, Phase 2 contains some ambiguity in the choice of the entering variable and in the coefficient on which pivoting occurs. Whether or not Phase 2 can cycle depends on how this ambiguity is resolved. No matter how this ambiguity is resolved, cycling is so rare that early commercial codes omitted anticycling rules.

For certain classes of optimization problems, a large fraction of the pivots are degenerate. In an assignment problem, for instance, many degenerate pivots can occur between successive nondegenerate pivots. Even for these problems, cycling is virtually unheard of in practice.

18.17. PHASE 1

With our presentation of Phase 2 now complete, we turn our attention to **Phase 1**, which performs these two functions:

- It determines whether or not a linear program has a feasible solution.
- If the linear program has a feasible solution, it constructs a basic feasible system with which to initiate Phase 2.

Phase 1 is presented for linear programs whose decision variables are nonnegative. In a later section, the entire discussion will be adapted to linear programs that include free variables.

Phase 1 can be organized in several different ways. No matter how it is organized, Phase 1 is a bit intricate. Program 18.6 will be used to illustrate Phase 1.

Program 18.6: Maximize $\{4p + 1q + 2r\}$, subject to the constraints

$$\begin{aligned} -1p + 1q + 2r &\geq 6, \\ 1p - 3.5q - 3r &= -10, \\ -2p - 4q \qquad &\leq 0, \\ p \geq 0, \quad q \geq 0, \quad r &\geq 0. \end{aligned}$$

Sketch of Phase I

The version of Phase 1 that we present here consists of three steps. To preview it, we describe its application to Program 18.6.

- Step 1 creates a basic system for Program 18.6, but its basic solution will not be feasible for Program 18.6.
- Step 2 inserts an "artificial" variable, y, such that the basic solution is feasible for all values of y that are sufficiently large. Step 2 uses the simplex method to reduce y to zero.
- Step 3 removes y and creates a basic feasible system with which to initiate Phase 2.

These three steps will be described in general terms, which apply to any linear program whose decision variables are nonnegative. Program 18.6 will be used to illustrate each step.

Step 1 (Setup)

Step 1 of Phase 1 is long-winded, but it is straightforward and familiar. Step 1 introduces an equation that defines z as the objective value, it converts inequalities to equations, and then it executes Gauss-Jordan elimination.

Step 1 of Phase 1:

(a) Introduce an equation that equates z to the objective value.

(b) Leave the nonnegativity constraints on the variables as they are. Convert all other inequality constraints to equations by inserting slack or surplus variables.

(c) Pivot, aiming to create a basic variable for each nontrite equation that lacks one. While doing so, keep $-z$ basic. Part (c) will terminate in one of three ways.

- If it terminates with an inconsistent equation, the equations can have no solution. In this case, stop; the linear program is infeasible.
- If it terminates with a feasible basis, proceed directly to Phase 2.
- If it terminates with a basic system whose basic solution violates one or more of the nonnegativity constraints on the variables, proceed to Step 2 of Phase 1.

Illustration of Step 1

To illustrate Step 1, we apply it to Program 18.6. To execute part (a), we introduce an equation that defines z as the value of the objective. To execute part (b), we subtract a nonnegative "surplus" variable s_1 from the first constraint, and we add a nonnegative "slack" variable s_3 to the third constraint. Parts (a) and (b) organize Program 18.6 in a familiar format, namely, as:

Program 18.$\bar{6}$: Maximize $\{z\}$, subject to the constraints

$$\begin{array}{rrrrrrrl} 4p + & 1q & + 2r & & & -z = & 0, & (18.14.0)\\ -1p + & 1q & + 2r - & s_1 & & = & 6, & (18.14.1)\\ 1p - & 3.5q & - 3r & & & = & -10, & (18.14.2)\\ -2p - & 4q & & & + s_3 & = & 0, & (18.14.3)\\ \end{array}$$

$$p \geq 0, \quad q \geq 0, \quad r \geq 0, \quad s_1 \geq 0, \quad s_3 \geq 0. \qquad (18.14.4)$$

As is usual, we think of $-z$ as a variable in Program 18.$\bar{6}$. The variable $-z$ is basic for Equation (18.14.0), and s_3 is basic for Equation (18.14.3). Equations (18.14.1) and (18.14.2) lack basic variables.

Part (c) of Step 1 pivots to create basic variables for Equations (18.14.1) and (18.14.2). We could pivot on any nonzero coefficients in these equations. The spreadsheet in Table 18.10 executes two such pivots. Evidently, we have elected to pivot on the coefficient of s_1 in Equation (18.14.1) and then on the coefficient of p in Equation (18.15.2).

Like any pivots, those in Table 18.10 create no solutions and destroy none. Systems (14) and (16) have the same set of solutions. System (16) *recasts* Program 18.$\bar{6}$ as:

Program 18.$\bar{6}$: Maximize $\{z\}$, subject to the constraints

$$\begin{array}{rrrrrrrl} & 15q + & 14r & & & -z = & 40, & (18.16.0)\\ & 2.5q + & 1r & + s_1 & & = & 4, & (18.16.1)\\ p - & 3.5q - & 3r & & & = & -10, & (18.16.2)\\ & -11q - & 6r & & + s_3 & = & -20, & (18.16.3)\\ \end{array}$$

$$p \geq 0, \quad q \geq 0, \quad r \geq 0, \quad s_1 \geq 0, \quad s_3 \geq 0. \qquad (18.16.4)$$

Step 1 has constructed a basic system. Its nonbasic variables are q and r. Its basic solution sets

$$q = r = 0, \quad -z = 40, \quad s_1 = 4, \quad p = -10, \quad s_3 = -20.$$

Step 1 is now complete. This basic solution is not feasible because it equates p and s_3 to negative values. For this reason, Phase 1 proceeds to Step 2.

Table 18.10 Pivoting to create a basic system for Program 18.$\bar{6}$.

	A	B	C	D	E	F	G	H
1	equation	p	q	r	s1	s3	-z	RHS
2	(14.0)	4	1	2			1	0
3	(14.1)	-1	1	2	-1			6
4	(14.2)	1	-3.5	-3				-10
5	(14.3)	-2	-4			1		0
6								
7	equation	p	q	r	s1	s3	-z	RHS
8	(15.0)	4	1	2	0	0	1	0
9	(15.1)	1	-1	-2	1	0	0	-6
10	(15.2)	1	-3.5	-3	0	0	0	-10
11	(15.3)	-2	-4	0	0	1	0	0
12								
13	equation	p	q	r	s1	s3	-z	RHS
14	(16.0)	0	15	14	0	0	1	40
15	(16.1)	0	2.5	1	1	0	0	4
16	(16.2)	1	-3.5	-3	0	0	0	-10
17	(16.3)	0	-11	-6	0	1	0	-20

The Motivation for Step 2

The motivating idea for Step 2 is straightforward. The basic solution to system (16) is infeasible because it equates p and s_3 to negative values. The idea is to insert an **artificial variable** y such that this basic solution is feasible for all sufficiently large values of y. Having inserted y, use the simplex method to minimize y, aiming to reduce y to zero. This scheme is aptly dubbed, *coming in on a ray*.

To implement this idea, we insert the variable y with a coefficient of $+1$ on the right-hand sides of Equations (18.16.2) and (18.16.3), thereby replacing Program 18.$\bar{6}$ with Program 18.7, below.

Program 18.7: Minimize $\{y\}$, subject to the constraints

$$
\begin{array}{lr}
-15\,q - 14\,r \qquad\qquad -z = 40, & (18.17.0)\\
2.5\,q + 1\,r + s_1 \qquad\qquad = 4, & (18.17.1)\\
p - 3.5\,q - 3\,r \qquad\qquad = -10 + 1\,y, & (18.17.2)\\
-11\,q - 6\,r \qquad + s_3 = -20 + 1\,y, & (18.17.3)\\
p \geq 0,\quad q \geq 0,\quad r \geq 0,\quad s_1 \geq 0,\quad s_3 \geq 0,\quad y \geq 0. & (18.17.4)
\end{array}
$$

For the moment, regard y as a number and not as a decision variable. Inserting y changes the basic solution system of (17) to

$$q = r = 0, \quad -z = 40, \quad s_1 = 4, \quad p = -10 + y, \quad s_3 = -20 + y.$$

For values $y \geq 20$, this basic solution is feasible.

Program 18.7 *includes* the constraint $y \geq 0$. The objective of Program 18.7 attempts to reduce y to zero. If this can be accomplished, a feasible solution to Program 18.$\bar{6}$ will have been constructed. Alternatively, if the smallest value of y is positive, then Program 18.$\bar{6}$ can have no feasible solution.

The first pivot in Program 18.7 should occur on the coefficient of y in the equation for which s_3 is basic. This pivot produces a solution to the equations $s_3 = -20 + y$ and $s_3 = 0$. It sets $y = 20$, which is feasible. Subsequent pivots will reduce y.

Step 2 (Simplex Pivots)

Step 2 is now described in terms that apply generally and not just to Program 18.7. In this description, we put y on the left-hand side of the equations, along with the other variables. This switches the nonzero coefficients of y from $+1$ to -1.

Step 2 of Phase 1:

(a) Introduce the nonnegative variable y on the left-hand side of the linear program, with these coefficients.

- 0 in the equation for which $-z$ is basic.
- 0 in each equation whose right-hand-side value is nonnegative.
- -1 in each equation whose right-hand-side value is negative, except for the equation for which $-z$ is basic.

(b) Pivot on a coefficient of y that equals -1 and, among these coefficients, pivot on one in an equation whose right-hand-side value is most negative.

(c) Use the simplex method to minimize y. This application of the simplex method terminates in one of two ways.

- If the smallest value of y is positive, stop. In this case, the linear program is infeasible.
- If y has been reduced to zero, proceed to Step 3.

For our example, Program 18.7 has executed part (a), except that we need to shift y to the left-hand side of constraints (18.17.2) and (18.17.3). Part (b) calls for a pivot on the coefficient of y in the equation whose right-hand-side value equals -20. In Table 18.11, this pivot occurs on the boxed-in coefficient. Rows 25–29 show that this pivot produces a basic solution that has $y = 20$, which guarantees its feasibility for Program 18.7.

Phase 1 Simplex Pivots

Part (c) of Step 2 uses the simplex method to minimize y. How shall we accomplish this? Since y is itself a decision variable, we can minimize y directly on the spreadsheet in Table 18.11, without introducing any rows or columns. To see how, we first cast system (18) in the format of a dictionary. To do so, we multiply Equation (18.18.0) by -1, and then we shift the nonbasic variables q and r to the right-hand sides of the equations. This produces:

$$\begin{aligned}
z &= -40 + 15\,q + 14\,r, \\
s_1 &= \quad 4 - 2.5\,q - 1\,r, \\
p &= \quad 10 - 7.5\,q - 3\,r + 1\,s_3, \\
y &= \quad 20 - 11\,q - 6\,r + 1\,s_3.
\end{aligned}$$

We aim to perturb the basic solution to this system in a way that reduces the value of y. The variable y is basic for the last of the above equations. Evidently, perturbing the basic

Table 18.11 Creating a basic feasible solution for Program 18.7.

	A	B	C	D	E	F	G	H	I	J
19	equation	p	q	r	s1	s3	-z	y	RHS	ratio
20	(17.0)	0	15	14	0	0	1		40	none
21	(17.1)	0	2.5	1	1	0	0		4	
22	(17.2)	1	-3.5	-3	0	0	0	-1	-10	
23	(17.3)	0	-11	-6	0	1	0	-1	-20	
24										
25	equation	p	q	r	s1	s3	-z	y	RHS	ratio
26	(18.0)	0	15	14	0	0	1	0	40	none
27	(18.1)	0	2.5	1	1	0	0	0	4	
28	(18.2)	1	7.5	3	0	-1	0	0	10	
29	(18.3)	0	11	6	0	-1	0	1	20	

solution by setting q positive or by setting r positive reduces y. On the other hand, perturbing the basic solution by setting s_3 positive increases y. Note that q and r have *positive* coefficients in Equation (18.18.3). (These coefficients became negative when they were transferred to the right-hand side.) The entering variable for a Phase 1 pivot should have a positive coefficient in the equation for which y is basic. The pivot element should be selected, as usual, to keep $-z$ basic and to keep the basic solution feasible. We summarize:

In a **Phase 1 simplex pivot** for Form 1, the entering variable and pivot element are found as follows:

- The entering variable can be any nonbasic variable that has a positive coefficient in the equation for which y is basic.
- The leaving variable is selected by the usual ratios, which keep the basic solution feasible and keep $-z$ basic.

Phase 1 simplex pivots continue until one of these two conditions has been met:

- The basic solution has $y = 0$. In this case, a feasible solution to the original problem has been constructed.
- The basic solution has $y > 0$, and no nonbasic variable in the equation for which y is basic has a positive coefficient. In this case, y has been minimized. Since it is impossible to decrease y further, no feasible solution exists to the original problem.

The spreadsheet in Table 18.12 illustrates two Phase 1 simplex pivots. In each pivot, the entering variable had the most positive coefficient in the equation for which y is basic. In system (18), the entering variable is q because its coefficient in row 29 is most positive. In system (19), the entering variable is r because its coefficient in row 35 is most positive. For the second pivot, there is a tie for the smallest ratio. We had a choice. We have elected to pivot on the coefficient of r in Equation (18.19.3).

Preview of Step 3

The simplex pivots in Table 18.12 have made y nonbasic. The basic solution to system (20) sets $y = 0$. To initiate Phase 2, all we need to do is to eliminate y from system (20). Eliminating y returns us to Program 18.6, which now appears in the format:

Table 18.12. Two Phase 1 simplex pivots make y nonbasic.

	A	B	C	D	E	F	G	H	I	J
25	equation	p	q	r	s1	s3	-z	y	RHS	ratio
26	(18.0)	0	15	14	0	0	1	0	40	none
27	(18.1)	0	2.5	1	1	0	0	0	4	1.6
28	(18.2)	1	7.5	3	0	-1	0	0	10	1.333
29	(18.3)	0	11	6	0	-1	0	1	20	1.818
30										
31	equation	p	q	r	s1	s3	-z	y	RHS	ratio
32	(19.0)	-2	0	8	0	2	1	0	20	none
33	(19.1)	-0.33	0	0	1	0.333	0	0	0.667	***
34	(19.2)	0.133	1	0.4	0	-0.13	0	0	1.333	3.333
35	(19.3)	-1.47	0	1.6	0	0.467	0	1	5.333	3.333
36										
37	equation	p	q	r	s1	s3	-z	y	RHS	ratio
38	(20.0)	5.333	0	0	0	-0.33	1	-5	-6.67	none
39	(20.1)	-0.33	0	0	1	0.333	0	0	0.667	
40	(20.2)	0.5	1	0	0	-0.25	0	-0.25	0	
41	(20.3)	-0.92	0	1	0	0.292	0	0.625	3.333	

Program 18.6: Maximize $\{z\}$, subject to the constraints

$$\begin{array}{lllllr}
(16/3)p & & & - (1/3)s_3 - z = & -20/3, & (18.21.0)\\
-(1/3)p & & & + s_1 + (1/3)s_3 \;= & 2/3, & (18.21.1)\\
(1/2)p & + q & & - (1/4)s_3 \;= & 0, & (18.21.2)\\
-(11/12)p & & + r & + (7/24)s_3 \;= & 10/3, & (18.21.3)\\
\multicolumn{5}{c}{p \geq 0, \quad q \geq 0, \quad r \geq 0, \quad s_1 \geq 0, \quad s_3 \geq 0.} & (18.21.4)
\end{array}$$

(A close comparison of systems (20) and (21) would reveal small discrepancies that are due to round off. For cosmetic reasons, decimals have been abbreviated in Table 18.12. The number $-0.91666\ldots -11/12$ has been rounded to -0.92, for instance.)

Program 18.6 has now been written in three different formats. In its first appearance, system (14) lacked a basis. Its second appearance, system (16) had a basis but not a feasible basis. In its third appearance, system (21) has a feasible basis, which sets the stage for Phase 2.

Step 3 (Cleanup)

After y has been driven to zero, Step 3 removes y and creates a basic feasible system with which to initial Phase 2. Step 3 can be a bit more complicated than has been suggested by our example. Step 3 is now described in general terms.

Step 3 of Phase 1:

- **(a)** Remove y and its column of coefficients.
- **(b)** If y had been basic for an otherwise nontrite equation, pivot on any nonzero coefficient in that equation.
- **(c)** Proceed to Phase 2.

Table 18.13 Step 2 terminates with system (22), and Step 3 executes a pivot.

	A	B	C	D	E	F	G	H	I
43	equation	p	q	r	s1	s3	-z	y	RHS
44	(22.0)	-4.67	-20	0	0	4.667	1	0	-6.67
45	(22.1)	-0.33	0	0	1	0.333	0	0	0.667
46	(22.2)	0.333	2.5	1	0	-0.33	0	0	3.333
47	(22.3)	-2	-4	0	0	1	0	1	0
48									
49	equation	p	q	r	s1	s3	-z	RHS	
50	(23.0)	0	-10.7	0	0	2.333	1	-6.67	
51	(23.1)	0	0.667	0	1	0.167	0	0.667	
52	(23.2)	0	1.833	1	0	-0.17	0	3.333	
53	(23.3)	1	2	0	0	-0.5	0	0	

For our example, part (b) of Step 3 was unnecessary because the final pivot in Step 2 had made y nonbasic. The same example can be made to illustrate part (b). To do so, we return to Table 18.12. In system (19), there had been a tie for the smallest ratio. We could have pivoted on the coefficient of r in Equation (18.19.2). Executing that pivot produces system (22), which appears in rows (43)–(47) of Table 18.13. In system (22), the variable y is basic, but the basic solution sets $y = 0$, which causes Step 2 of Phase 1 to terminate.

Let's see what Step 3 accomplishes when it is applied to system (22). Part (a) removes y, eliminating the basic variable for Equation (18.22.3). Part (b) of Step 3 executes a pivot on any nonzero coefficient in Equation (18.22.3). In Table 18.13, we elected to pivot on the coefficient of p in Equation (18.22.3). This pivot is degenerate. It changes no right-hand-side values, and it produces a feasible basis with which to initiate Phase 2. By the way, pivoting on the coefficient of q in Equation (18.22.3) would produce system (21), with the final two equations interchanged.

Phase 1 Recap

The theme of Phase 1 is straightforward. The basic idea is to introduce an artificial variable y in a way that creates a basic feasible solution and then pivot so as to minimize y. If y can be driven to zero, a basic feasible solution to the original linear program will have been found. If y cannot be reduced to zero, the original linear program can have no feasible solution.

In the literature, the problem of solving a linear program is parsed in several ways into Phase 1 and Phase 2. There are several versions of Phase 1, but no matter how Phase 1 is accomplished, it entails simplex pivots and is a bit intricate.

Many texts describe a version of Phase 1 that begins by inserting an artificial variable in each constraint that lacks a basic variable and then uses the simplex method to minimize the sum of the artificial variables. The version that we have presented has these advantages:

- The pivot elements in Step 1 can be selected to "hot start" the simplex method, for example, with the optimal basis from the prior optimization run.
- Step 2 introduces only one artificial variable and no extra equations.

- This artificial variable is used to "come in on a ray," which is a trick worth learning.
- Step 2 adapts simplex pivots to minimize the value of a decision variable directly rather than through an objective function.

Our method allows for trite equations. When it is adapted to accommodate "free" variables, it provides accurate sensitivity analysis.

18.18. ACCOMMODATING FREE VARIABLES

In a linear program, a decision variable is said to be **free** if it is not constrained in sign. A free variable can take any value, positive, negative, or zero. Free variables do occur in applications. To place a linear program in Form 1, we must replace each free variable by the difference of two nonnegative variables. That is awkward, and it is unnecessary. Modern computer codes accommodate free variables. They deal with a linear program that is cast in a format dubbed "Form 2." A linear program is said to be written in **Form 2** if:

- Its objective is to maximize the quantity z or to minimize z.
- Each decision variable other than z either is constrained to be nonnegative or is free (unconstrained).
- All other constraints are equations.

Evidently, Form 2 differs from Form 1 by allowing one or more of its decision variables to be unconstrained as to sign.

This section adapts the prior discussion to Form 2. This adaptation has three advantages. It eases the task of entering the linear program, it speeds up the simplex method, and it produces accurate sensitivity analysis. To accommodate free variables, we generalize three concepts: a feasible basis, a feasible pivot, and a simplex pivot.

A Basic Feasible System

Throughout this section, we focus on Phase 2. In Form 2, one or more of the decision variables can be free. A basic system for Form 2 is said to be **feasible** if its basic solution equates each basic variable that is constrained to be nonnegative to a nonnegative value. Now, a basic solution is allowed to assign a negative value to a free variable.

A Feasible Pivot

The idea of a feasible pivot is to select the leaving variable so as to keep the basic solution feasible. In Form 1, the leaving variable was the first to decrease to zero when we perturb the basic solution by setting the entering variable positive. What happens if this perturbation decreases a free variable to zero? We don't want to remove it from the basis. To the contrary, we want to allow this free variable to have any value, positive, negative, or zero. That's easily accomplished, as follows:

> After a free variable becomes basic, compute **no ratio** for the equation for which it is basic. This keeps the free variable basic, allowing it to have any sign in the basic solution.

The nonnegative variables can pivot in, then out, then in again, over and over. By contrast, the free variables can pivot in but not out. No feasible pivot causes a free variable to leave the basis.

Form 2 introduces a new type of feasible pivot. In Form 1, we avoided pivoting on negative coefficients because we wanted the basic solution to stay nonnegative. But suppose that a free variable x is nonbasic. We have a choice of pivot elements! We could perturb the basic solution by equating x to a positive value. This leads us to a feasible pivot on a positive coefficient of x, using the rule that has just been described. We could also perturb the basic solution by equating x to a negative value. That leads to a **feasible** pivot whose coefficient of a free variable x is determined as follows:

- For each equation whose coefficient of x is negative, compute the ratio of its right-hand-side value to its coefficient of x, but compute no ratio for any equation whose basic variable is free.
- Pivot on the coefficient of x in an equation whose ratio is largest (least negative).

If you wish, concoct an example, and use a dictionary or a spreadsheet to check that the preceding rule selects the pivot element in a way that keeps the basic solution feasible.

A Simplex Pivot

For specificity, we focus on Phase 2 for a maximization problem. As before, a simplex pivot is a particular type of feasible pivot. Now there are two types of simplex pivot. In a **simplex pivot**, the entering variable can be any nonbasic variable whose reduced cost is positive, and, in addition, the entering variable can be any *free* variable x whose reduced cost is negative. Why the latter? The objective increases if we perturb the basic solution by equating this variable x to a negative value. A simplex pivot keeps the basic solution feasible because it is a feasible pivot. Each nondegenerate simplex pivot improves the objective value.

The simplex method allows a free variable x to become basic if its reduced cost is any number other than zero. Once a free variable becomes basic, the simplex method never lets it leave. The simplex method "sucks" free variables into the basis and keeps them there. It executes at most one pivot per free variable.

We've described the effect of free variables on Phase 2. Free variables have analogous effects on Phase 1. To see how free variables improve sensitivity analysis, you may refer to Problem 21 on page 634.

18.19. THE PERTURBATION THEOREM*

In Chapter 3, the Perturbation Theorem helped us to interpret the optimal solution to a linear program. The Perturbation Theorem describes the changes that can occur in the optimal solution to a linear program if its data are perturbed by small amounts. This theorem's conclusion is reiterated as follows, where it is assigned the label (P).

> (P) If the data of the linear program are perturbed by amounts that are sufficiently small, its optimal solution can change, but its tight constraints remain tight, and its slack constraints remain slack.

These "perturbations" can occur anywhere in the linear program—in the right-hand-side values, in the objective, and within the constraints—but condition (P) requires each perturbation to be below some threshold.

In Chapter 3, it was noted that the Perturbation Theorem does not apply to every linear program; it requires a "nondegeneracy condition," which was not specified there. This section presents the nondegeneracy condition and demonstrates its validity.

A Specific Format

To ease the exposition slightly, we will describe the nondegeneracy condition for a linear program that has been cast in the following format:

- The objective to maximize z.
- In addition to z, there are n decision variables, which are labeled x_1 through x_n.
- These n decision variables are constrained to be nonnegative.
- In addition to the nonnegativity constraints on decision variables, the linear program has m other constraints, each of which is an equation.

Thus, this linear program has been cast as a maximization problem in Form 1. This linear program has m equality constraints and n variables, each of which is constrained to be non-negative.

The Nondegeneracy Condition

A feasible solution to this linear program is now said to satisfy the **nondegeneracy condition** if this feasible solution is basic and if, in addition:

(a) This solution assigns positive values to exactly m of the x_j's.

(b) Each variable x_j that it assigns the value zero has a negative number as its reduced cost.

It is easy to check whether or not a basic optimal solution satisfies the nondegeneracy condition. To check for condition (a), count the number of positive variables and see whether or not it equals the number of equations. To check for condition (b), check that each variable that is assigned the value zero has a reduced cost that is negative.

If this linear program is feasible and bounded, the simplex method finds an optimal solution that is basic. The remainder of this section shows that this basic optimal solution satisfies condition (P) if and only if it satisfies the nondegeneracy condition. To achieve these results, we delve more deeply into linear algebra and numerical analysis than is typical.

Matrix Notation

The linear program under consideration has m equality constraints and n decision variables. Its decision variables have been labeled x_1 through x_n. With these decision variables arrayed into the $n \times 1$ vector x, this linear program can be described succinctly, in matrix notation, as:

Program 18.8: Maximize z, subject to the constraints

$$z = c\,x, \quad A\,x = b, \quad x \geq 0.$$

The data in Program 18.8's constraints form the $m \times n$ matrix A and the $m \times 1$ vector b, with A_{ij} as the coefficient of the variable x_j in the ith constraint and with b_i as the right-hand-side value of the ith constraint. The objective coefficients of the variables form the $1 \times n$ vector c, where c_j is the coefficient of x_j in the objective.

Necessity

The "necessity" half of our equivalence argument shows that if the nondegeneracy condition is violated, then (P) does not hold.

Lemma 18.1. If a basic optimal solution violates the nondegeneracy condition, then (P) is violated.

Proof. The number of nonzero x_j's in a basic solution to $A\,x = b$ cannot exceed the rank of A, and the rank of A cannot exceed the number m of rows in A. Thus, condition (a) requires that the matrix A has m as its rank. Our proof of Lemma 18.1 entails three cases.

Case 1: Let us first examine the case in which the rank of A is less than m. In this case, every basic solution violates condition (a) and each basic tableau must have at least one trite row. The basic optimal tableau has a trite row. If the right-hand-side value of that row is perturbed by a miniscule amount, the row becomes inconsistent, and the linear program becomes infeasible, which violates (P).

Case 2: Now, suppose that the rank of A equals m but condition (a) is violated. Each basis has m basic variables. Since condition (a) is violated, the basic optimal solution equates some basic variable x_j to zero. A miniscule decrease in the right-hand-side value of the row for which x_j is basic makes this solution infeasible. This perturbation causes a different basic solution to be optimal, and that different basic solution must equate a previously zero variable to a positive value. Hence, this miniscule perturbation causes a tight constraint to become slack, which violates (P).

Case 3: In the remaining case, condition (a) holds, but (b) does not. Hence, the reduced cost of a nonbasic variable x_j equals zero. Increase the objective coefficient c_j of this variable by a miniscule amount. For the perturbed data, the optimal solution must have x_j positive. Thus, a miniscule perturbation causes the constraint $x_j \geq 0$, previously tight, to become slack, which violates (P). This completes the proof. ◆

Lemma 18.1 shows that the Perturbation Theorem does not hold when the nondegeneracy condition is violated.

Sufficiency

It remains to show that the Perturbation Theorem does hold when the nondegeneracy condition is satisfied.

Lemma 18.2. If a basic optimal solution satisfies the nondegeneracy condition, then (P) is satisfied.

Proof. As noted in the proof of Lemma 18.1, condition (a) guarantees that the rank of A equals m. Let us consider an optimal basis found by the simplex method. Since m equals the rank of A, this basis consists of exactly m of the variables x_1 through x_n. This basis is used to define the matrix B and the vectors x_B and c_B as follows:

- The $m \times m$ matrix B consists of the columns of A that correspond to the basic variables, in their natural order.
- The $m \times 1$ vector x_B consists of the rows of x that correspond to the basic variables, in their natural order.
- The $1 \times m$ vector c_B consists of the columns of c that correspond to the basic variables, in their natural order.

Condition (a) guarantees that the columns of B are linearly independent, so B is invertible. The basic solution equates each nonbasic variable to zero, and condition (a) guarantees that it equates each basic variable to a positive value. The values of these basic variables satisfy

$$Bx_B = b, \quad \text{so} \quad x_B = B^{-1}b.$$

The shadow prices of the constraints form the $1 \times m$ vector y. Since the shadow prices equate the contribution of each basic variable to its opportunity cost, they satisfy

$$c_B = yB, \quad \text{so} \quad y = c_B B^{-1}.$$

What happens when small changes occur in the values of A, b, and c? Since B is invertible, any of several routine arguments[3] shows that small changes in B cause small changes in B^{-1} and, hence, in x_B, in y, and in the vector $c - yA$ of reduced costs.

Thus, the variables that were positive remain positive (guaranteeing feasibility), and the reduced costs that were negative remain negative (guaranteeing optimality). In brief, the basic solution remains feasible, and it remains optimal, so statement (P) holds, which completes the proof. ◆

Taken together, Lemmas 18.1 and 18.2 show that the nondegeneracy condition is necessary and sufficient for the conclusions of the Perturbation Theorem to be valid.

18.20. SPEED OF THE SIMPLEX METHOD*

Intuitively, you have ample reason to be skeptical about the speed of the simplex method. If a linear program has hundreds of equations and thousands of variables, its feasible region could have a host of extreme points, and the simplex method could entail astronomically many pivots.

Empirical Evidence

The empirical evidence is truly astonishing. To measure the size of a linear program, we (again) designate:

$$m = \text{the number of equality constraints in a Form 1 representation,}$$
$$n = \text{the number of decision variables in a Form 1 representation.}$$

It is reasonable to suspect that the number of pivots increases rapidly in the number m of constraints and in the number n of variables. The empirical evidence is as follows.

> For the vast majority of practical problems, the number of pivots needed for Phases 1 and 2 combined does not exceed $3\,m$.

This *is* astonishing. A bound of $3\,m$ pivots is very low indeed, and this bound is independent of the number n of decision variables.

[3] This can be proved using determinants or, more elegantly, using matrix norms and the "condition number" of the matrix B. A lucid development of the matrix norm approach culminates with Equation (1.86) on page 27 of H. R. Schwarz, *Numerical Analysis. A Comprehensive Introduction* (New York: John Wiley & Sons, Inc., 1989).

We recall from Chapter 17 that each pivot requires mn multiplications and divisions. Thus, nearly every practical problem can be solved with approximately 3 $m^2 n$ multiplications and divisions. This is a low-order polynomial; if m and n are doubled, the amount of work grows by the factor of 8 (or 2^3). Work increases as the *cube* of the size of the linear program.

There is, we should add, one class of practical linear programs for which the simplex method seems to require more than 3 m pivots. These are coupled multiperiod problems whose nonzero coefficients include a "staircase" structure. Loosely speaking, the simplex method seems to ripple up and down the stairs.

Worst-Case Behavior

We now sketch a class of linear programs for which the simplex method performs poorly. The **unit cube** in m-dimensional space is easy to describe and to interpret; it is the set of all solutions to the inequality system,

$$0 \leq x_k \leq 1, \quad \text{for } k = 1, 2, \ldots, m.$$

The extreme points of the unit cube are easy to identify; for each k, the variable x_k can take the value 0 or 1. Evidently, this unit cube has 2^m extreme points.

Figure 18.4 depicts the unit cube in three-dimensional space. Its arrows describe a pivot sequence that visits all eight of its extreme points. The pivot sequence depicted in the figure begins at extreme point (0, 0, 0), reaches extreme point (1, 0, 0) after the first pivot, and attains extreme point (0, 0, 1) after the seventh pivot.

It's not difficult to tilt and stretch the unit cube so that Rule 1 follows the pivot sequence that is suggested by the arrows in Figure 18.4. In 1972, Klee and Minty constructed a now-famous family of linear programs whose feasible regions are deformations of the unit cube in m-dimensional space. For these examples, Rule 1 visits *all* 2^m extreme points en route to the optimal solution.

For these problems, the amount of work grows exponentially in the number of equations and variables. When $m = 100$, the number of pivots is $2^{100} - 1$. That's a large number. At one *trillion* pivots per second, it would take about 40 *billion* years to execute 2^{100} pivots.

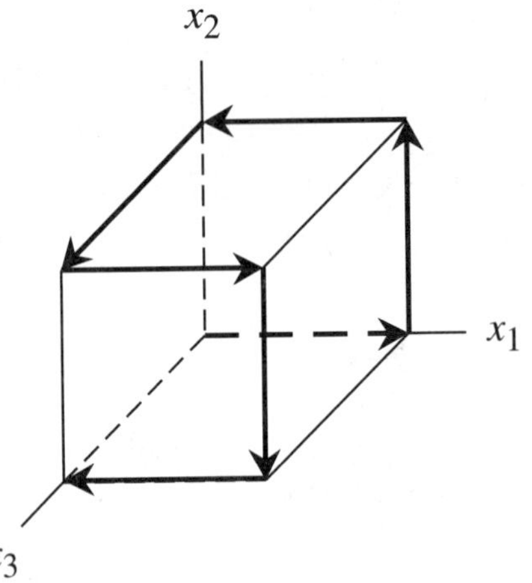

Figure 18.4 The unit cube in three-dimensional space, with a pivot sequence that visits all eight extreme points.

The simplex method requires 3 m pivots on typical problems, but it requires 2^m pivots on atypical problems. When it is good, it is very very good; when it is bad, it is awful. This conundrum has perplexed operations researchers and computer scientists for a half century.

The Expected Number of Pivots

Since the earliest days of the simplex method, researchers have sought to explain the bound of 3 m pivots on typical problems. Roughly speaking, this explanation would take the form of a theorem to the effect that the expectation of the number of pivots on a linear program whose data will be selected at random is a low-order polynomial of m and n. This theorem would have these components:

- A pivot rule that executes a sequence of feasible pivots.
- A probability law for drawing the data of linear programs of each given size.
- A calculation to the effect that the expectation, with respect to this probability law, of the number of pivots needed to solve a linear program is a low-order polynomial in the number m of its constraints and in the number n of its decision variables.

This probability law must draw the Klee-Minty-type examples with miniscule probability.

Suddenly, in the 1977–1984 time period, enormous progress was made on the probabilistic analysis of simplex-type methods. We mention two landmarks:

- K. Borgwardt published models whose expected number of pivots is a fifth-order polynomial in m and n.
- In 1983, M. Haimovich introduced an astoundingly simple model that requires an expectation of not more than n pivots to execute a version of Phase 2. He and others built models that execute Phase 1 and Phase 2 with an expected number of pivots that is a quadratic function of m and n.

To this date, however, no one has obtained a probabilistic model for which the expected number of pivots is a *linear* function of m and n.

18.21. THE COMPLEXITY OF LINEAR PROGRAMS*

In the previous section, we measured the size of a linear program in terms of the integers m and n, and we measured the work of the simplex method in terms of the number of pivots and the number of multiplications and divisions. Our attention will soon turn to solution methods that do not pivot. These methods require a more precise specification of the size of the linear problem.

This section and the next deal with linear programs whose data are integer-valued. A measure of the size of such a linear program is the integer L, where

$$L = mn + K,$$

where

$m =$ the number of equations in a Form 1 representation,

$n =$ the number of variables in a Form 1 representation,

$K =$ the number of bits needed to record each nonzero datum in the linear program.

The equation that defines L seems to "add apples and oranges." Why does L include the number K of bits that are needed to store the data? The answer lies in a "rounding lemma."

When a feasible solution reduces a certain "gap" below 2^{-2L}, this lemma shows that n pivots are guaranteed to "round" the feasible solution to an optimal solution.

An algorithm for linear programs is now said to be **polynomial** if the number of "operations" that is needed to "solve" each linear program is a polynomial in L. In this definition, the terms in quotes have the following meanings:

- To *solve* a linear program is to determine whether or not it is feasible, if feasible, whether or not it is bounded, and if feasible and bounded, to find an optimal solution to it.
- An *operation* is an arithmetic operation (including division) in infinite precision. For some algorithms, an *operation* may even include the taking of a square root or some other function.

This usage of "operation" takes license with standard complexity arguments. Strictly speaking, the number of bits required to perform all operations should be a polynomial in L.

Is the Simplex Method Polynomial?

Earlier, we mentioned the Klee and Minty examples, for which Rule 1 requires $2^m - 1$ pivots. Evidently, Rule 1 is not polynomial. Similar worst-case behavior has been established for other rules for resolving the ambiguity in the simplex pivot. No known version of the simplex method is polynomial. Many researchers speculate that *no* version of the simplex method is polynomial, but no one knows for sure.

The Conjecture, $P = NP$

A class of problems is said to be in ***NP*** if, given a candidate solution, we can confirm that it is a solution with a number of operations that is a polynomial in the size of the problem. Linear programs are in *NP*. (Checking that a candidate solution is feasible requires only mn operations; checking that a candidate is optimal does require polynomial work, but that fact may not be obvious.)

A class of problems is said to be in ***P*** if each member of the class can be solved with a number of operations that is a polynomial in the size of the problem. As indicated above, no known variant of the simplex method is polynomial.

An important, long-standing, and still unproven conjecture is that $P = NP$. Linear programs are in *NP*, so this conjecture would be *false* if no polynomial algorithm existed for linear programs. The conjecture $P = NP$ drove generations of researchers on a quest for a polynomial algorithm for linear programs.

The Ellipsoid Method

In 1979, L. Khachiyan published the **ellipsoid method**, which was the first polynomial-time algorithm for linear programming. The ellipsoid method solves linear programs with a number of operations that is proportional to $n^4 L$. (Here, the taking of a square root is counted as one operation.) The ellipsoid method employs a divide-and-conquer tactic that restricts the optimal solution to ever-shrinking ellipsoids. When the ellipsoid is small enough, this method rounds to an extreme point.

The ellipsoid is important to theory; it shows that linear programs are in P. Unfortunately, on practical problems, the ellipsoid method exhibits performance that is similar to its worst-case bound, $n^4 L$.

On typical problems, the simplex method requires 3 m pivots, and each pivot entails a number of operations that is proportional to mn. Thus, for typical problems, the number of operations required by the simplex method is proportional to mL, which is vastly superior to n^4L.

18.22. INTERIOR-POINT METHODS*

The quest persisted to find methods that would compete with the simplex method on typical problems and perform well in the worst case. In 1984, N. Karmarkar published a paper that contained stunning advances and ignited a firestorm of research. In this work, Karmarkar presented and analyzed a variety of **interior-point** algorithms that avoid the extreme points entirely. One of these algorithms employs "projective transformations"; this method requires a number of iterations that is proportional to nL, with work per iteration that is proportional to $n^{2.5}$ when his "partial updating" scheme is employed. Karmarkar also claimed computational speed that beats the simplex method on large linear programs.

Eventually, it was discerned that Karmarkar's fast computation times tended to be for his "affine scaling" algorithm, which is not a polynomial-time algorithm and which may not converge. It turns out that the affine scaling algorithm had been published in 1967 by I. Dikin.

Currently, interest in projective transformations and affine scaling seems to have faded, but the methods that Karmarkar introduced have persisted. The search has been for algorithms that solve *all* linear programs, that have low worst-case work bounds, and that beat the simplex method on large practical problems.

Today there are two leading contenders in this search. One is the "path-following" algorithm introduced J. Renegar in 1988, with significant contributions by other researchers. This method requires a number of iterations that are proportional to $\sqrt{n}L$ (as opposed to nL for projective transformations), with work per iteration that's proportional to $n^{2.5}$. The path-following method is easy to understand, and it is reminiscent of work done in the 1960s by A. Fiacco and G. McCormick.

The second major contender is the self-dual homogeneous method introduced in 1994 by Y. Ye, M. Todd, and S. Mizumo. Their method is novel, but it recalls facets of duality that were known early in the history of linear programming.

Empirically, these two methods—and several others—solve large practical linear programs within 35 iterations and with work per iteration that is proportional to $n^{2.5}$. Empirically, the simplex method requires 3 m iterations, with work per iteration that's proportional to mn. This connotes that for large enough problems, the interior-point methods are faster. Interior-point methods enjoy a second advantage: They solve nonlinear programs of enormous size, and they do so quickly. Interior-point methods lie beyond the scope of this text.

18.23. REVIEW

The simplex method provides a complete analysis of a linear program. The simplex method can be parsed in several ways into Phases 1 and 2, but no matter how that is accomplished:

- *Phase 1* determines whether or not the linear program has a feasible solution. If this linear program is feasible, Phase 1 constructs a basic feasible solution for it.
- *Phase 2* determines whether a feasible linear program is bounded or unbounded. If this linear program is bounded, Phase 2 constructs a basic solution that is optimal.

The version of Phase 1 that is presented in this chapter uses a single artificial variable y to "come in on a ray." Phase 1 pivots so as to reduce y.

Let us recall from Chapter 17 that Gauss-Jordan elimination requires one pivot per equation. By contrast:

- On typical linear programs, the simplex method requires no more than 3 m pivots, m being the number of equations in a Form 1 representation. This is only three times the work needed to solve the linear equations themselves.
- On the other hand, examples exist on which the standard implementations of the simplex method require 2^m pivots.

Although the simplex method is very fast on typical problems, no known version of the simplex method is polynomial, in the worst case. Some *interior-point* algorithms do have polynomial worst-case work bounds. Evidence has accumulated that some of these interior-point methods outperform the simplex method on linear programs that have enormous numbers of constraints and variables.

At the heart of the simplex method lies the *feasible pivot*, which transforms one basic feasible system into another. The feasible pivot allows any nonbasic variable to enter the basis. Given any entering variable, the feasible pivot uses *ratios* to find a pivot element that keeps the basis feasible.

An easy way to understand feasible pivots is to write the basic system as a *dictionary*. The dictionary clarifies the effects of *perturbing* the basic solution by setting one nonbasic variable positive and adjusting the values of the basic variables so as to satisfy the equation system. The dictionary shows which variables increase and which decrease, and how ratios determine a pivot element that keeps the basis feasible.

When a linear program is cast in Form 1, each feasible basis corresponds to an *extreme point* of the feasible region. Each perturbation corresponds to an *edge* of the feasible region in which the entering variable becomes positive. Each nondegenerate feasible pivot shifts the basic solution to the extreme point at the other end of this edge.

Each *nondegenerate* simplex pivot changes the basis and improves the basic solution's objective value. Each *degenerate* simplex pivot changes the basis, but not the basic solution or the objective value. A sequence of degenerate pivots can cause the simplex method to *cycle* by returning to a basis that had been visited previously. In practice, cycling is rare. Cycling can be precluded in several ways; Bland's anticycling rule is particularly easy to implement.

The Solver add-in **=PIVOT(cell, array)** takes the drudgery out of executing pivots.

This chapter includes an economic interpretation of the simplex method. In that interpretation, each basis establishes a market for the resources—a breakeven or shadow price for each resource. These prices cause the contribution of each basic variable to equal its opportunity cost. If the contribution of a nonbasic variable exceeds its opportunity cost, an agent can earn a profit by setting that variable positive, buying and selling the requisite resources at their shadow prices. This agent maximizes profit by setting this variable to the largest possible value. Dong so reduces to zero the level of another variable, which pivots to a new basis, with a new set of shadow prices. These shadow prices hint at the "revised" simplex method, which runs faster and enables far better control of round-off error.

In Chapter 3, the shadow prices and reduced costs were introduced as a property of the optimal solution. But, as was just noted, every basic solution has its own shadow prices and reduced costs, and these prices guide the simplex method in its choice of entering variable.

Accommodating *free variables* speeds up the simplex method and provides accurate sensitivity analysis. The simplex method makes a free variable basic if its reduced cost is any number other than zero. Once a free variable becomes basic, the simplex method keeps it basic.

18.24. HOMEWORK AND DISCUSSION PROBLEMS

1. **(Feasible Pivots)** In system (1), execute a pivot on the coefficient of A in Equation (18.1.2). What goes wrong?

2. Consider this linear program: Maximize $\{A\}$, subject to the constraints

$$\begin{aligned} -A + \quad B &\le 1, \\ A + \quad B &\le 4, \\ -A + \quad B &\le 2, \\ A \ge 0, \quad B &\ge 0. \end{aligned}$$

(a) Solve this linear program by executing simplex pivots on a spreadsheet.

(b) Solve this linear program graphically, and use your graph to trace the progress of the simplex method.

3. Consider this linear program: Maximize $\{A + 1.5\,B\}$, subject to the constraints

$$\begin{aligned} A \qquad\quad &\le 4, \\ -A + \quad B &\le 2, \\ 2\,A + \quad 3\,B &\le 12, \\ A \ge 0, \quad B &\ge 0. \end{aligned}$$

(a) Solve this linear program by executing simplex pivots on a spreadsheet.

(b) Execute a feasible pivot that finds a second optimal solution to this linear program.

(c) Solve this linear program graphically, and use your graph to trace the progress of the simplex method.

(d) How many optimal solutions does this linear program have? What are they?

4. For the following linear program, construct a basic feasible system and specify its basis and basic solution.

Maximize $\{2\,B - \quad 3\,C\}$, subject to the constraints

$$\begin{aligned} A + \quad B - \quad C &= 16, \\ B + \quad C &\le 12, \\ 2\,B - \quad C &\ge -10, \\ A \ge 0, \quad B \ge 0, \quad C &\ge 0. \end{aligned}$$

5. **(Graphical Interpretation)** On a spreadsheet, solve Program 18.2 with B as the entering variable for the first simplex pivot. Use Figure 18.2 to interpret the progress of the simplex method.

6. **(Graphical Interpretation)** Each part of this problem refers to Table 18.2 and Figure 18.2.

(a) Table 18.2 indicates that the coefficient of B in Equation (18.1.1) equals zero. Use Figure 18.2 to interpret this graphically. Does your interpretation apply to the coefficient of A in Equation (18.1.3)?

(b) With A as the entering variable, no ratio was computed for Equation (18.1.4). If this ratio had been computed, it would have equaled $9/(-1) = -9$. Use Figure 18.2 to interpret this number.

(c) True or false: Program 18.2 has a feasible basis whose nonbasic variables are A and s_4.

(d) True or false: Program 18.2 has a feasible basis whose nonbasic variables are A and s_1.

(e) True or false: Program 18.2 has a feasible basis whose nonbasic variables are A and s_3.

7. Rule 1 picks the most positive entering variable for a simplex pivot on a maximization problem. State a simplex pivot rule that makes the largest possible improvement in the basic solution's objective value? Use Program 18.2 to illustrate your rule.

8. Apply the simplex method to Program 18.4 (on page 604), but use your last name to select the entering variable from those whose reduced costs are positive, as follows: Counting from the left, for the kth pivot, use the kth letter in your last name to select the pivot element. For instance, if the second tableau has two variables whose reduced costs are positive, choose the leftmost as the entering variable if the second letter of your last name is a, c, e, g, and so forth and choose the rightmost if the second letter in your last name is b, d, f, h and so forth. Copy Figure 18.3, and plot the path that your implementation of the simplex method traced.

9. This concerns the *minimization* problem whose Form 1 representation is given in the tableau that follows.

	A	B	C	D	E	F	G	H
1	equation	A	B	C	s1	s2	-z	RHS
2	(1.0)	1	5	2			1	0
3	(1.1)	2	-5	1	1			-5
4	(1.2)	2	-1	2		1		4

(a) It this a basic tableau? Is its basis feasible?

(b) To make short work of Phase 1, pivot on the coefficient of B in Equation (1.1). Then continue Phase 2 to optimality.

10. **(Degeneracy)** This problem concerns the variant of Program 18.1 in which the right-hand-side value of Equation (18.1.4) equals 0 rather than 9.

(a) On a spreadsheet, execute the simplex method, with B as the entering variable for the first pivot.

(b) Draw the analog of Figure 18.2 for this linear program. Trace the progress of the simplex method, and create the analog of Table 18.6.

11. **(An Unbounded Linear Program)** Draw the feasible region for Program 18.3. Apply the simplex method to Program 18.3, selecting B (and not A) as the entering variable for the first pivot. What happens? Interpret your result graphically.

12. **(Forcing Free Variables into Form 1)** This problem concerns the variant of Program 18.2 in which A is a free variable. Now, the constraint $A \geq 0$ is omitted. It's easy to see, graphically, that eliminating the constraint $A \geq 0$ has no effect on the optimal value.

(a) *Without* forcing this variant of Program 18.2 into Form 1, use Solver to find its optimal solution, and obtain a Sensitivity Report. Note that the Allowable Increase in the objective coefficient of A equals 1.

(b) To place this variant of Program 18.2 in Form 1, replace the free variable A by the difference $AP - AN$ of two nonnegative variables. Use Solver to compute its optimal solution, and obtain a Sensitivity Report.

(c) In part (b), the Allowable Increase in the objective coefficient of AP equals zero, as does the allowable increase on the objective coefficient of AN. Why? *Hint:* Would these perturbations present an arbitrage opportunity?

13. **(Redundant Inequalities)** Consider a Form 1 representation of a linear program. Suppose you needed to know whether or not a particular inequality constraint in this linear program was redundant. How could you do this? *Hint:* Use a linear program.

14. True or false: When the simplex method is executed, a variable can:

(a) Leave the basis at a pivot and enter at the next pivot. *Hint:* If it entered, to which extreme point would it lead?

(b) Enter at a pivot and leave at the next pivot. *Hint:* Maximize $\{2B + A\}$, subject to the constraints $3B + A \leq 3, \quad A \geq 0, \quad B \geq 0.$

15. True or false:

(a) A nondegenerate pivot can result in a degenerate basic system.

(b) A degenerate pivot can result in a nondegenerate basic system.

16. True or false: For a linear program in Form 1, feasible pivots are the only pivots that keep the basic solution feasible.

17. The simplex method has been applied to a maximization problem in Form 1 (so that all variables other than $-z$ are constrained to be nonnegative). At some point in the computation, the tableau that is shown below has been encountered; in this tableau, u, v, w, and x denote numbers.

	A	B	C	D	E	F	G
1	A	B	C	D	E	-z	RHS
2	-2	-1	u			1	12
3	1	v		1			w
4	2	3	x		1		3

State conditions on u, v, w, and x such that:

(a) The basic solution to this tableau is the unique optimal solution.

(b) The basic solution to this tableau is optimal but is not the unique optimal solution.

(c) The linear program is unbounded.

(d) The linear program has no feasible solution.

18. **(Nonnegative Column)** For a maximization problem in Form 1, the following tableau has been encountered. In it,* stands for an unspecified data element.

	A	B	C	D	E	F	G	H
1	A	B	C	D	E	F	-z	RHS
2	-3		*		*		1	*
3	4	1	*		*			*
4	2		*	1	*			*
5	0		*		*	1		*

Prove there exist no values of the unspecified data for which it is optimal so set $A > 0$. *Hint:* If a feasible solution exists with $A > 0$, show that it is profitable to decrease A to zero and increase the values of B, D and F in a particular way.

19. **(Nonpositive Row)** For a maximization or a minimization problem in Form 1, the following tableau has been encountered. In it,* stands for an unspecified data element.

	A	B	C	D	E	F	G	H
1	A	B	C	D	E	F	-z	RHS
2	*		*		*		1	*
3	-4	1	-2		-2/3			11
4	*		*	1	*			*
5	*		*		*	1		*

(a) Prove that B is basic in every feasible basis.

(b) Prove that deleting B and the equation for which it is a basis can have no effect either on the feasibility of this linear program or on its optimal value.

20. In a linear program, the variables x and y are said to have **identical columns** if the coefficient of x in each constraint equals the coefficient of y in that constraint and if the coefficient of x in the objective equals the coefficient of y in the objective. Assume that the variables x and y have identical columns in the initial tableau.

(a) Do their columns stay identical after any number of Gaussian operations?

(b) Suppose you pivot to make x basic for an equation. What can you say about y?

(c) True or false: A basic system uniquely identifies the basis.

(d) True or false: A dictionary uniquely identifies the basis.

(e) True or false: If two columns are identical for a linear program in Form 1, one of them can safely be discarded.

21. Solve Problem 13 in Chapter 3 (**A Farmer**) with the free variable AS replaced by the difference $A1 - A2$ of two nonnegative variables. Obtain a Sensitivity Report. Compare it with the one on page 91. Explain the differences.

Remark: Let us observe a difference between two-dimensional feasible regions and three-dimensional feasible regions:

- A bounded two-dimensional feasible region is a polygon. Each of its extreme points touches *exactly two* edges. If a basic feasible solution to its linear program is degenerate, an extra line passes through its extreme point, and this line's constraint is redundant.
- A bounded three-dimensional feasible region is a polyhedron. Each of its extreme points touches *at least three* edges, perhaps more. For instance, the apex of a pyramid touches four edges. In three dimensions, we can encounter degeneracy without having any redundant constraints.

22. (**A Degenerate Pivot, with No Redundant Constraints**) This concerns the linear program: Maximize $\{0.1\,A + 1\,B + 0.1\,C\}$ subject to the constraints $A + B \leq 1$, $B + C \leq 1$, $A \geq 0$, $B \geq 0$, $C \geq 0$.

(a) On a spreadsheet, use simplex pivots with Rule 1 to solve this linear program. Did a degenerate pivot occur?

(b) True or false: The simplex method terminates when it first encounters a basic solution that is optimal.

23. (**Shadow Prices**) The spreadsheet and dialog box in Table 18.7 show how to compute the shadow prices and reduce costs for the basis for Program 18.4 that corresponds to the extreme point (35, 0, 15) in Figure 18.3 (on page 605).

(a) Do the same calculation for the basis that corresponds to the extreme point (0, 10, 20). Which perturbation(s) improve the objective? Which pivot(s) improve the objective?

(b) Repeat part (a) for extreme point (0, 34, 4).

24. (**An Example of Cycling**) Apply Rule 1 (on page 612) to the maximization problem in Table 18.9 (on page 612). *Hint:* The variables x_1 through x_6 will enter in numerical order, and on the sixth iteration, when x_6 enters, Table 18.9 will reappear.

25. (**Bland's Anticycling Rule**) Apply Bland's rule (on page 613) to the maximization problem in Table 18.9 (on page 612). Did it cycle?

26. (**Charnes's Anticycling Rule**) Charnes's idea is to resolve the ambiguity for the leaving variable by perturbing the right-hand-side values of the constraints in miniscule, nonlinear way. To describe his perturbation scheme, cast the linear program in Form 1 and begin with a basic feasible system. Do not perturb the equation for which $-z$ is basic. Number the other nontrite equations 1 through m, and perturb the right-hand-side value of the kth such equation by adding ε^k to it, where ε is a tiny positive number. For the example in Table 18.9, this perturbation is depicted below. The quantity on the right-hand side of Equation (18.13.1) equals $0 + \varepsilon$, for instance.

	A	B	C	D	E	F	G	H	I	J	K	L	M
1	equation	x1	x2	x3	x4	x5	x6	x7	-z	RHS	ε	ε^2	ε^3
2	(13.0)	0.75	-10	0.2	-6				1	0			
3	(13.1)	0.25	-4	-0.4	9	1				0	1		
4	(13.2)	0.5	-6	-0.2	3		1			0		1	
5	(13.3)			1				1		1			1

For the perturbed version of system (13), no feasible pivots will be degenerate because the right-hand-side values will stay positive.

(a) On a spreadsheet, use Charnes's scheme with Rule 1 to maximize z. *Remark:* Columns K, L, and M can help you to figure out what to pivot upon, but these columns needn't be kept track of because are identical to columns G, H, and I, respectively.

(b) There's a sense in which Charnes's scheme is lexicographic. Can you spot it? Can you state it?

27. (**Maximizing a Decision Variable**) Alter Program 18.1 so that its objective is to maximize B, but its constraints are unchanged. Adapt the simplex method to accomplish this directly, that is, *without* introducing an equation that defines z as the objective value. Execute your method on a spreadsheet.

28. (**Phases 1 and 2**) Consider this linear program: Maximize $\{2\,A + 6\,B\}$, subject to the constraints

$$\begin{aligned} 2\,A - \quad 5\,B \qquad\quad &\le -3,\\ 4\,A - \quad 2\,B + \quad 2\,C &\le -2,\\ 1\,A + \quad 2\,B \qquad\quad &\le \;\; 4,\\ A \ge 0, \quad B \ge 0, \quad C &\ge 0. \end{aligned}$$

(a) On a spreadsheet, execute Step 1 of Phase 1, as described on page 615.

(b) On the same spreadsheet, execute Step 2 of Phase 1. Continue Phase 1 until termination.

(c) If Phase 1 constructs a feasible solution to the linear program, execute Phase 2 on the same spreadsheet.

29. (**Phase 1**) In Step 2 of Phase 1, would any harm be done by giving y a coefficient of -1 in every equation other than the one for which $-z$ is basic?

30. (**Phase 1**) True or false: In a Phase 1 simplex pivot, the entering variable can have nonpositive coefficients in all rows other than the one for which $-z$ is basic.

31. (**Phase 1**) True or false: Part (b) of step 3 of Phase I can be avoided by selecting y as the leaving variable if it is basic for the equation whose ratio is smallest.

32. (**Free Variables in Phase 2**) In Program 18.2, allow s_2 to be a free variable. Does this affect the pivot element in Table 18.2? in Table 18.3? Execute the simplex method for this variant of Program 18.2. Adapt Figure 18.2 in a way that explains what happens.

33. (**Free Variables in Phase 2**) Alter Program 18.4 by allowing s_3 and s_4 to be free variables. What happens when you apply the simplex method? Why? *Hint:* No work is needed.

34. (**Phase 1 for Form 2**) For a linear program that is written in Form 1, the description of the Phase 1 simplex pivot appears in the boxed-in rectangle on page 618. Adapt this description to Form 2.

35. (**Nonpositive Variables**) In theory, at least, a linear program can have a variable that is constrained in sign to be nonpositive. Consider a linear program whose constraints include $x_6 \le 0$.

(a) Does Form 1 accommodate this constraint directly, that is, without a change of variables?

(b) Does Form 2 accommodate this constraint directly, that is, without a change of variables? If so, how?

(c) True or false: If a linear program has two identical columns, one of them is superfluous.

36. (**The Klee-Minty Examples**) A three-variable Klee-Minty example is a maximization problem whose Form 1 representation has this initial tableau.

	A	B	C	D	E	F	G	H	I
1	equation	A	B	C	s1	s2	s3	-z	RHS
2	(1.0)	100	10	1				1	0
3	(1.1)	1			1				1
4	(1.2)	20	1			1			100
5	(1.3)	200	20	1			1		10000

(a) On a spreadsheet, solve this linear program by executing the simplex method with Rule 1.

(b) Identify the bases that you encounter with Figure 18.4 (on page 626).

(c) *Guess* what a four-variable analog of this example looks like.

Chapter 19

Duality

Note: **Sections 1–11 require only Chapter 3.**

19.1. PREVIEW

The word "dual" suggests a pairing, and in this chapter, we will see that each linear program is paired with another. Each linear program in such a pair will be said to be the "dual" of the other.

Duality has been with us all along. When the simplex method solves a linear program, its shadow prices form an optimal solution to the dual of this linear program. This chapter presents a discussion of duality and probes its implications.

19.2. WHAT CAN YOU LEARN FROM THIS CHAPTER?

Is duality a mathematical curiosity, or is it important? The answer is the latter, for duality plays three key roles:

- From the *modeling* perspective, duality enlarges the scope of linear programming. It expands the set of decision problems that can be formulated as linear programs.
- From a *mathematical* perspective, the duality that emerges from linear programming has become one of the more potent tools in linear algebra, and its generalizations play important roles in analysis.
- From a *computational* perspective, duality is the driving force behind the simplex method, its generalizations, and several of its variants.

Being an introduction, this text focuses on modeling, making brief forays into the mathematics of duality and hinting as to its significance for computation.

The Basics

The first 11 sections of this chapter cover the basics of duality and do not require Chapter 18. They can be read immediately after Chapter 3.

To build intuition, we begin by revisiting the Recreational Vehicle problem. Its goal is to operate a production facility for one week in a way that maximizes profit. You will see that its dual linear program is the problem of renting the same facility for the same period, at minimum expense. You will see how these two linear programs are linked and how the shadow prices for either prescribe the optimal solution to the other.

Next, you will learn how to take the dual of any linear program. Then you will learn of two tests for optimality. In particular, you will see that feasible solutions to a linear program and its dual are optimal if they satisfy a famous condition that's known as "complementary slackness."

Modeling

After the basics have been covered, this chapter presents four examples of situations in which duality models decision making. These four examples are independent of each other, and they can be read selectively. Briefly, they are as follows:

- *Data envelopment:* This example concerns a collection of "units," each of which consumes certain inputs and produces certain outputs. A unit is said to be "enveloped" if there exists a linear combination of the other units that consumes less of each input and produces more of each output. From this example, you can see how duality relates envelopment to efficiency.
- *General equilibrium:* This example concerns an economic model of an "economy" that is in "general equilibrium." This economy consists of multiple agents, each of whom acts selfishly. Goods are traded at prices that are determined within the model. From this example, you can see how a linear program and its dual demonstrate the existence of a general equilibrium and construct one.
- *Arbitrage:* This example concerns the "no-arbitrage" hypothesis of financial economics, which is—roughly speaking—that no financial asset can earn a profit without risking the possibility of a loss. You will see how duality uses the no-arbitrage hypothesis to prescribe a probability distribution over the states of nature such that each asset earns zero as the expectation of its profit.

- *Interdiction:* In a network, a "cut" is a set of arcs whose removal makes it impossible to send any flow from the network's "source" node to its "sink" node. You will see how duality relates cuts to the quantity of goods that can flow from the source node to the sink node.

Math

Sections 19.12 and 19.14 of this chapter present the mathematics of duality as it relates to linear programs. Some readers may wish to start with these sections; others may wish to avoid them.

Section 19.12 demonstrates that each linear program is paired with another, that each linear program in a pair is the dual of the other. This section demonstrates that a linear program and its dual have the same optimal value, and moreover, that the shadow prices for either form an optimal solution to the other.

Section 19.14 relates duality to a theorem of the alternative, one that predates linear programming by a half century. This theorem of the alternative has many important implications, and it generalizes well beyond linear programming, but this introductory text can only hint at the generalizations.

Computation

The computational facet of duality has already been glimpsed. Specifically, in Section 18.13 of Chapter 18, we observed that each basis has its own set of shadow prices and that these shadow prices guide the simplex method as it pivots from basis to basis. Each set of shadow prices turns out to be a "trial" solution to the dual linear program. The simplex method pivots in search of shadow prices that are feasible for the dual linear program, and it stops when it finds them.

This chapter hints at pivoting schemes that solve a linear program and its dual without optimizing anything. These pivot schemes aim for feasible solutions that satisfy complementary slackness. They solve equilibrium problems as well as optimization problems.

19.3. RECREATIONAL VEHICLE REDUX

By now, the Recreational Vehicles problem, introduced in Chapter 2, must be numbingly familiar. It has popped up repeatedly, and here it is again, but this time, it will be used to introduce duality.

Despite its familiarity, let us review it briefly. A manufacturing facility consists of five shops that can make three types of vehicle. Table 19.1 lists the shops, the capacity of each shop, the vehicles, the contribution of each vehicle that is manufactured, and the time needed in each shop to make each type of vehicle. For instance, making one Standard model vehicle consumes 3 hours of capacity in the Engine shop, 1 hour of capacity in the Body shop, and 2 hours of capacity in the Standard Finishing shop.

The Recreational Vehicle Company wishes to operate this facility in a way that maximizes its weekly contribution (profit). In Chapter 2, their decision problem was formulated as a linear program whose decision variables are S, F, and L. This linear program is reproduced here and is dubbed the "Production LP." The data in each of its constraints are taken from a row of Table 19.1, as are the data in its objective.

Table 19.1 Shop capacities (in hours per week), manufacturing times (in hours per vehicle), and contribution of each vehicle that is made.

		Manufacturing times		
Shop	Capacity	Standard	Fancy	Luxury
Engine	120	3	2	1
Body	80	1	2	3
Standard Finishing	96	2		
Fancy Finishing	102		3	
Luxury Finishing	40			2
Contribution		\$840	\$1120	\$1200

Production LP: $z^* = \text{Max}\{840\,S + 1120\,F + 1200\,L\}$, subject to

$$
\begin{array}{llllr}
E: & 3\,S\,+ & 2\,F\,+ & 1\,L & \leq 120, \\
B: & 1\,S\,+ & 2\,F\,+ & 3\,L & \leq 80, \\
SF: & 2\,S & & & \leq 96, \\
FF: & & 3\,F & & \leq 102, \\
LF: & & & 2\,L & \leq 40, \\
 & S \geq 0, & F \geq 0, & L \geq 0. &
\end{array}
$$

In a linear program, the constraints divide themselves into two groups, which we now identify. A **sign** constraint is a constraint on the sign of a variable, and a **nonsign** constraint is any constraint in a linear program that is not a sign constraint.

Each nonsign constraint in the Production LP has been assigned the label that appears to its left. These labels will have two interpretations, one familiar, the other new. The familiar interpretation is as the shadow price of the indicated constraint; for example, E is the shadow price of the constraint on the Engine shop capacity. The new interpretation will be as a decision variable in a second linear program.

In the Production LP and in general, each nonsign constraint is paired with a decision variable in a second linear program. Each nonsign constraint and the variable with which it is paired are now said to be **complementary** to each other. For instance, the constraint $3\,S + 2\,F + 1\,L \leq 120$ and the decision variable E are complementary to each other; the constraint $1\,S + 2\,F + 3\,L \leq 80$ and the decision variable B are complementary to each other; and so forth.

In Chapter 2, we used Solver to find the optimal solution to the Production LP, along with its shadow prices. This optimal solution is

$$S = 20, \quad F = 30, \quad L = 0,$$

the optimal value z^* equals \$50,400 per week, and the shadow prices of its five nonsign constraints are, respectively,

$$E = 140, \quad B = 420, \quad SF = 0, \quad FF = 0, \quad LF = 0.$$

A second interpretation will soon be made of this optimal solution and of its shadow prices.

19.4. BIDDING FOR THE RESOURCES

The Production LP provides meaningful results from the perspective of the manufacturer who wishes to make the best use of the available resources. A second linear program provides equally meaningful results from the perspective of an outsider who wishes to bid for the use of the same resources. To view the situation faced by this outsider, we pose:

Problem A (Renting the Recreational Vehicle Facility)

The Recreational Vehicle Company has agreed to rent its Recreational Vehicle facility to you for one week according to the following terms:

- You must offer them a price for each unit of capacity of each shop.
- You must set each price high enough that they have no economic motive to withhold any capacity of any shop from you.

What prices should you set, and what will it cost you to rent the facility for one week?

The Recreational Vehicle Company can earn \$50,400 by operating this facility for one week. You must set your prices high enough that they have no motive to withhold any capacity from you. Intuitively, it seems clear that you must pay at least \$50,400 to rent their entire capacity. But must you spend more than \$50,400? And what prices should you offer? To answer these questions, we will build a linear program.

Decision Variables

The decision variables in this linear program are the prices that you will offer. By agreement, you must offer five prices, one per shop. Let us designate these prices as:

E = the price (\$/hour) you offer for each unit of Engine shop capacity.
B = the price (\$/hour) you offer for each unit of Body shop capacity.
SF = the price (\$/hour) you offer for each unit of Standard Finishing shop capacity.
FF = the price (\$/hour) you offer for each unit of Fancy Finishing shop capacity.
LF = the price (\$/hour) you offer for each unit of Luxury Finishing shop capacity.

A second meaning has now been assigned to the symbols E, B, SF, FF, and LF. They are the decision variables in the linear program that is under construction. Can you guess why this ambiguity has been introduced?

Renting the Capacity

Let us compute the cost to you of renting the entire capacity of the Recreational Vehicle facility. The Engine shop has a capacity of 120 hours. The cost to you of renting the entire capacity of the Engine shop is 120 E. Similarly, the cost of renting the capacity of the Body shop capacity is 80 B, and so forth. The total cost that you will pay to rent every unit of every shop's capacity is given by

$$\{120\,E + 80\,B + 96\,SF + 102\,FF + 40\,LF\}.$$

You wish to minimize this expression, which is your rental bill, subject to constraints that keep the Recreational Vehicle Company from withholding any capacity from you.

Leaving Resources Idle

The Recreational Vehicle Company need not make full use of the capacity of any of its shops. That fact constrains the prices that you can offer. For instance, the capacity constraint on the Engine shop is the inequality,

$$3\,S + 2\,F + 1\,L \leq 120.$$

Can you offer a price E that is negative? No. If you did, the Recreational Vehicle Company would not rent you any of the capacity of its Engine shop. Instead, it would leave those resources idle. You must offer a price E that is nonnegative. The decision variable E must satisfy the constraint $E \geq 0$.

Each shop's capacity constraint is a "≤" inequality. For this reason, each of the prices that you offer must be nonnegative. In other words, the decision variables must satisfy the constraints,

$$E \geq 0, \quad B \geq 0, \quad SF \geq 0, \quad FF \geq 0, \quad LF \geq 0.$$

Producing Vehicles

The Recreational Vehicle facility can be used to manufacture vehicles. Your prices must be high enough that manufacturing each type of vehicle becomes unprofitable. For each decision variable in the Production LP, we will construct a **complementary** constraint in the dual, and this constraint will make it unprofitable for the Recreational Vehicle Company to set that variable positive.

Let us begin with the Standard model vehicle. The Recreational Vehicle Company would earn \$840 for each Standard model vehicle that it made. To keep it from making any Standard model vehicles, you must offer at least \$840 for the bundle of resources needed to make one Standard model vehicle. From a *column* of Table 19.1, we see that making one Standard model vehicle requires 3 hours in the Engine shop, 1 hour in the Body shop, and 2 hours in the Standard Finishing shop. Thus, making Standard model vehicles becomes unprofitable if the prices you offer satisfy

$$S: \qquad 3\,E + 1\,B + 2\,SF \geq 840.$$

Similarly, your prices must make it unprofitable for the Recreational Vehicle Company to make any Fancy model vehicles. You assign to the variable F the complementary constraint

$$F: \qquad 2\,E + 2\,B + 3\,FF \geq 1120.$$

In the same way, you assign to the variable L the complementary constraint

$$L: \qquad 1\,E + 3\,B + 2\,LF \geq 1200.$$

The data in each of these three constraints have been taken from a column of Table 19.1.

A Price-Setting Linear Program

We have displayed the objective and constraints of your linear program. This linear program is assembled here and is dubbed the "Pricing LP."

Pricing LP: $z_* = \text{Min}\,\{120\,E + 80\,B + 96\,SF + 102\,FF + 40\,LF\}$, subject to

$$\begin{array}{llllllll}
S: & 3\,E + & 1\,B + & 2\,SF & & & \geq 840, \\
F: & 2\,E + & 2\,B + & & 3\,FF & & \geq 1120, \\
L: & 1\,E + & 3\,B + & & & 2\,LF & \geq 1200, \\
 & E \geq 0, & B \geq 0, & SF \geq 0, & FF \geq 0, & LF \geq 0. &
\end{array}$$

The Pricing LP calculates the prices that minimize the cost of renting the facility for one week, subject to constraints that make it unprofitable for the Recreational Vehicle Company to withhold any capacity from the renter.

Each nonsign constraint in the Pricing LP has been assigned the label that appears to its left. As usual, each label has two interpretations—as the shadow price for its constraint and as a decision variable in the "dual" linear program.

The optimal solution to the Pricing LP is easily found from Solver. This optimal solution is

$$E = 140, \quad B = 420, \quad SF = 0, \quad FF = 0, \quad LF = 0.$$

The optimal value z_* to the Pricing LP is 50,400 dollars per week. And the shadow prices of its three constraints are respectively,

$$S = 20, \quad F = 30, \quad L = 0.$$

Compare the Production LP with the Pricing LP. Their optimal values are identical, and the shadow prices of each linear program are optimal values of the complementary variables in the other! Had you guessed this?

19.5. DUAL LINEAR PROGRAMS

The relationship between the Production LP and the Pricing LP is no accident. It illustrates a general principle of "duality." We will see that each linear program is paired with another, that each linear program in the pair is the other's "dual." Let us recall that a linear program is **feasible** if it has at least one feasible solution. Also, a feasible linear program is **bounded** if its objective cannot be improved without limit. The relationship between a linear program and its dual is highlighted as follows:

Duality Theorem. Each linear program is paired with another. If either linear program in a pair is feasible and bounded, then:

- The other linear program is feasible and bounded, and both linear programs have the same optimal value.
- The shadow prices for either linear program form an optimal solution to the other.

In other words, the properties that we have observed for the Production LP and the Pricing LP hold in general. Later in this chapter, the Duality Theorem will be proved.

The simplex method was presented as a way of solving *a* linear program, but it actually solves two linear programs—the one under attack and its dual. These two linear programs have the same optimal value. The shadow prices for each linear program form an optimal solution for the other.

Is duality a surprise? Yes and no. We will see that duality is, in essence, a reinterpretation of the shadow prices and opportunity costs that were studied in Chapter 3.

Precisely how does one take the dual of a particular linear program? We have not yet answered this question. The Pricing LP and the Production LP suggest the general ideas but omit some important detail. To develop your intuition for that detail, we return once more to the Recreational Vehicle problem.

19.6. RECREATIONAL VEHICLES REVISITED

Let us alter, slightly, the situation faced by the Recreational Vehicle Company and see what changes occur in the prices you must pay to rent their facility.

Problem B (Renting the Recreational Vehicle Facility, revised)

The Recreational Vehicle example is as before, with this exception: The Recreational Vehicle Company must use all 102 hours of the capacity of its Fancy Finishing shop because letting this shop's capacity be underutilized would cause irreparable damage to an expensive tool. The Recreational Vehicle Company remains willing to rent its facility to you subject to the same terms as in Problem A, but with the added requirement you must make full use of the Fancy Finishing shop. Now, what prices will you set, and what will it cost you to rent their facility for one week?

To begin an analysis of Problem B, we reconsider the optimization problem faced by the Recreational Vehicle Company. Previously, in the Production LP, the constraint on the Fancy Finishing shop had been the inequality, $3\,F \le 102$. The unique optimal solution to the Production LP had set $F = 30$, so this constraint had been slack. Now, the Recreational Vehicle Company must increase F from 30 to 34 because $(3)(34) = 102$. As a consequence, they will be able to make less efficient use of the facility. It will be less profitable.

Changes to the Production LP

To learn how much the profit decreases, we rerun the Production LP with the inequality constraint $3\,F \le 102$ replaced by the equation $3\,F = 102$. Its optimal solution becomes

$$S = 12, \quad F = 34, \quad L = 0,$$

its optimal value becomes $z^* = \$48{,}160$, and the shadow prices of its constraints become

$$E = 0, \quad B = 840, \quad SF = 0, \quad FF = -186\ 2/3, \quad LF = 0.$$

Forcing the capacity of the Fancy Finishing shop to be fully utilized has decreased the profitability of the facility from 50,400 to 48,160 dollars per week. The shadow price for the Fancy Finishing shop has become negative.

Changes to the Pricing LP

If you rent the Recreational Vehicle facility, you must make full use of the capacity of the Fancy Finishing shop. This stipulation makes the Recreational Vehicle facility less desirable to you as a rental property. But how does it affect the prices that you offer?

To see how these prices change, we observe that the Fancy Finishing shop's capacity constraint has become the equation,

$$FF: \qquad 3\,F = 102,$$

and FF is still a variable in the Pricing LP. But must you still offer them a price FF that is nonnegative? No. You need not do that because they dare not leave any capacity of that shop unutilized. The price FF becomes a *free variable* in the Pricing LP; it can have any sign, positive, negative, or zero. We have re-solved the Pricing LP with FF as a free variable. It reports an optimal solution of

$$E = 0, \quad B = 840, \quad SF = 0, \quad FF = -186\ 2/3, \quad LF = 0,$$

it reports an optimal value $z_* = \$48{,}160$, and it reports shadow prices of

$$S = 12, \quad F = 34, \quad L = 0,$$

exactly as predicted by the Duality Theorem.

Recap

Problem B has caused these changes in the Production LP and in the Pricing LP:

- In the Production LP, the constraint on the capacity of the Fancy Finishing shop switched from a "≤" inequality to an equation.
- In the Pricing LP, the complementary variable *FF* switched from being nonnegative to being free.

This suggests a relationship between the "sense" of a constraint and the "sense" of its complementary variable.

19.7. COMPLEMENTARY VARIABLES AND CONSTRAINTS

To state this relationship precisely, we introduce some vocabulary. Each nonsign constraint has one of three **senses**; it can be an equation, a "≤" inequality, or a "≥" inequality. Similarly, each decision variable has one of three **senses**: it can be constrained to be nonnegative, nonpositive, or free. (A variable is free if it can be positive, negative, or zero.)

The Crossover Table

Problems 19.1 and 19.2 illustrate rules for building the constraints of the dual of any linear program. These rules are:

- To each nonsign constraint in the linear program, assign a **complementary** variable in its dual.
- To each decision variable in a linear program, assign a **complementary** constraint in its dual.
- To determine the sense of each complementary decision variable and constraint from the **crossover table** (Table 19.2).

When taking the dual of a maximization problem, read the crossover table from left to right; when taking the dual of a minimization problem, read the crossover table from right to left.

Let us illustrate the use of the crossover table. The original Production LP is a maximization problem. When taking its dual, we read the crossover table from left to right. The nonsign constraints in the Production LP are "≤" inequalities, so line 1 of the crossover

Table 19.2 The crossover table: the senses of complementary constraints and variables.

Line	Maximization problem	Minimization problem
1.	ith nonsign constraint $\le$ RHS value	ith variable ≥ 0
2.	ith nonsign constraint $=$ RHS value	ith variable is free
3.	ith nonsign constraint $\ge$ RHS value	ith variable ≤ 0
4.	jth variable ≥ 0	jth nonsign constraint $\ge$ RHS value
5.	jth variable is free	jth nonsign constraint $=$ RHS value
6.	jth variable ≤ 0	jth nonsign constraint $\le$ RHS value

table shows that the complementary decision variables in its dual are nonnegative. The decision variables in the Production LP are nonnegative, and line 4 shows that the complementary nonsign constraints in its dual are "$\geq$" inequalities. In Problem B, one nonsign constraint in the Production LP became an equation, and line 2 shows that this equation's complementary variable is free.

Similarly, the original Pricing LP was a minimization problem whose variables were nonnegative and whose nonsign constraints were "$\geq$" inequalities. By reading lines 1 and 4 from right to left, we see that its dual is a maximization problem with "$\leq$" constraints and nonnegative variables.

A Memory Aid

Must you memorize the crossover table? No. To see that lines 1 and 4 of the crossover table are "natural," use this memory aid:

- In any linear program, it's *natural* that the decision variables be nonnegative.
- In a maximization problem, it's *natural* that the nonsign constraints be less-than-or-equal-to inequalities, as in

  $$\text{Maximize } \{3\,x\}, \text{ subject to } 2\,x \leq 6 \text{ and } x \geq 0.$$

  (The sense of the constraint $2\,x \leq 6$ is "natural" because the linear program would be unbounded if its inequality went the other way.)
- In a minimization problem, it's *natural* that the nonsign constraints be greater-than-or-equal-to inequalities, as in

  $$\text{Minimize } \{6\,y\}, \text{ subject to } 2\,y \geq 3 \text{ and } y \geq 0.$$

  (If the sense of the constraint $2\,y \geq 3$ was reversed, this linear program would be unbounded.)
- *Natural* constraints have complementary variables that are nonnegative, as prescribed by rows 1 and 4.

In brief, the natural senses are those in rows 1 and 4. If a constraint is *unnatural* (reversed), its complementary variable is nonpositive (rows 3 and 6). And if a constraint is an equation, its complementary variable is free in sign (rows 2 and 5).

Another Memory Aid

Here's another way to avoid memorizing the crossover table. Remember that the simplex method terminates with shadow prices that equal optimal values of the dual variables.

- In a maximization problem, the shadow price of a "$\leq$" constraint (such as $2\,x \leq 6$) is nonnegative because increasing its right-hand-side value can only increase the optimal value. Since the shadow prices are optimal values of the dual variables, this justifies line 1.
- In a minimization problem, the shadow price of a "$\geq$" constraint (such as $2\,y \geq 3$) is nonnegative because increasing its right-hand-side value can only increase the optimal value. Since the shadow prices are optimal values of the dual variables, this justifies line 4.

Thus, lines 1 and 4 are easy to justify from the signs of the shadow prices. The other lines follow from them.

19.8. A RECIPE FOR CONSTRUCTING THE DUAL

This section presents a recipe for constructing the dual of any linear program. To illustrate this recipe, we will use it to take the dual of the Pricing LP. A detached-coefficient tableau for the Pricing LP appears as Table 19.3.

The recipe for taking the dual has three steps. We will describe each step and then use Table 19.3 to illustrate it algebraically. Later in this section, we will illustrate the recipe a second time, on a spreadsheet.

Three Steps

The first step specifies the decision variables in the dual linear program and the sign constraint on each of them.

Step 1: For each nonsign constraint in the original linear program, create a complementary decision variable for the dual, and use the crossover table to determine the sense of this dual variable from the sense of the nonsign constraint to which it is complementary.

The Pricing LP has three nonsign constraints. In Table 19.3, these nonsign constraints have been assigned the labels S, F, and L. The Pricing LP is a minimization problem, so we read the crossover table from right to left when taking its dual. Each nonsign constraint in the Pricing LP is a "$\geq$" inequality, so line 4 shows that its dual variables are nonnegative. In brief, Step 1 states that:

$$S \geq 0, \quad F \geq 0, \quad L \geq 0.$$

The second step reverses the sense of optimization and specifies the objective of the dual.

Step 2: If the original linear program was a maximization problem, its dual is a minimization problem, and conversely. The coefficient of each dual variable in its objective equals the right-hand-side value of the constraint to which this variable is complementary.

The Pricing LP is a minimization problem, so its dual is a maximization problem. In Table 19.3, we form the objective of the dual by multiplying each dual variable by the right-hand-side value in its row. Step 2 states that the goal of the dual linear program is to

$$\text{Maximize } \{840\, S + 1120\, F + 1200\, L\}.$$

The third step creates a constraint of the dual that is complementary to each decision variable in the original linear program.

Step 3: To each decision variable x in a the original linear program, create a complementary constraint whose sense, RHS value, and coefficients are determined as follows:

(a) The sense of this constraint is determined by the sense of x via the crossover table.

Table 19.3 A detached-coefficient tableau for the Pricing LP.

	E	B	SF	FF	LF		RHS
S:	3	1	2			$\geq$	840
F:	2	2		3		$\geq$	1120
L:	1	3			2	$\geq$	1200
min:	120	80	96	102	40		

(b) The right-hand-side value of this constraint equals the coefficient of x in the objective.

(c) The left-hand side of this constraint multiplies the coefficient of x in each constraint by that constraint's complementary dual variable, and takes the sum.

Step 3, though wordy, is easy to grasp. To illustrate it, we consider the decision variable E in the Pricing LP. Since E is a nonnegative variable in a minimization problem, line 1 of the crossover table shows that its complementary constraint is a "$\leq$" inequality. From the column of coefficients of E in Table 19.3, we see that this complementary constraint is

E: $\qquad\qquad 3S + 2F + 1L \leq 120.$

Similarly, the columns of coefficients of B in Table 19.3 specify the complementary constraint

B: $\qquad\qquad 1S + 2F + 3L \leq 80.$

In the same way, the columns of coefficients of SF, FF, and LF in Table 19.3 specify the complementary constraints

SF: $\qquad\qquad 2S \qquad\qquad \leq 96,$
FF: $\qquad\qquad\qquad 3F \qquad \leq 102,$
LF: $\qquad\qquad\qquad\qquad 2L \leq 40.$

Together, these three steps specify the objective and constraints of the Production LP. When this recipe is applied to either the Pricing or the Production LP, it gives the other. Each is the other's dual.

The Dual on a Spreadsheet

Let us execute this recipe on a spreadsheet. Table 19.4 presents a standardized spreadsheet for the Pricing LP, along with its Solver dialog box. In Table 19.4, the five heavily outlined cells in row 2 are reserved for the values of the decision variables in the Pricing LP. In the table, two blank rows have been inserted between the constraints of the Pricing LP and its objective. We will use these two rows to create the dual linear program *without* repositioning the data in Table 19.4.

Table 19.5 presents the spreadsheet and Solver dialog box that result from applying this recipe to the linear program in Table 19.4. In Table 19.5, cells A3, A4, and A5 record

Table 19.4 Standardized spreadsheet and Solver dialog box for the Pricing LP.

	A	B	C	D	E	F	G	H	I	J
1	name		E	B	SF	FF	LF			
2		value								
3			3	1	2				>=	840
4			2	2		3			>=	1120
5			1	3			2		>=	1200
6										
7										
8			120	80	96	102	40			
9										
10			= SUMPRODUCT(C$2:G$2, C8:G8)							

Solver Parameters
Set Target Cell: $H$8
Equal To: ○ Max ● Min
By Changing Cells: $C$2:$G$2
Subject to the Constraints:
$C$2:$G$2 >= 0
$H$3:$H$5 >= $J$3:$J$5

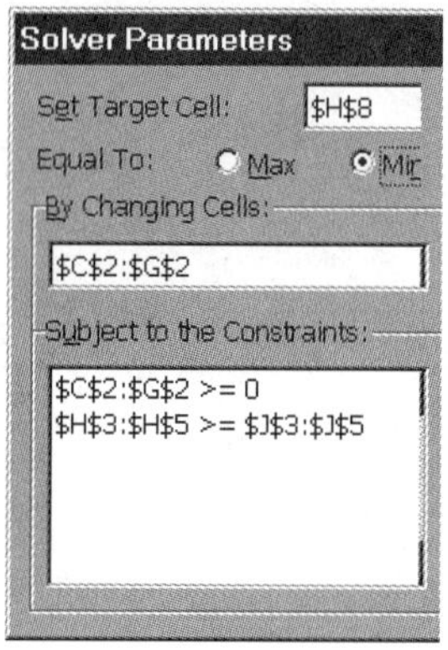

- Cells that describe inequality constraints have been shaded if these constraints are tight.

Candidate solutions to a linear program and its dual are now said to satisfy **complementary slackness** if each variable that is complementary to a slack constraint is assigned the value zero. The feasible solutions in Table 19.6 satisfy complementary slackness. To see that this is so, note that if a nonsign constraint in a row or column is slack (not shaded), then the complementary variable in the same row or column is shaded (equal to zero).

It is easy to verify (we'll do so later) that feasible solutions to a linear program and its dual are optimal if and only if they satisfy complementary slackness. We emphasize:

> **An Optimality Condition.** Feasible solutions to a linear program and its dual are optimal for their respective linear programs if and only if they satisfy complementary slackness.

A linear program can have more than one optimal solution. Its dual can have more than one optimal solution. Complementary slackness states conditions that must be satisfied by *each* pair of optimal solutions. Specifically:

- Suppose a constraint is slack in an optimal solution to a linear program. Then *every* optimal solution to its dual equates this constraint's complementary variable to zero.
- Suppose a variable is positive in an optimal solution to a linear program. Then *every* optimal solution to its dual causes this variable's complementary constraint to hold as an equation.

In 1964, C. E. Lemke and J. T. Howson created a sensation with a pivot method that aims directly for feasible solutions that satisfy complementary slackness. Their method solves linear programs, convex quadratic programs, and nonzero-sum two-person matrix games. Shortly after they introduced this method, Herbert Scarf adapted its ideas to construct an approximate solution to a "Brouwer fixed point," thereby enabling the computation of general equilibrium in an economy.

19.11. DATA ENVELOPMENT

In this section, duality is put to use. Here, a decision-making situation is formulated for solution by a linear program *and* by its dual. This decision problem concerns a type of input-output model that arises in organizations that serve the public.

Let us focus on an organization that contains several subdivisions or units, with these characteristics:

- All units consume the same inputs and produce the same outputs, but the quantities can differ from unit to unit.
- We can measure the amount of each input that each unit consumes and the amount of each output that each unit produces.
- The values of the outputs and the costs of the inputs are difficult to appraise.

Thus, measuring the inputs and outputs of each unit is relatively simple, but placing values on these inputs and outputs is not.

In 1978, Charnes, Cooper, and Rhodes introduced a quantitative technique called **data envelopment analysis** that distinguishes efficient units from inefficient units *without* requiring prices. They found a way in which the "data" (inputs and outputs) of efficient units "envelop" those of an inefficient unit. Settings in which their approach is fruitful include:

- Several medical offices
- A collection of branch banks
- A group of military units

To introduce data envelopment analysis, we describe a setting that is familiar to college students.

Academic Departments

A university contains a large number of academic departments that differ vastly in the contributions they make to the university and in the resources they consume. The university's provost (chief operating officer) is an advocate of programmed budgets and of cost-effectiveness measures. She suspects that some departments are strikingly inefficient. Intending to establish an "efficiency ratio" for each department, the provost plans to categorize and measure the inputs and outputs of each department. Each department's inputs include:

- the number of full-time equivalent faculty in the department,
- the number of square feet it occupies,
- the number of books and journals in the library that its students and faculty require,
- the volume of hazardous waste that its laboratories produce,

and so forth. Each department's outputs include:

- the total number of students taught each year,
- the number of Ph.D. degrees that its students earn each year,
- the number of research papers that its faculty publish per year,

and so forth.

An Illustration

The provost began with a highly aggregated model that has only 3 outputs and 2 inputs. Her staff measured the outputs and inputs of departments A, B, and C, which are presented in Table 19.7.

The provost hoped to place values on the outputs and costs on the inputs. Specifically, she aimed to assign values to the variables,

$$\begin{aligned} p_i &= \text{the annualized value per unit of output } i, \quad \text{for } i = 1, 2, 3, \\ q_j &= \text{the annualized cost per unit of input } j, \quad \text{for } j = 1, 2. \end{aligned}$$

Given these values, she would rank departments by an *efficiency ratio*, namely, by the ratio of the total value of its outputs to the total cost of its inputs. For instance:

$$\text{efficiency ratio of department A} = \frac{20\,p_1 + 3.25\,p_2 + 10\,p_3}{10\,q_1 + 15\,q_2}.$$

Table 19.7 Outputs and inputs of departments A, B, and C.

Department	Output 1	Output 2	Output 3	Input 1	Input 2
A	20	3.25	10	10	15
B	25	7	20	24	30
C	20	6	26	21	24

A department with a particularly low efficiency ratio would be a candidate for the provost's scrutiny.

A brouhaha developed as soon as the faculty got wind of this. The provost was accused of favoritism, of being arbitrary, of acting on whim, and of more serious sins against society. She found herself unable to forge a consensus about the costs of the inputs or the values of the outputs. Nonetheless, she suspected that certain departments might be relatively inefficient. At this point, she turned to you.

After some thought, you describe a department as **potentially efficient** if there are nonnegative prices, not all of which equal zero, such that this department's efficiency ratio is largest. The *potential* to be efficient is a weak test of cost-effectiveness. You suspect that a department that is inefficient for every price schedule should be a candidate for scrutiny by the provost. You ask yourself the question in:

Problem C (Efficiency Ratios)

For the data in Table 19.7, is there a department that fails to be potentially efficient?

You plan to build a linear program that determines whether or not a particular department is potentially efficient. This would seem to require a comparison of ratios, which is unwieldy, but a "scaling" trick avoids these comparisons. The effect of multiplying the value of each output by a positive constant c is to multiply every ratio by c, and this causes the largest ratio to remain the largest. Thus, a department is potentially efficient if and only if there exist nonnegative prices such that its efficiency ratio is at least 1 and such that the other departments' efficiency ratios are at most 1. When you linearize these ratio constraints in the usual way, you will obtain the constraints of a linear program.

A Linear Program

Program 19.1, determines whether or not department B is potentially efficient. Its first and third nonsign constraints require the efficiency ratios for departments A and C to be 1 or less. Its second nonsign constraint requires the efficiency ratio for department B to equal 1 or more. The constraints of Program 19.1 are satisfied, however, by equating all prices to zero. Evidently, Program 19.1 has a feasible solution whose objective is positive if and only if department B is potentially efficient.

Program 19.1: Maximize $\{q_1 + q_2 + q_3 + p_1 + p_2\}$, subject to the constraints

$$
\begin{aligned}
y_A: &\quad 20\,q_1 + 3.25\,q_2 + 10\,q_3 \le 10\,p_1 + 15\,p_2,\\
y_B: &\quad 25\,q_1 + 7\,q_2 + 20\,q_3 \ge 24\,p_1 + 30\,p_2,\\
y_C: &\quad 20\,q_1 + 6\,q_2 + 26\,q_3 \le 21\,p_1 + 24\,p_2,\\
&\quad q_1 \ge 0, \quad q_2 \ge 0, \quad q_3 \ge 0, \quad p_1 \ge 0, \quad p_2 \ge 0.
\end{aligned}
$$

Program 19.1 has a peculiar feature. It either has zero as its optimal value or is unbounded. We could add to Program 19.1 a constraint, such as

$$q_1 + q_2 + q_3 + p_1 + p_2 \le 10,$$

which would keep it from being unbounded. We omitted this constraint to simplify (slightly) the discussion that follows.

Table 19.8 contains a standardized spreadsheet for Program 19.1, along with its Solver dialog box. Evidently, the variables p_1 and p_2 have been shifted to the left-hand side of its constraints. Row 2 of this spreadsheet contains its changing cells, and column G contains the usual SUMPRODUCT functions.

Table 19.8 Standardized spreadsheet, Solver dialog box, and optimal solution to Program 19.1.

	A	B	C	D	E	F	G	H	I
1	name	q1	q2	q3	p1	p2			
2	value	0	0	0	0	0			
3	dept. A	20	3.25	10	-10	-15	0	<=	0
4	dept. B	25	7	20	-24	-30	0	>=	0
5	dept. C	20	6	26	-21	-24	0	<=	0
6		1	1	1	1	1	0		
7									
8		=SUMPRODUCT(B$2:F$2,B6:F6)							

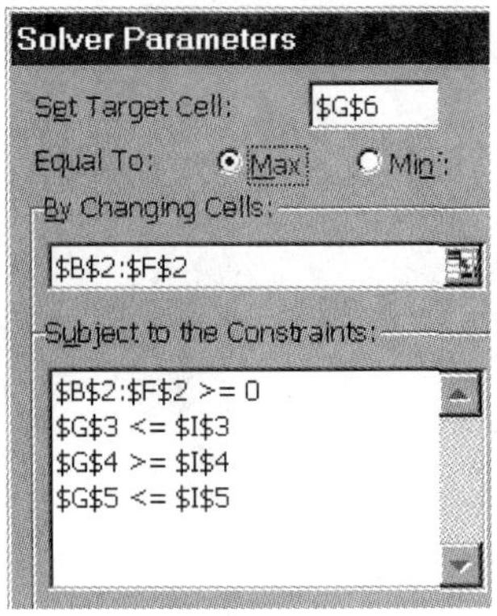

Table 19.8 also reports the optimal solution that Solver has found. Program 19.1 has 0 as its optimal value. Thus, no prices exist for which department B has the highest efficiency ratio. Department B is not potentially efficient.

For no reason that is (yet) apparent, you had the foresight to use Solver to obtain a sensitivity report, which includes the shadow prices of the nonsign constraints. These shadow prices are

$$y_A = 6.57, \quad y_B = -9.07, \quad y_C = 7.19,$$

when rounded off to three significant figures.

You report to the provost that no prices exist for which department B has the highest efficiency ratio. She is less pleased than you had hoped. She informs you that the conclusion that department B is not potentially efficient is a *negative* statement. It fails to provide her with any guidelines as to how she should coax department B to evolve. She encourages you to find a way to offer constructive advice.

Duality

After further thought, you see how duality might recast this negative statement as a positive one. You ask the provost two questions.

- Would multiplying the size of a department by a factor increase its inputs and outputs by roughly the same factor?
- Would the consolidation of two departments cause their inputs and outputs to add?

To both questions, she answers, "Essentially, yes. These measures are linear." At this point, you ask yourself the question posed in:

Problem D (Envelopment)

If a department is not potentially efficient, is there a nonnegative combination of the other departments that consumes less of each input and creates more of each output? If so, how can you find it?

To answer this question, you plan to investigate the dual of Program 19.1. The dual appears below as Program 19.1^d.

Program 19.1.[d] Minimize $\{0\, y_A + 0\, y_B + 0\, y_C\}$, subject to the constraints

$$
\begin{array}{lrcrcrl}
q_1: & 20\, y_A & + & 25\, y_B & + & 20\, y_C & \geq 1, \\
q_2: & 3.25\, y_A & + & 7\, y_B & + & 6\, y_C & \geq 1, \\
q_3: & 10\, y_A & + & 20\, y_B & + & 26\, y_C & \geq 1, \\
p_1: & -10\, y_A & - & 24\, y_B & - & 21\, y_C & \geq 1, \\
p_2: & -15\, y_A & - & 30\, y_B & - & 24\, y_C & \geq 1, \\
 & & & y_A \geq 0, & & y_B \leq 0, & y_C \geq 0.
\end{array}
$$

Program 19.1 is feasible and bounded, so the Duality Theorem demonstrates that Program 19.1[d] is too. The Duality Theorem also shows that the shadow prices are an optimal solution to Program 19.1[d]. You begin to wonder what you might divine from these shadow prices. After fiddling with the constraints of Program 19.1[d], it occurs to you (we omit the details) that the shadow prices prescribe a mixture of the other two departments that envelops department B. This mixture is the sum of r_A times department A and r_C times department C, where

$$r_A = \frac{y_A}{-y_B} = \frac{6.57}{-9.07} = 0.724 \quad \text{and} \quad r_C = \frac{y_C}{-y_B} = \frac{7.19}{-9.07} = 0.793.$$

On a spreadsheet, you find that the mixture of $r_A = 0.724$ times department A and $r_C = 0.793$ times department C produces these outputs and inputs:

$$
\begin{array}{rl}
20\, r_A + 20\, r_C = 30.3 > 25, & \text{(output 1)} \\
3.25\, r_A + 6\, r_C = 7.11 > 7, & \text{(output 2)} \\
10\, r_A + 26\, r_C = 27.6 > 20, & \text{(output 3)} \\
10\, r_A + 21\, r_C = 23.9 < 24, & \text{(input 1)} \\
15\, r_A + 24\, r_C = 29.9 < 30. & \text{(input 2)}
\end{array}
$$

Evidently, this mixture **envelops** department B; it obtains more of each output than does department B, and it consumes less of each input than does department B.

You present this result to the provost. Now, she's pleased. She notes that the mixture produces significantly more of outputs 1 and 3 while consuming about the same amounts of the inputs. She can use this information to coax department B to evolve toward a roughly equal mixture of the attributes of departments A and C.

The General Result

This example illustrates a general result, which appears as the following theorem. The proof of this theorem is omitted. This proof flows from duality in precisely the way that our example suggests.

Theorem 19.1 (Data Envelopment). Suppose each of several units consumes a fixed positive amount of each input and produces a fixed positive amount of each output. Then, for each particular unit, exactly one of the following is true:

(a) There exist nonnegative prices, not all of which are zero, such that this unit has the highest efficiency ratio.

(b) There exists a mixture (nonnegative combination) of the other units that obtains more of each output and consumes less of each input than does this unit.

Statement (a) is that the unit is potentially efficient. Statement (b) is that the unit is enveloped by a mixture of other units. The theorem states that exactly one of these statements must be true. Either a unit is potentially efficient, or that unit is enveloped by a mixture of the other units.

As stated, Theorem 1 is nonconstructive. More has been accomplished than it states. Specifically, we have constructed a linear program whose solution and shadow prices either determine prices for which a particular unit has the highest efficiency ratio or provide a mixture of the other units that envelops it.

A Theorem of the Alternative

A **theorem of the alternative** states that *exactly one* of two alternatives must occur. The preceding is a theorem of the alternative. In this case and in general, a theorem of the alternative can replace a negative statement by an equivalent statement that is positive. In our example, the negative statement is that there exist no prices for which a particular department has the highest efficiency ratio. The equivalent positive statement is that this department consumes more of each input and produces less of each output than does some mixture of the other departments.

In 1896, J. Farkas published a now-famous theorem of the alternative that is known as **Farkas's lemma**. It shows that a given system of linear inequalities has no solution if and only if a solution does exist to a different (dual) system of linear inequalities.

Farkas's lemma and duality are easy consequences of each other. As mathematical results, they are close kin and are of equal importance. A later section presents Farkas's lemma and shows how to obtain it as a direct consequence of duality. Farkas's lemma predates linear programming by a half century. Needless to say, Farkas used other means to prove it.

19.12. THE MATHEMATICS OF DUALITY

Until now, our discussion of duality has relied on examples, intuition, and spreadsheets. The mathematics has been circumvented. In this section, we start over. We now do the mathematics.

Since we are doing the math, why did we begin with an example? Examples can be easier to grasp than abstract discussions. For many purposes, understanding the examples is sufficient, in which case this section can be skipped or skimmed. On the other hand, for a deeper understanding of duality, beginning with examples can make the mathematics easier to follow.

Lmax

A linear program is given the acronym **Lmax** if it is written as a *m*aximization problem, if each of its variables is nonnegative, and if each of its nonsign constraints is a *less*-than-or-equal-to inequality. The Production LP is an example of Lmax.

Lmax is a **canonical form** for linear programs because every linear program can be wrestled into its format. To cast the Pricing LP in the format of Lmax, multiply its nonsign constraints by -1, multiply each coefficient in its objective by -1, and switch its objective from minimization to maximization.

A general description of Lmax requires mathematical notation. This notation may be familiar to you. It appeared in the discussion of activity analysis (on page 126 in Chapter 4) and in the economic perspective on the simplex method (on page 606 in Chapter 18).

In Lmax, the symbol n denotes the number of decision variables, which are labeled 1 through n. The symbol m denotes the number of nonsign constraints, which are labeled 1 through m. The symbol x_j denotes the value taken by the jth decision variable. Lmax is a linear program that is written in the following format. Once again, the ellipsis (three dots) means repeat the pattern.

Lmax: $z^* = \text{Maximize } \{c_1 x_1 + c_2 x_2 + \cdots + c_n x_n\}$, subject to the constraints

$$\begin{array}{ll}
y_1: & A_{11} x_1 + A_{12} x_2 + \cdots + A_{1n} x_n \le b_1, \\
y_2: & A_{21} x_1 + A_{22} x_2 + \cdots + A_{2n} x_n \le b_2, \\
\vdots & \quad \vdots \qquad\quad \vdots \qquad\qquad\quad \vdots \qquad \vdots \\
y_m: & A_{m1} x_1 + A_{m2} x_2 + \cdots + A_{mn} x_n \le b_m, \\
 & x_1 \ge 0, \ x_2 \ge 0, \ldots, \quad x_n \ge 0.
\end{array}$$

Clearly, Lmax describes a maximization problem whose decision variables are nonnegative and whose nonsign constraints are less-than-or-equal-to inequalities. The data in Lmax are the integers n and m and the numbers, A_{ij}, b_i, and c_j as i ranges from 1 to m and as j ranges from 1 to n. Evidently, c_j is the per-unit contribution of the jth decision variable, b_i is the right-hand-side value of the ith nonsign constraint, and A_{ij} is the coefficient of the jth decision variable in the ith nonsign constraint. The optimal value of Lmax is labeled z^*.

The symbols y_1 through y_m that appear to the left of the constraints will be given two interpretations. The y_i's denote the shadow prices for the constraints that they abut, and the y_i's will be the decision variables in the dual linear program. In Lmax, the ith constraint and the symbol y_i appear as

$$y_i: \qquad A_{i1} x_1 + A_{i2} x_2 + \cdots + A_{in} x_n \le b_i.$$

For each i, the ith constraint of Lmax and y_i are said to be **complementary** to each other.

Lmax drips with subscripts, but it is nothing new. The Production LP illustrates Lmax. It has $m = 5$ and $n = 3$. Its variables are labeled S, F, and L rather than x_1, x_2, and x_3. That example has $A_{11} = 3$, $A_{12} = 2$, $A_{13} = 1$, and so forth. For the Production LP, the A_{ij}'s form an array that is familiar from Table 19.1; this array has three columns (one per variable) and five rows (one per constraint).

The Dual of Lmax

Since Lmax is a canonical form for linear programs, defining the dual of Lmax defines the dual of every linear program. The **dual** of Lmax is defined to be the linear program that is presented here as Gmin. The data in Gmin are the same as those in Lmax. The acronym "Gmin" abbreviates *min*imization with *greater*-than-or-equal-to inequalities.

Gmin: $z_* = \text{Minimize } \{y_1 b_1 + y_2 b_2 + \cdots + y_m b_m\}$, subject to the constraints

$$\begin{array}{ll}
x_1: & y_1 A_{11} + y_2 A_{21} + \cdots + y_m A_{m1} \ge c_1, \\
x_2: & y_1 A_{12} + y_2 A_{22} + \cdots + y_m A_{m2} \ge c_2, \\
\vdots & \quad \vdots \qquad\quad \vdots \qquad\qquad\quad \vdots \qquad \vdots \\
x_n: & y_1 A_{1n} + y_2 A_{2n} + \cdots + y_m A_{mn} \ge c_n, \\
 & y_1 \ge 0, \ y_2 \ge 0, \ldots, \quad y_m \ge 0.
\end{array}$$

The decision variables in Gmin are y_1 through y_m. Each nonsign constraint in Gmin is paired with a variable in Lmax. The jth nonsign constraint of Gmin and the variable x_j appear as the pair

$$x_j: \qquad y_1 A_{1j} + y_2 A_{2j} + \cdots + y_m A_{mj} \ge c_j.$$

As before, x_j will have two interpretations: x_j is the shadow price of the jth constraint of Gmin, and x_j is the variable in Lmax that is **complementary** to the jth constraint of Gmin.

These definitions establish a complementarity between each variable in either linear program and a constraint in the other. Each variable in Lmax is complementary to a nonsign constraint in Gmin, and each nonsign constraint in Lmax is complementary to a variable in Gmin.

The optimal value of Gmin is labeled z_*. (As a memory aid, the "stars" in z^* and z_* are lifted up in the case of maximization, down in the case of minimization.)

If Gmin puzzles you, the subscripts may be obscuring the view. Gmin bears *exactly* the same relation to Lmax as the Pricing LP bears to the Production LP. The sense of optimization is reversed, the right-hand-side values become the objective coefficients, the objective-coefficients become the right-hand-side values, and the data in each constraint of Gmin are the column of coefficients of its complementary variable.

The Dual of Gmin

We've defined Gmin to be the dual of Lmax. This does *not* guarantee that Lmax is the dual of Gmin. We do know that Lmax is the linear program whose dual is Gmin. That Lmax is the dual of Gmin is a different statement, one that has not yet been proved.

Lemma 19.1. Lmax is the dual of Gmin.

Sketch of Proof. How could we prove Lemma 19.1? To prove it, we would need to place Gmin in the format of Lmax, take its dual, and see what emerges. To convert Gmin to a maximization problem, multiply each coefficient in its objective by -1. To convert its nonsign constraints to "$\leq$" inequalities, multiply them by -1. Then take the dual. This dual is a minimization problem with "$\geq$" constraints. To switch it to a maximization problem, multiply each coefficient in its objective by -1. To switch its nonsign constraints to "$\leq$" inequalities, multiply each of them by -1. When the details are implemented (and when the dust settles), the end result of this manipulation will be Lmax. ◆

Lemma 19.1 has two important implications. First, it justifies the use of the word "dual" by showing that Lmax is the dual of its dual. Second, it provides us with a second way in which to take the dual of a linear program: place it in the format of Gmin, in which case Lmax will be its dual.

The Crossover Table

We now know how to take the dual of two linear programs, Lmax and Gmin. In the context of the crossover table, we have justified the use of lines 1 and 4 in both directions. What about the other lines?

Lemma 19.2. The crossover table works.

Sketch of Proof. How could we prove Lemma 19.2? By reducing the other lines to lines 1 and 4 and looking at what happens when we take the dual. Consider line 2. We could replace an equality constraint in a maximization problem by a pair of inequalities, write them both as "≤" inequalities, use Lmax to take the dual, and observe that the difference between the pair's complementary variables is, effectively, a free variable. Similar manipulations work for lines 3, 5, and 6. ◆

Our earlier presentation of duality contained a logical flaw, which we had glossed over. We had used the crossover table to define the dual of a linear program. That gave us several different ways to find the dual of a particular linear program. It opened the possibility of inconsistency, of getting *different* linear programs by taking the dual in different ways. Lemmas 19.1 and 19.2 repair the logic. These lemmas show that, effectively, we get the same linear program no matter how we take the dual.

Slack and Surplus Variables

The nonsign constraints in Lmax and in Gmin are inequalities. Summation notation lets us write them compactly in the form

$$\sum_{j=1}^{n} A_{ij}\,x_j \le b_i \quad \text{and} \quad \sum_{i=1}^{m} y_i\,A_{ij} \ge c_j.$$

For current purposes, it will be convenient to convert these inequalities into equations. To convert the ith inequality in Lmax into an equation, insert a **slack variable** s_i, as in:

$$\sum_{j=1}^{n} A_{ij}\,x_j + s_i = b_i, \quad s_i \ge 0.$$

If the original inequality constraint holds strictly, then s_i is positive; if the original inequality holds as an equation, then s_i equals zero. In either case, s_i is nonnegative. This variable s_i "takes up the slack" in the original, inequality constraint. Similarly, to convert the jth inequality in Gmin into an equation, insert a **surplus variable** t_j from its left-hand side, as follows:

$$\sum_{i=1}^{m} y_i\,A_{ij} - t_j = c_j, \quad t_j \ge 0.$$

Including these slack and surplus variables rewrites Lmax and Gmin, as follows.

Lmax: $z^* = \max\{\sum_{j=1}^{n} c_j x_j\}$, subject to the constraints

y_i: $$\sum_{i=1}^{n} A_{ij}\,x_j + s_i = b_i, \quad i = 1, \ldots, m, \qquad (19.1)$$

$$x_j \ge 0, \quad j = 1, \ldots, n, \qquad (19.2)$$
$$s_i \ge 0, \quad i = 1, \ldots, m. \qquad (19.3)$$

Gmin: $z_* = \min\{\sum_{i=1}^{m} y_i b_i\}$, subject to the constraints

x_j: $$\sum_{i=1}^{m} y_i\,A_{ij} - t_j = c_j, \quad j = 1, \ldots, n, \qquad (19.4)$$

$$y_i \ge 0, \quad i = 1, \ldots, m, \qquad (19.5)$$
$$t_j \ge 0, \quad j = 1, \ldots, n. \qquad (19.6)$$

Thus, a feasible solution to Lmax assigns numerical values to the x_j's and the s_i's that satisfy the equations and inequalities in (19.1), (19.2) and (19.3). Similarly, a feasible solution to Gmin is a set of values of the y_i's and the t_j's that satisfy the equations and inequalities in (19.4), (19.5) and (19.6).

A Handy Relationship

The next lemma establishes a handy relationship between the objective values in Lmax and Gmin.

Lemma 19.3. Each pair of solutions to equations in (19.1) and (19.4) satisfies

$$\sum_{j=1}^{n} c_j x_j + \sum_{i=1}^{m} y_i s_i + \sum_{j=1}^{n} t_j x_j = \sum_{i=1}^{m} y_i b_i. \tag{19.7}$$

Proof. Multiply the ith equation in (19.1) by y_i and then sum over i. In the equation that results, shift each addend $y_i s_i$ to the right-hand side to get

$$\sum_{i=1}^{m} \sum_{j=1}^{n} (y_i A_{ij} x_j) = \sum_{i=1}^{m} (y_i b_i) - \sum_{i=1}^{m} (y_i s_i).$$

Similarly, multiply the jth equation in (19.4) by x_j and then sum over j. In the equation that results, shift each addend $t_j x_j$ to the right-hand side to get

$$\sum_{j=1}^{n} \sum_{i=1}^{m} (y_i A_{ij} x_j) = \sum_{j=1}^{n} (c_j x_j) + \sum_{j=1}^{n} (t_j x_j).$$

The double sums on the left-hand sides of the above two equations are equal to each other because they sum identical addends over identical ranges. Thus, the right-hand sides of these equations equal each other:

$$\sum_{j=1}^{n} (c_j x_j) + \sum_{j=1}^{n} (t_j x_j) = \sum_{i=1}^{m} (y_i b_i) - \sum_{i=1}^{m} (y_i s_i).$$

Rearranging the above verifies (19.7) and proves Lemma 19.3. ◆

The idea behind Lemma 19.3 is to multiply each constraint by its complementary variable and see what happens. The A_{ij}'s cancel out, and Equation (19.7) remains.

Lemma 19.3 states a fact about solutions to equations (19.1) and (19.4). These solutions need not satisfy inequalities (19.2), (19.3), (19.5), and (19.6). They need not be feasible.

Feasible Solutions to Lmax and Gmin

Each of the next four lemmas is a simple consequence of Lemma 19.3. The first of these lemmas concerns feasible solutions to Lmax and Gmin.

Lemma 19.4. Each pair of feasible solutions for Lmax and Gmin satisfies

$$\sum_{j=1}^{n} c_j x_j \leq \sum_{i=1}^{m} y_i b_i. \tag{19.8}$$

Proof. Feasible solutions to Lmax and Gmin satisfy expressions (19.1) through (19.6). Lemma 19.3 shows that they satisfy Equation (19.7). Feasibility guarantees that each addend $y_i\, s_i$ and $t_j\, x_j$ in Equation (19.7) is nonnegative. This verifies expression (19.8). ◆

Lemma 19.4 compares the objective values of feasible solutions to Lmax and Gmin. Lemma 19.4 shows that the objective value of each feasible solution to the minimization problem is at least as large as the objective value of each feasible solution to the maximization problem.

Weak Duality

A second implication of Lemma 19.3 relates the optimal values of Lmax and Gmin. Let us recall that a linear program is **feasible** if it has at least one feasible solution. Also, a feasible linear program is **bounded** if its objective cannot be improved without limit.

Lemma 19.5 (Weak Duality). Suppose Lmax and Gmin are feasible. Then both linear programs are bounded, and their optimal values z^* and z_* satisfy

$$z^* \le z_*. \tag{19.9}$$

Proof. Pick any feasible solution to Gmin. It and *each* feasible solution to Lmax satisfy (19.8). Maximize the left-hand side of (19.8) over all feasible solutions to Lmax to get

$$z^* \le \sum_{i=1}^{m} y_i\, b_i.$$

The preceding inequality holds for each feasible solution to Gmin. Minimize over all feasible solutions to Gmin and get (19.9), completing the proof. ◆

Lemma 19.5 states that the optimal value of Gmin is at least as large as the optimal value of Lmax. The name "weak duality" stems from the fact that in certain generalizations of linear programming the optimal value z^* of the maximization problem is below the optimal value z_* of the minimization problem. In those settings, the difference $(z_* - z^*)$ is known as the **duality gap**. For the case of linear programs, we will soon prove that $z_* = z^*$, and hence that the duality gap equals zero.

Unbounded and Infeasible Linear Programs

If Lmax is unbounded, we write $z^* = +\infty$ to record the fact that its objective can be made arbitrarily large. Similarly, if Gmin is unbounded, we write $z_* = -\infty$ to record the fact that its objective can be made arbitrarily small (negative).

Lemma 19.6. If a linear program is unbounded, its dual is infeasible.

Proof. Lmax is a canonical form. We cast the unbounded linear program in the format of Lmax. Its optimal value z^* equals $+\infty$. Aiming for a contradiction, we assume that its dual has a feasible solution. This feasible solution has a finite objective value, K. Lemma 19.4

shows that every feasible solution to Lmax has an objective value that does not exceed K and hence that $z^* \le K$. But we know that $z^* > K$ for every finite number K. This contradicts the assumption that the dual is feasible. That assumption must be false. Thus, the dual must be infeasible, which completes the proof. ◆

Thus, if a linear program is unbounded, its dual must be infeasible. If a linear program is infeasible, must its dual be unbounded? No! It can happen that both a linear program and its dual are infeasible. For an example, please refer to Problem 11, page 689.

Complementary Slackness

Earlier in this chapter, a complementary slackness condition was expressed verbally in the sentence, "If an inequality constraint is slack, then its complementary variable must equal zero." Insertion of slack and surplus variables measures the amounts by which the original inequality constraints were slack. This lets us restate the **complementary slackness** conditions for Lmax and Gmin as

$$t_j\, x_j = 0, \quad \text{for } j = 1, \dots, n, \tag{19.10}$$

$$y_i\, s_i = 0, \quad \text{for } i = 1, \dots, m. \tag{19.11}$$

Equations (19.10) and (19.11) are a crisp way of saying that if a nonsign constraint for either linear program holds strictly, then this constraint's complementary variable must equal zero.

The fourth simple consequence of Lemma 19.3 is:

Lemma 19.7. If the solutions to Equations (19.1) and (19.4) satisfy complementary slackness, then

$$\sum_{j=1}^{n} c_j\, x_j = \sum_{i=1}^{m} y_i\, b_i. \tag{19.12}$$

Proof. In Equation (19.7), use (19.10) and (19.11) to obtain (19.12). ◆

Lemma 19.7 states a condition under which candidate solutions to Lmax and Gmin have the same objective values for their respective linear programs.

Complementary Slackness and the Simplex Method

Complementary slackness is now related to the simplex method. In Chapter 18, we saw that each basis prescribes a set of shadow prices whose values are such that the contribution of each basic variable equals its opportunity cost. The next lemma observes that each basic solution and its shadow prices satisfy complementary slackness. This lemma also observes that the simplex method terminates when it identifies a feasible solution to the dual linear program.

Lemma 19.8 (Basic Solutions and Complementary Slackness). In Lmax, with slack variables included, consider any basis. For each i, designate as y_i this basis's shadow price for the ith constraint in Lmax. Given these values of y_1 through y_m, specify t_1 through t_n that satisfy Equation (19.4). Then:

(a) The complementary slackness conditions are satisfied.

(b) If the reduced costs of the nonbasic variables are nonpositive, the y_i's and t_j's are feasible for Gmin.

Proof. We first verify part (a). Let us consider any variable x_j that is basic. Equation (18.10) of Chapter 18 states that the opportunity cost of x_j equals its contribution, and equivalently, that the shadow prices satisfy

$$\sum_{i=1}^{m} y_i A_{ij} = c_j \quad \text{if } x_j \text{ is basic.} \tag{19.13}$$

By comparing (19.13) with (19.4), we conclude that $t_j = 0$. If x_j is nonbasic, then $x_j = 0$. In both cases, $t_j\, x_j = 0$, so that (19.10) holds.

Next, consider any slack variable s_i that is basic. Since s_i is basic, the opportunity cost of s_i equals its contribution, so the analog for s_i of (19.13) holds. The contribution of the slack variable s_i equals zero, s_i has a coefficient of $+1$ in the ith constraint of Lmax and s_i has coefficients of 0 in the other constraints of Lmax. Hence, the analog of (19.13) for s_i is

$$y_i = 0 \quad \text{if } s_i \text{ is basic.} \tag{19.14}$$

If s_i is nonbasic, then $s_i = 0$. In both cases, $y_i\, s_i = 0$, so that (19.11) holds. This proves part (a).

For part (b), we assume that the reduced costs of the nonbasic variables are nonpositive. The y_i's and t_j's satisfy (19.4). We must show that they satisfy the nonnegativity requirements in (19.5) and (19.6).

Consider t_j. If x_j is basic, equation (19.13) gives $t_j = 0$. Now consider the case in which x_j is nonbasic. Equation (11) of Chapter 18 holds with $\bar{c}_j \leq 0$, so that

$$\sum_{i=1}^{m} y_i A_{ij} \geq c_j \quad \text{if } x_j \text{ is not basic.} \tag{19.15}$$

In this case, Equation (19.4) guarantees $t_j \geq 0$. This verifies (19.6).

Consider y_i. The ith constraint of Lmax is a "$\leq$" inequality, so its shadow price y_i is nonnegative. In other words, (19.5) holds. This proves part (b). ◆

Part (a) of Lemma 19.8 indicates that the shadow prices for each basis are a "trial" solution to the dual linear program, one that satisfies complementary slackness. Part (b) shows that the shadow prices that cause the simplex method to stop form a feasible solution to Gmin. We emphasize:

> **Interpretation of Phase 2.** The simplex method preserves feasibility and complementary slackness. It pivots in search of a feasible solution to the dual linear program.

The Duality Theorem

The main goal of this section is to prove the Duality Theorem. This theorem shows that if a linear program is feasible and bounded, then so is its dual, and the two linear programs have the same optimal value. The proof that we present employs the simplex method, and it constructs optimal solutions to the two linear programs.

Theorem 19.2 (Duality Theorem). Consider a linear program that is feasible and bounded. Apply the simplex method with an anticycling rule to this linear program. This application of the simplex method terminates finitely. At termination, it finds an optimal solution and a set of shadow prices. These shadow prices are an optimal solution to the dual linear program, and the two linear programs have the same optimal value.

Proof. We prove this theorem for a linear program that is cast in the format of Lmax with slack variables included. Thus, its feasible solutions satisfy (19.1), (19.2), and (19.3). This linear program is feasible and bounded, so the simplex method must end with an optimal basic solution. Lemma 19.8 shows that this basic solution's shadow prices are feasible for the dual and that complementary slackness is satisfied. Lemma 19.7 shows that (19.12) holds. This demonstrates that the x_j's and the y_i's satisfy

$$z^* = \sum_{j=1}^{n} c_j x_j = \sum_{i=1}^{m} y_i b_i \geq z_*, \tag{19.16}$$

the last because the y_i's are feasible for Gmin, a minimization problem. Weak duality (Lemma 19.3) shows that $z^* \leq z_*$, so inequality (19.16) holds as an equation. This completes the proof. ◆

The Duality Theorem is the central result in linear programming. This theorem is sometimes called the **Strong** Duality Theorem to distinguish $z^* = z_*$ from the weaker result, $z^* \leq z_*$, which had been established in Lemma 19.5.

Optimality Conditions

Earlier, we had stated two optimality conditions, without proof. One was that feasible solutions to a linear program and its dual have the same objective value. The other optimality condition was that feasible solutions to a linear program and its dual satisfy complementary slackness. The Duality Theorem makes it easy to verify that these conditions are necessary and sufficient for feasible solutions to be optimal.

Theorem 19.3 (Optimality Conditions). Consider a feasible solution to Lmax and a feasible solution to Gmin. The following are equivalent:

1. These feasible solutions are optimal for their respective linear programs.
2. These feasible solutions satisfy complementary slackness.
3. These feasible solutions have the same objective value for their respective linear programs.

Proof. Let us consider a feasible solution $x = (x_1, x_2, \ldots, x_n)$ to Lmax and a feasible solution $y = (y_1, y_2, \ldots, y_m)$ to Gmin. Lemma 19.5 shows that Lmax is bounded. Hence, the Duality Theorem shows that $z^* = z_*$. Since Lmax is a maximization problem and since Gmin is a minimization problem, these feasible solutions have

$$\sum_{j=1}^{n} c_j x_j \leq z^* = z_* \leq \sum_{i=1}^{m} y_i b_i. \tag{19.17}$$

Equation (19.7) shows that both inequalities in (19.17) hold as equations if and only if these feasible solutions satisfy complementary slackness. And expression (19.17) shows that both feasible solutions are optimal if and only if their objective values equal each other. This proves Theorem 19.3. ◆

19.13. AN ECONOMY IN GENERAL EQUILIBRIUM

This section develops a model that plays a central role in economics. This model is of an "economy" that is in "general equilibrium." In this section, a linear program *and* its dual will be used to construct a general equilibrium, thereby demonstrating that an equilibrium exists.

The **economy** contains two types of agents, which are known as "consumers" and "producers." These agents are linked to each other through a "market" at which goods can change hands. The prices at which goods change hands are determined within the model. Thus, an economy differs from a typical linear program in two vital ways:

- There are multiple agents, each acting in his or her self-interest.
- The prices are endogeneous, which means that they are set within the model.

By contrast, in a linear program, there is a single decision maker, and that person has no influence on the prices of the goods that he or she buys or sells.

An Outline

The elements of this model are outlined now, with details to follow. In an economy, the consumers own all of the goods. Each consumer begins with an endowment of each good. Each consumer can buy or sell goods at the market. Each consumer sells the goods that he or she owns but does not wish to consume and buys the goods that he or she wishes to consume but does not own. Each consumer faces a budget constraint, namely, that the market value of the goods that the consumer buys cannot exceed the market value of the goods that the consumer sells.

The producers can operate technologies, each of which transforms one bundle of goods into another. If a producer operates a technology, the producer must buy its inputs at the market and must sell its outputs at the market. Producers aim to maximize the profit that they earn by operating technologies.

The market for a good is said to **clear** if the quantity of the good that is brought for sale is at least as large as the quantity that is demanded. (The model that we are developing allows for free disposal of unwanted goods.)

An economy is said to be in **general equilibrium** if the production quantities, the consumption quantities, and the market prices have these properties:

- Each agent maximizes his or her own welfare.
- The market for each good clears.

A great bulk of theoretical and applied economics rests on the assumption that an economy has a general equilibrium. In this section, duality is used to construct a general equilibrium.

A Simplification

The model of an economy that we will study is simplified in the following ways:

- We study a one-period planning problem.
- We assume that the production activities have constant returns to scale.
- We assume that the economy contains only *one* consumer.
- We assume that the consumer has a linear utility function on quantities consumed.

The results that we will obtain generalize easily to the cases of multiple periods, to decreasing returns to scale on production, and to decreasing marginal utility on consumption.

By contrast, our method does *not* generalize to multiple consumers. The assumption of a single canonical consumer is crucial. To obtain comparable results for the case of multiple consumers, we would need to switch tools from linear and nonlinear programming to fixed-point methods.

Goods and Technologies

A typical model of an economy is highly aggregated. The many commodities are grouped into general types of **good**, which might include capital, labor, land, steel, energy, foodstuff, and so forth. The symbol m denotes the number of goods in the model, and these goods are numbered 1 through m.

Typically, the many production processes are grouped into several different types of **technology**, which might include steel production, agriculture, steel capacity expansion, and so forth. The symbol n denotes the number of technologies in the model, and these technologies are numbered 1 through n.

These technologies are familiar. The same model is used here as in the discussion of activity analysis in Chapter 4. Each technology transforms one bundle of goods into another. Each technology can be operated at any nonnegative level. The production side of the economy is determined by the data,

$$A_{gt} = \text{the } \textit{net} \text{ output of good } g \text{ per unit level of technology } t,$$

where g ranges between 1 and m and where t ranges between 1 and n. As was the case in our earlier discussion of activity analysis, a positive net output is a quantity produced and a negative net output is a quantity consumed. If, for instance, good g is an input to every technology, then $A_{gt} < 0$ for each t.

Endowments and Utilities

In a model of an economy, the consumers own all of the assets; the producers own nothing. In our model, there is only one consumer, and that person owns all of the goods. For each good g, the model includes the datum

$$e_g = \text{the quantity of good } g \text{ that the consumer possesses at the beginning of the period.}$$

This quantity e_g is often called the consumer's **endowment** of good g. If, for instance, good 7 is steel, then e_7 equals the number of units of steel that the consumer possesses at the start of the period. Goods exist in nonnegative quantities, so these endowments (the e_g's) are nonnegative numbers.

In general, each consumer wishes to consume some goods (e.g., food and energy). Each consumer places a value (or utility) on the bundle of goods that he or she consumes. Our naïve model has only one consumer, and the total value that he or she places on consumption is assumed to be proportional to the level of consumption of each good. In brief, for each good g, the model includes the datum

$$u_g = \text{the value that the consumer places on the consumption of each unit of good } g \text{ during the period.}$$

This quantity u_g is often called the consumer's **utility** of good g. In our model, the utility can vary with the good but not with the quantity that is consumed.

The Central Issue

The model's data have now been specified. These data are the integers m and n and the numbers e_g, u_g and A_{gt}, as g ranges between 1 and m and as t ranges between 1 and n.

As indicated earlier, a general equilibrium occurs if each participant in the economy maximizes his or her own welfare, if each good changes hands at prices that are determined within the model, and if the market for each good clears. The motivating question in this section is now posed as:

Problem E (General Equilibrium)

For this model, does there exist a general equilibrium? If so, what does it look like, and how can we find it?

The Duality Theorem will be used to demonstrate that a general equilibrium exists and to construct one.

The Decision Variables

This model can be described in terms of three types of decision variable—the level at which each technology is operated, the amount of each good that the consumer consumes, and the market price of each good. Specifically, for $g = 1, \ldots, m$ and for $t = 1, \ldots, n$, the model includes the decision variables:

x_t = the level at which the producers operate technology t during the period.

z_g = the amount of good g that the consumer consumes during the period.

p_g = the market price of good g.

The production levels and the consumption levels are nonnegative; these decision variables must satisfy the constraints

$$x_t \geq 0, \quad t = 1, \ldots, n,$$
$$z_g \geq 0, \quad g = 1, \ldots, m.$$

The data introduced earlier and the decision variables introduced here will enable us to specify the conditions that are required for

- the market to clear,
- the consumer to be in equilibrium,
- the producers to be in equilibrium, and
- the economy to be in general equilibrium.

We shall explore each of these four conditions in turn.

Market Clearing

The **market** is the only place in this model where the consumers and producers can trade goods. They trade these goods at prices that are established within the model. As suggested earlier, the model that is under development allows for **free disposal** of each good, which means that any quantity of any good can be disposed of at the market, with no cost to any participant. Effectively, none of the goods in this economy are noxious.

With free disposal, the **market-clearing** requirement is that the aggregate supply of each good be at least as large as the aggregate demand for that good. Market clearing enforces a set of constraints, one for each good. The first step toward these constraints is to interpret:

$$\sum_{t=1}^{n} A_{gt}\, x_t = \text{the net output of good } g \text{ due to production during the period.}$$

This net output can be negative. Suppose, for instance, that good 2 is labor and that labor is an input to every technology. In this case, A_{2t} is negative for each t, and the net output of good 2 (labor) is negative whenever any technologies are employed. Moreover, the amount by which the net output of labor is negative equals the amount of labor that is employed.

The economy begins the period with an endowment e_g of good g. This lets us interpret:

$$e_g + \sum_{t=1}^{n} A_{gt}\, x_t = \text{the amount of good } g \text{ that is available for consumption during the period.}$$

The market-clearing constraints are that

$$z_g \leq e_g + \sum_{t=1}^{n} A_{gt}\, x_t, \quad \text{for each good } g.$$

Thus, each market-clearing constraint states that the amount of a good that is consumed during the period cannot exceed the amount that is made available for consumption during that period. The market-clearing constraints are inequalities because of our assumption of free disposal. If good g was noxious, then its market-clearing constraint would be an equation rather than an inequality.

Whether or not the market clears depends on the prices. If the price of a good is too low, the demand for that good may exceed the supply, in which case the market for that good would not clear.

Consumer's Equilibrium

In general, each consumer begins the period with an endowment (a nonnegative quantity) of each good. Each consumer can sell at the market the goods that he or she owns but chooses not to consume and can buy at the market the goods that he or she does not own but wishes to consume. Each consumer faces a **budget constraint**, namely, that the market value of the bundle of goods that the consumer consumes cannot exceed the market value of the consumer's endowment.

At a given set of prices, a **consumer's equilibrium** is any trading and consumption plan that maximizes the utility of the bundle of goods that the consumer consumes, subject to the consumer's budget constraint.

Our model has only one consumer, and the satisfaction that the consumer receives from each good is linear in the amount of that good that is consumed. For our model, a consumer's equilibrium is a solution to the linear program,

Program 19.2: $\underset{z*}{\text{Maximize}}\ \sum_{g=1}^{m} u_g z_g$, subject to the constraints

$$\sum_{g=1}^{m} p_q z_g \leq \sum_{g=1}^{m} p_g e_g,$$

$$z_g \geq 0, \qquad g = 1, \ldots, m.$$

In Program 19.2, placing "$z*$" under "Maximize" is a signal that the z_g's are its decision variables. The only decision variables in Program 19.2 are the consumption quantities, the z_g's. The prices (the p_g's) are fixed because the consumer has no direct effect on the prices.

The objective of Program 19.2 measures the consumer's level of satisfaction (utility) with the bundle of goods that he or she consumes. Program 19.2 includes the consumer's budget constraint, which keeps the market value of the bundle of goods that the consumer consumes from exceeding the market value of the consumer's endowment.

Producers' Equilibrium

In this model of an economy, the producers play rather simple roles. They own no assets. All they can do is to operate the technologies. Each producer who operates a technology:

- buys its inputs at the market,
- sells its outputs at the market,
- aims to make a profit.

In this model, capital is a good. If capital is an input to a technology, any producer who operates that technology must pay for the cost of the capital. We interpret

$$\sum_{g=1}^{m} p_g A_{gt} = \text{the net profit per unit level of technology } t.$$

This sum equals the revenue received from the outputs of the technology less the price paid for its inputs. Capital is an input, so this sum is positive if the producer earns an *excess* profit, that is, a profit that is above the market rate of return on capital.

At a given set of prices, a **producers' Equilibrium** is a set $x = (x_1, \ldots, x_n)$ of levels of the technologies that maximize the profit made by each producer. Our model has linear returns to scale. For it, a producers' equilibrium is described in the following lemma.

Lemma 19.9. At a given set of market prices, a set $x = (x_1, \ldots, x_n)$ of production levels is a producers' equilibrium if and only if it satisfies the constraints

$$x_t \geq 0, \quad \text{for } t = 1, \ldots, n,$$

$$\sum_{g=1}^{m} p_g A_{gt} \leq 0, \quad \text{for } t = 1, \ldots, n,$$

$$x_t \cdot \left(\sum_{g=1}^{m} p_g A_{gt} \right) = 0, \quad \text{for } t = 1, \ldots, n.$$

Proof. We establish necessity. The inequality $x_t \geq 0$ must be satisfied by any production plan. If the prices satisfied $\sum_{g=1}^{m} p_g A_{gt} > 0$, then no finite level x_t could maximize profit, so the second set of constraints is necessary. If the prices satisfy $\sum_{g=1}^{m} p_g A_{gt} < 0$, technology t operates at a loss and cannot be optimal to set $x_t > 0$, so the third set of constraints is necessary. Similar arguments, which we omit, establish sufficiency. ◆

The conditions in Lemma 19.9 are easy to express verbally. These conditions require that each production level be nonnegative, that no technology operates at a profit, where "profit" means profit in excess of the rate of return on capital, and that no technology operates if it would incur a loss. Today, these conditions seem natural and obvious. But they were unnoticed for decades after seminal work by Leon Walras (1884) on general equilibrium. Tjalling G. Koopmans published these conditions in 1951, early in the history of linear programming.

Koopmans' conditions imply that the existence of each technology imposes a constraint on market prices. To illustrate, suppose the economy is in equilibrium and then a new technology emerges, say, technology 9. If the prices that existed before this technology emerged violate the inequality $\sum_{g=1}^{m} p_g A_{g9} \le 0$, the prices will have to shift in order for a producers' equilibrium to be restored.

General Equilibrium

Taken together, the models of the producers, the consumers, and the market are known as the **economy**. The economy is said to be in **general equilibrium** if:

- Each producer is in equilibrium.
- Each consumer is in equilibrium.
- The market for each good clears.

Equilibrium prices are prices of the goods for which the economy can be in general equilibrium. A general equilibrium is sometimes called a **complete equilibrium**; these terms are synonyms.

A general equilibrium of an economy studies the interactions between all agents in that economy. By contrast, a **partial equilibrium** model explores the interactions within a fragment of the economy. In a partial equilibrium model, the prices of some goods are determined within the model, and the prices of other goods are fixed (exogenous). A partial equilibrium model does not verify whether the markets for exogenous goods clear.

A Linear Program

We are now poised to answer the questions posed in Problem E. For the case of a single (canonical) consumer, we will construct a general equilibrium from the optimal solutions to Program 19.3 and its dual.

Program 19.3: $u^* = \underset{z*,\ x*}{\text{Maximize}} \sum_{g=1}^{m} u_g z_g$, subject to the constraints

$$
\begin{aligned}
p_g: \qquad z_g - \sum_{t=1}^{n} A_{gt}\, x_t &\le e_g, \quad \text{for } g = 1, \ldots, m,\\
x_t &\ge 0, \quad \text{for } t = 1, \ldots, n,\\
z_g &\ge 0, \quad \text{for } g = 1, \ldots, m.
\end{aligned}
$$

The symbols $z*$ and $x*$ in Program 19.3 indicate that its decision variable are the consumption quantities (the z_g's) and the levels at which the technologies are operated (the x_t's). Program 19.3 sets the production levels and the consumption levels so as to maximize the total utility to the consumer of the bundle of goods that he or she consumes, subject to constraints that keep the consumption of each good g at or below its net supply. Program 19.3 is peculiar in that the producers are altruistic; they set their production levels so as to maximize the consumer's level of satisfaction, with no regard for their own welfare.

Is Program 19.3 feasible? Yes. The endowments (the e_g's) are nonnegative numbers, so it is feasible to equate each decision variable to zero. Program 19.3 enforces the market-clearing constraints. The facets of a general equilibrium that Program 19.3 omits are:

- The consumer's budget constraint.
- The market prices.
- The requirement that the producers maximize their profits.

On the other hand, the notation hints that the optimal values of the dual variables will be the market prices. A general equilibrium will be constructed from optimal solutions to Program 19.3 and its dual.

The Dual Linear Program

In Program 19.3, the market-clearing constraint on good g has been assigned the complementary dual variable, p_g. Each decision variable in Program 19.3 gives rise to a complementary constraint in its dual. This dual appears below as Program 19.3^d.

Program 19.3^d: $u^* =$ Minimize $\sum_{g=1}^{m} e_g p_g$, subject to the constraints

$$x_t: \qquad \sum_{g=1}^{m} p_g(-A_{gt}) \geq 0, \qquad \text{for } t = 1, \ldots, n,$$

$$z_g: \qquad p_g \geq u_g, \qquad \text{for } g = 1, \ldots, m,$$

$$p_g \geq 0, \qquad \text{for } g = 1, \ldots, m.$$

Each nonsign constraint in either linear program is labeled with the variable to which it is complementary.

Existence and Construction of a General Equilibrium

We *assume* that Program 19.3 is bounded and, equivalently, that the consumer cannot be made infinitely well off. A sufficient condition for Program 19.3 to be bounded is that there exists a good (such as labor) that is an input to each technology.

The central result of this section is to use Program 19.3 and 19.3^d to construct a general equilibrium. The construction of this equilibrium guarantees its existence.

Theorem 19.4 (General Equilibrium). Assume that Program 19.3 is bounded. Then:

(a) Program 19.3 and Program 19.3^d have optimal solutions and have the same optimal value.

(b) Each optimal solution $x = (x_1, \ldots, x_n)$ and $z = (z_1, \ldots, z_m)$ to Program 19.3 and each optimal solution $p = (p_1, \ldots, p_m)$ to Program 19.3^d form a general equilibrium. Moreover, if z_g is positive, then $p_g = u_g$.

Proof. Program 19.3 is feasible. By hypothesis, Program 19.3 is bounded, so it has an optimal solution $x = (x_1, \ldots, x_n)$ and $z = (z_1, \ldots, z_m)$. The Duality Theorem guarantees that Program 19.3^d is feasible and bounded and hence that it has an optimal solution $p = (p_1, \ldots, p_m)$. The Duality Theorem shows that these optimal solutions have the same objective value, u^* and, moreover, that they satisfy complementary slackness. The complementary slackness conditions include

$$x_t\left(\sum_{g=1}^{m} p_g A_{gt}\right) = 0, \qquad \text{for } t = 1, \ldots, n, \tag{19.18}$$

$$z_g(p_g - u_g) = 0, \qquad \text{for } g = 1, \ldots, m. \tag{19.19}$$

We will demonstrate that these values of the decision variables form a general equilibrium, with p_1 through p_m as the market prices.

To see that this is a producers' equilibrium, observe from the constraints of Program 19.3^d that no technology operates at a profit. Also, system (19.18) guarantees that no technology is used if it operates at a loss. Thus, the production quantities are a producer's equilibrium.

System (19.19) states that, if z_g is positive, then $p_g = u_g$, as asserted in the theorem. Let us rewrite system (19.19) as

$$u_g z_g = p_g z_g, \quad \text{for } g = 1, \ldots, m. \tag{19.20}$$

It remains to show that this is a consumer's equilibrium and, equivalently, that the consumer's budget constraint is satisfied. The fact that Programs 19.3^d and 19.3 have the same optimal value u^* couples with system (19.20) to give

$$u^* = \sum_{g=1}^{m} p_g e_g = \sum_{g=1}^{m} u_g z_g = \sum_{g=1}^{m} p_g z_g. \tag{19.21}$$

Equation (19.21) demonstrates that the market value $\sum_{g=1}^{m} p_g e_g$ of the consumer's endowment equals the market value $\sum_{g=1}^{m} p_g z_g$ of the bundle of goods that the consumer consumes. In other words, this is a consumer's equilibrium, and the proof is complete. ◆

This chapter began by establishing a duality between production quantities and prices. Production LP set optimal levels of the production quantities. Its dual, the Pricing LP, set optimal levels of the prices. The General Equilibrium Theorem deepened this duality between production quantities and prices. It shows that optimum (equilibrium) levels of production and consumption are found from one linear program and optimum prices from its dual. Taken together, these two linear programs solve the optimization problems faced by all participants in the economy. This amply illustrates the power of linear programming.

The key to this theorem has been our assumption of a single (canonical) consumer. With one consumer, the content of this theorem remains valid for the case of decreasing marginal returns on production and consumption. For this more general case, the Lagrange multipliers that Solver reports *are* the market prices.

19.14. FARKAS'S LEMMA AND THEOREMS OF THE ALTERNATIVE

In 1896, Farkas published a marvelously insightful result that is now known as Farkas's lemma. His result is that a certain set of linear inequalities has a solution if and only if a different set of linear inequalities has no solution. Farkas's lemma and the Duality Theorem are easy consequences of each other.

In this section, we establish Farkas's lemma as a consequence of the Duality Theorem. To develop Farkas's lemma, we examine a linear program that appears below as Program 19.4.

Program 19.4: Maximize $\{0x_1 + 0x_2 + \cdots + 0x_n\}$, subject to the constraints

$$A_{11}x_1 + A_{12}x_2 + \cdots + A_{1n}x_n \le b_1, \tag{19.22.1}$$

$$A_{21}x_1 + A_{22}x_2 + \cdots + A_{2n}x_n \le b_2, \tag{19.22.2}$$

$$\vdots$$

$$A_{m1}x_1 + A_{m2}x_2 + \cdots + A_{mn}x_n \le b_m, \tag{19.22.m}$$

$$x_1 \ge 0, \quad x_2 \ge 0, \ldots, \quad x_n \ge 0. \tag{19.22.m + 1}$$

Program 19.4 specializes Lmax by having zero as its objective function. Program 19.4 is feasible if and only if Lmax is feasible. Every feasible solution to Program 19.4 has 0 as its objective value. Thus, if Program 19.4 is feasible, then its optimal value equals zero.

The dual of Program 19.4 specializes Gmin by having 0's as its right-hand-side values. The dual of Program 19.4 appears below, where it is dubbed Program 19.4^d.

Program 19.4^d: Minimize $\{y_1 b_1 + y_2 b_2 + \cdots + y_m b_m\}$, subject to the constraints

$$y_1 A_{11} + y_2 A_{21} + \cdots + y_m A_{m1} \geq 0, \qquad (19.23.1)$$
$$y_1 A_{12} + y_2 A_{22} + \cdots + y_m A_{m2} \geq 0, \qquad (19.23.2)$$
$$\vdots \qquad \vdots \qquad \vdots \qquad \vdots \qquad \vdots$$
$$y_1 A_{1n} + y_2 A_{2n} + \cdots + y_m A_{mn} \geq 0, \qquad (19.23.n)$$
$$y_1 \geq 0, \quad y_2 \geq 0, \ldots, \quad y_m \geq 0. \qquad (19.23.n+1)$$

Program 19.4^d is definitely feasible; its constraints are satisfied by equating each of its decision variables to zero.

Lemma 19.10. Exactly one of the following alternatives holds:

(a) Program 19.4 has a feasible solution.

(b) Program 19.4^d has a feasible solution whose objective value is negative.

Proof. To prove this lemma, we examine the case in which Program 19.4 has a feasible solution and the case in which Program 19.4 has no feasible solution.

Case 1: Suppose that Program 19.4 has a feasible solution. This feasible solution has 0 as its objective value. Since Program 19.4 is an instance of Lmax, Lemma 19.4 shows that each feasible solution to its dual, Program 19.4^d, has objective value that is greater than or equal to 0. Thus, if (a) holds, then (b) cannot.

Case 2: Suppose that Program 19.4 has no feasible solution. Program 19.4^d has a feasible solution whose objective value equals zero, namely, the solution $y_i = 0$ for each i. Aiming for a contradiction, we assume that Program 19.4^d has no feasible solution whose objective value is negative. In this case, Program 19.4^d has 0 as its optimal value. Consequently, the Duality Theorem shows that Program 19.4 also has 0 as its optimal value. But this contradicts the hypothesis that Program 19.4 has no feasible solution. This contradiction demonstrates that Program 19.4^d has a feasible solution whose objective value is negative. In brief, if (a) does not hold, then (b) must hold. ◆

Lemma 19.10 is a direct consequence of the Duality Theorem. Lemma 19.10 is called a **theorem of the alternative** because it demonstrates that exactly one of two alternatives occurs. Theorems of the alternative are important because they replace negative statements by equivalent positive statements. When Lemma 19.10 is rephrased, it becomes:

Theorem 19.5 (Farkas's Lemma). Exactly one of the following alternatives holds:

(a) There exists a solution to system (19.22).

(b) There exists a solution to system (19.23) for which

$$y_1 b_1 + y_2 b_2 + \cdots + y_m b_m < 0. \qquad (19.24)$$

Proof. Condition (a) is that Program 19.4 has a feasible solution. Condition (b) is that Program 19.4^d has a feasible solution whose objective value is negative. Hence, Farkas's lemma is a rewording of Lemma 19.10. ◆

Proving Farkas's lemma through the Duality Theorem pays a handsome dividend. The crossover table adapts Lemma 19.10 and Farkas's lemma to systems of constraints whose senses differ from those in (19.22).

Let us illustrate. Suppose that constraints (19.22.1) through (19.22.m) were equations rather than inequalities. The crossover table would delete constraint (19.23.m + 1) from Program 19.4[d] because the variables that are complementary to equations are free variables. With these changes, Lemma 19.10 and Theorem 19.5 remain true; their proofs remain valid, *exactly* as written. This proves:

Corollary 19.1. There exists a solution to the system

$$\begin{aligned} A_{11}x_1 + A_{12}x_2 + \cdots + A_{1n}x_n &= b_1, \\ A_{21}x_1 + A_{22}x_2 + \cdots + A_{2n}x_n &= b_2, \\ \vdots \qquad \vdots \qquad\qquad \vdots \quad &\quad \vdots \\ A_{m1}x_1 + A_{m2}x_2 + \cdots + A_{mn}x_n &= b_m, \\ x_1 \geq 0, \quad x_2 \geq 0, \ldots, \qquad x_n &\geq 0, \end{aligned}$$

if and only if there exists no solution to the system

$$\begin{aligned} y_1A_{11} + y_2A_{21} + \cdots + y_mA_{m1} &\geq 0, \\ y_1A_{12} + y_2A_{22} + \cdots + y_mA_{m2} &\geq 0, \\ \vdots \qquad \vdots \qquad\qquad \vdots \quad &\quad \vdots \\ y_1A_{1n} + y_2A_{2n} + \cdots + y_mA_{mn} &\geq 0, \\ y_1b_1 + y_2b_2 + \cdots + y_mb_m &< 0. \end{aligned}$$

In precisely this way, quite a variety of theorems of the alternative can be proved via the Duality Theorem.

19.15. ARBITRAGE

In financial economics, "arbitrage" describes an investment opportunity that can earn a profit without any risk of a loss. This section links arbitrage to duality. To construct this link, we must account for the time value of money. To keep things simple, we focus on investing for a single period that has a fixed length of, say, six months.

A Risk-Free Asset

The economy is assumed to include an investment opportunity (asset) that is guaranteed to return $(1 + r)$ dollars at the end of the period for each dollar invested in that asset at the start of the period. This number r is known as the **risk-free** interest rate for the period. In models of the U.S. economy, the risk-free interest rate for a given period is often taken to be the interest rate of a U.S. Treasury bill or note for that period.

Risky Investments

In addition to the risk-free asset, the economy is assumed to include other investment opportunities (assets) whose returns are uncertain. Their returns depend on the state of the economy at the end of the period.

Table 19.9 The current price, the end-of-period price, and the net profit of each of three risky assets under each of four states of the economy.

	A	B	C	D	E	F	G	H	I
1									
2	r =	0.03				= B6 - B$11*(1 + $B$2)			
3									
4	**asset**	**1**	**2**	**3**		**asset**	**1**	**2**	**3**
5	state	end-of-period price				state		net profit	
6	a	107	215	305		a	4	9	-4
7	b	109	200	321		b	6	-6	12
8	c	95	212	305		c	-8	6	-4
9	d	107	197	293		d	4	-9	-16
10									
11	price =	100	200	300					

Table 19.9 illustrates this type of investment model. Columns B, C, and D of this table describe three risky assets, which are labeled 1, 2, and 3. Row 11 records the start-of-period price of each these assets. For instance, cell C11 reports that the current price of asset 2 is \$200. Cells B6 through D9 record the end-of-period price of each risky asset under each of four states of the economy, which are labeled *a* through *d.* If state *c* occurs, asset 2's price will be \$212, for instance.

Cell B2 of Table 19.9 records the risk-free interest rate for the current period, which is 3%. Columns G, H, and I report the "net profit" of assets 1, 2, and 3 under each state of the economy. The **net profit** of an asset is defined as its end-of-period price less the return that would be earned by investing its start-of-period price in the risk-free asset. For instance, cell G6 records a net profit of 4 because $4 = 107 - (100)(1 + 0.03) = 107 - 103$. Dragging the function in cell G6 across to column I and down to row 9 computes the net profit of each asset for each state of the economy.

Evidently, the net profit of each asset can be positive or negative, depending on the state of the economy at the period's end. If state *c* occurs, asset 2 returns more than an equivalent investment in the risk-free asset, but assets 1 and 3 earn less.

Omitted from Table 19.9 are the probabilities that states *a* through *d* will occur at the end of the period. Values of these probabilities will eventually emerge, as an indication that no arbitrage opportunity exists.

A Portfolio

In a general one-period model, there are n risky assets, and a **portfolio** is an n-tuple $x = (x_1, x_2, \ldots, x_n)$ of investment levels, where x_j denotes the number of units of asset j that the investor buys at the start of the period and sells at the end of the period. The model in Table 19.9 has $n = 3$.

An important part of the model of financial economics under development is the assumption that the investor can select investment levels (the x_j's) that take *any* values, positive, negative, or zero.

Longs and Shorts

Positive values of x_j are called "longs"; negative values are "shorts." The current market price of asset 3 is $300 per unit, so setting $x_3 = -2.2$ amounts to selling $660 worth of asset 3 that you do not own at the start of the period and buying it back at the end of the period.

On the stock market, for instance, a "short sale" occurs if you borrow stock, sell it, and thereby incur the obligation to return the stock by buying it later on, hoping that its price will decrease in the interim.

In this model of financial economics, the *same* data are assumed to apply to "longs" and to "shorts." Thus, for the data in Table 19.9, setting $x_2 = -1$ earns a net profit of $9 if state d occurs but incurs a net cost of $6 if state c occurs.

The Role of the Risk-Free Asset

The opportunity to invest in the risk-free asset does not appear in Table 19.9—not explicitly, that is. It is present implicitly, as is the opportunity to borrow at the risk-free rate. To see why, we interpret the net profit of investing in one unit of a risky asset as follows:

- At the start of the period, borrow the price of one unit of a risky asset at the risk-free rate. Use this loan to buy one unit of this risky asset.
- At the end of the period, sell this risky asset and pay off the loan, including the accumulated interest,
- What remains is the "net profit," which varies with the state and can be negative.

For instance, to set $x_2 = 1$ is to borrow $200 at the risk-free rate at the start of the period, use this loan to purchase one unit of asset 2 at the start of the period, sell this unit at the end of the period, and repay the loan and its interest, which has grown (at the risk-free rate) to $\$200\,(1 + r) = \$206$. This scenario earns a (random) net profit that equals the excess of the end-of-period price of one unit of asset 2 over $206.

An Arbitrage Opportunity

In the context of our one-period investment model, an **arbitrage opportunity** is said to exist if some portfolio satisfies these two conditions:

- For every state of the economy that can occur at the end of the period, the net profit of this portfolio is nonnegative.
- For at least one state of the economy that can occur at the end of the period, the net profit of this portfolio is positive.

If an arbitrage opportunity exists, any investor who prefers more money to less will elect to invest in it.

An equivalent way in which to describe an arbitrage opportunity is as follows. An arbitrage opportunity is a portfolio whose *rate* of return is at least as large as the risk-free rate under every state of the economy and whose rate of return exceeds that of the risk-free rate under at least one state of the economy.

The No-Arbitrage Tenet

The **no-arbitrage** tenet of financial economics is that the prices of the assets preclude arbitrage opportunities. The economic argument in support of the no-arbitrage tenet is this: If an arbitrage opportunity did arise, it would be discovered quickly, investors would invest heavily in portfolios that achieve it, and their actions would adjust the prices of the assets in ways that eliminate the arbitrage opportunity.

The no-arbitrage tenet implies that if a risky asset has a positive net profit under some state of the economy, it must have a negative net profit under another state of the economy. Do you see why? Similarly, if a risky asset has a negative net profit under some state of the economy, it must have a positive net profit under another state. Again, do you see why?

A Theorem

The no-arbitrage tenet applies not just to individual assets but to portfolios. Its application to portfolios gives rise to Theorem 19.6 where hypothesis recapitulates the prior discussion. Its conclusion is that there exists no arbitrage opportunity if and only if there exists a probability distribution over the states, with each probability positive, for which each risky asset earns at the risk-free rate.

Theorem 19.6 (Arbitrage). Suppose that:

- The economy includes a risk-free asset.
- Exactly one of m states of the economy will occur at the end of the period, these states being numbered 1 through m.
- The economy includes n risky assets, which are numbered 1 through n. The $m \times n$ array A is known, where

  A_{ij} = the net profit (excess over the risk-free return) earned per unit investment in asset j if state i occurs at the end of the period.

- The investment levels can take any values, including negative ones.

Then, exactly one of the following conditions holds.

(a) **(Arbitrage Opportunity)** There exists an arbitrage opportunity.

(b) **(Profit-Free Probabilities)** There exists a probability distribution over the states of the economy such that each probability is positive and such that each asset has zero as the expectation of its net profit.

For the data in Table 19.9, does an arbitrage opportunity exist? No. To the contrary, we will see that each asset has zero as the expectation of its profit if the probability distribution over the states is given by

$$p(a) = 9/79, \quad p(b) = 28/79, \quad p(c) = 31/79, \quad p(d) = 11/79.$$

Obviously, these probabilities are positive. On a spreadsheet, it is easy to check (as we shall) that these probabilities cause the expected profit of each asset to equal zero.

A Linear Program for Arbitrage Opportunity

A linear program will characterize the case in which an arbitrage opportunity exists. Its dual will be used to analyze the case in which no arbitrage opportunity exists. Program 19.5 is the first of these linear programs. Its decision variables describe a set $x = (x_1, x_2, \ldots, x_n)$ of investment levels, where

$$x_j = \text{the number of units of asset } j \text{ purchased at the start of the period.}$$

The investment levels x_1 through x_n can take any values, negative values denoting "short" sales. The variable y_i in Program 19.5 equals the net profit (excess over the risk-free return) earned by portfolio $x = (x_1, x_2, \ldots, x_n)$ if state i occurs at the end of the period. An arbitrage opportunity exists if there exists a portfolio such that y_i is nonnegative for each i and such that y_k is positive for some k.

Program 19.5: Maximize $\sum_{i=1}^{m} y_i$, subject to the constraints

$$q_j: \qquad y_i = \sum_{j=1}^{n} A_{ij}\, x_j \quad \text{for } i = 1, 2, \ldots, m,$$

$$y_i \geq 0 \qquad \text{for } i = 1, 2, \ldots, m.$$

In brief, an arbitrage opportunity exists if and only if Program 19.5 has a feasible solution whose objective value is positive.

A Linear Program for No Arbitrage

Program 19.5 does have a feasible solution whose objective value equals 0 because its constraints are satisfied by equating each variable to zero, that is, by setting $0 = x_1 = \cdots = x_n = y_1 \cdots = y_m$. Thus, no arbitrage opportunity exists if and only if Program 19.5 has 0 as its optimal value.

With regard to the three statements that follow, we have just seen that the first two of them are equivalent. The Duality Theorem of linear programming shows that the final two are equivalent. Hence, all three are equivalent:

- No arbitrage opportunity exists.
- Program 19.5 has 0 as its optimal value.
- The dual of Program 19.5 has 0 as its optimal value.

The dual of Program 19.5 appears next, where it is labeled Program 19.5^{d}. The data in each constraint of Program 19.5^{d} are the coefficients of its complementary variable. The sense of each constraint in Program 19.5^{d} is determined from the crossover table.

Program 19.5^{d}: Minimize $\{0\}$, subject to the constraints

$$x_j: \qquad \sum_{i=1}^{m} q_i(-A_{ij}) = 0, \quad \text{for } j = 1, 2, \ldots, n, \tag{19.25}$$

$$y_i: \qquad q_i \geq 1, \quad \text{for } i = 1, 2, \ldots, m.$$

Every feasible solution to Program 19.5^{d} has 0 as its objective value. Consequently, Program 19.5^{d} has a feasible solution if and only if no arbitrage opportunity exists.

A Probability Distribution

We've seen that no arbitrage opportunity exists if and only if Program 19.5^{d} has a feasible solution. The decision variables in Program 19.5^{d} are positive, but they cannot be interpreted as probabilities because their sum is at least m.

From any feasible solution $q = (q_1, q_2, \ldots, q_m)$ to Program 19.5^{d}, we aim to construct a probability distribution over the states. With $K = q_1 + q_2 + \cdots + q_m$, we define $p = (p_1, p_2, \ldots, p_m)$ by

$$p_i = q_i/K \quad \text{for } i = 1, 2, \ldots, m.$$

Since q_i is positive, so is p_i, and the p_i's sum to 1. Thus, $p = (p_1, p_2, \ldots, p_m)$ is a probability distribution that assigns a positive probability to each state. Dividing Equation (19.25) by K shows that

$$\sum_{i=1}^{m} p_i A_{ij} = 0 \quad \text{for } j = 1, 2, \ldots, n, \tag{19.26}$$

and hence that each asset j has 0 as the expectation of its profit, given this probability distribution.

In brief, from any feasible solution to Program 19.5^d, it's easy to construct a probability distribution over the states for which each probability is positive and for which each risky asset has 0 as its expected net profit.

Proof of Theorem 19.6. A proof of Theorem 19.6 is nearly complete. We've seen that no arbitrage opportunity exists if and only if Program 19.5^d has a feasible solution. From each feasible solution to Program 19.5^d, we've seen how to compute a probability distribution over the states, with all probabilities positive, such that each asset has 0 as the expectation of its net profit.

All that remains is to begin with positive probabilities p_1 through p_m that satisfy Equation (19.26) and construct a feasible solution to Program 19.5^d. To do so, set $q_i = p_i/L$ for each i, where the number L equals the smallest of the probabilities. These q_i's are at least 1, and they are easily seen to satisfy Equation (19.25). This completes a proof of Theorem 19.6. ◆

An Illustration

The investments in Table 19.9 are now used to illustrate Program 19.5^d. Table 19.10 replicates columns F through I of Table 19.9. Cells J6 through J9 of Table 19.10 record the values of the decision variables $q(a)$ through $q(d)$. Solver has minimized the value of $q(a)$ subject to constraints that each decision variable is greater than or equal to 1 and such that system (25) is satisfied. Column J records the optimal solution that Solver has found, and column K records the resulting probabilities.

Table 19.10 Spreadsheet computation of a probability distribution for which each asset has 0 as the expectation of its profit.

	F	G	H	I	J	K
3						
4	**asset**	**1**	**2**	**3**	**variable**	**probability**
5	state		net profit			
6	a	4	9	-4	1	9/79
7	b	6	-6	12	3.111111	28/79
8	c	-8	6	-4	3.444444	31/79
9	d	4	-9	-16	1.222222	11/79
10		0.0000	0.0000	0.0000		
11						
12		= SUMPRODUCT(G6:G9, $J6:$J9)				
13						
14					= J9/SUM(J$6:J$9)	

Solver minimized the value in cell J6 with cells J6:J9 as changing cells subject to the constraints

J6:J9 >= 1

G10:I10 = 0

Solver has found a feasible solution to Program 19.5[d], so Theorem 19.6 shows that no arbitrage opportunity exists. The probabilities in column K are those that we displayed earlier. The probabilities in column K appear as fractions rather than as decimals. As an aside, we note that Excel has been used to compute these fractions. To arrange for Excel to compute them:

- First, select cells K6 through K9.
- Then, on the Excel Format menu, scroll down to the Cells. . . entry and click.
- Next, on the window that pops up, click on the Number tab, select the Fraction entry, and choose the "Up to Three Digits" option.

Critique

Let us critique this model of investment. Its no-arbitrage tenet is robust and alluring. This tenet concerns a choice between two investment opportunities. The first alternative earns zero as its net profit under every state of nature. The second earns a nonnegative net profit under every state of nature and earns a positive net profit under at least one state of nature. For any decision maker who prefers more money to less, the choice is clear. The second **dominates** the first in the strongest of the senses described in Chapter 9. Results that rest on this sort of dominance are virtually unassailable.

But what about the investment alternatives postulated by Theorem 19.6? A small-scale investor can invest at the risk-free rate, but he cannot borrow at the risk-free rate. He incurs transaction costs and may not be able to sell short without extra restrictions (e.g., margin requirements) and expenses. He may also face tax consequences that may vary from asset to asset.

On the other hand, the prices of assets are determined largely by institutional investors who manage big portfolios. They can negotiate tiny transaction costs, and they can "short" at rates that are only slightly less advantageous than the long rates. They manage large tax-deferred retirement portfolios for which trades do not incur taxes. In brief, large institutional investors who manage retirement funds do face choices that satisfy the hypothesis of Theorem 19.6, at least approximately.

But is the risk-free asset actually free of risk? The **real** return on an investment equals its nominal return over a period *less* the increase in the cost of living over the same period. Many investors are concerned with the real return of their portfolios over a period of a decade or two. Long-term U.S. Treasury bonds and notes that pay fixed dividends fail to account for inflation. They cannot serve as a risk-free asset for an investor whose concern is real return over a lengthy period.[1]

Theorem 19.6 seems to violate a fundamental tenet of financial economics, which is that investors demand higher expected profits for riskier investments. The theorem states that there exists a probability distribution over the states of nature for which each *risky* asset has zero as its net profit. The theorem doesn't state that these are the objective probabilities, the ones that will occur. But many users of this theorem presume that they are, which implies that investors are not being rewarded for taking risks. Put another way, Theorem 19.6 seems to characterize those risks that can be eliminated by "diversification," namely, by investing in diversified portfolios.

---

[1]A fairly new U.S. Treasury security called TIPS (short for Treasure Inflation-Protected Security) may eventually provide a risk-free rate of return for tax-deferred accounts. This security bears a dividend equal to a nominal return (e.g., 3.25% per year until maturity) *and* has a principal that grows at the rate of inflation. The U.S. tax code makes TIPS unattractive for taxable investment accounts. As of this writing, so few TIPS have been issued that they may be illiquid unless held to maturity.

19.16. MINIMUM CUT

With the Duality Theorem in hand, we return to the max-flow problem that first appeared in Chapter 4. The focus there had been on maximizing the flow from the "source" node to the "sink" node of a network. Here the focus will be on a different problem. Roughly speaking, the "min-cut" problem is to remove from the network the least amount of capacity such that no flow can occur from the source to the sink on the arcs that remain.

In this section, a pair of dual linear program will be used to solve the max-flow problem and the min-cut problem. This analysis will prove the "max-flow min-cut" theorem.

An Example

Figure 19.1 reproduces the max-flow problem that was studied in Chapter 4. The arc pointing into node 1 has infinite capacity, as does the arc pointing away from node 8. Each of the other arcs has finite capacity, which is displayed next to its arrow. Arc (6, 8) has 12 as its capacity, for instance. Each arc's capacity is the upper bound on its flow.

Figure 19.1 illustrates the type of network on which the max-flow problem occurs. In this instance and in general, all but two of the arcs have finite capacities. The two arcs that have infinite capacities are special. These two arcs—and only they—touch one node apiece. One of these arcs points toward the **source** node, and the other points away from the **sink** node. In any feasible solution to the max-flow problem, the quantities that flow on these two special arcs must equal each other. (Flow in equals flow out.).

The **max-flow problem** is that of causing the largest possible quantity to flow on the infinite-capacity arc that points toward the source while conserving flow at each node and keeping the flow on each arc from exceeding its capacity.

Cuts and Their Capacities

In a max-flow problem, a **cut** is any set of nodes that includes the source but does not include the sink. There are many cuts. In Figure 19.1, for instance, one cut consists of the set $C = \{1, 3\}$ of nodes.

The **capacity** of a cut C equals the sum of the capacities of the arcs (m, n) for which m is in C and n is not in C. The capacity of the cut C is denoted $K(C)$ and can be written

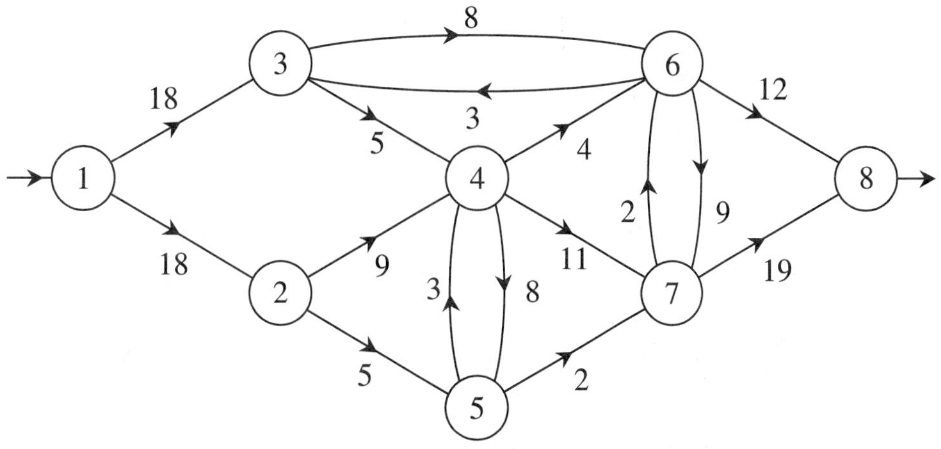

Figure 19.1 A max-flow problem, with node 1 as the source and node 8 as the sink.

succinctly as

$$K(C) = \sum_{(m,n):\, m \in C,\, n \notin C} U_{mn}.$$

The nasty-looking notation in this equation sums the capacity U_{mn} over each arc (m, n) that points from a node m that is in C and to a node n that is not in C. In Figure 19.1, for instance, the cut $C = \{1, 3\}$ has

$$K(C) = U_{12} + U_{34} + U_{36} = 18 + 5 + 8 = 31$$

because the arcs (1, 2) and (3, 4) and (3, 6) point from nodes in C to nodes that are not in C.

Having defined cuts and their capacities, we can pose the question on which this section is focused. It is:

Problem F (Minimum Cut)

For the max-flow problem, how can we find a cut C whose capacity $K(C)$ is smallest?

There is nothing linear about a cut. Either an arc is in the cut or it is not in the cut. Nevertheless, we will construct a minimum cut from the optimal solution to a linear program. This linear program is the dual of the max-flow problem.

Algebraic Description of the Max-Flow Problem

In Chapter 4, we built the max-flow problem directly on a spreadsheet (Table 4.9, on page 119). This avoided some algebra. We will need to construct the dual of the max-flow problem. That will be easier if we first describe the max-flow problem algebraically.

For a general description of a max-flow problem, we first re-label the nodes, if necessary, so that node 1 is the source and node 8 is the sink. A general description is as follows:

- Exactly two arcs have infinite capacity. These two arcs, and only they, touch one node apiece. One of these infinite-capacity arcs points toward node 1 (the source), and the other points away from node 8 (the sink). The symbols x_{s1} and x_{8d} denote the flows on the infinite-capacity arcs that point toward node 1 and away from node 8, respectively.
- The symbol E denotes the set of arcs whose capacities are finite. Each arc (i, j) in E has positive capacity U_{ij}, and x_{ij} denotes the flow on this arc.

Figure 19.1 illustrates this notation. In Figure 19.1, the set E contains 16 arcs. The figure has $U_{13} = 18$ and $U_{24} = 9$, for instance. This notation lets us cast the max-flow problem as Program 19.6.

Program 19.6: $z^* =$ Maximize $\{x_{s1}\}$, subject to the constraints

y_i:	$\sum_j x_{ij} - \sum_j x_{ji} = 0,$	for each node i,
z_{ij}:	$x_{ij} \le U_{ij},$	for each arc (i, j) in E,
	$x_{ij} \ge 0,$	for each arc (i, j).

In Program 19.6, the variable y_i is complementary to the flow-conservation constraint for node i. This flow-conservation constraint is to be read as "The sum of the flows on those arcs that point away from node i minus the sum of the flows on those arcs that point into node i equals zero." In Figure 19.1, the flow-conservation constraint for node 3 is

$$y_3: \qquad x_{34} + x_{36} - (x_{13} + x_{63}) = 0,$$

for instance.

The Dual of the Max-Flow Problem

The dual of Program 19.6 appears below as Program 19.6[d]. After specifying Program 19.6[d], we justify it.

Program 19.6[d]: $z^* =$ Minimize $\sum_{(i,j)\in E} z_{ij} U_{ij}$, subject to the constraints

$$\begin{array}{lrl} x_{ij}: & y_i - y_j + z_{ij} \geq 0, & \text{for each arc } (i, j) \text{ in } E, \\ x_{s1}: & -\ y_1 \qquad \geq 1, & \\ x_{8d}: & y_8 \qquad \geq 0, & \\ & z_{ij} \geq 0, & \text{for each arc } (i, j). \end{array}$$

In Program 19.6[d], the y_i's are free (unconstrained in sign) because they are complementary to equality constraints. The z_{ij}'s are nonnegative because they are complementary to "$\leq$" constraints in a maximization problem. The objective of Program 19.6[d] multiplies each dual variable by the right-hand-side value of the constraint to which it is complementary and takes the sum.

The nonsign constraints in Program 19.6[d] may be puzzling. These constraints are now justified. Let us first construct the constraint in Program 19.6[d] that is complementary to the variable x_{36}. The variable x_{36} is nonnegative, so this constraint is a "$\geq$" inequality. The variable has 0 as its coefficient in the objective of Program 19.6, so the right-hand-side value of this constraint equals 0. The variable x_{36} makes these three appearances in the nonsign constraints of Program 19.6:

- With a coefficient of $+1$ in the flow-conservation constraint for node 3, whose complementary variable is y_3.
- With a coefficient of -1 in the flow-conservation constraint for node 6, whose complementary variable is y_6.
- With a coefficient of $+1$ in the capacity constraint for arc (3, 6), whose complementary variable is z_{36}.

Thus, the constraint that is complementary to x_{36} is $y_3 - y_6 + z_{36} \geq 0$. A similar pattern holds for each variable x_{ij} whose arc (i, j) is in E. Replacing 3 by i and 6 by j produces the constraint $y_i - y_j + z_{ij} \geq 0$. This justifies the first group of nonsign constraints in Program 19.6[d].

The constraint that is complementary to x_{s1} is $-y_1 \geq 1$ because x_{s1} appears with a coefficient of -1 in the flow-conservation constraint for node 1 and with a coefficient of $+1$ in the objective of Program 19.6. The constraint that is complementary to x_{8d} is $y_8 \geq 0$ because x_{8d} appears with a coefficient of $+1$ in the flow-conservation constraint for node 8. This justifies the nonsign constraints of Program 19.6[d].

A Preliminary Result

Our analysis of Programs 19.5 and 19.5[d] begins with Lemma 19.11. This lemma relates each cut C to each feasible solution to Program 19.6.

Lemma 19.11. Each cut C and each feasible solution to Program 19.6 satisfy

$$x_{s1} + \sum_{(n,m)\in E:n\notin C, m\in C} x_{nm} = \sum_{(m,n)\in E:m\in C, n\notin C} x_{mn}. \tag{19.27}$$

Remark: Equation (19.27) is not as nasty as it may appear. Its left-hand side sums the flows on all of the arcs that point into the cut C, and its right-hand side sums the flows on all arcs that point out of C. In Figure 19.1, for instance, consider the cut $C = \{1, 3\}$. For this cut, Equation (19.27) asserts that

$$x_{s1} + x_{63} = x_{12} + x_{34} + x_{36}.$$

Proof. The statement, "Flow in equals flow out," applies to each individual node and, by summation, to any set of nodes. Applying this statement to the cut C verifies Equation (19.27). ◆

The Max-Flow Min-Cut Theorem

Theorem 19.7 constructs a cut whose capacity is smallest from any optimal solution to Program 19.6^d. This theorem also proves that the capacity of the smallest cut equals the maximum value of the flow, thereby proving the **max-flow min-cut theorem** of Ford and Fulkerson.

The method that we will use to prove Theorem 19.7 rests on duality. Like nearly every result that entails duality, this one has an "easy half" and a "not-so-easy half." Both halves employ Lemma 19.11.

Theorem 19.7 (Max-flow Equals Min-cut Capacity). Program 19.6^d is feasible and bounded, and hence has an optimal solution. From any optimal solution to Program 19.6^d, let C^* consist of those nodes i that have $y_i \le -1$. Then C^* is a cut, and

$$K(C^*) = \min\{K(C) : C \text{ a cut}\} = z^*,$$

where z^* is the maximum flow, that is, the optimal value of Program 19.6.

Proof. We do the easier half first. Program 19.6 is feasible and bounded. Let the set x of flows be an optimal solution to Program 19.6, so that $z^* = x_{s1}$. Let C be any cut. To see what (19.27) implies, first consider any arc (m, n) in E with $m \in C$ and $n \notin C$; we have $x_{mn} \le U_{mn}$, so (19.27) gives

$$x_{s1} + \sum_{(n,m)\in E:n\notin C, m\in C} x_{nm} \le K(C).$$

For each arc (n, m) with $n \notin C$ and $m \in C$, we have $x_{nm} \ge 0$, so the above gives $x_{s1} \le K(C)$. This inequality holds for each cut C. Minimize over C, and recall that $z^* = x_{s1}$.

$$z^* = x_{s1} \le \min\{K(C) : C \text{ a cut}\} \tag{19.28}$$

The harder part will entail (19.27) and complementary slackness. Program 19.6 is feasible and bounded. Hence, its dual is feasible and bounded, and the Duality Theorem shows that Programs 19.5 and 19.5^d have the same optimal value, z^*. Theorem 19.3 shows that each pair of optimal solutions to Programs 19.5 and 19.5^d satisfy complementary slackness. Let the set x of flows be optimal for Program 19.6, and let (y, z) be an optimal solution to Program 19.6^d. Complementary slackness assures us that:

$$\text{If } z_{ij} > 0, \quad \text{then } x_{ij} = U_{ij}. \tag{19.29}$$

$$\text{If } y_i - y_i + z_{ij} > 0, \quad \text{then } x_{ij} = 0. \tag{19.30}$$

The constraints of Program 19.6[d] include $y_1 \le -1$ and $y_8 \ge 0$. The statement of Theorem 19.7 designates $C^* = \{i : y_i \le -1\}$. This set C^* is a cut because C^* contains the source (node 1) but not the sink (node 8).

Aiming to apply (19.27), we first consider any arc (m, n) with $m \in C^*$ and $n \notin C^*$. From $y_m \le -1$ and $y_n > -1$, we get $y_m - y_n < 0$, so the constraint $y_m - y_n + z_{mn} \ge 0$ in Program 19.6^d guarantees $z_{mn} > 0$. Hence, (19.29) shows that $x_{mn} = U_{mn}$. Thus, Equation (19.27) gives

$$x_{s1} + \sum_{(n,m)\in E:n\notin C^*,m\in C^*} x_{nm} = K(C^*). \tag{19.31}$$

Let us now consider any arc (n, m) in E with $n \notin C^*$ and $m \in C^*$. Since $y_n > -1$ and $y_m \le -1$, we have $y_n - y_m > 0$, which assures us that $y_n - y_m + z_{nm} > 0$. Hence, (19.30) guarantees $x_{nm} = 0$. Thus, substituting in Equation (19.31) gives $x_{s1} = K(C^*)$. This and expression (19.28) complete the proof. ◆

In Chapter 4, we had constructed a minimum cut from the optimal solution to Program 19.6; this cut C^* consisted of each node i whose shadow price was -1 or less. The Duality Theorem demonstrates that these shadow prices are an optimal solution to Program 19.6[d]. Theorem 19.7 shows that these shadow prices prescribe a cut whose capacity is smallest, and, moreover, that the capacity of the smallest cut equals the maximum value of the flow.

Interdiction

At the start of this section, we alleged that we would solve the problem of removing from the network the least amount of capacity that makes it impossible for any flow to occur from the source node to the sink node. Did we accomplish what we set out to do?

An **interdiction** problem is now presented in which we change the definition of a cut. A **new-style cut** is defined to be any set B of arcs whose removal makes it impossible for any flow to occur from the source to the sink on the arcs that remain. In Figure 19.1, one new-style cut B consists of these four arcs:

$$B = \{(3, 6), (4, 6), (4, 7), (5, 7)\}$$

The **capacity** $\hat{K}(B)$ of each new-style cut B is defined to be the sum of the capacities of the arcs in B. The new-style cut B that is displayed above has $\hat{K}(B) = 8 + 4 + 11 + 2 = 25$. The interdiction problem is to find a new-style cut B^* whose capacity $\hat{K}(B^*)$ is smallest. This interdiction problem is what we had *alleged* we would solve.

Figure 19.1 illustrates what remains to be done. In its network, the maximum value of the flow is 25 units. Theorem 19.7 shows how to construct an old-style cut C^* that has $K(C^*) = 25$. We need to prove that each new-style cut B has $\hat{K}(B) \ge 25$. Although we haven't done this, we nearly have. Problem 24, on page 691, completes the discussion by showing how to transform new-style cuts into equivalent old-style cuts.

19.17. REVIEW

Every linear program is paired with another, which is called its dual. When you take the dual of a linear program:

- The sense of optimization reverses.
- Each nonsign constraint in the linear program gives rise to a complementary decision variable in its dual.
- The right-hand-side value of each nonsign constraint becomes the coefficient of that constraint's complementary variable in the objective of the dual.

- Each decision variable in the linear program gives rise to a complementary constraint in its dual.
- The objective coefficient of each decision variable becomes the right-hand-side value of its complementary constraint.
- Each variable's column of coefficients become the data in its complementary constraint.
- The crossover table determines the senses of the complementary variables and constraints.

In this chapter's preview, we asserted that duality is important to modeling, computation, and mathematics. Let us review what we have learned about each of these facets of duality.

Modeling

Table 19.11, summarizes five models into which duality has provided insight. In the chapter on game theory, two other models are added to this list.

In three of these models, the decision variables in one linear program are quantities, and the decision variables in the other are prices. In retrospect, that is not surprising. If the decision variables in a linear program are quantities and if its constraints allocate resources, its shadow prices (dual variables) are breakeven prices for marginal resources.

The duality exhibited in Table 19.11 generalizes—with qualifications—to the case of decreasing marginal return, but it fails to generalize in the case of increasing marginal return.

Mathematics

We have demonstrated that each linear program is the dual of its dual. We have also proved the Strong Duality Theorem. If a linear program is feasible and bounded, the Strong Duality Theorem tells us that its dual is feasible and bounded and that the two linear programs have the same optimal value. Our proof has rested on the simplex method and has showed that the shadow prices for each basic optimal solution to a linear program are optimal values of the complementary decision variables in the dual linear program. Thus, when you apply the simplex method to a linear program, you are constructing optimal solutions to it and to its dual.

We have seen that feasible solutions to a linear program and its dual are optimal for their respective linear programs if and only if they satisfy either of two conditions. One of these

Table 19.11 Models into which duality has provided insight.

Model	Decision variables in one linear program	Decision variables in the dual linear program
Allocation of scarce resources	Production levels that maximize profit	Prices that rent the facility at least cost
Units with common inputs and outputs	Prices for which a designated unit has the highest efficiency ratio	A nonnegative combination of the other units that dominates the designated unit
An economy that is in general equilibrium	Production and consumption quantities that maximize the consumer's welfare while clearing the market	Prices that create a producers' equilibrium and satisfy the consumer's budget constraint
A collection of risky assets	Levels of investment in these assets that create an arbitrage opportunity	A probability distribution over the states for which each risky asset has zero as the expectation of its profit
Flow from the source node in a network to the sink node	Flow levels that maximize the flow from source to sink	A least-capacity arc set whose removal makes source-to-sink flow impossible

conditions is that their objective values equal each other; the other is that they satisfy complementary slackness.

Complementary slackness has proved to be useful. Complementary slackness has been crucial to our analysis of the general equilibrium model and of the max-flow model, for instance.

We've also seen how duality proves Farkas's lemma and a variety of theorems of the alternativc, which are useful in themselves. They generalize readily, and these generalizations have broad application, but they lie a bit beyond the scope of this text.

Computation

From the computational viewpoint, duality has shed light on the simplex method. The shadow prices guide the selection of its pivot element. Pivoting stops when the shadow prices become feasible for the dual linear program.

Let us close by mentioning two pivot schemes that duality motivates. The **dual simplex method** pivots in a way that keeps the shadow prices feasible for the dual linear program. It pivots in search of a basic solution that is feasible for the linear program itself. Barring degeneracy, each of its pivots improves the objective value of the dual linear program.

The **complementary pivot method** begins by using an artificial variable to produce basic feasible solutions for the primal and the dual that satisfy complementary slackness. Each of its pivots preserves complementary slackness and keeps both basic solutions feasible. In each pivot, the entering variable is complementary to the one that just left. Pivoting stops when the artificial variable becomes nonbasic. At that point, it has satisfied the optimality condition—feasible solutions that satisfy complementary slackness. The complementary pivot method solves a variety of optimization problems, and it also computes equilibria.

19.18. HOMEWORK AND DISCUSSION PROBLEMS

1. This problem concerns the linear program:
Maximize $\{1\,x_1 - 2\,x_2 + 3\,x_3\}$, subject to the constraints

$$\begin{aligned} 4\,x_1 \qquad\qquad - 5\,x_3 &\le 6, \\ -7\,x_1 + 8\,x_2 \qquad\qquad &= -9, \\ 10\,x_2 - 11\,x_3 &\ge 12, \\ x_1 \ge 0, \qquad\qquad x_3 &\le 0. \end{aligned}$$

(a) Take the dual of this linear program.

(b) Re-cast the minimization problem that you obtained in part (a) as a maximization problem.

(c) Take the dual of the maximization problem you obtained in part (b).

(d) Convert the minimization problem you obtained in part (c) into a maximization problem. Is it familiar? If so, why?

2. This problem concerns the linear program:
Maximize $\{-4\,x_1 + 5\,x_2 - 6\,x_3\}$, subject to the constraints

$$\begin{aligned} - 2\,x_2 + 3\,x_3 &\le 4, \\ 2\,x_1 \qquad\qquad - 4\,x_3 &\le -5, \\ - 3\,x_1 + 4\,x_2 \qquad\qquad &\le 6, \\ x_1 \ge 0, \quad x_2 \ge 0, \quad x_3 &\ge 0. \end{aligned}$$

(a) Take the dual of this linear program.

(b) True or false: No linear program is its own dual. *Hint:* A square array B of numbers is said to be *skew symmetric* if $B_{ij} = -B_{ji}$ for each i and j.

3. Take the dual of the Pricing LP, but interpret B as a free variable and the constraint $B \geq 0$ as a nonsign constraint. What happens?

Remark: A feasible linear program is said to have a **bounded feasible region** if there exists some number K such that no feasible solution equates any decision variable to a value that exceeds K in absolute value. Intuitively, it might seem "normal" for a feasible region to be bounded. But the next problem suggests that bounded feasible regions occur about half the time.

4. This problem is couched in terms of the original version of Lmax, which has inequality constraints. Consider data for which Lmax is feasible and has a bounded feasible region. Show that its dual, Gmin, has an unbounded feasible region. *Hints:*

- Since Lmax is feasible and bounded, it has an optimal solution x, and Gmin has an optimal solution y.
- Since Lmax also has a bounded feasible region, the linear program that maximizes $\{x_1 + x_2 + \cdots + x_n\}$ subject to the constraints in Lmax has an optimal solution $\hat{x}$ and its dual has an optimal solution $\hat{y}$.
- See what you can say about $y + \theta\hat{y}$ where θ is a positive number.

Matrix notation: When Lmax and Gmin are cast in matrix notation, they become this pair of this linear programs.

Lmax: Maximize $\{c\,x\}$, subject to the constraints $A\,x \leq b$ and $x \geq 0$.

Gmin: Minimize $\{y\,b\}$, subject to the constraints $y\,A \geq c$ and $y \geq 0$.

The data in Lmax and Gmin are the $1 \times n$ vector c, the $m \times n$ matrix A, and the $m \times 1$ vector b. The decision variables form the $n \times 1$ vector x and the $1 \times m$ vector y.

5. **(Convexity and Linear Programs)** Suppose that Lmax is feasible and bounded when its right-hand-side vector equals b and when its right-hand side vector equals $\hat{b}$. Let x and $\hat{x}$ denote basic optimal solutions for right-hand-side vectors b and $\hat{b}$ respectively. Suppose that x and $\hat{x}$ have the *same* set of tight constraints in their respective linear programs. Hence, for instance, if $x_j = 0$, then $\hat{x}_j = 0$.

(a) Let the y be an optimal solution to the dual of Lmax when its right-hand side vector is b. Is y optimal when the right-hand-side vector equals $\hat{b}$? *Hints:* Does y stay feasible? Do y and $\hat{x}$ satisfy complementary slackness?

(b) Now, suppose the right-hand-side vector of Lmax is a vector $\tilde{b}$ that lies in the line segment between b and $\hat{b}$. In other words, there exists a number c that is between 0 and 1 such that $\tilde{b}_i = c\,b_i + (1 - c)\hat{b}_i$ for $i = 1, \ldots, m$. Set $\tilde{x}_j = c\,x_j + (1 - c)\hat{x}_j$ for $j = 1, \ldots, n$. Is $\tilde{x}$ feasible for the right-hand side vector $\tilde{b}$? Is y feasible for the dual? Do they satisfy complementary slackness? Are they optimal?

(c) True or false: The set of right-hand-side vectors b for which a particular set of constraints is tight is a convex set.

6. **(Strong Complementary Slackness)** This problem concerns the following pair of dual linear programs:

Program E: Maximize $\{c\,x\}$, subject to $A\,x = b$ and $x \geq 0$.

Program E^d: Minimize $\{y\,b\}$, subject to $y\,A - t = c$ and $t \geq 0$.

Suppose that both linear programs are feasible and hence that both linear programs have optimal solutions. Let x be optimal for Program E, let (y, t) be optimal for Program E^d, and let z^* be their optimal value. Complementary slackness guarantees $x_j t_j = 0$ for each j. This problem asks you to verify this **strong complementary slackness** condition: If every optimal solution to Program E has $x_k = 0$, then there exists an optimal solution to Program E^d that that has $t_k > 0$. To verify this, you will examine a third linear program, which is:

Maximize $\{v_k\}$, subject to the constraints

w:	$A\,v = b,$
θ:	$-c\,v \leq -z^*,$
	$v \geq 0.$

(a) Is this third linear program feasible? Suppose it has a feasible solution whose objective value is positive. In this case, is strong complementary slackness satisfied?

For parts (b) and (c), suppose that the optimal value of this third linear program equals zero.

(b) True or false: Every optimal solution to Program E has $x_k = 0$.

(c) Take the dual of the third linear program. From that dual, construct an optimal solution to Program E^d in which t_k is positive. *Hint:* There will be two cases to consider. If θ is positive, try w/θ. If $\theta = 0$, try $y + w$.

(d) Have you proved strong complementary slackness? Why?

7. **(Data Envelopment)** Suppose that cash is one of the inputs to each unit. The value of cash is easy to establish; one dollar is worth one dollar. Does the data envelopment model remain valid? Justify your answer.

8. **(Data Envelopment)** Adapt Program 19.1 to determine whether or not department A is potentially efficient. Solve it. Do the same for department C. In each case, adapt Program 19.1, if necessary, to find prices for which the department has the highest efficiency ratio.

9. **(Data Envelopment)** This concerns Program 19.1. You may use the fact that the optimal value of Program 19.1 equals zero and the fact that the data in Table 19.7 are positive numbers, but you may not use the specific values of these data.

(a) What can you say about the optimal value of Program 19.1^d?

(b) In each feasible solution to the Program 19.1^d, prove that y_B is negative.

(c) From each feasible solution to Program 19.1^d, construct a mixture of departments A and C that envelops department B. Prove that it does so.

(d) Is the enveloping mixture unique?

10. **(Data Envelopment)** For the data in Table 19.7 (on page 652), find the mixture of departments A and C that maximizes x where each output of this mixture exceeds that of department B by at least x and where the input of this mixture is exceeded by that of department B by at least x.

11. **(Infeasible Linear Programs)** This problem concerns the linear program:
Maximize $\{A + B\}$, subject to the constraints

$$A - B \leq -1,$$
$$-A + B \leq -1,$$
$$A \geq 0, \quad B \geq 0.$$

(a) Take the dual of this linear program.

(b) Is this linear program feasible? Is its dual feasible?

12. **(Producer's Equilibrium)** In Lemma 19.9, which of the following statements is true:

(a) No producers' equilibrium can exist if, for some technology t, the prices satisfy

$$\sum_{g=1}^{m} p_g A_{gt} > 0.$$

(b) Suppose that the prices satisfy

$$\sum_{g=1}^{m} p_g A_{gt} \leq 0 \quad \text{for } t = 1, \ldots, n.$$

Then setting $x_t = 0$ for each t is a producers' equilibrium.

(c) Suppose that the prices satisfy the condition in part (b). Then a set of nonnegative production levels is a producers' equilibrium if and only if $x_t = 0$ for each t such that

$$\sum_{g=1}^{m} p_g A_{gt} = 0.$$

13. **(General Equilibrium)** Theorem 19.4 shows that, in the general equilibrium, the market price p_g of each good g that the consumer actually consumes equals the consumer's marginal utility u_g for that good. Why must this be so?

14. **(General Equilibrium with One Noxious Good)** Suppose that good 13 is noxious and so cannot be thrown out. Assume that $u_{13} < 0$ and that $e_{13} = 0$. What changes occur in Programs 19.3 and 19.3^d? What changes occur in Theorem 19.4? Why?

15. **(General Equilibrium with Production Capacities)** Suppose that each technology t has a finite production capacity, C_t, so that the levels at which the technologies are operated must satisfy the constraints $x_t \leq C_t$ for $t = 1, \ldots, n$. The consumer owns all of the assets in the economy, including the production capacities. What changes, if any, occur in Programs 19.3 and 19.3^d? What changes, if any, occur in Theorem 19.4? Why?

16. **(General Equilibrium with Decreasing Marginal Returns)** This concerns the variant of the general equilibrium model in which the data u_g and A_{gt} are replaced by functions that exhibit decreasing marginal return. Specifically, for each good g and for each technology t:

$$u_g(z_g) = \text{the (total) utility to the consumer of consuming } z_g \text{ units of good } g.$$
$$A_{gt}(x_t) = \text{the net output of good } g \text{ due to setting technology } t \text{ at level } x_t.$$

Now for each g and t, the function $u_g(z_g)$ is a concave differentiable function of z_g, and the function $A_{gt}(x_t)$ is a concave differentiable function of x_t. You are to *guess* the answers to parts (a)–(c) and to use verbal (that is, nonmathematical) arguments to justify your guesses.

(a) Adapt Lemma 19.9 to this more general situation.

(b) Adapt Program 19.3 to this more general situation.

(c) Adapt Theorem 19.4 to this more general situation. *Hint:* Do not attempt to generalize Program 19.3^d.

17. **(General Equilibrium with Increasing Marginal Returns?)** For the one-generic-consumer model with constant or decreasing differential marginal returns, there can be multiple equilibria, but they have the same objective value (utility to the consumer), and the market price of each good that the consumer consumes equals its marginal value to the consumer. Guess which of these properties hold in the case of increasing differentiable marginal returns (economies of scale)?

18. **(Farkas)** Adapt Lemma 19.10 and Theorem 19.5 to the variant of system (19.22) in which the nonnegativity constraints in expression (19.22.m + 1) are omitted. Does the same proof work?

19. **(Zero-Sum Two-Person Matrix Games)** Chapter 16 illustrated the game that is now described. You must choose one of m rows of a matrix A. Simultaneously, I must choose one of its columns. If you choose row i and I choose column j, then I pay you the amount A_{ij} in the ijth position of the matrix A. We know the matrix. You wish to maximize the expectation of your earnings. I wish to minimize the expectation of my loss. Show that this game has an equilibrium in randomized strategies, and show how to compute it. *Hint:* Review Program 16.1 of Chapter 16, and plan to use duality.

20. **(Arbitrage)** Alter the investment problem whose data are in Table 19.9 so that asset 3 cannot be sold short.

(a) What changes occur, if any, in Program 19.5?

(b) What changes occur, if any, in the dual of Program 19.5?

(c) Can this change in the investment problem introduce an arbitrage opportunity?

21. **(Arbitrage)** For the investment problem whose data are in Table 19.9, find the range on the risk-free interest rate r for which no arbitrage opportunity exists. For a value of r that is slightly outside this range, find an arbitrage opportunity.

22. **(Arbitrage)** This problem concerns the following three investment opportunities:

- A risk-free asset with an interest rate $r = 0.06$ for one period.

- Asset S, which can be purchased at the start of the period at a unit price of \$100. Either state u or state d will occur at the end of the period. If state u (short for up) occurs, this asset can be sold at a unit price of \$140. If state d occurs, this asset can be sold at a unit price of \$90.
- An option to purchase one unit of asset S at a price of \$125 at the end of the period. The price of this option is \$$C$ per unit, payable at the start of the period. This option will be worthless if state d occurs; each unit of it will be worth \$15 = \$140 − \$125 if state u occurs.

(a) Find the probability distribution over the states for which asset S has zero as its expected net profit.

(b) Find the value (values?) of C for which no arbitrage opportunity exists.

(c) Construct a portfolio consisting of quantities of the risk-free asset and of asset S whose payoff at the end of the period duplicates the payoff of the option under each state. How much is invested in this portfolio at the start of the period? Compare your answer with the value of C that you computed in part (b). Interpret your answer.

23. **(Arbitrage)** For the investment problem whose data are in Table 19.9, suppose that the central bank incurs transaction costs and passes them along to borrowers and depositors. For this reason, the risk-free rate of 3% no longer applies. Instead, the central bank pays 2.9% on deposits and charges 3.1% on loans.

(a) Alter Program 19.5 accordingly.

(b) Alter the dual of Program 19.5 accordingly.

(c) What changes, if any, occur in Theorem 19.6?

24. **(Interdiction on a Max-Flow Network)** This problem concerns the *new-style* cut B that was introduced on page 685, at the end of Section 19.16. To relate the max-flow to $\hat{K}(B)$, build a new network in this (peculiar) way: For each arc (i, j) in E, add a new node labeled i–j. Replace each arc (i, j) in E by arcs $(i, i\text{–}j)$ and $(i\text{–}j, j)$ whose capacities are $2\,K(1)$ and U_{ij}, respectively. In Figure 19.1, for instance, arc (3, 6) is replaced by arc (3, 3–6) whose capacity equals $72 = 2 \times 36$ and by arc (3–6, 6) whose capacity equals 8. In the new network, each i–j type node has *one* arc pointing away from it.

(a) Argue that the new network has the same max-flow z^* as did the original network.

(b) Argue that $z^* = \min\{\hat{K}(B) : B \text{ is a new-style cut}\}$.

Appendix: Note on Excel

Chapter 2 gets you started with Excel and Solver. It also tells you how to get Solver up and running. This appendix supplements that discussion by describing:

- What is on the CD-ROM that accompanies this text and how you can use it.
- A few quirks of Excel and how you can get around them.

This appendix begins by introducing a program called the Windows Explorer.

WINDOWS EXPLORER

Windows Explorer will be used to locate folders on your computer and to transfer some Excel Add-Ins from the CD-ROM that accompanies this text to your Excel Library.

Where Is Windows Explorer?

To open Windows Explorer, click on your Start button. A menu will pop up. One of the items on that menu is labeled Programs. Shift the mouse pointer to Programs, then click. A second menu will pop up. One of the items on that menu will be Windows Explorer. Click on Windows Explorer. A window entitled Exploring – (C:) will appear. Table A.1 displays this window for a particular computer.

What Is This Window All About?

The window displayed in Table A.1 is easy to use, once you get the knack. The key is to understand the roles played by its three white boxes.

- The small white box at the top left contains a particular folder's name and icon. (In Table A.1, it's the (C:) folder.)
- The wide white box at the right lists the contents of the folder whose name is in the box at the top left. (In Table A.1, the wide white box lists the contents of the (C:) folder.)
- The skinny white box to the left contains the folder/subfolder structure by which information in your computer is organized. Typically, only a fragment of it will be in view.

We will soon use these boxes.

What's on the CD-ROM?

To find out what's on the CD-ROM that accompanies this text, insert that CD in your disc drive. Then:

- Open Windows Explorer, as indicated earlier.

Table A.1 The Windows Explorer window.

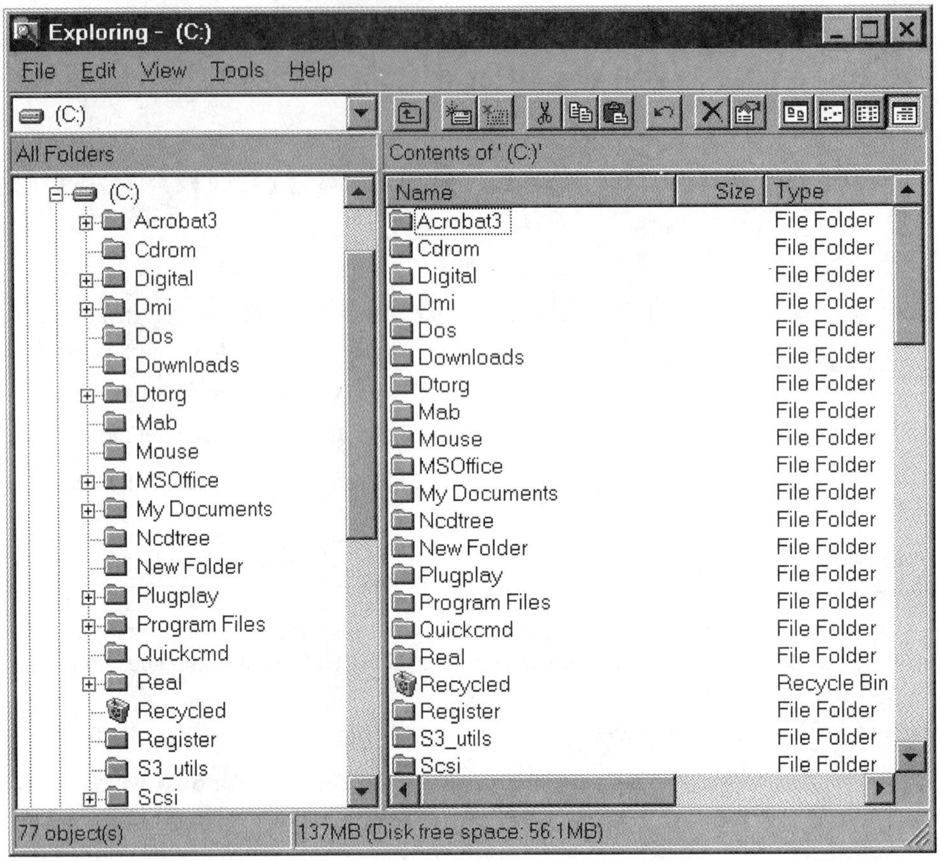

- Scroll down the skinny window to the left until the disc drive's icon is in view.
- Move the mouse pointer onto the disc drive's icon and then single-click it.

Table A.2 displays the result. The icon on which you clicked now appears in the small white box at the top left, and that folder's contents appear in the large white block to the right. Evidently, the CD-ROM contains six items. These items are listed next, with a thumbnail sketch of each.

1. A folder labeled "chapter aides." This folder contains the Excel spreadsheets that appear in the various chapters of this text. You may find it handy when you read the chapters or work the problems.

   Note: The "computer aides" folder is 4.8 million bytes long. Those of its spreadsheets that employ Add-Ins must be accessed directly from your C: drive. You could copy the entire folder and paste it onto your C: drive. Alternatively, to save space, you can copy-and-paste any chapter's subfolder.

2. An Excel Add-In labeled "or-tools." This Add-In contains three user-defined functions that were written by Kenneth Canfield '03, a Yale undergraduate. They are:

   a. The (normal loss) function **=NL(q, μ, σ)** that computes the expectation of the excess of D over q, where D is a random variable whose distribution is normal, with μ as its mean and σ as its standard deviation. (Chapter 11 describes this function.)

Table A.2 The Windows Explorer's display of the contents of the CD-ROM accompanying this text.

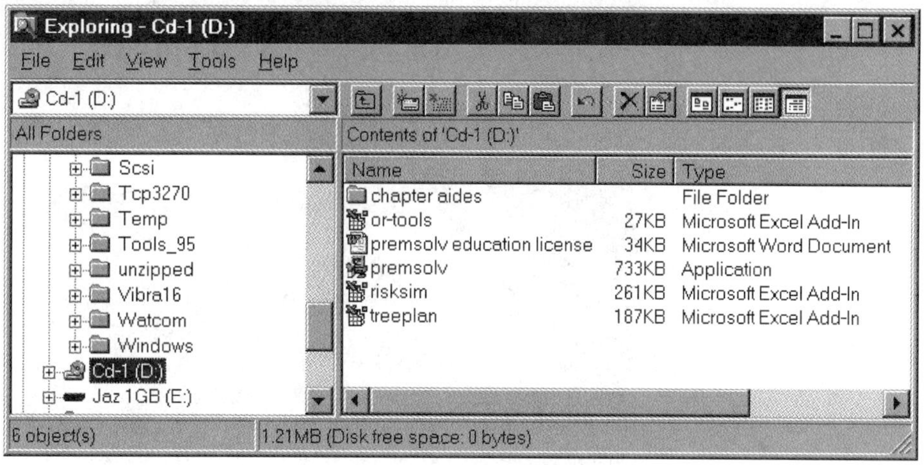

b. The function **=LQP(ρ, c)** that computes the expectation of the number of customers who are waiting for service to begin in a queue that has exponential interarrival times, exponential service times, c servers, and ρ as its utilization rate. (See Chapter 14 for a description.)

c. The function **=PIVOT(cell, array)** that executes a pivot on the coefficient in the cell of the array (matrix). (Chapter 17 explains this function.)

3. A Word file labeled "premsolv education license." It is the licensing agreement for Premium Solver for Education.

4. An Application labeled "premsovl." It executes Premium Solver for Education.

5. An Excel Add-In labeled "risksim." It identifies RiskSim, which was written by Michael Middleton. (Chapter 14 tells how to use RiskSim to construct simulations and run them.)

6. An Excel Add-In labeled "treeplan." It identifies TreePlan, which was also written by Michael Middleton. (Chapter 9 tells how to use it to draw decision trees and roll them back.)

Our next goal is to show you how to copy the three Excel Add-Ins (items 2, 5, and 6 above) into your Excel Library folder. Before you can do that, however, you must locate your Excel Library folder.

Where Is My Excel Library Folder?

The answer depends on how your computer's hard-drive memory is organized. You can use Windows Explorer to find out. To locate the Excel Library folder in your PC:

- Click on your Start button. On the menu that pops up, shift the mouse pointer to Find. Then, on the submenu that pops up, click on the "Files or Folders" entry. A dialog box entitled "Find: All Files" will pop up.
- In that dialog box's "Named" entry, type Library. Then click on Find Now.

For a typical Office 97 memory configuration, this results in Table A.3. The "folder"-shaped icons in this table identify two paths to the Excel Library folder. Evidently, the shorter

Table A.3 Windows Explorer after locating the Excel Library folder.

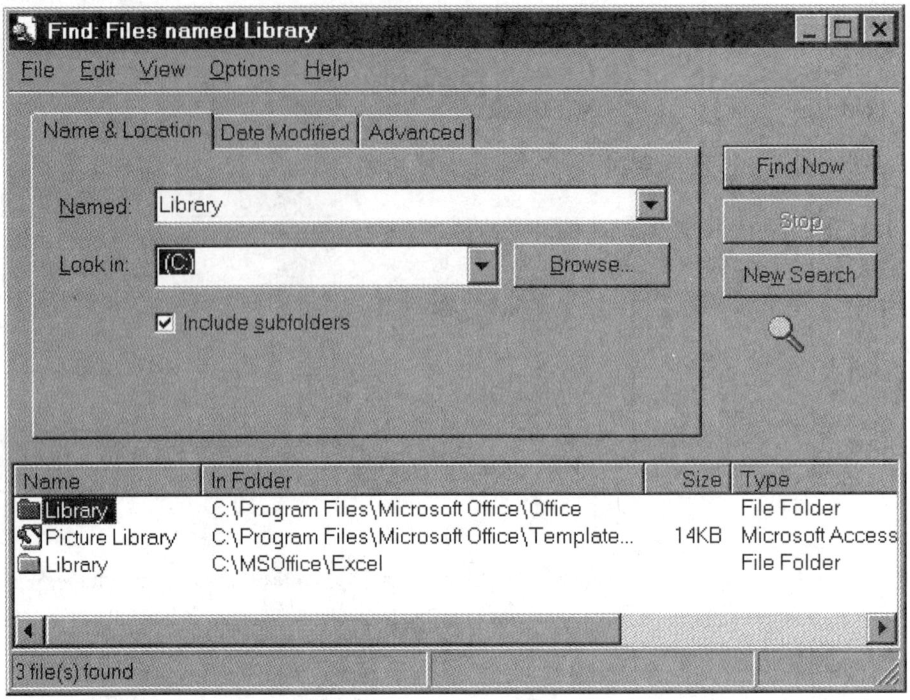

of these paths is

C:\MSOffice\Excel

which means that the Library folder is a subfolder of the Excel folder, which is itself a subfolder of the MSOffice folder, which is a subfolder of the C: folder.

How Can I Copy the Add-Ins into My Excel Library Folder?

Once you have found a path to your Excel Library folder, Windows Explorer makes short work of copying stuff into it. The next step (which will soon be described) causes your Windows Explorer dialog box to resemble Table A.4.

Described next is a procedure that constructs the window in Table A.4 for the file structure in that window. If the path to your Library folder differs, you will need to adapt this procedure, which is:

- With the book's CD-ROM in your disc drive, open Windows Explorer.
- In the skinny window to the left, put the mouse pointer on the icon for the MSOffice subfolder of (C:) folder, and then *double-click* the mouse pointer. The subfolders of the MSOffice folder will pop into view. Shift the mouse pointer onto the icon for the Excel subfolder, and then double-click again. The Library subfolder will pop into view. (You will need to adapt this step if the path to your Excel Library folder differs from that in Table A.4.)
- Next, scroll down the skinny box to the left until the icon for your disc drive is visible, and *single-click* on that icon. This drive's icon will pop up in the small white rectangle at the top left, and the wide white rectangle to the right will list its contents.

Table A.4 Windows Explorer with the Excel Library folder visible in the left window and the contents of the CD-ROM in the right window.

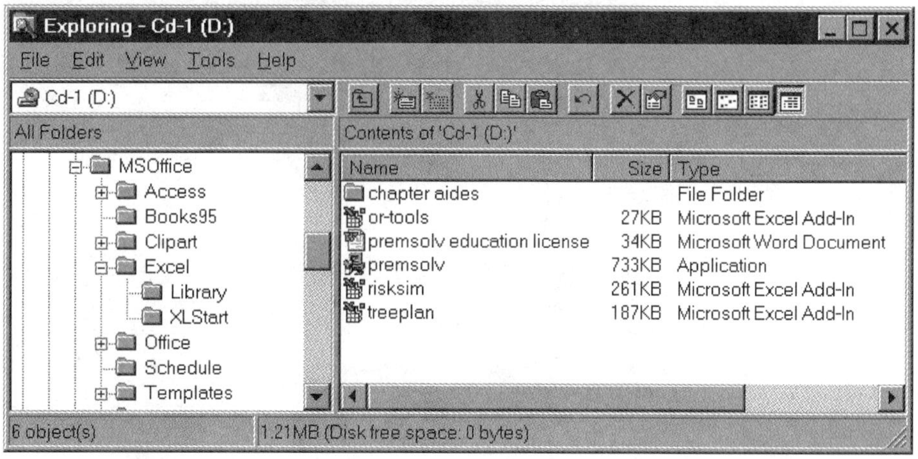

- Scroll back up until the Excel Library folder is in view. Your Windows Explorer box should now resemble Table A.4.

In Table A.4, the icon for the Excel Library folder is visible in the left window, and the icons for contents of the CD are in the right window. All that remains is to drag from right to left, as follows:

- To copy the Add-In labeled or-tools into the Excel Library folder, put the mouse pointer on its icon, click it, and then drag it leftward onto the Excel Library folder before releasing.
- Repeat for Add-Ins whose icons are labeled risksim and treeplan.

To check that you have succeeded, move the mouse pointer to the Library folder, and then single-click it. That folder's icon will pop into the small window at the top left of Windows Explorer, its contents will appear in the wide white window to the right, and they should include the three recently installed Add-Ins.

The Excel Tools Menu

The Add-Ins are now in your Excel Library, but they are not yet installed on your Excel Tools menu. To get them onto your Excel Tools menu, do these steps:

- Open Excel. Put the mouse pointer on its Tools menu, and then click it. Scroll down to the Add-Ins entry and click again. A window named Add-Ins will appear.
- Note the list of Add-Ins that are available to you. This list does not (yet) include the items you just placed into the Excel Library. In the Add-Ins window, click on the Browse button.
- In the "Browse" window, see the subfolders of the C: folder. To make the Browse window look like Table A.5, double-click on MSOffice, double-click on Excel, and then double-click on Library.
- Select the icon for an item in your library, such as or-tools. Then click on the OK button.
- Repeat for the other items.

Table A.5 The Browse menu, poised to install items to the Tools menu.

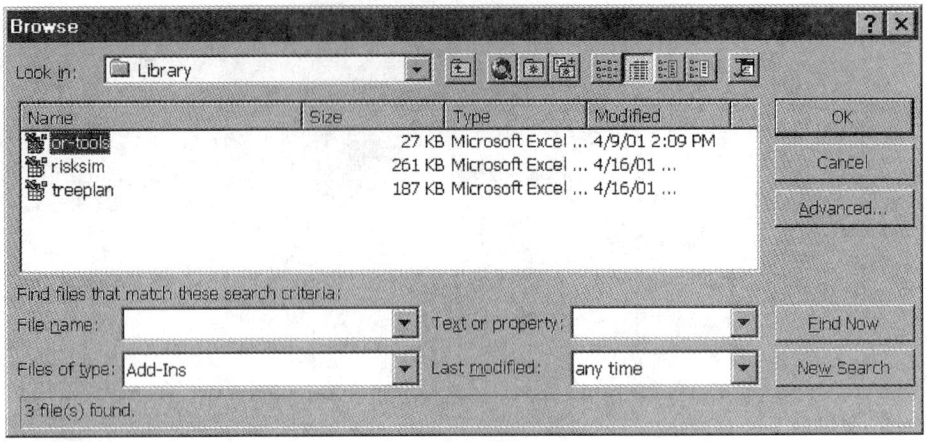

- Finally, on the Excel Tools Add-Ins submenu, check that or-tools, risksim and treeplan are listed, and check off their boxes if they are not already checked off.

What about Installing Premium Solver for Education?

As was explained in Chapter 2, plain old Solver works pretty well, but Premium Solver for Education has added capabilities as well as fewer bugs. You may use either.

You cannot use Premium Solver for Education, however, unless Solver itself has been installed and activated. Chapter 2 tells how to get Solver up and running. That done, installing Premium Solver for Education is a breeze. To do so, put the text's CD-ROM in your disc drive, select that drive's directory, double-click on the icon marked premsolv, and follow instructions, correcting the path to your Excel Library if necessary.

ROUGH SPOTS IN EXCEL

Excel is remarkably user-friendly, but it does have a few quirks. This section identifies some rough spots, and it tells how to deal with them.

How Can I Create a Chart from My Spreadsheet?

Excel charts come in many varieties, and there are many ways in which to create each type of chart. To become facile with charts, you will need to experiment. Described here is one way in which to create a chart. In these instructions, Step 4 is the place where you will need to try things out.

1. Select the cells that contain the data that you want to chart.
2. In Excel's Insert menu, click on Chart, and use the sequence of "dialog" boxes to create the type of chart you desire, without worrying much about how it looks. You can fix it later (in Step 4, below).

   Warning: In a so-called Line chart, the data on the x-axis will be equally spaced. If you want the x-axis to measure something, use the so-called XY (Scatter) chart.

3. When your chart appears, adjust its height or width to the size you want by moving the mouse pointer to an edge, waiting for it to turn into a double-headed arrow, and then stretching or shrinking that edge.
4. Put the mouse pointer inside the chart and double-click it. Black squares (like so ■) will appear along the edges of the chart. Then, to get your chart looking the way you want it to, follow this recipe:
   - Step a: Click on a feature that you want to fix—on the *x*-axis, on the *y*-axis, on a data series, on the plot area, or whatever.
   - Step b: On the Excel Format menu, the feature you selected in Step a will appear as a menu item! (If you selected the *x*-axis in Step a, a "Selected axis" option will appear on the Format menu.) Click on that menu item. A new menu will pop up. Use that menu to make the feature look the way you want it to look.
   - Repeat Steps a and b for each feature that you wish to fix.

How Can I Add Lines, Arrows, Text Boxes, or Greek to My Chart?

To insert arrows and text boxes, put the mouse pointer inside the chart and then double-click it to surround the chart with black squares (like so ■). Then, on the Excel View menu, click on Toolbars, check off the Drawing tool, and then use the Drawing tool bar to put in the stuff you want.

Alternatively, copy your chart, paste it onto a PowerPoint slide, and add what you need there. PowerPoint gives you better control of lines, arrows, and text.

Adding Greek or equations to Excel spreadsheets isn't easy. Avoid it if you can. One method that works is to open a Word document, use the Equation Editor to create the Greek or the equation you want, copy it, shift back to Excel, and paste it. Then nudge it into proper position. Finally, use the Object entry on the Excel Format menu to remove the ugly lines that surround it. That's hard work. A similar procedure adds equations and Greek letters to PowerPoint slides.

How Can I Print an Excel Worksheet the Way I Want It to Look?

In Excel, you can click on the File menu and then click on Print. But that may not print exactly what you want. Here are some options.

- To include row and column headings, click the File menu, click the Page Setup option, select the Sheet tab, check the Row and Columns heading, and click on OK.
- To omit the gridlines, click the File menu, click the Page Setup option, select the Sheet tab, remove the check on the Gridlines, and click on OK.
- To print the formulas, but not their values, on the Tools menu, click Options. On its View tab, click Formulas, then click OK. Then print.
- To re-scale a worksheet (onto one printed page, say), on the File menu click Page Setup. In its Scaling box, use *Adjust to* or *Fit to* to tailor your worksheet to the number of pages in the width and height dimensions that you wish. Then click on OK.

Can I Do Matrix Operations within Excel?

You can. Excel lets you add and multiply matrices and take the transpose. It even computes the inverse. But the way in which Excel does matrix operations includes a pair of booby traps that you should know about.

Table A.6 A matrix multiplication executed by the MMULT function.

	A	B	C	D	E	F	G	H
1								
2		4	3	2	1		4	1
3		3	1	3	1		3	0
4		-1	-2	-3	-4		2	1
5							1	0
6		30	6					
7		22	6					
8		-20	-4					

In Excel's language, a matrix operation is done by an **array formula**. To specify an array formula, you perform this sequence of steps.

1. Select the array of cells in which the result of the computation is to appear.
2. Type the array formula, but *do not* hit the **Enter** key.
3. Instead, hit **Ctrl+Shift+Enter**, which means that you should depress the **Ctrl** and the **Shift** key and, while they are depressed, hit the **Enter** key.

Table A.6 contains the result of an array computation, one that multiplies the 3 × 4 matrix in cells B2:E4 by the 4 × 2 matrix in cells G2:H5 and places the resulting 3 × 2 matrix in cells B6:C8. This was accomplished by the following procedure.

1. Select cells B6:C8.
2. Type **=MMULT(B2:E4, G2:H4)**
3. Hit **Ctrl+Shift+Enter**

The matrix product appears in the cells that you have selected. This sounds easy, but here is something worth remembering:

Booby trap #1: When Excel performs a matrix operation, it fails to interpret a blank as a zero. In the above calculation, for instance, suppose cell H3 contained a blank rather than a zero. Excel would report #VALUE! in cells B6:C8. Yuck! When doing matrix operations, you must replace all blanks by zeros.

Suppose you no longer want the above matrix product on your spreadsheet. How do you get rid of it? Easy. Select the entire array that you wish to delete and then hit the **Del** key. In particular, to delete the matrix product in Table A.6, select B6:C8 and then hit the **Del** key. The array will disappear, as it should.

But suppose you hit the **Backspace** key instead of the **Del** key. That normally works but not this time. If you hit the **Backspace** key, Excel will treat you to this message:

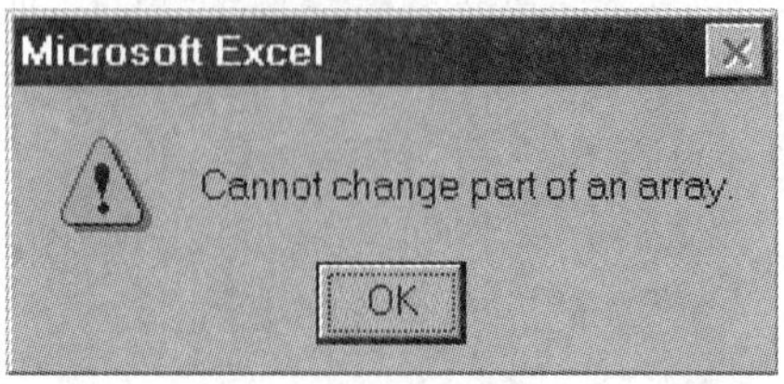

Table A.7 An inverse computed by the MINVERSE function.

	A	B	C
1			
2		1	2
3		4	5
4			
5		-1 2/3	2/3
6		1 1/3	- 1/3

Having no choice, you click on the **OK** button. The message disappears. You move the Mouse pointer somewhere else, and click it. The message reappears. *Swell—a loop*!

The same loop would occur if you had actually tried to delete a portion of an array. The **Esc** key gets you out of the loop.

Booby trap #2: If you get a "Can't change part of an array" message, click on ***OK***. *Then, to get out of the loop, hit the* **Esc** *button. If that doesn't work, try to select the entire array, and then hit the* **Esc** *button.*

Once you know how to deal with these booby traps, matrix operations can be very handy.

How Can I Compute the Inverse of a Matrix?

The array function **=MINVERSE(array)** computes the inverse of any square array (matrix) that has an inverse. In Table A.7, the 2 × 2 array (matrix) in cells B2:C3 has been inverted; its inverse is the 2 × 2 array in cells B5:C6. This was accomplished by the following procedure.

1. Select cells B5:C6.
2. Type **=MINVERSE(B2: C3)**
3. Hit **Ctrl+Shift+Enter**

If cell C3 had contained the number 8 rather than 5, the matrix would have no inverse, in which case Excel would report #NUM! in cells B5 through C6.

Can I Take the Transpose of a Matrix?

The array function **=TRANSPOSE(array)** takes the transpose of any array (matrix). But why would you ever want to take the transpose of a matrix? If your variables are lined up in a row, you may find yourself wanting to print them in a column or to do some computations that require them to be in a column. Take the transpose.

What Are the MATCH and OFFSET Functions? What Good Are They?

Excel is, in effect, a programming language, and like any other programming language, it has inherent advantages and disadvantages. In the main, Excel instructions are easy to write and easy to read. Excel lets you manipulate blocks of cells without having to use "subscripts," for instance.

Table A.8 Illustrating MATCH and OFFSET functions.

	A	B	C	D	E	F	G	H	I	J	K	L	M
1		travel times					scheduled trips					scheduled trips	
2		(in days)				trip	leave	on	for		from	to port	arrive
3		A	B	C		nmbr	port	day	port		port #	#	on day
4	A	0	2	3		1	A	1	B		1	2	3
5	B	2	0	1.5		2	B	1	C		2	3	2.5
6	C	3	1.5	0		3	C	3	B		3	2	4.5
7						4	A	4	B		1	2	6
8	return date =			12		5	A	6	C		1	3	9
9						6	C	9	A		3	1	12
10													
11					=MATCH(G9,A$4:A$6,0)								
12					=MATCH(I9,B$3:D$3,0)						=H9+OFFSET($A$3,K9,L9)		

But there is a price to pay, and you will pay it when you need to use the MATCH and OFFSET functions. Table A.8 illustrates this. In the table, cells A3:D6 specify the travel times between ports A, B, and C. Columns G, H, and I describe six trips and list the departure day of each. Let's calculate the day at which each trip ends. The final trip leaves port C on day 9 and heads for port A. It gets there three days later, on day 12. That was easy to do by "eyeball." To get Excel to do it, we will use both the MATCH and the OFFSET functions.

Cells A4, A5, and A6 identify cities A, B, and C with rows 1, 2, and 3 of the array of travel times. Column K assigns to each trip the *number* of the port from which that trip originates. Trips that begin in port A are assigned the number 1, for instance. The MATCH accomplishes this. For instance, cell K9 contains the function

=MATCH(G9, A$4:A$6, 0)

This function searches the one-dimensional array A4:A6 for an exact "match" with the entry in cell G9. If it finds a match, it records in cell G9 the row in which the match occurred, which is row 3. If it finds no match, it will record #N/A in cell G9. Dragging this function up column K records the number of the port from which each trip originates.

Similarly, column L assigns to each trip the column number of the port at which that trip ends. For instance, cell L9 contains the function

=MATCH(I9, B$3:D$3, 0)

whose value is 1 because the final trip is to port 1.

Evidently, the MATCH function has three arguments, which are separated by commas. They are:

leftmost—the value sought.

center—the range over cells to search. This range *must* be a row or column.

right—the number 0 if an exact match is required.

For instance, the function **=MATCH("A", A4:A6, 0)** returns 1 because A is the first entry in this list.

You may have guessed that the OFFSET function identifies an entry in a two-dimensional array. This function has three arguments, which are separated by commas. Its arguments are:

left—the "anchor" or reference cell.

center—the number of rows below the reference cell.

right—the number of rows to the right of the reference cell.

When applied to the above spreadsheet, the function **=OFFSET(A3, 2, 3)** returns 1.5 because 1.5 is the content of the cell that lies two rows below cell A3 and three cells to its right. Column M uses the OFFSET function to compute the arrival time of each trip.

RECAP

This book intends to be self-contained as to Excel. Chapter 2 got you started. The appendix helps you install the information on the book's CD-ROM. It introduces you to a potent tool, the Windows Explorer. It also acquaints you with a few peculiarities of Excel, and it shows you how to deal with them.

Index

Software License

Decision Support Services ("DSS") grants you the right to use this copy of the enclosed computer program (the "Software") on a single computer by one person at a time. The Software is protected by United States copyright laws and international treaty provisions. Therefore, you must treat the Software just like any other copyrighted material (e.g. a book or musical recording), except that: You may store the Software on a hard disk or network server, *provided* that only one person uses the Software on one computer at a time, and you may make one copy of the Software solely for backup purposes. You may not rent or lease the Software, but you may transfer it on a permanent basis if the person receiving it agrees to the terms of this license. This license agreement is governed by the laws of the State of California.

Limited Warranty

THE SOFTWARE IS PROVIDED "AS IS" WITHOUT WARRANTY OF ANY KIND. THE ENTIRE RISK AS TO THE RESULTS AND PERFORMANCE OF THE SOFTWARE IS ASSUMED BY YOU. DSS warrants only that the CD or diskette on which the Software is distributed and the accompanying written documentation (collectively, the "Media") is free from defects in materials and workmanship under normal use and service for a period of ninety (90) days after purchase, and any implied warranties on the Media are also limited to ninety (90) days. SOME STATES DO NOT ALLOW LIMITATIONS ON THE DURATION OF AN IMPLIED WARRANTY, SO THE ABOVE LIMITATION MAY NOT APPLY TO YOU. DSS's entire liability and your exclusive remedy as to the Media shall be, at DSS's option, either (i) return of the purchase price or (ii) replacement of the Media that does not meet DSS's limited warranty. You may return any defective Media under warranty to DSS or to your authorized dealer, either of which will serve as a service and repair facility.

EXCEPT AS PROVIDED ABOVE, DSS DISCLAIMS ALL WARRANTIES, EITHER EXPRESS OR IMPLIED, INCLUDING BUT NOT LIMITED TO IMPLIED WARRANTIES OF MERCHANTABILITY AND FITNESS FOR A PARTICULAR PURPOSE, WITH RESPECT TO THE SOFTWARE AND THE MEDIA. THIS WARRANTY GIVES YOU SPECIFIC RIGHTS, AND YOU MAY HAVE OTHER RIGHTS WHICH VARY FROM STATE TO STATE.

IN NO EVENT SHALL DSS BE LIABLE FOR ANY DAMAGES WHATSOEVER (INCLUDING WITHOUT LIMITATION DAMAGES FOR LOSS OF BUSINESS PROFITS, BUSINESS INTERRUPTION, LOSS OF BUSINESS INFORMATION, AND THE LIKE) ARISING OUT OF THE USE OR INABILITY TO USE THE SOFTWARE OR THE MEDIA, EVEN IF DSS HAS BEEN ADVISED OF THE POSSIBILITY OF SUCH DAMAGES. BECAUSE SOME STATES DO NOT ALLOW THE EXCLUSION OR LIMITATION OF LIABILITY FOR INCIDENTAL OR CONSEQUENTIAL DAMAGES, THE ABOVE LIMITATION MAY NOT APPLY TO YOU. In states that allow the limitation but not the exclusion of such liability, DSS's liability to you for damages of any kind is limited to the price of one copy of the Software and Media.

U.S. Government Restricted Rights

The Software and Media are provided with RESTRICTED RIGHTS. Use, duplication or disclosure by the Government is subject to restrictions as set forth in subdivision (b)(3)(ii) of The Rights in Technical Data and Computer Software clause at 252.227-7013. Contractor/manufacturer is Decision Support Services, 2105 Buchanan Street, #1, San Francisco, CA 94115.

THANK YOU FOR YOUR INTEREST IN DECISION SUPPORT SERVICES

Software License

Limited Warranty

U.S. Government Restricted Rights